THERMODYNAMICS AND PHASE TRANSFORMATIONS

THE SELECTED WORKS OF Mats Hillert

Scientific Editors:

John Ågren, KTH, Stockholm
Yves Bréchet, INPG, Grenoble
Christopher Hutchinson, Monash University, Melbourne
Jean Philibert, Université Paris Sud
Gary Purdy, Mc Master University, Hamilton

17, avenue du Hoggar
Parc d'activités de Courtabœuf, BP 112
91944 Les Ulis Cedex A, France

Illustration de couverture : Divergent pearlite in Fe-C-Mn.
Photo Chris Hutchinson.

Imprimé en France

ISBN: 2-86883-889-8

Tous droits de traduction, d'adaptation et de reproduction par tous procédés, réservés pour tous pays. La loi du 11 mars 1957 n'autorisant, aux termes des alinéas 2 et 3 de l'article 41, d'une part, que les "copies ou reproductions strictement réservées à l'usage privé du copiste et non destinées à une utilisation collective", et d'autre part, que les analyses et les courtes citations dans un but d'exemple et d'illustration, "toute représentation intégrale, ou partielle, faite sans le consentement de l'auteur ou de ses ayants droit ou ayants cause est illicite" (alinéa 1er de l'article 40). Cette représentation ou reproduction, par quelque procédé que ce soit, constituerait donc une contrefaçon sanctionnée par les articles 425 et suivants du code pénal.

© EDP Sciences 2006

A short biography of Mats Hillert

Mats Hillert was born in Gothenburg, Sweden on 28th of November 1924. He was the youngest of the three sons of Anna and Hildding Hillert. As they were very lively the young Hillert brothers were well known in the local neighbourhood.

Mats grow up in Gothenburg and went to high school at Realläroverket, later named Vasa Läroverk. He earned his B.S. in chemical engineering at Chalmers Technical University in Gothenburg in 1947 with a bachelor's thesis on diffusion of radioactive Ag in Ag_2HgJ_4 under the supervision of Karl Erik Zimen, a former student of Carl Wagner whose oxidation theory connected electrical conductivity with diffusion.

After his military service Mats started his research career as a scientist at the Swedish Institute for Metals Research in 1948. Mats was assigned to look into the use of internal friction but was left with a high degree of freedom. On the side he could thus start to work on the use of radioactive isotopes to study the distribution of various elements in metals. He also contacted Gudmund Borelius, who was professor of physics at Royal Institute of Technology, KTH, wanting to register for graduate studies. He was told that he first had to take the undergraduate courses for physicists that were not included in the chemistry curriculum. For a few years he spent most of his spare time on those studies, which was possible because his girl friend still went to school in Gothenburg. At work Mats came in contact with Sten and Helfrid Modin, two Swedish metallographers and early pioneers in the use of electron microscopy to study the microstructure of steels. Sten Modin inspired Mats to add yet another undergraduate course at KTH and thus participated in the lectures of Axel Hultgren who was professor in physical metallurgy at the nearby KTH. Mats says that his life-long deep interest in phase transformations and thermodynamics started with a scientific argument after one of these lectures. Mats felt that Hultgren's reasoning on the effect of Ni on carburization of steel was thermodynamically incorrect and told Hultgren his opinion. Hultgren, the grand old man of Swedish physical metallurgy, became impressed of this young scientist who showed such a remarkable intuition in thermodynamics and suggested that one should resolve the issue by an experiment. Of course the experiment showed that Mats was correct and it eventually led to a joint publication of Hultgren and Hillert in 1953. At that time Hultgren had developed the concept of paraequilibrium, which also concerned the effect of alloying elements on transformations in steel. This contact with Hultgren resulted in a number of manuscripts centred around the concept of paraequilibrium and isoactivity lines in phase diagrams and their applications to transformations in steels. Mats assignment on internal friction was simply forgotten by that time.

Eric Rudberg, who was the director of the Swedish Institute for Metals Research, encouraged Mats to apply for a grant from the Sweden-America Foundation. With his wife since 1951, Gerd, he thus moved to Boston and MIT in 1953 and started his studies under the supervision of the legendary teacher Morris Cohen. At MIT he also met Carl Wagner who taught thermodynamics and gave Mats valuable advice on some of his manuscript, which had all been rejected. He was then able to publish some, some were published later but one was not published until the present volume. His topic for the Master's thesis was an experimental study of the thermodynamics of the Ag-Al system. Inspired by one of his fellow students, Larry Kaufman, who was modelling the thermodynamic properties of the Fe-Ni system, Mats applied a similar model to the Ag-Al system and that work was the starting point for his life-long interest in the thermodynamics of alloys. Mainly by the efforts of Kaufman, that procedure has since developed into what is now called CALPHAD. From another fellow student, Eric Kula, he learned about the effect of fine inclusions on grain growth, which inspired him to start developing a mathematical theory for grain growth. After a Master's degree Mats was invited to stay for the Doctorate and while spending a semester to prepare for the qualifying exam he could not help worrying about the strange behaviour of Au-Ni alloys that one of his room mates, Ervin Underwood, was studying. He connected it with the phenomenon later called spinodal decomposition, which was the main research interest of Borelius at KTH. It lead him to developing a theory for that phenomenon. Fortunately, this topic fitted into one of Cohen's research programs and Mats could finish his degree after adding an experimental study of spinodal decomposition. He left MIT with a Doctor of Science in 1956 and returned to the Swedish Institute of Metal Research in Stockholm. A coincidence deserves mention. Another of Mats' room mates was John Hilliard who brought a copy of his thesis with him when he left for GE in Schenectady in 1956. There he met John Cahn and together they completed the theory of spinodal decomposition and thus opened up a new field.

During the Mats' stay in the US professor Hultgren had retired and Curt Amberg, a well known industrial researcher and an expert in heat treating, had been appointed as the new professor in physical metallurgy. However, he became seriously ill and passed away 1959. Mats was then appointed as a temporary replacement and started teaching in physical metallurgy. The coming years he published a large number of papers, some of them ground breaking, and when the position as professor in physical metallurgy was announced, Mats applied and was appointed full professor 1961. He remained at that position until his retirement 30 years later.

As a professor Mats taught several generations of Swedish metallurgists the most up to date knowledge in the fundamentals of physical metallurgy. He launched the graduate teaching on the subject and acted as the main supervisor of more than 30 doctors which are now active in Swedish industry or academia worldwide.

Mats Hillert is a fellow of the Royal Swedish Academy of Engineering Sciences and the Royal Swedish Academy of Sciences. He is also a fellow of ASM International and a Fellow of Met. Soc. AIME (TMS). His list of awards is impressive and includes, to mention only a few, R.F. Mehl medalist (Met. Soc. AIME), Bakhuis Roozeboom Gold Medal (Royal Acad. Netherland), Acta Metallurgica Gold Medal, Murakami Gold Medal (Japan Inst. Metals, Japan), Björkén award (Uppsala University, Sweden), Hume-Rothery Award (TMS, USA).

Mats Hillert at the TMS fall meeting in 1991.

Foreword

For over half a century, Mats Hillert has contributed greatly to the science of materials. He is widely known and respected as an innovator and an educator, a scientist with an enormous breadth of interest and depth of insight. In acknowledgment of his many contributions, a conference was held in Stockholm in December 2004 to mark his eightieth birthday.

This volume was conceived prior to, and publicly announced during the conference. The difficult choice of twenty-four papers from a publication list of more than three hundred was carried out in consultation with Mats. He also suggested or approved the scientists who would be invited to write a brief introduction to each paper.

A brief reading of the topics of the selected papers and their introductions reveals something of their range and depth. Several early selections (for example, those on "The Role of Interfacial Energy during Solid State Phase Transformations", and "A Solid-Solution Model for Inhomogeneous Systems") contained seminal material that established Mats as a leading figure in the study of phase transformations in solids. Others established his presence in the areas of solidification and computational thermodynamics. A review of his full publication list shows that he has consistently built upon those early foundational papers, and maintained a dominant position in those fields. Although many of his contributions have been of a theoretical nature, he has always maintained a close contact with experiment, and indeed, he has designed numerous critical experiments.

This volume represents a judicious sampling of Mats Hillert's extensive body of work; it is necessarily incomplete, but it is hoped and expected that it will prove useful to students of materials science and engineering at all levels, and that it will inspire the further study and appreciation of his many contributions.

The editors are very grateful to Arcelor Research for their financial support and to EDPSciences who have lavished much care in the production of this volume.

John ÅGREN
Yves BRECHET
Christopher HUTCHINSON
Jean PHILIBERT
Gary PURDY

List of contributors

Hubert I. Aaronson[1],
Department of Materials Science and Engineering
Carnegie Mellon University, USA

John Ågren
Department of Materials Science
KTH (Royal Institute of Technology), Sweden

Michael J. Aziz
Division of Engineering and Applied Sciences,
Harvard University, USA

Yves Bréchet
Laboratoire de Thermodynamique et Physico-Chimie Métallurgiques,
Institut National Polytechnique de Grenoble, France

John W. Cahn
Materials Science and Engineering Laboratory
NIST, USA

Masato Enomoto
Department of Materials Science,
Ibaraki University, Japan

Paul C. Fife
Department of Mathematics,
University of Utah, USA

Hasse Fredriksson
Professor in Casting of Metals
KTH (Royal Institute of Technology), Sweden

Ola Hunderi
NTNU
Trondheim, Norway

Christopher R. Hutchinson
Department of Materials Engineering,
Monash University, Australia

Gerhard Inden
Max-Planck-Institut für Eisenforschung
Düsseldorf, Germany

Larry Kaufman
Chair, CALPHAD Advisory Board

Jack S. Kirkaldy
Brockhouse Institute for Materials Research,
McMaster University, Canada

---

[1] Hubert Aaronson passed away on December 13, 2005.

Dmitri V. Malakhov
Department of Materials Science and Engineering
McMaster University, Canada

Thaddeus B. Massalski
Department of Materials Science and Engineering
Carnegie Mellon University, USA

John E. Morral
Department of Materials Science and Engineering
Ohio State University, USA

Jean Philibert
Université Paris-Sud
Orsay, France

Gary Purdy
Department of Materials Science and Engineering
McMaster University, Canada

Nils Ryum
NTNU
Trondheim, Norway

Peter W. Voorhees
Department of Materials Science and Engineering
Northwestern University, USA

Pavel Zieba
Institute of Metallurgy and Materials Science
Polish Academy of Sciences, Poland

Contents

A short biography of Mats Hillert iii

Foreword v

List of contributors vii

Chapter 1
– Introduction to "Nuclear Reaction Radiography" 1
– Original paper 5

Chapter 2
– Introduction to "Paraequilibrium" 7
– Original paper 9

Chapter 3
– Introduction to "Pressure-Induced Diffusion and Deformation During Precipitation, Especially Graphitization" 25
– Original paper 27

Chapter 4
– Introduction to "The Role of Interfacial Energy during Solid State Phase Transformations" 51
– Original paper 53

Chapter 5
– Introduction to "Thermodynamics of Martensitic Transformations" 87
– Original paper 89

Chapter 6

- Introduction to "The Growth of Ferrite, Bainite and Martensite" 111
- Original paper 113

Chapter 7

- Introduction to "A Solid Solution Model for Inhomogeneous Systems" 159
- Original paper 165

Chapter 8

- Introduction to "The Formation of Pearlite" 177
- Original paper 179

Chapter 9

- Introduction to "On the Theory of Normal and Abnormal Grain Growth" 231
- Original paper 233

Chapter 10

- Introduction to "Grey and White Solidification of Cast Iron" 245
- Original paper 247

Chapter 11

- Introduction to "The Role of Interfaces in Phase Transformations" and "Diffusion and Interface Control of Reactions in Alloys" 257
- Original paper 261

Chapter 12

- Introduction to "The Regular Solution Model for Stoichiometric Phases and Ionic Melts" 295
- Original paper 299

Chapter 13

- Introduction to "Diffusion Controlled Growth of Lamellar Eutectics and Eutectoids in Binary and Ternary Systems" 309
- Original paper 311

Chapter 14

- Introduction to "On the Theories of Growth During Discontinuous Precipitation" 321
- Original paper 323

Chapter 15

- Introduction to "The Effect of Alloying Elements on the Rate of Ostwald Ripening of Cementite in Steel" 337
- Original paper 339

Chapter 16

- Introduction to "The Uses of Gibbs Free Energy-Composition Diagrams" 347
- Original paper 351

Chapter 17

- Introduction to "A Treatment of the Solute Drag on Moving Grain Boundaries and Phase Interfaces in Binary Alloys" 401
- Original paper 403

Chapter 18

- Introduction to "Chemically Induced Grain Boundary Migration" 417
- Original paper 419

Chapter 19

- Introduction to "An Analysis of the Effect of Alloying Elements on the Pearlite Reaction" 427
- Original paper 429

Chapter 20

– Introduction to "Thermodynamics of the Massive Transformations" and "Massive Transformations in the Fe-Ni system" 447

– Original paper 449

Chapter 21

– Introduction to "On the Nature of the Bainite Transformation in Steel" 471

– Original paper 473

Chapter 22

– Introduction to "Solute Drag, Solute Trapping and Diffusional Dissipation of Gibbs Energy" 479

– Original paper 481

Credits 507

1 Introduction to "Nuclear Reaction Radiography"

published in *Nature* (1951)

Originality is a hallmark of Mats Hillert's work. He was often the first to recognize a problem or formulate a solution to an existing problem. There is no better example of his being first than his short paper in Nature on Nuclear Reaction Radiography, in which he proposed a variation of autoradiography to determine the distribution of boron in steels. Instead of using the decay products of radioisotopes to mark a film, he demonstrated that alpha particles resulting from the transmutation of boron into lithium could be used to mark film and indicate the location of boron atoms.

The article was published in 1951, just four years after he graduated from Chalmers University with a B.S. in Chemical Engineering. Five years later he obtained a Ph.D. from M.I.T. His approach was referenced in the 1950's in the biological [1] and chemical [2] literature from the 1950's, but wasn't referenced in metallurgical literature until Barbara Thompson's article [3] appeared in 1960. Unfortunately the reference to his paper was omitted from much later work which focused on analyzing nuclear reaction autoradiography data and identifying both the location of boron in steel [4–6] and the mechanism of boron hardenability [7]. Instead the articles referred to Thompson's paper as the earliest important reference. However, this oversight doesn't change the fact that his article planted the seed from which the later work grew and flourished.

John E. Morral

References

[1] W.M. Dugger, Jr., Autoradiography with plant tissue, *The Botanical Review* 23 (1957) 351.

[2] U. Schindewolf, Chemical analysis with neutron reactions, *Angew. Chem.* 70 (1958) 181.

[3] B.A. Thompson, Determining boron distribution in metals by neutron activation, *Trans. Met. Soc. AIME* 218 (1960) 228.

[4] J.D. Garnish and J.D.H. Hughes, Quantitative analysis of boron in solids by autoradiography, *J. Material Science* 7 (1972) 7.

[5] W.F. Jandeska and J.E. Morral, The distribution of boron in austenite, *Metall. Trans. A* 3 (1972) 2933.

[6] T.B. Cameron and J.E. Morral, A model describing quantitative neutron autoradiography for boron in iron, in *Neutron Radiography*, ed. by J.P. Barton and P. von der Hardt (D. Reidel Pub. Co., 1983, Dordrecht, Holland) p. 807.

[7] J.E. Morral and T.B. Cameron, Boron hardenability mechanisms, in *Boron in Steel*, ed. by S.K. Banerji and J.E. Morral (TMS, 1980, Warrendale, PA.) p. 19.

Mats Hillert in 1951.

Transcribed from Nature, vol. 168, (1951), pp 39,40

Nuclear Reaction Radiography

AUTORADIOGRAPHY has been employed for many years for the purpose of determining the distribution of an element in a given material. The essential requirement for the use of this method is that the element in question itself should emit a radiation which can be registered on a photographic film pressed against the surface of the sample. To make an element betray its presence in thins way, one provides a certain fraction of marked atoms by adding a small amount of radioactive isotopes of the same element.

I have investigated another method, which may be termed nuclear reaction radiography, since it makes use of the property of the non-radioactive element in question of reacting with an external particle radiation in such a way that a new radiation is born in the process, which is then registered on the photographic film. To use this method, the photographic film must be insensitive to the primary particle radiation used for exciting the reaction. This requirement would seem to restrict the species of available primary radiations at present to one kind, namely, neutrons. Another requirement is that the element to be detected should have a much higher probability of reacting with the primary radiation than other elements present in the material under investigation. This can be achieved in some cases by the choice of neutrons with suitable energy, making use of the fact that different reactions have different resonance energies. Finally, the radiation produced in the reaction must be of such a nature that it can be properly registered on the film.

Boron is one of those elements where the ordinary autoradiographic method fails, since no suitable radioisotope is available. Instead, conditions would seem to be favourable for the use of nuclear reaction radiography in this case. The reaction which suggests itself is:

$$_5^{10}B + _0^1n \rightarrow \; _3^7Li + _2^4\alpha$$

For thermal energy neutrons, the cross-section has the very high value of 3,000 barns. The α-particle is emitted with an energy of 1.5 MeV., which is a very favourable value for photographic trace production. The range of these particles is about 5 μ in the emulsion, and in iron it is considerably less than this. Hence, the method should be capable of yielding very high resolution. I have investigated this method, using Kodak Autoradiographic plates. These photographic plates give very good resolution because (i), the emulsion is very fine grained, (ii) the emulsion can be stripped directly from the glass plate and mounted directly on the sample under investigation. A further advantage with this material is that the sensitivity to γ-radiation is very small, whereas the sensitivity to α-radiation is high.

The method has been tested on an experimental alloy made up of 98 per cent iron and 1 per cent boron. A small sample of this was ground, polished and etched, and given a thin protective coating of collodion in order to keep the surface free from tarnish in the subsequent work (otherwise it will rust very readily). The collodion film was less than 1μ thick. The freshly stripped, wet emulsion was placed on top of this and was allowed to dry in position, which made it adhere firmly to the sample. The sample was then placed in a light-tight container, where it was irradiated for 16 min. with neutrons from a cyclotron. A block of lead, inserted between the sample and the cyclotron, served to reduce as much as possible the amount of γ-radiation reaching the emulsion. The sample was surrounded by paraffin wax in order to obtain a favourable proportion of thermal neutrons. After irradiation, the emulsion was developed and compared with the metal surface in an ordinary microscope. The results are illustrated by the accompanying reproductions. It is evident that the emulsion stripped form the sample reproduces the distribution of boron quite accurately, and that it is possible to work at quite considerable magnification.

I am grateful to the Nobel Institute for Physics in Stockholm, where Mr. S. Thulin kindly carried out the cyclotron irradiation, and to Mr. L. Erwall of the Division of Physical Chemistry at the Royal Institute of Technology, Stockholm, who placed the Kodak Autoradiographic plate material at my disposal; this emulsion undoubtedly played an important part in the success of the experiment.

MATS HILLERT
Swedish Institute for Metal Research,
   Stockholm.
   April 12.

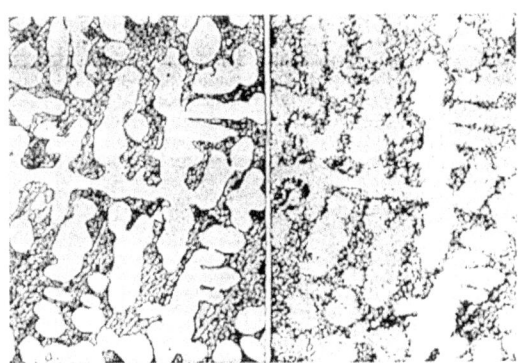

(a)   Photomicrograph of the metal surface. Light areas: iron; darker parts: eutectic of iron and Fe_2B. x 100
(b)   Reaction radiograph of the same surface as that shown in (a). Dark spots are produced by α-particles, showing position of reacting atoms of boron. x 100

2 Introduction to "Paraequilibrium"

published as an *Internal Report*, Swedish Institute for Metals Research (1953)

Paraequilibrium is a terminology introduced by Hultgren [1] to refer to the products of austenite decomposition that have the same concentration of alloy element as the parent austenite. As was indicated by Zener [2], if the diffusion of alloy element occurred to the full extent, the transformation would cause a larger reduction in Gibbs free energy, but the reaction can proceed as long as the reduction in free energy is sufficiently large even before the diffusion of alloy element occurs. Indeed, transformations occur in this way in alloy steels. In both this paper and a paper published the previous year, entitled "The use of isoactivity lines in ternary phase diagrams" [3], Hillert examined in detail the thermodynamic conditions for the formation of paraphases, *i.e.* the transformation products in which the alloy element concentration is inherited from the parent austenite.

In paraequilibrium, equilibrium is established only with respect to a component that has a greater diffusivity and thus is rendered mobile (*e.g.* carbon) compared to other component species (*e.g.* Fe and M, where M denotes alloy element). It is distinguished from the ordinary equilibrium in which equilibrium is established with all components and which Hultgren called orthoequilibrium [1]. In this paper, utilizing the understanding that a carbon isoactivity line cannot intersect a carbon component ray (line of constant ratio of Fe to M concentrations) more than once [4], Hillert was able to show that the paraequilibrium ($\alpha + \gamma$) two-phase field lies within that of orthoequilibrium. Next, he discussed the growth of a paraphase (*e.g.* paracementite) when the alloying element was not totally immobile, that is, when local equilibrium of substitutional element was maintained. Such chemical conditions were called false paraequilibrium or quasi-paraquilibrium conditions [5]. It was then proposed that, in steady-state, true and false paraequilibrium obtain when $D_M/v < 0.4d$ and $> 10d$, respectively, where d is the distance between the abutting atom planes of parent and growing phases, v is the growth rate, and D_M is the diffusivity of alloy element in the immediate vicinity of the boundary in austenite. Furthermore, he discussed the region in the two-phase field in which false paracondition is prevalent, *i.e.* the diffusion of alloying element is no longer necessary for the transformation to proceed. Finally, the effect of alloying elements on the growth of pearlite was discussed using isoactivity lines.

This paper includes essentially all the ingredients now used in the discussion of austenite decomposition reactions. His thoughts have been developed by a number of researchers and by himself as well. The thermodynamic condition of paraequilibrium was given in a more explicit form by Hillert [6], and Gilmour *et al.* [7]. The upper limit of false paracondition was called the envelope of zero partition by Purdy *et al.* [8]. Coates [9] called the growth mode controlled by alloy element diffusion partition local equilibrium (PLE) mode, and the one controlled by carbon diffusion Negligible partition local equilibrium (NPLE) mode.

More than half a century has elapsed since this paper was published. One of the major issues in physical metallurgy or phase transformation in steels is what chemical conditions, *e.g.* true or false paraequilibrium, prevail, or if they evolve, how the transition occurs [10–17]. This paper thus threw a much needed and powerful light on the question, not only for physical metallurgists, but also for steel engineers. Recently, it was proposed that the use of the term orthoequilibrium should be confined, and full equilibrium and full local equilibrium be used instead when discussing the state of equilibrium in the bulk and the chemical condition at a moving boundary, respectively [18].

Since this paper was not published in an archive journal, researchers did not have ready access to it[1]. With this book, many people can acquaint themselves with the original thoughts that brought forth the most important principles governing austenite decomposition in alloy steels.

Masato ENOMOTO

---

[1] It remained as an internal report, typewritten and reproduced with the usual tools of that time, with hand written corrections from the author. As such, with its poor appearance according to modern standards, it can be considered as an *historical monument* or a piece of *archives*, delivered without "restoration".

References

[1] A. Hultgren, *Trans. ASM* 39 (1947) 915.

[2] C. Zener, *Trans. AIME* 167 (1946) 550.

[3] M. Hillert, *Jernkont Ann.* 136 (1952) 25.

[4] M. Hillert, *Acta Metall.* 3 (1955) 34.

[5] M. Hillert, Phase Equilibria, Phase Diagrams and Phase Transformations – Their Thermodynamic Basis (Cambridge University Press, Cambridge, 1998).

[6] M. Hillert, in *Phase Transformations*, ed. H.I. Aaronson (1970, ASM, Metals Park, OH) p. 187.

[7] J.B. Gilmour, G.R. Purdy and J.S. Kirkaldy, *Metall. Trans.* 3 (1972) 1455.

[8] G.R. Purdy, D.H. Weichert and J.S. Kirkaldy, *Trans. TMS-AIME* 230 (1964) 1025.

[9] D.E. Coates, *Metall.Trans.* 3 (1972) 1203; 4 (1973) 1077.

[10] H.I. Aaronson and A.D. Domian, *Trans.TMS-AIME* 236 (1966) 781.

[11] J.R. Bradley and H.I. Aaronson, *Metall. Trans. A* 12 (1981) 1729.

[12] M. Hillert, in *Proc. Int. Conf. on Solid→Solid Phase Transformations*, ed. by H.I. Aaronson, R.F. Sekerka, D.E. Laughlin and C.M. Wayman (AIME, Warrendale, PA, 1982) p. 789.

[13] M. Enomoto and H.I. Aaronson, *Metall. Trans. A* 18 (1987) 1547.

[14] K. Oi, C. Lux and G.R. Purdy, *Acta Mater.* 48 (2000) 2147.

[15] M. Hillert, *Scripta Mater.* 46 (2002) 447.

[16] C.R. Hutchinson, A. Fuchsman and Y. Brechet, *Metall. Trans. A* 35 (2004) 1211.

[17] A. Phillion, H.S. Zurob, C.R. Hutchinson, H. Guo, D.V. Malakhov, J. Nakano and G.R. Purdy, *Metall. Trans. A* 35 (2004) 1237.

[18] M. Hillert and J. Agren, *Scripta Mater.* 50 (2004) 697.

Paraequilibrium

by Mats Hillert.

I. Definition of the concept.

Metastability is a concept that has been used for a very long time. The prefix meta means a restriction. Thus a metastable phase is not quite stable. It is stable against small internal fluctuations in composition--- it has internal stability. However, it can be transformed into two quite new phases, which then have lower Gibbs free energy together than the metastable phase. (This paper is limited to reactions and equilibrium states under constant pressure and temperature. All spontaneous reactions must then involve decrease in the Gibbs free energy.)

The concept metastability can be used in a more precise manner. Thus an equilibrium between two phases can be characterized as metastable with respect to a certain third phase. Then formation of this phase is attended by a decrease of the Gibbs free energy. One of the two phases in the previous equilibrium, the metastable one, will at last disappear.

Quite generally, metastability implies full stability until the nucleation difficulties for a new phase are overcome.

Hultgren[1] has introduced a new prefix, "para", that implies an even greater restriction. Two phases, which are in paraequilibrium with each other, will maintain their compositions only on account of the low mobility of some components. Therefore, after a sufficiently long time the two phases have adjusted their compositions so that an equilibrium state has been attained. Hultgren calls this an orthoequilibrium when it is a really stable state as well as when it is a metastable equilibrium.

The reason, why paraequilibrium is called equilibrium at all, is that the two phases are in real equilibrium with respect to a mobile component. Thus the paraphases have the same chemical activity for this component. Consequently, paraequilibrium can appear in systems with at least three components, one of them mobile. In Hultgren's investigations the mobile component has been carbon and the others iron and various alloying elements. At isothermal decomposition of austenite below the A_1 temperature cementite and ferrite are formed. If the temperature is low enough this splitting into two phases takes place solely by diffusion of carbon. The ratio alloying element/iron is thus the same in the paracementite and the paraferrite as in the original austenite. In the isothermal section of the ternary diagrams all these three phases lie on the straight line, that goes from the C corner. Such a line is called a component ray [2].

-2-

II. Phase Boundaries of the Para-State

It is well known that the transition from a metastable state (e.g. cementite + austenite) to a stable state (e.g. graphite + austenite) is accompanied by an increase of the two-phase region. The stable phase boundary (austenite/austenite + graphite) lies inside the metastable one-phase region (austenite). By comparison with the metastable case Hultgren[3] has drawn the conclusion that the transition from para- to ortho-state is accompanied by an increase of the two-phase region. That this really is so will be shown here by means of isoactivity lines.

Isoactivity lines have been used very little. However, their properties have recently been investigated.[4] By definition an isoactivity line is the collection of all points in an isothermal section of a ternary diagram, which have the same activity for a certain component. The broken line NLMO in Fig. 1 is an isoactivity line for component C. It has been proved that an isoactivity line for a certain component can not intersect the component ray to this component more than once. This rule will now be utilized.

Suppose that a suitable alloy consists of one homogeneous phase at a high temperature. The composition is such (e.g. K in Fig. 1) that the sample separates into two phases at a lower temperature. If the conditions are fulfilled for paraequilibrium, this separation involves the diffusion of one component only, here called C, the mobile component. The composition of the phases obtained must all the time lie on the component ray e-C, that goes through K. Thus the stable state represented by L and M (the ortho-state) cannot be realised. The para-state is obtained when diffusion of C has been completed and cannot consist of S and T but must therefore consist of two phases on the component ray e-C, which have equal C-activity. Thus they must lie on the same isoactivity line for C. But according to reference [4] it is not possible for an isoactivity line to intersect the component ray more than once. However, this rule is applicable only to an equilibrium diagram. The explanation must be that the two para-phase compositions lie in the two-phase region of the ordinary equilibrium diagram (e.g. a and b in Fig. 1). That these two points have the same C-activity is plain from the fact that they lie on extrapolations of one and the same isoactivity line. The extrapolations are drawn into the two-phase region from the one-phase parts of the isoactivity line.

It is obvious that the phase boundaries of the para-state do not coincide with those of the ortho-state but lie inside. Thus Hultgren's conclusion has been confirmed. In Fig. 1 the dotted lines represent the para-boundaries.

Fig. 2 shows an isothermal section through a ternary diagram in which three phases appear. The diagram is essentially in agreement with that of Fe-C-Mn. The dotted lines show the phase boundaries for the paracondition "only C mobile". The three-phase triangle of the ortho-state is thus reduced to a line. The vertical section e - C through the ternary diagram appears at the bottom of Fig. 2. Here too the dotted lines represent the "para-diagram". As will be seen, this entirely resembles a purely binary diagram. As we have seen earlier, the para-condition occasions an increase of the one-phase regions into the two-phase region. It is interesting to note that the para-condition can also cause the growth of a two-phase region into another two-phase region.

III. Requirements for Ortho- and Para-states.

The pre-requisite for the para-state is that only one component is mobile. No absolute immobility exists in nature. The para-state is therefore a borderline case which can never be attained exactly. In this Chapter an examination will be made of the conditions under which a deviation from the ortho-state can be obtained and therewith an approach to the para-state, and under which a practical attainment of the para-state may be expected.

At ortho-equilibrium between two adjoining phases there is no sudden change in any of the activities of the components at the phase boundary. The appearance of a sudden activity change implies a deviation from ortho-state and approach to para-state. The question is thus under what conditions noticeably sudden changes in activity can arise.

We consider a ternary system, Fe-C-M, in which an alloying element, carbon, diffuses very easily. Our sample is in a supersaturated state, austenite super-saturated with carbon (γ_0 in Fig. 3), and a new phase, cementite, is therefore deposited.

Assume in the first place that the alloying element M diffuses fairly easily, and that adjoining phases are always in ortho-equilibrium. If the alloying element M prefers cementite to austenite, as is apparent in Fig. 3, a concentration of M takes place in the separating cementite. The latter may then obtain, for example, the composition c_1 in Fig. 3 at a certain stage in the separation. The adjoining austenite, with which c_1 is in ortho-equilibrium, has the composition γ_1. If M had been somewhat less mobile, the composition of the cementite at the same stage of separation would have lain nearer to the component ray e - C, e.g. c_2, and in equilibrium with γ_2. As the degree of mobility of M decreases, the composition of the cementite would approach nearer and nearer to the limit value c_3 on the component ray e - C. The composition of the adjoining austenite approaches closer and closer to the limit value γ_3. The distance bet-

-4-

ween the composition of the cementite in question and the limit value c_3 is a measure of the diffusion of M away from the growing cementite, i.e. over fairly great distances in the specimen. The separation takes place as rapidly, however, as the very mobile carbon is able to diffuse over such distances. With even moderately low mobility of M, therefore, the limit value c_3 can be quite well attained. But it can never be attained exactly. The different stages of separation at such a low mobility of M can be discribed by the following illustration. The first formation of cementite occurs from the austenite with M-content as indicated by the component ray e - C in Fig. 3. Since the cementite, according to the equilibrium diagram 3, has a higher M-content than the austenite with which it is in equilibrium, this first formation of cementite obtains an elevated M-content and is surrounded by an austenitic layer with reduced M-content. The following formation of cementite thus occurs from an austenite of lower M-content, and will therefore be poorer in M than the cementite first formed. It is therefore able to render its neighbouring layer of austenite still poorer in M-atoms than did the cementite first formed, and so on. The forcing down of the M-content in the adjoining layer of austenite gives rise to diffusion of M-atoms from more distant austenite. This diffusion, which counteracts the reduction in M-content, increases as the M-content is further reduced. The successive reduction in M-content both in the separating cementite and in the adjoining austenite therefore takes place more and more slowly. Finally the separation approaches in some degree a state of continuity. This is attained when the M-content of the cementite has fallen so far that the ratio between the M- and Fe-contents is the same in the cementite as in the original austenite. This implies a somewhat lower absolute content of M in the cementite than in the austenite, for the carbon content is higher. Henceforth we shall indicate the M-content without consideration of the carbon content, and by $[M]$ therefore, we mean the ratio between the M-content and the sum of the M- and Fe-contents. Fig. 4 illustrates the state of continuity. The M-atoms which are here "lacking" in the austenite near the front thus correspond to a concentration of M-atoms in the early formed cementite.

We consider the simplified case that the reaction face is plane and that this face moves at a constant speed v. Assume that Fick's law is valid for diffusion of M in the austenite:

$$\frac{\partial [M]}{\partial t} = D_M \cdot \frac{\partial^2 [M]}{\partial x^2} \qquad (1)$$

-5-

The distribution of M in the austenite, when the state of continuity has been reached, is then

$$[M] = [M]_0 - ([M]_0 - [M]_3) e^{-\frac{v}{D_M}(x-vt)} \qquad (2)$$

$[M]_0$ is the original M-content, i.e. that in γ_0. $[M]_3$ is the M-content at the growing face, i.e. that in γ_3. Equation 2 has been obtained under simplified assumptions. Yet it permits certain conclusions of general applicability to be drawn. It is seen from equation 2 that the lower the mobility of the alloying element M in relation to the speed of growth v of the cementite, the greater is the contraction of the layer of austenite which has altered M-content in front of the face. Finally the change in the M-content in the neighbourhood of the face becomes so rapid that there must be an appreciable change in M-content and M-activity from atomic plane to atomic plane. The greatest discontinuity must lie at the interface, i.e. between the last atomic plane of the cementite and first of the austenite. Ortho-equilibrium can then no longer be realised and an approach to para-state must have taken place. If we imagine a gradual decrease in the mobility of M, the composition of the cementite now changes from a point very near c_3 towards c_4 which represents the para-state. The composition of the adjoining austenite is at the same time displaced from a point very near γ_3 towards the paracomposition γ_4. In Fig. 3 the whole displacement of the compositions has been marked with thick lines. c_3 lies, as does c_4, on the component ray e - C, and therefore, in the same way as c_4, has the same proportion between the M- and Fe-contents as γ_0. To distinguish between c_3 and c_4 we shall in the continuation call c_3 the false para-composition since it is in ortho-equilibrium with its adjoining austenitic layer, γ_3. Utilizing equation 2 we can make a quantitative calculation of the conditions required for distinct deviation from ortho-state and for an approximate approach to para-state. The first condition can, for example, be written $[M] = [M]_0 - ([M]_0 - [M]_3) \cdot 0,9$ for $x = vt + d$, where d is the distance between the last atomic plane of the cementite and the first of the austenite. The second condition can be written $[M] = [M]_0 - ([M]_0 - [M]_3) \cdot 0,1$ for $x = vt + d$. Equation 2 gives

$$e^{-\frac{v}{D_M} \cdot d} = 0,9 \text{ and } 0,1 \text{ respectively.}$$

Condition for ortho: $\dfrac{D_M}{v} > 10\,d$

Condition for para: $\dfrac{D_M}{v} < 0,4\,d$

When using these formulas it should be noted that it is not certain that the

−6−

diffusion constant D_M has the same value immediately in front of the growing interface as in the undisturbed austenite.

IV. Rate of Decomposition of Austenite.

It is known that the majority of alloying elements increase the hardenability of steel by postponing the disintegration of the austenite. It has been thought possible that one reason for this may be that these alloying elements diffuse slowly in comparison with carbon, and that it must therefore take longer for steel to be converted to, for example, pearlite, when not only the very mobile carbon but also a slow alloying element is to be distributed between the cementite and ferrite. It has been pointed out by Zener[5] that the eutectoid reaction would admittedly cause a greater reduction in the Gibbs free energy if the element were allowed to diffuse, but that the reaction has no reason to await this diffusion if it in any case causes a sufficient reduction in the free energy. An investigation will now be made as to whether isoactivity lines can enlighten us on this important problem.

A: Nucleation of cementite.

On separation of cementite from austenite supersaturated with carbon the changes in carbon activity can be quite closely estimated by means of the activity coefficient in austenite. For our purpose it is satisfactory to assume that the carbon activity in austenite is proportional to the carbon content. Let point γ_o in Fig. 5 represent the supersaturated austenite. γ_k and c_k represent the austenitic and cementitic compositions when separation is complete. Then the ratio of the carbon activity for the equilibrium state $\gamma_k + c_k$ (here called λ_C^k) to that for the original state γ_o (here called λ_C^o) is roughly the same as the ratio of the carbon contents in austenite. According to the diagram $\dfrac{\lambda_C^k}{\lambda_C^o} = 0{,}9$

It has recently[6] been pointed out that it is not quite necessary for a nucleus to have the same composition as the precipitate at equilibrium. If the carbon contents in the cementite and the austenite are taken into consideration, $\dfrac{\lambda_C^t}{\lambda_C^o} = 0{,}98$ is found. λ_C^t is the carbon activity in the cementite that has the thermodynamically most favourable nuclear composition. The same relation must also apply to the Cr-activity of the nucleus. In Fig. 5 the isoactivity line for Cr in austenite has been drawn which passes through γ_o. It has been assumed that the Cr-activity depends only on the Cr-content. The C axis represents the Cr-activity zero. A nucleus with Cr-activity 2 per cent lower than γ_o must clearly lie very close to the continuation of the isoactivity line drawn through γ_o, i.e. very close to c_o. Of particular interest is the fact that if the composition of the supersaturated austenite had been for instans γ_p, the thermodynamically most favourable nucleus would not have been cementite at all, but carbide of composition close to it. This is true on condition that equal work of surface tension and of volume is involved in the nucleation of the carbides $(Fe, Cr)_3C$ and $(Cr, Fe)_7C_3$. On this assumption it is even possible to calculate for which sort of carbide the nucleation is thermodynamically most favourable on separation from a given parent austenite. It is found that cementitic nuclei are most favourable only if the composition of the parent austenite lies below an arc such as that in Fig. 5 passing through the two-phase region austenite + cementite. If the composition of the parent austenite lies above this arc, the nucleation of $(Cr, Fe)_7C_3$-carbide is more favourable. This is due to the higher chromium and carbon content of this carbide. It has been considered that a nucleus must be of cementitic structure if it is to function as nucleus for pearlite. If the nuclei first formed are of another sort of carbide, the pearlitic separation may be delayed. We have now seen that this possibility may occur even if the supersaturated austenite lies within the two-phase region austenite + cementite. Previously [5] account has only been taken of the possibility that the alloying element is unevenly distributed due to a faulty structure in the space lattice, so that certain areas of the austenite, such as the grain boundaries, possibly lie within the area of existence of such a carbide.

B. Growth of a Phase.

When a cementitic nucleus has become considerably larger than the critical size, its growth takes place at a great speed. The neighbouring austenite does not have time to fully adjust its composition by conversion of unchanged austenite situated further away. At a very great rate of growth no appreciable diffusion of the alloying element can take place over great distances in the sample. The newly formed cementite then obtains the same alloy content, counted in proportion to iron, as the original austenite. It thus lies in the same component ray as the original austenite. As we have already seen (Chapter III), however, we should still reckon that it may be surrounded by an austenitic layer of changed composition (γ_3 in Fig. 3). In that case it has false para-composition. For the continued growth of cementite a diffusion of carbon is then required from γ_0 to γ_3. The isoactivity line for carbon through γ_3 has been so extrapolated that it intersects the ray e - C. If the composition γ_0 of the original austenite lies at the point of intersection S or to the left of it, carbon cannot diffuse up to the surface of the cementite. The rate of growth of the cementite therefore falls so much that the surrounding austenite has time to adjust its alloy content, i.e. in such a way that the composition of the cementite no longer lies in the component ray. The rate of growth of the cementite will be decided by the rate of diffusion of the alloying element! If the composition of the original austenite lies to the right of S, the alloying element will have a weaker effect. The above argument also holds good for separation of ferrite.

-8-

At temperatures low enough to give real paracementite this arresting action by alloying elements ceases.

C. Growth of Eutectoid.

We have seen above that an alloying element can arrest a growth of a phase. In the growth of a eutectoid the circumstances are different since the diffusion then takes place at right angles to the direction of growth of the eutectoid. Thus in the growth of pearlite or bainite the diffusion of carbon takes place from the austenitic regions, which will become ferrite, to the regions which will become cementite. In Fig. 6, which is an isothermal section at about $600°C$ through the constitution diagram of Fe-C-Ni, α_1 and c_1 indicate the composition of the ferrite and cementite, if the alloying element does not have time to diffuse away, i.e. false paracomposition. The neighbouring austenitic layers have compositions γ_α^e and γ_c^e. It is the difference in carbon activity between γ_α^e and γ_c^e which causes the diffusion of carbon at right angles to the direction of growth. It should be evident from the diagram that this driving activity difference is considerably smaller than for pure carbon steel (γ_α^o, γ_c^o). With a sufficient alloy content the difference in activity may even completely disappear. γ_α and γ_c then lie on the same isoactivity line. The growth of the eutectoid can then not be more rapid than that the alloying element has time to divide between the ferrite and cementite. The rate of growth of the eutectoid will be determined by the rate of diffusion of the alloying element! Fig. 7 shows part of the diagram in the case of a carbide-forming alloying element. Here γ_c^e and γ_α^e are displaced downward and approach γ_α^o and γ_c^o. It is seen, however, from the isoactivity lines that the difference in activity between γ_α^e and γ_c^e is always greater than between γ_α^o and γ_c^o, which holds for unalloyed steel. Thus for these alloying elements we find no obstructive effect on the growth of pearlite, but only on its nucleation. From Figs. 8 and 9, which hold for Mn and Si, we find that the latter elements have the same influence on the growth of the eutectoid as Ni. The carbide-forming elements are thus the only ones for which we cannot find any arresting action on the growth of the eutectoid. On the other hand we have found a possible arresting action on the rate of nucleation of cementite, and thereby of pearlite. The nucleation of bainite should not be influenced by this action. It is considered that the separation of bainite commences with a ferritic nucleus, and the mechanism described at the beginning of this Chapter under A should therefore not be of any importance.

Summary.

The recently introduced concept paraequilibrium is studied by means of isoactivity lines. New view points of the decomposition rate of austenite have been found.

REFERENCES

1. Hultgren, A.   Trans. A. S. M. __39__ (1947) p 915.

2. Rudberg, E.   Jernkontorets Annaler __136__ (1952) p 91.

3. Hultgren, A.   " __135__ (1951) p 403.

4. Hillert, M.   "Isoactivity Lines" Acta Met. __3__ (1955) p. 34

5. Zener, C.   Trans. A. I. M. E. __167__ (1946) p 550.

6. Hillert, M.   "Nuclear composition – a factor of interest in nucleation" Acta Met. __1__ (1953) p. 764

Text to the figures.

Fig. 1   Isothermal section of a simple ternary diagram. The dotted lines are the "para"-phase boundaries.

Fig. 2. Ternary diagram. The dotted lines constitute the "para"-diagram.

Fig. 3. Separation of cementite from austenite γ_o. $c_1 - c_4$ show the composition of cementite for different grades of low mobility of the alloying element M. $\gamma_1 - \gamma_4$ show the composition of the adjoining austenite.

Fig. 4. Concentration of the alloying element M in the neighbourhood of a moving interface between cementite and austenite.

Fig. 5. Principal appearance of the diagram Fe-C-Cr at $900°C$.

Fig. 6. Part of the diagram Fe-C-Ni at $600°C$ with isoactivity lines for carbon in the supercooled austenite.

Fig. 7. Part of the diagram Fe-C-Cr at $600°C$.

Fig. 8. Part of the diagram Fe-C-Mn at $600°C$.

Fig. 9. Part of the diagram Fe-C-Si at $600°C$.

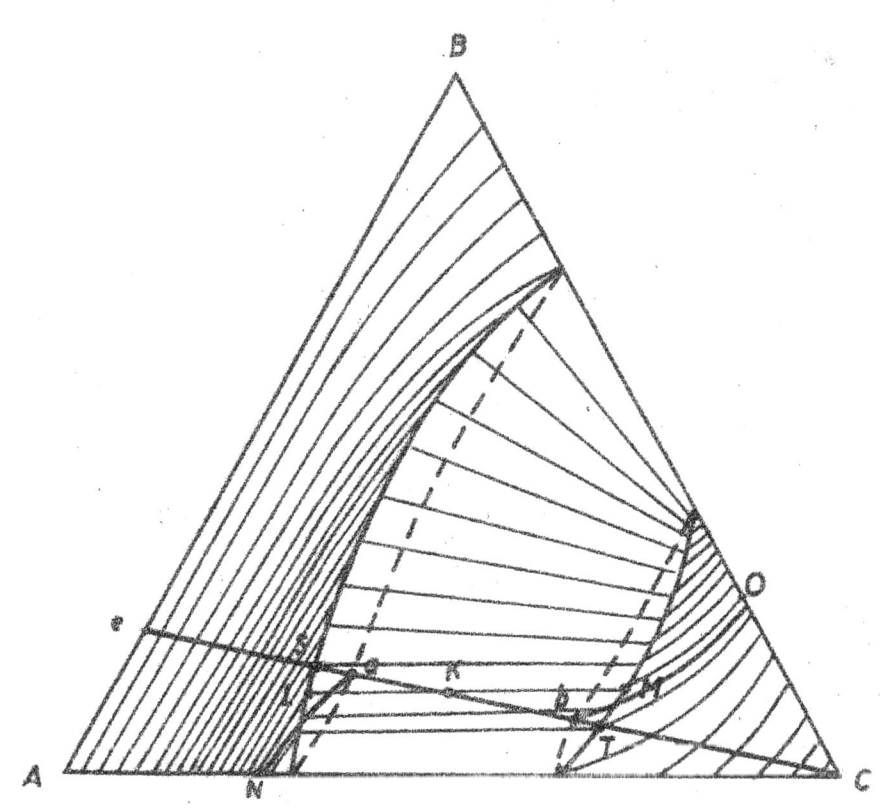

Fig.1

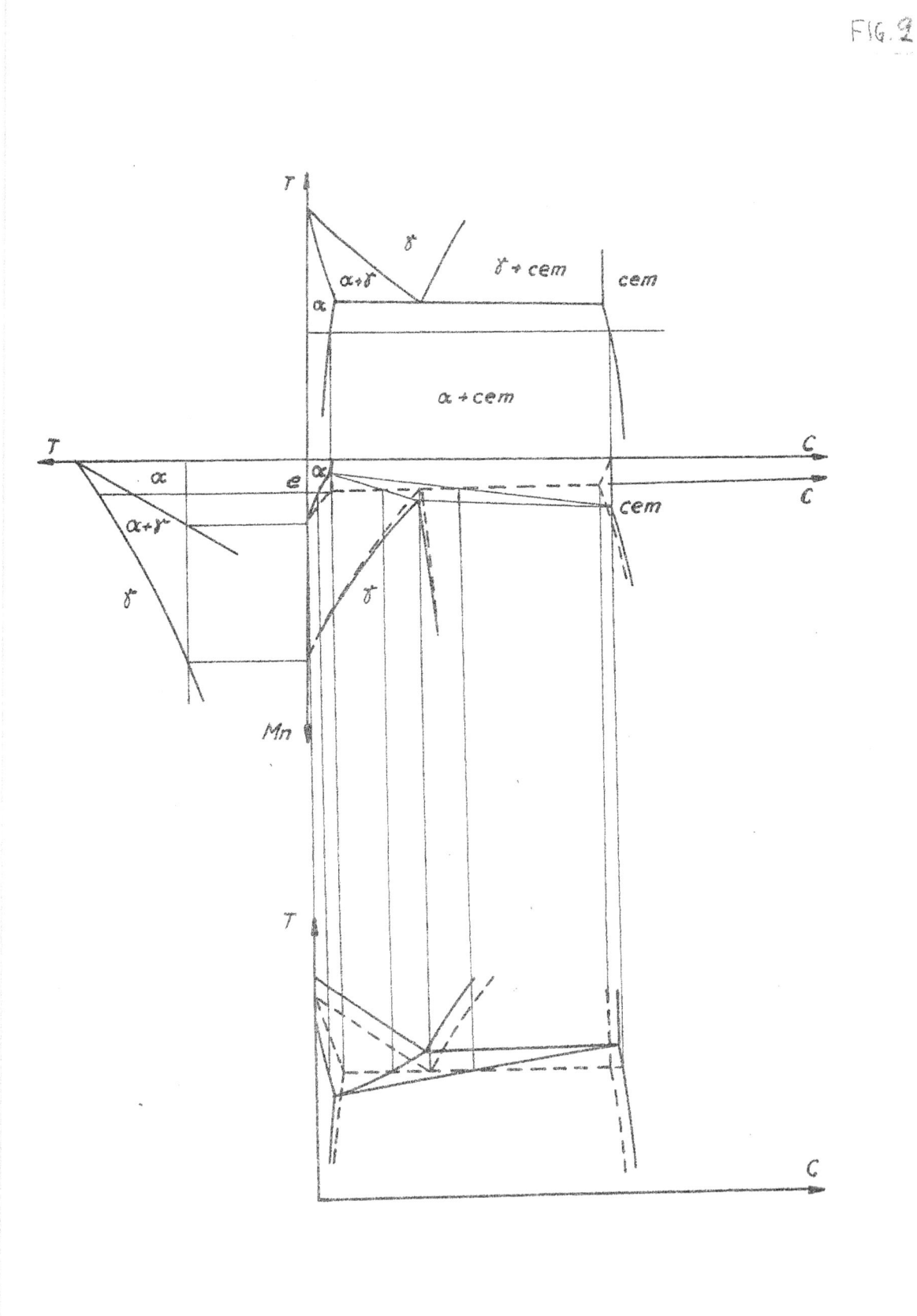

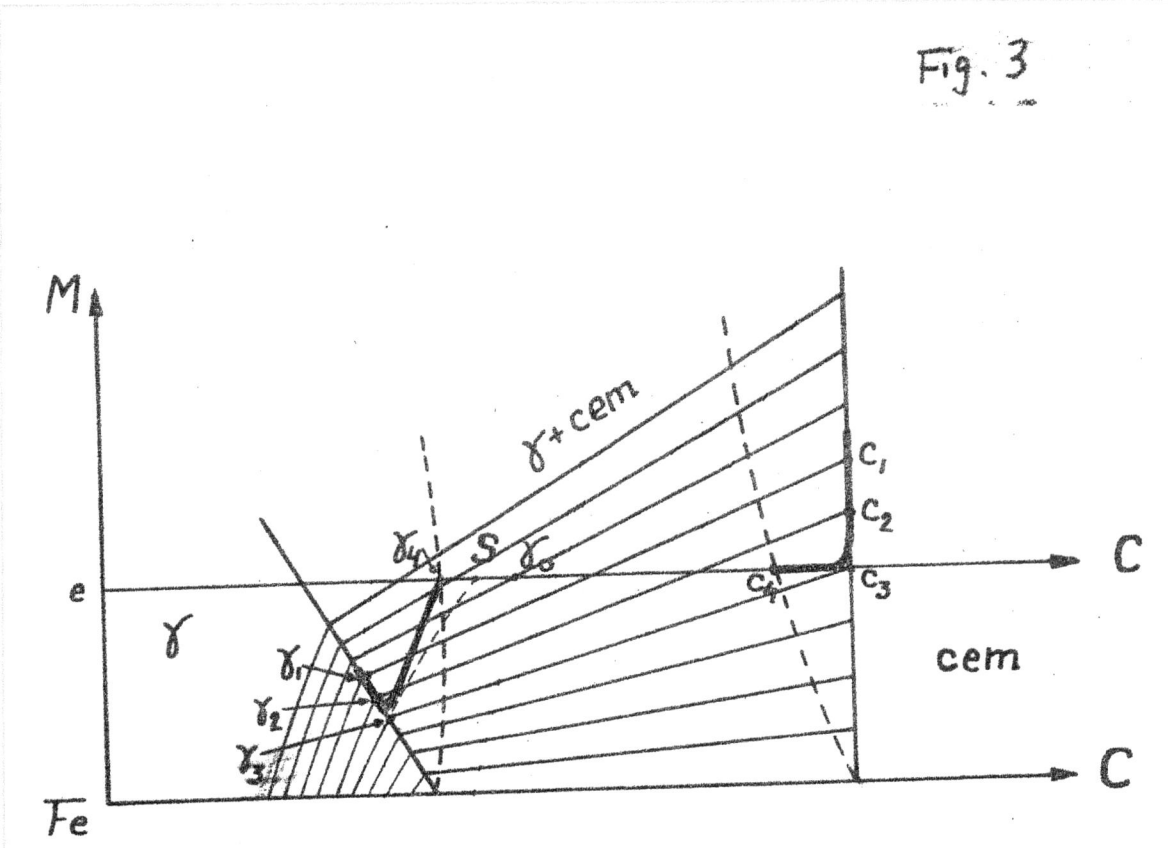

Fig. 3

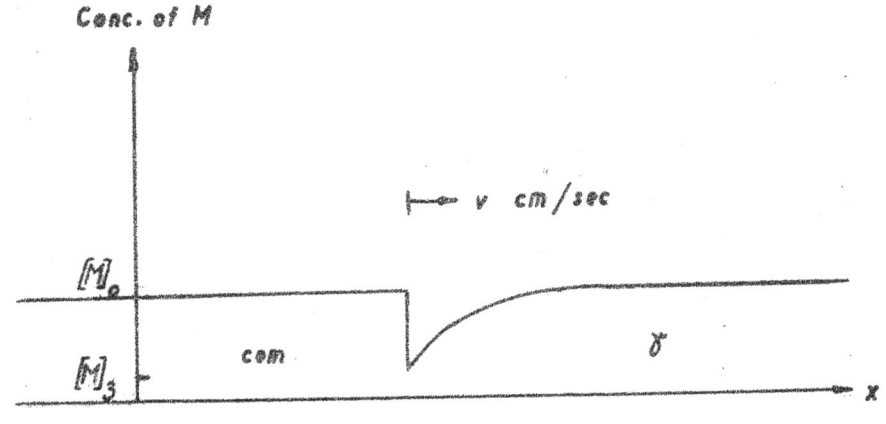

FIG. 4

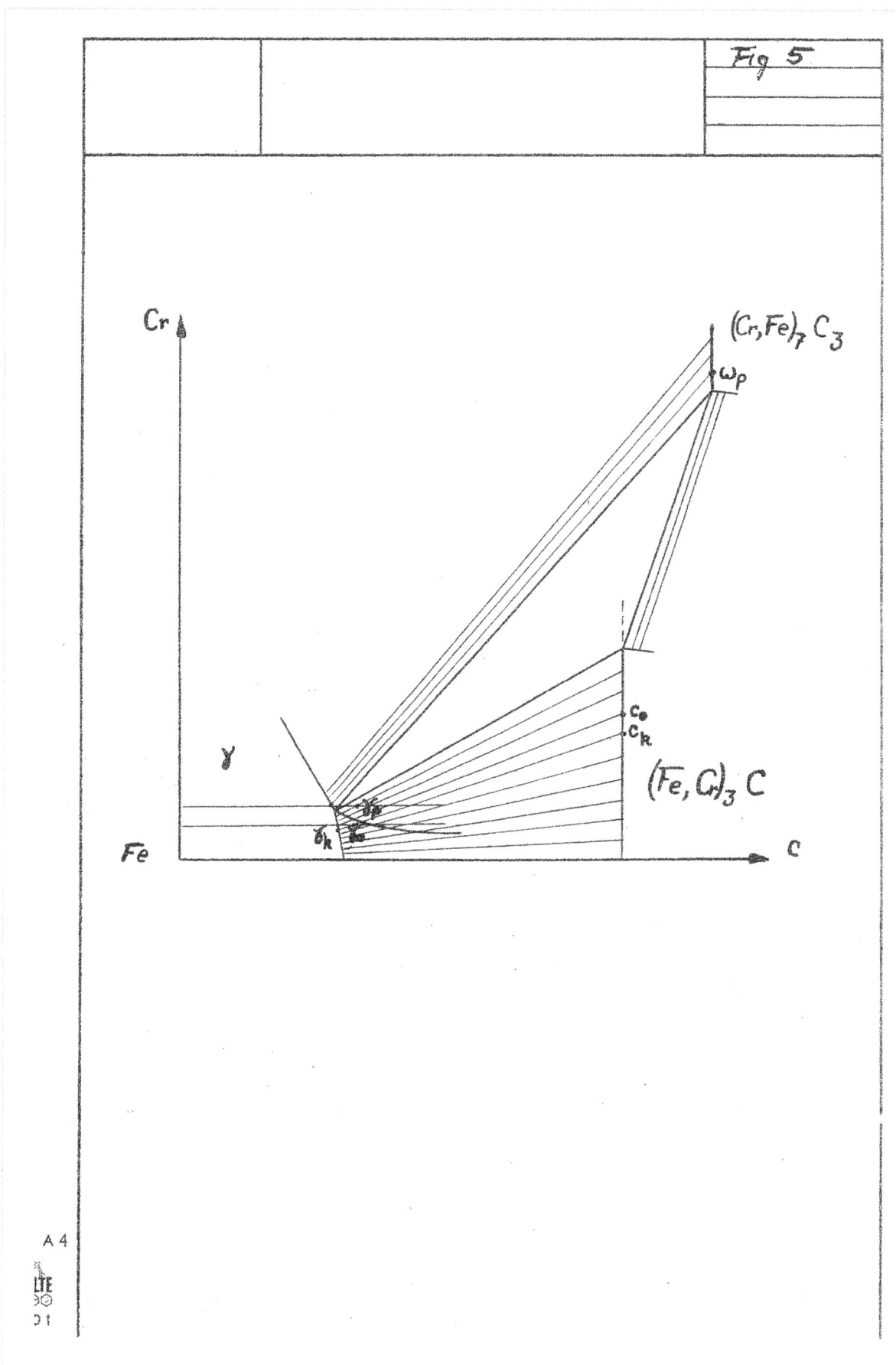

Fig 5

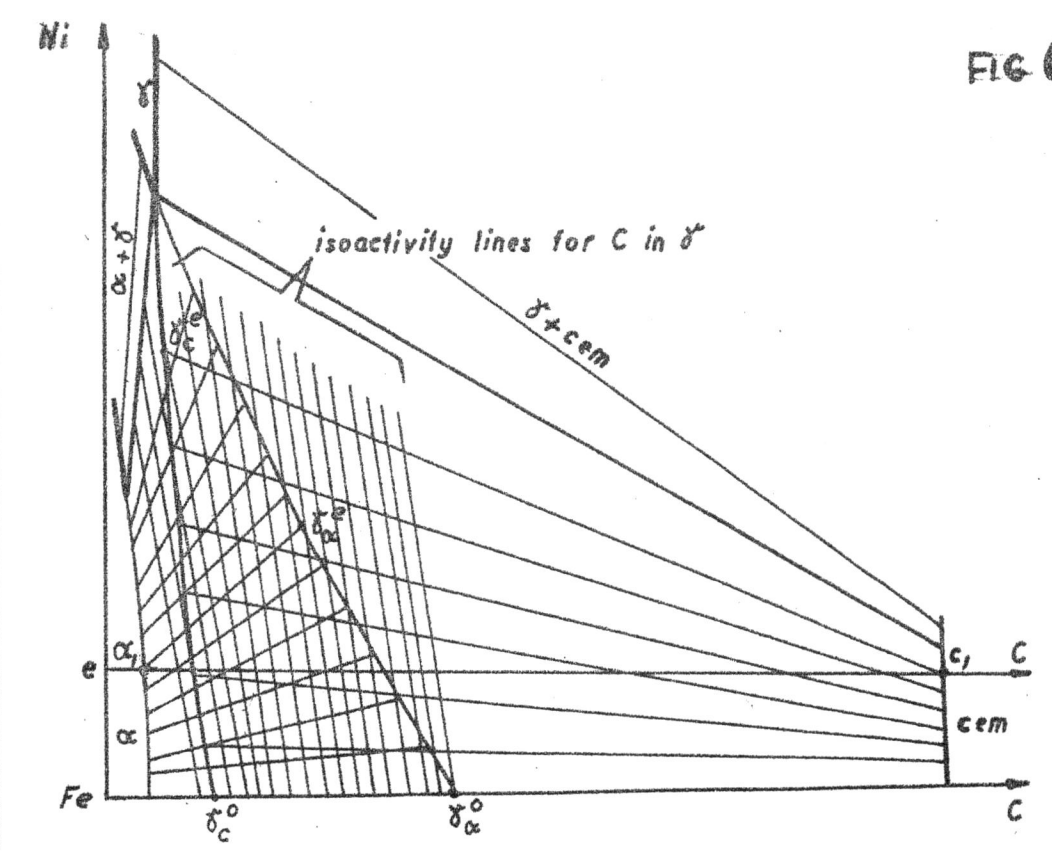

FIG 6

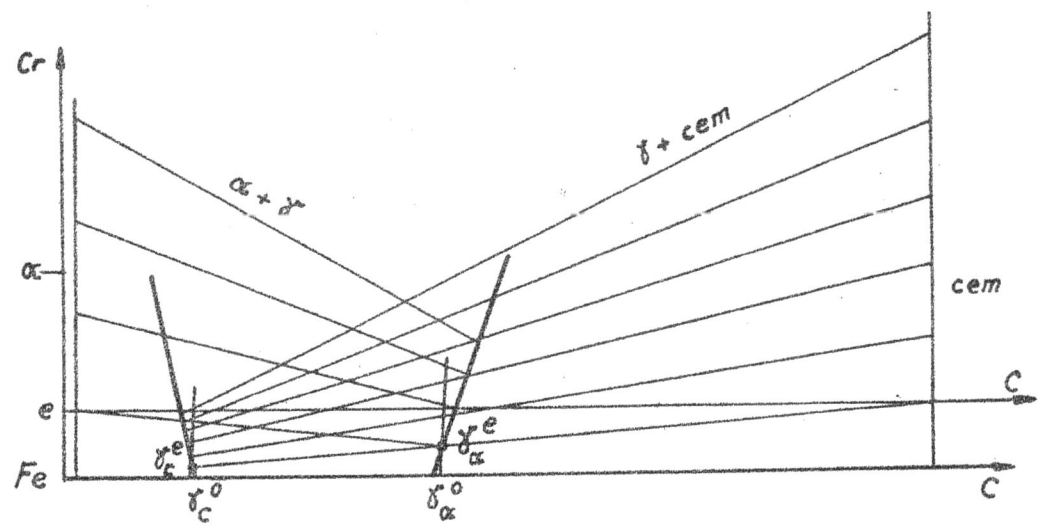

FIG 7

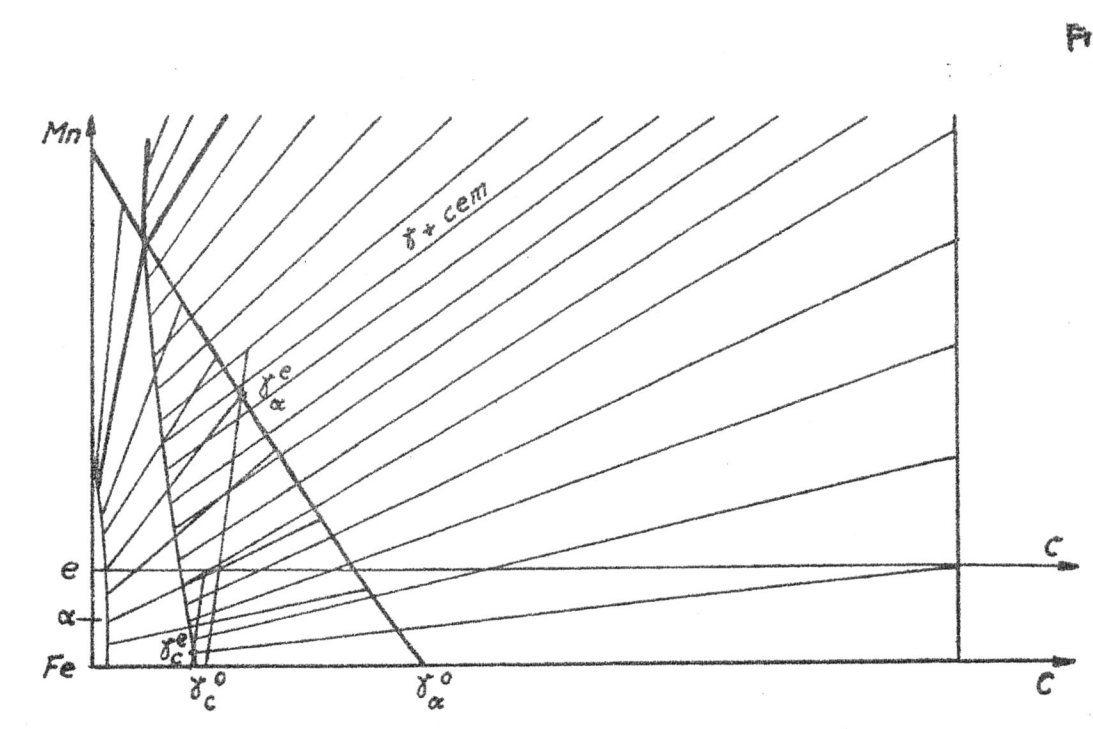

FIG. 8

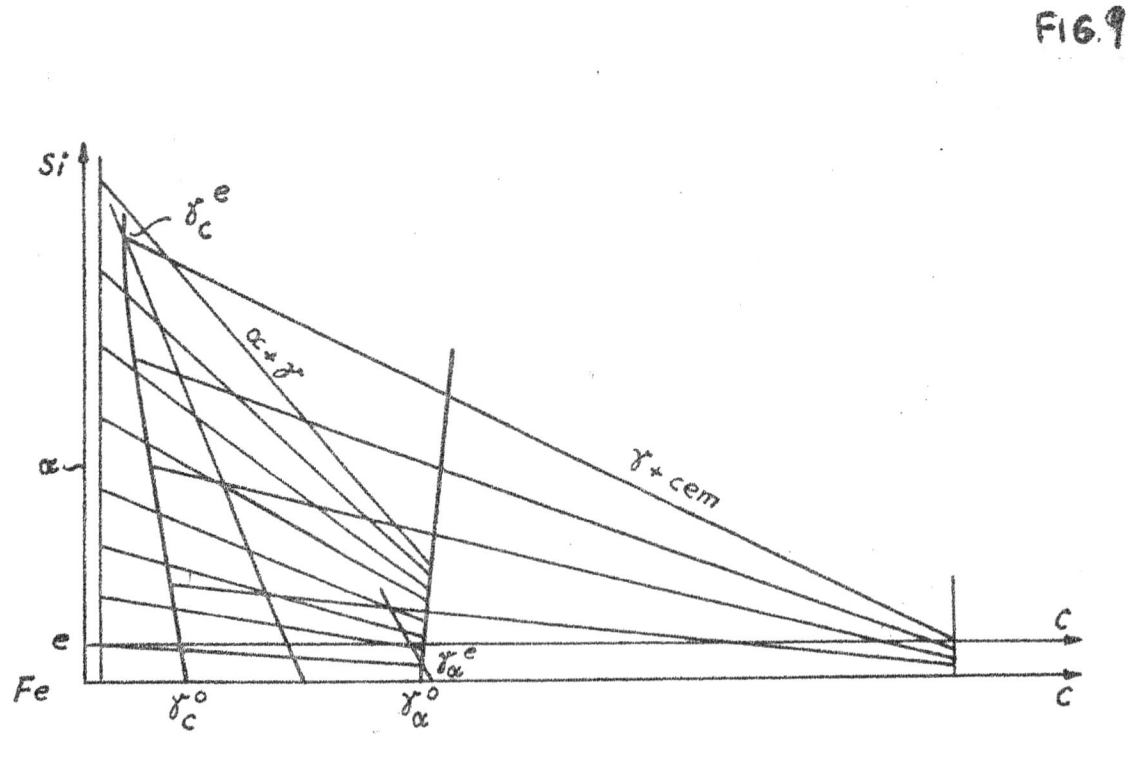

FIG. 9

3 Introduction to "Pressure-Induced Diffusion and Deformation during Precipitation, Especially Graphitization"

published in *Jernkontorets Annaler* (1957)

When assessing a paper's impact, it is crucial to place the work in the context of the field at the time it was published. This is particularly relevant to Mats' paper since so very much of what is discussed in it is now used routinely to model the phase transformations' kinetics. More remarkable is the fact that, while the pioneering insights contained in the paper apply to phase transformations in broad classes of materials, they were originally developed *not as an end in themselves but* to describe an important phase transformation in iron-carbon alloys.

The context for this paper is aptly provided by the then recently published work of Clarence Zener. As stated in a footnote, Mats' paper was based upon a lecture given by him only four years after Zener's article appeared in the *Journal of Applied Physics* in 1949. Thus, while it was agreed within the community that the local equilibrium assumption provided a good approximation of the concentrations at the interface of a growing particle, the manner in which to account for effects such as capillarity and stress on the motion of interfaces was not clear.

In contrast with Zener's approach, the model for particle growth was framed in terms of activities, instead of concentrations. This insight allowed him to easily include the effects of capillarity through the pressure dependence of the activity. The particle-size dependence of the growth rate could then be modeled by expressing the pressure jump at the interface in terms of particle size using the Laplace-Young equation. This is now the standard approach that is taught at the undergraduate level in most materials science and engineering departments. Once capillarity was included in the theory for particle growth, Mats noted that there was a critical particle size where "Smaller particles will have a tendency to be dissolved, larger particles will grow". He correctly associated this with the critical particle size for nucleation. However, unlike the well known Gibbs formulation for homogeneous nucleation in which the critical radius for nucleation is determined by finding the particle size that is in unstable equilibrium with the matrix, Mats' result follows from a treatment of the dynamics of particle growth. He proceeded further to provide an expression for the critical radius of nucleation (Eq. 30) that is valid for the non ideal thermodynamics of the Fe-C system. Given the understanding of the role of capillarity in phase transformations at the time, this was a remarkable achievement.

Mats goes on to include the effects of both elastic and plastic deformation on particle growth. He recognizes at the outset that including shear stress was beyond the scope of this work. Properly accounting for the energy of shear stresses in a thermodynamic description of a crystalline solid would remain an active area of research for some 50 years to come. Mats had the insight to realize that stress can be generated by the difference in molar volumes between the particle and matrix as well as the diffusion process itself. The result is a theory for particle growth that accounts not only for elastic stress, but also for plastic stress. Mats employed a particularly clever approach of coupling the strain rate to the trace of the stress and through this to the pressure dependence of the activity. Although the resulting differential equation for the time dependence of the radius of the particle could not be solved analytically, the various limits discussed in the paper illuminated the basic physical phenomena predicted by the model. This remains one of the few theories for particle growth during creep.

Mats Hillert's 1957 paper is truly pioneering. Even at this early stage in Mats' career it is clear that he was destined to become a giant in the field of physical metallurgy. I suspect that if the paper had been published in a more widely circulated journal the impact of this work would have been even more substantial. Nevertheless, his ideas endure.

Peter. W. Voorhees

Pressure-Induced Diffusion and Deformation during Precipitation, Especially Graphitization

by

MATS HILLERT

Communication from the Swedish Institute for Metal Research

Introduction

548. 526: 669. 112. 247

Most processes in physical metallurgy are treated as if they took place under uniform temperature and pressure. Metals are good thermal conductors and the assumption of uniform temperature is therefore justified in most cases. Often the assumption of uniform pressure, on the other hand, does not seem justified in view of the high mechanical strength of solid metals. In the present paper it will be shown how pressure differences are originated during solid-state precipitation processes, and the effect of such differences on the course of the process will be examined in some detail. Only binary systems will be considered in order to simplify the treatment as much as possible.

Formulation of the Problem

It is well known that the interfacial energy σ causes a pressure difference between the two sides of a curved interface according to the equation $\Delta P = \sigma(1/r_1 + 1/r_2)$, where r_1 and r_2 are the principal radii of curvature of the interface. A precipitating particle will thus be under a higher pressure than the surrounding matrix. Another cause of pressure differences is the increase in volume, which accompanies many precipitation processes. A region where such a process starts will tend to expand and a high pressure is automatically created if the surrounding material is mechanically strong.

Considerable pressure differences can be created by the precipitation process even if it does not cause any net volume change. It is true that the overall effect of such a process can be described simply as an exchange between different regions of equal amounts of the two components. However, such a balanced exchange can usually not be accounted for by a simple diffusion mechanism, because different components usually

---

This paper is based on a lecture given at the meeting of the Swedish Metallographers' Association in May 1953.

have different mobilities. This is well known for interstitial solutions, for instance the Fe-C system, where the carbon atoms are much more mobile than the iron atoms. The same has been found for substitutional solutions, where the phenomenon has been called the Kirkendall effect. An additional transportation mechanism is thus necessary in order to balance the amounts of material in the two directions. This fact is often overlooked, probably because the additional mechanism is trivial in the case of one-dimensional diffusion experiments. This case will first be discussed.

Consider two plates of different metals welded together and placed with the more mobile metal under the sluggish one. If the temperature is so high that there is a mutual miscibility, the mobile atoms from the lower plate will diffuse upwards and the sluggish atoms will slowly *fall* downwards as the amount of the mobile component under them decreases. If the temperature is lowered again very slowly, the solubility is decreased and the reaction is reversed. The mobile atoms now diffuse downwards and the sluggish atoms are simply *lifted* up.

The equivalent mechanism in a two- or three-dimensional process is much more complicated. The difference is that a plate of sluggish material in the one-dimensional case can move up or down without changing its width, whereas the material in a ring in the two-dimensional case cannot move away from or toward the center without changing the circumference of the ring. The ring will thus oppose to such a movement and mechanical stresses are automatically formed. The same will happen in the three-dimensional case.

Mechanical stresses may affect the transportation of material in a number of ways. Elastic or plastic deformation directly results in a movement of bulk material. Mobile and sluggish material are thus moved in the same direction by this mechanism and the direction might be against the diffusional flow of the mobile component and in the same direction as the slower diffusion of the sluggish component. The net flow of the mobile component is thus decreased and that of the sluggish component is increased. The mechanical deformation will thus tend to balance the flow of the two components and the stresses will automatically develop into such a state that the deformation exactly compensates for the unbalance due to the difference in diffusion rates.

Exact calculations of the mechanical deformation can be very difficult and only some simple cases will be treated in this paper. The results will be applied to the growth process of graphite in steel and cast iron.

Another effect of mechanical stresses is to change the driving force for the diffusion process in such a way that diffusion from a low-pressure region to a high-pressure region is slowed down and diffusion from a high-pressure region to a low-pressure region is accelerated. This effect will diminish the unbalance due to the diffusion and consequently diminish the need for mechanical deformation. Our first step will therefore be to evaluate this effect.

In order to simplify the equations the symbol P will be used all through this paper to denote the overpressure, i.e. the difference between the actual pressure and the normal atmospheric pressure.

Effect of Pressure on Activity and Diffusion

Fick's first law, which holds for diffusion in isobaric systems, can be written

$$J_A = -D_A \frac{1}{V_m} \frac{\partial x_A}{\partial l} \qquad (1)$$

J_A is the flux of the component A, D_A is the diffusion coefficient, V_m is the molar volume, x_A is the mole fraction and l is the length coordinate. This equation can be mathematically transformed into

$$J_A = -D_A^a \frac{\partial a_A}{\partial l} \qquad (2)$$

where a_A is the activity of component A and D_A^a is the "activity diffusion coefficient". The two diffusion coefficients are related by

$$D_A^a = D_A / V_m \left(\frac{\partial a_A}{\partial x_A} \right)_P \qquad (3)$$

It will now be suggested that equation (2) holds also for diffusion in non-isobaric systems if the following definition of the relative activity a_A at a hydrostatic pressure P is used

$$a_A = \lambda_A^P / \lambda_A^0 \qquad (4)$$

where λ_A^P is the absolute activity at the hydrostatic pressure P and λ_A^0 is the absolute activity for the pure component A at atmospheric pressure. This suggestion is equivalent to suggesting that the driving force for diffusion is the chemical potential gradient in non-isobaric systems as well as in isobaric systems.

Two neighboring phases, α and β, can be treated as being in local equilibrium if the chemical reaction at their interface is fast enough, especially in comparison with the diffusion process. This is usually assumed in physical metallurgy and the A activity in the two phases at their interface can then be denoted by a single symbol a_A^0 at atmospheric pressure and by another symbol a_A^P under a certain stress condition. This condition may very well involve different hydrostatic pressures P^α and P^β in the two phases. The pressure difference can be caused partly by interfacial energy, partly by shear stresses. It is possible that shear stresses can affect the activities directly to some extent but this effect cannot be deduced from simple thermodynamics and will be neglected here. The effect of the pressures P^α and P^β can be deduced in the following way.

In view of equation (4) one can write the Gibbs–Duhem relation for each of the two phases under isothermal conditions

$$\left.\begin{array}{l} x_A^\alpha \, d(RT \ln a_A^\alpha) + x_B^\alpha \, d(RT \ln a_B^\alpha) = V_m^\alpha \, dP^\alpha \\ x_A^\beta \, d(RT \ln a_A^\beta) + x_B^\beta \, d(RT \ln a_B^\beta) = V_m^\beta \, dP^\beta \end{array}\right\} \quad (5)$$

V_m^α and V_m^β are the molar volumes of the two phases and can usually be treated as constants for solid phases. Equation (5) can easily be integrated as long as the variation of the compositions of the phases, (x_A^α, x_B^α) and (x_A^β, x_B^β), is reasonably small. By integrating between the pressure 0 and P^α or P^β one obtains for the two phases

$$\left.\begin{array}{l} \left(\dfrac{a_A^P}{a_A^0}\right)^{x_A^\alpha} \cdot \left(\dfrac{a_B^P}{a_B^0}\right)^{x_B^\alpha} = \exp \dfrac{P^\alpha V_m^\alpha}{RT} \\ \left(\dfrac{a_A^P}{a_A^0}\right)^{x_A^\beta} \cdot \left(\dfrac{a_B^P}{a_B^0}\right)^{x_B^\beta} = \exp \dfrac{P^\beta V_m^\beta}{RT} \end{array}\right\} \quad (6)$$

By solving for a_A^P and a_B^P one obtains

$$\left.\begin{array}{l} a_A^P = a_A^0 \exp \dfrac{P^\alpha V_m^\alpha x_B^\beta - P^\beta V_m^\beta x_B^\alpha}{RT(x_B^\beta - x_B^\alpha)} \\ a_B^P = a_B^0 \exp \dfrac{P^\beta V_m^\beta x_A^\alpha - P^\alpha V_m^\alpha x_A^\beta}{RT(x_A^\alpha - x_A^\beta)} \end{array}\right\} \quad (7)$$

In the case of an almost pure component A under a pressure P^α one has $x_A^\alpha \simeq 1$ and $x_B^\alpha \simeq 0$ and thus

$$a_A^P = a_A^0 \exp \dfrac{P^\alpha V_m^\alpha}{RT}$$

and for almost pure B

$$a_B^P = a_B^0 \exp \dfrac{P^\beta V_m^\beta}{RT} \quad (8)$$

In the following we shall consider the growth of a spherical particle of a phase β in an infinite matrix of a phase α. Zener [1] has considered the diffusion of material to and from such a particle assuming that the diffusion coefficient is independent of concentration and apparently also assuming that there is no Kirkendall effect and no differences in volume. His results show that the so-called steady-state solution is a very good approximation, especially if the supersaturation of the matrix is low. This solution will therefore be used here. The flow of a component B in the direction toward the spherical particle, according to this solution, is

$$Q_B = 4\pi r^2 J_B = 4\pi r^2 D_B^\alpha \frac{d a_B}{d r} = 4\pi r D_B^\alpha (a_B^\infty - a_B^r) \tag{9}$$

If the diffusion coefficient is not a constant (it may very well vary with composition and stress condition), equation (9) would still hold if some kind of mean value is used. The quantities a_B^r and a_B^∞ in equation (9) are the activities of B at the surface of the particle with the radius r and in the matrix far away from the particle. The surface layer of the particle, which has just been formed, cannot contain any shear stresses and it is thus under a purely hydrostatic pressure P^β. The activity a_B^r is therefore given exactly by equation (8) if the particle consists mainly of one component, as is the case for ferrite or graphite. Otherwise equation (7) must be used and information is then also required about the pressure P^α in the surrounding matrix. The stresses in the matrix far away from the growing particle are very small and a_B^∞ is therefore very close to the activity at normal atmospheric pressure.

Volume Changes during Precipitation

The calculation will be limited to the case of two components. Let Q_A and Q_B be the flow (in mole/sec) of A and B atoms to the spherical particle of β phase. A negative value thus indicates a flow away from the particle. Let $-dv^\alpha$ be the volume element of the matrix adjacent to the particle, which is transformed in the time dt, and let dv^β be the volume element of the particle, which thereby is formed. All the volumes are measured in cm³ at atmospheric pressure. It follows immediately that

$$Q_A dt = \frac{x_A^\alpha dv^\alpha}{V_m^\alpha} + \frac{x_A^\beta dv^\beta}{V_m^\beta} \tag{10}$$

$$Q_B dt = \frac{x_B^\alpha dv^\alpha}{V_m^\alpha} + \frac{x_B^\beta dv^\beta}{V_m^\beta} \tag{11}$$

or

$$\frac{dv^\alpha}{dt} = -V_m^\alpha \frac{Q_B x_A^\beta - Q_A x_B^\beta}{x_A^\alpha - x_A^\beta} \tag{12}$$

$$\frac{dv^\beta}{dt} = V_m^\beta \frac{Q_B x_A^\alpha - Q_A x_B^\alpha}{x_A^\alpha - x_A^\beta} \tag{13}$$

The change in volume due to the transformation at the interface between the particle and the matrix is thus

$$\frac{dv^\alpha}{dt} + \frac{dv^\beta}{dt} = \frac{d\Delta v}{dt} = \frac{Q_A(V_m^\alpha x_B^\beta - V_m^\beta x_B^\alpha) + Q_B(V_m^\beta x_A^\alpha - V_m^\alpha x_A^\beta)}{x_A^\alpha - x_A^\beta} \quad (14)$$

or

$$\frac{d\Delta v}{dt} = \frac{Q_A(V_m^\alpha x_B^\beta - V_m^\beta x_B^\alpha) + Q_B(V_m^\beta x_A^\alpha - V_m^\alpha x_A^\beta)}{V_m^\beta(Q_B x_A^\alpha - Q_A x_B^\alpha)} \frac{dv^\beta}{dt} \quad (15)$$

If the change in volume is negative, the particle is growing in a hole in the matrix, which tends to be larger than the particle. There is thus a possibility for the formation of pores. This case will not be further treated here.

If the change in volume is positive, the particle tends to be larger than the hole. It is therefore growing under a pressure which affects the diffusion of the two components in accordance with equations (7), (8) and (9). The pressure also supplies an additional volume, Δv, for the particle by means of mechanical deformation of the particle itself and of the matrix. A quantitative calculation of this volume is very difficult. It is necessary to represent the mechanical properties of the material with some simple analytical expression. Three different expressions were chosen in this paper and the calculations are presented in Appendix I. It should be emphasized that these calculations deal with the volume gained by an increase of the hydrostatic pressure in a spherical particle by the value P_0, if the particle was surrounded by an infinite matrix which was stress free before the pressure increase took place. The pressure in the spherical particle must have had the value $2\sigma/r$ before the pressure increase because of the interfacial energy and its value is consequently increased to $P^\beta = P_0 + 2\sigma/r$. When applying the results of Appendix I to an actual case of precipitation one must also remember that volume changes in the matrix due to its loss of the mobile component by diffusion are neglected. This may be a serious approximation in some cases and a more rigorous calculation is then required. In the case of graphitization, however, this approximation seems to be quite justified, because the total contraction of the matrix due to the loss of carbon is only about a third of the volume of the graphite which is formed. In the case of pearlite it is an extremely good approximation because the matrix does not change its composition except in a zone very close to the growing pearlite colony.

In Appendix II three calculations of the energy required for the deformation are presented. They are based on the same three analytical expressions for the mechanical properties.

The volume change of the particle due to compression is rather small and the actual volume of the particle, which in the following will be denoted by V, can therefore with good accuracy be substituted by the volume at atmospheric pressure, v^β.

Growth Rate of a Precipitating Particle

It has already been mentioned that the diffusion flows, Q_A and Q_B, are pressure dependent. In order to calculate the rate of growth of the particle from equation (13) it is thus necessary to know the magnitude of the pressures acting in the particle and the surrounding matrix. They can be calculated by combining equation (14) or (15) with one of the equations (11 A), (12 A) and (21 A), provided that the mechanical properties of the material can be sufficiently well described by one of the three models treated in Appendix I.

Elastic and ideally plastic deformation

When the diffusion of component A is very slow the terms containing Q_A in equation (15) can be neglected and the equation transforms into

$$\frac{d \Delta v}{dt} = \frac{d v^\beta}{dt} \frac{V_m^\beta x_A^\alpha - V_m^\alpha x_A^\beta}{V_m^\beta x_A^\alpha} \tag{16}$$

which can be integrated

$$\Delta v = v^\beta \frac{V_m^\beta x_A^\alpha - V_m^\alpha x_A^\beta}{V_m^\beta x_A^\alpha} + \Delta v_0 \tag{17}$$

The integration constant Δv_0 is the original volume of the hole in the matrix where the particle is growing. We shall assume $\Delta v_0 = 0$, which implies that there were no pores in the original material, where the precipitation could start. The pressure P_0 can then be calculated easily by elimination of Δv and v^β between this equation and (11 A) or (12 A) if the ideally plastic or elastic model is chosen. It is interesting that a constant value of P_0 is found.

The pressure P_0 can stay constant even when the diffusion of A cannot be neglected, provided that the particle is so large that the interfacial energy does not cause any appreciable difference between P_0 and P^β. If for instance the approximate steady-state solution is applied, both Q_A and Q_B are proportional to the radius r of the particle, according to equation (9). The radius r can therefore be eliminated from (15) and a constant value of P_0 can be calculated from

$$\Delta v = v^\beta \frac{Q_A(V_m^\alpha x_B^\beta - V_m^\beta x_B^\alpha) + Q_B(V_m^\beta x_A^\alpha - V_m^\alpha x_A^\beta)}{V_m^\beta(Q_B x_A^\alpha - Q_A x_B^\beta)} \tag{18}$$

and (11 A) or (12 A).

The calculation of the pressure P_0 and of the growth rate by inserting the P_0 value in equation (13) by means of equations (9) and (7) or (8) is thus comparatively simple in these cases. The growth rate, measured as dr^2/dt, is here found to be constant because v^β in equation (13) is proportional to r^3 and Q_A and Q_B are proportional to r.

Deformation by creep

If the creep model of deformation is chosen to represent the mechanical properties of the material, the pressures can be found by combining equations (14) and (21 A). By using the steady-state solution (equation 9) one obtains

$$\frac{D_B^\alpha(a_B^\infty - a_B^r)(V_m^\beta x_A^\alpha - V_m^\alpha x_A^\beta)}{x_A^\alpha - x_A^\beta}$$

$$= \frac{D_A^\alpha(a_A^r - a_A^\infty)(V_m^\alpha x_B^\beta - V_m^\beta x_B^\alpha)}{x_A^\alpha - x_A^\beta} + \frac{1}{2}r^2\left(\frac{3n}{2s}(P^\beta - 2\sigma/r)\right)^{\frac{1}{n}} \quad (19)$$

Equation (13) can be transformed into

$$\frac{dr^2}{dt} = 2 V_m^\beta \frac{D_A^\alpha(a_A^r - a_A^\infty)x_B^\alpha + D_B^\alpha(a_B^\infty - a_B^r)x_A^\alpha}{x_A^\alpha - x_A^\beta} \quad (20)$$

These two equations relate the growth rate dr^2/dt to the size of the particle, r, because they both contain P^β as a parameter. It is, however, impossible to eliminate P^β between them and the best method is therefore in general to calculate dr^2/dt and r for a series of different P^β values and make a plot. It is obvious that the growth rate is not independent of the size r. However, different stages of the growth may be distinguished and the growth rate is constant during two of these, as now will be shown.

Consider the case where one component is much less mobile than the other, $D_A^\alpha \ll D_B^\alpha$. For small values of r the left hand side of equation (19) would be much larger than the right hand side, unless the pressure is so high that $a_B^\infty - a_B^r$ is very small. The pressure can therefore with good accuracy be calculated from $a_B^\infty = a_B^r$, where a_B^r is given by a_B^P in equation (7) or (8). The pressure P^β can therefore be calculated from the value of a_B^∞ using one of these equations. To simplify the following treatment we shall assume that the β phase contains so small amounts of component A that equation (8) can be used. Then

$$a_B^\infty = a_B^0 \exp\frac{P^\beta V_m^\beta}{RT} \quad \text{or} \quad P^\beta = \frac{RT}{V_m^\beta}\ln\frac{a_B^\infty}{a_B^0} \quad (21)$$

The pressure in the particle P^β is thus constant as long as the particle is sufficiently small. Three different stages can here be distinguished.

STAGE 1. At sufficiently low values of r the second term on the right hand side of equation (19) can be neglected and by elimination of D_B^α between (19) and (20) one obtains

$$\frac{dr^2}{dt} = \frac{2 D_A^\alpha(a_A^r - a_A^\infty)}{x_A^\alpha/V_m^\alpha - x_A^\beta/V_m^\beta} \quad (22)$$

PRESSURE-INDUCED DIFFUSION AND DEFORMATION 75

The value of a_A^r is given by equation (8) and is consequently dependent on the pressure P^α in the matrix close to the particle. This pressure is given as $(1-n)P_0$ by equation (23 A) and considering the interfacial energy one obtains $P^\alpha = (1-n)(P^\beta - 2\sigma/r)$. Consequently, the pressure P^α and the growth rate will be zero when the radius of the particle, r, has the value $2\sigma/P^\beta$. This is the critical size in the nucleation of the β phase. Smaller particles have a tendency to be dissolved, larger particles will grow. The pressure P^α and the growth rate are exceedingly low in the beginning but are rapidly increasing as the particle grows. The pressure has reached half the maximum value already when the particle has grown to a size twice the critical size. At larger sizes the effect of the interfacial energy can be neglected and the first stage, which was dominated by the interfacial energy, is succeeded by a second stage.

STAGE 2. Equation (22) is still valid but a_A^r is constant because P^α is constant. The growth rate, if measured as dr^2/dt, is then also constant. The rate determining process is here pressure induced diffusion of the sluggish component.

STAGE 3. Above a certain value of r the second term will dominate over the first term on the right hand side of equation (19). Elimination of D_B^α between (19) and (20) now gives

$$\frac{dr^2}{dt} = \frac{V_m^\beta x_A^\alpha}{V_m^\beta x_A^\alpha - V_m^\alpha x_A^\beta} r^2 \left(\frac{3n}{2s}(P^\beta - 2\sigma/r)\right)^{\frac{1}{n}} \qquad (23)$$

The growth rate is thus constant during this stage if measured as $d\ln r/dt$. The rate determining process is here the mechanical enlargement of the hole by creep.

STAGE 4. As r increases to even higher values, the pressure P^β will decrease from the constant value given by equation (21) toward the atmospheric pressure. The growth rate is therefore finally given by

$$\frac{dr^2}{dt} = \frac{2 D_B^\alpha (a_B^\infty - a_B^0) V_m^\beta x_A^\alpha}{x_A^\alpha - x_A^\beta} \qquad (24)$$

Consequently, it again becomes constant if measured as dr^2/dt. The rate determining process during this fourth stage is diffusion of the mobile component.

This section has indicated that the growth process of a precipitating particle is rather complicated when creep plays an important rôle. In order to examine such a growth process in more detail, the next section will be concerned with the phenomenon of graphitization in steel. This occurs after long annealing times at relatively high temperatures and

MATS HILLERT

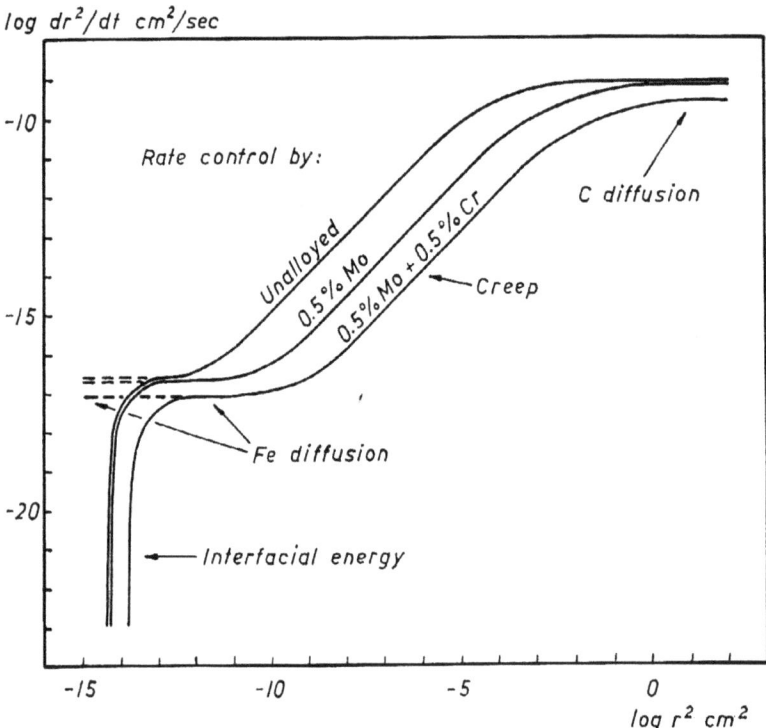

Fig. 1. Growth rate at 500°C of a single spherical graphite particle in three steels versus the size of the particle. As the particle grows the rate is first controlled by interfacial energy (stage 1), later by pressure induced diffusion of iron (stage 2) and creep (stage 3) and finally by diffusion of carbon (stage 4).

creep is therefore likely to be the dominating deformation mechanism. The following calculations are based on a considerable number of experimental data, some of which are of questionable accuracy. The final results might consequently be correct only to the order of magnitude. It is interesting that the four stages of the growth process mentioned in this section are clearly visible in the diagram with the calculated growth-rate values for graphite (Fig. 1).

Rate of Growth of Graphite in Steel

The growth of a spherical particle of graphite in an infinite matrix of ferrite containing fine and evenly distributed particles of cementite will now be considered. This is presumably a case where the mechanical properties of the matrix is best described by the creep model and equations (19) and (20) will consequently be applied. Three different steels will be considered and the following quantities are needed.

Concentrations and activities

The subscripts A and B stand for iron (Fe) and carbon (C) and the superscripts α and β stand for ferrite (α) and graphite (gr). Thus

$$x_B^\alpha = x_C^\alpha \cong 0, \qquad x_B^\beta = x_C^{gr} \cong 1, \qquad x_A^\beta = x_{Fe}^{gr} \cong 0,$$

$$x_A^\alpha = x_{Fe}^\alpha \cong 1, \qquad a_B^\infty = a_C^{cem+\alpha}, \qquad a_B^r = a_C^P = \exp\frac{P^{gr} V_m^{gr}}{RT},$$

$$a_A^\infty = a_{Fe}^\alpha \cong 1, \qquad a_A^r = a_{Fe}^P \cong \exp\frac{P^\alpha V_m^\alpha}{RT}, \qquad V_m^\beta = V_m^{gr} = 5{,}3 \text{ cm}^3/\text{mole},$$

$$V_m^\alpha = V_m^{ferrite} = 7{,}1 \text{ cm}^3/\text{mole}.$$

The activity of carbon for cementite and ferrite in equilibrium at atmospheric pressure can for a pure iron-carbon alloy be described by an equation [2]

$$a_C^{pure\ cem+\alpha} = 0{,}0353 \, \exp\frac{6900}{RT} \tag{25}$$

The effect of an alloying element is described by an approximate formula [3]

$$a_C^{cem+\alpha}/a_C^{pure\ cem+\alpha} = \exp\left(-k(1-x)-y\right) = \exp\left(-y(4F-3)\right) = \exp(wg) \tag{26}$$

k is the slope of the tie-line in the ternary diagram, on which the composition of the alloyed steel is located. y is the alloy content of the *ferrite*, measured as mole fraction. x is the carbon content of the ferrite, which is very low and can be neglected here. The slope k can be expressed by means of y and the partition coefficient F for the alloying element between cementite and ferrite. F is the ratio of the mole fractions of the alloying element in the two phases. For comparison of the effect of different alloying elements it is more convenient to use the weight per cent, w, instead of the mole fraction y. The related coefficient g in equation (26) can be computed from F.

Kuo and Hultgren [4] have studied the distribution of several elements between cementite and ferrite just below the eutectoid temperature. According to their results F is 6.5 and g is 0.13 for Mo. For Cr F is 22 and g is 0.92. Since there is no information available about the variation of these coefficients with temperature, they will here be assumed to be constant.

Creep data

The following values of the constants which determine the creep behavior will be used at 500°C.

unalloyed steel	$s = 10^{4.6}$ atm, sec^n	$n = 0.2$
0.5 % Mo	$s = 10^{4.9}$ atm, sec^n	$n = 0.2$
0.5 % Mo + 0.5 % Cr	$s = 10^{4.9}$ atm, sec^n	$n = 0.2$

The percentages refer to the alloy content of the ferrite in the steels. The values of s and n are in rather good agreement with experimental creep data which, however, show a considerable scatter due to differences in manufacturing and heat treatment of the steels. [5] The values of the growth rate when creep is the rate-controlling mechanism are therefore expected to be correct only to the order of magnitude.

The exact values of s and n have here been chosen so as to demonstrate the difference in action between Mo and Cr. Cr is known to have a much smaller influence on the creep strength than Mo and it was therefore here assumed that Cr does not improve the strength at all.

Diffusion coefficients

The value of D^a at atmospheric pressure has been estimated for diffusion of carbon in ferrite [6] by means of equation (3). According to the result

$$D^a_{C\ in\ \alpha} = 0.0094\ \exp\left(-\frac{36\ 700}{RT}\right)\ \text{mole C/sec, cm, unit activity} \qquad (27)$$

There is no information available about the variation of D^a or D with pressure for interstitial solutions and it will therefore be assumed that D^a is independent of pressure. It will also be assumed that it is independent of alloying elements, although there is some information available showing a variation.

The diffusion coefficient for iron in ferrite is not known directly but theoretically it is intimately related to the self-diffusion coefficient, which has been determined by Birchenall and Mehl. [7] In view of equation (3) we shall write

$$D^a_{Fe\ in\ \alpha} = 2.3 \cdot 10^3\ \exp\left(-\frac{73\ 200}{RT}\right)\frac{1}{V^\alpha_m}\ \text{mole Fe/sec, cm, unit activity} \qquad (28)$$

The interfacial energy for the ferrite-graphite interface is not known. Its exact value is not essential for the present treatment, however, and we shall quite arbitrarily use the reasonable value $\sigma = 500$ erg/cm².

Result

Fig. 1 shows the result of the calculation of the growth rate at 500°C for the three steels, based on equations (19) and (20). The four straight portions correspond to the four stages mentioned earlier.

The pressure in the graphite particle during the first three stages is given by equation (21)

$$P^{gr} = \frac{RT}{V^{gr}_m}\ \ln a^{cem+\alpha}_C \qquad (29)$$

STAGE 1. CONTROL BY INTERFACIAL ENERGY. The critical size, where the growth rate is zero, is

$$r_{critical} = \frac{2\sigma}{P^{gr}} = \frac{2\sigma V^{gr}_m}{RT\ \ln a^{cem+\alpha}_C} \qquad (30)$$

STAGE 2. RATE CONTROL BY PRESSURE INDUCED DIFFUSION OF FE. The value of the pressure in the ferrite close to the graphite particle is here $(1-n)P^{gr}$ according to equation (23 A) in the Appendix. The iron activity at the interface is then given by equations (8) and (29)

$$a_{Fe}^r = \exp\frac{(1-n)P^{gr}V_m^\alpha}{RT} = \left[a_C^{cem+\alpha}\right]^{(1-n)\frac{V_m^\alpha}{V_m^{gr}}} \cong a_C^{cem+\alpha} \qquad (31)$$

because the exponent is reasonably close to unity. Equation (22) finally yields the growth rate

$$\frac{dr^2}{dt} = 2V_m^\alpha D_{Fe}^a(a_C^{cem+\alpha}-1) \qquad (32)$$

STAGE 3. RATE CONTROL BY CREEP. The pressure P^{gr} is still given by equation (29) and equation (23) therefore gives

$$\frac{dr^2}{dt} = r^2\left(\frac{3nRT}{2sV_m^{gr}}\ln a_C^{cem+\alpha}\right)^{\frac{1}{n}} \qquad (33)$$

STAGE 4. RATE CONTROL BY DIFFUSION OF C. Equation (24) gives

$$\frac{dr^2}{dt} = 2V_m^{gr}D_C^a(a_C^{cem+\alpha}-1) \qquad (34)$$

Discussion

It is a common experience that the probability for spontaneous formation of a nucleus in a homogeneous, solid material is exceedingly low in most cases. Instead, the reaction usually starts at inhomogeneities such as interfaces. The retarding effect from the interfacial energy of the interface around a nucleus is much less in such preferred sites. The steep parts of the curves in Fig. 1, which represent the first stage where the interfacial energy is dominating, is therefore of theoretical interest only. In a real case of precipitation it seems more justified to substitute this part by an extrapolation of the second stage to smaller sizes. This extrapolation is respresented by dashed lines in Fig. 1. The growth-rate curves in Fig. 1, modified in this manner, were used for a calculation of the time of growth of a graphite particle from the end of the nucleation process to the size $r = 10^{-3}$ cm. The result for the three steels is listed in the last column of Table I. The addition of Mo thus increases the time of growth considerably (in our example by a factor of 25), because the strengthening of the matrix slows down the creep process. The further addition of Cr increases the time still more (in our example by another factor of 20), because the decreased carbon activity of the matrix means a decrease of the driving force

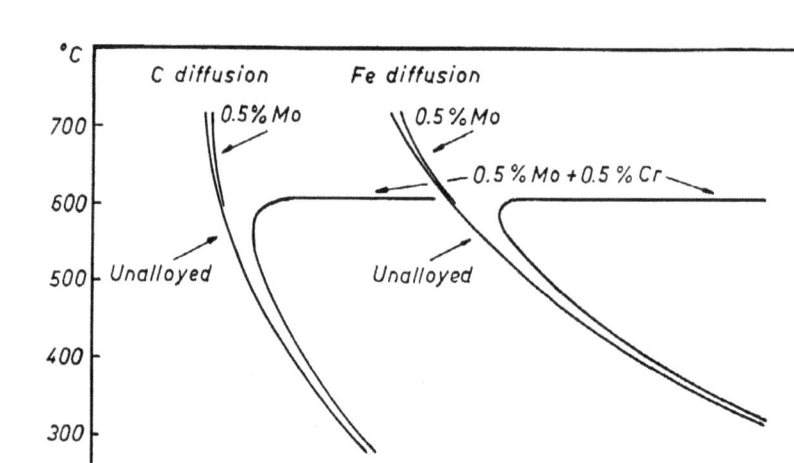

Fig. 2. The growth rate in stage 2 (iron diffusion) and in stage 4 (carbon diffusion) versus temperature. The growth rate is increasing to the left.

for all the stages. If we had assumed some strengthening effect of Cr on the ferrite, an even larger factor would have been found.

Table I also contains the results obtained by assuming that there is only one rate controlling process during the whole growth, namely diffusion of iron according to equation (32) or diffusion of carbon according to equation (34). The calculated values can be compared with the following experimental findings.

TABLE I. *Time in hours for the growth of a single spherical graphite particle to $r = 10^{-3}$ cm at $500°C$*

Steel	Rate controlling mechanism		
	Fe diffusion	C diffusion	Diffusion and creep
unalloyed	$1.4 \cdot 10^7$	0.4	$4 \cdot 10^2$
0.5 % Mo	$1.6 \cdot 10^7$	0.5	$1 \cdot 10^4$
0.5 % Mo + 0.5 % Cr	$3.7 \cdot 10^7$	1.1	$2 \cdot 10^5$

Dulis and Smith [8] have shown that graphitization is a relatively fast process in a pure iron-carbon alloy. It is a much slower process in the commercially used molybdenum alloyed steels. Furthermore, it is well known that these commercial steels get a greatly improved resistance against graphitization by the further addition of chromium.

The last column of Table I shows that these facts can be well accounted for by the suggested mechanism involving both pressure induced diffusion and creep. On the other hand, the assumption that the rate-controlling process is only carbon diffusion or iron diffusion results in a quite insufficient variation of the growth rate with the alloy content. The first two columns of Table I show a variation by a factor of 3 compared with 500 for the mechanism involving pressure induced diffusion and creep.

The change of the graphitization tendency with temperature is demonstrated by Fig. 2, where the growth rates according to the second stage (Fe diffusion, equation 32) and according to the fourth stage (C diffusion, equation 34) are plotted. The value according to the mechanism involving pressure induced diffusion and creep must always lie between these values. It is noticed from this plot that chromium is more effective the higher the temperature. Above a certain critical temperature, which is 610°C for 0.5% Cr, cementite becomes stable and already formed graphite could thus be dissolved again by a heat treatment above this temperature. The critical temperature can be calculated from equations (25) and (26) with $a_C^{cem+\alpha} = 1$.

Rate of Growth of Graphite in Cast Iron

The process of graphitization in the austenitic temperature range of cast iron, which has solidified white, can in principle be calculated in the same way as graphitization in steel. However, the information about the creep properties at such high temperatures is very limited and no quantitative calculation can therefore be presented in this paper. Only some qualitative conclusions will be drawn.

The growth process of graphite in cast iron must have the same general characteristics as in steel. The creep strength is much lower due to the higher temperature, however, and it is therefore quite possible that the fourth rate-controlling process, the diffusion of carbon, takes over the rate control at a smaller particle size in cast iron than in steel. If this is true the experimental measurements of the growth rate in cast iron may have dealt mainly with the fourth stage and should then fit the parabolic rate law (equation 24). This has actually been found by Brown and Hawkes. [9] Birchenall and Mead [10] have pointed out that this may also be the case with the data presented in a paper by Burke and Owen. [11] The data in a recent paper by Owen and Wilcock [12] fit the parabolic rate law at reasonably large particle size. At smaller size these data seem to fit better to a linear rate law. In view of Fig. 1 this would indicate that the rate control at this smaller size has not quite shifted from the creep process to the diffusion of carbon. The linear rate law would give the slope 1/2 when plotted with the logarithmic scales used in Fig. 1.

The nature of the rate-controlling mechanism in graphitization has recently been discussed in a number of papers. From a comparison of experimentally determined activation energies for graphitization with activation energies for different diffusion processes Bunin and Danilchenko

[13] as well as Burke and Owen [11] concluded that the rate-controlling process could not be the diffusion of carbon. Instead the diffusion of iron or silicon was suggested. However, Birchenall and Mead [10] have argued that the activation energy of a complex process like graphitization should not be compared with activation energies of simple diffusion processes. Instead they made a calculation of the rate of growth of graphite at several temperatures assuming different rate-controlling processes and compared the results directly with experimental values. The growth rates based on iron diffusion were too low by a factor of about 10,000 and they consequently concluded that this process could not be the rate-controlling one. The calculation based on carbon diffusion gave values too high by a factor of 5 to 10. The accuracy of the calculation was of about this magnitude and the agreement was therefore considered good. From this calculation Birchenall and Mead suggested that the diffusion of carbon is the rate-controlling process. In view of Fig. 1, however, it is now tempting to suggest that the experimental growth-rate data hold for such a particle size that the process is not quite controlled by carbon diffusion. The deformation by creep still plays some rôle. This would explain the disagreement by the factor of 5 to 10 in the growth rates as well as the tendency toward a linear rate law at small particle size, which was discussed earlier.

The fact that the early growth is controlled by diffusion of iron and creep, which give much lower growth rates than the diffusion of carbon, will result in an *apparent* incubation time, if the measured growth rates are extrapolated to zero size. Burke and Owen [11] have found a considerable incubation time and the interesting question is how much of this time is actually due to a delay in the nucleation process and how much is due to the bad method of extrapolation.

It is obvious that the apparent incubation time, i.e. the time it takes for a graphite particle to grow to a size large enough to be detected in microscope would be considerably shorter if the material originally contained pores where the graphite particles could start to grow without displacing any iron. The growth at small particle size would then be very fast instead of being extremely slow in accordance with Fig. 1.

Summary

1. It is suggested that the diffusion flow of a component in non-isobaric systems is proportional to the activity gradient.

2. An expression for the pressure dependence of activity is derived.

3. The difference in volume between a precipitate and the transformed part of the matrix is calculated. It is concluded that the precipitate is growing under a hydrostatic pressure if this difference is positive.

4. Formulae are derived for the volume supplied by mechanical deformation due to such a pressure and for the deformation energy. Three

PRESSURE-INDUCED DIFFUSION AND DEFORMATION 83

different assumptions about the mechanical behavior of the matrix are used, namely the elastic, the ideally plastic and a creep model.

5. The hydrostatic pressure and the growth rate of a spherical precipitate are calculated. The usual kinetic law, $dr^2/dt = $ constant, is found for the elastic and the ideally plastic models, when the interfacial energy is negligible.

6. For the creep model it is found that there are shifts in rate-controlling mechanism at certain particle sizes and at the same time a shift in kinetic law. When the rate-controlling mechanism is diffusion the kinetic law is $dr^2/dt = $ constant, and when the control is shifted to a creep mechanism the law is changed to $d \ln r/dt = $ constant.

7. The creep model is applied to the growth rate of graphite in steel and the effect of the alloying elements Mo and Cr is calculated. Comparison with experiments shows good agreement.

8. It is suggested that diffusion of carbon is the rate-controlling process for the growth of graphite in cast iron when the graphite particles are large. At a smaller particle size the necessary deformation of the austenite by creep plays some rôle.

Tryckinducerad diffusion och deformation under utskiljning, speciellt grafitisering

SAMMANFATTNING

Utskiljningsprocesser i fasta legeringar ger upphov till olikformigt tryck dels på grund av den volymsförändring som åtföljer processen dels på grund av olika rörlighet hos legeringsämnena. Inverkan av sådant tryck på utskiljningens förlopp har i allmänhet försummats. Vid närmare betraktande blir det emellertid tydligt att detta tryck är av vital betydelse för processen. I föreliggande uppsats har två olika mekanismer behandlats varigenom olikformigt tryck kan påverka transporten av legeringsämnena under utskiljningen, nämligen dels förändring av den drivande kraften för diffusion och dels mekanisk deformation.

Det föreslås att aktivitetsgradienten för en viss komponent kan betraktas som den drivande kraften för diffusionen av denna komponent i ett prov med olikformigt tryck. Ett uttryck för variationen av denna aktivitet med trycket kan härledas ur Gibbs-Duhems ekvation.

Den volym som ställs till förfogande för tillväxten av en sfärisk partikel under en utskiljningsprocess genom mekanisk deformation av omgivande material har uppskattats under antagande av tre olika mekaniska modeller för detta material, nämligen den idealt plastiska, den idealt elastiska och den exponentiella krypmodellen.

Under hänsynstagande till ovannämnda effekter har det hydrostatiska trycket i en sfärisk partikel och dennas tillväxthastighet beräknats. Den vanliga paraboliska tillväxtlagen $dr^2/dt = $ konstant erhölls med den plastiska och den elastiska modellen.

Med krypmodellen erhölls en förändring av den hastighetsreglerande mekanismen vid viss partikelstorlek och samtidigt en förändring av tillväxtlagen. Vid

Jernkont. Ann. 141 (1957): 2

84   MATS HILLERT

tillväxten av grafit i stål vid temperaturer under A_1 är enligt dessa beräkningar diffusionen av järn undan den växande grafiten hastighetsbestämmande när en grafitpartikel är mycket liten. Den paraboliska tillväxtlagen gäller då. Vid ökad partikelstorlek blir transporten av järn på grund av mekanisk deformation genom krypning viktigare och tillväxtlagen blir logaritmisk $d \ln r/dt = $ konstant. Vid ännu större partikelstorlek begränsas tillväxthastigheten av koldiffusionen fram till grafiten och tillväxtlagen blir åter parabolisk.

Effekten av legeringsämnen på tillväxthastigheten för grafit i stål har undersökts genom att deras inverkan på kolaktiviteten och den mekaniska hållfastheten betraktats. Effekten av Mo tycks huvudsakligen bero på dess förmåga att öka motståndet mot krypning medan för Cr ändringen av kolaktiviteten är viktigare.

Jämförelse mellan dessa beräkningar och experimentella mätningar på olegerat, Mo-legerat och Mo-Cr-legerat stål visar god överensstämmelse.

Även tillväxten av grafit vid aducering av vitt gjutjärn diskuteras. Kvantitativa beräkningar är här svårare men det föreslås att koldiffusionen är hastighetsbegränsande när grafitpartiklarna är av mätbar storlek. Vid mindre storlek torde mekanisk deformation genom krypning spela en viss roll. Den inkubationstid som man konstaterar experimentellt torde delvis bero på en sådan förändring i den hastighetsreglerande mekanismens natur.

References

1. C. ZENER, *J. Appl. Phys. 20 (1949) p. 950*.
2. M. HILLERT, *Acta Metallurgica 2 (1954) p. 11*.
3. M. HILLERT, *Acta Metallurgica 3 (1955) p. 34*.
4. K. KUO & A. HULTGREN, *Jernkont. Ann. 135 (1951) p. 449*.
5. See for instance Compilation of Available High-temperature Creep Characteristics of Metals and Alloys. Philadelphia 1938. (ASTM Sp. Techn. Publ. No. 37.)
6. M. HILLERT & R. D. SHARP, *Jernkont. Ann. 137 (1953) p. 785*.
7. C. E. BIRCHENALL & R. F. MEHL, *J. Metals 2 (1950) Trans. AIME vol. 188, p. 144*.
8. E. J. DULIS & G. V. SMITH, *Trans. Am. Soc. Metals 46 (1954) p. 1318*.
9. B. F. BROWN & M. F. HAWKES, *Trans. Am. Foundrymen's Ass. 59 (1951) p. 181*.
10. C. E. BIRCHENALL & H. W. MEAD, *J. Metals 8 (1956) Trans. AIME vol. 206, p. 1004*.
11. J. BURKE & W. S. OWEN, *J. Iron & Steel Inst. 176 (1954) p. 147*.
12. W. S. OWEN & J. WILCOCK, *J. Iron & Steel Inst. 182 (1956) p. 38*.
13. K. P. BUNIN & N. M. DANILCHENKO, *Doklady Akad. Nauk SSSR 72 (1950) p. 889*.

Appendix I

Calculation of the Volume Δv

The volume Δv will be calculated which is made available for a spherical particle with the radius r by means of a hydrostatic pressure P_0 in the particle. The particle is surrounded by an infinite matrix and the three principle stresses in the matrix at the distance ϱ from the center of the particle will be denoted by $\sigma_1(\varrho)$, $\sigma_2(\varrho)$ and $\sigma_3(\varrho)$, where $\sigma_2(\varrho) = \sigma_3(\varrho)$ and $\sigma_1(\varrho)$ is in the radial direction. It will be assumed that the elastic modulus E and Poison's ratio v have the same values in the particle as in the matrix.

Ideally plastic material

This model assumes that there is a certain yield strength, S, which is independent of the degree of deformation. The particle will be surrounded by a region of plastically deformed matrix if the pressure P_0 is high enough. Within this region

$$\sigma_2(\varrho) - \sigma_1(\varrho) = S \tag{1A}$$

By considering the forces on a volume element between ϱ and $\varrho + \Delta\varrho$ and using the relation (1A) one finds

$$\Delta\sigma_1(\varrho) = 2S\frac{\Delta\varrho}{\varrho} \tag{2A}$$

By integrating (2A) and using the boundary value $\sigma_1(\varrho) = -P_0$ at $\varrho = r$ one obtains

$$\sigma_1(\varrho) = -P_0 + 2S \ln\frac{\varrho}{r} \tag{3A}$$

The radius R of the plastically deformed region can be found by the criterion $\sigma_1(\varrho) + \sigma_2(\varrho) + \sigma_3(\varrho) = 0$, which together with (1A) and the equality $\sigma_2(\varrho) = \sigma_3(\varrho)$ gives

$$\sigma_1(R) = -\tfrac{2}{3}S \tag{4A}$$

Equation (3A) thus gives

$$\ln\frac{R}{r} = \frac{P_0}{2S} - \frac{1}{3} \tag{5A}$$

Let V be the actual volume of the particle and of its hole in the matrix (approximately equal to the volume of the particle at atmospheric pressure v^β), let v_1 be the volume increase of the hole in the matrix due to elastic deformation of the matrix outside $\varrho = R$, let v_2 be the volume increase due to elastic compression of the plastically deformed region between r and R and let v_3 be the volume decrease of the particle due to compression caused by the hydrostatic pressure P_0. It is easy to show that

$$V = \tfrac{4}{3}\pi r^3 \tag{6A}$$

$$v_1 = 4\pi R^3 \varepsilon_2(R) = V\frac{S}{E}(1+\nu)\left(\frac{R}{r}\right)^3 \tag{7A}$$

$$v_3 = P_0 v^\beta \frac{3(1-2\nu)}{E} \tag{8A}$$

The hydrostatic pressure $P(\varrho)$ in the plastically deformed region of the matrix is obtained from (1A) and (3A)

$$P(\varrho) = -\tfrac{1}{3}[\sigma_1(\varrho) + \sigma_2(\varrho) + \sigma_3(\varrho)] = P_0 - \tfrac{2}{3}S - 2S\ln\frac{\varrho}{r} \tag{9A}$$

Jernkont. Ann. 141 (1957): 2

Thus

$$v_2 = \int_r^R 4\pi\varrho^2 \, d\varrho \cdot P(\varrho) \frac{3(1-2\nu)}{E} = V\frac{S}{E}\left[(2-4\nu)\left(\frac{R}{r}\right)^3 - 3(1-2\nu)\frac{P_0}{S}\right] \quad (10\,A)$$

Equations (6 A), (7 A), (8 A) and (10 A) now give

$$\frac{\Delta v}{v^\beta} = \frac{v_1 + v_2 + v_3}{v^\beta} = \frac{V}{v^\beta}\left[\frac{S}{E}3(1-\nu)\left(\frac{R}{r}\right)^3 - \frac{P_0}{E}3(1-2\nu)\right] + \frac{P_0}{E}3(1-2\nu)$$

$$\cong \frac{S}{E}3(1-\nu)\left(\frac{R}{r}\right)^3 = \frac{S}{E}3(1-\nu)\exp\left(\frac{3P_0}{2S}-1\right) \quad (11\,A)$$

utilizing (5 A) and $V \cong v^\beta$.

Elastic model

Equation (5 A) shows that there is no plastic deformation if $P_0 \leq \frac{2}{3}S$. For this case there is only elastic deformation and (11 A) transforms into

$$\frac{\Delta v}{v^\beta} = \tfrac{9}{2}(1-\nu)\frac{P_0}{E} \quad (12\,A)$$

Deformation by creep

The rate of expansion of the spherical hole in the matrix is here of interest. Let r be the radius of the hole and dr the change in the time dt. Consider two concentric shells in the matrix with the radii ϱ_1 and ϱ where ϱ_1 is equal to the radius r of the hole at the time t and ϱ is any value larger than ϱ_1. By neglecting the volume change due to the change in the state of compression, one finds for the time interval $t - dt$ to t

$$4\pi\varrho^2 \frac{d\varrho}{dt} = 4\pi\varrho_1^2 \frac{d\varrho_1}{dt} \quad (13\,A)$$

where $d\varrho$ and $d\varrho_1$ are the mechanical movement outwards of the two shells during this interval. For the radial strain rates of the material in the two shells one finds

$$\dot\varepsilon(\varrho) = \frac{2}{\varrho}\frac{d\varrho}{dt} = \left(\frac{\varrho_1}{\varrho}\right)^3 \frac{2}{\varrho_1}\frac{d\varrho_1}{dt} = \left(\frac{\varrho_1}{\varrho}\right)^3 \dot\varepsilon(\varrho_1) \quad (14\,A)$$

It has been found that the equation

$$S = s(\dot\varepsilon)^n \quad (15\,A)$$

where s and n are two constants, in many cases holds quite well for secondary creep. It might therefore be a good approximation for the mechanical behavior of the matrix, especially for the material rather close to the expanding hole, where the strain is very large. For the material in the two shells one obtains the relation

$$S(\varrho) = \left(\frac{\varrho_1}{\varrho}\right)^{3n} S(\varrho_1) \qquad (16\,A)$$

Equation (2 A) now gives for the difference in radial stress at the distances ϱ and $\varrho + \Delta\varrho$ from the center

$$\Delta\sigma_1(\varrho) = 2\,S(\varrho)\frac{\Delta\varrho}{\varrho} = 2\frac{\Delta\varrho}{\varrho^{1+3n}}\varrho_1^{3n} S(\varrho_1) \qquad (17\,A)$$

Integration and the use of the value $\sigma_1(\varrho) = -P_0$ at $\varrho = \varrho_1$ at the time t yields

$$\sigma_1(\varrho) = -P_0 + \frac{2\,S(\varrho_1)}{3\,n}\left[1-\left(\frac{\varrho_1}{\varrho}\right)^{3n}\right] \qquad (18\,A)$$

For large values of ϱ, σ_1 is small compared with P_0 and therefore

$$S(\varrho_1) \cong \frac{3\,n\,P_0}{2} \qquad (19\,A)$$

if n is not very small. Equations (18 A) and (19 A) give

$$\sigma_1(\varrho) = -P_0\left(\frac{\varrho_1}{\varrho}\right)^{3n} \qquad (20\,A)$$

The increase of the volume Δv which is made availabe for the new phase due to the deformation of the matrix is thus during the time interval $t-dt$ to t

$$d\Delta v = d\left(\frac{4}{3}\pi\varrho_1^3\right) = 4\pi\varrho_1^2 d\varrho_1 = 2\pi\varrho_1^3\frac{2}{\varrho_1}\frac{d\varrho_1}{dt}dt = 2\pi\varrho_1^3\dot\varepsilon(\varrho_1)dt$$

$$= 2\pi\varrho_1^3\left(\frac{S(\varrho_1)}{s}\right)^{\frac{1}{n}}dt = 2\pi\varrho_1^3\left(\frac{3\,n\,P_0}{2s}\right)^{\frac{1}{n}}dt \qquad (21\,A)$$

where ϱ_1 can be substituted by r. Some room for the new phase is also provided by the transformation of a part of the matrix. The rate of growth of the new phase, i.e. the hole, is thus at the time t

$$\frac{d(\frac{4}{3}\pi r^3)}{dt} = \frac{dV}{dt} \cong \frac{dv^\beta}{dt} = \frac{v^\beta}{\Delta v}\frac{d\Delta v}{dt} = 2\pi r^3\left(\frac{3\,n\,P_0}{2s}\right)^{\frac{1}{n}}\frac{v^\beta}{\Delta v} \qquad (22\,A)$$

The hydrostatic pressure in the matrix can also be derived

$$P(\varrho) = -\frac{1}{3}(\sigma_1(\varrho)+\sigma_2(\varrho)+\sigma_3(\varrho)) = -\sigma_1(\varrho)-\frac{2}{3}S(\varrho)$$

$$= P_0\left(\frac{r}{\varrho}\right)^{3n} - \frac{2}{3}\left(\frac{r}{\varrho}\right)^{3n}\cdot\frac{3\,n\,P_0}{2} = (1-n)\,P_0\left(\frac{r}{\varrho}\right)^{3n} \qquad (23\,A)$$

88   MATS HILLERT

Appendix II

Calculation of the Deformation Energy

Ideally plastic model

The elastic energy of a system can easily be calculated from the well-known expression for the energy density

$$w = \frac{1}{2E}[\sigma_1^2 + \sigma_2^2 + \sigma_3^2 - 2\nu(\sigma_1\sigma_2 + \sigma_2\sigma_3 + \sigma_3\sigma_1)] \tag{24 A}$$

The calculation is straight-forward and for the elastic energy in the spherical particle, in the plastically deformed shell of the matrix and in the elastically deformed part of the matrix one obtains

$$W_1 = \frac{4}{3}\pi r^3 \frac{3(1-2\nu)}{2E} P_0^2 \tag{25 A}$$

$$W_2 = \frac{4}{3}\pi r^3 \frac{1}{E}\left[\frac{1}{3}(5-7\nu)\left(\frac{R}{r}\right)^3 S^2 - \frac{3}{2}(1-2\nu)P_0^2 - (1-\nu)S^2\right] \tag{26 A}$$

$$W_3 = \frac{4}{3}\pi r^3 \frac{1}{E}\frac{1}{3}(1+\nu)\left(\frac{R}{r}\right)^3 S^2 \tag{27 A}$$

By adding these expressions and substituting the value of R/r given by equation (5 A) one finds for the total elastic energy

$$\frac{dW_{el}}{dV} = \frac{1-\nu}{E} S^2 \left[\frac{\Delta v}{v^\beta} \frac{2E}{3(1-\nu)S} - 1\right] \tag{28 A}$$

An exact calculation of the plastic work is more difficult, and for the sake of simplicity we shall neglect the changes in volume in the plastically deformed part of the matrix due to the remaining elastic strains. The relation between the mechanical enlargement $d\varrho$ of a shell with the radius ϱ and the enlargement dr of the hole is then obtained from equation (13 A).

$$\varrho^2 d\varrho \simeq \varrho_1^2 d\varrho_1 = r^2 dr \frac{\Delta v}{v^\beta} \tag{29 A}$$

The plastic work per unit volume is found by an integration of the product $S d\varepsilon$ where $d\varepsilon = 2(d\varrho/\varrho)$ (see equation 14 A). The plastic work on the matrix when the radius of the particle increases from r to $r + dr$ is thus

$$dW_{pl} = \int_r^R 4\pi\varrho^2 d\varrho \cdot S \cdot 2\frac{d\varrho}{\varrho} = 4\pi r^2 dr \frac{\Delta v}{v^\beta} \cdot 2S \int_r^R \frac{d\varrho}{\varrho}$$

$$= 4\pi r^2 dr \frac{\Delta v}{v^\beta} \cdot 2S \ln \frac{R}{r} \tag{30 A}$$

Jernkont. Ann. 141 (1957): 2

PRESSURE-INDUCED DIFFUSION AND DEFORMATION 89

The volume increase of the particle is $4\pi r^2 dr$. By inserting the value of R/r from equation (5 A) and eliminating P_0 by means of (11 A) one obtains

$$\frac{dW_{pl}}{dV} = \frac{\Delta v}{v^\beta} \cdot \frac{2}{3} S \ln\left[\frac{\Delta v}{v^\beta} \frac{E}{3(1-\nu)S}\right] \tag{31 A}$$

Elastic model

Equation (31 A) is now transformed into

$$\frac{dW_{el}}{dV} = \frac{E}{9(1-\nu)}\left(\frac{\Delta v}{v^\beta}\right)^2 \tag{32 A}$$

Deformation by creep

Here we shall again neglect the elastic strains and only calculate the plastic work.

$$\frac{dW_{pl}}{dt} = \int_{\varrho_1}^{\infty} 4\pi \varrho^2 d\varrho \cdot S(\varrho) \frac{d\varepsilon(\varrho)}{dt} = 4\pi \varrho_1^3 P_0 \cdot \frac{1}{2} \frac{d\varepsilon(\varrho_1)}{dt}$$

$$= 4\pi \varrho_1^3 \cdot P_0 \cdot \frac{1}{\varrho_1} \frac{d\varrho_1}{dt} = 4\pi r^2 \frac{dr}{dt} \cdot \frac{\Delta v}{v^\beta} P_0 \tag{33 A}$$

$$\frac{dW_{pl}}{dV} = \frac{\Delta v}{v^\beta} \cdot P_0 \tag{34 A}$$

4 Introduction to "The Role of Interfacial Energy during Solid State Phase Transformations"

published in *Jernkontorets Annaler* (1957)

Over his career, Mats Hillert has made many profound contributions to the classical theory of phase transformations in solids. This one was among the first. Upon its publication, this paper was quickly recognized as representing not one, but a series of state-of the-arts advances. The scope of the work is astonishing; the subjects covered in this single publication could well form the basis for a textbook on phase transformations, with an emphasis on ferrous materials. The work is specific to diffusional transformation processes in binary systems, but touches on the martensite/bainite debate and several other still-current issues.

Consider that the sections include:

I. The difference in pressure caused by interfacial energy.
II. The effect of pressure on the activities and equilibrium compositions.
III. Nucleation.
IV. Grain growth.
V. Formation of a new phase:
Approximate calculation;
Rate of edgewise growth of a plate;
Shape of a growing plate;
Growth rate of a needle;
Growth of a plate along a grain boundary.
VI. Formation of two new phases:

Approximate treatment;

More accurate calculation.

VII. Comparison of separate and cooperative growth.
VIII. The nature of bainite.

Summary.
Appendices:
Concentration dependency of the diffusion coefficient;
Diffusion through ferrite;
Mechanical deformation.

On reading this "index" (which arose from the original paper) one is immediately struck by the scope of the subject matter and by the way in which the theoretical treatments are grounded in a deep interest in, and knowledge of, ferrous metallography.

The list of topics only hints at the depth to which these subjects are taken. The treatment of nucleation in section III, for example, allows for the first time the determination of the composition and size of a classical critical nucleus for the case where the partial molar volumes of the two components are different (*e.g.* the case of an interstitial solute). The treatment of grain growth (IV) shows how the driving force for grain growth can be evaluated, again in the case of a highly mobile interstitial solute.

Section V, on the formation of a new phase, is one of two most often cited. The detailed analysis of the edgewise growth of plates and needle crystals by volume diffusion is an elegant extension of Zener's work, who gave approximate solutions to these and other phase growth problems of in his celebrated 1946 paper [1]. Like Zener, Hillert assumed a local equilibrium at the precipitate-matrix interface. Unlike Ivantsov [2], who had found an exact solution for the growth of the isoconcentrate paraboloid, both Zener and Hillert included capillarity in their treatments, which, as a result, defined a critical radius of curvature at the tip (r_c) for which the diffusion gradient at the growing tip (and the tip velocity) would vanish. For larger values of the tip radius r, the analysis yielded values of the growth velocity V which increased to a maximum at $r = 2r_c$, decreasing for still larger values. Hillert's analyses also contained several quite reasonable approximations, including the fitting of an exponential diffusion profile in advance of the growing precipitate tips. At the end of this section, he noted that the diffusion-controlled growth of a needle will be faster than that of a plate, all other factors being equal, and then proceeded to a discussion of the likely effects of anisotropy of interfacial free energy on the choice of plate versus needle growth. The "Zener-Hillert" equation(s) have stood the test of time, and have been consistently utilized in the analysis of precipitate growth kinetics.

Section VI, on the cooperative growth of two phases, again builds on one of Zener's approximate analyses, this time concerning the growth of pearlite in steel. Several other researchers, notably Scheil [3] and Brandt [4], had also contributed to the development of an analytical model of the steady process, and Brandt had provided a first-order estimation of the shape of the growth front. Hillert offered a complete and elegant solution to the problem of volume-diffusion controlled growth of pearlite (including the precise shapes of the growth fronts) for a range of spacings. These varied, ranging from critical spacing (for which both phases have uniform curvature equal to the critical value for nucleation, and at which the growth rate vanishes) to that corresponding to a maximum in the growth rate (at twice the critical spacing) and beyond (his Figure 6). Like his treatment of single phase plate/needle lengthening, Hillert's analysis of cooperative growth yielded a relationship between a rate of growth and a characteristic length; in each case, the maximum growth rate corresponds to twice the critical spacing or radius.

In section VII, he compared the maximum rates of "separate and cooperative growth" for systems supersaturated with respect to two phases, and concluded that the rate of transformation will generally be greater if the two phases cooperate. However, he argued that separate growth would be favored if one of the phases has a low interfacial energy. This leads naturally into the section VIII, on the nature of bainite, which he considers to be, like pearlite, a product of eutectoid decomposition.

Section VII therefore also contains an early expression of Hillert's arguments in support of a diffusional (as opposed to diffusionless [1,5]) mechanism of bainite formation. He noted that surface relief, as demonstrated by Ko and Cottrell [6] is not exclusive to martensitic transformations, but is also found for higher temperature Widmanstätten ferrite. Finally, he suggested that a product of transformation of higher carbon austenite, termed "inverse bainite" should form, with cementite as the leading phase, and provided preliminary metallographic evidence for its existence.

This paper represents Mats Hillert's early and incisive contribution to the "classical" phase transformation theory. It effectively closed the book on the subject of pearlite growth by volume diffusion of carbon, and significantly advanced the theory of Widmanstätten precipitate growth. This paper also paved the way for many of his later contributions, several of which are included in the current volume of selected works.

It is clear from this paper alone that we owe an enormous debt of gratitude to Mats Hillert for his seminal contributions, as so many of these have formed the basis for our current understanding of the nature of phase transformations.

Gary PURDY

References

[1] C. Zener, *Trans. AIME* 167 (1946) 550.

[2] G.P. Ivantsov, *Dokl. Akad. Nauk SSSR* 58 (1947) 567.

[3] E. Scheil, *Z. Metallkunde* 37 (1946) 123.

[4] W.H. Brandt, *J. Appl. Phys.* 16 (1945) 139.

[5] H.K.D.H. Bhadeshia and D.V. Edmonds, *Acta Metall.* 28 (1980) 1265.

[6] T. Ko and S.A. Cottrell, *J. I. S. I.* 172 (1952) 307.

The Role of Interfacial Energy during Solid State Phase Transformations

by

MATS HILLERT

Communication from the Swedish Institute for Metal Research

Introduction

The role of interfacial energy during the grain growth process and during the very first stage of phase transformations, the so-called nucleation process, has long been recognized and many quantitative treatments have been published on this subject.[1,2] On the other hand, the effect of interfacial energy on the kinetics of phase transformations has been largely neglected. An exception is the treatment by Zener of the growth rate of pearlite.[3] By a simple computation, based on dimensional arguments, Zener gave an explanation and a quantitative description of the variation with temperature of the interlamellar spacing of pearlite.

The present paper will be limited to binary systems and will deal mainly with solid state transformations. The effect of interfacial energy on the composition of two crystals separated by a curved interface will be considered. The derived equations will first be applied to nucleation and grain growth. The main portion of the paper, however, will deal with the growth of a new phase and of an aggregate of two new phases from a parent phase. Special attention will be paid to the isothermal transformation of austenite and the nature of bainite.

I. The Difference in Pressure Caused by Interfacial Energy

It is customary to consider the effect of interfaces on different processes as caused by the extra energy required for the formation of a certain amount of interface. It is equally justified, however, to consider the pressure difference between the two sides of a curved interface as the driving force. This pressure difference is related to the interfacial energy σ by

$$\Delta P = \sigma \left(\frac{1}{r_1} + \frac{1}{r_2} \right) \tag{1}$$

where r_1 and r_2 are the two principal radii of curvature of the interface. This method has been chosen for the present paper, since it allows a more detailed treatment of many processes. A previous paper,[4] which considered the effect of non-uniform pressure on precipitation, was mainly concerned with pressure differences caused by volume changes. Such differences will here be neglected.

If the curvature varies along an interface, the pressure must vary at least within one of the two crystals. This may cause mechanical deformation. It will be assumed, however, that the strength of the material is so high that this deformation can be neglected for our present purpose. It will, furthermore, be assumed that the heat conductivity is so high that there are no temperature gradients and also that local chemical equilibrium is everywhere established. The rate of a process is then controlled by the diffusion of material. In order to simplify the mathematical treatment, the diffusion coefficient D will be treated as independent of composition and the interfacial energy σ of a certain interface will be treated as a constant along the whole interface.

The notation adopted by Wagner[5] has been chosen but the symbol P will be used to denote the overpressure, i.e. the difference between the actual pressure and the normal atmospheric pressure.

II. The Effect of Pressure on the Activities and Equilibrium Compositions

The activity of a component in a binary system varies with composition and pressure. Consequently,

$$d\,RT \ln a_A(x_A, P) = \left(\frac{\partial RT \ln a_A(x_A, P)}{\partial x_A}\right)_P dx_A + \left(\frac{\partial RT \ln a_A(x_A, P)}{\partial P}\right)_{x_A} dP \quad (2)$$

For relatively low pressures it seems reasonable to assume that the two partial derivatives have the same values as under normal pressure. The value of the latter derivative is obtained from the expression for Gibbs' free energy

$$dG = VdP - SdT + RT \ln a_A \cdot dn_A + RT \ln a_B \cdot dn_B \quad (3)$$

This is an exact differential and consequently,

$$\left(\frac{\partial RT \ln a_A}{\partial P}\right)_{T, n_A, n_B} = \left(\frac{\partial V}{\partial n_A}\right)_{P, T n_B} = \overline{V}_A \quad (4)$$

By integration
$$RT \ln \frac{a_A(x_A, P)}{a_A(x_A, 0)} = \overline{V}_A P \quad (5)$$

When chemical equilibrium is established between two phases, α and β, each under its own hydrostatic pressure, P^α and P^β, one obtains

$$RT \ln \frac{a_A^\beta (x_A^\beta, O)}{a_A^\alpha (x_A^\alpha, O)} = \overline{V}_A^\alpha P^\alpha - \overline{V}_A^\beta P^\beta \qquad (6)$$

$$RT \ln \frac{a_B^\beta (x_A^\beta, O)}{a_B^\alpha (x_A^\alpha, O)} = \overline{V}_B^\alpha P^\alpha - \overline{V}_B^\beta P^\beta \qquad (7)$$

For the sake of simplicity we shall in the following presume that the pressure is normal in one of the phases, e.g. $P^\beta = 0$.

In the next two sections it will be found that equations (6) and (7) are very convenient for showing the effect of pressure on different processes by means of free-energy diagrams. On the other hand, they are not suited for numerical calculations. The following method is more convenient in such cases. When two phases in a binary system are in equilibrium, their compositions (x_A^α, x_B^α) and (x_A^β, x_B^β) and the activity of the two components a_A and a_B have certain values. These values depend on the pressures in the two phases P^α and P^β in accordance with the Gibbs–Duhem relation.

$$x_A^\alpha \,\mathrm{d} \ln a_A + x_B^\alpha \,\mathrm{d} \ln a_B = V_m^\alpha \,\mathrm{d} P^\alpha / RT \qquad (8)$$

$$x_A^\beta \,\mathrm{d} \ln a_A + x_B^\beta \,\mathrm{d} \ln a_B = V_m^\beta \,\mathrm{d} P^\beta / RT \qquad (9)$$

The variation of the molar volumes V_m^α and V_m^β with pressure and composition is usually small enough to be neglected for our present purpose. In a previous paper[4] it was assumed that the variation of the equilibrium compositions could also be neglected. Equations (8) and (9) could then be integrated, yielding the equation

$$RT \ln \frac{a_B^P}{a_B^O} = \frac{P^\alpha V_m^\alpha / x_A^\alpha - P^\beta V_m^\beta / x_A^\beta}{z^\alpha - z^\beta} \qquad (10)$$

where $x_B/x_A = z$, and a_B^P and a_B^O denote the activity of B at equilibrium under the pressures P^α and P^β and under normal pressure, respectively.

In the present paper cases will be considered where the equilibrium compositions vary markedly. Equations (8) and (9) then yield the equation

$$(z^\alpha - z^\beta) RT \,\mathrm{d} \ln a_B = \mathrm{d} P^\alpha V_m^\alpha / x_A^\alpha - \mathrm{d} P^\beta V_m^\beta / x_A^\beta \qquad (11)$$

which cannot be integrated unless information on the thermodynamic properties of the two phases is available.

In the following we shall consider the growth of a phase α in a parent phase β and we shall assume that this process takes place without any volume change. There will then be no reason to expect any change of the pressure in the parent phase and, consequently, $dP^\beta = 0$. Furthermore, we shall treat V_m^α/x_A^α and V_m^β/x_A^β as constants and denote these quantities by V_A^α and V_A^β since they are the volumes which contain one mol of A atoms. The concentration measured as c_B mol B/cm^3 is then proportional to z mol $B/$mol A and the relation $z = c_B \cdot V_A$ holds. Finally, we shall limit ourselves to cases where the growing α phase is well defined as to its composition. In that case z^α and x_A^α are constants. For the parent phase, supposed to be under normal pressure, one could write $a_B = x_B \cdot f$ or $\ln a_B = \ln x_B + \ln f$. However, usually f is not a constant and thus one could as well write

$$\ln a_B = \ln z + \ln f_1 \tag{12}$$

which is more convenient for our present purpose. In a small range of composition, e.g. for very dilute solutions, we can treat f and f_1 as constants. Equation (12) can be used to cover a somewhat larger range of composition if we let

$$\ln f_1 = a + bz \tag{13}$$

where a and b are two constants. Combining equations (12) and (13) yields $d\ln a_B = (b + 1/z) \cdot dz$ which can be used for the integration of (11). One thus obtains

$$P^\alpha V_A^\alpha/RT = z^\alpha \ln z^{\beta\alpha}/z_0^{\beta\alpha} + (z^\alpha b - 1)(z^{\beta\alpha} - z_0^{\beta\alpha}) - \tfrac{1}{2} b (z^{\beta\alpha^2} - z_0^{\beta\alpha^2}) \tag{14}$$

where $z_0^{\beta\alpha}$ and $z^{\beta\alpha}$ are the equilibrium compositions of the β phase when the α phase is under normal pressure $(P=0)$ and under P^α, respectively.

When the change in composition $z^{\beta\alpha} - z_0^{\beta\alpha}$ is small, equation (14) can be approximated by

$$P^\alpha V_A^\alpha/RT = (z_0^{\beta\alpha} - z^{\beta\alpha})[-z^\alpha/z_0^{\beta\alpha} + 1 + b(z_0^{\beta\alpha} - z^\alpha)] \tag{15}$$

or

$$P^\alpha = RT k^{\beta\alpha}(z_0^{\beta\alpha} - z^{\beta\alpha})/V_A^\beta = RT k^{\beta\alpha}(c_0^{\beta\alpha} - c^{\beta\alpha}) \tag{16}$$

where

$$k^{\beta\alpha} = [1 + b(z_0^{\beta\alpha} - z^{\beta\alpha}) - z^\alpha/z_0^{\beta\alpha}] V_A^\beta/V_A^\alpha \tag{17}$$

According to measurements by R. P. Smith,[6] $b = 6.6$ for carbon in austenite at 800 and 1 000°C. We shall later apply this value to temperatures below A_1, where no direct measurements have been made.

III. Nucleation

Consider a spherical nucleus of a new phase surrounded by a parent phase with a homogeneous composition (x_A^β, x_B^β). At the critical size the nucleus can be said to be in unstable equilibrium with the parent phase

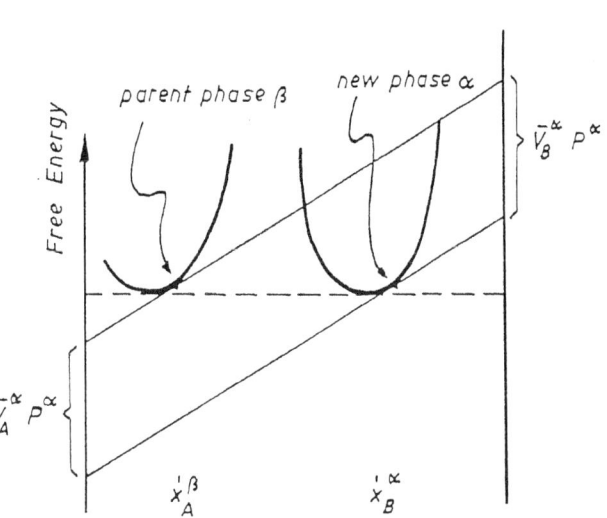

Fig. 1. Free energy diagram demonstrating the composition and size of a nucleus of a new phase.

and it must therefore satisfy equations (6) and (7). If we assume for the sake of simplicity that $\bar{V}_A^\alpha = \bar{V}_B^\alpha = V_m^\alpha$ and $P^\beta = 0$, we obtain

$$RT \ln \frac{a_A^\beta(x_A^\beta, 0)}{a_A^\alpha(x_A^\alpha, 0)} = RT \ln \frac{a_B^\beta(x_A^\beta, 0)}{a_B^\alpha(x_A^\alpha, 0)} = V_m^\alpha P^\alpha \qquad (18)$$

The composition of the critical nucleus can thus be obtained in the free-energy diagram by drawing a tangent to the free-energy curve of the new phase parallel to the tangent through the point which represents the parent phase, as shown in Fig. 1. This fact was pointed out in a previous paper.[7] The distance between these tangents will be $V_m^\alpha P^\alpha$ and the radius of the critical nucleus is thus easily obtained from the relation $V_m^\alpha P^\alpha = 2\sigma V_m^\alpha / r$.

It should be noticed that our conclusion is correct only when the interfacial energy σ is independent of the composition of the nucleus. Otherwise the pressure P^α would depend on the composition and the basic equation (2) would not hold.

As the nucleus grows, r increases and the pressure P^α decreases. Assuming that local chemical equilibrium is established at the surface of the nucleus during the growth process, one can calculate the compositions of the two phases at this surface for any size by drawing two parallel tangents, one for each free-energy curve, with the distance $2\sigma V_m^\alpha / r$. The distance will thus decrease as the size r grows and finally the two tangents will coincide to give the well-known double-tangent, which denotes the stable compositions of the two phases (dotted line in Fig. 1).

When $\bar{V}_A^\alpha \neq \bar{V}_B^\alpha \neq V_m^\alpha$, the tangents should not be drawn parallel to each other, but in such a way that the distance between the intersections on the axes gives the ratio $\bar{V}_A^\alpha / \bar{V}_B^\alpha$.

Jernkont. Ann. 141 (1957): 11

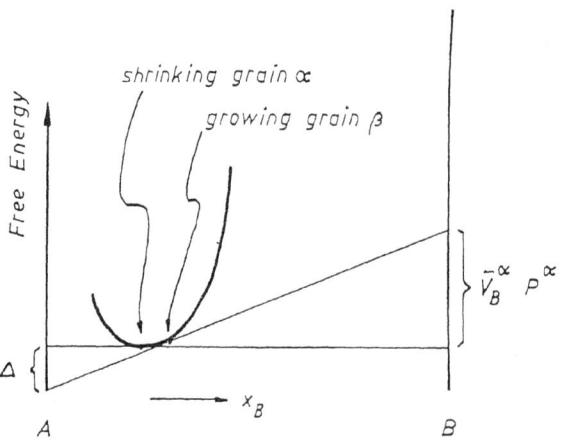

Fig. 2. Free energy diagram demonstrating the different compositions of two adjacent grains during grain growth.

IV. Grain Growth

It is generally accepted[1] that grain growth in a single-phase material is caused by the tendency of a curved grain boundary to move in the direction toward the center of curvature. The grain boundary area and, consequently, the total grain boundary energy, is decreased by such a movement and this fact is usually considered as the cause of the movement. In some cases, however, it may be more convenient to consider the pressure difference between the two sides of a curved boundary as the cause. Equation (5) shows that the activities of the components A and B are higher on the compressed side of a grain boundary and, consequently, the atoms will have a tendency to move over to the other side.

It is apparent that the two conditions for local equilibrium (6) and (7) cannot be satisfied if the two tangents in Fig. 1 belong to the same free-energy curve, hence local equilibrium cannot be established in this case where a curved interface separates two grains, α and β, of the same phase. In a binary system where one component is much more mobile than the other, e.g. Fe–C, it seems reasonable to suppose that the mobile component is in local equilibrium. Fig. 2 illustrates the position of the two tangents in this case when B is the mobile component. It is shown that a certain displacement of the composition can be expected. At the grain boundary, the growing grain β has a higher B content than the shrinking grain α. The ratio of the activities of the sluggish component A on the two sides of the boundary is

$$RT \ln a_A^\alpha / a_A^\beta = \overline{V}_A^\alpha P^\alpha + \Delta \tag{19}$$

where Δ is the distance between the intersections of the tangents with the A axis. If the displacement of the composition is small, one obtains

SOLID STATE PHASE TRANSFORMATIONS

$$\Delta = \overline{V}_B^\alpha P^\alpha x_B^\alpha / x_A^\alpha \tag{20}$$

and $\quad R T \ln a_A^\alpha / a_A^\beta = (\overline{V}_A^\alpha P^\alpha x_A^\alpha + \overline{V}_B^\alpha P^\alpha x_B^\alpha)/x_A^\alpha = V_m^\alpha P^\alpha / x_A^\alpha \tag{21}$

Hence, this is the driving force for the A atoms of the shrinking grain to rearrange themselves into the lattice of the growing grain.

V. Formation of a New Phase

Suppose that a new phase α is formed with a plane front facing the parent phase β and that the two phases have different compositions. If the rate of formation is controlled by diffusion, the plane front will advance with a decreasing rate according to a parabolic law

$$\frac{v}{D} = \frac{\text{constant}}{\sqrt{Dt}} \tag{22}$$

provided the front remains plane.[8] It has been pointed out, however, that a plane front is not stable unless the parent phase is supersaturated only at the front.[9, 10] This case can be realized at the solidification of a melt under external cooling. It cannot be realized if the reaction is so slow that all of the parent phase is undercooled before the transformation occurs. In that case any little roughness on a plane front will be further exaggerated as the transformation proceeds. Thus, if a small region of the front is only slightly ahead of the rest it will grow with a higher speed because of its favoured position with respect to the supersaturated parent phase. It will therefore grow even further ahead of the rest of the front and its speed will continue to increase. A needle with an increasingly sharp edge is developed by this process. Fig. 3 seems to be a two-dimensional example of such a mechanism.

The sharpness is finally limited by the interfacial energy. If the needle becomes too sharp the rate of growth of its extreme end will decrease because the favoured position is more than balanced by the interfacial energy, hence the sharpness will automatically be reduced again. The shape of the needle close to its edge will thus automatically be adjusted in such a way that it continues to grow with a constant maximum velocity and without further changing this shape. This velocity and shape will now be calculated.

A. *Approximate calculation*

The maximum growth velocity of a needle and the related curvature of its end can be estimated by the method applied by Zener in a treatment of the growth rate of pearlite.[3]

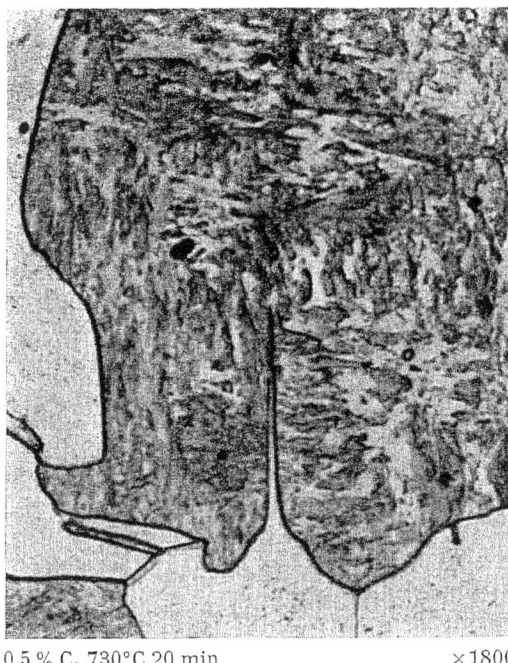

0.5 % C, 730°C 20 min   ×1800

Fig. 3. The formation of a plate from a growing crystal of ferrite.

Consider such a small part of the end that its shape can be described by a constant value of the radius of curvature r. This curvature results in a pressure difference between the growing phase and the parent phase and the equilibrium composition of the latter phase is changed approximately in accordance with equation (16). The difference in concentration of the parent phase at the growing phase interface and far away would be $c_0^{\beta\alpha} - c_\infty^\beta$ if there were no pressure difference, i.e. if the front was plane. With the radius r the concentration difference is $(c_0^{\alpha\beta} - c_\infty^\beta)(1 - r_c/r)$ where r_c is a constant, which can be identified with the radius of curvature that would completely eliminate the concentration difference. This critical radius r_c will thus give a zero growth rate and can be estimated by means of equation (16). The effective diffusion distance is approximately proportional to the radius of curvature, according to Zener, and it can be represented by $a \cdot r$. The diffusion can now be estimated

$$J = -D \frac{dc}{dx} \cong D \frac{(c_0^{\beta\alpha} - c_\infty^\beta)(1 - r_c/r)}{a r} \tag{23}$$

This diffusion makes possible the growth of the new phase with a certain rate v.

$$J = v(c_\infty^\beta - c^\alpha) \tag{24}$$

Combination of equation (23) and (24) gives

$$\frac{v}{D} = \frac{1}{a} \frac{c_0^{\beta\alpha} - c_\infty^\beta}{c_\infty^\beta - c^\alpha} \frac{1}{r} \left(1 - \frac{r_c}{r}\right) \tag{25}$$

It can easily be shown by a derivation that the maximum growth rate is obtained at $r = 2r_c$.

The above calculation has the advantage of simplicity but it is difficult to judge how good the approximations are and to estimate the value of the constant a.

Exactly the same expression as (25) is obtained for the edgewise growth of a plate but the values of a are of course different in the two cases. Zener's method of calculation gives no information on the relative magnitude of these values and as a consequence it cannot be used for a comparison of the growth rates of a needle and a plate. A more accurate calculation must be based on the variation of the concentration in the whole of the parent phase and on the shape of the whole needle. An exact treatment appears very complicated but the method will be outlined in principle in the next section and an approximate calculation will be made with the aid of some simplifying assumptions.

B. Rate of edgewise growth of a plate

Consider a plane front advancing in the x-direction. As already described any roughness will be exaggerated on further growth and finally a new shape of the front is developed which is stable. At first we shall limit the treatment to variations of the shape in only one direction, the y-direction, and the stable shape is then a plate rather than a needle. We shall neglect all diffusion through the new phase. Then the following equation holds for the diffusion in the parent phase β of the component which must be removed in order for the α phase to advance:

$$\frac{1}{D} \frac{\partial c^\beta}{\partial t} = \frac{\partial^2 c^\beta}{\partial x^2} + \frac{\partial^2 c^\beta}{\partial y^2} \tag{26}$$

We want to find a solution, $c^\beta(t, x, y)$, which gives a constant growth rate v. Such a solution can be considered as a function of only two variables, $u = x - vt$ and y. It can be determined by separation of the variables

$$c^\beta(u, y) = g(u) \cdot h(y) \tag{27}$$

Equation (26) then yields

$$\frac{1}{g}\left(\frac{d^2 g}{du^2} + \frac{v}{D}\frac{dg}{du}\right) = -\frac{1}{h}\frac{d^2 h}{dy^2} = -b^2 \tag{28}$$

where b^2 is the separation constant. The solution of (28) is

$$g = e^{\lambda u} \quad \text{where} \quad \lambda = \frac{-v}{2D}\left(1 + \sqrt{1 + 4b^2 D^2/v^2}\right) \tag{29}$$

$$h = \cos by \tag{30}$$

The constant b can have any value and the complete solution is thus

$$c^\beta - c^\beta_\infty = \int_0^\infty A(b)\, e^{\lambda u} \cos by\, db \tag{31}$$

where c^β_∞ is the concentration of the parent phase far away from the front. The problem is now to determine the function $A(b)$.

Consider a thin volume element from $x = -\infty$ to $x = +\infty$. The rate of decrease of the component considered is $v(c^\beta_\infty - c^\alpha)$ but it must also be equal to the loss by diffusion in the y-direction which we shall denote by $D \cdot F$, where

$$D \cdot F = -\int_{-\infty}^{\infty} D\, \frac{\partial^2 c}{\partial y^2}\, du = -D \int_{u_0}^{\infty} \frac{\partial^2 c^\beta}{\partial y^2}\, du = D \int_{u_0}^{\infty}\int_0^\infty b^2 A(b)\, e^{\lambda u} \cos by\, du\, db$$

$$= D \int_0^\infty \frac{b^2}{-\lambda} A(b)\, e^{\lambda u_0} \cos by\, db \tag{32}$$

The diffusion inside the α phase has been neglected here. u_0 denotes the varying u value along the front. The following equation is thus obtained for the shape of the plate, i.e. for the variation of u_0 with y.

$$\frac{v}{D}(c^\beta_\infty - c^\alpha) = \int_0^\infty \frac{b^2}{-\lambda} A(b)\, e^{\lambda u_0} \cos by\, db \tag{33}$$

The left-hand side does not vary along the front and the radius of curvature of the front can thus be obtained from this equation as

$$r = -\frac{F_{u_0}[1 + (F_y/F_{u_0})^2]^{3/2}}{F_{yy} - 2F_{y u_0} F_y/F_{u_0} + F_{u_0 u_0} F_y^2/F_{u_0}^2} \tag{34}$$

where F_y etc. denote the partial derivatives of F with respect to the variables represented by the indices. The radius of curvature finally determines the change in composition from the equilibrium value along the front according to equation (16)

$$c_0^{\beta\alpha} - c^{\beta\alpha} = \frac{P^\alpha}{RT k^{\beta\alpha}} = \frac{\sigma^{\alpha\beta}}{RT k^{\beta\alpha} r} \tag{35}$$

The concentration $c^{\beta\alpha}$ in the parent phase along the front is thus determined by the function $A(b)$ using equations (32) and (34) but may also be calculated directly from equation (31). The problem is to find such a function $A(b)$ that these two ways to calculate $c^{\beta\alpha}$ give the same result. It may be possible to solve this problem by successive approximations but only a first approximation will be attempted here.

Let the origin be at the edge of the plate. In the plane $u=0$ the parent phase must be symmetric around $y=0$, must also reach the value c_∞^β at $y=\pm\infty$ and go through a maximum at $y=0$. We can thus represent the composition in the plane $u=0$ with the following expression in the approximate calculation

$$c^\beta - c_\infty^\beta = K \cdot e^{-\pi y^2/16\, m^2} \tag{36}$$

where K and m^2 are two constants which will be determined by the boundary conditions.

At $u=0$ the exponential factor in (31) is unity and this equation is simply a Fourier representation of the above function

$$\int_0^\infty A(b)\cos by\, db = K \cdot e^{-\pi y^2/16\, m^2} \tag{37}$$

$A(b)$ can thus be determined as

$$A(b) = \frac{4}{\pi} K m\, e^{-4 b^2 m^2/\pi} \tag{38}$$

Inserting (38) in (33) yields

$$\frac{v}{D}(c_\infty^\beta - c^\alpha) = \frac{4}{\pi} K m \int_0^\infty \frac{b^2}{-\lambda} e^{-4 b^2 m^2/\pi} db \tag{39}$$

This integral cannot be solved exactly because λ is a function of b. However, for reasonably small m one can apply the approximation $-\lambda \simeq b$, obtaining:

$$K \cong 2\, \frac{v m}{D}(c_\infty^\beta - c^\alpha) \tag{40}$$

At the very edge of the plate $du_0/dy = 0$ and (34) is simplified to

$$\frac{1}{r} = -\frac{F_{yy}}{F_{u_0}} = \int_0^\infty \frac{b^4}{-\lambda} A(b)\, db \bigg/ \int_0^\infty b^2 A(b)\, db \tag{41}$$

Inserting (38) in (41) yields

$$\frac{1}{r} = \frac{32}{\pi^2} m^3 \int_0^\infty \frac{b^4}{-\lambda} e^{-4 b^2 m^2/\pi} \, db \qquad (42)$$

This integral also cannot be solved exactly but the approximation $-\lambda = b$ now gives

$$\frac{1}{r} \cong \frac{1}{m} \qquad (43)$$

The growth rate can now be calculated if equation (35) is applied at the origin

$$c_0^{\beta\alpha} - c_\infty^\beta - K = \frac{\sigma^{\alpha\beta}}{RT k^{\beta\alpha} r} \qquad (44)$$

By inserting the respective expressions for K and r (40) and (43) in (44), we finally obtain

$$\frac{v}{D} = \frac{1}{2} \frac{c_0^{\beta\alpha} - c_\infty^\beta}{c_\infty^\beta - c^\alpha} \frac{1}{m} \left(1 - \frac{m_c}{m}\right) \qquad (45)$$

where the constant m_c is given as

$$m_c = \sigma^{\alpha\beta} / RT k^{\beta\alpha} (c_0^{\beta\alpha} - c_\infty^\beta) \qquad (46)$$

It can be shown by derivation that the maximum growth rate is obtained at $m = 2 m_c$. The constant m_c can be considered as the critical radius of curvature r_c giving the zero growth rate, since $m \cong r$.

Equation (45) is very similar to equation (25) derived by Zener's method. For complete agreement, the proportionality constant a in Zener's method must have the value 2 for a growing plate.

The maximum growth rate can be determined from (44) without the use of the approximate expressions for K and r, (40) and (43). A numerical calculation of the integrals (39) and (42) was performed for a series of m values and $(-v/D)_{max}$ was determined by inserting the results in (44). A deviation of only 10 % from equation (45) was found, thus showing that the approximate expressions for K and m appear to be quite valid.

C. *Shape of a growing plate*

If the expression (38) for $A(b)$ is inserted in (33) the following equation for the shape is obtained

$$\int_0^\infty \frac{b^2}{-\lambda} e^{\lambda u_o - 4 b^2 m^2/\pi} \cos by \, db = \frac{v}{D} (c_\infty^\beta - c^\alpha) \qquad (47)$$

Again this integral cannot be solved exactly. Without claiming any high degree of accuracy we shall make an approximate calculation assuming that λ can be approximated here by $-v/D$. By inserting (38) in (31) and integrating, one obtains

$$c^\beta - c^\beta_\infty = K \cdot \exp\left(-\frac{vu}{D} - \frac{\pi y^2}{16 m^2}\right) \tag{48}$$

If the calculation is continued using this expression, one finds the following equation for the shape close to the edge of the plate,

$$-\frac{v}{D} u_0 = \pi y^2/16 m^2 - \ln(1 - \pi y^2/8 m^2) \cong 3 \pi y^2/16 m^2 \tag{49}$$

and for the maximum growth rate,

$$\left(\frac{v}{D}\right)_{max} = \frac{R T k^{\beta\alpha}}{12 \sigma^{\alpha\beta}} \frac{(c_0^{\beta\alpha} - c^\beta_\infty)^2}{c^\beta_\infty - c^\alpha} \tag{50}$$

for

$$m^2 = \left(\frac{3\sqrt{\pi}\,\sigma^{\alpha\beta}}{R T k^{\beta\alpha}}\right)^2 \cdot \frac{c^\beta_\infty - c^\alpha}{(c_0^{\beta\alpha} - c^\beta_\infty)^3} \tag{51}$$

This growth rate is $\frac{2}{3}$ of the rate found in the previous section. It is apparent that the growth rate is not very sensitive to the approximations applied in carrying out the calculations.

D. Growth rate of a needle

If the growing phase develops into a needle, the diffusion in the parent phase is best described using cylindrical coordinates

$$\frac{1}{D}\frac{\partial c^\beta}{\partial t} = \frac{\partial^2 c^\beta}{\partial x^2} + \frac{\partial^2 c^\beta}{\partial \varrho^2} + \frac{1}{\varrho}\frac{\partial c^\beta}{\partial \varrho} \tag{52}$$

If this differential equation is solved in the same way as (26), one obtains

$$c^\beta - c^\beta_\infty = \int_0^\infty A(b)\, e^{\lambda u} J_0(b\varrho)\, db \tag{53}$$

where $J_0(b\varrho)$ is a Bessel function. It may be possible to carry out a calculation similar to that in section B. The simpler method in section C appears to be quite good, however, and will therefore be applied instead.

By comparison with (48) we shall assume that the concentration in the parent phase can be represented by

$$c^\beta - c^\beta_\infty = K \cdot \exp\left(-\frac{vu}{D} - \frac{\varrho^2}{l^2}\right) \tag{54}$$

where K and l^2 are constants. If one considers a volume element from $u = -\infty$ to $+\infty$ and neglects the diffusion in the new phase, i.e. between $u = -\infty$ and $u = u_0$, one now obtains

$$F = -\int_{u_0}^{\infty} \left(\frac{\partial^2 c^\beta}{\partial \varrho^2} + \frac{1}{\varrho}\frac{\partial c^\beta}{\partial \varrho}\right) du = \frac{v}{D}(c_\infty^\beta - c^\alpha) \qquad (55)$$

The shape of the needle is obtained by inserting (54) in (55), giving

$$-\frac{v}{D} u_0 = \varrho^2/l^2 - \ln(1 - \varrho^2/l^2) = 2\varrho^2/l^2 \qquad (56)$$

and the maximum growth rate is finally calculated as

$$\left(\frac{v}{D}\right)_{max} = \frac{R T k^{\beta\alpha}}{8\,\sigma^{\alpha\beta}} \frac{(c_0^{\beta\alpha} - c_\infty^\beta)^2}{c_\infty^\beta - c^\alpha} \qquad (57)$$

for

$$l^2 = 2\left(\frac{12\,\sigma^{\alpha\beta}}{R T k^{\beta\alpha}}\right)^2 \cdot \frac{c_\infty^\beta - c^\alpha}{(c_0^{\beta\alpha} - c_\infty^\beta)^3} \qquad (58)$$

A comparison of the corresponding expression for the plate, equation (50), shows that the maximum growth rate of a needle is 1.5 times greater than the growth rate of a plate. One should not expect this ratio to be quite exact on account of the approximate nature of the calculations but the order of magnitude may be correct. In general, then, we can expect a growing phase to develop into a needle and by a repetition of the same process sidewise dendrites can form. This is also the normal process for the solidification of melts. It should be noted, however, that the surface energy $\sigma^{\alpha\beta}$ is included in the expressions for the growth rate.

Up to now we have assumed that the surface energy has a constant value but it is well known that it generally varies for different crystallographic planes. One should thus expect to find that the growth rate is highest in a certain crystallographic direction. For the same reason a needle should not have an exactly circular cross section but a more polygonal-like shape which is energetically favoured. If one crystallographic plane has appreciably lower interfacial energy than all the others, which can occur in noncubic crystals, one should expect the formation of plates rather than needles. According to the approximate calculations above, the condition necessary for this to take place is that the favoured crystallographic plane have an interfacial energy less than $\frac{2}{3}$ of the value for any other plane. It is not inconceivable that this is the case with the hexagonal plane of graphite in an iron-carbon melt, thus explaining the formation of graphite flakes during the solidification of grey cast iron rather than needles or dendrites.

Jernkont. Ann. 141 (1957): 11

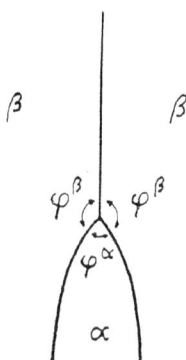

Fig. 4. The preferential growth of an α crystal along a β/β grain boundary.

For solid-state phase transformations it is quite usual for the new phase to form as plates coherent with the parent phase. They are often called Widmanstätten plates. The shape of these crystals as well as the fact that they form coherently rather than incoherently can be explained by the above discussion if the coherent interfaces have an energy lower than $\frac{2}{3}$ of the value for the most favoured incoherent interface. This condition may be satisfied since it is usually assumed that coherent interfaces have even lower energies than this critical limit.

It must be emphasized that the discussion above is based on the assumption that local equilibrium is everywhere established, including all interfaces. This can not always be the case as was shown for example in section IV. Explanations of the shape and crystallographic directions of dendrites as well as of the formation of thin, coherent plates have in fact been proposed based on the assumption that equilibrium is not established at the interfaces. Furthermore, one should not neglect the role of the nucleation process.

E. Growth of a plate along a grain boundary

It is a well-known observation that the formation of a new phase often starts in the grain boundaries and spreads along them. A grain boundary net work is thus formed. This cannot be fully explained by preferred nucleation in the grain boundaries. There must also be a higher rate of growth along the grain boundaries than at right angles to them. The approximate calculation in section V A can easily be modified to cover this case.

At the point of intersection of a growing α grain with two β grains, the grain boundaries will form certain angles with each other which can be calculated from the grain boundary energies (Fig. 4). Equation (25) was derived for the growth rate normal to the boundary of the new phase and for the edge of a growing plate this direction coincided with the actual direction of growth. When a plate is growing in a grain boundary, however,

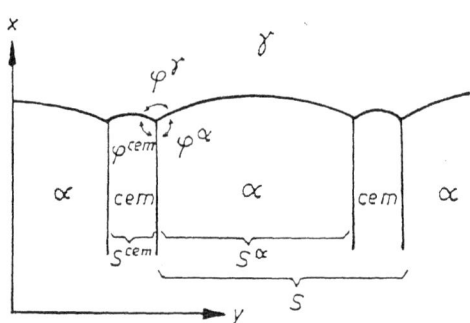

Fig. 5. The edgewise growth of pearlite.

the two directions do not coincide but differ by the angle $\varphi^\beta - \frac{\pi}{2}$. The rate of growth in the direction of the grain boundary is thus

$$\frac{v}{D} = \frac{1}{a \sin \varphi^\beta} \frac{c_0^{\beta\alpha} - c_\infty^\beta}{c_\infty^\beta - c^\alpha} \frac{1}{r} \left(1 - \frac{r_c}{r}\right) \tag{25 b}$$

where r is the radius of curvature of the α grain close to the edge. The growth rate is thus higher along a grain boundary because of the factor $\sin \varphi^\beta$, which is always less than 1, but probably also because of the factor a, which may be much less at the sharp edge of a plate growing in a grain boundary than at the smooth edge of a plate growing inside a grain.

VI. Formation of Two New Phases

For several transformations the end product consists of two new phases. This is the case, for example, for eutectic and eutectoid transformations. The two phases can then grow into the parent phase each independently as described in the previous section. Such a mechanism will be called in the following "separate growth". On the other hand, there is a possibility that each of the two phases can accelerate the others growth if they grow as an aggregate. This will be called in the following "cooperative growth". The formation of pearlite is an example of such a cooperative process. The two new phases, ferrite and cementite, are here formed as alternating lamellae (Fig. 5). We shall consider the growth of such an aggregate in the direction of the edges, the x direction, letting the y direction be normal to the lamellae. Let S^α and S^{cem} be the thicknesses of the two kinds of lamellae. Then S, the sum of $S^\alpha + S^{cem}$ is the period by which the structure is repeated in the y-direction and is called the interlamellar spacing. Like the growth rate of a separate phase which varies with the radius of curvature of its tip, the growth rate of the pearlitic aggregate will vary with the spacing S. The approximate treatment of this case by Zener[3] will first be outlined.

A. Approximate treatment

During the growth of pearlite carbon must be transported from the edge of each ferrite lamella to the edges of the neighbouring cementite lamellae. As before, we shall assume that this diffusion goes only through the parent phase, here the austenite. If the interfaces ferrite-austenite and cementite-austenite were plane, the concentration difference which provides the driving force for the diffusion would be $c_0^{\gamma\alpha} - c_0^{\gamma\text{cem}}$, where $c_0^{\gamma\alpha}$ and $c_0^{\gamma\text{cem}}$ denote the equilibrium concentrations according to the phase diagram for austenite in contact with ferrite and cementite, respectively. It was suggested by Zener that the real concentration difference could be represented approximately by $(1 - S_c/S)(c_0^{\gamma\alpha} - c_0^{\gamma\text{cem}})$ considering that the interfaces are curved. S_c is a constant that can be identified with a small, critical lamellar spacing for which there will be no concentration difference at all.

Furthermore, Zener suggested that the effective distance for the diffusion up to the edge of a cementite lamella is proportional to its thickness and that the proportionality constant a should be close to 1. The diffusion up to a cementite lamella is thus obtained as

$$J = -D\frac{dc}{dx} \cong -D\frac{c_0^{\gamma\alpha} - c_0^{\gamma\text{cem}}}{a S^{\text{cem}}}\left(1 - \frac{S_c}{S}\right) \qquad (59)$$

This diffusion makes possible an edgewise growth of the lamella with a velocity v given by

$$J = -v(c^{\text{cem}} - c^{\alpha}) \qquad (60)$$

The ratio between the thicknesses of the two kinds of lamellae is determined by the original composition of the austenite, c_∞^γ, which exists far away from the reaction front. Neglecting the volume change which accompanies the reaction, one finds

$$S^\alpha(c_\infty^\gamma - c^\alpha) = S^{\text{cem}}(c^{\text{cem}} - c_\infty^\gamma) = \frac{S^{\text{cem}} S^\alpha}{S}(c^{\text{cem}} - c^\alpha) \qquad (61)$$

Combination of (59), (60) and (61) gives

$$\frac{v}{D} = \frac{1}{a}\frac{S^2}{S^\alpha S^{\text{cem}}}\frac{c_0^{\gamma\alpha} - c_0^{\gamma\text{cem}}}{c^{\text{cem}} - c^\alpha}\frac{1}{S}\left(1 - \frac{S_c}{S}\right) \qquad (62)$$

The maximum growth rate $v_{\max}$ is found at $S = 2 S_c$ and from the arguments in section V concerning the shape of a growing plate Zener suggested that this would be the actual growth rate and spacing of pearlite. This suggestion is accepted in the present paper although it is realized that there is an important difference in the two cases. The curvature of the edge of a growing plate r can be adjusted continuously whereas the spacing of pearlite S can only change by an increase in the number of lamellae and the latter process may require an appreciable deviation from the optimum condition $S = 2 S_c$.

B. *More accurate calculation*

As in the treatment of the separate growth of a phase the calculation must now be based on the variation of concentration in the whole parent phase. The separation constant b^2 cannot have an arbitrary value in the present case, however, since the structure is periodic in the y-direction with the period S. Only those b values are permitted which satisfy the relationship

$$b_n = n \cdot \frac{2\pi}{S} \tag{63}$$

where n is an integer. The integrals obtained in section VB are thus transformed into summations. The variation of the concentration in the austenite, for example, is thus (cf. equation 31)

$$c^\gamma - c^\gamma_\infty = \sum_0^\infty A_n e^{\lambda_n u} \cos b_n y \tag{64}$$

As before the problem of determining the coefficients A_n may be solved rigorously by successive approximations but in this paper only a first approximate calculation will be presented. As shown in section V this calculation could be based on a reasonable assumption about the variation in the y-direction. Brandt,[11] for example, assumed a sinusoidal variation, i.e. $A_n = 0$ for all $n > 1$, while Scheil[12] used a parabolic variation.

Contrary to the previous calculations, which treated the separate growth of a phase, an alternative method can be applied for the present calculation which seems to have certain advantages. Experimental work has often shown that the interface between pearlite and austenite is fairly plane. We shall here assume that it is so plane that $\exp \lambda_n u$ can be approximated by the value 1 along the whole front. Combination of equations (33) and (61) then yields

$$\sum_0^\infty \frac{A_n b_n^2}{-\lambda_n} \cos b_n y = S^{\text{cem}} \cdot \frac{v}{DS} \cdot (c^{\text{cem}} - c^\alpha) \quad \text{for } -\frac{S^\alpha}{2} < y < \frac{S^\alpha}{2}$$

$$= -S^\alpha \cdot \frac{v}{DS} \cdot (c^{\text{cem}} - c^\alpha) \quad \text{for } \frac{S^\alpha}{2} < y < S - \frac{S^\alpha}{2} \tag{65}$$

The following values of the Fourier coefficients in this representation are then obtained

$$\frac{A_n b_n^2}{-\lambda_n} = \frac{4v}{DS b_n}(c^{\text{cem}} - c^\alpha) \sin n\pi S^\alpha/S \quad \text{for } n > 0 \tag{66}$$

When calculating the growth rate of pearlite we must utilize the fact that the angles φ^α, φ^γ and φ^{cem} between the three interfaces at a ferrite-austenite-cementite junction are determined by the interfacial energies. Providing that the ferrite-cementite interfaces between the lamellae are parallel with the x-direction, an integration along the front from the central axis of a ferrite lamella to the junction will yield

$$-\cos \varphi^\alpha = \int_0^{\frac{1}{2}S^\alpha} \frac{1}{r} \,\mathrm{d}y = \int_0^{\frac{1}{2}S^\alpha} \frac{P^\alpha}{\sigma^{\alpha\gamma}} \,\mathrm{d}y$$

$$= \frac{RT k^{\gamma\alpha}}{\sigma^{\alpha\gamma}} \int_0^{\frac{1}{2}S^\alpha} (c_0^{\gamma\alpha} - c^{\gamma\alpha}) \,\mathrm{d}y$$

$$= \frac{RT k^{\gamma\alpha}}{\sigma^{\alpha\gamma}} \int_0^{\frac{1}{2}S^\alpha} (c_0^{\gamma\alpha} - c_\infty^\gamma - \sum_0^\infty A_n \cos b_n y) \,\mathrm{d}y \qquad (67)$$

which gives

$$-\frac{2\sigma^{\alpha\gamma} \cos \varphi^\alpha}{S^\alpha RT k^{\gamma\alpha}} = c_0^{\gamma\alpha} - c_\infty^\gamma - A_0 - \frac{2}{S^\alpha} \sum_1^\infty \frac{A_n}{b_n} \sin n\pi S^\alpha/S \qquad (68)$$

In the same way one obtains by an integration along the edge of a cementite lamella

$$-\frac{2\sigma^{cem\,\gamma} \cos \varphi^{cem}}{S^{cem} RT k^{\gamma\,cem}} = c_0^{\gamma\,cem} - c_\infty^\gamma - A_0 + \frac{2}{S^{cem}} \sum_1^\infty \frac{A_n}{b_n} \sin n\pi S^\alpha/S \qquad (69)$$

A_0 and A_n for $n > 0$ can now be eliminated by a combination of (66), (68) and (69), and using the approximation $-\lambda_n \simeq b_n$ one obtains

$$\frac{v}{D} = \pi^3 \frac{S^\alpha S^{cem}}{S^2 \sum_1^\infty \frac{1}{n^3} \sin^2 n\pi S^\alpha/S} \frac{c_0^{\gamma\alpha} - c_0^{\gamma\,cem}}{c^{cem} - c^\alpha} \frac{1}{S} \left(1 - \frac{S_c}{S}\right) \qquad (70)$$

where

$$S_c = \frac{-1}{RT(c_0^{\gamma\alpha} - c_0^{\gamma\,cem})} \left(\frac{2S\sigma^{\alpha\gamma} \cos \varphi^\alpha}{S^\alpha k^{\gamma\alpha}} - \frac{2S\sigma^{cem\,\gamma} \cos \varphi^{cem}}{S^{cem} k^{\gamma\,cem}} \right) \qquad (71)$$

The maximum growth rate is again obtained at $S = 2S_c$. The angles, φ^α, φ^γ and φ^{cem} in the expression for S_c are determined by the interfacial energies

$$\frac{\sigma^{\alpha\gamma}}{\sin \varphi^{cem}} = \frac{\sigma^{cem\,\gamma}}{\sin \varphi^\alpha} = \frac{\sigma^{\alpha\,cem}}{\sin \varphi^\gamma} \qquad (72)$$

Jernkont. Ann. 141 (1957): 11

They can therefore be eliminated, giving the result

$$S_c = \frac{\sigma^{\alpha\,\text{cem}}}{RT(c_0^{\gamma\alpha} - c_0^{\gamma\,\text{cem}})} \left(\frac{S}{S^\alpha k^{\gamma\alpha}} - \frac{S}{S^{\text{cem}} k^{\gamma\,\text{cem}}}\right) \left[1 + \frac{\sigma^{\alpha\gamma^2} - \sigma^{\text{cem}\gamma^2}}{\sigma^{\alpha\,\text{cem}^2}} \cdot \frac{S^{\text{cem}} k^{\gamma\,\text{cem}} + S^\alpha k^{\gamma\alpha}}{S^{\text{cem}} k^{\gamma\,\text{cem}} - S^\alpha k^{\gamma\alpha}}\right] \tag{73}$$

In the following, we will use the value 1 for the bracketed expression above which is a good approximation if $-S^{\text{cem}} k^{\gamma\,\text{cem}}$ and $S^\alpha k^{\gamma\alpha}$ have similar values. This is the case for ordinary pearlite where the values at 650°C are less than 10 % apart. It is also a good approximation if the two interfacial energies $\sigma^{\alpha\gamma}$ and $\sigma^{\text{cem}\gamma}$ have similar values. The only interfacial energy remaining in the expression for S_c is thus $\sigma^{\alpha\,\text{cem}}$. It should be noted, however, that local equilibrium at the $\alpha/\gamma/\text{cem}$ junctions can be established only if each one of the energy values $\sigma^{\alpha\,\text{cem}}$, $\sigma^{\text{cem}\gamma}$ and $\sigma^{\alpha\gamma}$ is less than the sum of the other two. If, for example, $\sigma^{\text{cem}\gamma} > \sigma^{\alpha\gamma} + \sigma^{\alpha\,\text{cem}}$, the ferrite phase will wet the γ/cem interfaces resulting in tendency for the ferrite lamellae to break the contact between the cementite lamellae and the austenite and thus the lamellar mode of growth will be disturbed.

There is a great similarity between equation (70) and the equation which was derived by Zener's method, (62). The requirement for complete agreement is that

$$a = \frac{1}{\pi^3} \left(\frac{S^2}{S^\alpha S^{\text{cem}}}\right)^2 \sum_1^\infty \frac{1}{n^3} \sin^2 n\pi S^\alpha/S \tag{74}$$

and the value of Zener's proportionality constant a can thus be calculated. It is evident that a depends on the ratio between the thicknesses of the lamellae, and hence on the original carbon content of the austenite. A calculation shows that $a = 0.54$ when $S^\alpha = S^{\text{cem}} = \tfrac{1}{2} S$ and $a = 0.72$ when $\tfrac{1}{7} S^\alpha = S^{\text{cem}} = \tfrac{1}{8} S$. Zener's suggestion that a should be close to unity is thus confirmed.

Because $-\lambda_n = (v/2D)(1 + \sqrt{1 + 4 b_n^2 D^2/v^2})$ and $b_n = 2\pi n/S$, the approximation $-\lambda_n = b_n$ implies that the magnitude 1 is neglected in comparison with $4\pi n D/vS$. This approximation can now be tested. With $S^\alpha/S = \tfrac{7}{8}$, the numerical factor in (70) becomes 12.6 and at the maximum growth rate one obtains

$$4\pi n D/vS = 4\pi n \cdot 2(c^{\text{cem}} - c^\alpha)/12.6(c_0^{\gamma\alpha} - c_0^{\gamma\,\text{cem}})$$
$$= 2n(c^{\text{cem}} - c^\alpha)/(c_0^{\gamma\alpha} - c_0^{\gamma\,\text{cem}}) \tag{75}$$

The value of this expression is much larger than 1 even for $n = 1$ and the use of $-\lambda_n = b_n$ in the derivation of equation (70) thus appears justified.

The growth rate can be calculated using the more accurate equation (14) for P if equation (67) is integrated numerically. Such a calculation was performed for the temperature 650°C assuming that the interfacial

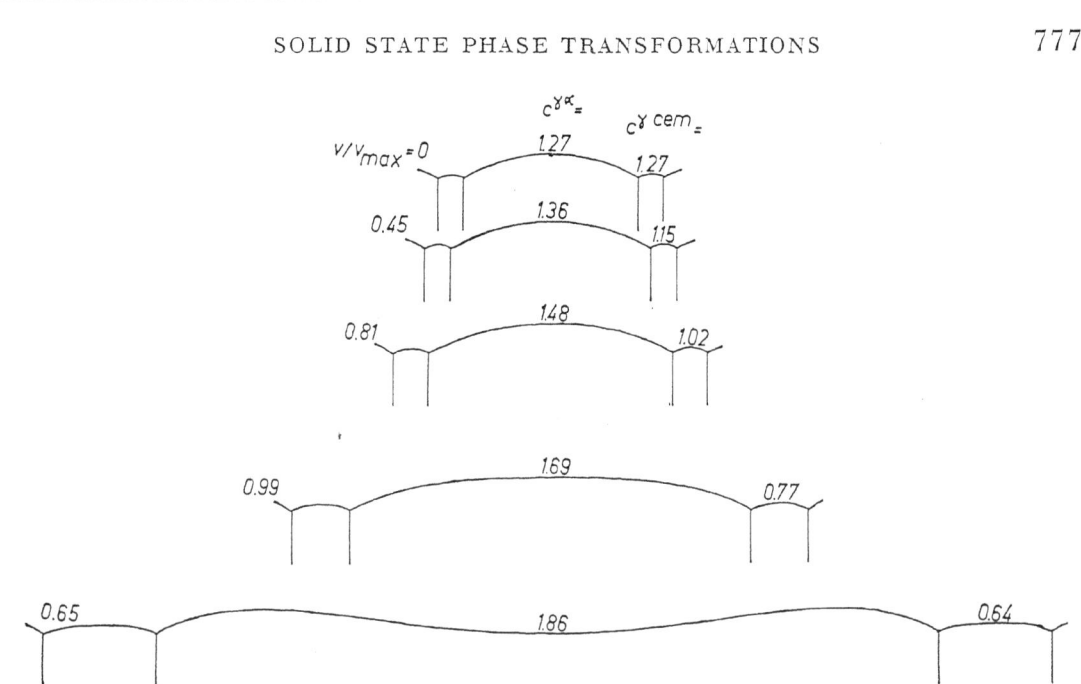

Fig. 6. Calculated shape and growth rate of pearlite with different interlamellar spacings.

energies are equal. The result agreed quite well with equation (70). At the same time the shape of the interface between austenite and pearlite was calculated by another numerical integration, the results of which are presented in Fig. 6 for a series of S values. Relative values of the growth rate are given as well as the carbon content in wt % in the austenite at the center of each lamella. It is seen that the interface really is quite plane when v is in the neighbourhood of v_{max} and consequently the new method of calculation which has been applied here may have been rather accurate. Moreover, it is a great advantage that no assumption is made about either the variation of the concentration along the pearlite-austenite interface or the shape of this interface. It was thus possible to calculate these values. The interface shape has previously only been calculated by Brandt.[11] However, he has not taken the interfacial energies into account at all and, as a consequence, the shape calculated by him may differ very much from the real shape. Brandt's curve should rather be considered as the "ideal" shape for which his assumption about a sinusoidal variation of the concentration is valid. How closely this "ideal" shape resembles the real shape, i.e. the validity of Brandt's assumption, largely depends, in fact, on the values of the interfacial energies.

It seems natural that ferrite and austenite should, if possible, arrange themselves in such a way that $\sigma^{\alpha cem}$ has a low value since this would increase the growth rate according to equations (70) and (73). In such a case the pearlite interface might be even flatter than Fig. 6 demonstrates, and the present method of calculation would be even better.

Scheil, Brandt and Zener all found that experimental growth rate values for pearlite are higher than their calculations predicted. Furthermore,

Zener's calculation predicted the correct interlamellar spacing only if an unreasonably high value of $\sigma^{\alpha\,cem}$ was assumed. The present calculation of the growth rate and spacing yields almost the same values as the earlier treatments and it would therefore be of considerable interest to test the approximations common to all the calculations. Some of them are discussed in the Appendix but the material presented is not extensive enough to allow any conclusion in regard to the discrepancy between measured and calculated values.

VII. Comparison of Separate and Cooperative Growth

When a system is supersaturated with respect to two phases, these can form either by separate or cooperative growth. The cooperative mode of formation is favoured to the exclusion of separate growth only if it results in the highest growth rate. If, on the other hand, either of the two new phases grows separately with maximum speed there is no reason for it not to dominate the transformation. It is therefore of interest to compare the derived growth rate equations.

The maximum growth rate of a separately growing plate of ferrite is obtained from equations (45) and (61).

$$\frac{v_{max}^{\alpha}}{D} = \frac{1}{8\,r_c} \frac{S}{S^{cem}} \frac{c_0^{\gamma\alpha} - c_\infty^{\gamma}}{c^{cem} - c^{\alpha}} \tag{76}$$

and the maximum rate for coooperative growth is obtained from equation (62)

$$\frac{v_{max}^{\alpha+cem}}{D} = \frac{1}{4\,a\,S_c} \frac{S^2}{S^{\alpha}\,S^{cem}} \frac{c_0^{\gamma\alpha} - c_0^{\gamma\,cem}}{c^{cem} - c^{\alpha}} \tag{77}$$

where the constants r_c and S_c are given by (46) and (73). Thus

$$\frac{v_{max}^{\alpha}}{v_{max}^{\alpha+cem}} = \frac{a}{2} \left(\frac{c_0^{\gamma\alpha} - c_\infty^{\gamma}}{c_0^{\gamma\alpha} - c_0^{\gamma\,cem}} \right)^2 \left(1 - \frac{S^{\alpha}\,k^{\gamma\alpha}}{S^{cem}\,k^{\gamma\,cem}} \right) \frac{\sigma^{\alpha\,cem}}{\sigma^{\alpha\gamma}} \tag{78}$$

The constants $k^{\gamma\alpha}$ and $k^{\gamma\,cem}$ are given by equation (17). For the transformation of eutectoidal austenite at 650°C, one finds

$$\frac{v_{max}^{\alpha}}{v_{max}^{\alpha+cem}} = 0.5 \frac{\sigma^{\alpha\,cem}}{\sigma^{\alpha\gamma}} \tag{79}$$

In the same way one finds $\quad \dfrac{v_{max}^{cem}}{v_{max}^{\alpha+cem}} = 0.02 \dfrac{\sigma^{\alpha\,cem}}{\sigma^{cem\,\gamma}} \tag{80}$

Although based on a number of approximations, these two equations do indicate that the rate of transformation will be higher if the two phases cooperate, providing that the interfacial energies have similar values. It may then be only a question of time as to whether a separately growing plate of one of the phases gives rise to a cooperatively growing aggregate of the two phases or whether its further growth is stopped by a cooperative aggregate nucleated elsewhere in the specimen. The cooperative growth of a lamellar structure will be favoured in comparison with separate growth also in the case of a needle-like crystal since this grows with only 50 % greater velocity than a plate according to our previous approximate calculations.

In order for the separate growth of one of the phases to take place preferentially it must have a low interfacial energy. Equations (79) and (80) yield the critical conditions $\sigma^{\alpha\gamma} < 0.5\,\sigma^{\alpha\,cem}$ and $\sigma^{cem\,\gamma} < 0.02\,\sigma^{\alpha\,cem}$ for the case considered. As previously mentioned, energies of coherent interfaces are generally considered to be very low. Consequently, the above criterion for separate growth may be satisfied for coherently growing crystals of one of the phases. Thus the edgewise growth of a Widmanstätten plate should continue undisturbed even if the second new phase nucleates in contact with it. The cooperatively growing aggregate which may be formed after such a nucleation grows too slowly to overtake the leading edge of the Widmanstätten plate.

VIII. The Nature of Bainite

In the metallography of steel one often uses the terms "proeutectoid" and "eutectoid" to describe different structures. The two new terms, "separate" and "cooperative", which have been defined in an earlier section, coincide approximately with the older terms but were introduced here because they offer a better phenomenological description of the phase transformation.

A low carbon content promotes the formation of proeutectoid ferrite, i.e. separately growing ferrite, which is quite natural and can be explained by equation (78) since the ratio $(c_0^{\gamma\alpha} - c_\infty^{\gamma})/(c_0^{\gamma\alpha} - c_0^{\gamma\,cem})$ increases as the carbon content of the austenite, c_∞^{γ}, decreases. By the same reason high carbon content promotes the formation of proeutectoid cementite. These proeutectoid structures can be of two types. The grain boundary type nucleates at the austenitic grain boundaries and grows preferentially along them, forming a more or less even lining around the original austenitic grains. The growth rate away from the boundaries is quite low for this type. The Widmanstätten type developes into thin plates growing rapidly into the center of the austenite grains. In a preceding section the variation in growth rate of these two types was shown to be explained by the difference in interfacial energy for incoherent and coherent interfaces.

Two types of eutectoid structure are usually encountered also, namely pearlite and bainite, however, the difference in the mode of formation of

780   MATS HILLERT

these aggregates is still unresolved. The following discussion in this section will indicate that the interfacial energy plays an important role in this case also and the difference between pearlite and bainite will be explained in terms of the new concepts "separate" and "cooperative".

S. Modin[13] has studied very carefully the structures in a low carbon steel (0.18 wt % C) after interrupted isothermal transformation. The observed structures can be described in the following way with the new terminology. At temperatures just below A_1 the transformation started with separate formation of ferrite, largely as groups of parallel and rather thick Widmanstätten plates. Later on the intervening spaces cooperatively transformed mainly into lamellar pearlite. However, some larger particles of cementite were also formed along the sides of the Widmanstätten plates of ferrite. It appears as if the cooperative process does not at once assume the lamellar arrangement which would give a high growth rate, but is more irregular at the beginning, resulting in a coarser structure. At lower transformation temperatures, the Widmanstätten plates became thinner and more closely spaced. The intervening spaces were thus thinner and it seemed to be increasingly difficult for the cooperative process to assume the lamellar arrangement before such a space was completely transformed. Fig. 7 shows a case where some intervening spaces have transformed completely to the coarse, cooperative structure and others have partly transformed into the lamellar structure, pearlite. At a low enough temperature the lamellar, cooperative structure is practically absent and the intervening spaces have transformed completely into the coarse irregular structure. The resulting structure is then identical to so-called upper bainite. As a consequence, this bainite should not be considered as a eutectoid structure formed by a single process but as resulting from an initial separate formation of Widmanstätten ferrite plates followed by a cooperative process, not greatly different from the high temperature structure consisting of coarse Widmanstätten ferrite plates and lamellar pearlite.

The appearance of the bainitic structure just described is largely determined by the initial formation of ferrite plates, since the succeeding cooperative process is restricted to the intervening spaces. However, the cooperative process could also start on the outside of the last plate in a group of parallel plates and could there develop without restriction. Fig. 8, taken from a paper by H. and S. Modin,[14] demonstrates that quite a lamellar, cooperative structure could thus form at an angle to the plane of the ferrite plates. An examination of the whole series of micrographs in the same paper gives the impression that the number of separately formed parallel plates is quite small in the bainitic aggregates which form in the interior of the austenite grains at lower temperatures, as shown, for example, in Fig. 9. It is thus possible that the separate growth process is of less importance and cooperative, sidewise growth plays a more important role here.

Jernkont. Ann. 141 (1957):11

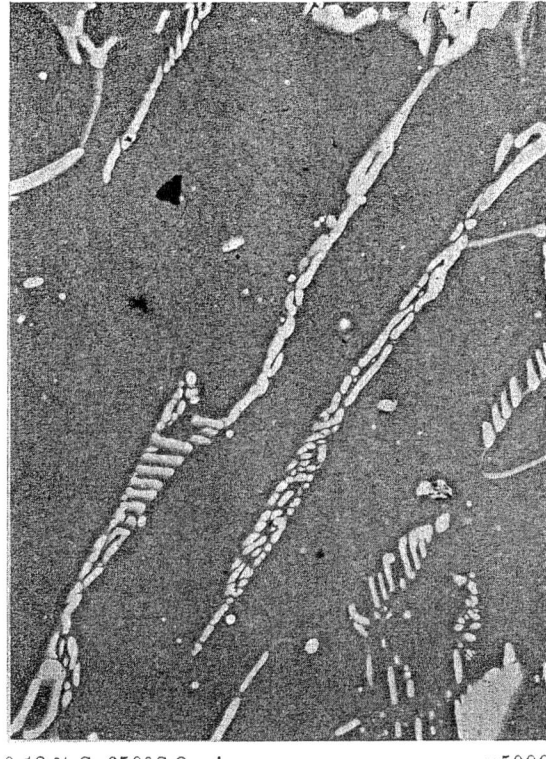

0.18 % C, 650°C 2 min   ×5000

Fig. 7. Formation of pearlite between Widmanstätten ferrite plates, resulting in lamellar and more irregular structure. (Courtesy by S. Modin)

If the above description of the nature of bainite formation is correct, there should also exist a similar structure which forms by the initial appearance of Widmanstätten cementite, followed by a cooperative process. We shall call this postulated structure "inverse bainite". It is well known that groups of Widmanstätten cementite can form in high-carbon steels. The question is simply then, can these plates form into such a fine structure that the cooperative process following does not have time to develop into the ideal lamellar structure before the transformation is completed in the intervening spaces? The chance to find such a structure should increase with the carbon content. A high carbon alloy was therefore prepared from a very pure iron by adding 1.7 % carbon.

A thin plate, 0.5 mm thick, was cut from the ingot, austenitized for 30 minutes at 1100°C, isothermally transformed in a lead bath at 450°C for 3 seconds and finally quenched in brine. The microstructure was examined in an electron microscope and evidence for postulated inverse bainite was found. Fig. 10 gives an example of this structure. Very thin and closely spaced Widmanstätten plates of cementite (light in Fig. 10) have formed on the upper side of the grain boundary cementite. Later on the intervening spaces have started to transform cooperatively giving rise to large ferrite plates (dark in Fig. 10). The parts of the intervening spaces

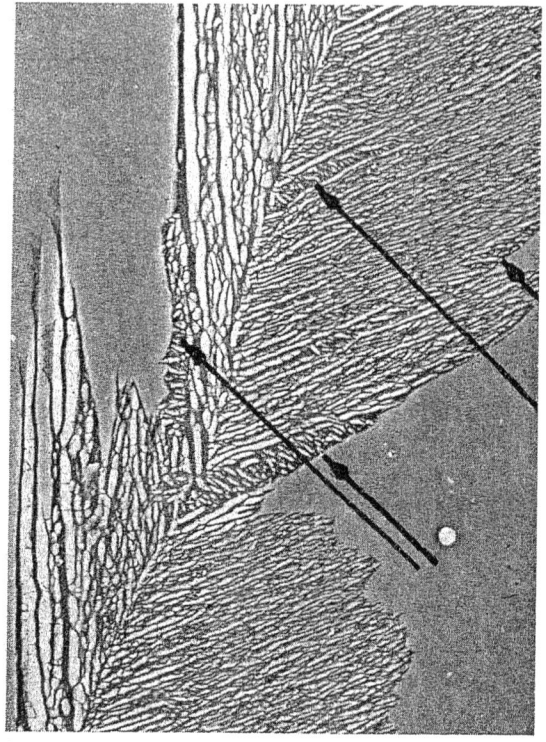

0.86 % C, 500°C 6 sec   ×3000

Fig. 8. Bainite growing from a grain boundary, mainly by an initial formation of ferrite plates, followed by a cooperative reaction in the intervening spaces. Arrows show examples of sidewise cooperative growth. (Courtesy by H. and S. Modin)

0.86 % C, 450°C 25 sec   ×5000

Jernkont. Ann. 141 (1957): 11

Fig 9. Bainite forming in the interior of an austenite grain. Arrows show examples of sidewise cooperative growth. (Courtesy by H. and S. Modin)

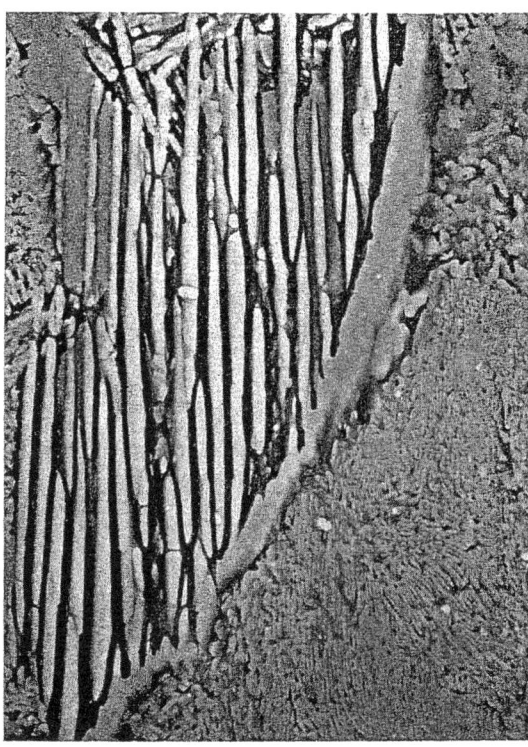

Fig. 10. Inverse bainite and pearlite growing from a grain boundary.

1.7 % C, 450°C 3 sec   ×12000

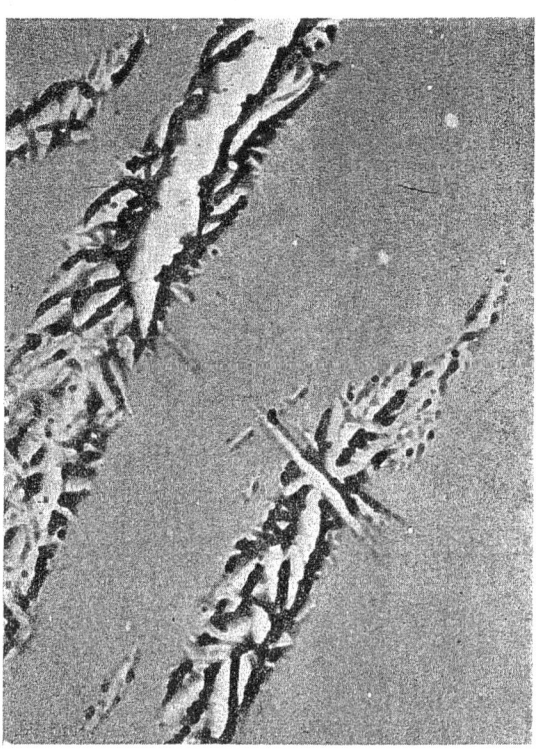

Fig. 11. Inverse bainite forming in the interior of an austenite grain.

1.7 % C, 450°C 3 sec   ×12000

Jernkont. Ann. 141 (1957): 11

which have not yet been reached by this cooperative process still contain austenite (grey in Fig. 10). The amount of cementite which must have formed at the same time as the ferrite has apparently deposited on the sides of the Widmanstätten plates. The resulting structure thus resembles pearlite in that it consists of alternating parallel plates of cementite and ferrite although the mode of formation are quite different. A comparison with the pearlite in the lower part of Fig. 10 shows that inverse bainite is much coarser than pearlite formed at the same temperature.

The resemblance between the inverse bainite in Fig. 10 and the bainite in Fig. 8 is striking although the cooperative process does not give rise to any large cementite lamellae in bainite, supposedly because the volume fraction of cementite is much less than that of ferrite. The resemblance is perhaps even more striking in Figs. 9 and 11 which show bainite and inverse bainite in the interior of austenite grains.

In addition to the difference in appearance between pearlite and bainite, there have been established the following three differences, each one of which considered so fundamental that a hypothesis about the different natures of pearlite and bainite has been based on it.

(1) Cottrell and Ko[15] found that the formation of bainite causes a relief effect on the surface of a specimen as does martensite but not pearlite. They therefore suggested that bainite forms by a shear mechanism as well as martensite.

(2) Mehl et al.[16] found that the bainitic ferrite has the same orientation in relation to the parent austenite as proeutectoid ferrite, whereas the pearlitic ferrite has a different orientation.

(3) Hultgren[17] found that in some of the alloy steels examined by him the alloy element will distribute itself between the two phases of pearlite according to the equilibrium diagram, whereas such a distribution was never observed during the formation of bainite.

These three observations are not in conflict with the description of the bainite formation given above. The relief effect does not necessarily indicate any close relationship to martensite formation since it has recently been found that proeutectoid Widmanstätten ferrite also gives such a relief effect.[18] The relief effect should thus be expected from our picture of the nature of bainite. It should also be expected that the bainitic ferrite and Widmanstätten ferrite have the same orientation because they are identical according to our picture. The difference in the distribution of the alloy elements, finally, might partly reflect the difference in the temperature of formation and partly the difference in coarseness of pearlite and bainite.

Summary

The interfacial energy of a curved interface changes the compositions and activities of the two phases. These changes are calculated and the derived equations are applied to different processes.

Jernkont. Ann. 141 (1957): 11

It is demonstrated how the composition and size of a critical nucleus can be evaluated directly from a free energy diagram, and also how the composition of the new phase changes as it grows.

The driving force for grain growth in a single-phase alloy is calculated.

The diffusion-controlled growth of a new phase in a supersaturated matrix is considered. It is described how a growing crystal develops into a plate or needle, growing with constant speed and shape. Equations are derived for these quantities by two methods, one based on dimensional arguments, the other on the two- and three-dimensional diffusion equation.

The diffusion-controlled growth of two new phases is also considered and the shape of the interface between growing pearlite and austenite is determined. Calculated values of the growth rate and interlamellar spacing are essentially in agreement with earlier calculations. Some approximations are examined in order to explain the discrepancy between calculated and measured data.

The theoretical growth rate of a single phase and of two phases in cooperation are compared. It is found that a single phase has the highest growth rate if its interfacial energy is low enough. The nature of bainite formation is thus explained in terms of the interfacial energy.

The existence of a bainitic structure is postulated, where cementite is the leading phase. Some experimental evidence of this inverse bainite is presented.

Ytenergins roll under fasomvandlingar i fasta tillståndet

SAMMANFATTNING

Ytenergin hos gränsytor mellan två faser ger upphov till en tryckskillnad som beror av ytans krökning. Tryckskillnaden i sin tur förskjuter jämviktsvärdena hos fasernas sammansättningar och aktiviteter. Dessa ändringar beräknas och de härledda ekvationerna utnyttjas i samband med olika processer.

Ytenergin spelar en avgörande roll vid kärnbildningen av en ny fas ur en övermättad lösning. Med hjälp av de härledda ekvationerna visas hur sammansättningen och storleken hos en kritisk kärna kan bestämmas direkt ur ett fri-energi-diagram och även hur den nya fasens sammansättning ändras när den växer.

Ett uttryck för den drivande kraften vid korntillväxt i en en-fasig legering härledes.

Den diffusionskontrollerade tillväxten av en ny fas i en övermättad lösning diskuteras och en beskrivning ges av uppkomsten av platt- och nålformiga kristaller, som tillväxer med oförändrad hastighet och form. Dessa storheter beräknas med två olika metoder, den ena baserad på dimensionella betraktelser, den andra på den tre- resp. tvådimensionella diffusionsekvationen.

Den diffusionskontrollerade tillväxten av två nya faser under ömsesidigt samarbete diskuteras också och den yttre formen hos perlit som växer i austenit beräknas. Beräknade värden för perlitens tillväxthastighet och lamellavstånd överensstämmer i stort sett med tidigare beräkningar. Några av de approximationer, som är gemensamma för alla dessa beräkningar, diskuteras i ett försök att förklara den dåliga överensstämmelsen mellan beräknade och experimentellt bestämda värden.

Vid jämförelse mellan de teoretiska tillväxthastigheterna för en ensam fas och för två faser under samarbete visar det sig att den ensamma fasen växer fortast om den har tillräckligt låg ytenergi. Detta faktum ger en ny bild av bainitens natur. Om slutsatserna är riktiga, bör det finnas en bainitisk struktur där cementiten och ferriten har bytt roller. Denna struktur har döpts till invers bainit och elektronmikroskopiska bilder presenteras, som tyder på att den verkligen existerar.

Appendix

Concentration dependency of the diffusion coefficient

Measurements[19] of the diffusion coefficient of carbon in austenite have been made at temperatures above A_1 and can be represented fairly well by the expression

$$\ln D = -\left(\frac{37\,300}{RT} + 0.7\right) + z\left(\frac{130\,000}{RT} - 30.5\right) \qquad (A\,1)$$

Consequently, there is a considerable variation with composition, z. In order to test the effect of such a variation on the growth rate and interlamellar spacing, a new approximate calculation will be performed here.

It will be assumed that the diffusion distance is $\frac{1}{2} S$ for all the carbon atoms from the edge of a ferrite lamella to the edge of the neighbouring cementite lamella. Upon application of a stationary state approximation, one obtains

$$v = 2 D_0 \frac{S^\alpha - S^{cem}}{S^\alpha S^{cem} \ln S^\alpha/S^{cem}} \frac{e^{Az^{\gamma\alpha}} - e^{Az^{\gamma cem}}}{A(z^{cem} - z^\alpha)} \qquad (A\,2)$$

if $D = D_0 \cdot e^{Az}$, D_0 and A being constants. On the other hand

$$v = 2 D \frac{S^\alpha - S^{cem}}{S^\alpha S^{cem} \ln S^\alpha/S^{cem}} \frac{z^{\gamma\alpha} - z^{\gamma cem}}{z^{cem} - z^\alpha} \qquad (A\,3)$$

if D is constant.

The growth rate v at 650°C was evaluated from equation (A2) using the diffusion data in (A1), and from (A3) considering the constant D as the value given by (A1) at the eutectoidal composition. It was found that equation (A2) gave a higher value by a factor of 2.3. However, equation (A3) was found to give essentially the same result if the constant D value was chosen from (A1) at the mean composition of the austenite at the front, i.e. at $z = \frac{1}{2}(z^{\gamma\alpha} + z^{\gamma\text{cem}})$. It seems reasonable that the assumption of a constant D value is valid also in the previous calculations if this D value is chosen.

The growth rate v was calculated from equation (A2) for different S/S_c ratios and its maximum value was determined graphically. The maximum occurs very close to $S/S_c = 2$ as in Zener's theory and the variation of D thus seems to have little influence with regard to the interlamellar spacing.

Diffusion through ferrite

In all the calculations described in this paper, the diffusion rate through ferrite and cementite has been neglected. Following a suggestion by Onsager,[20] Fisher[21] made an approximate calculation which showed that the diffusion rate through ferrite at 630°C gives a growth rate 7 times as high as the diffusion through austenite and thus some doubt was cast on the validity of the previous calculations. However, Mehl[22] argued that the cementite lamellae should be pointed at the front if the diffusion through ferrite predominates and that this is not observed experimentally.

There remains the possibility that the diffusion through cementite is as fast as that through ferrite, resulting in lamellae of even thicknesses. A comparison of the activity diffusion coefficients[23] indicates that this is not the case, and Mehl's conclusion thus seems justified. Hence, there are two alternatives; either the diffusion through austenite predominates and Fisher's comparison is in error or there is a third mechanism, for example boundary diffusion along the interfaces, which dominates over the diffusion through austenite as well as ferrite.

An examination of Fisher's calculation suggests that he has probably used too low a value for the diffusion coefficient through austenite. If we choose the value at the mean composition at the pearlite front in accordance with the previous section, the diffusion through austenite increases by a factor of 2.8. Moreover, in his comparison Fisher has not used identical values for the amount of carbon diffusion necessary for a certain amount of growth, the correction of this error gives a factor of 2.5. The combined effect of these two corrections gives a factor of 7, indicating that the diffusion through ferrite is of the same order of magnitude as through austenite. Because of all the extrapolations used in our numerical calculations, it is possible that the diffusion through ferrite is even less important. Consequently, there does not seem to be any reason at present to distrust the calculations based on the diffusion through austenite, nor to accept the existence of a third, dominating mechanism. The discrepancy between observed and calculated growth rates may be explained partly by extrapolation errors, and partly by an increase in the diffusion coefficient in austenite due to its continuous deformation, as suggested by Scheil.[12]

Mechanical deformation

There are at least three causes of deformation present during the growth of pearlite. First, the reaction involves a volume increase and the austenite surrounding a growing pearlite nodule must be deformed to give room for the growth. Secondly, cementite occupies a larger volume per iron atom than ferrite, causing the austenite to flow sidewise at the advancing pearlite front. Thirdly, the curved interfaces result in pressure differences, which may cause deformation in all three phases.

Only the first mechanism will be treated in detail here. In an earlier paper,[4] an expression was derived for the energy required for the deformation of a parent phase around a growing spherical particle. It was assumed that the parent phase behaved as an ideally plastic material with a constant yield strength S. Applying the same assumption here, we obtain

$$\frac{dW}{dV} = \frac{1-\nu}{E} S^2 \left[\frac{\Delta v}{v^\beta} \frac{2E}{3(1-\nu)S} - 1 \right] + \frac{\Delta v}{v^\beta} \frac{2}{3} S \ln \left[\frac{\Delta v}{v^\beta} \frac{E}{3(1-\nu)S} \right] \quad (A\,4)$$

by combining equations 28 A and 31 A in the above reference.[4] The factor $\Delta v/v^\beta$ is the relative increase in volume, amounting to about 3 % for pearlite. Poisson's ratio ν is about 0.3 and the elastic modulus E is about 2×10^6 kg/cm². The main difficulty is to choose a reasonable value for the yield strength S. If we take the values $S = 500$ kg/cm² at 700°C and $S = 800$ kg/cm² at 600°C, in accordance with data presented in *Metals Handbook*,[24] equation (A 4) shows that the mechanical work is 0.15 cal/g pearlite at 700°C and 0.25 cal/g pearlite at 600°C. These values should be compared with the free energy of the transformation which is about 0.6 cal/g pearlite at 700°C and 3 cal/g pearlite at 600°C. It appears that the mechanical deformation is quite important at temperatures close to A_1 but not in the lower temperature range. Consequently, the neglect of mechanical deformation in the previous calculations may have resulted in too high growth rate values and too small spacings at high temperatures but not at low temperatures.

The effect just considered depends on the relative change in volume $\Delta v/v^\beta$ and, consequently, it does not directly affect Zener's criterion $S = 2S_c$. It is not inconceivable, however, that the second cause of deformation listed above may change this criterion to give better agreement between theory and experiment. This would be the case, for example, if the yield strength of very small regions is higher than for larger regions, as the examinations of so-called metal whiskers suggest.

Jernkont. Ann. 141 (1957): 11

References

1. Metal Interfaces. ASM, Cleveland 1952.
2. H. K. Hardy & T. J. Heal, Nucleation-and-growth processes in metals and alloys. *The Mechanism of Phase Transformations in Metals. Inst. of Metals, London 1956, p. 1–46.*
3. C. Zener, Kinetics of the decomposition of austenite. *Trans. Am. Inst. Min. Met. Engrs. 167 (1946) p. 550.*
4. M. Hillert, Pressure-induced diffusion and deformation during precipitation, especially graphitization. *Jernkont. Ann. 141 (1957) p. 67.*
5. C. Wagner, Thermodynamics of Alloys. Addison-Wesley, Cambridge, Mass. 1952.
6. R. P. Smith, Equilibrium of iron-carbon alloys with mixtures of $CO-CO_2$ and CH_4-H_2. *J. Am. Chem. Soc. 68 (1946) p. 1163.*
7. M. Hillert, Nuclear composition—a factor of interest in nucleation. *Acta Met. 1 (1953) p. 764.*
8. C. Zener, Theory of growth of spherical precipitates from solid solution. *J. Appl. Phys. 20 (1949) p. 950.*
9. J. W. Rutter & B. Chalmers, A prismatic substructure formed during solidification of metals. *Can. J. Phys. 31 (1953) p. 15.*
10. C. Wagner, Theoretical analysis of diffusion of solutes during the solidification of alloys. *J. Metals 6 (1954) p. 154.*
11. W. H. Brandt, Solution of the diffusion equation applicable to the edgewise growth of pearlite. *J. Appl. Phys. 16 (1945) p. 139.*
12. E. Scheil, Über die Berechnung der eutektischen Kristallisationsgeschwindigkeit, dargestellt an Beispiel des Perlits. *Z. Metallkunde 37 (1946) p. 123.*
13. S. Modin, The isothermal transformation of austenite in two carbon steels with 0.50 and 0.18% C. An electron microscopic investigation. *Jernkont. Ann. (to be published).*
14. H. Modin & S. Modin, Pearlite and bainite structures in a eutectoid carbon steel. An electron microscopic investigation. *Jernkont. Ann. 139 (1955) p. 481.*
15. T. Ko & S. A. Cottrell, The formation of bainite. *J. Iron Steel Inst. 172 (1952) p. 307.*
16. G. V. Smith & R. F. Mehl, Lattice relationships in decomposition of austenite to pearlite, bainite and martensite. *Trans. Am. Inst. Min. Met. Engrs. 150 (1942) p. 211.*
17. A. Hultgren, Isothermal transformation of austenite. *Trans. Am. Soc. Metals 39 (1947) p. 915.*
18. A. P. Miodownik, Discussion. *The Mechanism of Phase Transformations in Metals. Inst. of Metals, London 1956, p. 319.*
19. C. Wells, W. Batz & R. F. Mehl, Diffusion coefficient of carbon in austenite. *J. Metals 2 (1950) p. 553.*
20. L. Onsager, Discussion. *Cornell Conference on Solid State. Aug. 1948.*
21. J. C. Fisher, Eutectoid decomposition. *Thermodynamics in Physical Metallurgy. ASM, Cleveland 1950, p. 201. 241.*
22. R. F. Mehl & W. C. Hagel, The austenite:pearlite reaction. *Progress in Metal Physics. Vol. 6, London 1956, p. 74.*
23. M. Hillert & R. D. Sharp, Diffusion of carbon in the different phases of carbon steel. *Jernkont. Ann. 137 (1953) p. 785.* [In Swedish.]
24. Metals Handbook. ASM, Cleveland 1948.

5 Introduction to "Thermodynamics of Martensitic Transformations"

in *Acta Metallurgica* (1958)

and "Thermodynamics of Martensitic Transformations"

published in *Martensite-A Tribute to Morris Cohen* (1992)

I have had the pleasure of associating with Mats for over 50 years, since 1953 when we were both students of Morris Cohen in the Department of Metallurgy at M.I.T. The subject paper was prepared by Mats and myself for Morris Cohen's Festschrift 80th birthday celebration. Our paper reviews the methods for employing thermodynamic data to describe martensitic transformations of parent phases to daughter phases of the same compositions. In the early 1950's this was not a very popular idea, especially at MIT where the practicing faculty included Carl Wagner, John Chipman and John Elliot who were giants in the fields of experimental and theoretical chemical as well as process metallurgy. By contrast, the graduate students in physical metallurgy were researching athermal and isothermal martensitic transformations, spinodal decomposition, aging reactions, tempering as well as pearlitic and bainitic reactions in steels. Mats Hillert's arrival from Sweden exposed this active research group to the ideas of Hultgren and Johannson and broadened the scope that we envisioned for thermodynamics as an important tool in our research. Mats also brought a very tolerant view toward consideration of a wide variety of ideas that others held in approaching these topics. The first few pages (Figures 1–5) present the salient experimental and theoretical features of the classical (isothermal) and martensitic (constant composition) description of the thermodynamics of the iron-nickel system. Pages 43 and 44 outline the method advanced to permit explicit numerical calculations of the equilibrium phase diagrams and the driving force for diffusionless transformations. All these calculations were done by hand by suffering though all of the missed minus signs! Nevertheless, the framework for the "Computational Thermodynamics" was conceived, pioneered by the CALPHAD, Thermo-Calc and SGTE groups that Mats and his students developed for the world in the past twenty years. Figure 6 published by Kubaschewski and von Goldbeck (Kuba's future wife Ortrude) in 1949 shows the same features displayed in Figure 3 for iron-nickel alloys. However they completely ignored the composition at which the FCC and BCC phases have the same Gibbs energy! This $T_0 - x_0$ relation, essential in the transformation behavior that we were observing experimentally, was of no interest to the classical thermodynamicists (Kubaschewski and von Goldbeck) who carried out the meticulous research required to generate Gibbs energy vs. Composition curves for the BCC and FCC phases in Fe-Ni alloys at 527 °C and 327 °C. Figures 6–11 and Table 1 illustrate other examples of martensitic and classical thermochemical data that are related explicitly. Included are examples of $T_0 - x_0$ curves for hydrated nitrates that were called "EGC" curves by Harry Oonk in 1968 (Figure 9). Harry Oonk was completely unaware of the way metallurgists were looking at phase diagrams. It is interesting to note that in 1953, when I told Carl Wagner about Zener's Equation 2, he shook his head, cleared his throat and informed me that in 1908 Van Laar showed that the correct description for the binary Fe(i)-Ni(j) system is given by Equations 6 and 7, where G_{Ni}^{bcc}, the Gibbs free energy of BCC nickel, was an important descriptor of the FCC and BCC phases of the Fe-Ni system, even though BCC nickel is unstable.

Equations 8–15 expand of the Gibbs energy to multicomponent systems using the Redlich-Kister model.

Mats Hillert, Bosse Sundman and their associates have developed a wide variety of models that describe solid and liquid oxide, carbide, ionic and composite phases that have been applied to a wide variety of multicomponent systems. Figure 12 discusses the differences between recent measurements of the variation of the enthalpy of transformation of ferrite and martensite to austenite in Fe-Ni alloys in order to evaluate the energy stored in the martensite by defects. Equations 16–18 reviews the description developed by Fisher and Zener for the energy difference between ordered tetragonal martensite and ferrite.

In the following section Mats revisited the formation of martensite as an adiabatic process, with much the same result as the one he obtained 35 years earlier (reference 30). Equations 19–28 can be used to calculate the difference between the A_d (the lowest temperature at which austenite can form from martensite which is being deformed) and M_d (the highest temperature at which martensite can be formed from austenite which is being deformed). The A_d and M_d temperatures bracket the T_0 temperature. For a Fe-30Ni alloy, the value of $A_d - M_d$ is calculated as 95C in agreement with experiment. Typically, Mats states that "one did not have any right to expect such an agreement" and then goes on to further discuss the questions these findings raise.

In the final sections of this paper Mats discusses two additional relevant problems of current interest: the interaction of diffusion with martensitic transformations which could develop into a bainitic transformation, and the relation of martensitic growth conditions and Paraequilibrium in iron-base alloys.

These final sections illustrate how far the field has developed in the 1953–1992 period. Mats Hillert and his students played key roles in these developments.

Larry KAUFMAN

Thermodynamics of martensitic transformations*

The thermodynamics of solid-state phase transformations is considered in this note and it will be shown that in many problems it is inconvenient to base the attack on Gibbs or Helmholtz free-energy functions

$$G = U - TS + PV \qquad (1)$$

$$F = U - TS \qquad (2)$$

with their established properties. It is better to start from the two basic principles of thermodynamics, which give

$$dU - q + w = 0 \qquad (3)$$

and $\quad dS > \dfrac{q}{T} \quad$ (natural changes) $\quad$ (4a)

or $\quad dS = \dfrac{q}{T} \quad$ (reversible changes) $\quad$ (4b)

The quantity q is the heat absorbed by the system and w is the work done by the system. All spontaneous processes must satisfy relation (4a) whereas equilibrium conditions can be calculated from (4b). Combination gives

$$dU - T\,dS + w \leq 0 \qquad (5)$$

In many cases the work w can be expressed as $P\,dV$. When this holds, equation (5) is equivalent to $dG \leq 0$ in the special case of constant temperature and pressure, since then

$$dG = dU - T\,dS - S\,dT + V\,dP + P\,dV$$
$$= dU - T\,dS + w \leq 0 \qquad (6)$$

If instead the temperature and volume are kept constant, equation (5) becomes equivalent to $dF \leq 0$, since

$$dF = dU - T\,dS - S\,dT$$
$$= dU - T\,dS + w \leq 0 \qquad (7)$$

The usefulness of the functions G and F is based on the fact that many transformations take place under constant temperature and pressure or temperature and volume. It is not unusual for solid-state transformations, however, that neither the pressure nor the volume is kept constant and, furthermore, the work w cannot be expressed as $P\,dV$ when there are shear stresses, for instance. It is customary[1] in such cases to evaluate w explicitly and to use one of the following criteria for natural and reversible changes.

$$dG + w \leq 0 \qquad (8)$$
$$dF + w \leq 0 \qquad (9)$$

neglecting the difference between G and F, which is usually very small for condensed phases. Estimates of w have for instance yielded values around 65 cal/mol for the martensitic transformation in Fe–C alloys.[1,2]

The above procedure does not seem to have any advantages. It is more straightforward to use equation (5) directly. The condition for equilibrium should thus be written

$$T = \frac{dU + w}{dS} \qquad (10)$$

Theoretically this equation is of general validity and there are no restrictions concerning the pressure, volume, temperature or any other parameter being kept constant. However, if a system is thermally isolated from its surroundings, i.e. if $q = 0$, equation (10) cannot be used to determine T, because $dU + w$ and dS are both zero. The original equations (3) and (4) must then be applied.

Krisement, Houdremont and Wever[2] have suggested that martensitic transformations should be considered as adiabatic processes because of the high velocity of propagation. The quantity q is then zero and consequently

$$dU + w = 0 \qquad (11)$$

They state quite correctly that the equilibrium condition for such a process is

$$dS = 0 \qquad (12)$$

If the transformation of a region containing 1 mol. is considered, these equations can be written

$$O = U^M(T_1) - U^A(T_0) + w = [U^A(T_1)$$
$$- U^A(T_0)] + [U^M(T_1) - U^A(T_1)] + w \qquad (13)$$

$$O = S^M(T_1) - S^A(T_0)$$
$$= [S^M(T_1) - S^A(T_0)] + [S^M(T_1) - S^A(T_1)] \qquad (14)$$

The superscripts A and M denote the parent phase and the transformation product. T_0 is the temperature of the parent phase before the transformation and T_1 is the temperature of the new phase immediately after its formation. If the value of w is known and also U and S of the two phases at different temperatures, T_0 and T_1 can be calculated from equations (13) and (14). At this point, however, Krisement et al. argue that dS is the difference between two large quantities and cannot be evaluated accurately enough. Without any further explanation they instead introduce the condition

$$w = F^A(T_1) - F^M(T_1) = T_1[S^M(T_1)$$
$$- S^A(T_1)] - [U^M(T_1) - U^A(T_1)] \qquad (15)$$

and they base their calculation of T_0 and T_1 on this equation and equation (13). It will be shown here that this procedure leads to erroneous results and all calculations should therefore be based on the two original equations (13) and (14).

In a reasonably small temperature range it is a good approximation to consider the specific heat of the parent phase c, and the difference in internal energy and entropy of the two phases, $U^M(T) - U^A(T)$ and $S^M(T) - S^A(T)$, as independent of temperature. Denote these differences by ΔU and ΔS. Equations (13) and (14) then give

$$\Delta U + w = U^A(T_0) - U^A(T_1)$$
$$= -c \cdot (T_1 - T_0) \qquad (16)$$

$$\Delta S = S^A(T_0) - S^A(T_1)$$
$$= -c \cdot \ln T_1/T_0 \qquad (17)$$

These equations can be solved for T_0 and T_1. One obtains

$$T_1 = \frac{\Delta U + w}{c\,[\exp \Delta S/c - 1]} = T_i \cdot \frac{\Delta S}{c\,[\exp \Delta S/c - 1]}$$

$$\cong T_i \cdot \frac{1}{1 + \Delta S/2c} \quad (18)$$

$$T_0 = \frac{\Delta U + w}{c\,[1 - \exp -\Delta S/c]} = T_i \cdot \frac{\Delta S}{c\,[1 - \exp -\Delta S/c]}$$

$$\cong T_i \cdot \frac{1}{1 - \Delta S/2c} \quad (19)$$

where T_i denotes the equilibrium temperature under isothermal conditions, which can be calculated directly from equation (10).

$$T_i = \frac{\Delta U + w}{\Delta S} \quad (20)$$

The relation (15) suggested instead of (14) by Krisement et al. yields an expression for T_1 identical to this expression for T_i, whereas our result indicates that T_1 is situated approximately as much above T_i as T_0 is situated below T_i.

The significance of T_0 and T_1 is the following. Assuming that a martensitic transformation takes place adiabatically and reversibly, the parent phase must be cooled to T_0 in order to transform spontaneously to the new phase. In the first moment this new phase will have the temperature T_1 but it will soon be cooled to T_0 by its surroundings. Once the reaction has taken place it will be necessary in order for the reaction to reverse to have the new phase and consequently the whole specimen heated to the temperature T_1. The parent phase which is then regenerated will aquire the temperature T_0 in the first moment. This hysteresis can be calculated directly from equation (16).

$$T_1 - T_0 = -\frac{\Delta U + w}{c} = T_i \frac{-\Delta S}{c} \quad (21)$$

In an actual case it cannot be expected that all the work, w, done by the martensitic transformation can be recovered on the reverse reaction. The actual hysteresis might therefore be much higher. It can be calculated from the first parts of equations (18) and (19) using different values of w in the two equations.

It has been shown[3,4] that the hysteresis is appreciably decreased if the transformations are studied under plastic deformation of a specimen, supposedly because the necessary work w is decreased almost to zero by this technique. The difference between the w values for the primary reaction and for its reverse would then be negligible. However, there would still be a remaining hysteresis according to equation (21). The following comparison with experiments can be made. Kaufman and Cohen[4] found that the remaining hysteresis is about 100° for Fe–Ni alloys with about 30 at. per cent Ni. With the values $T_i = 450°K$, $-\Delta S = 1.5$ cal/mol.°K and $c = 7$ cal/mol.°K, equation (21) yields an hysteresis of 95°, in good agreement with the experimental value.

Usually martensitic transformations are treated as isothermal processes. The work w can then be computed from experimental data on the M_s temperature by identifying this temperature with T_i, given by equation (20). The result from such calculations is about 350 cal/mol. for Fe–C alloys[2] and is thus in conflict with the previously mentioned w values of only 65 cal/mol. obtained from direct estimates of w. Krisement et al. introduced the adiabatic concept in order to resolve this conflict. When identifying the experimental M_s temperature with T_0 in their adiabatic treatment, they obtained w values around 80 cal/mol. The agreement with the estimated w values of 65 cal/mol. was so good that the conflict seemed to be resolved. However, if the calculation of w from T_0 is based on the proper equation (18), w values of about 200 cal/mol. are obtained instead of 80. It thus appears impossible to eliminate the discrepancy in the w values merely by introducing the adiabatic concept. On the other hand, the remaining discrepancy is no reason for abandoning the adiabatic treatment. In fact, it is quite possible that a value of 65 cal/mol. yields the correct criterion for the continuation of the growth of an already nucleated plate of martensite whereas a value of 200 cal/mol. determines the start of the formation of such a plate. There is no reason to expect that a single value of w should apply to all stages of the formation of a plate.

Metallografiska Institute   M. HILLERT
Stockholm, Sweden.

References

1. M. COHEN, E. S. MACHLIN and V. G. PARANJPE *Thermodynamics in Physical Metallurgy* p. 242. American Society of Metals (1949).
2. O. KRISEMENT, E. HOUDREMONT and F. WEVER *Rev. Metall.* **51**, 401 (1954).
3. J. B. HESS and C. S. BARRETT *Trans. Amer. Inst. Min. (Metall.) Engrs.* **194**, 645 (1952).
4. L. KAUFMAN and M. COHEN *Trans. Amer. Inst. Min. (Metall.) Engrs.* **206**, 1393 (1956).

* Received May 31, 1957.

CHAPTER 4

Thermodynamics of Martensitic Transformations

Larry Kaufman, ManLabs, Division of Alcan Aluminum Corporation
Mats Hillert, Royal Institute of Technology (Sweden)

"In order for a phase change to proceed spontaneously at a given temperature and pressure, it must be accompanied by a decrease in free energy. The martensitic transformation is no exception to this generalization. Austenite (the parent phase) will not decompose into martensite unless the free energy of austenite is larger than the free energy of martensite. This excess in free energy of the austenite over that of the martensite may be regarded as the 'driving force' behind the martensite transformation" [Cohen et al. (Ref 1)]. The driving force is defined as the "excess in chemical free energy per mole of austenite over that of martensite of the same composition." The austenite-martensite transformation does not start just below T_0, where the driving force becomes positive because of barrier conditions [Cohen, 1949 (Ref 2)], "but at a sufficiently lower temperature where enough free energy is available from the transformation to provide for the strain energy that is an inevitable product of the reaction" [Cohen, 1951 (Ref 3)].

Figures 1 to 5 [Kaufman and Cohen, 1958 (Ref 4)] provide some specific examples that illustrate the relations between the "driving force" for martensitic transformations, T_0, and the conventional thermochemical descriptions of isothermal equilibria in binary systems. Figure 1 illustrates the electrical resistance changes that occur when an Fe-30Ni alloy and an Au-47.5Cd alloy are heated and cooled at rates which are sufficiently fast to ensure that the composition of the parent and daughter phases do not change during heating and cooling.

The abrupt changes in resistance define the M_s temperature where the high-temperature parent phase begins to transform into the low-temperature daughter phase on cooling. Similarly, at A_s the low-temperature

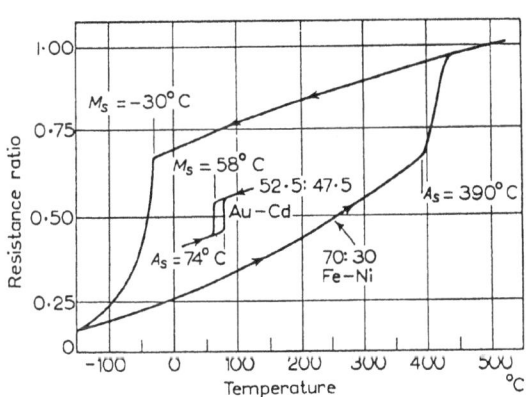

Fig. 1 Electrical resistance changes during the cooling and heating of an Fe-Ni alloy and an Au-Cd alloy, illustrating the hysteresis between the martensite reaction on cooling and the reverse transformation on heating

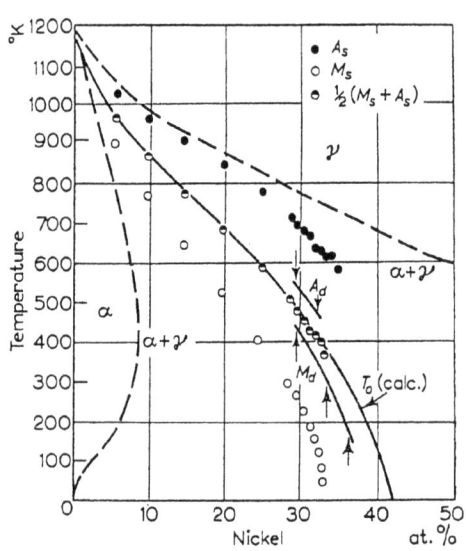

Fig. 2 *Experimental and theoretical determination of T_0 in the Fe-Ni system*

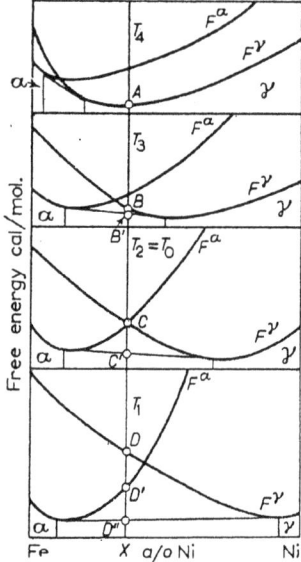

Fig. 3 *Schematic representation of chemical free energy versus composition for the alpha and gamma phases in the Fe-Ni system at four temperatures*

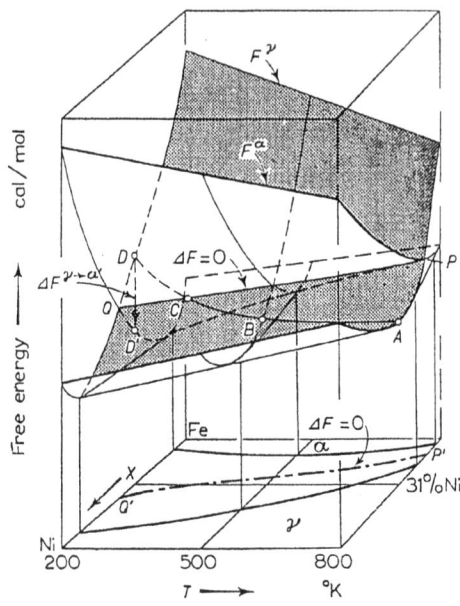

Fig. 4 *Schematic representation of F^γ and F^α surfaces as a function of α and T*

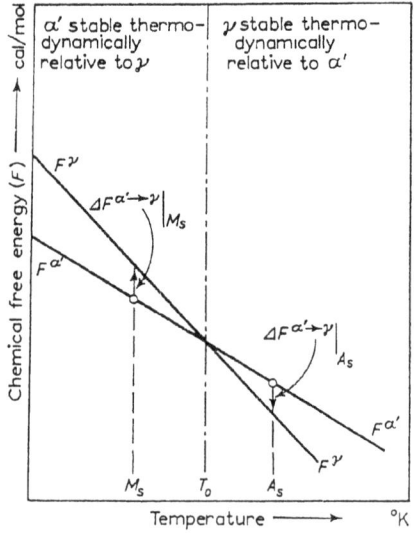

Fig. 5 *Schematic representation of $F^{\alpha'}$ and F^γ versus temperature for an iron-base alloy*

daughter phase begins its transformation on heating without any change in composition. Figure 1 shows a substantial hysteresis between A_s and M_s in the Fe-30Ni case and a much smaller gap in Au-47.5Cd. The size of the hysteresis range will be discussed below.

Thermodynamics of Martensitic Transformations

In Fig. 2, the M_s and A_s temperatures for Fe-Ni alloys are displayed along with the equilibrium phase boundaries between the α (bcc) and γ (fcc) fields in the Fe-Ni system. The latter are illustrated by dashed lines.

Figures 3 and 4 provide schematic representations of the chemical free energies of bcc and fcc phases versus temperature and composition in the Fe-Ni system. Figure 3 can be used to illustrate the free energy of the fcc and bcc phases in an Fe-Ni alloy similar to that shown in Fig. 1, where T_4 corresponds to 390 °C. At $T_2 = T_0$ (approximately 190 °C), the chemical free energies of the fcc and bcc phases in this alloy are equal. Figure 3 illustrates the relation between isothermal equilibrium, designated by the common tangent (which equilibrates the chemical potentials of iron and nickel atoms in the fcc and bcc phases at the phase boundaries), and the chemical driving force for martensitic transformations in terms of T_0 and the entropy difference.

Figures 4 and 5 provide different views of the information displayed in Fig. 3 to illustrate the continuity of free energy curves with temperature and the chemical free energy difference between the bcc and fcc phases as a function of temperature. Definition of this difference is the main subject of this chapter. Nevertheless, it is important to view definition of the chemical driving force (as the free energy difference shown in Fig. 5 has come to be known) in the overall framework of the free energy curves and surfaces shown in Fig. 3 and 4. Thus we see that the free energy surfaces shown in Fig. 4 give rise to isothermal equilibrium (defining the phase boundaries) shown in Fig. 3 as well as the driving force and T_0 shown in Fig. 5.

CALCULATION OF THE CHEMICAL DRIVING FORCE

It appears that Johansson (Ref 5) was the first to attempt a detailed calculation of Gibbs energy curves like those in Fig. 3 for the Fe-C system. Unfortunately, Johansson was forced to make several assumptions about details in the equations describing the Gibbs energy of the austenite, ferrite, and martensite in order to fill gaps in the information available to him. Subsequent experimental studies showed that some of these assumptions were poor, so that his numerical results are not in keeping with current opinions (Ref 3, p 644-645).

Zener (Ref 6) published the next serious attempt to calculate the driving force for martensitic transformations and the position of T_0 as a function of carbon concentration in the Fe-C system. It would appear that Zener (Ref 6, p 585, 595) was not aware of Johansson's prior work. Moreover, Zener did not attempt to describe the individual Gibbs energy curves for the austenite and ferrite as in Fig. 3, but only their difference! In his approach, the austenite and ferrite were treated as dilute solutions and the difference in the excess entropy of solution between the austenite and ferrite was assumed to be zero (precipitating a very caustic discussion by Darken [Ref 6, p 546]) so that the difference in the standard free energy of solution and the difference in the standard heat of solution of carbon in austenite and ferrite were *equal* and *constant*. Under these conditions:

$$\Delta G^{\alpha \to \gamma} = G^{fcc} - G^{bcc}$$
$$= (1-x)\Delta G_{Fe}^{\alpha \to \gamma} + x(\Delta H_s^{\alpha \to \gamma} + RT \ln \beta) \quad (1)$$

where G is the Gibbs energy, x is the atom fraction of solute (i.e., carbon in the present case), $\beta = 3$, $\Delta G_{Fe}^{\alpha \to \gamma}$ is the Gibbs energy difference between fcc and bcc forms of pure iron, and $\Delta H_s^{\alpha \to \gamma}$ is the difference in the enthalpy of solution of carbon in the fcc and bcc phases. As a consequence of the dilute solution approximation applied by Zener (Ref 6), it follows that:

$$\Delta H_s^{\alpha \to \gamma} = RT \ln (x^{bcc}/\beta x^{fcc}) \quad (2)$$

where x^{bcc} and x^{fcc} are the carbon concentrations at the $\alpha/(\alpha + \gamma)$ and $\gamma/(\alpha + \gamma)$ equilibrium phase boundaries. The coefficient β in Eq 1 and 2 is equal to unity for substantial solutions (i.e., nickel or chromium in iron) and is equal to the ratio of interstitial sites to host atom sites in the bcc lattice divided by the same ratio in the fcc lattice. Thus, for carbon in iron, $\beta = 3/1 = 3$. Zener was apprised of the appropriate value for this ratio, while in Johansson's time it was thought that $\beta = 6/1 = 6$, leading to errors in his computation. In the dilute solution approach employed by Zener (Ref 6):

$$\Delta G_{Fe}^{\alpha \to \gamma} \cong RT \ln (1 - x^{bcc})/(1 - x^{fcc})$$
$$\cong RT (x^{fcc} - x^{bcc}) \qquad (3)$$

Examination of Eq 1 to 3 discloses that this dilute solution approach could readily be applied to alloy systems once the Gibbs energy difference for pure iron was established (Ref 5) and the phase boundaries, x^{fcc} and x^{bcc}, determined. Thus, Eq 3 could be employed to check the range of validity of the dilute solution calculation (Eq 2) used to compute $\Delta H_s^{\alpha \to \gamma}$ and Eq 1 used to calculate Gibbs energy difference between the fcc and bcc phases of a fixed composition. Zener (Ref 6) employed this method to compute T_0 as a function of carbon concentration and, by further calculating $\Delta G^{\alpha \to \gamma}$ as a function of temperature and composition, discovered that the experimentally observed M_s in Fe-C alloys occurred at a temperature below T_0 where the driving force, $\Delta G^{\alpha \to \gamma}$, was equal to 290 cal/mol. Of course, closer examination of this calculation discloses that Eq 1 to 3 were used by Zener (Ref 6) well beyond the composition and temperature range over which they apply. Moreover, for the Fe-C case there is a distinct difference between martensite, which becomes increasingly tetragonal and ordered as the carbon content increases and the temperature is lowered, and ferrite.

Nevertheless, the simplicity of Zener's 1946 approach attracted wide attention. Cohen et al. (Ref 1, 3) and Fisher (Ref 7) attempted to broaden the range of Zener's dilute solution approach by using the activity measurements of carbon in ferrite and austenite performed by Smith (Ref 8) to replace the approximations embodied in Eq 2 and 3. This permitted a slight relaxation of the dilute solution restriction, although the need to extrapolate the high-temperature activity data to low temperatures remained. In addition, an attempt was made to calculate the Gibbs energy difference between ferrite and martensite. Cohen et al. (Ref 1, 3) made this accommodation by calculating the strain energy stored in the lattice when the bcc structure is distorted to become tetragonal, while Fisher (Ref 7), following Zener (Ref 6), described the Gibbs energy of martensite as being less than that of ferrite of the same composition at low temperatures due to ordering of carbon atoms in martensite.

Thus Zener (Ref 6), Cohen et al. (Ref 1-3), and Fisher (Ref 7) all concentrated on calculation of T_0 and the driving force for martensitic transformations. They did not consider the broader problem of the individual free energy curves for the α and γ phases as Johansson (Ref 5) did. By contrast, Kubaschewski and von Goldbeck (Ref 9) carried out a thermochemical study of the Fe-Ni system by measuring the activity of iron in the austenite and coupling this information with calorimetric data for bcc alloys.

The results were used to calculate the Gibbs energy curves for the bcc and fcc phases at 527 and 327 °C shown in Fig. 6. These curves, which are similar to those illustrated schematically in Fig. 3, were then employed to calculate the equilibrium phase boundaries shown in Fig. 2 by the common tangent rule. Kubaschewski and von Goldbeck (Ref 9) paid no attention to the points of equal Gibbs energy or the "driving force"

Thermodynamics of Martensitic Transformations 45

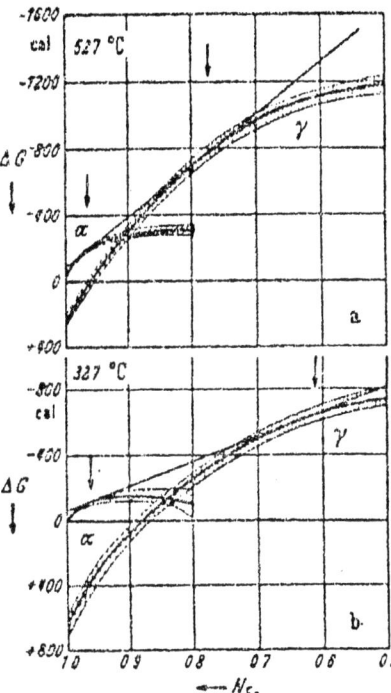

Fig. 6 *Free energy curves for α and γ Fe-Ni alloys at (a) 527 °C and (b) 327 °C for evaluating phase boundaries*

$(G^\gamma - G^\alpha)$ implied by their results! In this instance, consideration of the characteristics of martensitic transformations in Fe-Ni alloys would have provided an independent means for evaluating their experimental results and calculations.

Kaufman and Cohen (Ref 4, 10) provided an analysis of the thermodynamics of fcc and bcc alloys in the Fe-Ni system covering isothermal equilibria (Fig. 3) and martensite transformations (Fig. 5). Figure 7 displays the numerical values for the Gibbs energy difference between the bcc and fcc phases shown in Fig. 5. The individual equations describing the Gibbs energy of the fcc and bcc phases as a function of composition and temperature that were used to generate the Gibbs energy difference shown in Fig. 7 were also used to compute the phase boundaries between the $\alpha/(\alpha + \gamma)$ and $\gamma/(\alpha + \gamma)$ fields. Moreover, once the Gibbs energy difference is defined as a function of temperature, the enthalpy difference can be calculated.

Figure 8 shows the results of calorimetric measurements of the enthalpy change performed by Scheil and coworkers (Ref 11, 12) as a function of temperature and composition. The latter are compared in Table 1 with the results calculated by Kaufman and Cohen (Ref 4, 10) at M_s and A_s. The good level of agreement shows how a general treatment of the system combining information from isothermal equilibria and martensitic transformations can result in a broader picture of the system than that which can be obtained when only one kind of phenomena is considered. The "common tangent" or equilibration of chemical potentials method for defining isothermal equilibrium phase boundary compositions as illustrated in Fig. 3 and 6 were established by Gibbs in the 19th century. However, the concept of the equal Gibbs energy composition (i.e., point C at $T_2 = T_0$ in Fig. 3) and the T_0 versus composition curve illustrated in Fig. 2 and 5 seems to be of much more recent vintage. It appears to have originated with Zener (Ref 6) and was exploited by Cohen (Ref 1-3) and his students in considering martensitic transformations in metallic systems. In principle, however, the T_0-x concept could readily be applied in considering any transition in a binary or multicomponent system in which the temperature is altered rapidly enough to preclude diffusion. Thus in the case of a rapidly cooled liquid, the T_0-x curve defining the conditions under which liquid and solid of the same composition have equal Gibbs energy might have special significance.

It is interesting to note that Oonk (Ref 13-16), working in the field of heterogeneous equilibria in organic compounds, has developed a formalism for the calculation of phase

46   MARTENSITE

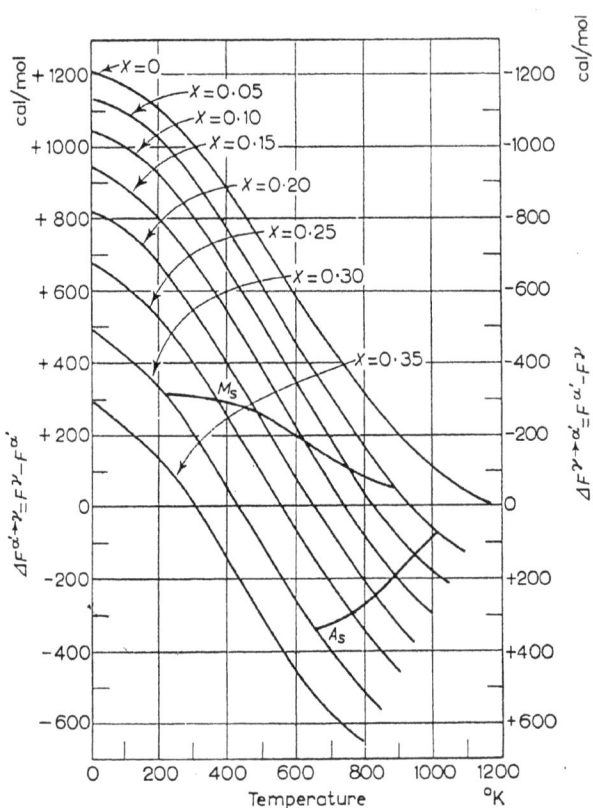

Fig. 7 Chemical free energy change accompanying the martensitic transformation in the Fe-Ni system

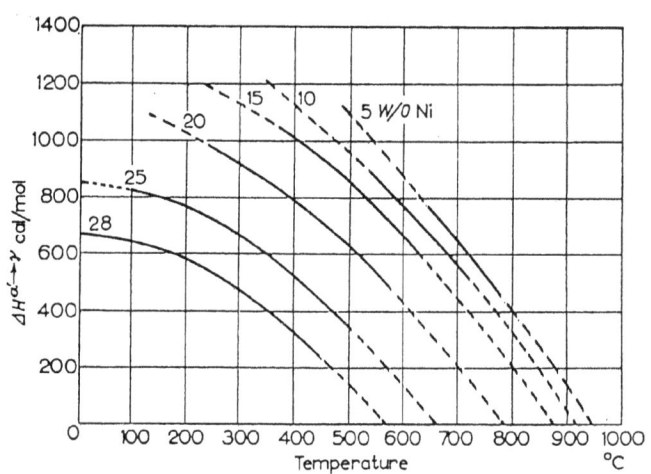

Fig. 8 Enthalpy change accompanying the martensitic transformation in Fe-Ni alloys measured as a function of temperature (Ref 11)

Thermodynamics of Martensitic Transformations

Table 1 Calculated and Observed Heat Effects Accompanying the $\alpha' \rightarrow \gamma$ and $\gamma \rightarrow \alpha'$ Reactions in Fe-Ni Alloys at the A_s and M_s Temperatures

Atomic fraction Ni	A_s, K	$\Delta H^{\alpha \rightarrow \gamma}$, cal/mol		M_s, K	$\Delta H^{\gamma \rightarrow \alpha'}$, cal/mol	
		Calc.	Obs.		Calc.	Obs.
0.05	1020	+574	+540	890	−1018	−840
0.10	950	+720	+630	750	−1160	−980
0.15	890	+748	+650	610	−1144	−1080
0.20	835	+714	+520	480	−1000	−1000
0.25	765	+669	+380	365	−810	−820
0.30	660	+587	...	235	−593	...

Note: $\Delta H^{\alpha \rightarrow \gamma}$ positive signifies heat absorbed by the martensite-to-austenite transformation, and $\Delta H^{\gamma \rightarrow \alpha'}$ negative signifies heat evolved by the austenite-to-martensite transformation.

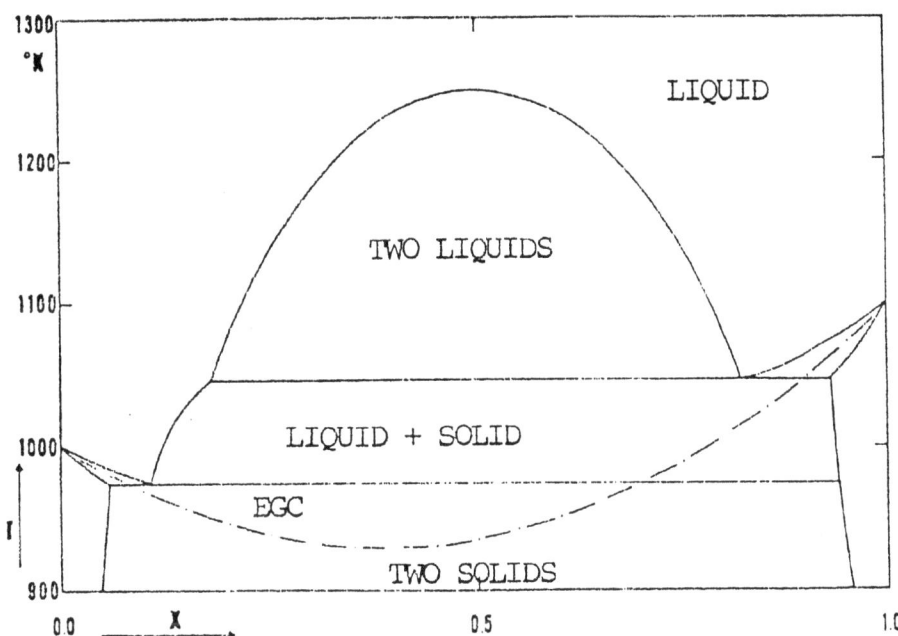

Fig. 9 Phase diagram showing incomplete solid as well as incomplete liquid miscibility. Dashed curve: equal-G curve

diagrams based upon the "equal Gibbs free energy curve," or ECG. The line of reasoning pursued by Oonk was developed without any knowledge of the use of the T_0-x idea discussed above until 1980, when he had occasion to meet one of the current authors at the ninth CALPHAD meeting in Montreal. Oonk cites the work of Hasselblatt (Ref 17) on $Cd(NO_3)_2 \cdot 4H_2O$-$Ca(NO_3)_2 \cdot 4H_2O$ and α-bromzimtaldehyd-α-chlorzimtaldehyd as the source for the EGC concept. Figure 9 (Ref 13) illustrates the EGC concept in

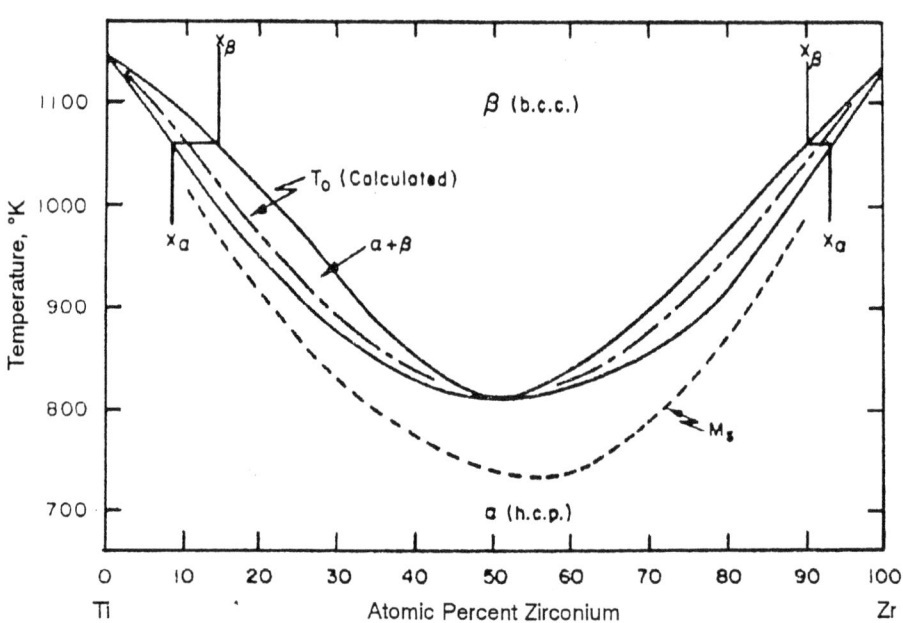

Fig. 10 Ti-Zr phase diagram

Oonk's development. Figures 10 and 11 (Ref 18) illustrate the T_0-x (or EGC) curves for the bcc/hcp martensitic transformation in the Ti-Zr system. It should be evident that the T_0-x or EGC curves are of general interest and have significance in dealing with a wide range of phenomena. Accordingly, subsequent sections of this chapter will review current methods and resources for calculating the T_0 versus x curves for binary and multicomponent alloys before returning to a discussion of the driving force for martensitic transformations.

CALCULATION OF THE TEMPERATURE-COMPOSITION COORDINATES AT WHICH PARENT AND DAUGHTER PHASES HAVE EQUAL GIBBS ENERGIES

From a practical viewpoint, the Ti-Zr case shown in Fig. 10 and the Fe-Ni case shown in Fig. 2 represent two different levels of difficulty in the location of the T_0-x or EGC curve. Figure 9, showing Oonk's hypothetical case, presents yet a third level of difficulty. In the Ti-Zr case, the approximate location of the T_0-x curve is trivial since it must be within the two-phase field. The narrower the two-phase field, the easier it is to locate the T_0-x curve, even without formal computations. In the Fe-Ni case, the two-phase $\alpha+\gamma$ field is broad, and restricting the T_0-x curve to the two-phase field does not help to define it very well. The case shown in Fig. 9 at temperatures below 970 K does not even display a two-phase liquid plus solid field, making location of the EGC from the equilibrium phase diagram uncertain. Similarly, location of the austenite-ferrite T_0-x curve in the Fe-C system from the equilibrium diagram below 1000 K would be difficult because of the intrusion of cementite (or graphite) on the phase diagram.

To overcome such difficulties, explicit descriptions of the Gibbs energies of the parent and daughter phases have been

Thermodynamics of Martensitic Transformations

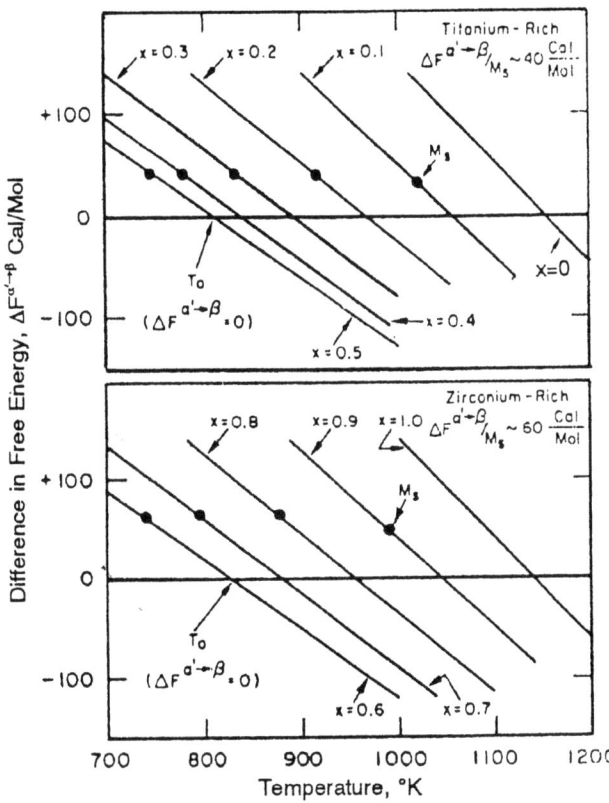

Fig. 11 Chemical driving force for martensitic transformations in Ti-Zr alloys as a function of composition and temperature

developed which can be applied over a wide range of temperature and composition in order to describe the curves illustrated for the Fe-Ni system in Fig. 3 to 5. The CALPHAD journal, published quarterly since 1976 by Pergamon Press, serves as a compendium of such information. In general, the Gibbs energy of a substitutional fcc solid solution in a binary system is given by:

$$G^{fcc} = x_i G_i^{fcc} + x_j G_j^{fcc} + {}^E G^{fcc}$$
$$+ RT(x_i \ln x_i + x_j \ln x_j) \; cal/mol \qquad (4)$$

where x_i and x_j are the atom fractions of elements i and j, respectively, T is the temperature in Kelvin, and R is the gas constant. In this formulation, G_i^{fcc} and G_j^{fcc} are the Gibbs energies of the fcc forms of pure i and pure j (which are temperature dependent), and ${}^E G^{fcc}$ is the excess Gibbs energy of mixing of the fcc, which is dependent on temperature but is defined to be equal to zero for the cases where $x_i = 0$ and $x_i = 1$. For a multicomponent substitutional alloy:

$$G^{fcc} = \Sigma x_i G_i^{fcc} + RT \Sigma x_i \ln x_i + {}^E G^{fcc} \qquad (5)$$

Similarly, for the case of a bcc substitutional alloy:

$$G^{bcc} = \Sigma x_i G_i^{bcc} + RT \Sigma x_i \ln x_i + {}^E G^{bcc} \qquad (6)$$

Thus, the Gibbs energy difference between the fcc and bcc phases in such an alloy is:

$$G^{fcc} - G^{bcc} = \Delta G^{bcc \to fcc} = \Sigma x_i \Delta G_i^{bcc \to fcc}$$
$$+ \Delta^{EG bcc \to fcc} \qquad (7)$$

In order to estimate the difference in the excess Gibbs energy of mixing in a multicomponent alloy system from data on the component binary systems, various approximations have been employed. In the absence of ternary or higher order effects, these methods for synthesizing the properties of a multicomponent solution from the properties of the component binary systems can be employed to calculate the last term in Eq 7. Two commonly used formulations are the Kohler Equation:

$$EG_{fcc} = \Sigma x_i x_j (x_i + x_j)^{-1} (x_i A_{ij} + x_j A_{ji}) \quad (8)$$

and the Redlich-Kister Equation of order k:

$$^E G^{fcc} = \Sigma x_i x_j (x_i - x_j)^k A_k[ij] \quad (9)$$

When each of the binary systems is symmetrical, i.e., $A_{ij} = A_{ji}$, Eq 8 is equivalent to a Redlich-Kister Equation of order k = 0. If this is not the case, then differences exist between the Kohler approximation (Eq 8) and the Redlich-Kister approximation (Eq 9). Clearly, Eq 9 can accommodate higher-order compositionally dependent terms in the description of the excess Gibbs energy. In either case, the A_{ij}, A_{ji}, and $A_k[ij]$ parameters are temperature dependent and are derived from a variety of phase diagrams and thermochemical information. If the corresponding equation for the excess Gibbs energy of the bcc phase is designated in terms of the coefficients B_{ij}, B_{ji}, and $B_k[ij]$, then the final term in Eq 7 in terms of the Kohler Equation would be given as:

$$\Delta^{EGbcc \to fcc} = \Sigma x_i x_j (x_i + x_j)^{-1}$$
$$[x_i(A_{ij} - B_{ij}) + x_j(A_{ji} - B_{ji})] \quad (10)$$

and on the basis of the Redlich-Kister Equation of order k as:

$$\Delta^{EGbcc \to fcc} = \Sigma x_i x_j (x_i - x_j)^k (A_k[ij] - B_k[ij]) \quad (11)$$

Examples of the A and B parameters as well as the Gibbs energy differences between the fcc and the bcc forms of the pure metals for many binary systems are given in Volumes 1 to 3 of the CALPHAD journal (Ref 19-24).

Generalization of Eq 4 to 11 to cover interstitial sites has been carried out in terms of a vacancy model (Ref 25, 26) in which there are substitutional elements $M_1, M_2...M_n$ and interstitial elements $N_1, N_2...N_n$. The concentration variables are specified in terms of site fractions y_m ($M = M_1, M_2...M_n$) and y_n ($N = N_1, N_2...N_n$), with y_{va} defining the fraction of vacant interstitial sites. In this formulation, the site fraction of substitutional elements is defined as the fraction of substitutional lattice sites occupied by a given substitutional element. Thus, if i represents a substitutional element,

$$y_i = x_i / \sum_M^m x_M \quad (12)$$

where the x_i values are the atom fractions and the denominator is equal to unity if no interstitial atoms are present! On this basis, the site fraction for an interstitial element j is given by:

$$y_j = (a/c) x_j / \sum_M^m x_M \quad (13)$$

where a/c is the ratio of substitutional sites to interstitial sites in a particular lattice. This ratio is equal to unity in austenite and one-third (a/c = 1/3) for ferrite. Finally, the fraction of vacant interstitial sites is defined as:

$$y_{va} = 1 - \sum_N^n y_N \quad (14)$$

These definitions lead to specification of the Gibbs energy per mole of substitutional atoms in Eq 15 as:

$$G = \sum_M^m y_M \left(\sum_N^n y_N G_{MNc/a} + y_{va} G_M \right) + RT \sum_M^m y_M$$

Thermodynamics of Martensitic Transformations

$$\ln y_M + RT(c/a) \sum_{N}^{n} (y_N \ln y_N + y_{va} \ln y_{va}) +$$

$$\sum_{i}^{m-1} \sum_{j>i}^{m} y_i y_j (\sum_{N}^{n} y_N L_{ij}^N + y_{va} L_{ij}^{va}) + \sum_{k}^{n-1} \sum_{l>k}^{n} y_k y_l \sum_{M}^{m}$$

$$y_M L_{kl}^M + \sum_{k}^{n} y_k y_{va} \sum_{M}^{m} y_M L_{kva}^M \quad (15)$$

Equation 15 can be applied to the fcc or bcc structure with the appropriate value of c/a, where M, i, and j denote substitutional elements and N, l, and k refer to interstitial elements. In addition, G_M is the Gibbs energy of the pure element M in the structure under consideration and $G_{MN_{c/a}}$ is the Gibbs energy of the compound $MN_{c/a}$ (corresponding to the case where there are no vacant interstitial sites). Thus, for ferrite in the Fe-C system, bcc FeC_3 (while for austenite, fcc FeC) would correspond to the compound of interest. The symbols L_{ij}^N and L_{ij}^{va} refer to the regular solution interactions between i and j in a lattice where all interstitial sites are filled with N-atoms (L_{ij}^N) or with vacancies (L_{ij}^{va}). Finally, L_{kl}^M and L_{kva}^M represent the interactions between interstitial elements in a metal M or interstitial elements and a neighboring vacancy in a metal M, respectively. The Gibbs energy per mole of atoms can be obtained from Eq 15 through multiplication by the factor:

$$\sum_{M}^{m} x_M.$$

When no interstitials are present, this factor is equal to unity, as is y_{va}. Since the remaining y_N values are equal to zero, the y_M values can be replaced by atom fractions. Under these conditions, only the terms containing G_M, $y_M \ln y_M$, and L_{ij}^{va} contained in Eq 15 survive, and Eq 15 is reduced to Eq 5 or 6.

Computer-based methods can be used to generate the Gibbs energy of substitutional and interstitial phases for multicomponent alloys based on equations such as Eq 7 or 15 in order to define the T_0 versus composition curves and the driving force at the M_s temperature. Since data bases capable of describing wide ranges of alloys are being developed, the opportunities for carrying out such computations should increase in the future.

CALCULATION OF THE GIBBS ENERGY DIFFERENCE BETWEEN FERRITE AND MARTENSITE

In the Fe-Ni case cited as an example early in this chapter, the martensite, formed on cooling, and the equilibrium ferrite phase have identical bcc structures. This situation simplifies consideration of the driving force for martensitic transformations as compared to Fe-C, where the martensite is tetragonal while the equilibrium low-temperature ferrite is bcc. Even Fe-Ni may offer some complications if the energetics of the martensitic bcc phase, formed on cooling austenite, is compared with that of the ferrite phase, formed by equilibrium decomposition of austenite. In comparing the overall energetics of these processes, particularly in the case where substantial transformation of the parent phase into the daughter phase occurs, the energy of defects stored in the daughter phase could differ.

Although this quantity of energy is generally thought to be small (i.e., of the order of 1 to 10 cal/mol) and negligible (Ref 4), Chang et al. (Ref 27) have reported on a series of experiments which suggests that the energy of defects stored in the daughter phase can be of the order of 100 cal/mol. These experiments consisted of annealing bcc Fe-Ni martensite (α') in the ($\alpha+\gamma$) field (see Fig. 2) for times up to 1000 h. This treatment results in the growth of equilibrium fcc (γ) and equilibrium bcc (α). The enthalpy of transformation from α to γ was then determined in the duplex ($\alpha+\gamma$) samples by DSC and by ascertaining the exact amounts and composition of the α phase. The latter deter-

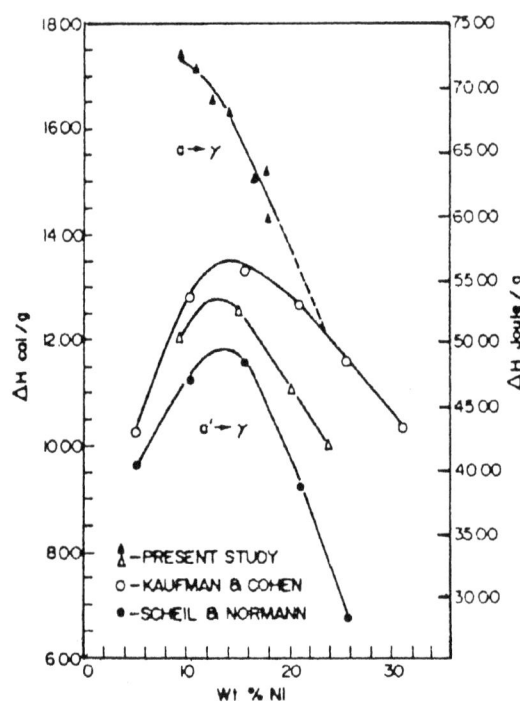

Fig. 12 Variation of $\Delta H^{\alpha \to \gamma}$ and $\Delta H^{\alpha' \to \gamma}$ as a function of nickel content in Fe-Ni alloys (Ref 27)

mination was carried out by means of magnetic measurements. Comparison of the enthalpy of the α→γ transformation with that for α'→γ shown in Fig. 12 leads to the conclusion that dislocations stored in the martensite increase its energy by 100 cal/mol above ferrite of the same composition.

The validity of this interesting result depends critically on the magnetic techniques employed to ascertain the amount and composition of the ferrite (α) formed during annealing in the α'→γ field. The measurement of the enthalpy of the α'→γ transformation by Chang et al. (Ref 27) shown in Fig. 12 presents no difficulties as the compositions of the martensite and austenite are known and are identical. However, the amount and composition of the equilibrium α formed during annealing can be questioned. It is indeed curious to examine the experimental enthalpy difference for the α→γ transition shown in Fig. 12 as a function of nickel content in light of the fact that this difference is equal to 210 cal/mol (3.76 cal/g) for the equilibrium transformation in pure iron.

Notwithstanding the foregoing interesting set of experiments (related to the energy of defects stored in the martensite as a result of the transformation from austenite to martensite and the comparison of the energy of the resulting structure with that of a similar structure containing no defects), it is important to consider the difference in energy between martensite and the low-temperature equilibrium phase when differences in crystal structure exist. The most practical example of such a case is the Fe-C situation where the ferrite has the bcc structure while the martensite is bct. As noted earlier, this problem has been considered by Johansson (Ref 5), Zener (Ref 6), Cohen and coworkers (Ref 1-3), Fisher (Ref 7), and again by Kaufman, Radcliffe, and Cohen (Ref 28). The latter discussion, which follows the developments of Fisher (Ref 7) and Zener (Ref 6), describes the Gibbs energy difference between the bct martensite (α') and the bcc ferrite (α) by means of:

$$\Delta G^{\alpha \to \alpha'} = -50{,}000 \, (x^2 p^2 / (1-x) + 0.7x \, T\phi[p]) \quad \text{(cal/mol)} \quad (16)$$

where x is the atomic fraction of carbon in the noncarbon alloy, p is the order parameter, and $\phi[p]$ is a numerical function of the order parameter p as a ratio of the temperature T to the ordering temperature T_c. The latter is related to the carbon content by:

$$T_c = 28{,}000x \, (1-x) \quad (K) \quad (17)$$

On this basis, the tetragonal martensite (α') has a lower Gibbs energy than ferrite due to the ordering of carbon atoms. The Gibbs energy difference between austenite and martensite is just the sum:

$$\Delta G^{\gamma \to \alpha'} = \Delta G^{\gamma \to \alpha} + \Delta G^{\alpha \to \alpha'} \quad (18)$$

Thermodynamics of Martensitic Transformations

A similar scheme for calculating the driving force for martensite transformation could be developed for other cases in which the daughter phase is similar to, but not exactly the same as, a low-temperature equilibrium phase.

FORMATION OF MARTENSITE AS AN ADIABATIC PROCESS

It is well known that the martensite transformation in steels and in many other materials can take place very rapidly. As a consequence, the transformation cannot be truly isothermal. An estimate of the diffusion distance for heat, based on the expression D/v, where D is the thermal diffusivity and v is the speed of propagation of the martensitic transformation, may yield a value of less than $0.1\ \mu m$.

It may be interesting to examine the consequences of treating the martensitic transformation as an adiabatic process as was first done by Krisement, Houdremont, and Wever (Ref 29). Before going into any details of their theory, we should emphasize that it can hardly be used in order to reveal the true mechanism of the martensite transformation. The reason stems from the fact that the driving force for the adiabatic reaction is less than for the corresponding isothermal reaction. If the atomic mechanisms of the two reactions are the same, we should thus expect the reaction to start in the isothermal mode before the driving force is large enough for the adiabatic mode. By the isothermal mode we simply mean that the reaction proceeds at such a slow rate that the temperature uniformity is maintained throughout the system. The fact that martensite has been observed to grow very rapidly can be explained by assuming that the real barrier to the formation of a martensite unit is to be found in the initial stage, or nucleation stage. We thus arrive at the picture that the nucleation barrier is surmounted by a slow, isothermal reaction. Then the resistance to the reaction decreases and the speed can increase toward adiabatic conditions because the driving force for that mode of reaction is now sufficient. In passing, we note that this reasoning suggests an experiment aimed at studying isothermal growth of martensite by using a specimen with a concentration gradient where the growth progresses into regions of less favorable composition.

The theoretical treatment of the adiabatic mode of martensite formation suggested by Krisement et al. (Ref 29) was worked out in detail by Hillert (Ref 30). A reversible, adiabatic change requires two conditions:

$$dU + w = 0 \qquad (19)$$

$$dS = 0 \qquad (20)$$

w is usually identified with the mechanical work done during the reaction, but for the present purpose we may generalize it and include other types of resistance to the reaction, such as surface and dislocation energies. Many such effects were considered by Cohen et al. (Ref 1) in their analysis of the thermodynamics of the martensitic transformation.

By applying the conditions of Eq 19 and 20 to a hypothetical system that can change homogeneously from austenite of temperature T_1 to martensite of temperature T_2, we find:

$$U^a(T_2) - U^a(T_1) + U^m(T_2) - U^a(T_2) + w = 0 \qquad (21)$$

$$S^a(T_2) - S^a(T_1) + S^m(T_2) - S^a(T_2) = 0 \qquad (22)$$

For a reasonably small temperature range, it might be a good approximation to consider the heat capacity, c, of the austenite and the difference in U and S between the two phases, ΔU and ΔS, as independent of temperature. We thus find:

$$\Delta U + w = -c(T_2 - T_1) \quad (23)$$

$$\Delta S = -c \ln T_2/T_1 \quad (24)$$

Solving for T_1 and T_2:

$$T_2 = T_i \frac{\Delta S/c}{\exp \Delta S/c - 1} \quad (25)$$

$$T_1 = T_i \frac{\Delta S/c}{1 - \exp -\Delta S/c} \quad (26)$$

$$T_2 - T_1 = -T_i \Delta S/c \quad (27)$$

where

$$T_i = \frac{\Delta U + w}{\Delta S} \quad (28)$$

In this case, T_i is the equilibrium temperature between austenite and martensite under isothermal conditions. Let us first neglect w. We can then identify T_i with T_0, the temperature of equal Gibbs energy. It is evident that T_1 and T_2 would then bracket T_0. Under adiabatic conditions, T_0 does not define the equilibrium temperature, but the austenite must be cooled further before the transformation can start. At the lower temperature, T_1, the system can transform to martensite, which will be at a higher temperature, T_2. If one tries to heat this martensite, it would immediately transform back to the austenite of the lower temperature, T_1.

When applying this treatment to a real system, one must first realize that the system does not transform homogeneously. Accepting that only a small portion is transformed to martensite, we understand that it will quickly cool from T_2 to T_1, and the reverse reaction will not occur unless the entire system is heated back to T_2. The temperature difference $T_2 - T_1$ is thus a hysteresis and it is tempting to compare this value with the experimental hysteresis observed between the start of the martensitic reaction, M_s, and the start of the reverse reaction, A_s. Before making such a comparison, we should discuss the resistance w.

The work done on the formation of martensite may not be recovered on the reverse reaction. It is even likely that there is a mechanical resistance to both the reactions, which would mean that w changes sign. Two different w values should thus be used in Eq 28, yielding different T_i values to be used in Eq 25 and 26.

It is well known that the martensitic transformation can be promoted by mechanical deformation. Kaufman and Cohen (Ref 10) examined this phenomenon in detail in the Fe-Ni system and defined new critical temperatures, M_d and A_d. They found that the starting temperatures for the formation of martensite and the reverse reaction move much closer to each other under plastic deformation, but they were not able to make M_d and A_d overlap.

There are two obvious ways in which plastic deformation can promote the martensitic transformation. It can either provide an additional driving force, i.e., change the quantity ΔU, or it can decrease the resistance w. Either way, it should be interesting to compare the experimental value of $A_d - M_d$ with the adiabatic hysteresis obtained by assuming that w is negligible or has the same value for the two reactions, i.e., obtained from Eq 27. For an Fe-30Ni alloy, a value of $T_2 - T_1 = 95$ K was obtained by Hillert (Ref 30), in good agreement with the experimental value of $A_d - M_d$. Of course, one did not have any right to expect such an agreement. First of all, it remains to be proved that the martensitic reaction on plastic deformation is rapid enough to be an adiabatic process. If this is actually the case, one may ask why it is not possible to promote the nucleation of martensite by plastic deformation to such an extent that martensite forms before the driving force for the adiabatic reaction has turned positive.

Finally, it should be mentioned that the adiabatic hysteresis should probably be about twice as large as calculated here, because the first adiabatic martensite to form will heat the surrounding austenite and thus prevent further growth unless the austenite is initially at an even lower temperature than T_1, as given by Eq 26. This complication will be dealt with in another connection (Ref 31, 32).

INTERACTION OF DIFFUSION WITH THE MARTENSITIC TRANSFORMATION

In the preceding section, it was argued that the critical event in the nucleation of martensite is a slow process and that a thermodynamic treatment should be based on the assumption of isothermal conditions. Of course, there is a limit to how slow a martensitic reaction is allowed to be, because the partitioning of individual elements between the two phases would change the reaction into a diffusion-controlled transformation. This fact will probably make the martensitic mode of transformation difficult at high temperatures. However, in substitutional iron-base alloys such as Fe-Ni, the temperature is relatively low and the rate of diffusion is so low that the diffusion-controlled $\gamma \to \alpha$ transformation cannot compete with the martensitic transformation.

On the other hand, if the interface between the γ and α crystals is of an incoherent type that migrates by individual atomic jumps, there may be a considerable atomic mobility in the interface as revealed by a high boundary diffusivity. The alloying element may then partition between the two phases and, if long-range diffusion is prevented due to the low temperature, the reaction cannot continue unless the temperature is lowered into or close to the α one-phase field. Evidently, this reasoning does not apply to martensite with its glissile interface, but it may apply to the massive diffusionless transformation (Ref 31). This may be the reason why diffusionless solid-state transformations, observed inside a two-phase field, are generally of the martensitic type.

In an interstitial iron-base alloy such as Fe-C, the diffusivity is much higher, and the possible movements of carbon should be seriously considered. On the short-range scale there may be a tendency for carbon to jump into a neighboring site. In the martensite crystal there are three families of interstitial sites, one of them coming from the interstitial sites in the parent austenite and the other two being created as the austenite is deformed by the Bain distortion. In the first instance, all the carbon atoms are situated in sites of the first family. This is an ordered situation, and it was first pointed out by Zener (Ref 6) that this may actually be a metastable situation. However, above a critical line in the phase diagram, disordering should take place, and it is then an important question whether the transformation is slow enough to allow this rearrangement to occur at the interface and thus increase the driving force for the martensitic transformation. This may be the case during the nucleation event in low-carbon steels. In order to make predictions of this effect, it is necessary to have a good thermodynamic treatment of the order-disorder reaction in bcc Fe-C alloys.

On the long-range scale, there is a tendency for carbon to segregate into a parent austenite. This factor would decrease the driving force for a further transformation and may result in a slow-growing diffusion-controlled bainitic transformation. It is easy to imagine conditions where all slowly formed martensitic nuclei will thus develop into bainite. A question that has been discussed (Ref 33) is whether such a slow-growing nucleus may gradually accelerate and develop into martensite.

RELATION OF MARTENSITE GROWTH CONDITIONS AND PARAEQUILIBRIUM CONDITIONS IN IRON-BASE ALLOYS

In the preceding section, we discussed the possibility that a slowly formed martensitic nucleus may develop into bainite due to the partitioning of carbon. On the other hand, it was concluded that substitutional alloying elements would be too sluggish to give the same effect. Let us now consider a ternary alloy with one substitutional and one interstitial alloying element, such as Fe-Ni-C. A mixed behavior can then be expected, giving partitioning of carbon but not of nickel. The existence of this type of transformation was first established by Hultgren (Ref 34), who found by chemical analysis that the ferritic constituent of bainite inherits the alloy content of the parent austenite. As far as iron and nickel are concerned, the bainitic transformation may thus be a purely martensitic transformation. Hultgren was interested in the driving force for the diffusion of carbon, and he introduced the concept of paraequilibrium to describe the local equilibrium conditions at the interface between austenite and ferrite. He hypothesized that there would be complete equilibrium with respect to carbon, but no movements of iron and nickel with respect to each other. This type of equilibrium can be evaluated if the ordinary Gibbs-energy functions of the two phases are known. Two equations are required. Naturally, the carbon equilibrium is defined:

$$G_C^\alpha = G_C^\gamma \quad (29)$$

The other equation defines equilibrium with respect to a mixture of iron and nickel in a fixed ratio, x_{Fe}/x_{Ni}. It can be written in two convenient ways (Ref 35, 36):

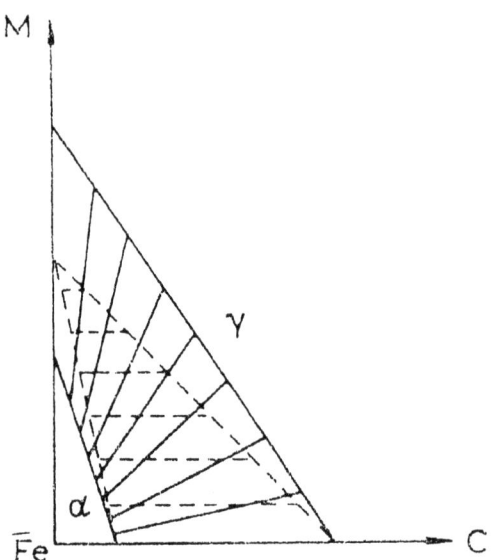

Fig. 13 Paraequilibrium in the ternary Fe-M-C system

$$\left[\frac{G_m - (x_C G_C)}{1 - x_C}\right]^\alpha = \left[\frac{G_m - (x_C G_C)}{1 - x_C}\right]^\gamma \quad (30)$$

$$\frac{x_{Fe}}{x_{Fe} + x_{NI}}(G_{Fe}^\alpha - G_{Fe}^\gamma) = \frac{x_{Ni}}{x_{Ni} + x_{Fe}}(G_{Ni}^\alpha - G_{Ni}^\gamma) \quad (31)$$

The paraequilibrium phase boundaries fall inside the ordinary phase boundaries in the phase diagrams, as illustrated schematically by the dashed lines in Fig. 13 (Ref 37). The conclusion is that the driving force for diffusion of carbon inside either one of the two phases will be higher under paraequilibrium conditions than under full equilibrium conditions at the interface, and the effect of the alloying element to increase the hardenability will thus be less if paraequilibrium conditions are established.

Figure 13 demonstrates the relation between the paraequilibrium concept and the equal-Gibbs-energy concept, discussed earlier, because the point where the paraequilibrium phase boundaries meet on the ordinate axis is the T_0 point.

The paraequilibrium concept has been applied many times in discussions of the γ→α transformations. However, it was early realized that a growing phase may inherit the alloy content of the parent phase even if full equilibrium conditions are established at the interface (Ref 38). In any case, one should always test for paraequilibrium conditions not only by chemical analysis but also by checking that $D/v < d$, where D is the volume diffusion constant for the alloying element, v is the speed of the interface, and d is the jump distance in diffusion.

In addition, when the interface is of such a type that it migrates by individual atomic jumps, the paraequilibrium conditions may never be fulfilled (Ref 39). This may apply to the ferrite/austenite interface in pearlite at all temperatures, but not to the ferrite/austenite interface in bainite at low temperatures. This may be the reason why bainite can form in alloyed steels under conditions where pearlite is completely prevented. This is an analogue to the advantage of the martensitic transformation to the massive one.

ACKNOWLEDGMENT

This work has been sponsored by the National Science Foundation under Grant DMR79-11916.

REFERENCES

1. M. Cohen, E.S. Machlin, and V.G. Paranjpe, Thermodynamics of the Martensitic Transformation, in *Thermodynamics in Physical Metallurgy*, American Society for Metals, 1950, p 242-270
2. M. Cohen, Retained Austenite, *Trans. ASM*, Vol 41, 1949, p 35-94
3. M. Cohen, The Martensite Transformation, in *Phase Transformations in Solids*, John Wiley & Sons, 1951, p 588-660; appendix by V.G. Paranjpe
4. L. Kaufman and M. Cohen, Thermodynamics and Kinetics of Martensitic Transformations, *Prog. Met. Phys.*, Vol 7, 1958, p 165-245
5. C.H. Johansson, Thermodynamishe begrundete der Vorgange bei der Austenite-Martensit-Umwandlung, *Arch. Eisenhüttenwes.*, Vol 11, 1937, p 241-251
6. C. Zener, Equilibrium Relations in Medium Alloy Steels, *Trans. AIME*, Vol 167, 1946, p 513-534; discussion, p 546, 585, 595
7. J.C. Fisher, The Free Energy Change Accompanying the Martensite Transformation in Steels, *Trans. AIME*, Vol 185, 1949, p 688-690
8. R.P. Smith, Equilibrium Mixtures of Iron-Carbon Alloys with Mixture of CO-CO_2 and CH_4-H_2, *J. Am. Chem. Soc.*, Vol 68, 1946, p 1163
9. O. Kubaschewski and O. von Goldbeck, The Thermodynamics of Iron-Nickel Alloys, *Trans. Faraday Soc.*, Vol 45, 1949, p 958
10. L. Kaufman and M. Cohen, The Martensitic Transformation in the Iron-Nickel System, *Trans. AIME*, Vol 206, 1956, p 1393-1401
11. E. Scheil and W. Normann, *Arch. Eisenhüttenwes.*, Vol 30, 1959, p 751
12. E. Scheil and E. Saftig, *Arch. Eisenhüttenwes.*, Vol 31, 1960, p 623
13. H.A.J. Oonk, The Use of the Equal-G Curve in Interpretation of Isobaric Binary Phase Equilibrium Diagrams, *Recueil*, Vol 87, 1968, p 1345-1358
14. H.A.J. Oonk and A. Sprenkels, Isothermal Binary Liquid-Vapor Equilibrium and the ECG-Method, *Z. Phys. Chem. Neue Folge*, Vol 75, 1971, p 225-233
15. A.C.G. Van Gerdem, C.G. de Kruif, and H.A.J. Oonk, Properties of Mixed Crystalline Organic Material Prepared by Zone Leveling—Experimental Determination of EGC for the System p-Dichlorobenzene and p-Dibromobenzene, *Phys. Chem. Neue Folge*, Vol 107, 1977, p 167-173
16. H.A.J. Oonk, *Phase Theory—The Ther-*

modynamics of Heterogeneous Equilibria, Elsevier, 1981

17. M. Hasselblatt, *Z. Phys. Chem.*, Vol 83, 1913, p 1-40
18. L. Kaufman, The Lattice Stability of Metals—I, Titanium and Zirconium, *Acta Metall.*, Vol 7, 1959, p 575-587
19. L. Kaufman, Proceedings of the Fourth CALPHAD Meeting, *CALPHAD*, Vol 1, 1977, p 7-89
20. L. Kaufman and H. Nesor, Coupled Phase Diagrams and Thermochemical Data for Transition Metal Binary Systems—I and II, *CALPHAD*, Vol 2, 1978, p 55-108
21. L. Kaufman, Coupled Phase Diagrams and Thermochemical Data for Transition Metal Binary Systems—III, *CALPHAD*, Vol 2, 1978, p 117-146
22. L. Kaufman and H. Nesor, Coupled Phase Diagrams and Thermochemical Data for Transition Metal Binary Systems—IV, *CALPHAD*, Vol 2, 1978, p 295-318
23. L. Kaufman and H. Nesor, Coupled Phase Diagrams and Thermochemical Data for Transition Metal Binary Systems—V, *CALPHAD*, Vol 2, 1978, p 325-348
24. L. Kaufman, Coupled Phase Diagrams and Thermochemical Data for Transition Metal Binary Systems—VI, *CALPHAD*, Vol 3, 1979, p 45-76
25. M. Hillert and M. Waldenstrom, Isothermal Sections of the Fe-Mn-C System in the Temperature Range 873K-1373K, *CALPHAD*, Vol 1, 1977, p 97-132
26. C. Chatfield and M. Hillert, A Thermodynamical Analysis of the Fe-Mo-C System Between 973 and 1273K, *CALPHAD*, Vol 1, 1977, p 201-223
27. H. Chang, S. Sastri, and B. Alexander, A Comparison of Enthalpy Change ΔH Between $\alpha' \rightarrow \gamma$ and $\alpha \rightarrow \gamma$ Transformations in Fe-Ni Alloys, *Acta Metall.*, Vol 28, 1980, p 925-932
28. L. Kaufman, S.V. Radcliffe, and M. Cohen, Thermodynamics of the Bainite Reaction, in *Decomposition of Austenite by Diffusional Processes*, V.F. Zackay and H.I. Aaronson, Ed., Interscience, 1962, p 313-352
29. O. Krisement, E. Houdrement, and F. Wever, Contribution à la termodynamique de la transformation austenite-martensite dans les alliages fer-carbone, *Rev. Metall.*, Vol 51, 1954, p 401-409
30. M. Hillert, Thermodynamics of Martensitic Transformations, *Acta Metall.*, Vol 6, 1958, p 122-124
31. M. Hillert, Thermodynamics of the Massive Transformation, *Metall. Trans.*, Vol 15A, 1984, p 411-419
32. M. Hillert, Paraequilibrium and Other Restricted Equilibrium, in *Alloy Phase Diagrams*, L.H. Bennett, T.B. Massalski, and B.C. Giessen, Ed., Elsevier, 1983
33. M. Hillert, Diffusion and Surface Control of Reactions in Alloys, *Metall. Trans.*, Vol 6A, 1975, p 5-19
34. A. Hultgren, Isothermal Transformation of Austenite, *Trans. ASM*, Vol 39, 1947, p 915
35. M. Hillert, Calculation of Phase Equilibria, in *ASM Seminar on Phase Transformations*, American Society for Metals, 1968, p 181-218
36. J.B. Gilmour, G.R. Purdy, and J.S. Kirkaldy, Thermodynamics Controlling the Proeutectoid Ferrite Transformations in Fe-C-Mn Alloys, *Metall. Trans.*, Vol 3, 1972, p 1455-1464
37. M. Hillert, The Role of Interfaces in Phase Transformations, in *Mechanism of Phase Transformations in Crystalline Solids*, Institute of Metals of London, 1969, p 231-247
38. M. Hillert, "Paraequilibrium," internal report, Swedish Institute for Metals Research, Stockholm, 1953
39. M. Hillert, An Analysis of the Effect of Alloying Elements on the Pearlite Reaction, in *Proc. Int. Conf. Solid-Solid Phase Transformations*, H.I. Aaronson et al., Ed., TMS-AIME, 1982, p 789-806

6 Introduction to "The Growth of Ferrite, Bainite and Martensite"

published as an *Internal Report Swedish Institute for Metals Reserch* (1960)

This paper is an internal report from the Swedish Institute for Metallurgy dating from 1960. Although never "officially" published in a journal, it has been very influential as a kind of "underground literature" that people would copy and forward to each other. It seems most appropriate to give this report a wider distribution *via* this collection of selected papers. It does not fall into the categories of papers that stabilize a field, but rather in the category which fertilizes it by the wealth of new ideas, still worthy of exploring and pursuing, that it contains.

The paper is very much in the "Hillert style", eminently recognisable with its systematic use of free energy diagrams, its preference for simple approximate solutions of the Zener type for diffusion problems, as well as its use of metallography to support statements on the mechanism, and, back from the 60's, its precursory use of numerical solutions when analytical ones are no longer available.

This paper contains a number of experimental investigations using tools which have now become classics: salt bath quenching and tempering of martensite to identify the temperature of its formation, and decarburisation experiments to investigate transformation kinetics. These "classic experiments" can nowadays be complemented by modern tools such as local microprobe chemical analysis to measure directly the carbon content of ferrite. They could perhaps be complemented by Synchrotron experiments to measure *in situ* the Carbon content while the austenite to ferrite transformation takes place. The decarburisation experiments have recently been used to investigate the concentrations at the migrating phase interface in ternary alloys and to discriminate between the various possibilities (Local equilibrium with negligible partitioning (LENP) or Paraequilibrium (PE)).

A number of theoretical ideas are introduced in this paper, which have irrigated the research on modelling phase transformations in ferrous systems for the last 45 years. One of the most original ideas deals, within the same model, with both the kinetics of phase transformations and the limiting temperatures (T_0, B_s, WB_s, M_s, M_d ...). The possibility for allowing for a discontinuity in chemical potentials across the migrating interface for both Carbon and Iron and the introduction of a finite interface mobility are at the root of all the models for phase transformations which try to refine on the original Zener treatment of the problem. Another idea still to be fully explored is the difference between incoherent interfaces, which are assumed to be totally diffusion controlled, and coherent interfaces, where a finite mobility is introduced. These ideas appear progressively throughout the paper; *via* the analysis of edgewise growth of Widmanstätten ferrite, *via* the analysis of non symmetric allotriomorphic ferrite, *via* the analysis of decarburisation experiments. Solutions to these questions require reliable knowledge of thermodynamic and diffusion data; the initial motivation for the Calphad philosophy, embodied in the THERMOCALC and DICTRA databases which were another of Mats Hillert's contributions to Physical Metallurgy.

But the most original part of the paper is the so called "generalised growth model". This formidable achievement contains, as nuclei, most of the ideas that have since been developed in the modelling of ferrous phase transformations. The main ideas behind this model are i) to allow for a discontinuity of chemical potentials at the interface for both Carbon and Iron atoms, and ii) the existence of a critical driving force to put the interface in motion. The model has four unknowns: the Carbon concentrations on either side of the interface and the velocity and radius of curvature of the growing needles. It leads to three equations expressing: i) the diffusion of carbon in front of the growing ferrite, ii) the mass balance, and iii) the transfer of carbon through the interface under the driving force. As it stands, a whole range of solutions are possible, among which, with some arbitrariness, and following Zener's suggestion, one can select the fastest growing one. Hillert goes further and explores the whole range of possibilities. The paper presents the growth rates for various initial saturations, as a function of the tip radius. Similarly, the Carbon content of the ferrite is calculated as a function of the tip radius. These two figures (10 and 11 in the paper), due to the structure of the equations, exhibit a "catastrophe" in a mathematical sense, *i.e.* a continuous variation of the initial super saturation leads to a discontinuous, and even singular, variation of the selected growth rate. It is tempting to interpret these singularities as transitions between different growth modes, Widmanstätten ferrite and Martensite. The analysis is performed in terms of a control parameter which would be the carbon supersaturation (the mobility of the interface can also be a possible control parameter). Hillert then extends this analysis, taking the temperature as the experimental parameter. The qualitative sequence of growth modes can be interpreted in this way, but again a quantitative prediction would require thermodynamic and kinetic information versus temperature, which were not reliable enough at the time. Very lucidly, Hillert also points out that the nucleation step, lacking in the present description, has to be included in a comprehensive model.

This paper is fascinating since it raises a number of questions, some of them still unsolved after 50 years: How can one deal with ternary elements? How can one incorporate in a consistent manner solute drag effects? How can one analyse the selected solution (if any selection criterion is indeed operative)? How can one couple the structure of the interface with the departures from equilibrium? How can one calculate *a priori* the interface transfer coefficient?

One can safely say that this paper has been one of the most influential "unpublished papers" of physical metallurgy, and no doubt, reprinting it in this volume will give a new impetus to the many ideas it has put forward.

Yves BRÉCHET and Jean PHILIBERT

The Growth of Ferrite, Bainite and Martensite

by M Hillert

Abstract

Quantitative measurements of the edgewise growth rate of Widmanstätten ferrite and bainite are presented and shown to fit a theoretical growth model involving diffusion of carbon in austenite as the rate-controlling process. There are some indications of a discontinuity in iron activity at the moving ferrite-austenite interface. A microscopic study of grain boundary ferrite indicates the existence of a similar discontinuity at coherent interfaces of such precipitate. Microscopic evidence of a carbon gradient in austenite indicates that the movement of incoherent ferrite-austenite interfaces is mainly diffusion controlled. A quantitative study of decarburization of steel supports this conclusion.

A generalized growth model is developed, allowing discontinuities in iron as well as carbon activity at the moving interface. The numerical calculation results in diffusion controlled growth at low supersaturations, in martensitic and diffusion-controlled growth at medium supersaturations and in only/martensitic growth at high supersaturations.

Swedish Institute for Metal Research
Stockholm 1960

1. Introduction

When formed at low temperatures, bainite has an appearance very like that of martensite and it is sometimes even impossible to distinguish them microscopically. Moreover, when bainite was first recognized as a special kind of structure, one already suggested that its ferrite constituent was supersaturated with carbon[1,2,3]. In this respect also, it was thus believed to resemble martensite. As a consequence, bainite has been regarded as a structure closely related to martensite and it is not surprising that the theories about the mechanism of formation of bainite have been based on this relation. The upper temperature limit for the formation of bainite, B_s, has thus been compared to the upper temperature limit for the formation of martensite, M_s, and both of them have been related to the temperature of no free energy change for a diffusionless transformation, T_o[4,5]. In this connection, diffusion of carbon during growth of bainite has not entered into the discussion. In connection with the rate of formation of bainite, on the other hand, diffusion has sometimes been regarded as essential and the favorite hypothesis has here been that the dominating diffusion process occurs inside the ferrite phase. The reason for this belief may partly stem from the idea of supersaturation. If there is an excess amount of carbon in the ferrite phase which has to leave this phase and precipitate as carbide particles, in order to allow further growth, diffusion through ferrite should undoubtedly be of great importance no matter where the carbides form. Another reason may have been that the diffusion coefficient for carbon in ferrite is much larger than in austenite, if compared at temperatures above the eutectoid point, and in view of the activation energy values it could be expected that the difference should be even more pronounced at low temperatures. The diffusion of carbon through austenite thus seemed quite too slow to be of any importance at the temperatures of bainite formation.

3.

Recently, the edwise growth rate of bainite was measured microscopically using a hot stage[6, 7], and the activation energy for this process was found to be quite close to that of diffusion in ferrite but much lower than for diffusion in austenite. These experiments thus lend some support to the hypothesis that the rate-controlling process for bainitic growth is diffusion through ferrite.

On the other hand, it has become increasingly clear in recent times, that there exist many similarities between bainite and proeutectoid ferrite of the Widmanstätten morphology, and it seems natural that the mechanism of bainitic growth should be studied from the high-temperature side rather than from the low-temperature side. As a consequence, first one should try to find out how the growth rate of Widmanstätten ferrite is controlled and, secondly, one should examine how far down in temperature the same mechanism can be applied. As a first step along these lines, a theoretical model for the edgewise growth of Widmanstätten ferrite was developed[8]. An experimental test of the theoretical growth equation has now been carried out at high temperatures where the transformation product is Widmanstätten ferrite, as well as at low temperatures where the product must be classified as bainite. The result of these experiments will be presented in the present paper. The results seem to indicate that a single growth rate equation can be applied for all the temperatures examined, confirming the close relationship between Widmanstätten ferrite and bainite. In fact, there seems to be a deviation from chemical equilibrium at the moving interface in both these cases which is of the same order of magnitude as in the case of martensite. As a consequence it seems reasonable to attempt to describe the growth of martensite with the same growth rate equation. In order to achieve this, one must dispose of the usual assumption of local equilibrium with respect to carbon at the moving interface. An attempt in this direction will be presented in the final section. The result indicates a close relationship between ferrite

4.

and martensite and the possibility of a transition from one to the other during growth.

2. Theoretical growth model for edgewise growth

The theoretical model which was developed in a previous paper[8] had to take into account the fact that Widmanstätten ferrite, as well as bainitic ferrite has a characteristic orientation relationship to the matrix austenite and forms on characteristic habit planes. These facts are generally taken as indications of coherency between the precipitate particles and the matrix and this view was accepted. What consequences such a coherency could have on the growth process is uncertain, however. Coherency may, for instance, be necessary in order to make the transfer of the sluggish iron atoms from the austenite lattice into their correct positions in the growing ferrite lattice so rapid that this process does not limit the growth rate at low temperatures. In the theoretical model it was simply assumed that the transfer of atoms, iron as well as carbon, over the interface is rapid enough to allow local chemical equilibrium to be established between the two phases. The growth rate should then be controlled by how rapid carbon can be transported away from the growing ferrite, and a growthrate equation was derived assuming that the main part of this transportation takes place by ordinary volume diffusion in the austenite phase. The only direct effect of coherency on this equation is through the interfacial energy of the ferrite-austenite interface, σ. The derivation followed a more approximate treatment carried out by Zener[4] for the edgewise growth of a plate and it was not able to take into account any variation of σ along the interface. Due to the coherency, σ is expected to have a low value on the sides of a Widmanstätten plate but a normal value at the edge. The correct growth rate would be given by the equation if only the correct mean value of σ could be chosen. It is quite uncertain how this choice should be made but it appears reasonable to expect that the correct value of σ should be considerably lower than the normal value of 600 ergs/cm^2 which has been found for incoherent interfaces between ferrite and austenite[9].

By combining eqs. 45 and 46 in reference 8 and using eq. 17, which yields a value of about 1 for the quantity $k^{\gamma a}$, one obtains

$$\frac{8\sigma}{R} = (c_o^{\gamma a} - c_\infty^\gamma)^2 \Big/ \frac{vc_\infty^\gamma}{DT} \qquad (1)$$

where R is the gas constant, $c_o^{\gamma a}$ the equilibrium carbon content of austenite in contact with ferrite, c_∞^γ the original carbon content, v the edgewise growth rate and T the absolute temperature.

This equation could be tested by plotting experimental growth rate data as $\sqrt{vc_\infty^\gamma/DT}$ versus $c_o^{\gamma a} - c_\infty^\gamma$. A straight line through the origin should then be obtained and from the slope a value of σ could be calculated.

3. Measurements of growth rate

The growth of Widmanstätten ferrite has been studied microscopically by Johanson[9] who was able to follow the growth by examining a series of specimens, austenitized and isothermally heat treated for different lengths of time. The length of the largest plates growing out from grain boundaries was measured in each specimen and plotted graphically versus the holding time. The points could be well represented by a straight line as long as there was no interference with the opposite grain boundary or with other plates. A linear growth rate was thus obtained in accordance with the growth-rate equation. The range of temperature (750-650°C) and carbon content (0,25 - 0,55%) was not large enough to allow a further test of the equation, but it is interesting to note that Johanson did not find any noticeable effect of 1% silicon or nickel on the growth rate, after correcting for the shift of the A_3-line by the method given by Zener[10].

6.

In the present study the same experimental technique has been adopted but the range of temperature and carbon content of the steels have now been extended enough to include bainite in addition to Widmanstätten ferrite. The chemical composition of the steels and the observed growth-rate values are given in Table I together with the same information from Johanson's work. Johanson used synthetic steels (upper part of Table I) and in the present study commercial plain carbon steels were employed (lower part of Table I).

The quantity $c_o^{\gamma\alpha}$ must be known in order to allow a test of the growthrate equation. Above the eutectoid temperature, it can be read from the iron-carbon phase diagram but at lower temperatures one must use an extrapolation of the A_3-line. Since there seemed to be no reliable method available for calculating this extention, a straight line was simply drawn through the eutectoid point and a point chosen at 5 % at 400°C. This extrapolation seems reasonable at the higher temperatures but gives very large values at low temperatures. Consequently, at present there appears to be no way of testing the theoretical growth-rate equation for temperatures below say 400°C.

The value of D has been determined by Wells, Batz and Mehl[11] at temperatures and compositions where austenite is stable. For the present purpose these data must also be extrapolated and the equation derived in an earlier paper [8] was used for this purpose. It must be emphasized that the extrapolations of measured data which had to be used for $c_o^{\gamma\alpha}$ and D are more uncertain the lower the temperature and may be quite wrong at 400°C.

The experimental values of D show a strong concentration dependency, whereas the growth-rate equation was derived assuming a constant D. As a consequence, a value of D must be chosen corresponding to some mean value of composition in the diffusion zone. For the present purpose this composition was chosen as $\frac{1}{2}(c_o^{\gamma\alpha} + c_\infty^\gamma)$.

7.

4. Test of growth-rate equation

The measured growth-rates are plotted in Fig. 1 as $\sqrt{vc_\infty^\gamma/DT}$ together with the data obtained by Johanson. The symbols shown in Table I were used in order to identify the steels and thin lines were drawn around all points from the same temperature.

All the points are fairly well gathered in a band directed towards the origin. If they are represented by a stright line through the origin and most weight is given to the date above 630°C this line would have a slope corresponding to $\sigma = 1000$ ergs/cm^2. This value is considerably higher than expected but it must be realized that it depends strongly on the methods of extrapolating $c_o^{\gamma\alpha}$ and D. If the straight line extrapolation of $c_o^{\gamma\alpha}$ had been drawn through 4% instead of 5% at 400°C, a σ value of about 600 ergs/cm^2 would have resulted. No conclusion should thus be drawn from its value. The mere fact that measured data from such a large range of supersaturation and temperature are so well gathered around a straight line through the origin may be taken as an indication that the equation is basically correct.

It is especially interesting to note from Table I that the growth rates <u>increase</u> as the temperature is lowered. This holds for all the observations above 400°C except the last ones on the eutectoid steel. Furthermore, it is evident from Fig. 1 that this increase can be accounted for in terms of diffusion in austenite, in spite of the fact that one has generally believed that the rate of this diffusion process would <u>decrease</u> rapidly as the temperature is lowered.

A close examination of the values indicates that the growth rate for most of the steels is approaching a maximum in the vicinity of 400°C and one could expect the values to decrease in a lower temperature range. This is in agreement with the low-temperature results obtained by hot-stage microscopy[6, 7].

8.

The method of interpreting this decrease in terms of an activation energy and comparing with values for diffusion does not seem justified in view of the above discussion. As a consequence, the hot-stage experiments cannot be used in support of the hypothesized rate control by diffusion in ferrite. In this connection, it might be interesting to note that the extrapolation formula for $D^{(8)}$ indicates that the activation energy for diffusion in austenite with 3% carbon may be around 20000 cal/mole, a value usually connected with diffusion in ferrite.

If one would dare to go one step further in extracting information about the growth mechanism from the experimental growth rates in spite of the uncertainties involved in the extrapolations of $c_o^{\gamma\alpha}$ and D, one should base the further analysis on the observation that the points from any particular temperature in Fig. 1 seem to scatter very little from a straight line which intercepts the abscissa. It even seems possible to make these lines parallel for all the temperatures. This observation could perhaps be interpreted in terms of the following growth-rate equation,

$$\sqrt{\frac{8\sigma}{R}}\sqrt{\frac{vc_\infty^\gamma}{DT}} = c_o^{\gamma\alpha} - c_\infty^\gamma - a = c^{\gamma\alpha} - c_\infty^\gamma \qquad (2)$$

The difference "a" between $c_o^{\gamma\alpha}$ in the previous growth-rate equation and $c^{\gamma\alpha}$ in the new equation is given by the intercepts on the abscissa in Fig. 1 and may be considered as representing a deviation from complete chemical equilibrium. The diffusion of carbon in austenite would then be driven by the concentration difference $c^{\gamma\alpha} - c_\infty^\gamma$, only. Except for this difference, a value of σ could be evaluated for the diffusional growth process as before. All the parallel lines give the same result, $\sigma = 230$ ergs/cm^2. This value seems to be low enough to lend some support for the suggested interpretation.

9.

The deviation from equilibrium, i.e. the intercept, increases towards lower temperatures, as shown in Table II, indicating that the interface becomes more sluggish. It does not indicate that carbon finds it increasingly difficult to escape from the ferrite. On the contrary, there seems to be less chance of carbon being caught by ferrite and thereby causing a supersaturation of this phase.

If the parallel lines, one for each temperature, are really of physical significance, one should be able to conclude that edgewise growth of Widmanstätten ferrite or bainite cannot occur until the supersaturation of austenite is higher than "a". If the values of $c^{\gamma\alpha}$, i.e. $c_o^{\gamma\alpha}-a$, are plotted in an iron-carbon phase diagram one would thus obtain a curve representing the upper temperature limit of Widmanstätten ferrite or bainite growth, Fig. 2. This line may be termed WB_s and in Fig. 2 it is compared with M_s, B_s and T_o. It should be noted that the old B_s-curve was determined on high chromium steels and on the supposition that bainite is not related to Widmanstätten ferrite. The position of the new WB_s-line seems quite reasonable. A direct determination of WB_s for hypereutectoid steels would be preferred but might prove difficult due to interference from cementite.

Although based on uncertain extrapolations, the extended analysis of the measured growth rates seems to yield reasonable results and support the modified growth rate equation. It would be highly desirable if justification for this modification could be obtained by some independent method. In particular, one would like to establish the existence of deviation from chemical equilibrium during the growth of ferrite in austenite. This seems to be possible and will be discussed in the next section.

5. Growth of grain-boundary ferrite

It is frequently found that crystals of ferrite which nucleate and grow along grain boundaries, will thicken into only one of the grains of austenite. The interface towards the other grain seems to be quite immobile. In Fig. 3 one crystal of ferrite has grown in one direction and another crystal in the opposite direction. This growth can certainly not be explained by a pure diffusion control and it might be interesting to evaluate how big a deviation from chemical equilibrium is involved. Furthermore, in view of a detailed microscopic study of ferrite which has recently been carried out[12], the immobile interface of a grain boundary crystal of ferrite is intimately related to the interfaces on the sides of a Widmanstätten plate.

A theoretical analysis can be based on the assumption that the movement of the mobile interface is diffusion-controlled and that perfect chemical equilibrium holds at that interface. The rate of interface movement will depend upon the diffusion of carbon to both sides, as illustrated in Fig. 4. The immobile interface is situated at $x = 0$ and the carbon concentrations, $c^{\gamma\alpha}$ and $c^{\alpha\gamma}$, are here supposed to differ from the equilibrium values $c_o^{\gamma\alpha}$ and $c_o^{\alpha\gamma}$. However, it will be assumed that there is no abrupt change in the activity of carbon, the carbon atoms being very mobile. If the two phases, α and γ, are approximated as ideal solutions, one thus has $c^{\gamma\alpha}/c^{\alpha\gamma} = c_o^{\gamma\alpha}/c_o^{\alpha\gamma}$. For convenience, this ratio will be represented by k. Following the procedure by Wagner[13] we can represent the variable concentrations in the three crystals by the expressions

$$c^I = c_{\infty}^{\gamma} + B^I \cdot (1+ \mathrm{erf}\frac{x}{2\sqrt{D^{\gamma} \cdot t}}) \qquad (3)$$

$$c^{II} = A^{II} + B^{II} \cdot \mathrm{erf}\frac{x}{2\sqrt{D^{\alpha} \cdot t}} \qquad (4)$$

11.

$$c^{III} = c_\infty^\gamma + B^{III} \cdot (1 - \text{erf} \frac{x}{2\sqrt{D^\gamma \cdot t}}) \qquad (5)$$

We shall apply the solution $\xi = \eta \cdot 2\sqrt{D^\gamma \cdot t}$ where η is a dimensionless growth constant. Let $\varphi = \sqrt{D^\gamma/D^\alpha}$. The boundary conditions at $x = 0$ and ξ give the following relations

$$c^{\gamma\alpha} = c_\infty^\gamma + B^I \qquad (6)$$

$$c^{\alpha\gamma} = A^{II} \qquad (7)$$

$$c_o^{\alpha\gamma} = A^{II} + B^{II} \cdot \text{erf } \eta\varphi \qquad (8)$$

$$c_o^{\gamma\alpha} = c_\infty^\gamma + B^{III} \cdot (1 - \text{erf } \eta) \qquad (9)$$

The balance of mass transport at $x = o$ and ξ yields respectively

$$D^\gamma \cdot B^I / 2\sqrt{D^\gamma t} = D^\alpha \cdot B^{II}/2\sqrt{D^\alpha t} \qquad (10)$$

$$(c_o^{\gamma\alpha} - c_o^{\alpha\gamma}) \cdot \frac{\eta\sqrt{D^\gamma}}{\sqrt{t}} = D^\gamma \cdot \frac{2}{\sqrt{\pi}} \cdot \frac{B^{III}}{2\sqrt{D^\gamma \cdot t}} \cdot \exp{-\eta^2} + D^\alpha \cdot \frac{2}{\sqrt{\pi}} \cdot \frac{B^{II}}{2\sqrt{D^\alpha \cdot t}} \cdot \exp{-\eta^2\varphi^2} \qquad (11)$$

If the values of B^I, A^{II} and B^{II} given by (6), (7) and (8) are inserted into (10), one obtains

$$\varphi \text{ erf } \eta\varphi = \frac{c_o^{\alpha\gamma} - c^{\alpha\gamma}}{c^{\gamma\alpha} - c_\infty^\gamma} \qquad (12)$$

After eliminating $c^{\alpha\gamma}$ by means of k and rearranging one can write (12):

$$\frac{c^{\gamma\alpha} - c_\infty^\gamma}{c_o^{\gamma\alpha} - c_\infty^\gamma} = 1/(1 + k\varphi \text{ erf } \eta\varphi) \qquad (13)$$

If the values of B^{II} and B^{III}, given by (8) and (9), are inserted into (11), one obtains

$$(c_o^{\gamma\alpha} - c_o^{\alpha\gamma}) \cdot \eta \cdot \sqrt{\pi} = \frac{(c_o^{\gamma\alpha} - c_\infty^\gamma)\exp{-\eta^2}}{1 - \text{erf } \eta} + \frac{(c_o^{\alpha\gamma} - c^{\alpha\gamma})\exp{-\eta^2\varphi^2}}{\varphi \text{ erf } \eta\varphi} \qquad (14)$$

and combination with (12) and (13) yields

$$\frac{c_o^{\gamma\alpha} - c_o^{\alpha\gamma}}{c_o^{\gamma\alpha} - c_\infty^{\gamma}} \cdot \eta \cdot \sqrt{\pi} = \frac{\exp{-\eta^2}}{1-\text{erf}\,\eta} + \frac{\exp{-\eta^2\phi^2}}{1+k\phi\,\text{erf}\,\eta\phi} \qquad (15)$$

The growth constant η can be calculated from (15) and $c^{\gamma\alpha}$ from (13). It may be instructive to study the approximate equations which are obtained for $D^\gamma \ll D^\alpha$, i.e. $\phi \ll 1$. Eq. (15) can then be written

$$\frac{c_o^{\gamma\alpha} - c_o^{\alpha\gamma}}{c_o^{\gamma\alpha} - c_\infty^{\gamma}} \cdot \eta \cdot \sqrt{\pi} = \frac{\exp{-\eta^2}}{1-\text{erf}\,\eta} + \frac{1}{1+\eta\ell} \qquad (16)$$

where ℓ stands for $k \cdot \phi^2$ which in turn is identical to $c_o^{\gamma\alpha} \cdot D^\gamma / c_o^{\alpha\gamma} \cdot D^\alpha$. The quantity ℓ which enters into eq. (16) can thus be considered as the ratio between the permeabilities of austenite and ferrite. If eq. (13) is approximated and rearranged, one can write

$$\frac{c_o^{\gamma\alpha} - c^{\gamma\alpha}}{c_o^{\gamma\alpha} - c_\infty^{\gamma}} = \frac{\eta\ell}{1+\eta\ell} \qquad (17)$$

and we can at once conclude that the deviation from equilibrium (measured by $c_o^{\gamma\alpha} - c^{\gamma\alpha}$, for instance) is less the lower the value of ℓ. In the present case ℓ is about 1/3, yielding rather small deviations from equilibrium. A calculation based on the exact eqs. (13) and (15) was carried out for 710°C and the steel compositions c_∞^γ = 0,21 and 0,59. The results were $c_o^{\gamma\alpha} - c^{\gamma\alpha}$ = 0,33 and 0,08, respectively. These values should be compared with the a-value of 0,3 obtained from the edgewise growth and given in Table II. It thus seems that the same deviation from equilibrium (a=0,3) which must be established before a Widmanstätten plate of ferrite can start to grow at 709°C, can be established at the immobile interface of grain boundary ferrite without being able to force this interface to move.

13.

It is admitted, however, that immobile interfaces were observed to be more common in the steel with 0,59% C than in the steel with 0,21%C. One may perhaps conclude that an increase of the deviation from equilibrium from 0,08 to 0,33 has made most of the interfaces move.

This analysis of the growth of grain boundary ferrite has established the existence of a deviation from equilibrium of the correct order of magnitude to justify the modification of the growth-rate equation suggested in Section 4.

Our theoretical evaluation of the magnitude of the deviation from equilibrium was based on the assumption that the movement of the mobile interface was diffusion-controlled. Until this assumption has been checked, one can only say that the deviation from equilibrium must be at least as large as the calculated values. Since the existence of an appreciable deviation has thus been established, one should now inquire whether it is not much larger than calculated and whether diffusion plays any role in determining the rate of growth. This could be tested by measuring the rate of thickening of the grain boundary ferrite and comparing with the value of the growth constant η_1, given by equation 15. This method would be dependent upon the extrapolated values of $c_o^{\gamma\alpha}$ and D, however.

A superior method would be to study directly the change in carbon distribution in the austenite, caused by the growing ferrite. If the growth is essentially diffusion-controlled, there should be an appreciable enrichment of carbon close to the ferrite and the concentration profile should be directly related to the width of the ferrite crystal but otherwise independent of the actual value of D. If there is no diffusion-control, the increase in carbon content should occur uniformly in all the austenite. Unfortunately, there is no good method available for carbon analysis on a scale small enough for the present purpose. The carbon content of austenite can be estimated metallographically, however, thanks to the concentration dependency of the martensitic transformation. For the study it was decided to utilize the variation of M_f with the carbon content. The heat treatment used in producing the microstructure in Fig. 3 was repeated with the difference that the final quench was made into a low-melting metallic bath instead of brine. After 5 minutes, all the martensite formed in the bath had become sufficiently tempered to yield a dark appearance after polishing and etching, and the specimen was then allowed to cool to room temperature. Any additional martensite which might form during this final cooling would be untempered and look white under the microscope, much like retained austenite.

When the metallic bath was kept at a temperature of $280°C$, it was observed that the centre of the austenite grains was cooled below its particular M_f temperature. The structure only showed tempered martensite. Close to the grain boundary ferrite, on the other hand, there were a large number of white, triangular areas, Fig. 5, indicating that here the austenite was not cooled below its particular M_f temperature. The variation of M_f in the specimen, which was thus established, should correspond to an appreciable variation in carbon content. The width of the zone containing the white areas was found to be a few times larger than the width of the grain boundary ferrite and this result compares favorably with the expected width of the zone enriched in carbon. One may thus conclude that diffusion plays an important role in the rate-control of the growth of grain boundary ferrite.

6. Decarburization experiments

Another method of checking the degree of diffusion-control during growth of ferrite from austenite involves decarburization of iron-carbon alloys in a wet hydrogen atmosphere, at a temperature between A_1 and A_3. A ferritic surface zone is formed during this process and the width of the zone, d, is related to the diffusion coefficient in ferrite, D^α, by

$$d^2 / D^\alpha \cdot t = \Delta c^\alpha / 2 \cdot (c_\infty - c_m^\alpha) \qquad (18)$$

where c_∞ is the original carbon content of the steel and c_m^α the average carbon content of the ferrite. The quantity Δc^α is the difference in carbon content over the ferritic zone and the degree of diffusion-control can be studied by evaluating Δc^α from decarburization experiments and comparing with the equilibrium solubility of carbon in ferrite, $c_o^{\alpha\gamma}$, which is fairly well known. Any difference $c_o^{\alpha\gamma} - \Delta c^\alpha$, which may be found, represents the total deviation from chemical equilibrium at the outer and inner sides of the ferritic zone.

The solubility of carbon in ferrite has been determined by various methods. The agreement is quite good as demonstrated by the following list, where the value of $c_o^{\alpha\gamma}$ at the eutectoid temperature, $727°C$ [14], has been evaluated from the most recent works:

Smith,	1946[15]	0,023 x
Ham,	1947[16]	0,023
Stanley,	1949[17]	0,019
Dijkstra,	1949[18]	0,022
Wert,	1950[19]	0,019

---

x) This value is obtained by a linear extrapolation through the A3-point of pure iron and the lowest experimental point. Smith himself gives the value 0,025, probably obtained by a non-linear extrapolation.

16.

If the value of Δc^α is evaluated from various studies of decarburization, published in the literature, and extrapolated to 727°C by a straight line through the A_3-point of pure iron, the following values of Δc^α are obtained,

Snoek,	1941[20]	0.012
Pennington	1946[21]	~ 0.027
Stanley,	1949[17]	≥ 0.080
Darken,	1950[22]	≫ 0.020
Weyerer,	1952[23]	≤ 0.017

The value of D^α determined by Stanley[17] was used for this evaluation.

The scatter among the values of Δc^α is quite too large to allow any comparison with $c_o^{\alpha\gamma}$. However, the observation by Pennington[21] that the rate of decarburization is dependent upon the silicon content of the steel may provide an explanation of the scatter, and in planning new decarburization experiments it was hoped that the 0.21 and 0.42% carbon steels with their low silicon content (see Table I) would yield reliable values. The experiments were performed at 750°C and the results are presented in Fig. 6. The straight lines have been drawn in agreement with the equation. Using the value $D^\alpha = 1.07 \cdot 10^{-8}$ cm^2/sec, obtained from the work by Stanley[17], one can calculate the value $\Delta c^\alpha = 0.020$ from the slope of both these lines. This corresponds to a Δc^α value of 0.023 at 727°C which compares favorably with the listed values of $c_o^{\alpha\gamma}$. Hence, the rate of decarburization of a high-purity steel seems to be mainly diffusion-controlled and there is no appreciable deviation from chemical equilibrium.

17.

7. Nature of deviation from equilibrium

From the results obtained in this investigation one may conclude that the movement of incoherent ferrite-austenite interfaces is purely diffusion-controlled, whereas the movement of coherent interfaces is partly interface-controlled, partly diffusion-controlled. Coherent crystals of ferrite do not grow until the carbon content at their interface/s deviates appreciably from equilibrium. Assuming that there is no abrupt change of the carbon activity over the interface, this implies that there must be sharp increase in iron activity over the interface as demonstrated by the free energy diagram in Fig. 7. The mechanism of transfer of the iron atoms over a coherent interface is not very well known. It seems reasonable, however, that a considerable driving force is necessary. A calculation of the value of the driving force per mole iron atom, $RT \ln a_{Fe}^{\gamma a}/a_{Fe}^{a\gamma}$, can be made from the position of the WB_s line in Fig. 2, using the relation $\ln a_{Fe}^{\gamma} = -c^{\gamma}/(1-c^{\gamma}) - 3.3[c^{\gamma}/(1-c^{\gamma})]^2$, obtained from the work by Smith[15]. c^{γ} is here the carbon content of austenite measured as mol fraction. The result is given in Fig. 8 and again depends upon the correctness of the extrapolated A_3-line. For comparison, the results obtained from the more conservative extrapolation of A_3 through 4% at 400°C are also given. For the higher temperatures the two curves agree and the value of the driving force seems reliable. For the low temperatures it seems impossible to tell what value between 100 and 500 cal/mol the driving force is approaching.

It is quite feasible that the interface will be able to move with any speed, once the necessary driving force is supplied and, hence, in effect be diffusion-controlled. The modification of the growth-rate equation in section 4 assumes that this is the case, otherwise a function of v would

18.

have been chosen instead of the constant "a". This is not in agreement with the model used by Cahn[24] in a treatment of the growth of pearlite, for instance. There the driving force is assumed to be proportional to the velocity. On the other, it is in agreement with the characteristics of the martensitic transformation in steels, where the interface moves with a very high speed. In view of the old hypothesis that there is a close relationship between martensite and bainite, and new metallographic observations[12] of a close resemblance between martensite and Widmanstätten ferrite, it is tempting to suggest that the same mechanism of interface movement operates in the two cases. The same driving force would then be needed. In the case of martensite, the driving force necessary for the formation has been calculated theoretically from the distance between M_s and T_o[25] and the results for Fe-C and Fe-Ni alloys are plotted in Fig. 8 for comparison. The values for Fe-C alloys fall quite close to each other and a very slight modification of the extrapolated A_3-line would make it possible to draw a common curve through the points calculated for ferrite and bainite (from the WB_s-line) and the points calculated for martensite (from the M_s-line). This curve would have the same general shape as the curve obtained for martensite in the Fe-Ni system. The necessary modification of the A_3-line is obtained from the old straight-line extrapolation through the eutectoid point and the point 5% at 400°C, by a slight displacement towards lower carbon contents at temperatures below about 550°C.

In view of the above result, it seems quite possible that there is a very close relationship between all three of Widmanstätten ferrite, bainite and martensite. In this connection it is interesting to note that Knapp and Dehlinger[26, 25] have calculated theoretically the interfacial energy between martensite and austenite and obtained a value of $\sigma = 200$ erg/cm^2, in close agreement with the value of $\sigma = 230$ erg/cm^2 obtained in the present study for the interface between Widmanstätten ferrite and austenite.

19.

8. Generalized growth model

So far we have assumed that there is equilibrium with respect to carbon at the ferrite-austenite interface. In principle, it is equally reasonable that there is an abrupt change of the carbon activity as well as the iron activity, although the former might be quite small at high temperatures where carbon atoms are very mobile. In the generalized case, we should thus consider the compositions of the two phases at the interface, $c^{\alpha\gamma}$ and $c^{\gamma\alpha}$, as two unknown and independent quantities and attempt to calculate both of them together with the growth rate v from a theoretical treatment of the growth process. The principles for such a calculation will now be discussed by means of the schematic free energy diagram in Fig. 9.

Consider a thin and infinitely long volume element, going through the edge of the growing plate and directed parallel to the growth direction. Let A be the cross section of this thin element. The carbon content of the element must change with a rate of $v \cdot A \cdot (c_\infty^\gamma - c_\infty^\alpha)$ if the plate is growing with the speed v under steady-state growth conditions. This change is caused by a diffusional flow of carbon, $\frac{dn}{dt}$, out of the element due to the difference in carbon activity at the ferrite-austenite interface, $a_c^{\gamma\alpha}$, and in the austenite at large distances, $a_{c\infty}^\gamma$. The quantity $RT \ln a_c^{\gamma\alpha}/a_{c\infty}^\gamma$ in Fig. 9 can thus be regarded as the driving force for this process and one has:

$$\frac{dn}{dt} = v \cdot A \cdot (c_\infty^\gamma - c_\infty^\alpha) \qquad (19)$$

For the movement of the coherent interface, there is a driving force of the size $\mathcal{E}$ available (see Fig. 9), from which one must subtract the effect of the pressure difference ΔP between the two phases caused by the interfacial energy σ of the curved interface. For the edge of a plate one has $\Delta P = \sigma/r$, r being the radius of curvature at the tip. There should be some relationship between the rate of the interface movement v and the force $\mathcal{E} - \Delta P \cdot V_m$. In the present case we shall assume that v=0 for all

$\epsilon - \Delta P \cdot V_m$ less than a critical calue ϵ_{crit}. In view of the previous results we shall assume that the interface can move with any speed, as soon as the force reaches the critical value, and, for a moving interface, we thus get the condition

$$\epsilon - \Delta P \cdot V_m = \epsilon_{crit} \tag{20}$$

where, in principle, ϵ_{crit} is obtained from Fig. 8. The fact that the curve calculated from the WB_s-line refers to one mole iron atoms whereas the other curves in Fig. 8 as well as the force $\epsilon - \Delta P \cdot V_m$ refer to one mole of atoms, both elements counted, makes only a small change.

We shall assume that the mechanism of interface movement in the present case is such that fragments of the austenite lattice are incorporated into the ferrite lattice without the order among the atoms being changed. The additional ferrite formed by such a process would have the same composition as the austenite unless there is some transfer of carbon atoms back into the austenite, due to the difference in carbon activity at the ferrite side of the interface, $a_c^{\alpha\gamma}$, and the austenite side, $a_c^{\gamma\alpha}$. The quantity $RT \ln a_c^{\alpha\gamma}/a_c^{\gamma\alpha}$ in Fig. 9 can be regarded as the driving force for this process and one has:

$$c^{\gamma\alpha} - c^{\alpha\gamma} = J_{transfer}/v \tag{21}$$

In order to carry out an actual calculation of the growth process, we must find some expressions for the quantities $\frac{dn}{dt}$, ϵ and $J_{transfer}$. In view of the results in reference 8 (particularly eq. 23), we shall adopt the relation

$$\frac{dn}{dt} = A \cdot D \cdot (c^{\gamma\alpha} - c_\infty^\gamma)/2r \tag{22}$$

which is based on the assumption that there is no diffusion occuring inside the ferrite. This is equivalent to assuming that there are no concentration differences in the ferritic part of the infinitely long volume element previously discussed, and hence, $c_\infty^\alpha = c^{\alpha\gamma}$.

The quantity ϵ can easily be calculated from the thermodynamics of the two phases. Assuming that they are both ideal solutions of carbon in iron, we can write

$$\epsilon/RT = c^{\alpha\gamma} \ln f_c^\gamma c^{\gamma\alpha}/f_c^\alpha c^{\alpha\gamma} + (1-c^{\alpha\gamma})\ln f_{Fe}^\gamma \cdot (1-c^{\gamma\alpha})/f_{Fe}^\alpha \cdot (1-c^{\alpha\gamma}) \tag{23}$$

21.

Very little is known about the transfer of atoms over an interface due to an activity difference. It is possible that the rate of such a process can be estimated from a consideration of the elemental process of diffusion. For the present purpose we shall simply assume that the rate of carbon transfer is proportional to the difference in chemical potential.

$$J_{transfer} = M \cdot RT \ln f_C^\alpha c^{\gamma\alpha} / f_C^\gamma c^{\gamma\alpha} \tag{24}$$

By inserting the expressions of $\frac{dn}{dt}$, ε and $J_{transfer}$ in eqs. (19), (20) and (21), we obtain

$$v = \frac{D}{2r} \cdot \frac{c^{\gamma\alpha} - c_\infty^\gamma}{c_\infty^\gamma - c^{\alpha\gamma}} \tag{25}$$

$$\frac{\sigma V_m}{rRT} = \frac{\varepsilon - \varepsilon_{crit}}{RT} = \ln f_{Fe}^\gamma / f_{Fe}^\alpha - \frac{\varepsilon_{crit}}{RT} - c^{\alpha\gamma}(\ln f_{Fe}^\gamma / f_{Fe}^\alpha + \ln f_C^\alpha c^{\alpha\gamma}/f_C^\gamma c^{\gamma\alpha})$$
$$-(1-c^{\alpha\gamma}) \ln(1-c^{\alpha\gamma})/(1-c^{\gamma\alpha}) \tag{26}$$

$$\frac{v}{MRT} = \frac{\ln f_C^\alpha c^{\alpha\gamma}/f_C^\gamma c^{\gamma\alpha}}{c^{\gamma\alpha} - c^{\alpha\gamma}} \tag{27}$$

Here we have three equations relating the four unknown quantities v, $c^{\gamma\alpha}$, $c^{\alpha\gamma}$ and r and thus yielding an infinite number of solutions. The same situation was met in the derivation of the previous growth rate equation[8]. In that case, the proposal by Zener[4] was adopted, suggesting that the curvature of the edge of a growing plate will automatically adjust itself to give the highest possible growth rate. The final growth rate equation, eq. (1) in this paper, was derived to give this maximum only. In the present case we want to calculate all the solutions for reasons later to become apparent.

22.

A numerical calculation was carried out for 600°C where the following values seem reasonable: $\ln f_{Fe}^{\gamma} / f_{Fe}^{\alpha} = 0,12$; $\varepsilon_{crit}/RT = 0,08$; $f_{c}^{\alpha}/f_{c}^{\gamma} = 66$; $D = 0,3 \cdot 10^{-8}$ cm^2/sec; $V_m = 7$ cm^3/mol. Finally, a value of 100 was chosen for the dimensional quantity $D/2\sigma V_m M$ which enters into the numerical calculation. This value implies that $M = 10^{-14} \frac{cm}{sek} \frac{erg}{mol}$ if $\sigma = 200$ erg/cm^2. There is no justification for this particular M value and the effect of changing this choice will be considered later on.

The result of the calculation is presented in Figs. 10 and 11. Each curve represents the infinite number of possible solutions for a certain alloy composition at 600°C. As an example, the lowest curve in Fig. 10 holds for a steel with the composition $c_{\infty}^{\gamma} = 0,0200$ which corresponds to a carbon content of 0,43 % by weight. It is shown that the highest growth rate is obtained when the edge of the growing plate has a radius of curvature of about 200 Å. Larger as well as smaller radii give lower growth rates. According to Zener's proposal, we can thus expect that a plate which starts to grow with any curvature at the edge, will eventually attain an r value of 200 Å and grow with the maximum possible rate. The modified growth rate equation (2) yields the same result for this alloy composition.

As we increase the supersaturation by choosing a steel with less carbon, the maximum growth rate increases, as expected, and the radius of curvature yielding this rate decreases. The optimum r values are about 120, 110 and 105 Å for $c_{\infty}^{\gamma} = 0,0090, 0,0070$ and $0,0040$, respectively, which corresponds to steels with a carbon content of 0,19, 0,15 and 0,086 % by weight. Fig. 11 demonstrates that there is a simultaneous decrease of the carbon content of the ferrite as we go from $c_{\infty}^{\gamma} = 0,0200$ to $0,0040$. As the supersaturation of the steel is further increased, the dotted curve in Fig. 11, which is drawn through the maximum points, demonstrates that the carbon content of the ferrite is beginning to increase and, at the same time, the optimum radius of curvature increases, as demonstrated by the dotted curve in Fig. 10.

23.

This phenomenon is connected with the appearance of a different class of solutions as a critical value of supersaturation is passed at $c_\infty^\gamma = 0.0095$ (0.20% carbon by weight). These solutions are first shown for the composition $c_\infty^\gamma = 0.0090$ at the upper left corner of Fig. 10 and at the right hand side of Fig. 11. They represent growth conditions yielding very high growth rates. In fact, the growth rate here approaches infinity as the radius of curvature approaches a limiting value of about 2000Å and the composition of the ferrite approaches the alloy composition $c_\infty^\gamma = 0.0090$. In this limit the solution represents martensitic growth and, accepting Zener's proposal, we should expect that a plate, which starts to grow according to any of the solutions on this curve, will automatically develop into martensite. On the other hand, there does not seem to be any possibility for a plate to develop into martensite if it has started to grow according to a point on the lower curve. Consequently, we could expect the formation of martensite as well as Widmanstätten ferrite or bainite at 600°C in this steel. However, the nucleation of martensite must also be considered and it is quite evident that nuclei, large enough to allow the formation of edges with a radius of curvature larger than 2000Å, are not very probable. It appears that martensite should not be expected until the supersaturation has been further increased. For a steel with $c_\infty^\gamma = 0.0040$ (0.086% carbon by weight), the limiting radius is as low as 80Å and the formation of martensite seems much more probable here.

Another critical degree of supersaturation is approached as the carbon content of the steel is lowered to $c_\infty^\gamma = 0.0020$ (0.043% by weight). Here the two curves meet in a point which is the maximum point of the lower curve. The development of a ferrite plate is now changed radically. If it starts to grow with a very sharp edge, for instance, it will gradually attain a higher growth rate as the radius of curvature increases toward a value of 200Å. Having reached this maximum on the lower curve, the plate finds a new possibility to increase the growth rate still further by following the upper curve. The radius of curvature will then start to decrease again and, as a limiting value of about 60 Å is approached, the growth rate approaches infinity. The growth characteristics of the plate of ferrite has thus changed to martensitic. Fig. 11 demonstrates that the carbon content of the ferrite plate during this whole process increases from a low value towards that of the steel, $c_\infty^\gamma = 0.0020$.

In view of the above argument it appears that the simultaneous formation of ferrite and martensite will no longer occur at the critical composition of $c_\infty^\gamma = 0.0020$. Instead, all nuclei will here lead to martensite. At even higher supersaturations this development of ferrite into martensite will occur even more readily, the two curves now being combined to yield a smooth curve as shown to the right in Fig. 10 and to the left in Fig. 11. The combination also yields another curve situated at the left side of Fig. 10. This curve grows smaller as the supersaturation increases further and finally it completely disappears at about $c_\infty^\gamma = 0.0006$ · (0.013% carbon by weight). These solutions thus exist in a narrow range of supersaturations only. Their physical significance is not clear to the present author.

The numerical calculation of the generalized case of edgewise growth was carried out for a constant temperature and the supersaturation was varied by allowing the alloy composition to vary. In practice, the supersaturation is usually varied by a variation of the temperature, the alloy composition being constant. A corresponding calculation was not undertaken because of the great uncertainty regarding the temperature dependence of the proportionality constant M in eq. (24). However, in principle one can expect the same sequence of events when the supersaturation is varied, irrespective of the method of variation. In view of the above calculation, one can thus define a number of curves in the phase diagram which, starting from high temperatures, are:

1. The upper temperature limit for incoherent growth of ferrite, identical to the solubility limit, A_3
2. The upper temperature limit for coherent growth of Widmanstätten plates, here called WB_s.
3. The temperature of no free energy change, T_o.
4. The upper temperature limit for the <u>growth</u> of martensite. This may possibly be intimately related to the so-called M_d curve, obtained for the formation of martensite during plastic deformation.
5. The upper temperature limit for the formation of martensite, M_s. This limit is probably caused by the process of nucleation rather than growth. It should be observed, however, that it is not necessarily identical to the temperature limit where the formation of a nucleus of martensite is energetically possible, the reason being that such a nucleus may very well develop into ordinary ferrite. Above the next temperature limit in this list, the nucleation of martensite is not successful until some point on the upper curve in Fig. 10 has been reached. Consequently, the critical nucleus for martensite seems to be of a <u>dynamic</u> rather than static nature in this range of supersaturation.
6. The lower temperature limit for the growth of non-martensitic ferrite. Below this curve all plates of ferrite automatically develop into martensite.

The calculated position of the last curve in this list depends upon the adopted value of the constant M, which was chosen by a pure guess in the present calculation. The choice of a higher M value would have resulted in a displacement of this curve towards lower temperatures and increase the range of supersaturations where growth of ferrite and martensite can both

26.

occur. Judging from experimental evidence, this range must be quite wide since ferrite, in the form of bainite, is often observed below the M_s temperature.

An experimental test of the position of the lower temperature limit of non-martensitic ferrite could possibly be carried out by allowing Widmanstätten ferrite to start forming at a high temperature and, on cooling, studying at what temperature it will continue as martensite. Unfortunately, such measurements will probably be severely disturbed by a preceding formation of other plates of martensite after cooling below the M_s temperature. However, in some cases such an experiment may be successful and Figs. 12 and 13 probably show an example. Fig. 12 shows a number of parallel plates of Widmanstätten ferrite formed at 630°C and etched by picral. Fig. 13 shows the same region after repolishing and etching in nital. Ferrite crystals with a certain orientation show a pronounced passivation in this etch[27] and thus appear unetched in the microstructure. This behavior is displayed by the set of parallel plates of ferrite but also by the adjoining plates of martensite formed on cooling from 630°C, indicating that ferrite has here continued to grow martensitic.

8. Summary

The edgewise growth rate of Widmanstätten ferrite and bainite was studied metallographically and the results were used to test a theoretical growth-rate equation, derived on the assumption that the growth is completely diffusion controlled. Some indications were obtained that the equation might be basically correct.

Although based on rather debatable extrapolations, a more detailed analysis of the growth rate data was performed. This indicated that the growth-rate equation should be modified to take into account a deviation from chemical equilibrium at the ferrite-austenite interface. When this was done, the experimental data agreed well with the equation. The analysis yielded a curve, here called WB_s and representing the upper temperature limit for the growth of Widmanstätten ferrite and bainite.

Support for the idea that there may exist a deviation from chemical equilibrium at a ferrite-austenite interface was obtained from a consideration of the immobile, and presumably coherent, interface sometimes observed on grain boundary ferrite. The same order of magnitude was obtained in the two cases.

A metallographic study of the distribution of carbon in the austenite matrix during growth of ferrite as well as a study of the rate of decarburization indicated that the movement of incoherent interfaces is mainly diffusion-controlled.

The deviation from chemical equilibrium at a coherent interface can be represented by an abrupt change in iron activity over the interface. The driving force necessary in order to make a coherent interface move, was calculated from the WB_s curve. It is suggested that the same driving force is needed for martensitic transformations.

28.

The principles are outlined for a more generalized treatment of the growth rate of ferrite from austenite by disposing of the usual assumption of local equilibrium with respect to carbon at the interface. A system of equations are obtained and solved numerically for a particular case. Solutions representing martensitic growth and ordinary ferritic growth are both obtained, indicating that the two kinds of growth may occur simultaneously. Below a certain temperature limit, the solutions indicate that a plate, which starts to grow as ordinary ferrite, will automatically develop into martensite.

Acknowledgment

Thanks are due to Mr. Sven Erik Johanson who kindly allowed the use of his data in this paper, to Mr. Bo Granath as a student at K. T. H. Stockholm, and to Mr. Raymond Ekvall as a trainee at the Swedish Institute for Metal Research, who both contributed to the success of the metallographic technique for showing the distribution of carbon in austenite, and to Mr. Nils Lange at the Swedish Institute for Metal Research for valuable technical assistance.

References

1. J. M. Robertson, J. Iron and Steel Inst. 119 (1929) p. 391
2. E. S. Davenport and E. C. Bain, Trans. A. I. M. E. 90 (1930) p. 117
3. F. Wever and H. Lange, Mitt. Kais. Wilhelm Inst. f. Eisenforschung, 14 (1932) p. 71
4. C. Zener, Trans. A. I. M. E., 167 (1946) p. 550
5. J. C. Fisher, "Thermodynamics in Physical Metallurgy" (ASM, Cleveland 1950) p. 201
6. K. Tsuya, J. Mech. Lab. Japan, 1 (1955) p. 1
7. K. Tsuya and T. Mitsuhashi, J. Mech. Lab. Japan, 1 (1955) p. 42
8. M. Hillert, Jernkontorets Annaler, 141 (1957) p. 757
9. S. E. Johanson, Thesis, K. T. H., Stockholm 1959
10. C. Zener, Trans. A. I. M. E., 167 (1946) p. 513
11. C. Wells, W. Batz and R. F. Mehl, J. Metals, 2 (1950) p. 553
12. M. Hillert, to be published
13. C. Wagner, "Diffusion" by Jost (Acad. Press, New York 1952) pages 69-75
14. R. P. Smith and L. S. Darken, Trans A. I. M. E., 215 (1959) p. 727
15. R. P. Smith, J. Am. Chem. Soc., 68 (1946) p. 1163
16. J. L. Ham, mentioned by Stanley, reference 17
17. J. K. Stanley, J. Metals, 1 (1949) p. 752
18. L. J. Dijkstra, J. Metals, 1 (1949) p. 252
19. C. A. Wert, J. Metals, 2 (1950) p. 1242
20. J. L. Snoek, Physica, 8 (1941) p. 734
21. W. A. Pennington, Trans. A. S. M., 37 (1946) p. 48
22. L. S. Darken, J. Metals, 2 (1950) p. 1023
23. H. Weyerer, Disserlation, Göttingen 1952
24. J. W. Cahn, Acta Met., 7 (1959) p. 18
25. L. Kaufman and M. Cohen, Progress in Metal Physics, 7 (1958) p. 165
26. H. Knapp and U. Dehlinger, Acta Met., 4 (1956) p. 289
27. A. Hultgren, A. Josefsson, E. Kula and G. Lagerberg, Jernkontorets Annaler 142 (1958) p. 165

Table I.

Experimental edgewise growth rates of Widmanstätten ferrite and bainite

Composition in %				Temp. °C	v µ/sec	$c_o^{\gamma\alpha} - c_\infty^\gamma$ %	Symbol used in Fig. 1
C	Si	Ni	Mn				
0,25	-	1,0	-	750 700 650	0,3 10 19	0,20) 0,70) 1,33)	■
0,29	-	-	-	700 650	16 22	0,78) 1,41)	▲
0,36	1,0	-	-	750 700 650	1,7 15 29	0,33) 0,84) 1,47)	●
0,46	1,0	-	-	700 650	9 20	0,74) 1,37)	◌
0,49	-	-	-	700 650	2,4 10	0,58) 1,21)	-
0,53	-	1,0	-	700 650	0,8 5	0,42) 1,05)	⊟
0,55	1,0	-	-	700 650	3,3 14	0,65) 1,28)	O
0,21	0,02	-	0,26	680 630 590 520	270 270 400 400	1,16) 1,80) 2,33) 3,24)	⬢
0,35	0,28	-	0,58	709 680 630 590 520 480 440 380	25 34 95 105 150 165 160 155	,65) 1,02) 1,66) 2,19) 3,10) 3,61) 4,14) 4,91)	
0,42	0,08	-	0,28	709 680 630 590 520 480 440 380	35 73 160 140 220 220 230 200	,58) ,95) 1,59) 2,12) 3,03) 3,54) 4,07) 4,84)	+
0.59	0,25	-	0,56	709 680 630 590 520 480 440 380	3 14 80 70 110 120 100 200	,41) ,78) 1,42) 1,95) 2,86) 3,37) 3,90) 4,67)	X

cont.

Table I. Cont.

Composition in %				Temp. °C	v μ/sec	$c_o^{\gamma a} - c_\infty^{\gamma}$ %	Symbol used in Fig. 1
C	Si	Ni	Mn				
0,81	0,26	–	0,23	630	30	1,20	⬢
				590	70	1,73	
				520	74	2,64	
				480	70	3,15	
				440	60	3,68	
				380	40	4,45	

33.

Table II.

Evaluated deviation from equilibrium carbon content during edgewise growth.

t °C	a % C
750	0.2
709	0.3
700	0.4
680	0.6
650	0.7
630	1.0
590	1.4
520	1.9
480	2.3
440	2.8
380	3.4

34.

Figure captions

Fig. 1. Edgewise growth rate data plotted according to theoretical growth rate equation. Straight lines drawn according to modified equation.

Fig. 2. Upper temperature limit for formation of Widmanstätten ferrite and bainite, WB_s, evaluated from Fig. 1. plotted together with M_s, B_s and T_o according to Fisher [5].

Fig. 3. Crystals of grain boundary ferrite with one immobile interface each, still outlining the position of the austenite-austenite grain boundary. 0.59 % carbon steel, austenitized, transformed for 20 min. at 710°C, quenched in brine. 600 x.

Fig. 4. Distribution of carbon during growth of ferrite crystal with an immobile interface at $x = 0$ and a mobile interface at $x = \xi$.

Fig. 5. White areas of untempered martensite and retained austenite revealing a high carbon concentration at the growing ferrite. 0.59% carbon steel, austenitized, transformed for 20 min. at 710°C, quenched to 280°C and tempered for 5 min. 600 x.

Fig. 6. Depth of decarburized ferritic zone plotted according to equation based on diffusion control.

Fig. 7. Schematic free energy diagram demonstrating suggested nature of deviation from equilibrium at coherent interface.

Fig. 8. Driving force for movement of coherent interface, calculated from WB_s line in Fig. 2, compared with driving force for martensite calculated from M_s.

35.

Fig. 9. Schematic free energy diagram demonstrating suggested nature of deviation from equilibrium at the interface according to the generalized growth model.

Fig. 10 Calculated growth rate as a function of the shape of the edge. r is the radius of curvature. Each curve holds for the particular alloy composition indicated (given as mole fraction of carbon). $MRT = 7 \cdot 10^{-4}$ cm/sec. $\sigma V_m/RT = 1.9 \cdot 10^{-8}$ cm.

Fig. 11 Calculated growth rate as a function of the carbon content of the growing plate of ferrite.

Fig. 12 Widmanstätten plates of ferrite formed at 630°C in 1/2 sec. Quenched. Etched in picral. 1200 x.

Fig. 13 Some area as Fig. 12. Etched in nital. 1200 x.

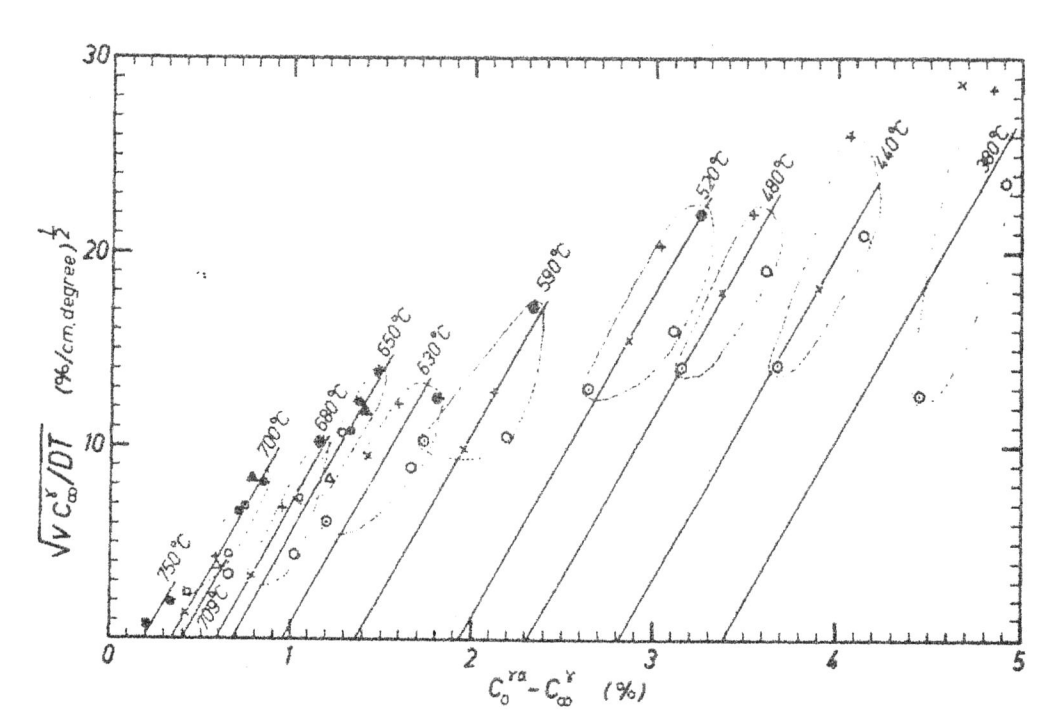

Fig. 1. Edgewise growth rate data plotted according to theoretical growth rate equation. Straight lines drawn according to modified equation.

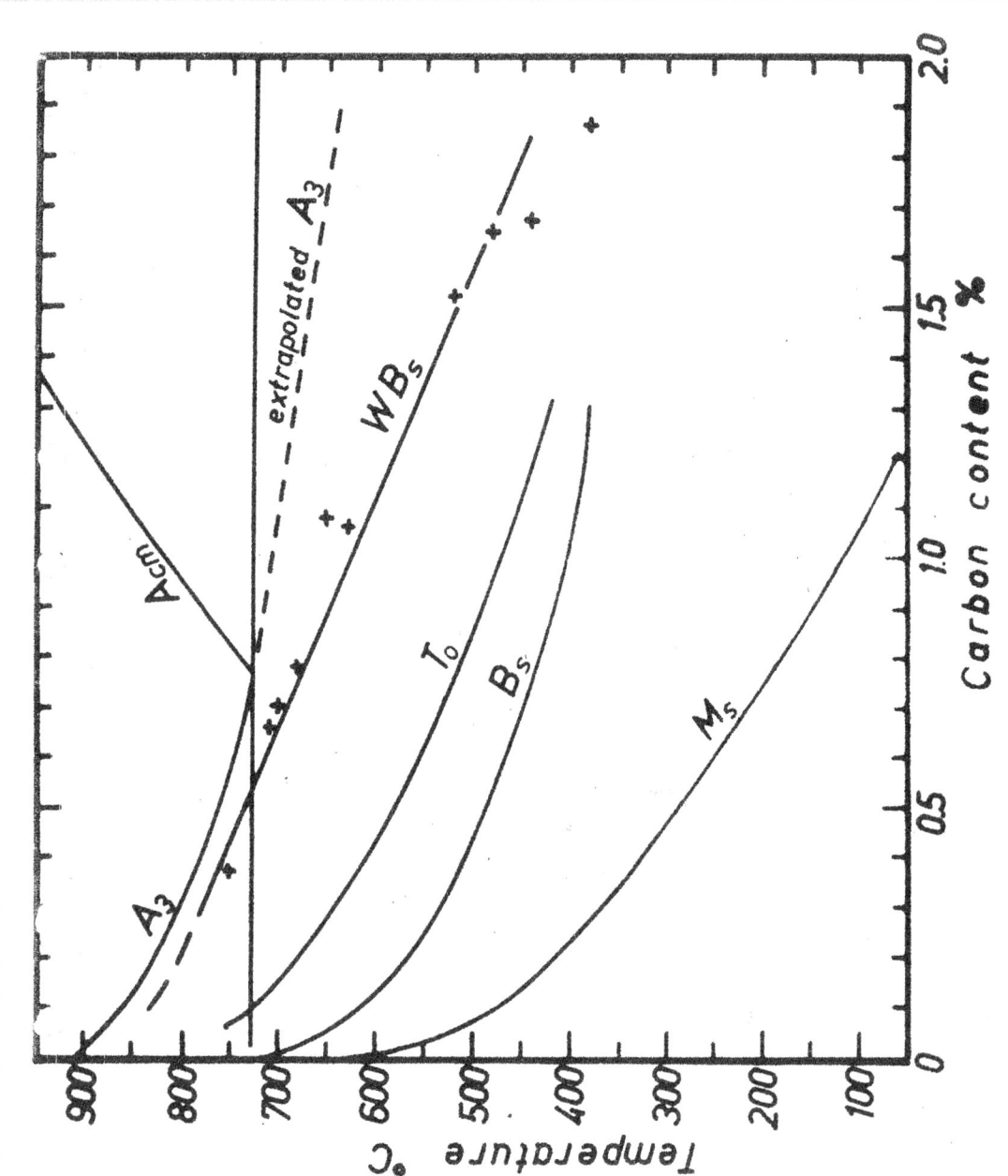

Fig. 2. Upper temperature limit for formation of Widmanstätten ferrite and bainite, WB_s, evaluated from Fig. 1, plotted together with M_s, B_s and T_o according to Fisher [5].

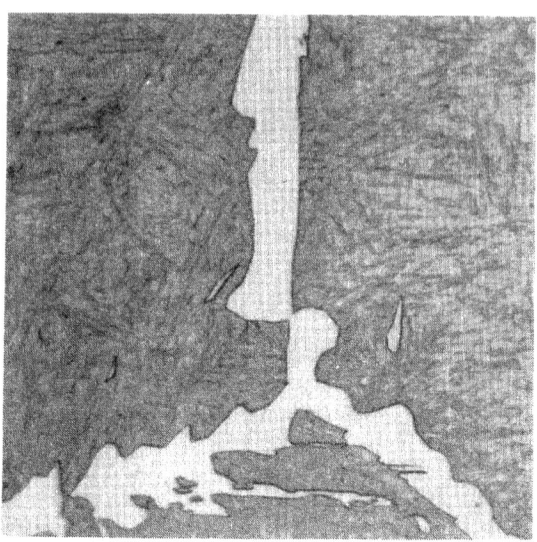

Fig. 3   Crystals of grain boundary ferrite with one immobile interface each, still outlining the position of the austenite-austenite grain boundary. 0.59% carbon steel, austenitized, transformed for 20 min. at 710°C, quenched in brine. 600x.

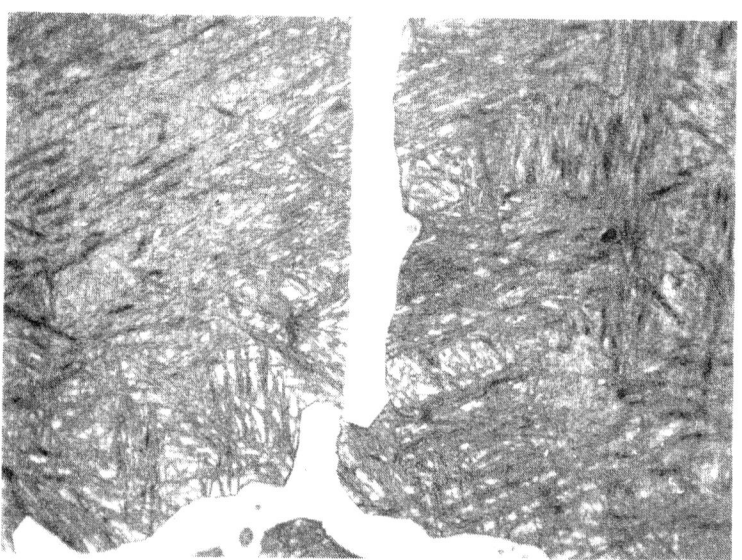

Fig. 5   White areas of untempered martensite and retained austenite revealing a high carbon concentration at the growing ferrite. 0.59% carbon steel, austenitized, transformed for 20 min. at 710°C, quenched to 280°C and tempered for 5 min. 600x.

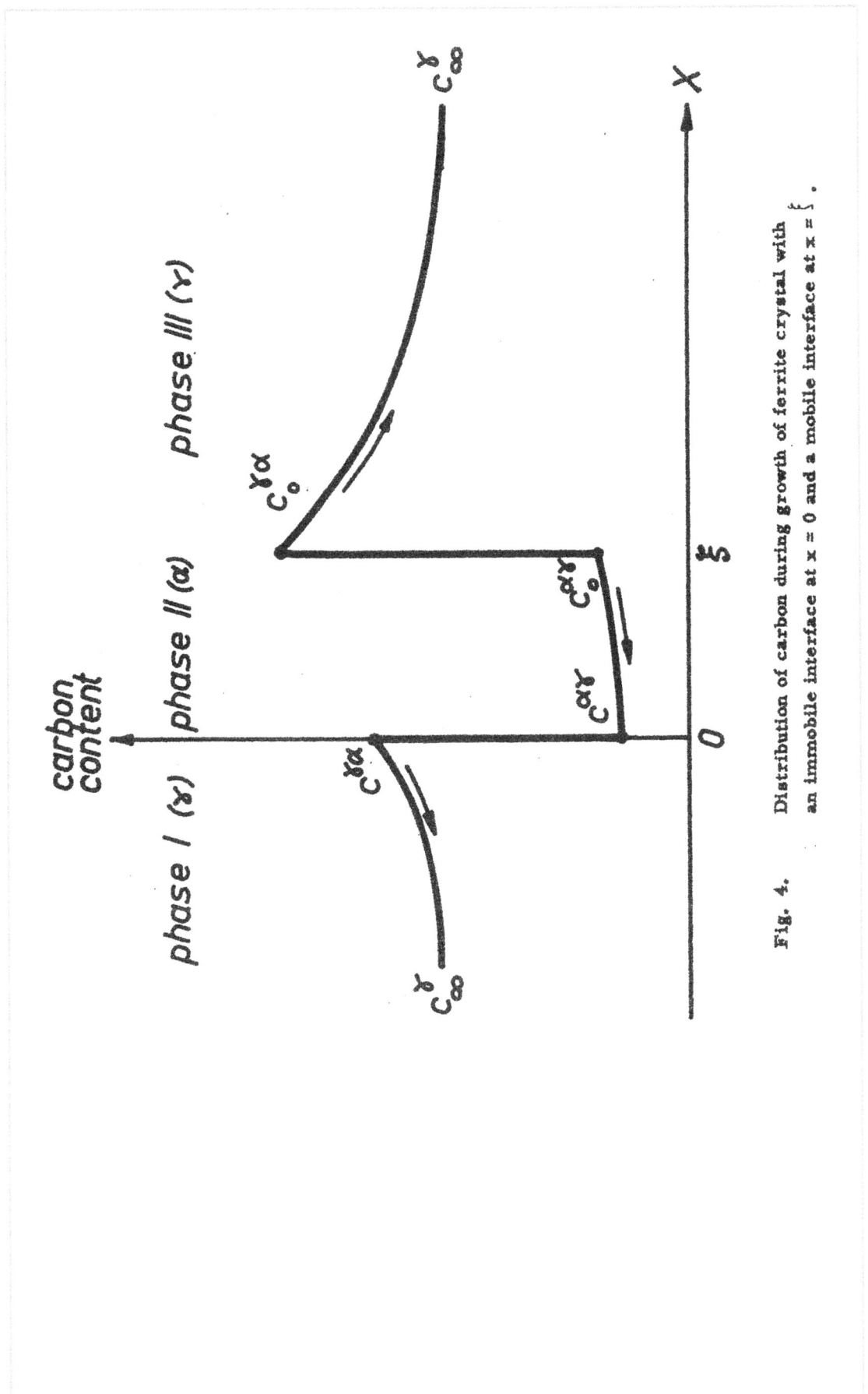

Fig. 4. Distribution of carbon during growth of ferrite crystal with an immobile interface at $x = 0$ and a mobile interface at $x = \xi$.

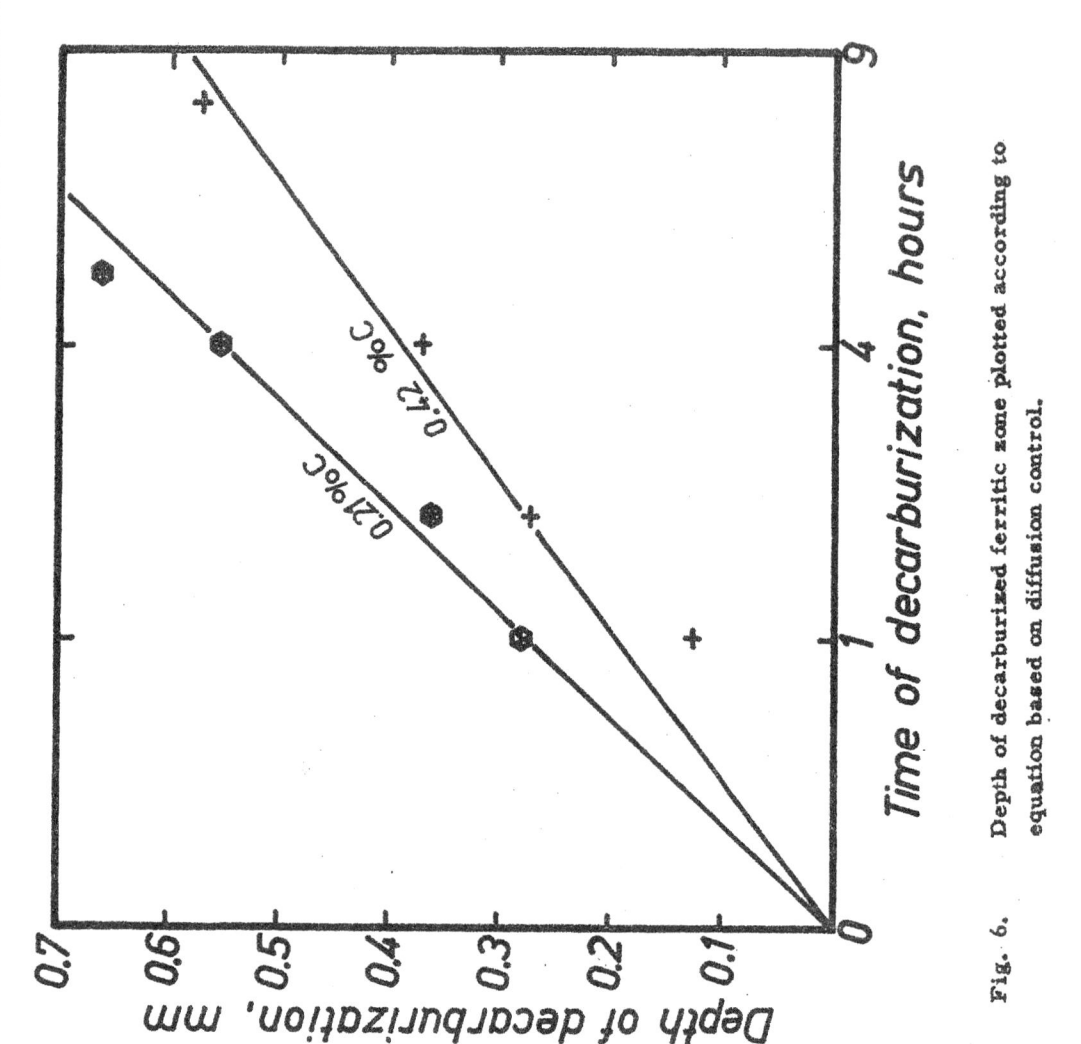

Fig. 6. Depth of decarburized ferritic zone plotted according to equation based on diffusion control.

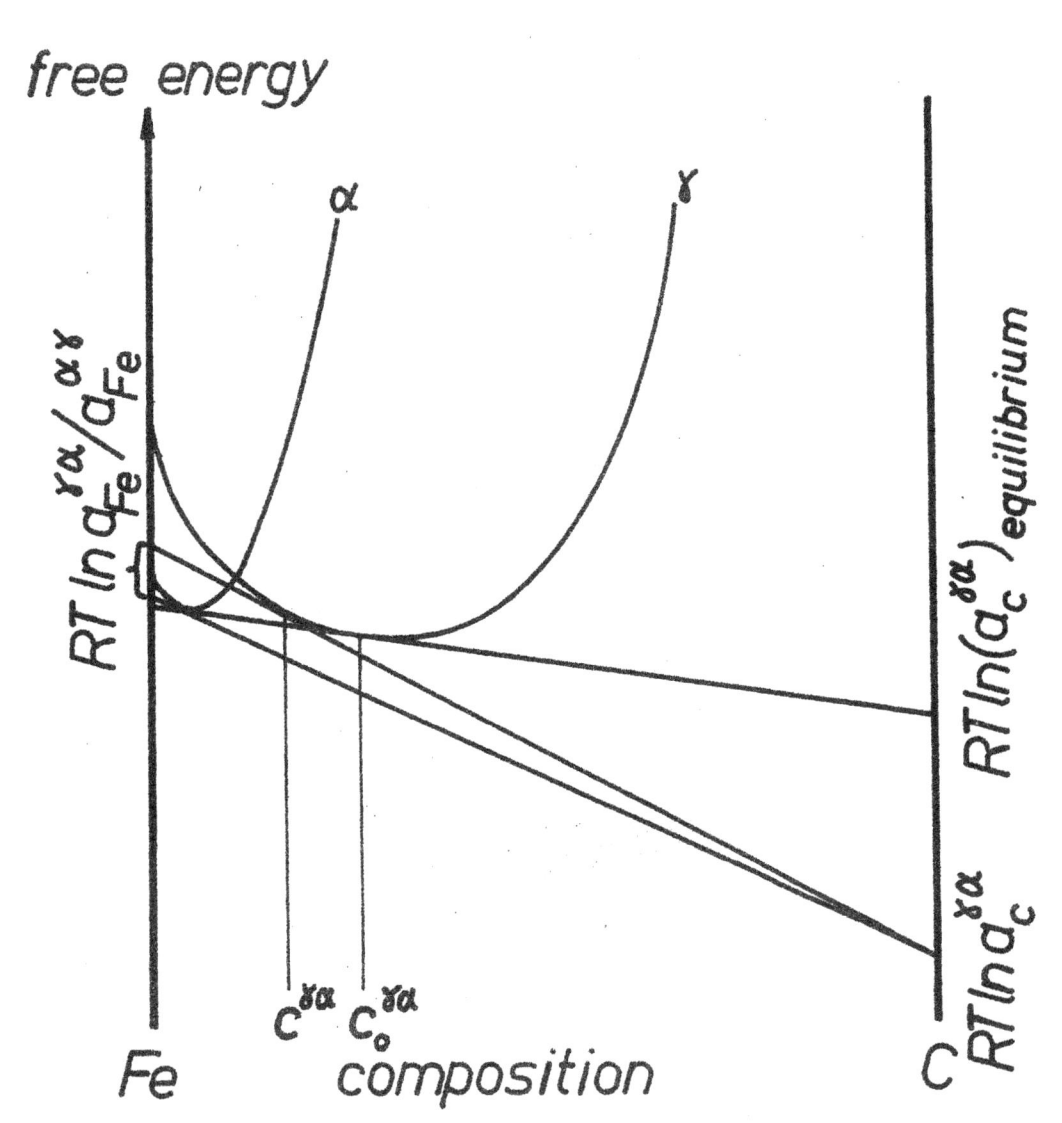

Fig. 7. Schematic free energy diagram demonstrating suggested nature of deviation from equilibrium at coherent interface.

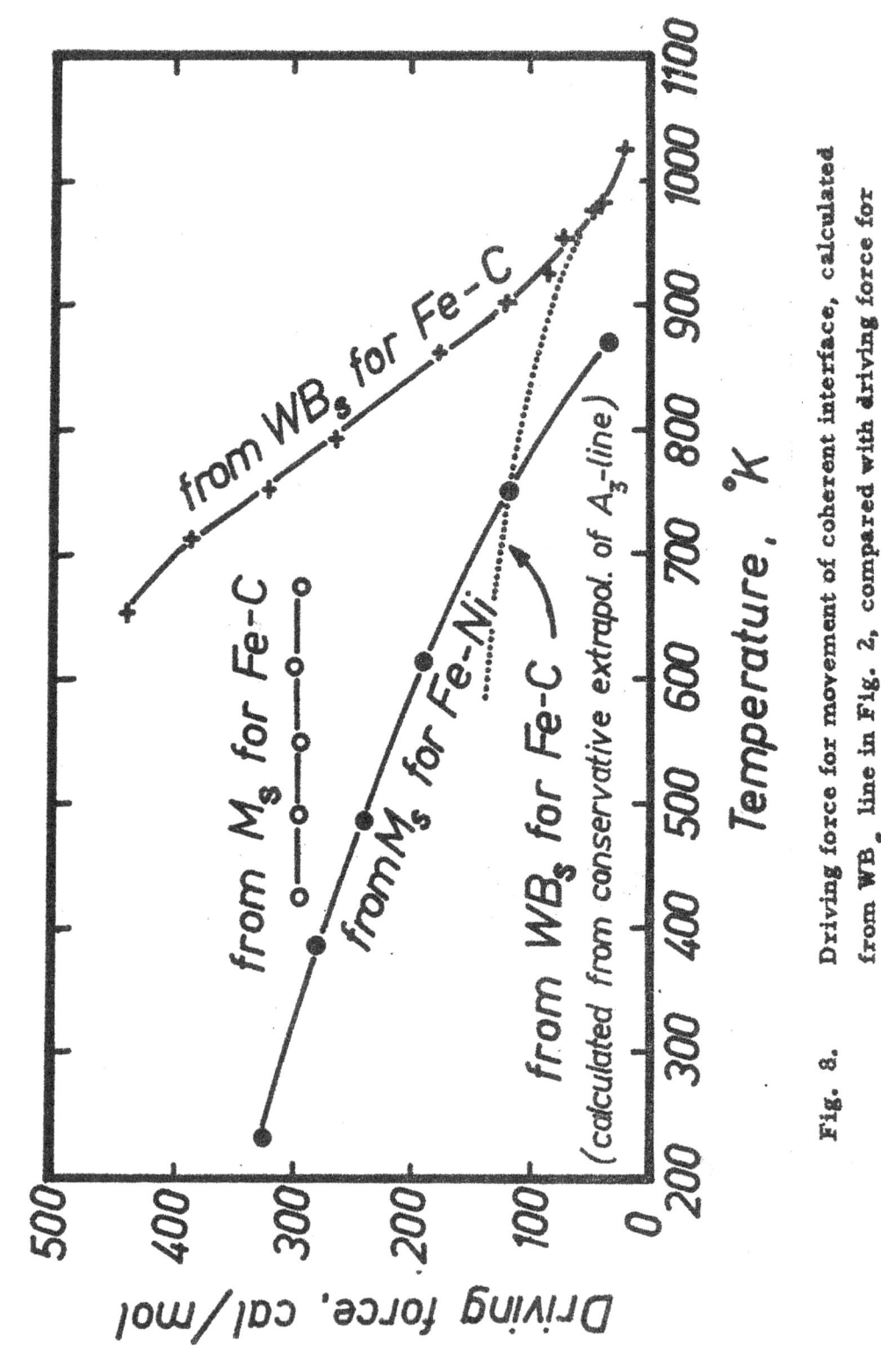

Fig. 8. Driving force for movement of coherent interface, calculated from WB_s line in Fig. 2, compared with driving force for martensite calculated from M_s.

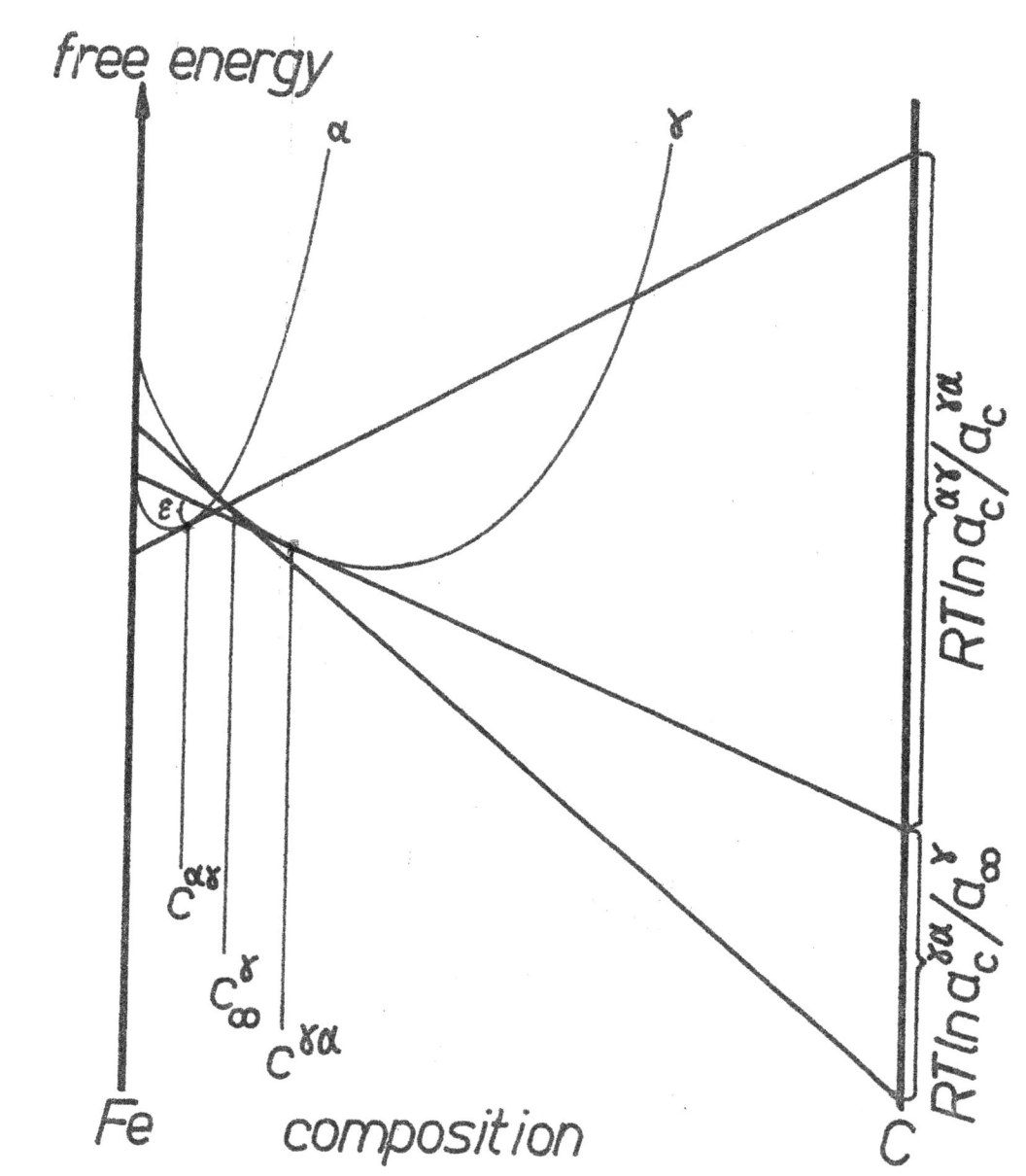

Fig. 9. Schematic free energy diagram demonstrating suggested nature of deviation from equilibrium at the interface according to the generalized growth model.

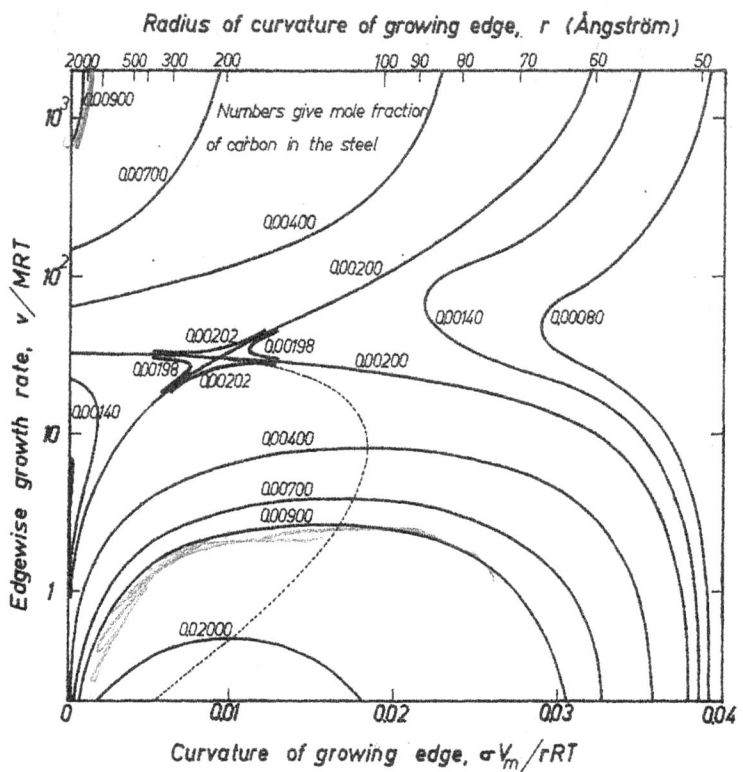

Fig. 10   Calculated growth rate as a function of the shape of the edge. r is the radius of curvature. Each curve holds for the particular alloy composition indicated (given as mole fraction of carbon). MRT = 7 · 10^{-4} cm/sec. $\sigma V_m/RT$ = 1.9 · 10^{-8} cm.

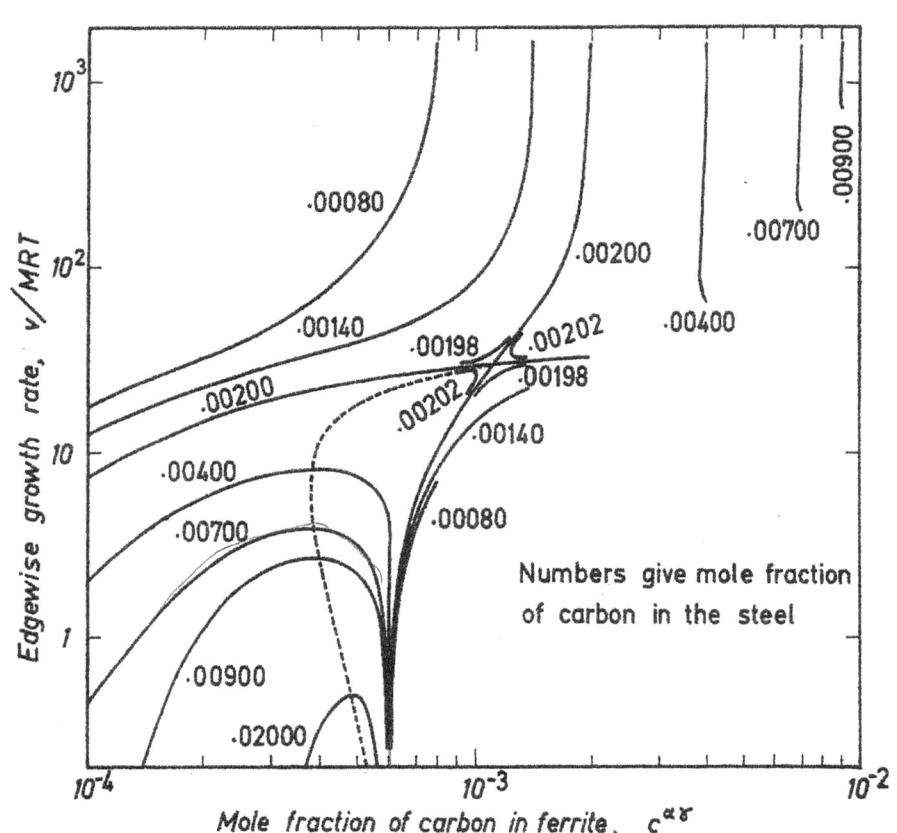

Fig. 11   Calculated growth rate as a function of the carbon content of the growing plate of ferrite.

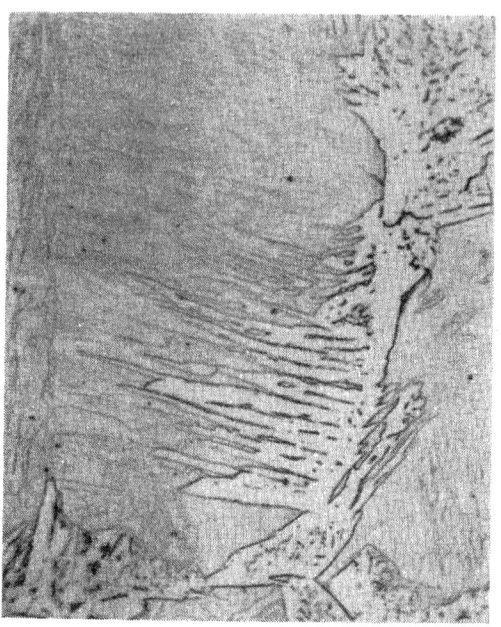

Fig. 12   Widmanstätten plates of ferrite formed at 630°C in 1/2 sec. Quenched. Etched in picral. 1200x.

Fig. 13   Same area as Fig. 12. Etched in nital. 1200x.

7

Introduction to "A Solid Solution Model for Inhomogeneous Systems"

published in *Acta metallurgica* (1961)

Although the modest title doesn't suggest it, this is a revolutionary paper about stability and the diffusional mechanisms and kinetics of alloy evolution, for both ordering and compositional phase separation, using a combination of trail-blazing theory, analysis, intuition, and a pioneering use of computer simulations. The first paragraph of the abstract describes a thermodynamic model from which interfaces and modulated structures are displayed as equilibria. The last paragraph seems like an afterthought. It is a single sentence "A kinetic treatment is developed from this model, allowing calculations to be made of changes from time to time in a thermodynamically unstable system". This kinetic treatment is about spinodal decomposition and ordering; kinetics is not in the title and is not mentioned again until the last part of Section 5. In his own language Mats Hillert then develops what today would be called an h-1 gradient system approach based on his free energy function, which includes a term for concentration gradients, to derive equations for diffusional evolution and equilibria of unstable alloys [1]. He connects the mathematical solutions with the then mysterious modulated structures seen by X-ray diffraction [2], and opens the modern use of structure factors from the diffuse scattering of X-ray for studying equilibrium and evolving microstructures [3]. He notices a myriad of unexpected aspects in the results of his calculations, and provides prescient explanations for them and new insights into the decades old controversy between the proponents of nucleation theory and the spinodal. This paper has stimulated so many lines of research that it would require a much longer commentary. I will give an idiosyncratic synopsis, for it greatly influenced my own work [4].

The 1950s was a period of great ideas and creativity in Physical Metallurgy. Pertinent to this paper was Darken's incorporation of thermodynamics into the formulation of diffusion in alloys [5]. At MIT, where Mats Hillert was studying for his doctorate, granted in 1956, the CALPHAD method was being created with Hillert's profound input. This paper's first equation concerns a free energy F in the CALPHAD modelling style, mean-field bond-counting with near-neighbour interaction energies of strength v, but with an important new twist: the solid solution is inhomogeneous, but in one direction only with an arbitrary composition, x_p for plane p. There is a separate term with the interplanar coordination number z as a factor for the inhomogeneity, now called the gradient energy term.

Requiring conservation of the components for the minimization of F, Hillert reinvents Lagrange multipliers[1], and arrives at a large system of nonlinear algebraic difference equations, equations (3) or (4), the solutions of which are periodic and are at stationary states, satisfying one of the necessary conditions for equilibrium. He recognizes that some of the equations' solutions will be global or local minima (which he calls true minima or stable and metastable states, respectively), and that others are unstable equilibria (which he calls saddles and also "critical nuclei", see below).

Hillert explores the solutions of these large systems of algebraic difference equations. In Section 3 he finds that the spinodal bounds the condition for which the solution for the homogeneous state is a local minimum. Inside the spinodal the homogeneous state is unstable, and he finds "periodic" solutions to a linearization of these equations, valid for small amplitude composition excursions[2]. In section 4 he displays (quasi)periodic solutions to these nonlinear systems of equations, obtained by numerical calculations on a mechanical adding machine at MIT, and then repeated in 1959 on an early vacuum tube computer in

---

[1] One explanation why no one at MIT told Hillert could be that this standard method was little known at MIT; until the 1980s it was absent from all the editions of Thomas' freshman calculus.
[2] It is a minor quibble that the solutions are only quasiperioidic unless the RHS of equation (10) is an integer.

Sweden[3]. Some results are shown in Figure 3 for $v > 0$ (phase separation), and in Figure 4 for $v < 0$ (ordering). The infinite wavelength cases show, respectively, an interface between coexisting phases and an (100) antiphase domain wall in an ordered *BCC* system (for which the global minimum would be a single domain).

Figures 1 and 2 display a new and amazingly complex interrelationship of the wavelength and amplitude of these periodic solutions to the average alloy composition and the sign of v, all calculated at a temperature of 0.9 of the critical temperature for this model. In Section 5 he calls these periodic solutions metastable rather than unstable states, reasoning fallaciously that although their free energy can be lowered with increasing wavelengths, a "complete reorganization of component atoms" is needed for this small change in wavelength.

To see whether a stationary solution, an infinite set of x_p, is a local minimum of F, one must examine F for all the neighbouring states, those for which none of the changes in the x_p are large, and include aperiodic changes. The periodic states, with slightly different wavelengths, are not neighbouring states with their "complete reorganization of component atoms" because some of the x_p will change by large amounts[4]. More formally, the matrix of second derivatives of F with respect to the x_p has to be shown to be positive definite. The infinitesimal variations likely to demonstrate instability are aperiodic ones, which bring two peaks closer together locally. Regardless of whether the periodic states are metastable or not to increasing wavelength, Hillert recognizes that his time-independent formulation is hampered by allowing only quasi-periodic solutions. The need to consider aperiodic solutions leads him, at last, to consider the time-dependent formulation of Section 6ff.

For the net diffusional flux between plane $(p-1)$ and p he considers three equations (11-13). All are discrete, rather than continuum, flux equations and contain Darken's thermodynamic factor, which for Hillert's model is proportional to $[kT - 2Zvx(1-x)]$. In each equation there is a term linear in the concentration difference, the coefficient of which would be proportional to the diffusion coefficient. At the spinodal, $kT = 2Zvx(1-x)$ (which implies $v > 0$), the Darken factor and the diffusion coefficient change sign, and are negative within the spinodal. Equations (11) and (13) contain an additional term, a third difference in concentration with factor z, and therefore a term resulting from the gradient energy. Hillert does not display the discrete versions of Fick's second law for the rate of concentration changes for each plane, easily obtained[5] by adding the fluxes to plane p from both planes $(p-1)$ and $(p+1)$, and equating this to the time derivative of x_p. Had he done so, the equations for the rate of concentration changes for each plane would have shown a standard diffusion term proportional to the second difference in concentration, but in addition the equations derived from equations (11) and (13) would have shown a fourth difference resulting from the gradient energy. These derived equations form large systems of ordinary differential difference equations, one equation for each plane.

For $v > 0$, the solution for the system derived from equation (12) is plane by plane alternation in x, similar to what is computed in Section 9 and displayed in Figure (14) for $v < 0$. The system based on equation (12) gives this "ordered" structure for $v > 0$ because when there is no gradient energy term ($z = 0$), the fastest growing length scale for separation into coexisting phases is a single plane. With $z = 0$, the interfaces between these single layer "phases" have no energy. To confuse matters more, for $v < 0$ the diffusion coefficient in Equation (12) is always positive, and leads to a homogeneous phase instead of the ordered phase. Although equation (12) is a discrete diffusion equation with a Darken factor based on the thermodynamics of the homogeneous phases of Hillert's model, its solutions do not evolve to the equilibria computed for this model.

Hillert had not solved this system, Equation (12), for small gradients and, without even seeing this problem, he remedies it by adding the gradient energy terms to create Equation (11). Now when $v > 0$ there is an energy penalty for gradients, and inside the spinodal the result is a phase separation on a larger scale, which becomes coarser with time. When $v < 0$ gradients lower the energy, and below the ordering transition, alternations in concentration on the atomic scale result. Solutions from either one of Hillert's equations (11) or (13) can display either spinodal decomposition or ordering. However such systems of equations are hard

---

[3] Hillert writes: "... there were no computers at MIT in 1955, at least none available to non-experts. There were machines for multiplication. You inserted the first number on a cylinder and for each time you turned the cylinder around the number was added to a register. If you turned 5 times it had been added 5 times, *i.e.*, multiplied by 5. Then you moved the cylinder on step to the left and for each new turn you added the number 10 times etc. Fortunately, there were electrical machines at MIT, called Frieden and Marchant so you did not have to turn by hand... After returning to Sweden I could use in 1959 a so-called mathematical machine, *i.e.* a computer operated with electron tubes. Then I could use the new diffusion equation and simulate what happens if you start with a fluctuation of some sort".

[4] With small, $\alpha \neq 0$, $\sin[(k+\alpha)p] - \sin[kp] = \sin[kp]\{\cos[\alpha p] - 1\} + \cos[kp]\sin[\alpha p]$ is not infinitesimal for all p.

[5] Hillert in 2004 gives a mistaken worry about mass conservation as his reason: "About Fick's 2nd law: By using the flux and not the 2nd law I did not lose any atoms. With the 2nd law I would have had to correct for the content of mass all the time".

to analyze, and can only deal with ordering when the structure can be described as homogeneous layers alternating in concentration.

The standard diffusion equation is a parabolic second-order partial differential equation (PDE). An easy conversion of Hillert's discrete system of ordinary differential equations (ODEs) to a PDE consists of letting the differences in concentrations between adjacent planes become gradients, second differences become second derivatives, etc., all multiplied by an appropriate power of the interplanar spacing. The PDE derived from equation (12) is the standard diffusion equation with a diffusion coefficient which contains Darken's thermodynamic factor. But when the diffusion coefficient is negative, this equation becomes ill-posed. It has no solution, since any concentration differences on an infinitesimally small scale will grow infinitely fast. The ODE resulting from Hillert's discrete formulation of the kinetics are never ill-posed; the interplanar spacing puts a bound on how small the distance scale can become. Inside the spinodal the resulting solution is growth in amplitude at a finite rate on the interplanar distance scale.

The gradient energy term in equations (11) or (13) adds a fourth-order term to the diffusion equation. When $\nu > 0$ this term regularizes the diffusion equation; it does not become ill-posed even when the diffusion coefficient is negative. Such a kinetic equation was the basis of the early treatments of spinodal decomposition [4]. On the other hand the equation is ill-posed whenever $\nu < 0$, regardless of the sign of the diffusion coefficient. Such a PDE cannot treat ordering. Thus, the reasons for adding the gradient energy term to the kinetic equation were quite different for the continuum and discrete formulations. Without the gradient term the solutions to the discrete formulation are deemed unphysical; with it, both ordering and spinodal decomposition can be treated. Without the gradient term the early continuum formulation is ill posed inside the spinodal; with it, spinodal decomposition, but not ordering, can be treated.

A method of solving the early stages of discrete kinetics for both ordering and phase separation in any dimension was soon proposed [6]. Using Fourier transforms of the concentration, Cook, de Fontaine and Hilliard found a continuum treatment of the evolution of structure in reciprocal (Fourier) space. Intensity increases near the origin signified phase separation into phases with different compositions, while intensity increases at particular reciprocal lattice positions identifiable as superlattice reflections indicated which ordered phase was developing [7].

Since x oscillates by large amounts from lattice site to lattice site the formulation of a continuum model for ordering ($\nu < 0$) can not be based on x itself, but on a properly defined order parameter [8] or the structure factor at a superlattice reflection [7]. This has been more difficult and has taken longer. Recently Hillert demonstrated his continuing concerns with this case, and described a continuum formulation which models separation into an ordered and disordered phase on the same lattice [9]. He used variables suggested by Novick-Cohen and myself [8]; a locally averaged concentration and an order parameter, which is a second difference of the local concentration. Both are slowly varying and lead to a system of two PDEs, one, for the average composition, is a fourth-order, while the other, for the ordering, is a second-order. The reduction from a fourth-order to a second-order PDE results from choosing as the variable a second difference in x.

In the 1961 paper, Hillert proposes Equation (13) to deal with the large gradients expected in the solutions for ordered systems, while stating squarely after the insertion of one factor in the step in the derivation that "the author has not been able to justify this choice by applying absolute reaction rate theory". Regardless, equation (13) is the key kinetic equation, and its solutions display qualitatively all the important new features. The quantitative changes from modifications from more sophisticated alloy solution are pale in comparison with changes produced by just adding a gradient energy term. This one equation describes not only both the separation into two phases and ordering, which are fast, but also at a later stage the much slower coarsening. Figures 5 and 9 show, for average x of respectively 0.5 and 0.35, inside the spinodal, and $\nu > 0$, schematic free energy landscapes connecting the homogeneous state with the metastable states (dashed curve), of which the free energies can be calculated.

The equation describes a rapid descent from the neighbourhood of the initial homogeneous state into the valley (the dashed curve) into a state of spinodal decomposition (or ordering) near some average wavelength. The coarsening is the slow movement down this valley to larger wavelengths. The mechanical analogue of a ball rolling rapidly down from a ridge to the valley and then slowly following the riverbed is a good example of how a single formulation can give two quite different kinetics. In recent years the continuum formulation of one-dimensional spinodal decomposition has been extensively studied as an interesting example of a "dynamic system" for which much profound new mathematics are being created [10]. The valley is called an "unstable attractor manifold" for this equation. The system moves rapidly to the attractor, but never reaches it (except in the unlikely event of a periodic solution). The attractor is a manifold because each part with a different number of particles is specified by different sets of solutions. It is a connected manifold, because there are borderline structures, as seen on the left in Figure 8, where the merging of two particles decreases their number. Because the solution is not periodic, this coarsening at longer times by combining the two peaks can be simpler than the rearrangements Hillert discussed for the periodic solutions in Section 5. Since interfaces in one dimension have no curvature, the coarsening is driven by long range attractions between these diffuse interfaces, and the coarsening rate becomes exponentially slow as the particle size increases [11]. In this paper Hillert understood and made clear what took many mathematicians years to prove.

Because the homogeneous solution is metastable, solutions of Equation (13) must start from an inhomogeneous composition. Hillert chose two types of concentration profiles for his initial data for computing Figures 7 (a localized perturbation) and 8 (random fluctuations, which are still widely used). Random thermal fluctuations at all times, introduced by Cook [12] into the PDE, are important, mainly in the early stages.

The B2 ordering of BCC alloys with average $x = 0.5$ and $v < 0$ is an example of a structure which can be described as the one-dimensional plane-by-plane alternations of concentration along $\langle 001 \rangle$. Figures 13 and 14 show how initial data affect the evolution. The resulting domain walls Hillert displays would be planar ones along (001). As mentioned above, this too was an early description of what has become a big field. Figure 15 shows some contrasting aspects of positive and negative v.

Nucleation is much discussed in this paper. It is mentioned in almost every section, and Section 8 is entitled "Homogeneous nucleation". The title of Hillert's 1956 thesis is "A theory of nucleation of solid metallic solutions". However, the main thrust of the thesis and the great significance of this paper are about clarifying the role of the spinodal, and his use of the word of "nucleation" has to be understood in the murky context of a long-standing polemic at the time.

The spinodal first appears as a limit of stability of gases and liquids in van der Waals thesis on the continuity of states between them. Soon after, Gibbs gave a set of necessary conditions for the stability of fluid phases with any number of components [13], which led to a generalization of the limiting conditions for stability, later all incorporated in the term spinodal. One of these necessary conditions for stability of a homogeneous fluid binary phase is that $\partial^2 F/\partial x^2 > 0$. When $\partial^2 F/\partial x^2 < 0$, which we term "inside the spinodal", the fluid phase is unstable. In the first half of the twentieth century estimations of temperature, composition and pressures for the spinodal by predecessors of CALPHAD methods was part of the practical science of estimating accessible ranges of undercooling or supersaturation [14]. Gibbs' role in spawning the spinodal was soon forgotten. Gibbs also discussed critical nuclei which would render a metastable phase unstable, and gave formulas for their work of formation [15], but that was ignored until 1926. The opening sentence and footnote in the first modern nucleation paper by Volmer and Weber [16] mentions that Gibbs had done this fifty years before but that, since it had been forgotten, it needed to be redone[6]. They then extended it to kinetics for single component droplet nucleation. This mention of Gibbs was also soon forgotten in the ensuing decades-long controversy between adherents of nucleation and spinodal theory about how new phases got started in clean systems. With our hindsight this controversy is hard to understand.

Lacking the clear distinctions that Gibbs had drawn, the debate led to attempts at a false conciliance in which the proponents of the spinodal adapted the language of nucleation. At the spinodal they claimed there would be copious homogeneous nucleation. Thereafter it became hard to settle the debate by most experiment. The debate seemed to have ended with studies of precipitation in Pb-Sn. Borelius, one of the main proponents of the spinodal, had calculated its location on the phase diagram[7] and claimed it marked the onset of the rapid changes in resistance he found and associated with precipitation from solid solution. Turnbull and Treaftis [17] used a microscope and showed clear evidence for heterogeneous nucleation on grain boundaries, followed by growth over tens of microns. They proved by calorimetry [18] that elements of the supersaturated solid solution were not unstable, but remained stable until reached by the growth front of precipitation. Borelius conceded, and when Hillert and I visited him in 1961, his first reaction was a profound apology to us for having been the cause of us wasting our time on the spinodal. He claimed that Turnbull's experiment had proven that the spinodal was a useless concept. After we insisted on going on with our story, he conceded that we might be right, but that what we laid out differed profoundly from what he had thought.

It is ironic that until Hillert's thesis, no one seems to have realized that the modulated structures seen by Lipson and Daniels with X-rays were the result of the spinodal. These structures simply did not fit in with nucleation theory, and their interpretation of the X-ray scattering data was conveniently ignored as controversial.

With this background, much of the use of the words "nucleation" and "solid" in the thesis and this paper can be understood, and the true greatness of Hillert's contribution to our understanding of the spinodal is revealed. To him "solid" only meant that the atoms are on a lattice, even after there has been ordering

---

[6] Volmer blames Ostwald for this on page 14ff in the historical introduction to his book Kinetik der phasenbildung, T Steinkopf, Dresden, 1939.

[7] He used his own determination of the phase diagram, which was off by more than 20 °C, and he assumed a single F would cover both the FCC lead-rich phase and tetragonal tin. I conceived and carried out a better test, using the miscibility gap in Au-Ni, and found no evidence of spinodal behaviour. This led to the concept of the coherent spinodal for solids.

and phase separation; there is no coherency stress. To him "nucleation" means finding states from which an ordering or phase separation can proceed with no further configurational free energy barriers. The processes revealed by the solution he computed inside the spinodal was "nucleation"; there were no further configurational free energy barriers to a phase change. Clearly the progress revealed in Figure 7 is not that of "growth" of the nucleus, his "Guinier Zone", inserted as initial data. Instead of growth he found many small new particles created in a spatially coordinated, approximately periodic, manner. The particles seem to grow in compositional amplitude and not in spatial extent. It is obvious that many new particles are created by the first one, but by neither growth nor random nucleation.

One-dimensional nucleation theory has a few applications. Hillert's problem is quite similar mathematically to that of double kink nucleation on a dislocation line under an applied stress and lying on a low index direction. Hillert's treatment came before and is superior to the early treatments [19]. It is difficult to extend Hillert's nucleation theory to higher dimensions. A planar nucleus starts out with infinite interfacial area, and would always have an infinite energy barrier (he does not mention that the energy barrier in the schematic Figure 10 is per unit area of these infinite plates), except when the system is inside the spinodal, or in the case of perfect wetting on a substrate [20]. Attempts to link the phenomena inside the spinodal with nucleation outside the spinodal are still actively studied. Hillert's one-dimensional model did reveal what goes on at early times inside the spinodal even in higher dimensional systems. It ended the controversy, and nucleation theory was unaffected, except near the spinodal.

For these and many more reasons, this modestly titled paper remains one of the most profound and influential papers in our field.

John W. CAHN

References

[1] W. Craig Carter, J.E. Taylor and J.W. Cahn, Variational methods for microstructural evolutions, *JOM* 49 #12 (1997) 30-36.

[2] V. Daniel and H. Lipson, *Proc. Roy. Soc.* 182 (1943) 378.

[3] M. Hillert, M. Cohen and B.L. Averbach, *Acta Metall.* 9 (1961) 536.

[4] J.W. Cahn, On Spinodal Decomposition, *Acta Metall.* 9 (1961) 795–801; Spinodal Decomposition, The 1967 Institute of Metals Lecture, *TMS AIME* 242 (1968) 166–180.

[5] L.S. Darken, *Trans. AIME* 175 (1948) 184.

[6] H.E. Cook, D. de Fontaine and J.E. Hilliard, *Acta Metall.* 17 (1969) 765.

[7] D. de Fontaine, *Solid State Physics* 34 (1979) 73.

[8] A. Novick-Cohen and J.W. Cahn, Evolution Equations for Phase Separation and Ordering in Binary Alloys, *J. Stat. Phys.* 76 (1994) 877–909.

[9] M. Hillert, *Acta Mater.* 49 (2001) 2491.

[10] G. Fusco, in *NATO Adv. Study Inst. on Dynamics of infinite dimensional system*, ed. by S.-N. Chow and J.K. (Hale), p. 115.

[11] J. Carr and R.L. Pego, *Commun. Pure and Appl. Math.* 42 (1989) 523–576.

[12] H.E. Cook, *Acta Metall.* 18 (1970) 297.

[13] J.W. Gibbs, *Collected Works*, Vol. 1, p. 100ff.

[14] G. Borelius, F. Larris and E. Ohlsson, *Arkiv Mat, Astr. och Fys.* 31A (1944).

[15] *Loc. cit.*, p. 252ff.

[16] M. Volmer and A. Weber, *Z. Phys. Chem.* 119 (1926) 277.

[17] D. Turnbull and H. Treaftis, *Acta Metall.* 3 (1955) 43.

[18] W. Desorbo and D. Turnbull, *Acta Metall.* 4 (1956) 495.

[19] F.R.N. Nabarro, Theory of Crystal Dislocations (Clarendon Press, Oxford, 1967) p. 738ff.

[20] J.W. Cahn, *J. Chem. Phys.* 66 (1977) 3667–3672.

A SOLID-SOLUTION MODEL FOR INHOMOGENEOUS SYSTEMS*

M. HILLERT†

A solid-solution model allowing compositional variations in one dimension is developed from the zeroth approximation of nearest-neighbor interactions. It predicts the existence of periodically modulated structures in ordering as well as in precipitation systems. It provides a method of calculating the diffuse structure of grain and domain boundaries; it indicates the existence of an activation energy for boundary movement; and it constitutes a new approach for treating homogeneous nucleation that is free of *a priori* assumptions about the critical nucleus.

A kinetic treatment is developed from this model, allowing calculations to be made of changes from time to time in a thermodynamically unstable system.

UN MODELE DE SOLUTION SOLIDE POUR DES SYSTEMES INHOMOGENES

L'auteur développe un modèle de solution solide, permettant des variations de composition dans une dimension, à partir de l'approximation de zéro des interactions entre atomes voisins les plus proches. Le modèle prévoit l'existence de structures périodiquement modulées dans des systèmes d'ordre, aussi bien que dans des systèmes à précipitation. Il fournit une méthode de calcul des structures diffuses des frontières des grains et des domaines; il indique l'existence d'une énergie d'activation pour le mouvement des frontières; enfin il permet d'aborder d'une manière nouvelle le problème de la germination homogène exempte de supposition *à priori* sur les germes critiques.

L'auteur développe en partant de ce modèle une étude cinétique permettant le calcul des changements avec le temps dans un système thermodynamiquement instable.

MODELL EINER FESTEN LÖSUNG FÜR INHOMOGENE SYSTEME

Aus der nullten Näherung des Modells der Wechselwirkung nächster Nachbarn wird ein Modell für eine feste Lösung entwickelt, daß eindimensionale Veränderungen der Zusammensetzung zuläßt. Es sagt das Auftreten von periodisch modulierten Strukturen bei Ordnungs- wie bei Ausscheidungssystemen voraus. Es führt zu einer Methode, die die diffuse Struktur von Korn- und Bezirksgrenzen zu berechnen gestattet, es deutet auf die Existenz einer Aktivierungsenergie für die Korngrenzenbewegung, und es stellt einen neuen Weg zur Behandlung homogener Keimbildung dar, der frei ist von *a-priori*-Annahmen über den kritischen Keim.

Das Modell wird zu einer kinetischen Behandlung ausgebaut, mit der sich zeitliche Veränderungen in einem thermodynamisch instabilen System berechnen lassen.

1. INTRODUCTION

Many theories have been developed for binary solid solutions, one of the simplest being the so-called zeroth approximation of the nearest-neighbor interaction model[1] which was first suggested by Gorsky[2] and further developed by Bragg and Williams[3]. A common feature of these theories is that they are restricted to solutions which are essentially homogeneous in composition, a fact that causes some limitations. The advantage of relaxing this restriction was recently pointed out by the present author, and a new solid-solution model was evolved based on the zeroth approximation.[4] This model allowed compositional variations in one direction only and was considered as a first attempt along new lines. Among other things, it led to the development of a new treatment of homogeneous nucleation. Later, the treatment of nucleation was extended to three dimensions by Cahn and Hilliard[5–7], and a self-consistent thermodynamic formalism was developed by Hart[8].

The original one-dimensional model still has some advantages, especially for systems which give modulated structures on transformation. The purpose of this paper is to describe the model, to discuss its applicability and to present new machine calculations based on the model. The conclusions that can be drawn with regard to the kinetics of transformation in a system with a miscibility gap as well as in an ordering system will be discussed in addition to the diffuse nature of grain boundaries and domain boundaries. The process of homogeneous nucleation will not be dealt with in any greater detail in view of the extended treatment developed by Cahn and Hilliard[7].

2. DERIVATION OF BASIC EQUATIONS

We shall consider variations in composition along

---

* This paper is based on a thesis submitted in partial fulfillment of the requirements for the degree of Sc.D. in Metallurgy at the Massachusetts Institute of Technology, Cambridge, Mass., in May 1956. The research was sponsored by the United States Atomic Energy Commission. Received July 1, 1960.

† Formerly Department of Metallurgy, Massachusetts Institute of Technology; now at the Swedish Institute for Metal Research, Stockholm.

a certain direction in a crystalline solid of the two components A and B, and hence we introduce a set of variables $\ldots x_{p-2}, x_{p-1}, x_p, x_{p+1}, x_{p+2} \ldots$ representing the composition (atom fractions) of a succession of atomic planes, perpendicular to the direction considered. First, we want to find an expression for the free energy of the total system in terms of these variables. This is most readily done by applying the zeroth approximation of the nearest-neighbor interaction model. One simply counts the number of A–A, A–B and B–B bonds in the system assuming random mixing of the atoms within any one plane. Substantially the same method was used by Becker[9] in developing his nucleation theory, the only difference being that he used only two variables, the composition of the matrix and that of the nucleus. In the present case, it is found[4] that the difference in free energy between the inhomogeneous state under consideration and a homogeneous state is given by the expression

$$\Delta F = -vm \sum_p [Z(x_p - x_a)^2 - z(x_p - x_{p+1})^2]$$
$$+ mkT \sum_p [x_p \ln x_p/x_a$$
$$+ (1 - x_p) \ln (1 - x_p)/(1 - x_a)] \quad (1)$$

where x_a is the average composition of the total system, v is the so-called interaction energy, m is the number of atoms in each atomic plane, Z is the total number of nearest neighbors to each atom and z is the number of nearest neighbors in one plane to a given atom in the next plane.

Next, we find the condition for equilibrium states. The thermodynamic criterion for stability of the system is that $\partial \Delta F / \partial x_p = 0$ for all x_p. We are dealing with a closed system, however, and care must be taken that the content of the two components in the total system is kept constant. This can be done by considering the variation of ΔF when atoms are exchanged between two neighboring planes, p-1 and p. It is found that

$$\partial \Delta F / \partial x_p = mkT \ln \frac{x_p(1-x_{p-1})}{x_{p-1}(1-x_p)} - 2mv[Z(x_p - x_{p-1})$$
$$+ z(x_{p+1} - 3x_p + 3x_{p-1} - x_{p-2})] \quad (2)$$

and in view of $\partial \Delta F / \partial x_p = 0$, one obtains the relation

$$x_{p+1} = x_{p-2} + \left(\frac{Z}{z} - 3\right)(x_{p-1} - x_p)$$
$$+ \frac{kT}{2zv} \ln \frac{x_p(1-x_{p-1})}{x_{p-1}(1-x_p)} \quad (3)$$

which must hold for all equilibrium states.

An alternative method can be used when one is only interested in states which contain some large region where all the planes have a common composition, x_∞, in addition to regions of varying composition. One can then let the pth plane exchange atoms with this large reservoir and a simpler relation results:

$$x_{p+1} = \frac{Z}{z} x_\infty + \frac{kT}{2zv} \ln \frac{1-x_\infty}{x_\infty}$$
$$- \left(\frac{Z}{z} - 2\right) x_p - x_{p-1} + \frac{kT}{2zv} \ln \frac{x_p}{1-x_p}. \quad (4)$$

Mathematical solutions to the difference equation (3) can be obtained by starting from any chosen set of values for x_{p-2}, x_{p-1} and x_p and calculating x_{p+1}, x_{p+2}, etc. and also x_{p-3}, x_{p-4}, etc. one by one. All solutions where any x is found to be higher than unity or less than zero must be discarded as being without any physical meaning. Two classes of physically significant solutions can be expected, one representing true minima in free energy, i.e. stable or metastable states, the other class representing saddle points on the many-dimensional free-energy surface in the $(\ldots x_{p-2}, x_{p-1}, x_p \ldots)$-space. This class represents critical nuclei. We are interested in both these classes and discuss them in the following sections.

3. SOLUTIONS FOR SMALL AMPLITUDES

The difference equation (3) can be solved explicitly only in the case of small variations in composition from the average composition x_a. The logarithmic term can then be approximated by the first term of the Taylor series expansion. One then obtains

$$x_{p+1} = x_{p-2} - x_{p-1} + x_p - 2M(x_{p-1} - x_p) \quad (5)$$

where M is a constant given by the expression

$$2M = \frac{kT}{2zv} \cdot \frac{1}{x_a(1-x_a)} + 2 - \frac{Z}{z}. \quad (6)$$

The solutions of equation (5) can be described by

$$x_p = C_1 + C_2 \cdot \beta^p + C_3 \cdot \beta^{-p} \quad (7)$$

where C_1, C_2 and C_3 are constants. The value of β is determined by inserting the expression for x_p in equation (5), which yields

$$\beta = M + \sqrt{(M^2 - 1)}. \quad (8)$$

For $|M| > 1$ one finds that β is a real number, which implies that x_p assumes infinite value as $p \to \infty$ or $-\infty$ unless C_2 and C_3 are both zero. The only solution of physical significance is then $x_p = C_1 = x_a$, which represents the homogeneous state. It is interesting to note that, according to equation (6), $M = 1$ occurs when $kT/2zvx_a(1-x_a) = 1$, which is the equation of the spinodal* curve for a system with positive v, assuming to the zeroth approximation.

---

* The spinodal is the locus of points within a miscibility gap where $\partial^2 F/\partial x_a^2 = 0$.

For an alloy outside the spinodal, there is thus only one solution with small compositional variations and this solution represents the homogeneous state.

For $|M| < 1$, which occurs inside the spinodal for positive ν, one finds that β is a complex number and the solution can be written

$$x_p = x_a + C_4 \cdot \sin p\phi \tag{9}$$

where $\phi = \operatorname{arctg} \sqrt{(1/M^2 - 1)}$, and C_4 is a constant. This solution represents a sinusoidal variation of small amplitude, and the wavelength expressed in number of atomic planes is

$$l = 2\pi/\operatorname{arctg} \sqrt{(1/M^2 - 1)}. \tag{10}$$

4. SOLUTIONS FOR LARGE AMPLITUDES

Solutions representing states with large compositional variations cannot be obtained explicitly but, as already mentioned, they can be calculated step by step. This work is very tedious and machine methods are necessary if one wants to calculate a large number of solutions. It was found that all the solutions represent periodic variations in composition. Each solution can thus be identified by its amplitude† and wavelength, a fact that is utilized in Figs. 1 and 2 where the results of a series of machine calculations are presented. Each one of the curves in such a diagram represents all the solutions for a certain alloy composition. These calculations were carried out with $Z/z = 2$ and $kT/2z\nu = +0.45$ and -0.45, respectively, which holds for a b.c.c. structure in the cube direction at a temperature $T = 0.9\ T_M$, where T_M

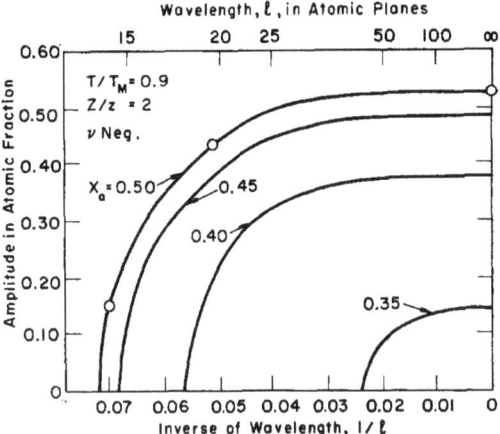

Fig. 2. Periodic states for different alloy compositions negative ν. States marked with circles are shown in Fig. 4.

denotes the maximum temperature of the miscibility gap or the ordering region in the phase diagram. The solutions represented by the three small circles in Figs. 1 and 2 are shown in Figs. 3 and 4. The physical significance of the solutions in Figs. 1 and 2 will now be discussed.

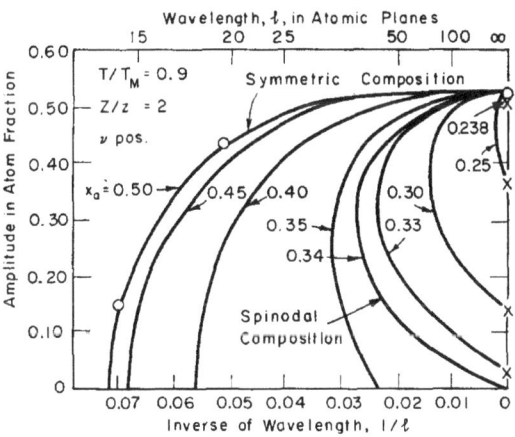

Fig. 1. Periodic states for different alloy compositions, positive ν. States marked with circles are shown in Fig. 3, states marked with crosses in Fig. 11.

† Difference between maximum and minimum compositions in the periodic variation.

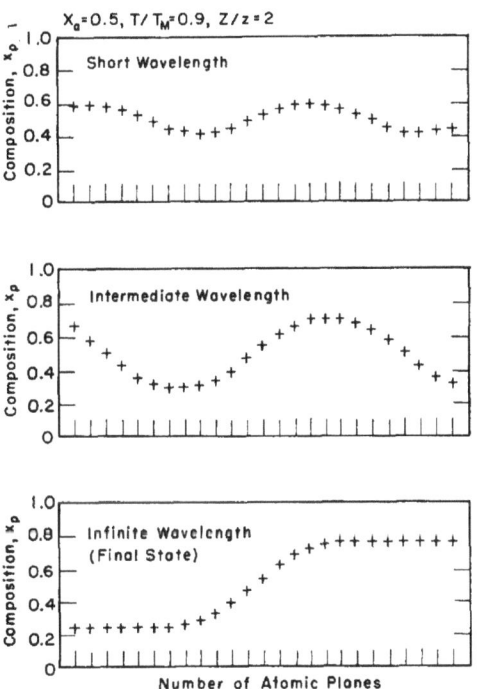

Fig. 3. Solutions to equation (3) showing periodic variation of composition. Positive ν in alloy of symmetric composition.

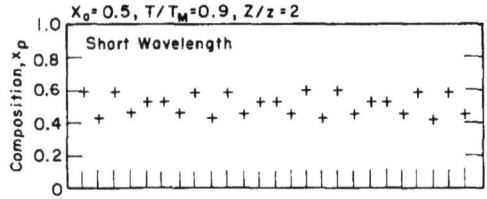
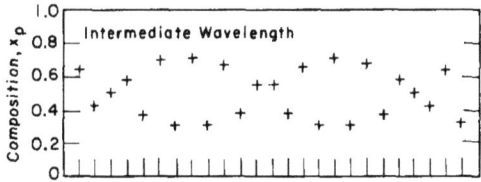
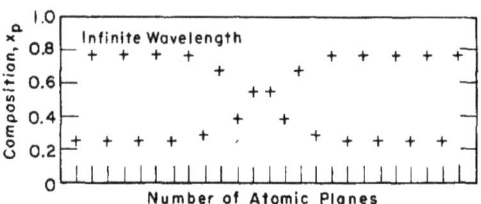

Fig. 4. Solutions to equation (3) showing periodic variation of composition. Negative ν in alloy of symmetric composition.

5. SYMMETRIC ALLOY WITH POSITIVE ν

It is possible to calculate the free energy for each state by applying equation (1), and Fig. 5 demonstrates schematically the situation for a symmetric alloy with positive interaction energy ν. In this diagram the wavelength, l, has been plotted along the x-axis in order to make the value $l = 0$ visible in the figure. In Figs. 1 and 2, on the other hand, $1/l$ was used in order to make the value $l = \infty$ visible. The curve for $x_a = 0.5$ has thus been replotted in the lower part of Fig. 5. The upper part of the figure shows not only the free energy of these states, which are solutions to equation (3), but also the free energy of nonequilibrium states. Each of these states is defined as the state of lowest free energy for the particular combination of wavelength and amplitude. In this way, a complete free-energy surface is obtained, which is useful in clarifying the nature of the particular states representing solutions to equation (3).

A symmetric alloy ($x_a = 0.5$) that has been homogenized above the miscibility gap and quenched to a temperature inside the gap can, momentarily at the new temperature, be represented by any point on the l-axis (Fig. 5). The shape of the free-energy surface demonstrates that this homogeneous state is stable with respect to fluctuations of a shorter wavelength than a value l_{crit}, which in fact is the value given by equation (10). Fluctuations with a longer wavelength, on the other hand, can increase in amplitude because the free energy will then be lowered, until the dashed line at the bottom of the valley on the free-energy surface is reached. This line thus represents stable or, rather, metastable states whereas the l-axis to the right of l_{crit} represents saddle points in the many-dimensional x_p-space, i.e. critical nuclei for homogeneous nucleation. Critical nuclei for the transformation are thus already present in the homogeneous state and the transformation can start spontaneously as soon as the alloy is cooled inside the miscibility gap.

If the system for $x_a = 0.5$ has reached a state on the dashed line in the bottom of the valley, the diagram shows that the free energy will decrease further if the system can move to the right along this line and the equilibrium state will finally be reached at infinite wavelength. As demonstrated by the lower part of Fig. 3, this state consists of two large regions of different compositions, one corresponding to each side of the miscibility gap. These regions are separated by a grain-boundary interface which is quite diffuse in contrast to the sharp interface suggested by Becker[9].

The calculation of the diffuse grain-boundary interface can be carried out by means of the simpler equation (4) because the necessary large reservoir is

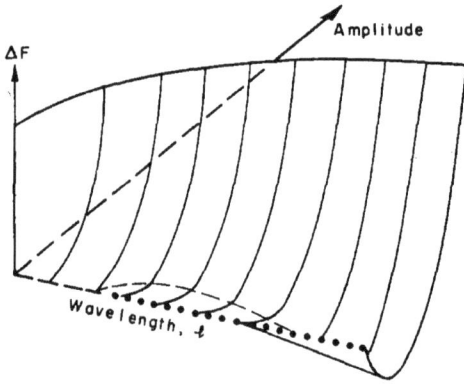
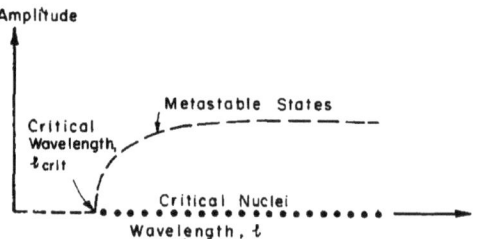

Fig. 5. Free energy of periodic states for symmetric composition, $x_a = 0.5$. Positive ν. Schematic diagram.

available here. One can use either of the two large regions having compositions corresponding to the sides of the miscibility gap. In fact, equation (4) has previously been derived and employed for this special purpose by Ono[10]. The development of the ideas concerning diffuse interfaces has recenly been reviewed by Cahn and Hilliard[5] and the subject will not be further discussed here.

All the states on the dashed line along the bottom of the valley in Fig. 5 are metastable, which implies that a system cannot spontaneously move to the right down the valley. For each increase in wavelength, a new nucleation event is necessary because a complete reorganization of the component atoms is needed if the wavelength of a perfectly periodic structure is to increase from one value to another. Essentially, such a system must go back up-hill to the l-axis in order to choose a longer wavelength and descend into the valley again. The l-axis can thus be considered as an activation barrier. A nucleation event of this kind seems very unlikely, and hence, one must inquire how the system can possibly increase its wavelength. It appears that an answer can be provided by the kinetic treatment to be described in Section 6.

Another important question is: What wavelength will first be developed on transformation? Any wavelength longer than l_{crit} can be expected on thermodynamic grounds. It seems that kinetic considerations should be applied to settle this question because the decisive factor might be the rate of development of the various metastable states. The rate should be higher, the larger the decrease in free energy accompanying the formation of a state; and it should be lower, the longer the diffusion distance for the component atoms. As a first attempt, one could thus study the expression $-\Delta F/l$ as a function of l. Fig. 6

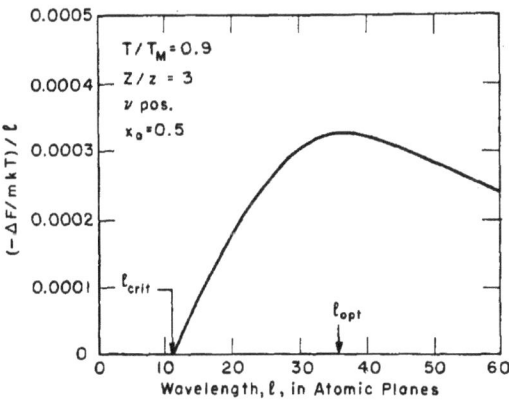

FIG. 6. Variation with wavelength l of ratio between driving force and diffusion distance in symmetric composition.

shows the result of a calculation for a specific case ($T = 0.9\ T_M$, $Z/z = 3$, $x_a = 0.5$), from which it may be concluded that a wavelength a few times longer than the critical wavelength can be expected in this instance. In the following, this will be referred to as the optimum wavelength, l_{opt}.

6. KINETIC TREATMENT

So far the discussion has been based on equation (3) which describes stable and metastable states and also critical nuclei. A more powerful method would be to calculate the changes in a system from time to time due to the diffusion or interchange of the component atoms. All that is needed is a diffusion equation describing the exchange of atoms between neighboring atomic planes. Such an equation should be derived using the absolute rate theory. In the original treatment,[4] the following equation was suggested for the case of small gradients.

$$J = -\frac{B}{a^2} x(1-x) \cdot \frac{\partial \Delta F}{\partial x_p}$$
$$= \frac{mB}{a^2} x(1-x)\left[\left(\frac{kT}{x(1-x)} - 2Z\nu\right)(x_{p-1} - x_p) \quad (11)\right.$$
$$\left. + 2z\nu(x_{p+1} - 3x_p + 3x_{p-1} - x_{p-2})\right]$$

where a is the distance between neighboring planes, and B is the atomic mobility.

Becker's[9] and Darken's[11] equations yield the expression:

$$J = \frac{mB}{a^2} x(1-x)\left(\frac{kT}{x(1-x)} - 2Z\nu\right)(x_{p-1} - x_p) \quad (12)$$

when applied in the same case. The difference is due to the neglect in equation (12) of the discontinuity in composition from plane to plane. The two equations become identical for $z = 0$, as could be expected.

Now we are interested in an equation for large gradients, It is evident that the value of $\partial \Delta F/\partial x_p$ must be taken from equation (2) but it is not clear what expression should take the place of $x(1-x)$. For the following calculations, the expression $\frac{1}{2}[x_p(1-x_{p-1}) + x_{p-1}(1-x_p)]$ was chosen although the author has not been able to justify this choice by applying the absolute rate theory. (It seems that one runs into trouble due to the shortcomings inherent in the zeroth approximation.) The resulting diffusion equation is:

$$J = \frac{mBkT}{2a^2}\{x_p(1-x_{p-1}) + x_{p-1}(1-x_p)\}$$
$$\times \left\{\ln\frac{x_{p-1}(1-x_p)}{x_p(1-x_{p-1})} + \frac{2\nu Z}{kT}\left[x_p - x_{p-1}\right.\right.$$
$$\left.\left. + \frac{z}{Z}(x_{p+1} - 3x_p + 3x_{p-1} - x_{p-2})\right]\right\}. \quad (13)$$

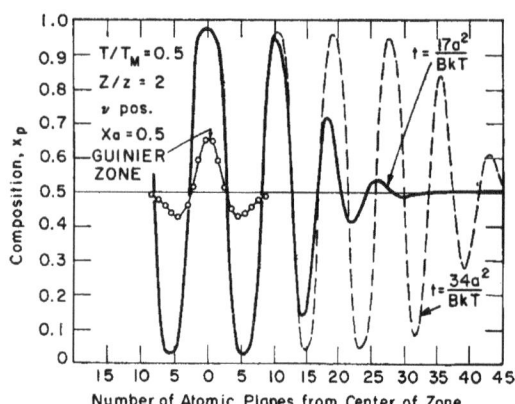

FIG. 7. Development of Guinier zone after different times, for symmetric composition, positive ν.

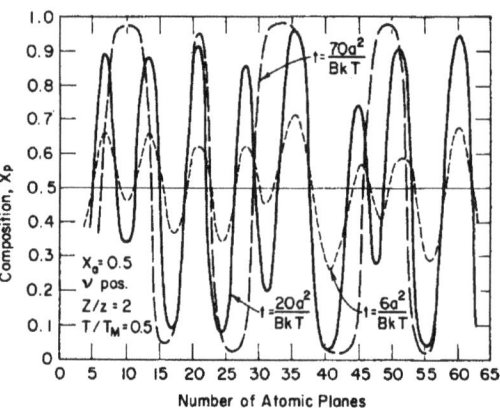

FIG. 8. Development of random fluctuations after different times, in symmetric composition, positive ν.

Starting from any distribution of the component atoms, the changes from time to time can now be calculated. Again, machine methods are applied to advantage. As an example, the development of a Guinier zone[12] was computed. The results after two different times are presented for $x_a = 0.5$ in Fig. 7, where the circles represent the original Guinier zone. It is shown that the presence of one zone will induce the formation of a whole series of zones and thus give rise to a periodic structure with a fairly uniform wavelength and amplitude, quite similar to the metastable states discussed in the previous sections. This consequence of a Guinier zone inside the spinodal curve was predicted by the present author[4] and also by Tiedema et al.[13] from a qualitative reasoning based on up-hill diffusion. An analogous situation seems to exist in the development of a grain-boundary groove on thermal etching. Here also, one reaches the conclusion that a whole series of waves may be expected outside the groove. More recent calculations by Mullins[14] have given the same result, but showed that the new waves will probably be too small to be detected.

It may not be quite realistic to assume that the homogeneous nucleation of an alloy starts with the formation of zones, well separated from one another, as suggested by Guinier[12]. Alternatively, one might visualize that there are many small fluctuations at the beginning, and equation (13) can then be used to calculate how such fluctuations will develop into a periodic structure. Fig. 8 shows the result of a calculation based on a "random" fluctuation which was chosen in such a way that it contained a wide spectrum of wavelengths. It is found that waves of different widths are at first developed. Later on, the small waves gradually disappear, causing a progressive growth of the *average* wavelength. As far as this calculation was carried, there was still a wide spectrum of wavelengths present and the growth of the average wavelength was still continuing. The kinetic treatment based on the diffusion equation (13) thus seems capable of providing answers to our questions about the transformation. The perfectly periodic states, on the other hand, which are obtained as solutions to equation (3), provide a description of a real system which may be approximately correct in many details but fails completely when the presence of a spectrum of wavelengths is important.

7. ASYMMETRIC ALLOY WITH POSITIVE ν

The lower part of Fig. 9 shows a curve from Fig. 1 representing the states for a slightly asymmetric alloy, (e.g. $x_a = 0.35$) now replotted on an l-scale. The upper part shows the corresponding free-energy surface. With increasing amplitude, for sufficiently large wavelengths, the points representing solutions to equation (3) first lie along a ridge on the free-energy surface (dotted line) and along the bottom of a valley (dashed line). This valley slopes down to the same equilibrium state at infinite wavelength that holds for the symmetric alloy already discussed. As a consequence, the final state always exhibits two large regions of the compositions represented by the sides of the miscibility gap and separated by the same diffuse grain boundary. The only difference between alloys of different compositions, x_a, at the same temperature is the relative size of the two regions. This result is quite natural and contains no new information. On the contrary, it must be expected from any model that is not basically wrong.

For the asymmetric composition, part of the valley is hidden behind the ridge and cannot be reached from

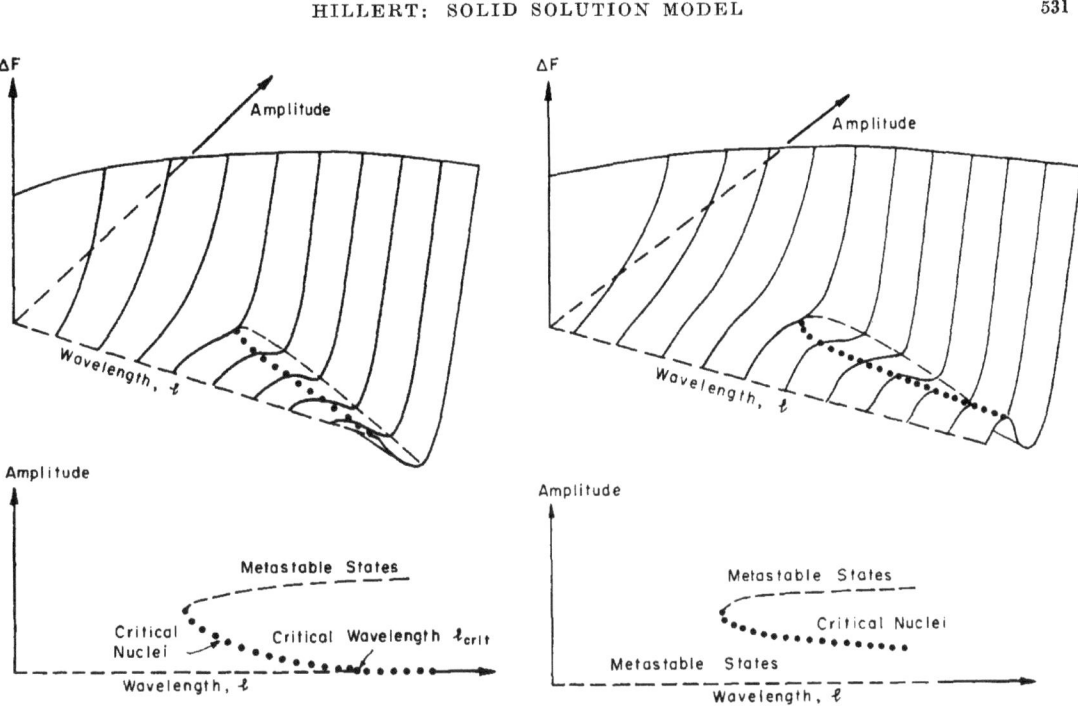

Fig. 9. Free energy of periodic states for asymmetric composition, $x_a \simeq 0.35$. Positive v. Schematic diagram.

Fig. 10. Free energy of periodic states for composition outside spinodal, $x_a \simeq 0.30$, positive v. Schematic diagram.

the l-axis without the system passing over this free-energy barrier. The states on the ridge thus represent critical nuclei. They all have shorter wavelengths than l_{crit}. All stable or metastable states with a wavelength longer than l_{crit}, on the other hand, can still be reached directly from the l-axis. For such wavelengths, the transformation can start without awaiting any nucleation event.

As the alloy composition is chosen closer and closer to the spinodal, l_{crit} approaches infinity and all the stable or metastable states become hidden behind the ridge on the free-energy surface, as illustrated by Fig. 10, which holds for an asymmetric composition outside the spinodal (e.g. $x_a = 0.30$ in Fig. 1). In this case, a nucleation event is necessary before the transformation can start. The lowest point on the ridge is situated at infinite wavelength ($1/l = 0$); this point represents the most probable critical nucleus for homogeneous nucleation in the alloy considered. For all the alloy compositions between the spinodal and the solubility limit, there is such a state and the right-hand axis in Fig. 1 can be regarded as the locus of all points representing such states. Their amplitudes can thus be read directly from this diagram, and it is found that the amplitude is highest for a composition at the solubility limit and approaches zero as the composition approaches the spinodal.

8. HOMOGENEOUS NUCLEATION

According to the preceding discussion, the present model can be used for a new treatment of the homogeneous nucleation process, which is free from all *a priori* assumptions about the critical nucleus regarding size, shape, composition, uniformity, etc. Instead, the critical nucleus is now treated as a state of the whole system, although a detailed calculation will reveal that the most probable critical nucleus is characterized by a localized, but somewhat diffuse clustering. An example is given in Fig. 11 where this state is shown for four alloy compositions. The same four states are represented by the crosses on the right-hand axis of Fig. 1. Far away from the localized clustering, the matrix composition remains unchanged. The simple equation (4) can thus be used for the calculation of the state if one chooses $x_\infty = x_a$. After finding the state, one can compute the free energy for nucleation from equation (1). Fig. 12 shows the result of such a calculation compared with the results yielded by Becker's nucleation theory[9] and by the more elaborate treatments of Hobstetter[15] and Scheil[16] when applied to compositional variations in one dimension.

According to the thermodynamic criterion for stability, a system inside the spinodal curve cannot have internal stability, and hence there is no free

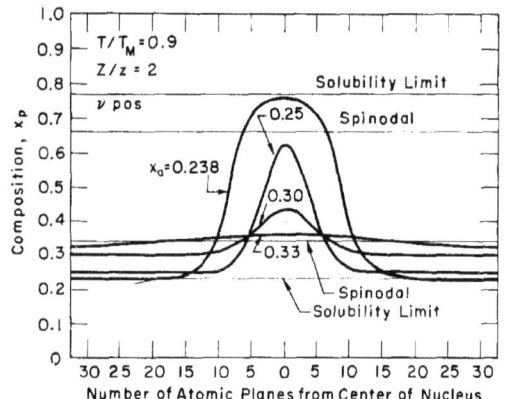

FIG. 11. Most probable critical nuclei for four alloy compositions outside the spinodal, marked with crosses, in Fig. 1.

energy of nucleation. This is borne out by the Borelius nucleation theory,[17] for instance, and was also realized by Becker[9]. However, as shown in Fig. 12, Becker's theory yields a free energy of nucleation for all compositions, and thus it incorrectly predicts internal stability for compositions inside the spinodal. It is interesting to note from Fig. 12 that the treatments by Hobstetter and Scheil correctly predict a vanishing free energy of nucleation inside the spinodal. However, this situation was not stressed by Hobstetter, apparently because he did not notice that the size of the critical nucleus approaches infinity as the alloy composition approaches the spinodal.[4] It is quite unexpected that the nucleus size should become infinite at the spinodal composition, considering that the nucleus size also approaches infinity as the composition approaches the solubility limit (i.e. the side of the miscibility gap). The smallest nucleus size is thus obtained at some intermediate composition. As is seen from Fig. 11, the same result now comes out of the present treatment, if nucleus size is identified with some measure of the size of the diffuse cluster. The nucleus size is connected with the diffuse nature of the boundary between cluster and matrix, and will be discussed again in Section 10.

The limitations of the new approach to homogeneous nucleation are, first, the restriction of compositional variations to one dimension only, and secondly, the application of the zeroth approximation. In principle, these limitations can be removed and the present treatment should be regarded as a first attempt only. It has been improved by Cahn and Hilliard[7] who succeeded in extending the treatment to three dimensions. Further details of the one-dimensional treatment of homogeneous nucleation will therefore not be presented here. It is interesting to note, however, that all the conclusions regarding homogeneous nucleation which can be drawn from the one-dimensional treatment seems to be confirmed by the more elaborate three-dimensional treatment.

A shortcoming of the present one-dimensional treatment is that an absolute value of the free energy of nucleation cannot be calculated because of the unknown parameter, m, designating the size of the atomic planes considered. However, this deficiency does not seem serious for the following reason. When calculating the free energy of nucleation outside the spinodal, Becker[9] found very low values even far away from the spinodal. He concluded that there would, in fact, be no efficient activation barrier for nucleation except for quite low supersaturations, i.e. for alloy compositions quite close to the solubility limit. For all other alloys, one should therefore expect the transformation to start spontaneously. This conclusion is now confirmed by the present treatment because it yields even lower values as demonstrated by Fig. 12. The absolute value of the free energy barrier is of little interest as long as it is too low to provide an effective obstacle for nucleation. Correspondingly, the old question, as to whether the free energy of nucleation goes to zero at the spinodal may be of little practical importance.

From the above discussion, it can be concluded that no sharp discontinuity in kinetic behavior is to be anticipated at the spinodal composition because a considerable part of the ridge in Figs. 9 and 10 may be so low that it constitutes no real barrier for the transformation. In particular, we now see that l_{opt} may have a finite value at the spinodal composition even though l_{crit} approaches infinity.

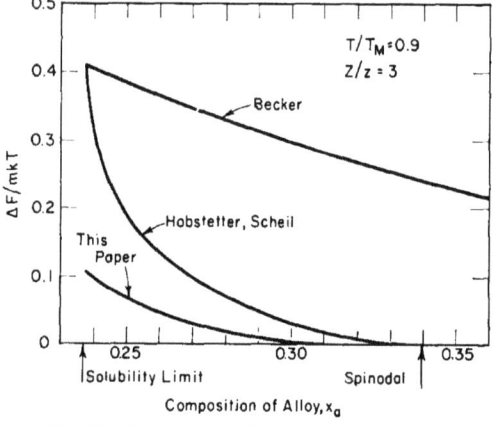

FIG. 12. Free energy of nucleation according to different theories.

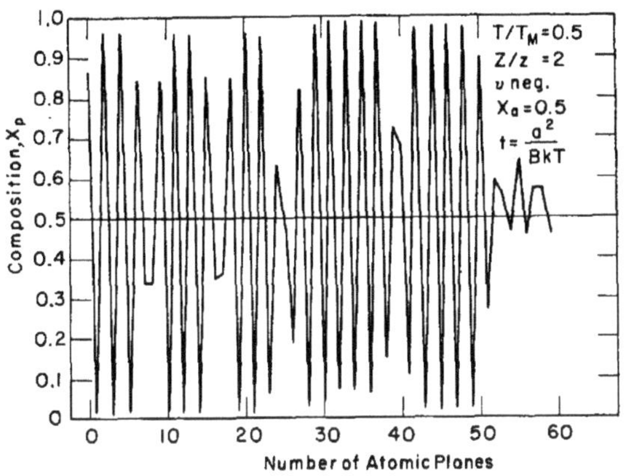

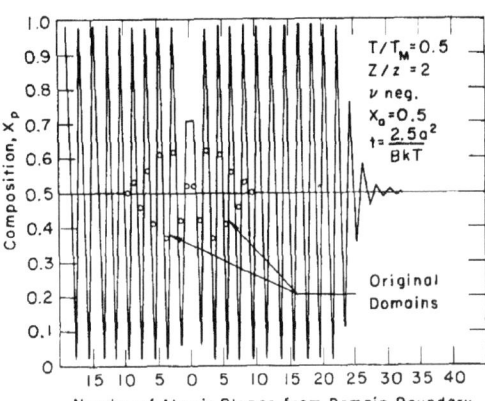

FIG. 13. Development of random fluctuations in symmetric composition, negative ν.

FIG. 14. Development of two small domains in symmetric composition, negative ν.

9. ALLOYS WITH NEGATIVE ν

In an ordering system, the value of the interaction energy ν is negative according to the nearest-neighbor interaction models. The zeroth approximation of this model has been applied most successfully to ordering systems of the CsCl structure, a well-known example being β-brass. The following discussion will therefore be concerned mainly with this structure, which gives the value $Z/z = 2$. Many of the calculations for positive ν presented in the preceding sections were also carried out with the same choice $Z/z = 2$ in order to allow comparison between positive and negative ν-values.

The possible states for a system with negative ν are presented in Fig. 2. The curve for the symmetric alloy ($x_a = 0.5$) is identical to the corresponding curve for positive ν in Fig. 1. The free-energy surface also resembles the corresponding diagram for positive ν, which is presented in Fig. 5. The kinetic treatment, on the other hand, reveals that there is a distinct difference, as demonstrated by Fig. 13 in comparison with Fig. 8. Although l_{crit} has the same value and the same significance in the two cases, the average wavelength (or domain size) at the first stage of transformation seems to be considerably larger when ν is negative than when it is positive. The difference in development of a zone is quite striking. If there is a small ordered region, the ordering will spread very rapidly by the growth of the original domain, as demonstrated in Fig. 14, and it does not induce the formation of other domains. A very large value of the wavelength will thus result. In the corresponding case for positive ν, the reaction will spread out relatively slowly and a whole series of zones are formed, resulting in a wavelength close to l_{crit} (Fig. 7). The difference is even more pronounced when one compares asymmetric compositions. As demonstrated by the shape of the curves in Fig. 2, l_{crit} is, for all compositions, the minimum wavelength when ν is negative. As a consequence, l_{opt} can never be less than l_{crit} and must approach infinity as l_{crit} does. The difference in this respect between systems with positive and negative ν is demonstrated schematically in Fig. 15.

With negative ν, the free-energy surface for asymmetric compositions also resembles Fig. 5 and, as a consequence, there is no theoretical free-energy barrier for nucleation in this instance. The states on the right-hand axis of Fig. 1, represent the most probable critical nuclei in the case of positive ν, whereas the states on the right-hand axis of Fig. 2 represent the final equilibrium states for negative ν. With positive ν, the degrees of separation in composition (i.e. the amplitude) in the final state is independent of alloy composition. The compositions

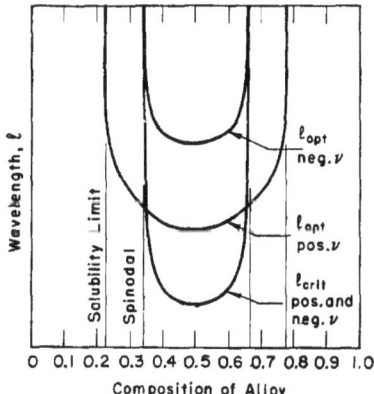

FIG. 15. Variation of optimum wavelength with alloy composition, positive and negative ν. Schematic diagram.

of the two regions always correspond to the sides of the miscibility gap. For negative v, the final degree of separation, or the amplitude, varies with composition as shown by the right-hand axis of Fig. 2. It is equivalent to saying that the degree of order depends on the alloy composition. This is a well known consequence of the zeroth approximation, but it is interesting that the present model relates this fact to the absence of critical nuclei.

For positive v, it was found that the diffuse grain boundary has the same structure, independent of alloy composition. For negative v, the domain boundaries are also diffuse, and the lower part of Fig. 4 gives an example. However, the structure of the diffuse domain boundary is dependent on composition due to the varying degree of order.

10. WIDTH OF DIFFUSE GRAIN AND DOMAIN BOUNDARIES

Points close to the lower right-hand corner in Figs. 1 and 2 represent states with wavelength approaching infinity and with small amplitude, implying a sinusoidal variation in composition or degree of order according to Section 3. The width of the diffuse boundaries of these states thus approaches infinity. For negative v, this suggests that the width of the diffuse domain boundary at *any* temperature approaches infinity as the composition approaches the side of the ordering region in the phase diagram. For positive v, it implies that the width of the diffuse grain boundary between the matrix and cluster in a critical-nucleus state (Section 8) approaches infinity as the composition approaches the spinodal curve in the phase diagram. This is equivalent to saying that the nucleus size approaches infinity. The width of the equilibrium grain boundary (the upper right corner in Fig. 1), on the other hand, is independent of composition and approaches infinity *only* at the peak temperature of the miscibility gap, where the amplitude of the equilibrium state approaches zero. This state can be realized only at the symmetric composition.

As mentioned in Section 5, one can use equation (1) to calculate the value of the interfacial free energy. This is true for domain boundaries as well as for grain boundaries. For diffuse boundaries with a small width, it can be expected that the value will depend on how the diffuse boundary profile is situated in the atomic structure. For instance, different results could be expected if one chose to place the steepest gradient of the profile exactly on an atomic plane or exactly between two neighboring planes. Numerical calculations indicate that this effect is usually too small to influence the interfacial free-energy value markedly. On the other hand, it may be important with regard to the movement of a domain boundary. In order to move one atomic distance, a domain boundary must pass from a preferred position in the structure through a position of higher free energy before it can reach a new preferred position. This effect should contribute an activation energy for domain-boundary movement.

11. APPLICABILITY OF ONE-DIMENSIONAL MODEL

Although not previously emphasized in this paper, it should be noticed that the present model is applicable to a certain class of transformations only. The model is based on the assumption that all changes in the system occur by an exchange of positions of component atoms. The initial and the final states thus contain exactly the same atomic sites. There is no generally accepted designation for this class of transformation but the name "exchange transformation" has been suggested.[4] This class includes ordering as well as precipitation. The formation of Guinier–Preston zones is an example. During recent years, the formation of modulated structures has attracted considerable interest.[18] This is also an exchange transformation and is closely related to the formation of Guinier–Preston zones.

Many of the features exhibited by modulated structures indicate that their formation can be described quite well by the model discussed in the present paper. In order to test the applicability of the model further, an experimental investigation of the formation of modulated structures in Cu–Ni–Fe alloys was undertaken and the results are reported in a separate paper. To test the model further, an experimental investigation of an ordering system should also be undertaken, with special attention focussed on kinetics and domain size.

The main limitation of the present model is its restriction to variations in one dimension only. Of course, it may be argued that at any one place in a system the compositional variations are largest in some direction and it is natural to expect these variations to develop, yielding a structure modulated in one direction only. However, it seems that such a structure would be quite unstable unless there were some additional factor. Probably, strains due to a difference in atomic size of the component atoms is such a factor. A structure, modulated in one dimension only, may be stable because this state minimizes the strain energy. Mathematically, this may mean that a modulation in one direction causes a decrease of the interaction energy v between neighbors in other directions. As a matter of fact, Manenc[19] has pointed out that structures modulated in one dimension are formed if there is a certain difference in lattice parameter. If the difference is too small, isotropic zones are formed. The one-dimensional model can only apply to the former case, of course. A three-dimensional model is needed for the latter case.

12. SUMMARY

Based on the so-called zeroth approximation, an expression is derived for the free energy of an inhomogeneous system with compositional variations in one direction. The thermodynamic criterion for internal stability yields a difference equation and the mathematical solutions of this equation are discussed.

One class of solutions represents stable or metastable states of the system and they show a periodic variation of composition in the direction considered, indicating that there is a thermodynamic reason for the formation of so-called modulated structures of the type found in the Cu–Ni–Fe system.

The other class represents critical nuclei, which are here defined as states of the whole system which the system must pass in order to transform from a metastable state to a state of lower free energy. The present approach allows the calculation of such homogeneous nucleation without the usual *a priori* assumptions concerning nucleus size, composition, shape, etc. The most probable critical nuclei can be characterized as a local but diffuse clustering in the homogeneous matrix. The dimension of this cluster (which can be identified with nucleus size in classical nucleation theory) is found to approach infinity as the alloy composition approaches the solubility limit and also as the alloy composition approaches the spinodal. The minimum size of the cluster is thus found at some intermediate alloy composition. The difference in composition at the center of the cluster and in the homogeneous matrix also depends on alloy composition, approaching zero as the alloy composition approaches the spinodal. This fact results in a vanishing free energy of nucleation at the spinodal even though the "nucleus size" approaches infinity at the same time.

For ordering systems, all solutions belong to the first class of metastable or stable states, implying that all homogeneous alloys with a composition inside an ordering region are inherently unstable and tend to transform spontaneously.

A calculation of the final stable state shows that the grain boundary between two compositions, corresponding to each side of the miscibility gap, is quite diffuse. The width of the grain boundary varies with temperature and approaches infinity at the peak temperature of the miscibility gap. For an ordering system, the domain boundaries are also diffuse but in this case there is, at any temperature, an alloy composition for which the boundary width approaches infinity. In this respect, a domain boundary does not resemble an equilibrium grain boundary but rather the boundary between cluster and homogeneous matrix in a critical-nucleus state.

A diffusion equation is derived, taking into account the discontinuity in composition from plane to plane in a crystalline solid. The equation is used for kinetic calculations of the transformation in unstable systems. This kinetic model predicts the formation of a wide spectrum of wavelengths (i.e. zone or domain sizes) during the first stage of transformation and a gradual growth of the wavelength later on. It is thus capable of explaining the main characteristics of the formation of modulated structures.

ACKNOWLEDGMENTS

The financial support of the Atomic Energy Commission is gratefully acknowledged. The author is indebted to his thesis supervisors, Professors Morris Cohen and B. L. Averbach, for frequent discussions and constant encouragement during the course of this work. Thanks are also due to Per Spiegelberg of the Swedish Institute for Metals Research who introduced the author to the machine methods of calculation.

REFERENCES

1. E. A. GUGGENHEIM, *Mixtures*. Oxford University Press (1952).
2. W. GORSKY, *Z. Phys.* **50**, 64 (1928).
3. W. L. BRAGG and E. J. WILLIAMS, *Proc. Roy. Soc.* A **145**, 699 (1934).
4. M. HILLERT, *A Theory of Nucleation of Solid Metallic Solutions*, Sc.D. Thesis, Mass. Inst. Tech. (1956).
5. J. W. CAHN and J. E. HILLIARD, *J. Chem. Phys.* **28**, 258 (1958).
6. J. W. CAHN, *J. Chem. Phys.* **30**, 1121 (1959).
7. J. W. CAHN and J. E. HILLIARD, *J. Chem. Phys.* **31**, 688 (1960).
8. E. W. HART, *Phys. Rev.* **113**, 412 (1959).
9. R. BECKER, *Z. Metallk.* **29**, 245 (1937).
10. S. ONO, *Mem. Fac. Engrs Kyushu Univ.* **10**, 195 (1947).
11. L. S. DARKEN, *Trans. Amer. Inst. Min. (Metall.) Engrs* **175**, 184 (1948).
12. A. GUINIER, *Acta Met.* **3**, 510 (1955).
13. T. J. TIEDEMA, J. BOUMAN and W. G. BURGERS, *Acta Met.* **5**, 310 (1957).
14. W. W. MULLINS, *J. Appl. Phys.* **28**, 333 (1957).
15. J. N. HOBSTETTER, *Trans. Amer. Inst. Min. (Metall.) Engrs* **180**, 121 (1949).
16. E. SCHEIL, *Z. Metallk.* **43**, 40 (1952).
17. G. BORELIUS, *Ann. Phys., Lpz.* **28**, 507 (1937).
18. A. GUINIER, *Solid State Phys.* **9**, 293 (1959).
19. J. MANENC, *Rev. Met.* **54**, 867 (1957).

8. Introduction to "The Formation of Pearlite"

published in *Decomposition of Austenite by Diffusional Processes'*, Eds. Zackay and Aaronson,
AIME (1962)

The following article on the "Formation of Pearlite" reminds us of the rich state of the micrographic art for carbon and alloy steels up to 1960 due to the superior skills of Axel Hultgren's group at the Swedish Institute for Metal Research, Stockholm. Mats' celebrated 1957 theory of the steady state Pearlite reaction [1], fully incorporating a detailed frontal shape and degenerate spacing based upon capillarity, is reviewed in terms of his Figure 14 in which the shape for $S/S_o = 2$, where spacing $S = S_o$ defines the frontal *free* velocity cut-off and the equality corresponds to the maximal frontal velocity in Zener's 1946 conjecture [2]. Concomitantly, he underlines his belief in a quasi-equilibrium state in which the interface is as flat as possible, carrying the implication that S/S_o should be > 2. In 1959 J.W. Cahn had theoretically examined the lamellar crystallization (or discontinuous precipitation) reaction [3], closely related to Pearlite in alloy steels, and conjectured that the stability criterion might be a maximum in the dissipation (rate), a proposition which was later positively tested by Solarzano and Purdy [4]. For a volume diffusion reaction this criterion leads to $S/S_o = 3$. This stability point also agrees with a Le Chatelier kinetic stability argument [5]. Mats, I believe, preferred to rely on his intuition in such matters, although he has recently included Onsager's Reciprocity theorem in his monograph on thermodynamics and phase transformations [6], the corollary of which for contrasting initial and boundary conditions such as *forced* velocity cellular solidification corresponds to a minimum in the dissipation [7]. The foregoing dichotomy in the dissipation presages a more general minimax stability point when two or more kinetic degeneracies allow it.

Here one can profitably refer to Ostwald's Step Rule, shown by Van Santen [8] to be a non-steady state corollary of the principle of minimum dissipation and applicable to isothermal time-dependent reaction sequences tending towards equilibrium. This is paraphrased such that "the reaction by using intermediate steps proceeds closest to reversible conditions, which minimizes the entropy production". The best known example is the sequential formation of localized high free energy G-P zones in Al-Cu. This is also exemplified by the very common lamellar coarsening sequences of Pearlites in the accompanying article of Spencer and Mack [9], and indeed pertains to all initiating free energy-rich fine structures followed by coarsening reactions. A very pertinent example within the venue of Hillert's contributions has to do with the intimate connection between the Pearlite structure in Au-Ni studied by Underwood [10], his associate at MIT, and Mats' other celebrated paper on clustering with diffuse interfaces contained in his 1956 MIT thesis [11]. While to Mats the contemporary connection may have been subconscious, when he encountered the degeneracy in the specification of an initiating, near stable clustering wavelength he was intuitively led to a plausible substitute for optimal dissipation in the negative ratio of the driving free energy to the wavelength, $-\Delta F/\lambda$ [12], which numerically yielded a maximum value near $\lambda = 3\lambda_c$ where λ_c is the cut-off wavelength. From another viewpoint, the Ostwald's Rule analogy is propitious, for all of Underwood's Pearlites in Au-Ni samples were templated by the yet unidentified nano-structured, elastic energy-rich substrate modulations [13], now clearly evidenced at all temperatures up to the critical point at 1080 K by the arrays of intragranular, octahedral near single crystal two-phase nodules observed by Underwood and Cahn [14]. Evidently, Van Santen's *necessary* high gradient energy initiating modulation sequences clash with, and thereby obviate Cahn's proposed strong critical point depressing, long range elastic energy effect which forces the effective diffusion coefficient to zero at the so-called "coherent spinodal", and furthermore dismisses the 4th order partial differential equation (PDE) formulation which is predicated upon an unnecessary flux constraint and a concomitant paradoxical "uphill" diffusion effect [15]. The correct local atomic misfit energetics based upon the Crum theorem and image forces was already established by Friedel [16] and Eshelby [17] and later clarified by Khachaturyan [18] and Hirth [19]. This generates a 2nd order PDE when formulated within a continuum version of Hillert's discrete free energy functional.

Although Mats and I have not completely exorcised all of our differences on these theoretical matters after a lengthy dialogue, I would like to assure him that I am still a loyal disciple and enthusiastically congratulate him on his incomparable contributions to empirical and computational science and his leadership in establishing the central importance of both discrete (Pearlitic) and diffuse (clustering) surface energy in predicting phase transformation morphologies.

Jack S. KIRKALDY

References

[1] M. Hillert, *Jernkontorets Ann.* 141 (1957) 757.

[2] C. Zener, *Trans. AIME* 167 (1946) 550.

[3] J.W. Cahn, *Acta Met.* 9 (1959) 18.

[4] I.G. Solarzano and G.R. Purdy, *Met. Trans.* 15A (1984) 1055.

[5] J.S. Kirkaldy, *Scripta Met.* 14 (1980) 531.

[6] M. Hillert, Phase Equilibria, Phase Diagrams, Phase Transformations (Cambridge, 1998) p. 99.

[7] I. Prigogine, Introduction Thermodynamics of Irreversible Processes (Thomas, Springfield, 1955).

[8] R.A. VanSanten, *J. Phys. Chem.* 88 (1984) 5768.

[9] C.W. Spencer and D.J. Mack, Decomposition of Austenite (Interscience, New York, 1962).

[10] E. Underwood, Sc.D. Thesis, MIT, 1954.

[11] M. Hillert, Sc.D. Thesis, MIT, 1956.

[12] M. Hillert, *Acta Met.* 9 (1961) 525.

[13] K. Janghorbian, J. Kirkaldy and G. Weatherly, *J. Phys.: Condensed Matter* 13 (2001) 8661.

[14] J.W. Cahn, Collected Works, *TMS*, 1998.

[15] J.W. Cahn, *Acta Met.* 9 (1961) 795.

[16] J. Friedel, *Adv. Phys.* 3 (1954) 446.

[17] J.D. Eshelby, 3 (1956) 79.

[18] A. Khachaturyan, Theory of Structural Transformations (Wiley, New York, 1983).

[19] J.P. Hirth, private communication, 2002.

The Formation of Pearlite

Mats Hillert

Swedish Institute for Metal Research, Stockholm, Sweden

I. Introduction	198
II. Orientation Relationships to Austenite	199
A. Accepted Theory of Pearlite Formation	199
B. Active Nucleus for Pearlite	201
C. Smith's Hypothesis Concerning Orientation Relationships	205
D. Generalization of Smith's Hypothesis	210
III. Nucleation and Growth of Pearlite	215
A. Edgewise Growth of Pearlite	215
B. Sidewise Growth of Pearlite	216
C. Development of a Pearlite Colony	221
D. Establishment of Cooperation	226
IV. Pearlite-Related Structures	228
A. Intermediate Structures	228
B. Comparison of Two Phase Structures	231
C. The Nature of Bainite	232
V. Interface and Orientation Effects	233
A. Effect of Grain Boundaries on the Growth of Pearlite	233
B. Effect of Twin Boundaries on the Growth of Pearlite	234
C. Effect of Orientation Relationships between Ferrite and Cementite	234
D. Effect of Lattice Orientation of Cementite	235
VI. Conclusions	236
References	236
Discussion	237

Abstract

The characteristic features of pearlite are discussed and compared with bainite. In view of recent microscopic evidence radical changes must be made in the old theory of pearlite formation. Cementite and ferrite act as equal partners in the formation of pearlite and either one can be the so-called "active nucleus."

There is no evidence of any sidewise growth of pearlite by a mechanism of repeated nucleation. Instead, the individual lamellae form by branching during edgewise growth.

198   DECOMPOSITION OF AUSTENITE

Each pearlite colony usually consists of only two interwoven crystals, one being ferrite, the other cementite. The real nucleation of pearlite is a gradual process involving a mutual arrangement of these two crystals, leading to a favorable geometry which results in a high growth rate. So-called "abnormal structure," "degenerate" or "divorced" pearlite, "semipearlite" and other coarse and irregular forms of pearlite are explained as the results of a low degree of cooperation between ferrite and cementite. Bainite may in part belong to this group of structures.

The ferrite and cementite constituents of a pearlite colony can have any orientation relationship to the matrix austenite, except for certain combinations which permit the formation of coherent interfaces. Cooperation between ferrite and cementite is then impossible and the favorable arrangement is destroyed.

Intermediate structures half way between pearlite and bainite can form when the orientation relationship is close to a critical combination.

In the old theory, bainite and pearlite were more or less considered to be mirror images, ferrite in one of them playing the same role as cementite in the other. The new theory also shows a high degree of symmetry but the mirror image of bainite is now a structure called inverse bainite, whereas pearlite lies in the center of the picture.

I. Introduction

Pearlite was one of the first structures in metals to be described in considerable detail, and over the years it has been the subject of a great many investigations. The wealth of information available in the literature was reviewed critically by Hull and Mehl in 1942[1] and they defined the purpose of their paper as: "(a) to subject past knowledge to a critical inspection, (b) to report the results of experiments that were designed to check the validity of old and new concepts, and (c) to rationalize well attested facts into a consistent theory of formation of pearlite."

TABLE I

Characteristic Features of α-Cementite Structures (according to Hull and Mehl[1])

	Bainite	Pearlite
Active nucleus	α	Cementite
First characteristic unit	Widmanstätten plate of α	Platelet of cementite
Orientation relationship of:		
primary importance	α/γ	cementite/γ
secondary importance		α/cementite
Effect of twin boundary	Stop	None
Effect of grain boundary	Stop	Stop

Indeed, these authors were able to rationalize available information and arrived at a theory which was not only self-consistent but also consistent with the accepted view regarding the formation of bainite. The main characteristics of the formation of these two structures are given in Table I. Comparison of their characteristics demonstrates that the two structures are regarded in many respects as mirror images, ferrite in one case playing the same role as cementite in the other. The only indication that the situation may not be quite so symmetrical is given by the information that a twin boundary stops the growth of bainite but has no effect on the growth of pearlite.

The general picture of the formation of pearlite thus given by Hull and Mehl still seems to be almost universally accepted today (compare Mehl and Hagel,[2] for instance) although a few critical voices have been heard in the meantime.[3-5] However, the time now seems ripe to subject the study of Hull and Mehl to the same test as they subjected the earlier work. The purpose of the present paper will thus be (a) to subject the theory of pearlite formation given by Hull and Mehl to a critical inspection, (b) to report the results of new experiments that were designed to check the validity of old and new concepts, and (c) to rationalize well-attested facts into a consistent theory.

II. Orientation Relationships to Austenite

A. Accepted Theory of Pearlite Formation

It is far beyond the scope of the present paper to review the wealth of experimental information available on the formation of pearlite or the development of theory in this field which led to the view of pearlite formation given in Table I. Reference is made instead to the excellent review given by Hull and Mehl[1] and to the more recent review by Mehl and Hagel.[2] For the present purpose it will suffice to point out in what respects the accepted picture of pearlite formation is logical and self-consistent.

In the eutectoid decomposition of austenite the two new phases ferrite and cementite can arrange themselves in two different ways, giving rise to two different structures, bainite and pearlite. Consequently it must be of primary interest to arrive at a picture of the eutectoid transformation of austenite which is capable of explaining the existence of two different structures. Since there is good evidence

200 DECOMPOSITION OF AUSTENITE

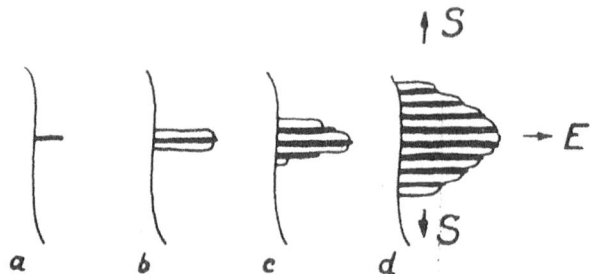

Fig. 1. Nucleation and growth of pearlite according to Hull and Mehl. Edgewise growth (E) and sidewise growth by repeated nucleation (S).

that bainite is nucleated by ferrite, it may seem quite natural to assume that pearlite is nucleated by cementite (see first line in Table I) although the evidence is not quite as good on this point. Furthermore, it is well known that the formation of nuclei with some special orientation relationship to the parent phase is often favored in comparison to randomly oriented nuclei. In view of the well-established fact that the nuclei for bainite are ferrite crystals of Widmanstätten nature, one could thus feel quite safe in assuming that pearlite is nucleated by a *platelet* of cementite in a Widmanstätten relationship to the parent austenite (see second line in Table I). This hypothesis is also very attractive from another point of view. First, it yields an explanation for the lamellar nature of pearlite and, second, it yields some understanding of the accepted fact that all parallel plates which form during sidewise growth by repeated nucleation exhibit the same lattice orientation* (see Fig. 1). Accepting this view, one must expect the orientation relationships between the two new phases and the parent phase to be of much importance during growth, at least in the sidewise direction. Thus one can see a natural explanation for the observation that "the case is never met with where any one zone of parallel lamellae (i.e., a pearlite colony) is not entirely localized in a single grain of austenite and is not arrested in its growth by the boundaries of this grain" (quoted from Jolivet[6]). As a matter of fact, one should expect the same effect at twin boundaries since it is not probable that the lattice orientation of cementite (or ferrite), which gives the correct orientation relationship to a particular grain

* Actually, since there are always a number of lattice orientations satisfying a certain Widmanstätten relationship due to its multiplicity, an additional hypothesis is required stating that all the plates chose the same orientation among these.

of austenite, could also yield the correct orientation relationship to a twin of this grain. The fact that no effect of twin boundaries had been reported was thus rather puzzling. Except for this detail, all the experimental facts appeared to be consistent with each other. As a consequence, if one should find, on re-evaluating the experimental evidence, that a mistake has been made on some point, it will probably prove necessary to revise the whole picture of the formation of pearlite.

B. Active Nucleus for Pearlite

The active nucleus of a pearlite colony may be defined as the first one of ferrite or cementite to form with the particular lattice orientation which will later be found in the pearlite colony. The question of which phase provides the active nucleus is thus intimately related to the question of orientation relationships.

Long before bainite was ever described, pearlite was known to be composed of two phases, ferrite and cementite. In view of this knowledge, Benedicks[7] made the natural suggestion that proeutectoid ferrite as well as cementite could act as "germs" for the formation of troostite (i.e., fine pearlite). Benedicks could support his suggestion by the experimental observation that this kind of pearlite forms preferentially in contact with proeutectoid ferrite or cementite. Later on, when bainite had been described and x-ray methods were available for orientation determinations, Mehl and co-workers[8-10] reported that bainitic ferrite has the same orientation relationship to the parent austenite as proeutectoid ferrite, whereas pearlitic ferrite appeared to have quite a different orientation relation with respect to the parent austenite. In the case of cementite, the x-ray methods did not yield such a definite answer but seemed to indicate that proeutectoid cementite and pearlitic cementite very well could have identical orientation relationships to the parent austenite. As a consequence one started to doubt the validity of Benedicks' suggestion. Upon microscopic inspection of the junction between proeutectoid ferrite and pearlitic ferrite, Hull and Mehl found a grain boundary between the two and reported that "the case is never observed, however, in which proeutectoid ferrite is continuous with ferrite of pearlite."

In view of this, Dubé[11] and Aaronson[12] suggested that the proeutectoid ferrite should only be considered as an "informal" nucleus, which provides a preferred site for the nucleation of cementite with the correct lattice orientation. This latter crystal, being the active nucleus for pearlite, is capable of nucleating ferrite with the lattice orientation characteristic of pearlitic ferrite and thus with a grain boundary toward the proeutectoid grain of ferrite. Furthermore, Hull and Mehl found that on quenching a partially transformed specimen with an insufficient rate of cooling, coarse pearlite will give rise to a rim of very fine pearlite (troostite), whereas proeutectoid ferrite gives rise to bainite. There was thus strong evidence in favor of the idea that bainite is nucleated by ferrite but pearlite is not.

Modin[3] pointed out that many micrographs published by Hanemann and Schrader in their *Atlas Metallographicus*[13] show that there often is no grain boundary between proeutectoid and pearlitic ferrite. After a thorough microscopic investigation Modin came to the same conclusion. Later on[14] Modin found in a particular case that the grain boundary was missing for about 30% of the pearlite units, a figure which seems quite too high to be ignored. This result is in serious conflict not only with the metallographic results of Hull and Mehl but also with the conclusions drawn from the x-ray measurements stating that pearlitic ferrite does not have the same lattice orientation as proeutectoid ferrite. As a consequence, Mehl and Hagel[2] found it difficult to accept Modin's result in spite of the high quality of his metallographic work. In this situation it appeared essential to test the validity of the earlier x-ray work which had been carried out on separate specimens, one containing proeutectoid ferrite of Widmanstätten character, another containing pearlite. A definite answer would be obtained by an orientation determination of a single colony of pearlite and the adjoining grains of ferrite. This determination could be carried out by means of a microbeam x-ray technique. However, metallographic techniques are also available and have been applied to this problem independently by Hillert[15] and Hultgren and Öhlin.[16]

Hillert used the characteristic propensity of nital to attack all ferrite crystals in a polished section except the ones with a cube axis closely perpendicular to the surface.[17] Unetched crystals of ferrite were thus seen in the microstructure of a hypoeutectoid steel. After etching in picral, most of them were found to consist of pearlite as well as adjoining proeutectoid ferrite, as illustrated in Figure 2. Hultgren[18] has refined an etching method which produces etch pits characteristic of the orientation of each grain of ferrite. Different lattice orientations can thus be distinguished microscopically by the application of polarized light. Hultgren and Öhlin applied this etch to partially transformed specimens in which the pearlite colonies were still quite small, the idea being that every pearlite colony found in a microsection should be close to the proeutectoid grain of ferrite which

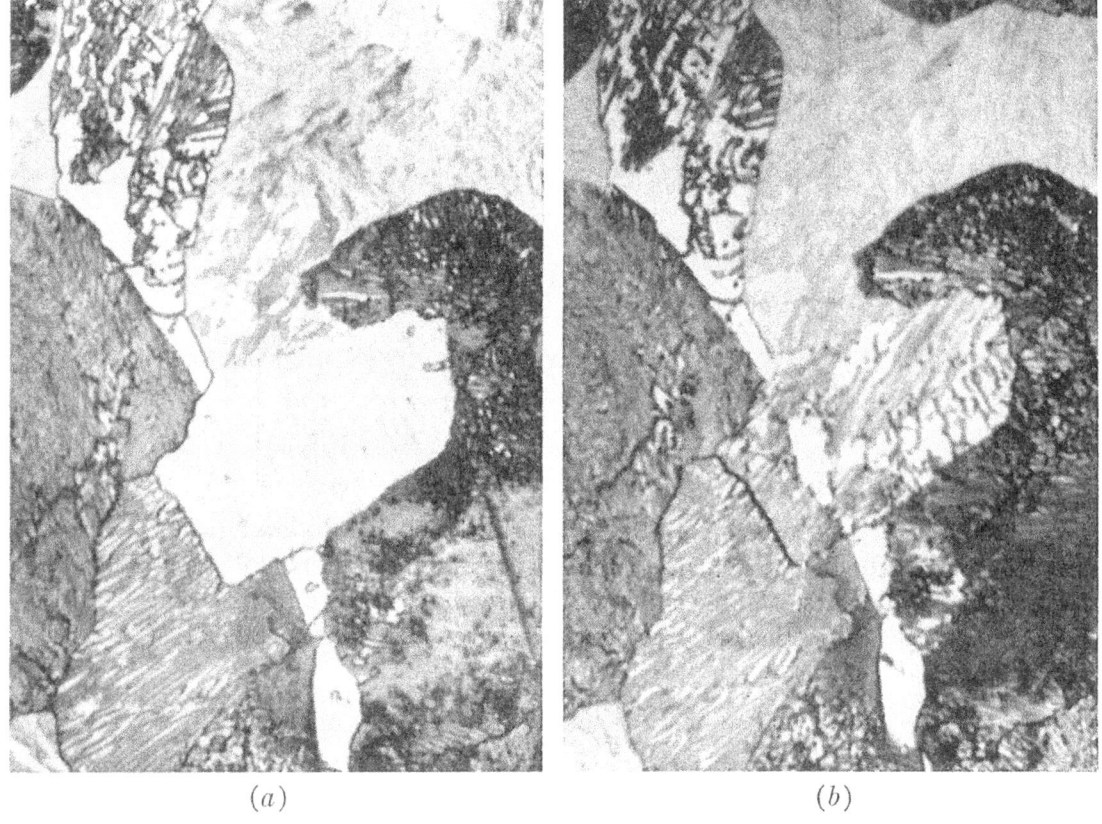

(a)   (b)

Fig. 2. Hypoeutectoid steel (0.6%C) partially transformed for 15 sec at 640°C, quenched. Row of proeutectoid grains of ferrite along austenite grain boundary (small white areas), pearlite (grey and black), martensite (light grey). (b) Etched in picral, shows attack on all grains of ferrite, proeutectoid and pearlitic. (a) Etched in nital, shows a passive grain of ferrite which is partly proeutectoid, partly pearlitic. 1800×.

had served as its nucleus, "informal" or "active." This investigation gave a very clear answer to the question concerning the role of proeutectoid ferrite. The ferrite constituent in 60–80% of all the pearlite colonies was found to have the same lattice orientation as some adjoining grain of proeutectoid ferrite, proving that these grains had been active, rather than informal, nuclei. Figure 3 shows a

204   DECOMPOSITION OF AUSTENITE

Fig. 3. Hypoeutectoid steel (0.6%C), slowly cooled to and partially transformed at 710°C, quenched. Etched in ammonium persulfate. Polarized light, crossed polars. Three pairs of ferrite precipitate and adjacent pearlitic ferrite of identical lattice orientation (white, grey and dark). 800×. (Courtesy Hultgren and Öhlin.)

microsection with three pearlite colonies. Each colony has an adjoining grain of ferrite with the identical lattice orientation.

Although proeutectoid cementite has always been considered able to serve as active nucleus for pearlite in a hypereutectoid steel, this opinion did not seem to be based on any experimental information until Modin[3] reported that inspection by polarized light revealed that pearlitic and adjoining proeutectoid cementite have the same lattice orientation. This point was again tested by Hultgren and Öhlin[16] who found this to be true for almost 100% of the pearlite colonies in a partially transformed hypereutectoid steel.

In view of these results, it appears necessary to conclude that Benedicks was correct when he suggested that pearlite can be nucleated by either ferrite or cementite. In hypereutectoid steels, cementite will normally form first and will then nucleate pearlite; in hypoeutectoid steels, ferrite will form first and then nucleate pearlite. Nicholson[5] arrived at the same conclusion by analyzing kinetic data for the formation of pearlite. His arguments have been criticized by Cahn,[19] however, and may not be quite valid although his conclusion now seems to be confirmed.

Accepting the fact that both ferrite and cementite may nucleate pearlite, an explanation of the earlier x-ray results must now be sought out. It seems that an answer can be provided by a hypothesis proposed by Smith.[4]

C. Smith's Hypothesis Concerning Orientation Relationships

Smith[4] suggested that a crystal of proeutectoid ferrite, formed at the grain boundary between two grains of austenite, would have a definite orientation relationship to one of them, resulting in a partially coherent interface. Ordinarily the lattice orientation of the ferrite crystal cannot at the same time be related to the other grain of austenite and an incoherent interface will thus form on this side. At a low degree of undercooling, growth will occur predominantly by the movement of the incoherent interface. The crystal of ferrite thus grows into the grain of austenite to which it bears no orientation relationship. At a higher degree of undercooling, the available free energy may be large enough to overcome the high elastic strain energy that opposes the movement of the coherent interface. The crystal of ferrite can then grow into the grain of austenite to which it is related. This growth will result in Widmanstätten forms.

Smith further proposed that the ferrite component of a pearlite unit, formed at a grain boundary, should also be related to one of the grains of austenite, whether nucleated before or after the cementite component. By analogy to discontinuous precipitation he suggested that the pearlite unit would only be able to grow by the advance of the incoherent ferrite-austenite interface, i.e., into the grain of austenite to which the lattice orientation of the ferrite is unrelated. The pearlitic ferrite found in a transformed grain of austenite should then bear a specific orientation relationship to a neighboring grain of austenite, this grain being the true parent grain for the crystal of ferrite which later developed into pearlite. This hypothesis is thus able to explain why an x-ray examination will yield different orientation relationships for pearlitic and Widmanstätten ferrite when referred to the matrix grain of austenite.* According to Smith, the lattice orientation of pearlitic ferrite in a particular matrix grain of austenite should be random except for the avoidance of certain orientations, which would give a coherent interface between ferrite and matrix austenite, preventing the formation of pearlite.

The reason for Smith's suggestion concerning the growth of pearlite was twofold. First, the mobility of an incoherent interface is high whereas a coherent interface sometimes may be rather immobile. Second, an appreciable increase of the diffusivity can be expected along an incoherent interface, allowing a rapid lateral diffusion and a high growth rate of a two-phase composite.

In view of the high potential value of Smith's hypothesis, an experimental study was carried out in order to test its validity in the Fe–C system.[20] The first part of this study concerns the formation of proeutectoid ferrite. The results are summarized in Figure 4, showing the main types of morphology. Intermediate types have also been found. It is suggested that all straight lines in Figure 4 represent partially coherent interfaces and are thus indications of an orientation relationship between ferrite and austenite, whereas the curved lines represent incoherent interfaces and thus indicate the absence of any close orientation relationship. Two kinds of coherent interfaces may be distinguished, the first kind being directed away from the original austenite grain boundary and thus giving rise to Widmanstätten plates, the second kind being approximately parallel to the original grain boundary and thus giving the impression of being quite immobile. It is suggested, however, that both kinds are crystallographically equivalent.

It is remarkable that the ferrite morphologies of type a, b, and c in Figure 4, which show straight interfaces toward both grains, are

---

* One may say that the earlier misinterpretation of the x-ray results stems from the fact that Mehl et al. used a *single crystal* of austenite in their study of pearlite. In such a specimen the matrix grain is also the parent grain because there is no other austenite grain present. Mehl could thus argue that all the proeutectoid ferrite in this particular case must bear a specific orientation relationship to the matrix grain and the fact that the pearlitic ferrite was found not to obey this relationship led him to conclude that cementite was the active nucleus of the pearlite. This conclusion may perhaps be justified for a single crystal specimen but it has no bearing on a polycrystalline specimen where each grain of austenite could easily be in contact with many crystals of proeutectoid ferrite nucleated in the neighboring grains of austenite.

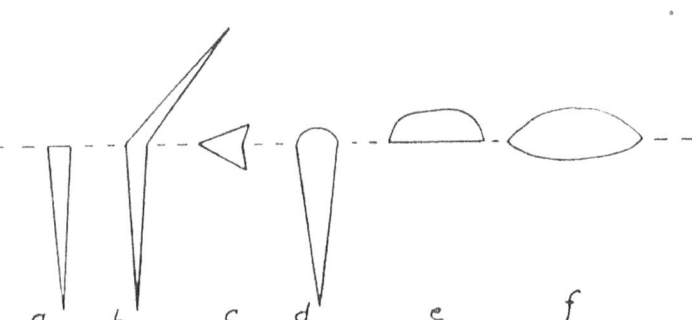

Fig. 4. Different morphologies of proeutectoid ferrite, observed at austenite grain boundaries.

found much more frequently than one would expect to find in a definite orientation relationship between the two neighboring grains of austenite. It thus appears that partially coherent interfaces may develop between ferrite and austenite, even when their orientation relationship deviates appreciably from the ideal condition, and it seems reasonable that the degree of coherency may vary from high to low depending on how well the two lattices fit together. A crystal of ferrite may thus be closely related to one grain of austenite, having a highly coherent interface toward this grain, and yet be somewhat related to the other grain and have an interface of some low degree of coherency toward that grain.

Smith only discussed morphologies of the types d and e (expecting e to form at low supersaturation and d at high), but a, b, c, d, and e may be regarded as confirmations of his hypothesis concerning a close relationship between proeutectoid ferrite and one of the grains of austenite, although the relationship to the other grain apparently may sometimes be good enough to allow the development of interfaces with some degree of coherency toward this grain also (a, b, and c).

When a specimen heat treated to give some proeutectoid ferrite was quenched with an insufficient rate through the temperature region where pearlite and bainite can form rapidly, the ferrite morphologies in Figure 4 were found to give rise to these two structures. In the cases where proeutectoid ferrite had grown incoherently into a grain of austenite (upward in d, e, and f and downward in f) further growth occurred on cooling, resulting in a rim of fine pearlite (d, e, and f in Fig. 5). On the other hand, in those cases where ferrite had advanced coherently into a grain of austenite (upward in b and c, downward in a, b, c, and d) further growth during cooling resulted

208   DECOMPOSITION OF AUSTENITE

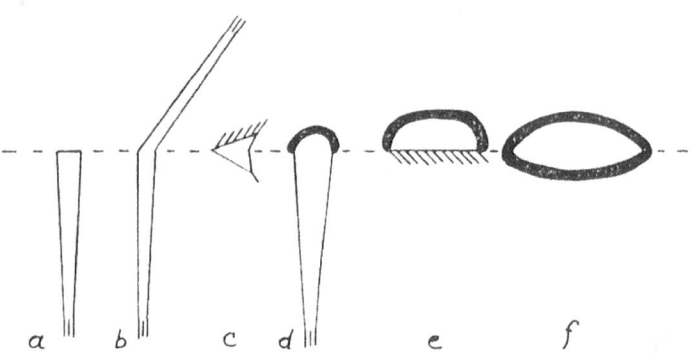

Fig. 5. Continued growth of proeutectoid ferrite during ineffective quench. Parallel lines represent acicular (bainitic) growth, dark rims represent troostite (fine pearlite).

in bainite. Among those cases, finally, where the high temperature transformation yielded coherent ferrite-austenite interfaces of a more or less immobile nature, further growth occurred occasionally and then resulted in bainite displaying a new habitus (upward in c and downward in e). It is believed that this occurs when the orientation relationship between ferrite and austenite deviates somewhat from the ideal condition. As a rule, the immobile interfaces developed at high temperature remain immobile on cooling and do not result in any low temperature structure (e.g., upward in type a). The sides of the proeutectoid plates obey the same rule. They do not move sidewise readily and they usually do not yield any low temperature structure on cooling.

The results presented in Figure 5 strongly support Smith's hypothesis concerning the orientation relationship and the nature of the interface between pearlitic ferrite and the matrix austenite. It appears that Smith's hypothesis can even be used as a means of deciding whether an interface is incoherent or not. Figure 6 shows a proeutectoid grain of ferrite where the shape does not give any clear indication of the nature of the interface. However, the structures formed during the slow quench indicate that the lattice orientation of the grain of ferrite is related to the lower grain of austenite, because there bainite has formed. No close relationship seems to have existed to the upper grain of austenite; the interface was here incoherent, allowing a thin rim of pearlite to form by a further advance of the incoherent interface.

The same types of morphology which were obtained by isothermal precipitation of ferrite followed by a slow quench may also be obtained

Fig. 6. Hypoeutectoid steel (0.6%C) partially transformed for 30 min at 710°C, inefficiently quenched. Bainitic growth into lower grain of austenite and pearlitic growth into upper grain during quench. 1800×.

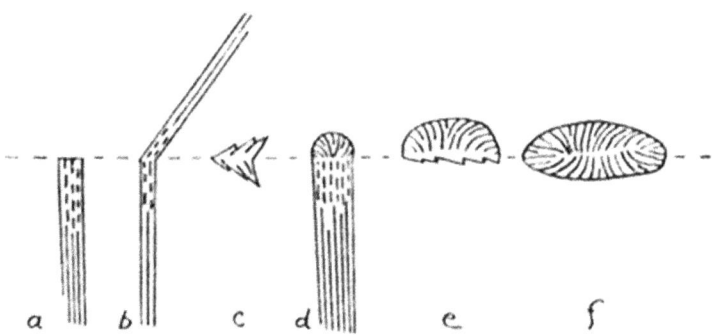

Fig. 7. Two-phase structures observed at austenite grain boundaries. Units with straight sides or jagged contours are bainite, units with rounded contours and curved lines are pearlite.

directly under isothermal conditions at a sufficiently low temperature, as demonstrated by Figure 7. By etching in nital it could be shown for all the types of morphology presented in Figures 4, 5, and 7 that it is ferrite of a uniform crystalline orientation which has grown into both the neighboring grains of austenite. Figure 8 gives an example where a ferrite crystal has developed as bainite of a characteristic direction in one grain of austenite, as bainite of another characteristic direction in another grain (compare Fig. 7b), and as pearlite into a third grain (compare Fig. 7d).

210   DECOMPOSITION OF AUSTENITE

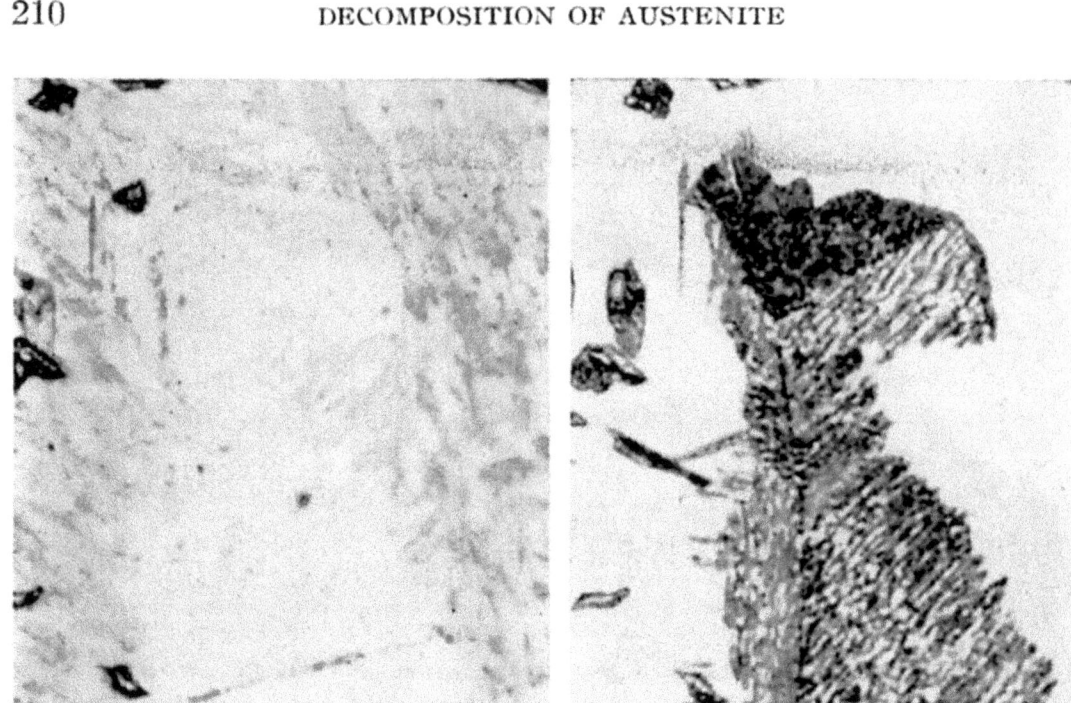

(a) (b)

Fig. 8. Hypoeutectoid steel (0.6%C) partially transformed for 6 sec at 443°C, quenched. (b) Etched in picral, shows bainite of two characteristic directions (striped constituent) and pearlite (dark constituent at the top). (a) Etched in nital, shows that all three have the lattice orientation of ferrite causing passivity. 1800×.

D. Generalization of Smith's Hypothesis

A microscopic study[20] of proeutectoid cementite in hypereutectoid steels has indicated that the kinds of morphology found are the same as those formed in the proeutectoid ferrite reaction. However, the facts that cementite always forms in rather small quantities and is strongly affected by alloying elements make the results more uncertain. It appears that Widmanstätten plates of a single grain of cementite may develop into two neighboring grains of austenite even more frequently than in the case of ferrite, indicating that partially coherent interfaces between cementite and austenite may form at even larger deviations from the ideal orientation relationship.

Hultgren and Edström[21] found that proeutectoid cementite develops a "porous" appearance in steels with a silicon content higher than 0.2%. It has been suggested[15] that the "pores" in this struc-

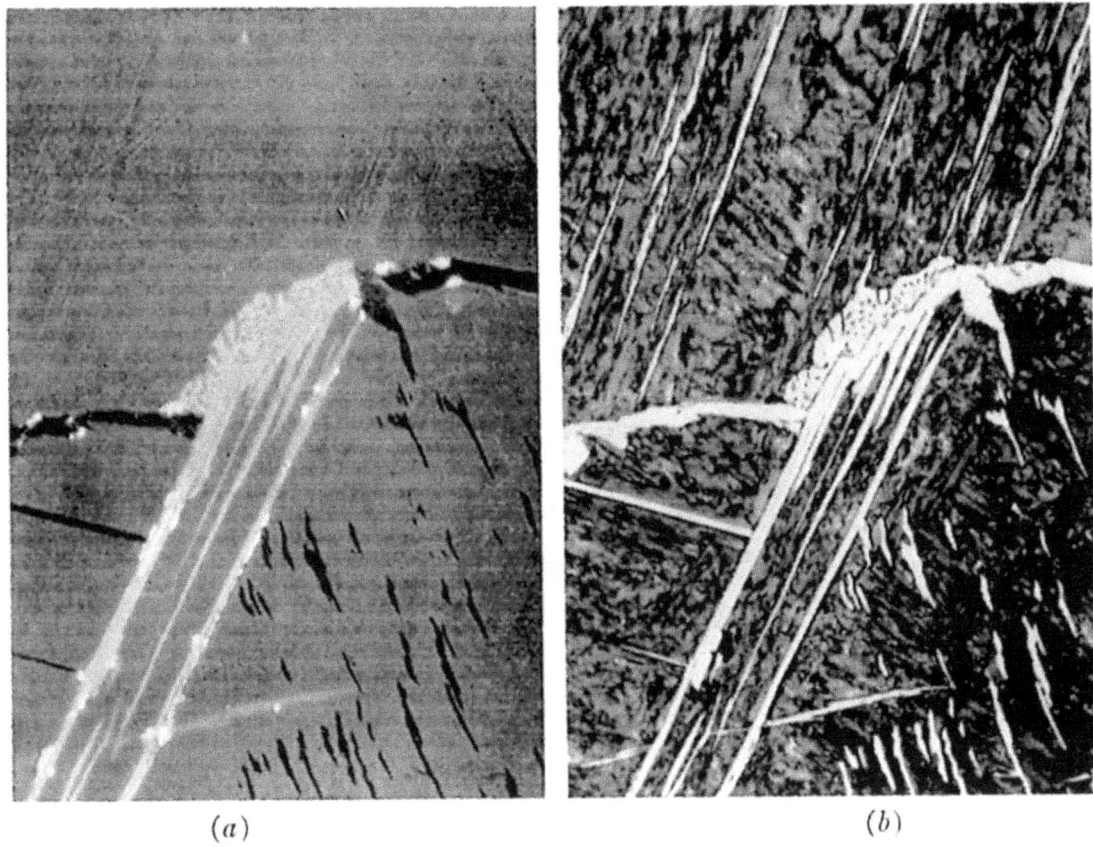

Fig. 9. Hypereutectoid steel with 0.25% Si reacted just above the eutectoid temperature, quenched. One crystal of proeutectoid cementite (appearing white under crossed polars in left picture) has developed acicularly into lower grain of austenite and pearlitically into upper grain. 1200×.

ture are ferrite which can form above the eutectoid temperature due to local enrichment in silicon during the growth of cementite, and Heckel and Paxton recently reached the same conclusion.[22] "Porous cementite" should thus be considered as a special kind of pearlite, the growth of which is totally dependent on the diffusion of the alloying element silicon, whereas the rate of formation of ordinary pearlite depends upon the rate of redistribution of carbon. Ordinary pearlite should thus be classified as para-pearlite and porous cementite as ortho-pearlite. This terminology will be adopted in the present paper.*

Thanks to the formation of ortho-pearlite above the eutectoid

---

* The use of the prefix "ortho" to denote a transformation where an alloying element partitions between the new phases according to the equilibrium conditions and the prefix "para" to denote a transformation where all the resulting phases inherit the alloying content of the present phase is due to Hultgren et al.[18]

212   DECOMPOSITION OF AUSTENITE

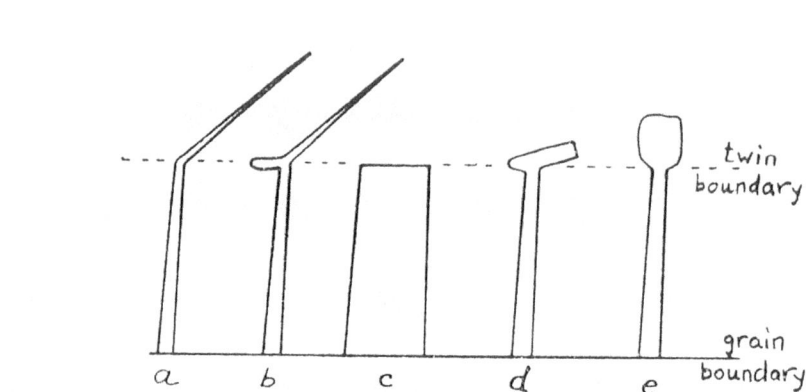

Fig. 10. Effect of austenite twin boundary on the growth of proeutectoid ferrite and cementite.

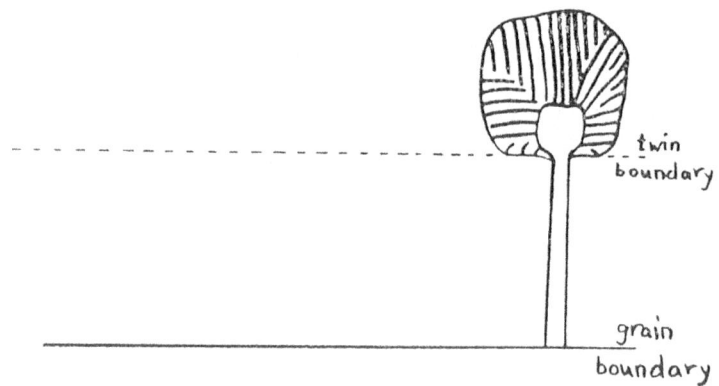

Fig. 11. Pearlite in an austenite twin nucleated by proeutectoid cementite which has developed as a Widmanstätten plate in the adjacent twin.

temperature, silicon- and aluminum-containing steels[15] are very suitable for studying orientation relationships in hypereutectoid steels. Figures 9a and 9b (Fig. 9a taken with polarized light) demonstrate that a single grain of cementite may develop as porous cementite into one grain of austenite and as Widmanstätten plates into the neighboring grain. Cementite thus behaves exactly like ferrite (cf. Fig. 7d). As a matter of fact, there are reasons to call the Widmanstätten structure in the lower grain of austenite in Figure 9 "inverse bainite," thereby emphasizing the close similarity with ordinary bainite but recognizing that Widmanstätten cementite, not ferrite, is the leading phase during the formation of this structure.[23] According to this terminology, Figure 9 demonstrates that a single grain of cementite develops into inverse bainite in one grain of austenite and ortho-pearlite in the other grain.

Further information about the orientation relationships can be obtained by studying the effect of twin boundaries on growing Widmanstätten plates. A number of different effects have been observed, as demonstrated in Figure 10. Again it is surprising to note that straight and presumably coherent interfaces may develop in both twins of austenite. This should be impossible according to the accepted orientation relationships between Widmanstätten ferrite or cementite and the matrix austenite. Figures 10a, 10b, and 10d show cases which have been observed for cementite; Figure 10c is for ferrite.[20] It is believed that the degree of coherency with the upper twin is quite low in Figure 10d and quite high in Figures 10a, 10b, and 10c.

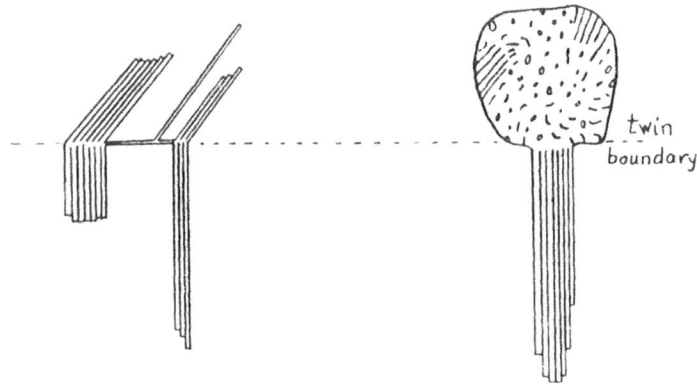

Fig. 12. Development of cementite into Widmanstätten plates (inverse bainite) and porous cementite (ortho-pearlite) at twin boundary.

On the other hand, the Widmanstätten character may be completely lost at the twin boundary and an irregularly shaped head is then formed on the thin Widmanstätten plate (Fig. 10e). This case has been observed for ferrite[20] as well as cementite.[16]

Hultgren and Öhlin[16] allowed the structure in Figure 10e to form in a hypereutectoid steel just above the eutectoid temperature and found that, after cooling below the eutectoid temperature, such a crystal of cementite develops into pearlite in the upper twin, i.e., in the twin to which it does not bear any orientation relationship (Figs. 11 and 13). On isothermal transformation of a silicon-containing hypereutectoid steel above the eutectoid temperature, Hillert[20] found combinations of inverse bainite and orthopearlite (Fig. 12) which were equivalent to the morphologies in Figures 10a, 10b and 10e.

In view of the experimental results described in this section, it appears that the same rules hold for cementite as those discussed for

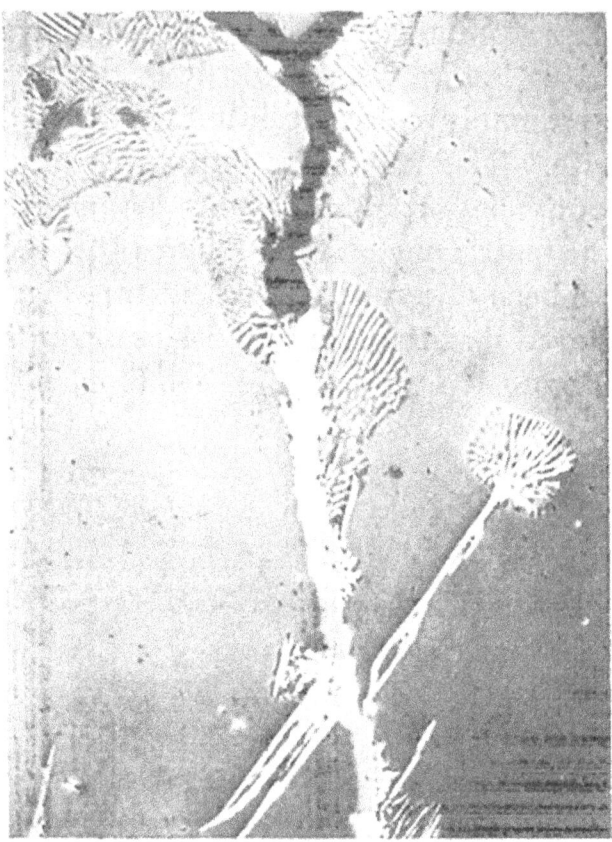

Fig. 13. Hypereutectoid steel slowly cooled to 710°C, quenched after a short holding time. Proeutectoid cementite, pearlite, martensite. A crystal of grain boundary cementite (appearing white under the crossed polars) has developed a Widmanstätten plate growing upwards in the right grain of austenite. Close to the middle of the right hand side of the picture, it has met an austenite twin boundary and developed as pearlite into the next twin (cf. Fig. 11). (Courtesy Hultgren and Öhlin.) 400×.

ferrite in the previous section. Smith's hypothesis should thus be generalized to read as follows:

"The ferrite and cementite constituents of pearlite can have any orientation relationships to the matrix austenite except for those which allow the formation of interfaces which are partially coherent with the matrix austenite."

The lattice orientations of pearlitic ferrite and cementite are thus random with respect to the matrix austenite except for the avoidance of some orientations.

Hultgren and Öhlin in their recent study of pearlite[16] independently reached the conclusion that Smith's hypothesis can be generalized to hold for the cementite constituent of pearlite as well as the ferrite. In order to check the importance of orientation relationships further, they particularly studied the formation of pearlite in a hypereutectoid steel where the individual grains of austenite were first isolated from each other by the formation of a continuous film of proeutectoid cementite. Apparently, in this case the lattice orientation of the ferrite constituent of pearlite cannot be directly related to any of the neighboring grains of austenite, as had been suggested by Smith.[4] Hultgren and Öhlin found that pearlite could form without any difficulty even under these circumstances, a fact that indicates that orientation relationships of the pearlite phases to austenite are of even less significance than suggested by Smith.

III. Nucleation and Growth of Pearlite

A. Edgewise Growth of Pearlite

We have concluded that the same rules hold for ferrite and cementite during the formation of pearlite. The two phases may thus be regarded as equal partners, a result which is particularly satisfactory in view of the theoretical calculations of the edgewise rate of growth of pearlite published in the literature,[23-27] all of which started from

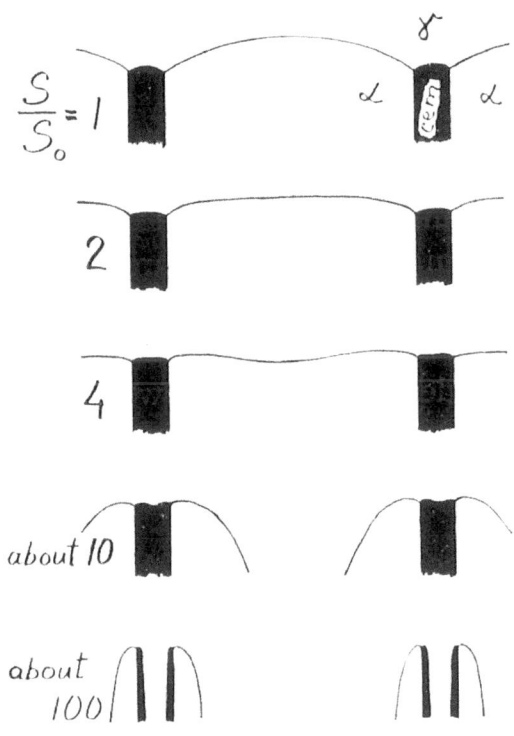

Fig. 14. Calculated profile of pearlite-austenite interface during transformation, as dependent upon the ratio of the actual spacing S to the minimum spacing S_0 which would give zero growth rate. Dark phase is cementite.

216   DECOMPOSITION OF AUSTENITE

this basic assumption. It is simply assumed that the growth rate is determined by the rate of carbon diffusion from the tips of the ferrite lamellae to the tips of the cementite lamellae. The results can be checked experimentally by comparing measured and calculated growth rates and interlamellar spacings. This problem is discussed by Cahn and Hagel.[28] However, the last calculation[23] also yielded some information about the shape of the interface between the growing pearlite and the matrix austenite which may be tested microscopically. Figure 14 shows the results of the calculation for various ratios of S/S_0, where S_0 is the minimum value of the interlamellar spacing S which would give a zero growth rate. The profiles for the first three ratios were actually calculated; the remaining two profiles represent estimated configurations. In view of these diagrams one could expect different profiles depending upon the actual spacing. However, according to a suggestion by Zener, the spacing will automatically adjust itself to a value of $S = 2S_0$ because this will give the highest growth rate. According to the diagram, the profile will then be fairly flat and in a real case one should expect to find an even flatter profile because the calculation uses the crude assumption that all interfacial energies involved are equal, whereas one can expect that a real system will find some way of attaining a low interfacial energy between the two growing phases.

A few experiments have been carried out in order to study the real shape of the profile of growing pearlite.[29-31] Quite surprisingly, a flat profile was not found. Schrader obtained a shape similar to the second from the bottom in Figure 14. Tardif obtained various shapes, and Fisher and Darken found a thin film of ferrite covering the tips of the cementite lamellae. Before discussing these results further, we shall turn to the sidewise mode of growth.

B. Sidewise Growth of Pearlite

Although it has become increasingly clear that a dominating part of the volume of pearlite forms by edgewise growth, it is still accepted that the sidewise mode of growth is essential at an early stage of the development of a pearlite colony[2] (Fig. 1). However, in view of the generalized Smith hypothesis, it is difficult to accept the fact that a lattice orientation of cementite which allows the formation of a platelet (Fig. 1a) will be suitable in a pearlite colony. Hultgren's micro-

FORMATION OF PEARLITE 217

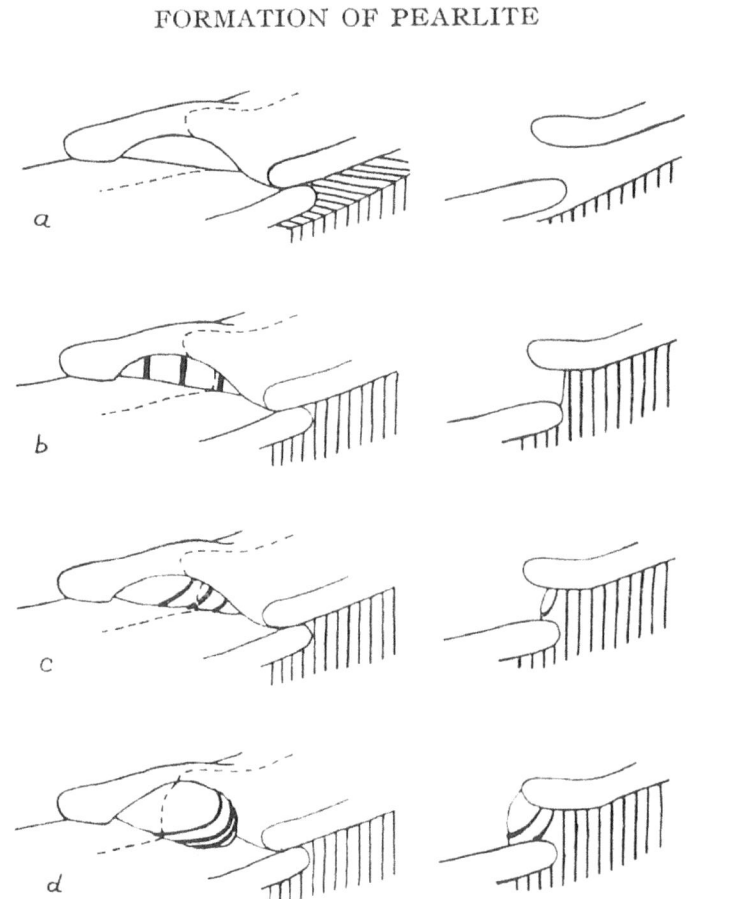

Fig. 15. Mechanism by which pearlite has been observed to grow from one grain of austenite (lower grain), through a hole in the grain boundary network of ferrite and to spread into the adjacent grain without using the repeated-nucleation processes. The four pictures to the left show successive stages of the process. The dotted sections, shown separately to the right, give a false impression of a repeated-nucleation process.

graph (Fig. 13) gives direct proof that a plate will not develop into pearlite unless it happens to cross a twin boundary and thereby *loses* its ability to grow as a plate.

The stages of development from Figures 1b to 1c and from Figures 1c to 1d have now been tested in two ways.[15,32] In the first method a hypoeutectoid steel was allowed to develop an almost complete network of ferrite just above the eutectoid temperature. The growth of pearlite across austenite grain boundaries was then studied by further reacting these specimens at a lower temperature. By means of a method of repeated sectioning, it was observed that such growth always takes place through holes in the ferrite network. These holes were often very small and allowed only a few lamellae to come through. A situation equivalent to Figure 1b or 1c was thus realized in the adjacent grain of austenite. A large number of such cases were studied,

218 DECOMPOSITION OF AUSTENITE

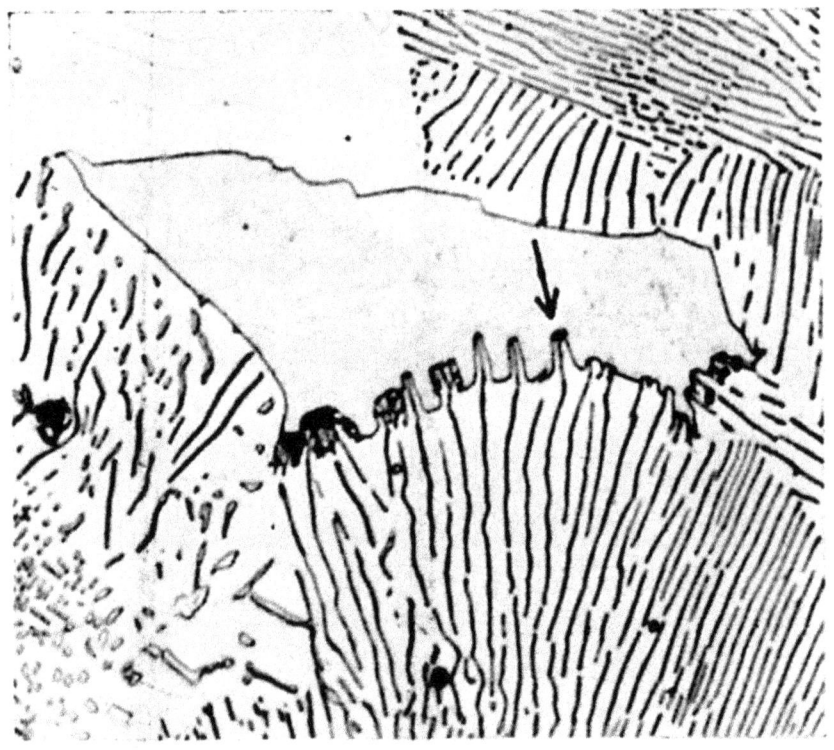

Fig. 16. Hypoeutectoid steel (0.6%C) isothermally reacted for 20 min at 713°C to form pearlite, held for 2 sec at 645°C to allow some further growth of the pearlite, quenched. In the first moment at 645°C the pearlite spacing corresponds approximately to $S/S_0 = 10$ and develops accordingly (cf. Fig. 14). Fine pearlite characteristic of 645°C develops gradually by branching, e.g., at arrow, not by nucleation of cementite in the "bays" between the finger-like outgrowths. 1800×.

but sidewise growth by nucleation of new lamellae was never observed. Instead, the number of lamellae was increased by some mechanism of branching. In a particularly instructive case (Fig. 15), the hole was almost at a right angle to the plane of the lamellae (see Fig. 15b, for instance). In this case the pearlite colony managed to advance through the hole by a change in direction of the old lamellae. By judging any one of the micrographs, one would conclude that sidewise growth had occurred through the hole by repeated nucleation of new lamellae (see right-hand side of Fig. 15d), but the three-dimensional picture obtained by a combination of all the microphotographs reveals the correct sequence of events (see left-hand side of Fig. 15d).

The second method of testing the sidewise mode of growth involves the formation of some coarse pearlite at a high temperature followed by a sudden drop of temperature and continued reaction at a lower level. The spacing of the coarse pearlite will then be many times

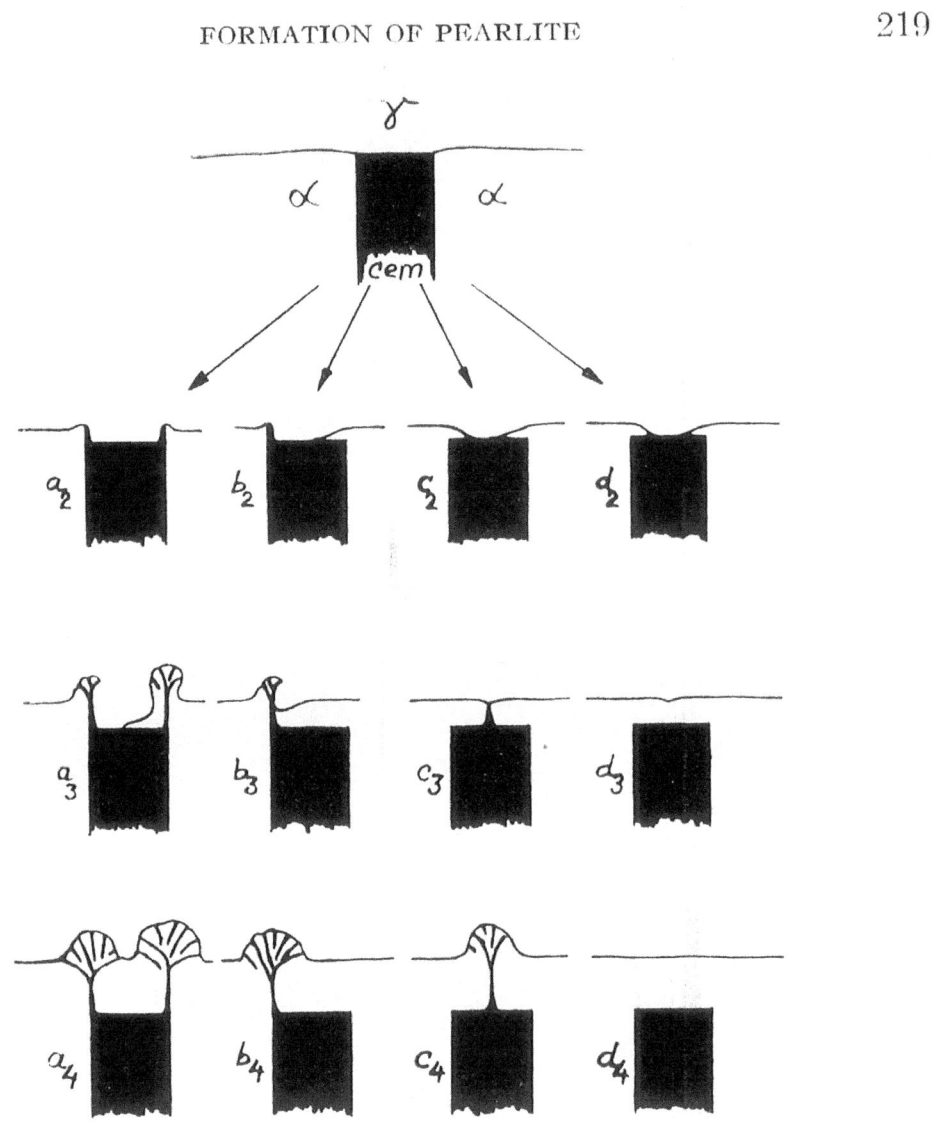

Fig. 17. Different possibilities for the further growth at low temperatures of the very coarse pearlite at the top of the figure. The sequence d_2–d_4 illustrates how cementite can become isolated from the austenite. The sequences a_2–a_4, b_2–b_4, and c_2–c_4 show how fine pearlite can develop.

larger than S_0 of the new temperature level, and one could expect that the coarse pearlite would develop a profile like those in the lower part of Figure 14, provided the calculations are basically correct. This seems to be the case, since a structure resembling that for $S/S_0 \sim 10$ in Figure 14 was obtained in this way (Fig. 16). Here the situation in Figure 1b is realized at each one of the old cementite lamellae. It is interesting to note that the large number of new lamellae that must form at the new temperature before the condition $S = 2S_0$ is fulfilled do not form at the sides of the old lamellae, but at the tip (at the arrow, for instance). In this experiment, also, the mechanism seems to be one of branching rather than repeated nucleation.

220) DECOMPOSITION OF AUSTENITE

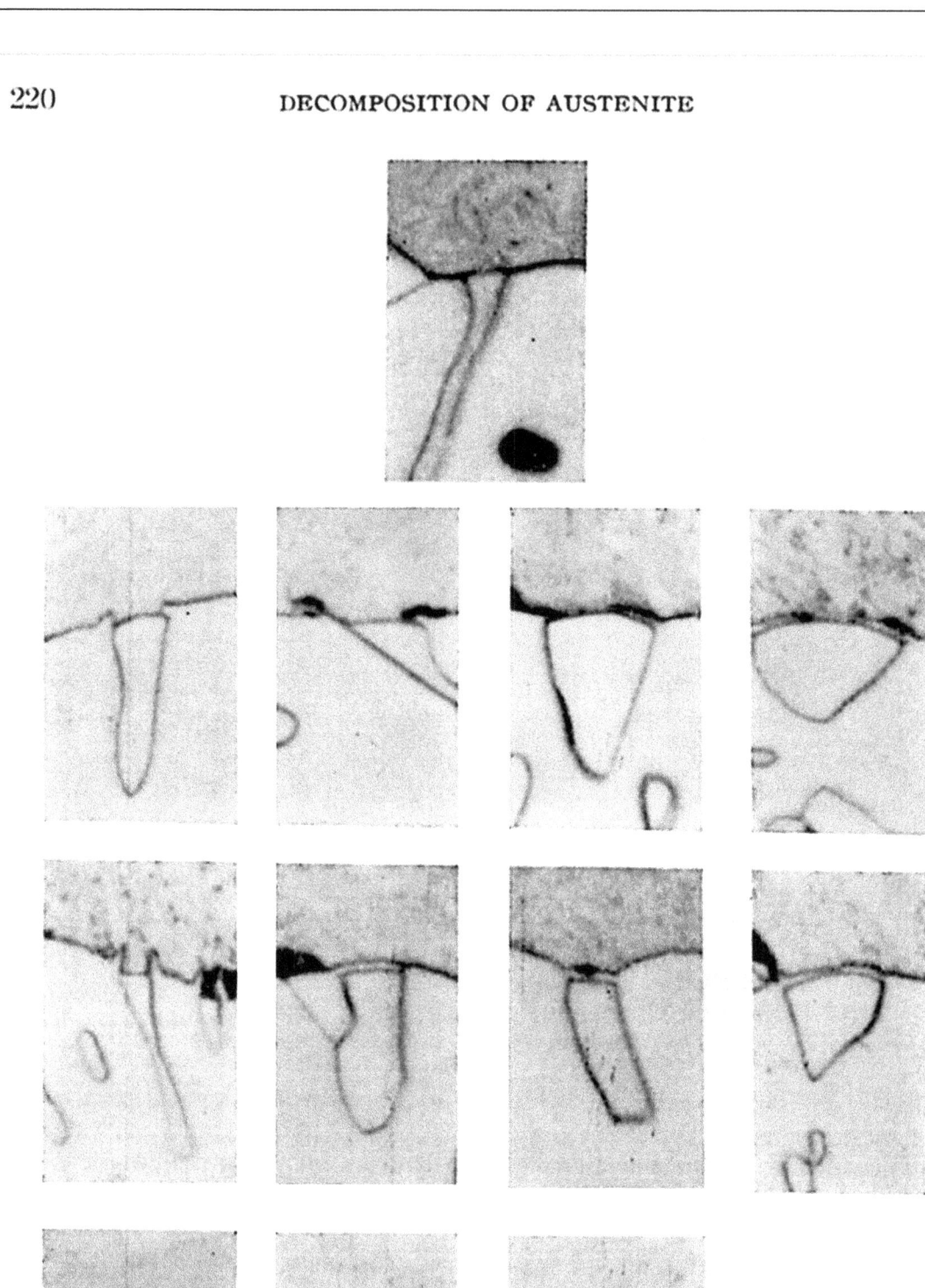

Fig. 18. Details from actual micrographs illustrating the different possibilities shown in Fig. 17. Very coarse pearlite formed just below the eutectoid temperature, some very fine pearlite (dark) formed during quench, martensite. 3000×.

A more drastic drop in temperature will produce an even higher value of S/S_0, and the profile at the bottom of Figure 14 can then be expected (Fig. $17a_2$). This has been found,[32] but other types shown schematically in Figure 17 were also observed (Fig. 18). The reason may be that the calculations are concerned with the steady state condition during growth, and after a temperature change it may take some time before this is again established. In the meantime, the process is very sensitive to local disturbances and may proceed in any direction. All different types of development observed have the common feature that they give no evidence of repeated nucleation. In all cases the appearance of fine pearlite can be explained by a branching mechanism. It is especially interesting to note that fine pearlite often develops at both corners of a coarse cementite lamella (Fig. $17a_2$–a_4). This result further emphasized the previous conclusion that ferrite and cementite are equal partners. What really matters in the development of pearlite is not one of the phases or the other but the lines of intersection between ferrite, cementite, and austenite. The formation of pearlite can be imagined as the result of the movements of those lines.

The results of Schrader, Tardif, and Fisher and Darken (quoted in Section IIA) may now be explained as due to an insufficient rate of cooling.[32] The structure observed by Schrader is similar to Figure 16. The structures obtained by Tardif resemble various types shown in Figure 17, and one of his micrographs actually shows a thick rim of fine pearlite (troostite) formed during the quench, proving that the quench was not efficient. The structure described by Fisher and Darken resembles that in Figures $17d_3$ and d_4.

From the above results it may be concluded that sidewise growth by repeated nucleation is at least a very rare phenomenon. The final answer to the question of whether or not it ever occurs can only be obtained by studying the complete three-dimensional structure of a pearlite colony. Such a study will also answer the question how a pearlite colony will form if the mechanism in Figure 1 is incorrect. These questions will be considered in the next section.

C. Development of a Pearlite Colony

Hillert and Lange[33] carburized an electrolytic iron, precipitated grain boundary cementite just above the eutectoid temperature, and completed the transformation of austenite just below this temperature.

222   DECOMPOSITION OF AUSTENITE

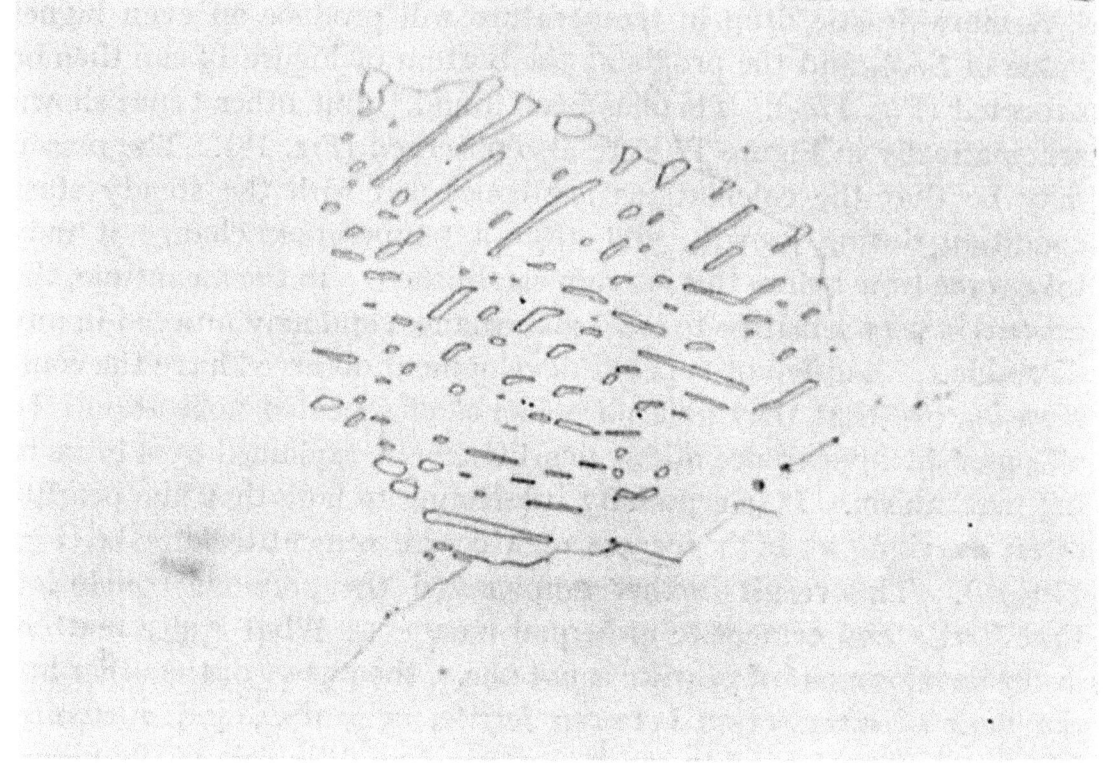

Fig. 19. Pearlite colony formed during slow cooling of carburized electrolytic iron showing strong tendency for abnormal structure. Pearlitic ferrite is part of the left grain of ferrite. 1200×.

Under such conditions, the steel transforms to abnormal structure (i.e., the cementite network is covered by a thick layer of ferrite which often fills the whole grain of austenite). In some large grains, there is a small volume at the center which transforms to pearlite. Such a colony was chosen situated at the intersection between three grains of ferrite far away from the grain boundary network (Fig. 19). The microstructure was repeatedly photographed after successive polishing and etching, a layer of about 1 μ being removed each time. This procedure was repeated 240 times until the colony had disappeared. The collection of micrographs represented a three-dimensional picture of the complete structure of the colony except for a small part which had been polished away before the colony was detected and chosen for this study. The micrographs were transferred to a motion picture film. By running this forward and backward one could move through space and study the true structure. It was thus found that all the cementite lamellae visible in the first micrograph were branches from one single stem of cementite which

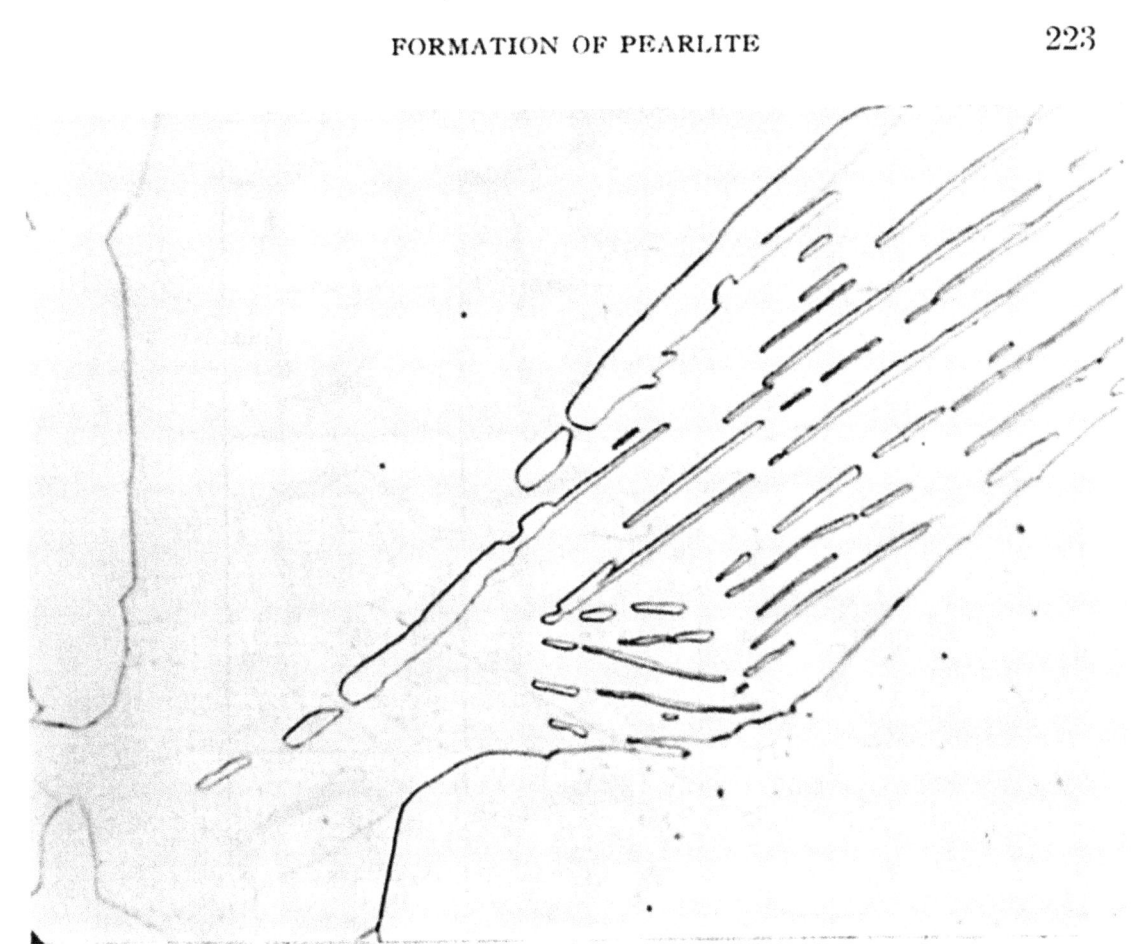

Fig. 20. A lower section through the pearlite colony in Fig. 19, showing an arm of cementite coming from the grain boundary cementite shown at the left-hand side of picture. 1200×.

had grown out from the cementite network. This fact was not apparant from any one micrograph. Figure 20 gives a fairly good view of the stem coming from the network cementite at the left-hand side of the picture and starting to branch just before reaching the center of the picture. In this particular microsection the stem is shown as a series of elongated crystals, but the next few sections revealed that they all joined together forming a continuous stem. The course of transformation of the austenite in this specimen seemed to be the following (cf. Fig. 21).

1. Cementite nucleates at the grain boundaries forming a network, thus displacing the austenite composition toward the hypoeutectoid side in the Fe-C diagram (Figs. 21a, 21b and 21c).

2. Ferrite nucleates at the interface between cementite and austenite (Fig. 21c) and grows along this interface isolating the cementite from the remaining austenite (Fig. 21d).

224   DECOMPOSITION OF AUSTENITE

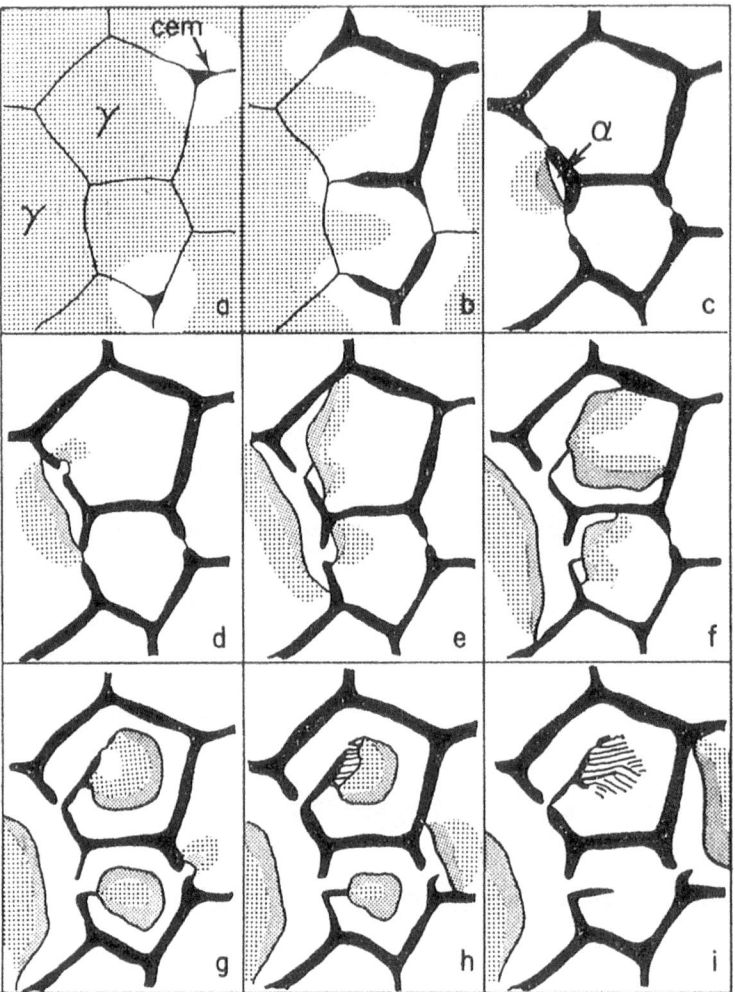

Fig. 21. Successive stages of transformation of hypereutectoid Fe-C alloy. Cementite (black), austenite with increasing carbon content (white, light grey, dark grey), ferrite (white).

3. When growing along the cementite-austenite interface, the ferrite may reach a hole in the network or a sharp edge of the cementite. Here, the ferrite may not be able to isolate the cementite completely from the austenite. An arm of cementite grows out from the network and retains contact with the austenite (Figs. 21d and 21e).

4. During the growth of ferrite, the austenite composition will move back toward the hypereutectoid side. The growth of the arm of cementite will then become more favored and finally the cementite starts to grow along the interface between ferrite and austenite, now isolating the ferrite from the austenite (Figs. 21f and 21g).

5. Owing to the small volume fraction of cementite, this phase has more difficulty than ferrite in forming a complete layer and thus

isolating the other phase. Consequently there is now a good chance that a structure forms which contains alternating units of cementite and ferrite, with both of them in contact with austenite. This may be regarded as the first stage of branching and has occurred in Figure 21g, although it is not clearly visible in the particular section chosen in Figure 21.

6. During further growth, more branching occurs until the spacing is close to the value characteristic of the temperature (Fig. 21h). If the austenite grain size is too small the transformation may be completed before 5 or 6 occurs. No lamellar pearlite is formed in such grains (the lower grain in Fig. 21, for instance).

These results lend further support to the author's conclusion that sidewise growth by repeated nucleation is not normally operative in the formation of pearlite. Instead, a pearlite unit is composed of two interwoven crystals, one ferrite and the other cementite. A pearlite unit should thus be regarded as a bi-crystal. It is interesting that in 1917 Forsman[34] arrived at essentially the same conclusion from a three-dimensional study of pearlite by repeated sectioning.

It is also interesting to compare the results of this three-dimensional study with the results of an earlier theoretical study.[23] In that connection the term "cooperation" was used to describe the interplay between the two phases in a eutectoid or eutectic structure, as distinguished from the "separate" growth of plates of one of the phases. From the present results it appears that the cooperation between ferrite and cementite does not start spontaneously after their nucleation. It would rather seem that the two phases are first "fighting" each other and that cooperation is only gradually developed. As the degree of cooperation increases the structure becomes finer and more regular and, finally, lamellar pearlite with a spacing characteristic of the temperature forms. This may be termed "ideal cooperation" because the two phases now help each other to attain the highest possible growth rate. As soon as this well-organized structure has developed somewhere, it will predominate during the remaining part of the transformation because of its high growth rate.

A recent study of the microstructure of white cast iron indicates that the concept "cooperation" is applicable to that case as well.[35] The cooperation was found not to start immediately after the two phases have been nucleated. Instead the cooperation develops gradually and the resulting structure becomes finer and more regular.

In concluding this section it should be emphasized that lamellar pearlite is not nucleated by either ferrite or cementite but is a result of a gradual process. The development of lamellar pearlite is not guaranteed after the nucleation of one phase or the other, nor after both of them are nucleated. As a consequence, it does not appear worthwhile to inquire about the nucleus of pearlite because it cannot be defined in any simple way. Of course, the concept "active nucleus" as defined in the first paragraph of Section II-B can still be used but it may not be very valuable.

D. Establishment of Cooperation

It has thus been concluded that the so-called "abnormal structure" in hypereutectoid steels, which was discussed in the preceding section, is a result of a low degree of cooperation. Another example is the equivalent structure in hypoeutectoid steels, which was discussed by Hanemann and Schrader[13] when they introduced the concept of "degenerate" pearlite (*entarteter Perlit*). In the new terminology, "degeneracy" or "divorcement" is the result of a very low degree of cooperation.

Grange[36] has described an irregular and rather coarse pearlitic structure which he calls "semipearlite." From his investigation of this structure he obtained the impression "that the lamellar growth is more difficult when the individual volumes of austenite about to transform are very small." His observations thus confirm the results quoted in the preceding section and his "semipearlite" should probably be considered as a structure formed with quite a low degree of cooperation between the two phases, ferrite and cementite.

Another example of structures formed with a low degree of cooperation occurs when the temperature is suddenly decreased during the growth of well-organized pearlite. One may say that the cooperation is destroyed by the temperature drop and it takes some time before it is again established. The coarse structures (coarse in comparison with the very fine lamellar pearlite characteristic of the new temperature) which form in the meantime were discussed in Section IIB and shown in Figures 17 and 18. The film of ferrite formed between austenite and the tip of cementite lamella in Figures 17d and 18d and described by Fisher and Darken is thus quite equivalent to the thick layer of ferrite covering the network cementite in hypereutectoid steels showing so-called "abnormal structure" or divorced pearlite (shown in Fig. 21d, for instance).

A very important question is what factors determine the rate of development of a high degree of cooperation, i.e., the rate of "nucleation" of lamellar pearlite. The rate seems to be quite low in pure Fe-C alloys. This should thus be considered the *normal* behavior although the resulting microstructure containing large quantities of divorced pearlite is unfortunately called "abnormal." The effect of alloying elements on the rate of development of cooperation is not very well known although there are a number of papers actually dealing with this question.[36,37] It was recently suggested[15] that those elements will increase the rate which favors the formation of so-called "porous cementite" or ortho-pearlite above the eutectoid temperature, whereas those elements will decrease the rate which opposes its formation. Silicon and aluminum belong to the former category and nickel to the latter.

The effect of temperature on the rate of development of cooperation is quite uncertain. It is true, of course, that less divorced pearlite forms at lower temperatures. However one should not look at the amount of divorced pearlite, represented by the absolute thickness of the ferrite layer, for instance, but rather compare the thickness with the spacing of lamellar pearlite formed at the same temperature.

The nature of the interface between ferrite and austenite or cementite and austenite will probably have a pronounced effect on the rate of development of cooperation. In view of the theoretical treatment of lamellar growth,[23] there are two important cases. (*a*) Even the high growth rate of well-organized lamellar pearlite will be less than the rate of separate growth of one of the phases, if this phase can form as Widmanstätten plates, directed away from the growth front. When the lattice orientation of a phase allows Widmanstätten plates of a proper direction to form, this phase will thus have no advantage of cooperation. It will refuse to cooperate and instead become the leading phase, a bainitic structure being the result. First, the second phase will eventually form between the plates, but the exact time and morphology of this later stage of the transformation is of little importance for the rate of formation of bainite as well as for its structure. (*b*) If one of the phases ferrite or cementite can form a low energy interface to the matrix austenite at a low angle to the growth front it may form a complete film covering the growth front, thus isolating its partner from the matrix austenite and making good cooperation impossible. These two effects may occur as results of coherency with the matrix austenite. They may be more important for the explanation of the generalized Smith hypothesis that lattice orientations allowing coherency make the formation of pearlite impossible than Smith's original suggestion which, in any event, concerns the case of discontinuous precipitation rather than of pearlite. Smith regarded the low mobility of coherent interfaces and the low diffusivity in them as decisive. In this connection, it should be pointed out that a reaction resembling discontinuous precipitation has been observed in hypereutectoid steels.[20] This structure, rather than pearlite, should be compared with discontinuous precipitation as observed in other systems.

IV. Pearlite-Related Structures

A. Intermediate Structures

If different degrees of coherency can be established between a growing phase and its matrix, as suggested by Smith and indicated by the microscopic study,[20] one may expect to find cases where the formation of pearlite has not been completely inhibited, but sufficient coherency remains to prevent really good cooperation from developing. Such structures have been found and may be regarded as structures intermediate between bainite and pearlite when ferrite is the coherent phase, and between pearlite and inverse bainite when cementite is the coherent phase.

Figure 22b shows a case of pearlite where the orientation relationship between ferrite and the matrix austenite apparently has deviated from the ideal relationship for coherency so much that pearlite may form. Here and there, however, some low degree of coherency has been developed, allowing the formation of platelike outgrowths of ferrite ahead of the pearlite front. Figure 22a shows the same area etched with nital and proves that the pearlitic ferrite and the protruding plates are parts of the same crystal of ferrite.

Figures 23 and 24 show two similar cases, but there the lattice orientation was such that the coherency which is established locally favors flat interfaces approximately parallel to the pearlite front.

In hypereutectoid chromium steels, pearlite may sometimes grow with a very jagged shape at low temperatures. The same phenomenon has recently been found to occur in a low-alloyed hypereutectoid steel[38] and it thus seems that the important factor is the high carbon

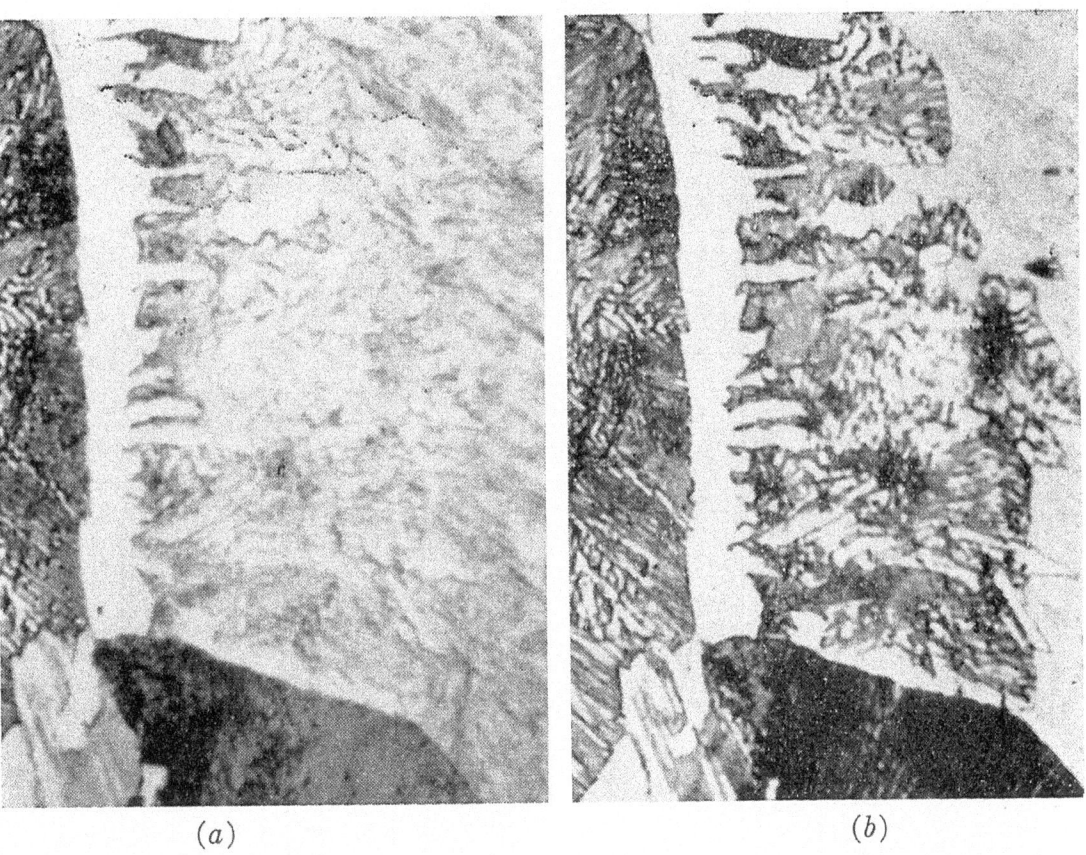

(a)   (b)

Fig. 22. Hypoeutectoid steel (0.6%C) transformed for 15 sec at 640°C. Ferrite, pearlite, martensite. (b) Etched in picral, shows a uniform attack on all grains of ferrite, proeutectoid, and pearlite. (a) Etched in nital, shows that pearlitic ferrite as well as platelike outgrowths of ferrite ahead of the pearlite have the particular lattice orientation causing passivity. 1800×.

content, not the alloying additions. It is thus tempting to suggest that this pearlite is the intermediate structure between ordinary pearlite and inverse bainite with cementite being the leading phase during edgewise growth. Some support for this suggestion is given by the electron microscopic study by Schrader[29] which revealed that there is a plate of cementite at the center of all the jagged irregularities protruding from the pearlite-austenite boundaries. Further support is given by Modin's observation that the same specimen may contain smooth pearlite nodules as well as jagged ones, indicating that the lattice orientation of one of the constituents, presumably cementite, is decisive.

Bainite, Widmanstätten ferrite, and Widmanstätten cementite are known to form with a surface relief effect. This may be a natural consequence of the coherency. Some surface relief effect should then occur for the intermediate structures also, for instance accompanying

TABLE II

Characteristic Features of α-Cementite Structures (Present Theory)

	Bainite	Intermediate structure	Pearlite	Intermediate structure	Inverse bainite
Active nucleus	α	α	α or cementite	Cementite	Cementite
First characteristic unit	Widmanstätten plate of α		Gradual development of α-cementite aggregate		Widmanstätten plate of cementite
Orientation relationship	Ideal α/γ	Imperfect α/γ	Random (except for the avoidance of particular orientations)	Imperfect cementite/γ	Ideal cementite/γ
Effect of twin boundary	Stop of bainitic growth		No effect (except at particular orientations)		Stop of bainitic growth
Effect of grain boundary	Stop of bainitic growth		No effect (except at particular orientations)		Stop of bainitic growth

FORMATION OF PEARLITE 231

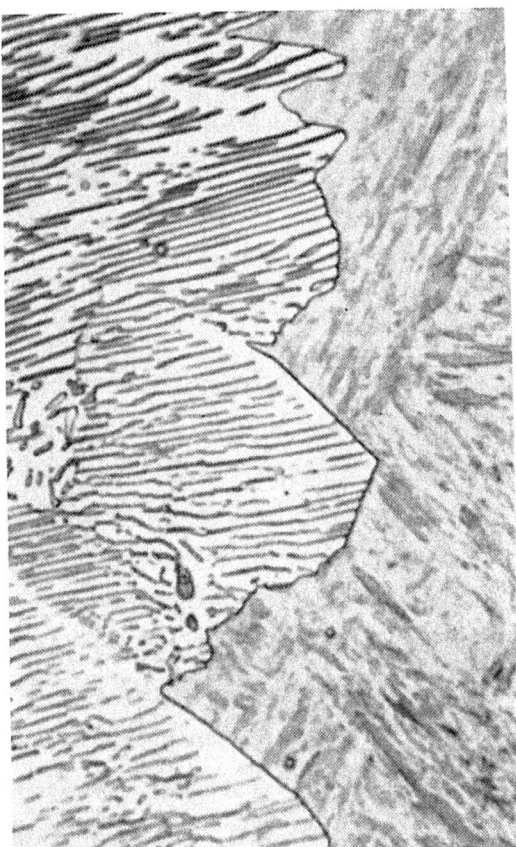
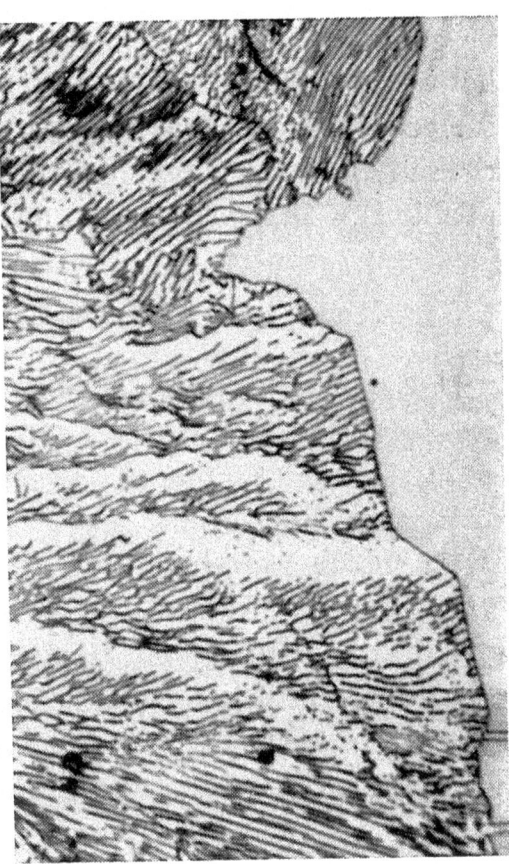

Fig. 23 (left). Hypoeutectoid steel (0.6%C) transformed for 10 min at 693°C. Pearlite with bands of ferrite, possibly caused by localized coherency between pearlitic ferrite and austenite. 1800×.

Fig. 24 (right). Hypoeutectoid steel (0.6%C) transformed for 32 sec at 690°C. Pearlite with bands of ferrite, possibly caused by localized coherency between pearlitic ferrite and austenite. 1800×.

the formation of the platelike outgrowths of ferrite in Figure 22. For hypoeutectoid steels this has not been tested but in the case of jagged pearlite it seems to have been found by Speich and Cohen,[39] although they regarded this structure as bainite and thus attributed the relief effect to the ferrite phase. Instead, it is now suggested that it is due to the cementite phase.

B. Comparison of Two Phase Structures

In view of the generalized Smith hypothesis the various two phase structures may now be rationalized by means of Table II. Table II is quite symmetrical and it thus seems that ferrite and cementite are quite equivalent to each other in the transformation of austenite, although they behave as equal partners only in pearlite. There may

232   DECOMPOSITION OF AUSTENITE

be some difference with regard to the lamellar nature of pearlite, however. This will be discussed in Section V-D.

According to Table II grain and twin boundaries should produce the same effects. Previously, a difference in their effects was reported (Table I). This point will be discussed in sections V-A and V-B.

C. The Nature of Bainite

During the formation of bainite, ferrite is the leading phase at the growing edges but for the sidewise growth, which starts at the sides of the primary plates, the coherency does not favor protruding tips of ferrite. Here, some degree of cooperation can be established between ferrite and cementite although the tendency of ferrite to form a complete film toward the matrix austenite, thus isolating the edges of cementite, will prevent the formation of a well-organized structure. The resulting structure will thus be rather coarse as compared to lamellar pearlite of the same temperature and it can be added to the list of coarse structures given in Section III-D.

It is well known[1] that sub-boundaries frequently develop in the ferrite constituent during growth of pearlite. Small changes in lattice orientation thus occur during growth. If the same phenomenon occurs during the sidewise growth from the primary plates in bainite, there is a possibility that the degree of cooperation may increase and finally lead to lamellar pearlite. It is uncertain whether this actually occurs, but it must be realized that the chances of observing this development will thus be higher, the larger the distance between the primary plates in bainite. At quite high temperatures, the primary plates of ferrite in low-carbon steels are very widely separated and pearlite will eventually fill the intervening spaces. It is possible, but has not been proved, that this pearlite has developed by sidewise growth from the primary plates. The resulting structure, which is ordinarily described as Widmanstätten ferrite and pearlite, should then be very closely related to upper bainite, where the intervening spaces are so narrow that only a coarse, two-phase structure forms between the primary plates. Cases have been observed where most of the intervening space is filled with a coarse, irregular two-phase structure, the remainder being lamellar pearlite[14,23]

In view of this reasoning it may be suggested[23] that bainite is a structure formed in two steps.

FORMATION OF PEARLITE 233

1. The separate formation of Widmanstätten ferrite, growing edgewise.

2. The cooperative growth of ferrite and cementite, growing sidewise from the primary plates. The degree of cooperation is low at the beginning but may increase with time until the intervening spaces are completely transformed.

It is suggested that this description applies to bainite formed at all temperatures, lower and upper bainite as well as the composite structure of Widmanstätten ferrite and lamellar pearlite. Some support for the suggestion that the mechanism of edgewise growth (point 1) is the same for all these structures was recently obtained from a kinetic study of the edgewise growth rate of Widmanstätten ferrite and bainite.[40]

V. Interface and Orientation Effects

A. *Effect of Grain Boundaries on the Growth of Pearlite*

It is very often observed that a pearlite unit forms at a grain boundary and grows into only one of the grains. This phenomenon is explained by the Smith hypothesis. On the other hand, if a growing pearlite unit reaches a new grain of austenite there should be only a very slight chance that the lattice orientation relationships to this grain allows coherency to be established and one should not expect to find any arrest in the growth of the pearlite unit, nor the formation of a new colony as suggested by Jolivet, Mehl, and co-authors. According to the experience of the present author, grain boundaries usually have no effect on the growth of pearlite. Sometimes an effect is observed, however, resulting in a new direction of the lamellae.[41,32] It may be suggested that this is caused by the effect of the high-energy grain boundary on the steady state conditions at the pearlite front.

On one occasion, another effect of a grain boundary was observed.[32] The growing pearlite hit upon a grain boundary (which may have been one of low energy) at a low angle and lost its lamellar character when passing into the new grain of austenite. It seems that the good cooperation was destroyed either due to the small angle of incidence or to some coherency effect. The resulting structure was one of low cooperation and could be added to the list of coarse structures in Section III-D.

234 DECOMPOSITION OF AUSTENITE

B. Effect of Twin Boundaries on the Growth of Pearlite

In principle, one should expect the same effect from a twin boundary as from a grain boundary. However, here the chances are much higher that one of the growing phases will bear an orientation relationship to the new "grain" of austenite, because a large number of parallel twins are often found in the austenite grains. It is thus possible that a crystal of ferrite nucleates at some favored point on a twin boundary bearing an orientation relationship to one of the twins only. It will then develop as Widmanstätten plates into that twin but may form pearlite into the other twin. On further growth the pearlite may hit upon a third twin with a lattice orientation identical to that of the first one. The ferrite constituent will then be unable to continue to cooperate with cementite and instead form Widmanstätten plates. On the other side of the third twin, cooperation may again be established, leading to lamellar pearlite. A case like this has been observed,[32] and the above mechanism may apply to other cases reported in the literature.[29,38] It is particularly interesting to note that Modin has found that in a hypereutectoid steel the effect of twin boundaries became very pronounced at low temperatures. This observation may be related to the increased effect of twin boundaries as preferred sites for nucleation at low temperatures.[42]

C. Effect of Orientation Relationships between Ferrite and Cementite

According to Table II, the lattice orientations of ferrite and cementite should have no influence on the growth of pearlite as long as some critical orientations are avoided which would allow coherency with the matrix austenite. It has already been emphasized, however, that a low interfacial energy between ferrite and cementite which may occur at some particular orientation relations would give a high growth rate. Furthermore, it is natural to expect an orientation relationship between these two phases due to the second phase being nucleated in contact with the first one to form. However, there does not seem to be any experimental proof so far of any orientation relationship between pearlitic ferrite and cementite.

From a theoretical point of view, the interesting question is not really whether any orientation relationship exists (or usually exists) between ferrite and cementite, but rather whether such a relationship is of any importance for the development or growth of pearlite.

FORMATION OF PEARLITE 235

There is some experimental evidence suggesting that such a relationship is relatively unimportant. First, it seems that a change of partners may occur when a growing pearlite unit reaches another crystal of ferrite. The pearlitic cementite may then establish cooperation with the new crystal of ferrite and leave the old ferrite behind.[32] Second, a pearlite unit has sometimes been found to be composed of two quite distinct crystals of ferrite[18,16]

D. Effect of Lattice Orientation of Cementite

Fisher[43] has found that plates of cementite, isolated from a specimen transformed to pearlite, had their c axis directed perpendicular to their flat sides, suggesting that the lamellar structure of pearlite depends upon the crystallographic properties of cementite. This is in conformity with a theoretical study which has shown that a eutectic or eutectoid structure should be rodlike, not lamellar, unless a particular interface of low energy could be formed.[44] It may thus be suggested that such interfaces are formed with the ab plane of cementite. However, it is frequently observed that a pearlite unit contains lamellae of two distinct directions (Figs. 19 and 20, for example) or even curved lamellae. As a consequence, Fisher's result cannot have general validity. Recent results by Modin[45] seem to indicate that, although most of the lamellae develop with the ab planes as sides, there are other cases. In most of these also the b axis but not the a axis seems to lie in the plane of the lamellae. It thus appears that there are several possible crystallographic directions of cementite allowing lamellar growth.

If the formation of pearlite were diffusion controlled and all the interfacial energies involved were independent of crystallographic direction, the geometry alone would determine the direction of lamellar growth. However, since cementite is highly anisotropic,[35] one can expect the system to choose a favorable crystallographic direction which only approximately coincides with the direction favored by the geometry. Furthermore, it is likely that the direction will remain the same even when the geometry changes slightly during the growth. A considerable change in geometry, on the other hand, may result in a new direction and again it is likely that a crystallographically favorable direction is chosen. It may also happen that there is no favorable crystallographic direction within the new range of orienta-

236   DECOMPOSITION OF AUSTENITE

tions demanded by the geometry after some growth has occurred. In such a case, the lamellar nature of pearlite may be lost more or less completely when the growth direction finally changes. Such cases have been observed,[32] and it may be suggested that the interfacial energy is then too high to allow lamellar growth in the new direction. It is not known whether the anisotropic nature of the cementite lattice is solely responsible for this effect or whether the lattice orientation with respect to ferrite or austenite has any influence.

One could perhaps expect that the lattice orientation of cementite is of great importance during the first development of a lamellar structure from the two crystals of ferrite and cementite. It would thus be worthwhile to examine the crystallographic directions shortly after the establishment of a high degree of cooperation.

VI. Conclusions

Past and present information about the formation of pearlite has been discussed and the previously accepted theory of the transformation of pearlite has been tested. This theory was found to be incorrect on essential points. A new attempt is made to rationalize available information and a new theory is obtained which shows that ferrite and cementite are very similar in their behavior with regard to the transformation of austenite. It is realized that all the experimental evidence used in this study may not stand the thorough test which will undoubtedly be undertaken in the years to come. It is hoped, however, that the new theory will provide a suitable basis for future research, which must aim at obtaining reliable information on small details, many of which may have been neglected or completely ignored up to now.

References

1. Hull, F. C., and R. F. Mehl, *Trans ASM*, **30**, 381 (1942).
2. Mehl, R. F., and W. C. Hagel, *Progr. in Metal Phys.*, **6**, 74 (1956).
3. Modin, S., *Jernkontorets Ann.*, **135**, 169 (1951).
4. Smith, C. S., *Trans. ASM*, **45**, 533 (1953).
5. Nicholson, M. E., *J. Metals*, **6**, 1071 (1954).
6. Jolivet, H., *J. Iron Steel Inst. (London)*, **2**, 95 (1939).
7. Benedicks, C., *J. Iron Steel Inst. (London)*, **2**, 352 (1905).
8. Mehl, R. F., and D. W. Smith, *Trans. AIME*, **113**, 203 (1934).
9. Mehl, R. F., and D. W. Smith, *Trans. AIME*, **116**, 330 (1935).

10. Smith, G. V., and R. F. Mehl, *Trans. AIME*, **150,** 211 (1942).
11. Dubé, A., thesis, Carnegie Inst. of Technology, 1948.
12. Aaronson, H. I., thesis, Carnegie Inst. of Technology, 1954.
13. Hanemann, H., and A. Schrader, *Atlas Metallographicus*, Bd 1., Verlag von Gebrüder Borntraeger Berlin 1927–33.
14. Modin, S., *Jernkontorets Ann.*, **142,** 37 (1958).
15. Hillert, M., *Värmländska Bergsm. Fören. Ann.* 1958, p. 29.
16. Hultgren, A., and H. Öhlin, *Jernkontorets Ann.*, **144,** 356 (1960).
17. Hultgren, A., A. Josefsson, E. Kula, and G. Lagerberg, *Jernkontorets Ann.*, **142,** 165 (1958).
18. Hultgren, A., et al., *Kgl. Svenska Vetenskapsakad. Handl.*, Ser. 4, **4,** No. 3 (1953).
19. Cahn, J. W., *J. Metals*, **9,** 140 (1957).
20. Hillert, M., *Jernkontorets Ann.*, to be published.
21. Hultgren, A., and O. Edström, *Jernkontorets Ann.*, **121,** 163 (1937).
22. Heckel, R. W., and H. W. Paxton, *Trans. AIME*, **218,** 799 (1960).
23. Hillert, M., *Jernkontorets Ann.*, **141,** 757 (1957).
24. Brandt, W. H., *J. Appl. Phys.*, **16,** 139 (1945).
25. Zener, C., Trans. *AIME*, **167,** 550 (1946).
26. Scheil, E., *Z. Metallk.*, **37,** 123 (1946).
27. Ljubov, B. J., *J. Tech. Phys.* (*U.S.S.R.*), **20,** 872 (1950).
28. Cahn, J. W., and W. C. Hagel, this volume, p. 131.
29. Schrader, A., *Arch. Eisenhüttenw.*, **25,** 465 (1954).
30. Tardif, H. P., see ref. 2.
31. Fisher, R. M., and L. S. Darken, in Möllenstedt, ed., *Proc. 4th Intern. Conference on Electron Microscopy*, Springer-Verlag, Berlin, 1960, p. 579.
32. Hillert, M., *Jernkontorets Ann.*, to be published.
33. Hillert, M., and N. Lange, Swedish Inst. for Metal Research, Stockholm, Sweden, unpublished research.
34. Forsman, O., *Jernkontorets Ann.*, **102,** 1 (1918).
35. Hillert, M., and H. Steinhäusser, *Jernkontorets Ann.*, **144,** 520 (1960).
36. Grange, R. A., *Trans. ASM*, **35,** 879 (1947).
37. Digges, T. G., *Trans. ASM*, **30,** 287 (1942).
38. Modin, S., *Jernkontorets Ann.*, to be published.
39. Speich, G. R., and M. Cohen, *Trans. AIME*, **218,** 1050 (1960).
40. Hillert, M., to be published.
41. Rathenau, G. W., and G. Baas, *Acta Met.*, **2,** 875 (1954).
42. Lyman, T., and A. R. Troiano, *Trans. ASM*, **37,** 402 (1946).
43. Fisher, R. M., *Am. Soc. Testing Materials*, Spec. Publ. No. 155, 49 (1953).
44. Frank, F. C., and K. E. Puttick, *Acta Met.*, **4,** 206 (1956).
45. Modin, H., *Jernkontorets Ann.*, to be published.

Discussion

Cyril Stanley Smith (*Univ. of Chicago, Illinois*): It is pleasing to see that my comments of some ten years ago on pearlite growth are beginning to take root. Dr. Hillert's proof of the continued connectivity of both ferrite and cementite

238 DECOMPOSITION OF AUSTENITE

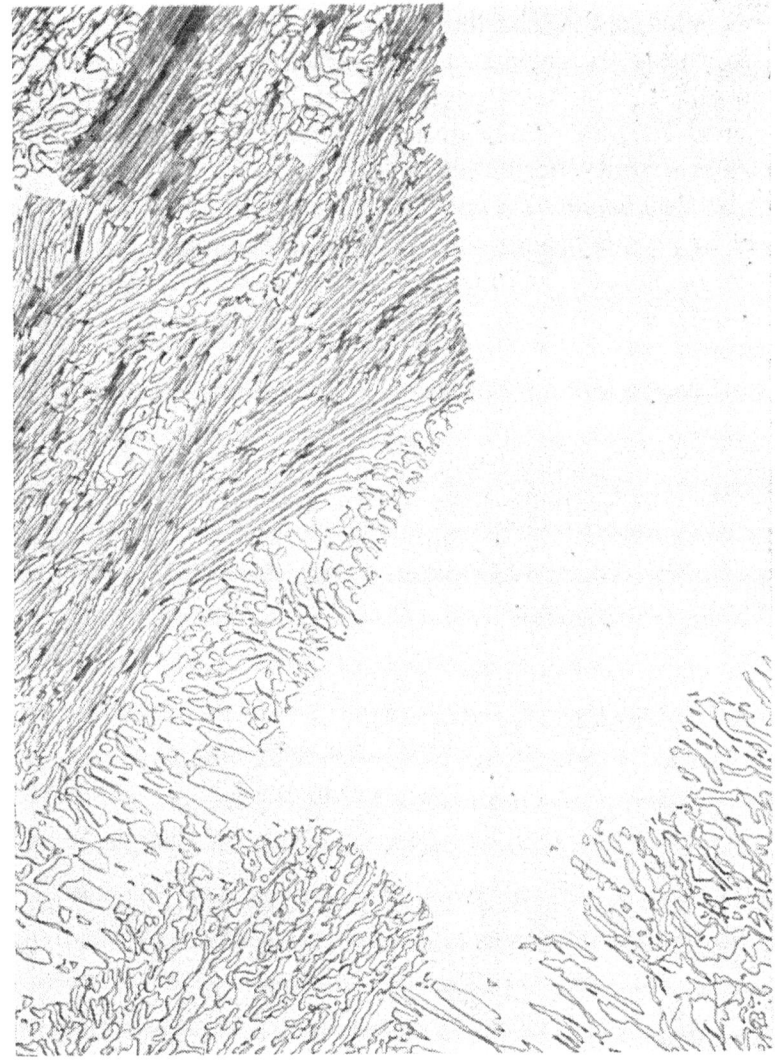

Discussion Fig. 1. Copper alloy with 11.8% aluminum. Furnace cooled from 850 to 550°C, held 12 hr at 550°C, and quenched. Etched with NH_4OH KOH and H_2O_2. 200×.

removes the last shred of support for the sideways-repeated-nucleation theory of pearlite growth.

It is surprising how rarely it has been stated clearly that since pearlite is a duplex structure the nucleus must itself be duplex. Pearlite is not nucleated either by ferrite or by cementite but by both together, and, even more important, there must be the proper kind of interface between the two new phases and the old one. Indeed, the critical aspect of nucleation in general is the formation of an interface of such a structure that it is transitional and so constituted that by its translation alone it produces the transformation.

It is generally true, as Dr. Hillert stated, that the orientation relationships between the ferrite and cementite in a pearlite colony are relatively unimportant, although such epitaxy does exist for it decreases the surface energy of the growing structure. In some systems the orientation relations are more rigidly defined

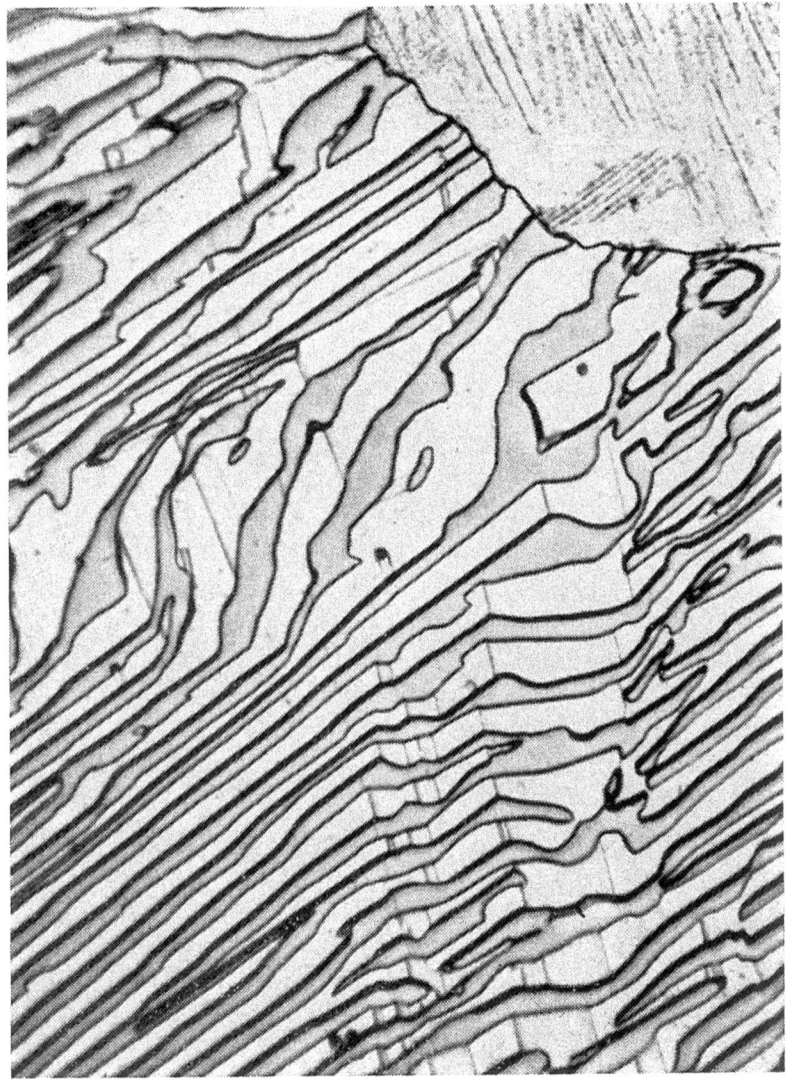

Discussion Fig. 2. Same specimen and etch as Discussion Fig. 1.   1500×.

because of some lucky lattice matching in a particular orientation. Such seems to be the case in the copper-aluminum eutectoid where there is a strong preference for alpha and delta to be in plane-sided plates definitely oriented. However, as Discussion Figure 1 (made by H. Hu at Chicago some years ago) shows, if the gross geometry calls for lateral growth the plates can change directions. They do so rather suddenly and the change is accompanied by the appearance of a twin boundary in order to allow edgewise growth in the desired direction while maintaining a low energy coherent interface. The photomicrograph (Discussion Fig. 2) shows a twin in alpha. Twins have also been observed in delta with similar results.

In the copper-aluminum eutectoid alloy the formation of a kind of bainitic structure sometimes occurs in competition with pearlite (Discussion Fig. 3). It has oriented plates of alpha growing into beta, which becomes enriched in aluminum and eventually nucleates delta more or less independently. Unlike pearlite, with its incoherent interface providing an easy route for diffusion, these

240 DECOMPOSITION OF AUSTENITE

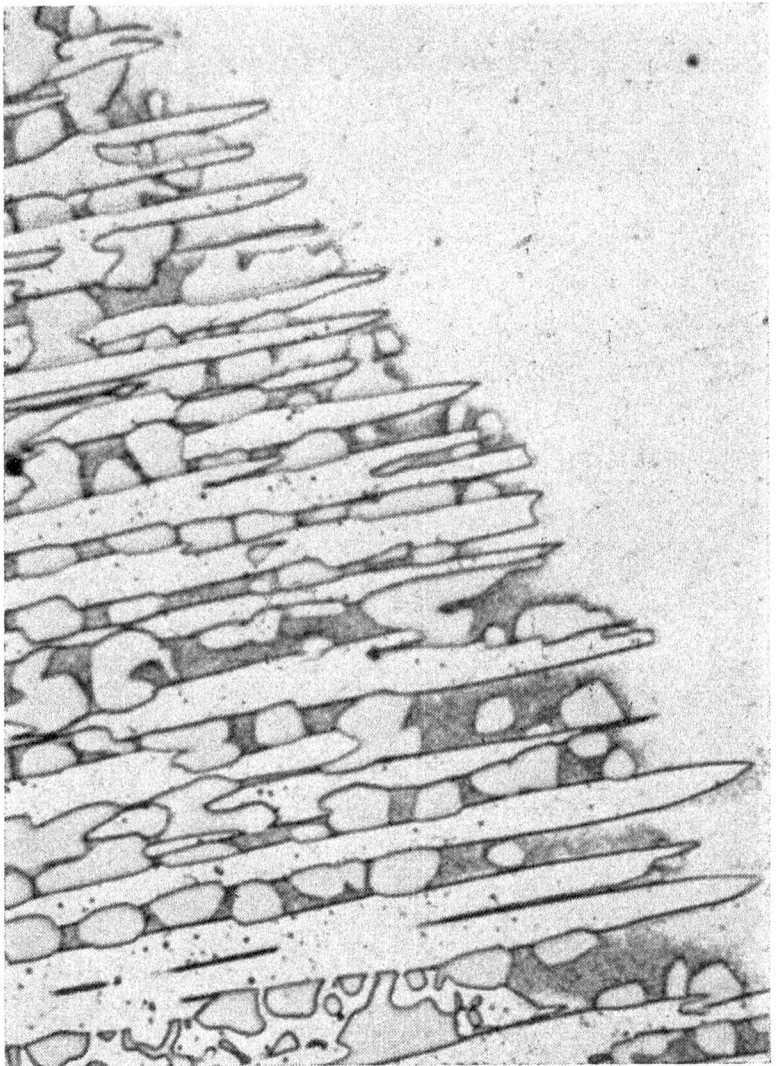

Discussion Fig. 3. Same specimen as Discussion Fig. 2. Etched electrolytically in 1% CrO_3. 1500×.

phases can only grow as the result of volume diffusion, and beta remains behind the advancing front for a considerable time. This is in complete concordance with Hillert's picture of bainite as being essentially martensite growing at a temperature above M_s because of the composition change produced in it by diffusion-controlled precipitation of a second phase.

Regarding Dr. Hillert's observation of the occasional coherent growth of ferrite into two adjacent austenite grains (Fig. 4b), could not this be simply due to the fact that there is always some degree of preferred orientation, resulting partly from chance but more often from the fact that several austenite grains may have had their origin in coherent nucleation from a single ferrite crystal?

Although sidewise nucleation does not occur in typical eutectoid structures, there are cases where it does. One of the most interesting is the two phase grains of alpha-kappa copper-silicon alloys [C. S. Smith, Trans. AIME, 137, 313 (1940)]. There is an extremely low energy interface between the octahedral

Discussion Fig. 4. Copper alloy with 5.15% silicon. Hot rolled at 820°C then annealed 22 hr at 725°C and quenched. Etched electrolytically in 1% CrO$_3$. 100×.

plane of the face-centered cubic alpha and the basal plane of hexagonal kappa, for the atomic arrangement and spacing at the interface plane are indistinguishable from those of a twin. If kappa is precipitated from supersaturated alpha it will nucleate more or less at random on any of the octahedral planes and there will be patches of various orientations interfering with each other (Discussion Fig. 4). If, however, such a duplex alloy is cold worked and then annealed, it recrystallizes into two phase grains looking much like twinned grains of alpha, and as the recrystallized grain grows either into the strained matrix or by the normal process of soap bubblelike grain growth into adjacent smaller grains, new plates of alpha and kappa will be formed alternately to keep the volumetric composition approximately correct. In the analogous case of grain growth in a single phase alloy, twins are nucleated at any of the possible directions with about equal frequency. In the copper-silicon case, however, the new plates in any

242 DECOMPOSITION OF AUSTENITE

Discussion Fig. 5. Same alloy as Discussion Fig. 4, annealed 22 hr, at 750°C, quenched, cold rolled 38% reduction, and annealed 53 hr at 725°C. Etched as Fig. 4. 150×.

given grain are always parallel to the pre-existing ones (Discussion Fig. 5). There must be some quite long range force which determines this. It is not composition gradient, for this would vary more or less at random depending on the position of the kappa and alpha plates in the adjacent grains. Diffusion seems to be mostly along the grain boundaries and the plates do not deviate from planarity near the boundary even though there must be some composition gradient which would produce local changes in the relative volumes of the two phases if the surface energy were not so highly anisotropic. Can there be a very long range strain within the alpha or kappa which is associated with or caused by the coherent interface?

This alloy also illustrates an important fact, that there can be grains which are crystallographically defined but which consist of more than one phase. If there is any orientation at which the interphase interface has a low energy, a two phase structure will always prefer to grow in this related fashion. Polyphase grains may be very common. They are just as crystallographically definite as is a twinned grain, and they are energetically favored in any crystalline matter containing more than one phase. Hu and the writer [*Acta Met.*, **4**, 638 (1956)] discussed alpha-beta brass as an example, while De Vore [*J. Geol.*, **67**, 211 (1959)], who explored geological structures, showed that the incremental surface energy necessarily varies as the volume fraction of a second phase increases, because contact necessarily occurs with boundaries of higher energy than those of which the phase originally nucleated.

Many years ago I pointed out that although there is a definable lowest energy configuration for a boundary of any orientation, in practice a boundary has little chance to reach equilibrium. When moving it always acquires a higher concentration of imperfections, for it is effectively under a negative hydrostatic pressure due to the volume change accompanying either grain growth or recrystallization. The effect of external hydrostatic pressure supports this view, and it may be that even simple mechanical stress enhances or retards grain boundary diffusion.

J. S. Kirkaldy (*McMaster Univ., Hamilton, Ontario*): The experimental material presented here supports the approach to solid-solid morphology subscribed to by B. H. Alexander (Discussion to article by R. F. Mehl and A. Dubé in *Phase Transformations in Solids*, Wiley, New York, 1951), viz., that "much can be learned from the liquid to solid transitions of pure metals and also of eutectic alloys." In the latter regard, the work of W. A. Tiller on polyphase solidification (*Liquid Metals and Solidification*, American Society For Metals, Cleveland, 1958, p. 276) is most pertinent and it is indeed found in that work that much is analogous and pertinent to the present solid-solid morphology problem.

While differing with Tiller in some detail I subscribe to his general viewpoint that the morphological development is essentially a problem in thermodynamics and kinetics and must invariably be so treated. Since ferrite and cementite are far from being structurally and thermodynamically equivalent I question the theoretical value of Hillert's attempt to classify eutectoid morphological variations within a simple framework of symmetry (cf. Table II). Major morphological transitions are primarily determined by changes in supersaturation, and the pearlite → bainite transition must be so regarded. As Hillert has observed, both structures may appear for the same supersaturation within a narrow range due to nucleus orientation differences, but this is thermodynamically of secondary importance.

The eutectoid transitions may be regarded as analogous to the cell to dendrite transition in dilute alloys and the lamellar to rod to globule transitions which occur in eutectic growth as the supersaturation increases. Furthermore, the observation that pearlitic carbide plates have a common nucleus (Fig. 19) supports the view that eutectoid structures are dendritic in nature, thus extending the analogy with eutectic structures. In general we note a parallel tendency with increasing supersaturation toward transformed states containing proportionately increased stores of available energy as surface tension and solute segregation. This is in line with the theoretical considerations in my paper (this volume, p. 39).

244 DECOMPOSITION OF AUSTENITE

Considerable attention is given to C. S. Smith's hypothesis that pearlite grows edgewise in such a way that an incoherent leading interface is maintained. (We will take the usual interpretation here that the mobility is an intrinsic property of the interface which satisfies an Arrhenius relation for which the activation energy is a function of relative orientation.) Since Hillert also subscribes to the principle that nature seeks the configuration of maximum growth rate (after Zener), one is led to suppose that maximum mobility and growth rate are here regarded as synonymous. It will be attempted to demonstrate in the following that this, however, is not always the case.

Let us assume for demonstrative purposes that interfacial reaction and its reaction products act independently to determine the orientation relationship of pearlite growth while diffusion controls the spacing. Let us consider pearlite as nucleated upon a strip of proeutectoid ferrite with one angular degree of freedom for outward direction of growth and let us assume in the most general case that the direction of minimum ferrite-cementite interfacial energy does not coincide with the direction of maximum mobility. We will measure the angle θ from this direction of maximum mobility. Now the rate equation will have the approximate form

$$v = A \phi \exp\{-Q/RT\}$$

where $\exp\{-Q/RT\}$ is proportional to the mobility of the interface and ϕ is the driving force or degree of advancement of the reaction. For maximum mobility the activation energy will be a minimum so let us represent it by the quadratic form $Q = B + \alpha\theta^2$. The degree of advancement ϕ can be represented by the linear form $\phi = C(1 + \beta\theta)$ provided that the directions of maximum mobility and maximum α-Fe$_3$C coherency do not lie too far apart. The rate now has the form

$$v = D(1 + \beta\theta) \exp\{-\alpha\theta^2\}$$

and for maximum rate

$$\frac{\theta\beta}{2\alpha} > 0$$

Thus the maximum growth rate corresponds to an angle different than that for maximum mobility ($\theta = 0$).

I have maintained, in any case, that there is no thermodynamic justification for a choice of maximum growth rate and have recommended the alternative optimal rate as a solution of

$$v = \phi(v)/\phi'(v)$$

This gives the optimal condition, $\theta = 0$, which corresponds to maximum mobility (as postulated by Smith). This is also the state of highest dynamic efficiency since more available energy is conserved as interfacial tension.

M. Hillert: In my paper I have tried to show that the crystallographic aspects, particularly orientation relationships, are far less important in the case of pearlite formation than previously believed. However, as discussed in Sections V-C and

FORMATION OF PEARLITE

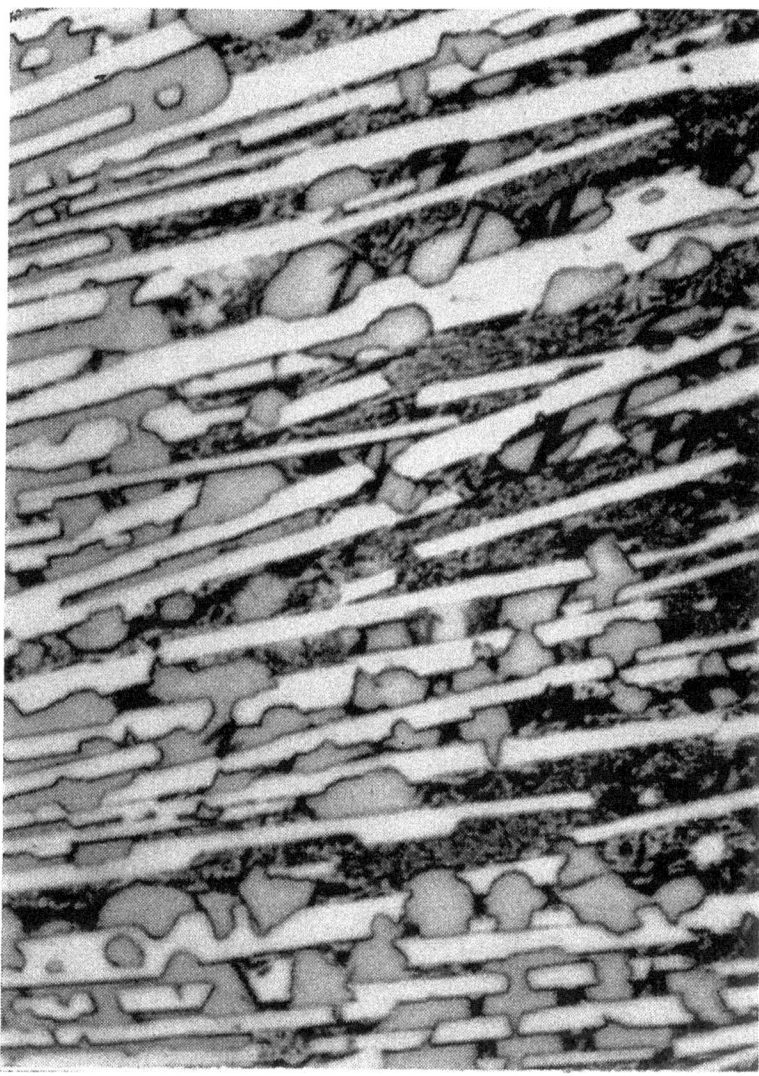

Discussion Fig. 6. Hypereutectic cast iron with 5% Mn and 0.5% Cr. Plates of cementite (white), irregularly shaped austenite (light grey) with some martensite (black), very fine eutectic structure formed on quenching (dark grey). 1800×.

V-D, there may be some influence with regard to the lamellar nature of pearlite. Dr. Smith discusses another eutectoid system, Cu-Al, where this influence seems to be somewhat more pronounced and a precipitation system, Cu-Si, where it is very pronounced.

It is certainly very interesting that the two kinds of eutectoid structure which can form in the Fe-C system, pearlite and bainite, also occur in many other eutectoid systems, e.g., Cu-Al as pointed out by Dr. Smith. When one looks for an explanation of this phenomenon, it may be wise to bear in mind that the same dualism exists in some eutectic systems as well. For example, Discussion Figure 6 shows a bainite-like structure formed in a white cast iron. The similarity with the bainite-like structure in Cu-Al presented by Dr. Smith in Discussion Figure 3 is striking. It thus appears likely that the important feature of the interface between the leading phase and the matrix is not primarily its

246 DECOMPOSITION OF AUSTENITE

coherent nature, nor the rate of interface diffusion. Instead, it has been suggested that the factor of primary importance is an anisotropic growth characteristic of the leading phase caused, for instance, by a low interfacial energy for a particular crystallographic plane (ref. 34 in my paper).

The morphologies represented by Figures 4a, 4b, and 4c in my paper are only expected when the crystal of ferrite is able to form partially coherent interfaces with both grains of austenite. Only a statistical study of the occurrence of these morphologies, combined with an assumption that neighboring grains of austenite are randomly oriented with respect to each other, can reveal what deviation in angle from the ideal orientation relationship is tolerated before coherency becomes impossible. Such a detailed study has not yet been undertaken, but it is my impression that these morphologies are much more frequent than usually expected. Dr. Smith's suggestion that neighboring grains of austenite are related to each other may be valid under certain experimental conditions and it appears essential that the statistical study is carried out under special precautions.

Dr. Smith raises a very important question concerning the possible difference between a moving interface and an interface at equilibrium. In this connection it may be pointed out that our knowledge regarding the magnitude of interfacial energies is mainly based on studies of interfaces at equilibrium but often applied to freshly formed or moving interfaces in attempts to test ideas concerning the growth mechanism of phase transformations. As an example, Zener (ref. 25 in my paper) derived an equation allowing him to predict the pearlite spacing but found that he had to use a very high energy of the ferrite-cementite interface in order to account for experimental data on the pearlite spacing. Should this result be taken as an indication of Zener's growth theory being in error or is it possible that the energy values of the interfaces involved in a phase transformation are in fact much higher than those we are used to finding when studying interfaces at equilibrium? It appears possible that the accumulation of lattice imperfections at a moving interface results in high values for this interface and the segregation of impurities to an equilibrium interface results in low values in that case.

Dr. Kirkaldy raises a number of interesting questions which need more experimental work to be settled. However, I should like to define my own standpoint in a number of cases because it seems to differ from Dr. Kirkaldy's.

I consider Table II in my paper as a rationalization of experimental observations. The fact that it displays a remarkable symmetry, to me indicates that the structural features of the individual phases, ferrite and cementite, are relatively unimportant with regard to the eutectoid transformation of austenite. I consider this as the normal behavior of a eutectoid reaction and believe that the structural features of a phase must be quite distinct in order to affect a eutectoid reaction considerably. Dr. Smith in this discussion draws attention to two cases where this effect is more evident than in the Fe-C system.

It should not be inferred from Table II that supersaturation is less important than the structural features listed. On the contrary, bainite has been placed to the left of pearlite and inverse bainite to the right in order to indicate that bainite is formed preferably to the left of the eutectoid composition in the Fe-C diagram and inverse bainite to the right. In spite of this fact, one can observe the simultaneous formation of two different morphologies within a wide range of

FORMATION OF PEARLITE 247

supersaturations. As an example, pearlite and bainite can both form in a hypoeutectoid steel at temperatures ranging from the eutectoid far down into the bainitic region. The normal behavior seems to be that a ferrite nucleus develops bainitically into one grain of austenite and, after the nucleation of cementite, as pearlite into the adjacent grain of austenite. As a consequence, I believe that it is incorrect to talk about a morphological transition pearlite → bainite caused by a change in supersaturation, although it is correct that the relative amounts of the two structures in a fully transformed specimen are sensitive to changes in supersaturation.

With regard to Dr. Kirkaldy's comparison of the transition pearlite → bainite to the lamellar → rod transition in eutectic growth which is caused by a radical decrease in the amount of one of the solid phases, I should like to draw attention to the fact that a similar lamellar → rod transition occurs in pearlite when the carbon content is decreased. The rodlike pearlite thus formed is not related to bainite any more than lamellar pearlite.

As pointed out in my paper, I do not subscribe to Smith's idea that the reason why pearlite grows with an incoherent interface is the high mobility of such an interface. In his theoretical calculation in the discussion, Dr. Kirkaldy considers the orientation dependency of the mobility of the pearlite-austenite interface. It should be pointed out that this calculation is based on the assumption that the growth rate is limited by the mobility of the interface. As far as I know there are no experimental indications that this would be the case in the Fe-C system under discussion. The experimental growth rate of pearlite is *not* lower than predicted from the rate of carbon diffusion. Furthermore, it is sometimes found that Widmanstätten-like protrusions develop from growing pearlite. Apparently, they are able to grow even *faster* than pearlite in spite of the coherent nature of their interfaces. As a consequence, it seems safe to conclude that at least the growth rate of pearlite is not limited by the interface mobility. For Widmanstätten ferrite and bainite, on the other hand, a theoretical analysis of growth rate data indicates that the interface mobility is more important (ref. 39 in my paper).

I quite agree with Dr. Kirkaldy that there is no thermodynamic justification for Zener's maximum growth rate criterion but there are kinetic arguments in favor of it. A thermodynamically justified criterion would be much preferred, of course, and Dr. Kirkaldy's contributions in this direction are most welcome.

9 Introduction to "On the Theory of Normal and Abnormal Grain Growth"

published in *Acta metallurgica* (1965)

Rather early in the 19th century came the realization that the size and crystallographic orientation of the grains in a polycrystalline material can have a pronounced effect on many materials properties, including also the production properties. A very striking effect was the discovery made by British metallurgists in the middle of the century of the flow stress dependence on the mean grain size in steels and also in other metallic materials. These observations led to the famous Hall-Petch equation (1951–1953). These observations prompted a rather large research activity into the formation and development of the grain structure in metals and alloys and the distinction between normal and abnormal grain growth was firmly established. The driving force for grain growth was accepted to be due to a reduction in the grain boundary area, and thus the total energy of the material and the mechanism behind grain growth stagnation was disclosed. Attempts to treat the grain growth process theoretically were made at this time, even though these first attempts were on a quantitative or semi-qualitative level.

The first quantitative theory of normal grain growth was made by P. Feltham. By assuming that a quasistationary grain size distribution is approached during grain growth and that this distribution was lognormal, he succeeded in developing a deterministic growth law which he solved to give the rate of growth of the mean grain size.

Mats Hillert submitted his investigation on normal and abnormal grain growth to Acta Metallurgica 19th June 1964. It is tempting to guess that two circumstances triggered Mats Hillert's work on these two closely connected subjects: an attempt to apply the Lifshitz and Slyozov's theoretical treatment of particle coarsening that had been recently published, and an attempt to shed more light on grain growth stagnation and abnormal grain growth. In addition, he was not completely satisfied with Feltham's growth equation. The first subject can possibly be looked upon as a purely theoretical undertaking, while the second one had strong ties over to applied physical metallurgy and was also of great industrial importance.

Lifshitz and Slyozov's treatment was a mean field treatment which in the coarsening reaction implies that the volume fraction of the coarsening phase approached zero. The intention to use the same approximation for normal grain growth is particularly daring, because in this case the fictious volume fraction is exactly equal to 1. In spite of this fact Mats Hillert made this assumption even though he did not state it explicitly. He introduced a critical grain of size $\overline{R_{cr}}$ with which all the grains in the material interact in the same way. He then derived a growth law which contained a proportionality constant which he, in the two-dimensional case, determined in an exact way by the use of von Neuman-Mullins law and argued that it is roughly twice as large in three dimensions. His growth equation predicted a constant negative growth rate for very small grains, which is in accordance with von Neumann-law while Feltham's equation predicted infinitely high negative values. Hillert did not realize until much later that Feltham solved his equation in an incorrect way.

Two important results emerged from Hillert's theoretical treatment of normal grain growth: 1) when the quasistationary state is reached the maximum grain size $\overline{D_{max}}$ in a section through a three-dimensional system is about $2\overline{D_s}$ where $\overline{D_s}$ is the mean apparent grain size in the section, and 2) the grain size distribution is given by an analytical expression; it is rather peaked and skewed in the right-handed direction. This distribution is now generally referred to as the Hillert distribution. Very few comparisons between experiments and Hillerts theory have been made. The simplest test would be to compare the grain size distribution observed experimentally with the Hillert distribution. In cases where comparison can be made, there is a clear difference between the two distributions. This discrepancy is most likely due to factors effecting grain growth that are not included in Hillerts treatment. It is likely that Hillert himself realized this fact and thus did not carry out such a comparison.

Following the introduction of computer simulations (around 1980), virtual experiments became possible and the simulation of grain growth by different techniques (*e.g.* Monte Carlo, Surface Evolver) became very popular. It now appears that in the idealized case (with uniform boundaries), the distributions observed are rather different from the Hillert's distribution. There have been quite a lot of speculations concerning the reason for this difference and on what point Hillert's theory of normal grain growth breaks down. Most likely it is related to the mean field assumption. In fact a grain size correlation effect has been observed in two (and one) dimensional systems (with uniform boundaries) that is fairly small but still has a pronounced effect both on the grain size distribution and also on the rate of grain growth. Hillert's assumption that grain growth could, in a similar way as coarsening (LSW-theory), be described as a mean field process was nearly correct, but the slight difference from mean field that existed in the system after quasistationary conditions were reached had a rather large effect on the process. Hillert's assumption was thus a very ingenious one and it is extremely difficult to see how it can be refined in a simple way.

After having finished his theoretical treatment of normal grain growth, Hillert turned to the effect of particles on grain growth and the important phenomenon of grain growth stagnation and abnormal grain growth. He introduced the effect of particles into his growth equation and derived an expression for the limiting grain size. However, he was unable to predict the approach to stagnation and the development of abnormal grain growth. This was probably not a surprise because these processes are not steady state processes. He thus left his deterministic growth law and instead introduced a new and very elegant approach: the defect model of grain growth. He first applied this model to normal grain growth in two dimensions and arrived at the conceptually important statement that normal grain growth could be considered as the shrinking of grains with less than 6 neighbours rather than the growth of grains with more than 6 neighbours. When the effect of particles is introduced into the defect model and combined with results obtained with the deterministic growth equation, he discovered the presence of two limiting grain sizes, one when normal grain growth is impossible and one when all grain growth, including abnormal grain growth, is impossible. The development of abnormal grains is of particular great industrial importance and Hillert discussed at length the conditions that must be satisfied for such a process to occur. This is obviously a problem that fascinated Mats Hillert, because during the following years he made several more penetrating studies on it.

Even though this publication is basically a theoretical one, in particular the first part of it, it is presented in such a lucid manner that it is fairly easy to follow even for people with a limited experience of theoretical thinking. In fact, Hillert included roughly the same presentation as lecture notes to his students at the same time as the Acta Metallurgica paper appeared. He also communicated at the same time the same ideas in approximately the same way to his Scandinavian colleagues in a publication (in Swedish) in the journal *Vuoriteollisuus-Berghanteringen* NR 2/1965.

Mats Hillert's publication "On the theory of Normal and Abnormal Grain Growth" is one of the most important publication on these topics. It is by far the most cited publication of all of Hillert's publications. From 1965 until 2005 it has been cited more than 830 times (Source: Sciences Citation Index). The paper initiated in the field of grain growth and was followed by a large number of papers by other authors. A reasonable comprehensive review of the field of grain growth was published by Mullins [1].

Nils RYUM and Ola HUNDERI

References

[1] W.W. Mullins, *Acta Mater.* 46 (1998) 6219–6229.

ON THE THEORY OF NORMAL AND ABNORMAL GRAIN GROWTH*

M. HILLERT†

A growth equation for individual grains in single-phase materials is suggested. It is used to calculate a rate equation for normal grain growth and the size distribution in the material. It predicts a maximum size of twice the average size. The theory is modified to take into account the effect of second-phase particles.

In an alternative treatment the array of grains is described in terms of a kind of defects introduced into a perfect array. The defects move through the array during grain growth. The rate of grain growth is calculated from the number of defects and their mobility. The defect concentration is predicted by comparing the two treatments. The defect-model predicts two grain size limits due to second-phase particles. Normal grain growth takes place below the lower limit. Abnormal grain growth can take place between the two limits if the material contains at least one very large grain. No grain growth can take place above the higher limit.

Several possible mechanisms for the development of abnormal grain growth are examined. An explanation is offered for the observation that most of the well-known cases occur as the second-phase is dissolving.

SUR LA THEORIE DES CROISSANCES GRANULAIRES NORMALE ET ANORMALE

L'auteur propose une équation rendant compte de la croissance des grains individuels dans un solide monophasé. Cette équation est utilisée pour calculer la vitesse de croissance normale et la distribution des tailles des grains dans la matière. Elle prédit une taille maximum correspondant à 2 fois la taille moyenne. La théorie a été modifiée pour tenir compte de la présence de particules d'une seconde phase.

Par ailleurs, la morphologie des grains est décrite en fonction d'un type de défauts introduits dans un réseau parfait. Ces défauts se déplacent dans le réseau pendant la croissance granulaire. La vitesse de croissance est calculée à partir du nombre de défauts et de leur mobilité. En comparant les deux modes de calcul, on peut prévoir la concentration en défauts: un modèle basé sur la distribution des défauts conduit à prévoir deux tailles de grains limites lorsqu'il reste des particules d'une seconde phase. La croissance granulaire normale se produit donc sous la taille limite inférieure. La croissance anormale se développe lorsque la taille est située entre les deux limites précitées à condition que le solide contienne au moins un grain très grand.

Plusieurs mécanismes possibles pour la croissance granulaire anormale sont examinés. Une explication pourrait être trouvée dans le fait que la plupart des cas de croissance anormale se produisent lorsque la seconde phase est en voie de dissolution.

ZUR THEORIE DES NORMALEN UND DES ANOMALEN KORNWACHSTUMS

Es wird eine Wachstumsgleichung für individuelle Körner in einphasigen Stoffen vorgeschlagen. Sie wird benutzt zur Berechnung einer Beziehung für normales Kornwachstum und für die Größenverteilung in dem Material. Sie sagt eine Maximalgröße von zweimal der durchschnittlichen Größe voraus. Die Theorie wird modifiziert, um dem Einfluß von Teilchen einer zweiten Phase Rechnung zu tragen.

In einer zweiten Behandlung wird die Anordnung von Körnern als eine Art von eingefügten Fehlern in einer idealen Anordnung beschrieben. Während des Kornwachstums wandern die Fehler durch die Anordnung. Die Geschwindigkeit des Kornwachstums wird berechnet aus der Zahl der Fehler und ihrer Beweglichkeit. Die Fehlerkonzentration wird durch einen Vergleich der beiden Verfahren vorausgesagt. Das Fehler-Modell sagt zwei Korngrößengrenzen auf Grund von Teilchen einer zweiten Phase voraus. Normales Kornwachstum findet unterhalb der unteren Grenze statt. Anomales Kornwachstum kann zwischen den beiden Grenzen stattfinden, wenn das Material wenigstens ein sehr großes Korn enthält. Oberhalb der oberen Grenze können die Körner nicht wachsen.

Für die Entstehung des anomalen Kornwachstums werden verschiedene in Frage kommende Mechanismen untersucht. Es wird eine Erklärung für die Beobachtung gegeben, daß die meisten wohlbekannten Fälle bei der Auflösung der zweiten Phase auftreten.

1. INTRODUCTION

By grain growth one usually understands the increase of the grain size in a single-phase material or of the matrix grain size in a material with second-phase particles. The sum of the individual grain sizes is constant and the increase in average grain size is thus connected with a disappearance of some of the grains, usually the smaller ones. In practice, one distinguishes between "normal" or "continuous" grain growth and "abnormal" or "discontinuous" grain growth. During normal grain growth, the size of the individual grains are relatively uniform. During abnormal grain growth, on the other hand, the differences in individual sizes increase by some of the grains growing rapidly. When they have consumed all the other grains, the remaining grains may again be of a relatively uniform size.

The theory of normal grain growth is based on the grain boundary interfacial free energy being the driving force. Several authors[1,2,3] have been able to deduce a parabolic growth law for the average grain

---

* Received June 19, 1964.
† Royal Institute of Technology, Stockholm 70, Sweden.

size, $\bar{R}^2 - R_0^2 = A\sigma t$, using slightly different arguments, a common feature being that a fixed distribution of shapes and relative sizes is assumed. C. S. Smith[4] suggested that there is a natural tendency toward such a fixed distribution. Apart from his discussion of the general topological rules governing the shapes of grains, there seems to be no theoretical discussion of how this distribution might look or whether there is a tendency toward such a distribution. Experimental data have often been interpreted in terms of a lognormal distribution. Accepting this distribution function, Feltham[5] was able to estimate the size of the rate constant A. The main interest of the experimenters has been focussed on the temperature dependence of the rate constant and on the time exponent in the rate law when written as R proportional $t^{1/2}$. Experimentally one has usually found exponent values appreciably less than 1/2.

The particle size of second-phase particles in a matrix may also increase by the larger particles growing at the expense of the smaller ones. This kind of grain growth is often called coalescense and is closely related to the first kind of grain growth, the main driving force being interfacial free energy in both cases. In the case of coalescence, Lifshitz and Slyozov[6] have recently been able to prove that there actually is a tendency toward a fixed distribution of relative sizes. They have calculated the distribution function in detail and also the rate constant in the growth law for coalescense, $r^3 = Kt$.

In the present paper it will be shown how the method of Lifshitz and Slyozov can be applied to the first kind of grain growth. The treatment will not be quite rigorous, however. The high complexity of the geometric shapes of the grains in a single-phase material as compared to the perfect spherical shapes of the second-phase particles in the case of coalescense makes some approximations necessary when formulating mathematically the growth law for individual grains. In an attempt to recover some of the finer details that may get lost by these approximations, an alternative model of grain growth will be discussed. In this model, the complicated geometric pattern of the grains in a single-phase material is described simply by a number of defects introduced into a perfect array of grains. This model is particularly interesting in connection with the retarding effect on the grain boundary movements due to second-phase particles, as treated by Zener.[7]

Grain growth in two-dimensional as well as three-dimensional systems will be considered. The three-dimensional case is the most interesting one from the practical viewpoint but the two-dimensional case seems to lend itself more readily to theoretical treatment.

2. CHOICE OF BASIC EQUATION

For coalescence, Greenwood[8] derived an equation for the change in size of an individual particle:

$$\frac{dr}{dt} = \frac{k}{r^2}\left(\frac{r}{r_{cr}} - 1\right) \quad (1)$$

The critical size r_{cr} is related to the average particle ones dissolve. Using equation (1) and assuming a constant total volume of the particles Lifshitz and Slyozov were able to show that the growth of the average particle size $\bar{r}$, will asymptotically approach the following expression:

$$\bar{r}^3 = \frac{4}{9}kt \quad (2)$$

They were also able to show that $r_{cr} = \bar{r}$ and to calculate the asymptotic distribution function for the particle sizes, finding that the radius of the largest particle should be 3/2 times as large as the average $\bar{r}$ when the asymptotic distribution has been attained.

In order to apply the method of Lifshitz and Slyozov to the case of grain growth in single-phase materials we need an expression similar to equation (1) for the change in size of an individual grain. We shall start with the generally accepted assumption that the velocity of a grain boundary is proportional to the pressure difference caused by its curvature:

$$v = M \cdot \Delta P = M\sigma \cdot \left(\frac{1}{\rho_1} + \frac{1}{\rho_2}\right) \quad (3)$$

The proportionality constant M may be regarded as the mobility of the grain boundary. ρ_1 and ρ_2 are the principal radii of curvature.

The size of each grain will be expressed by the radius R of an equivalent circle or sphere having the same area or volume, respectively. The net increase in size of a grain can, in principle, be calculated by integrating v around the grain. The net increase described by dR/dt is thus intimately related to an average value of v around the grain. We can put $dR/dt = g \cdot v_{\text{average}}$ where the value of the factor g depends upon the shape of the grain. For a circular grain it is unity, for ordinary grains it might be somewhat larger. Equation (3) can now be written

$$\frac{dR}{dt} = M\sigma \cdot g \cdot \left(\frac{1}{\rho_1} + \frac{1}{\rho_2}\right)_{\text{average}} \quad (4)$$

The curvatures will vary from grain to grain as well as around the periphery of each grain, due to the complex shapes of the grains in a single-phase material.

However, for our analysis we are only interested in a value of $g \cdot \left(\dfrac{1}{\rho_1} + \dfrac{1}{\rho_2}\right)$ averaged over all the grains of each size R. A central problem in the theory of grain growth is to find this average. In the present paper we shall not try to solve this complex geometric problem in a rigorous way but rather to make the simplest possible choice. Then we shall use it to show how the characteristics of grain growth can be calculated. Some justification for the choice will be given but the final justification may have to await an experimental test of the predicted characteristics of grain growth.

We want an expression of the correct dimension and with the characteristic feature that it is positive for large R but negative for small R. The critical size where the value of the expression goes through zero will be denoted by R_{cr}. The simplest choice seems to be

$$g \cdot \left(\dfrac{1}{\rho_1} + \dfrac{1}{\rho_2}\right)_{\text{average}} = \alpha \cdot \left(\dfrac{1}{R_{cr}} - \dfrac{1}{R}\right) \qquad (5)$$

where α is a dimensionless constant. Of course, the critical size R_{cr} might be related to the average grain size in some way but the exact relationship will not be discussed until later. Introducing (5) into equation (4) yields the following expression for the average growth rate of all the particles of size R.

$$\dfrac{dR}{dt} = \alpha M \sigma \cdot \left(\dfrac{1}{R_{cr}} - \dfrac{1}{R}\right) \qquad (6)$$

This is the basic equation on which our calculations will be based.

Assuming a lognormal distribution of the individual grain sizes, Feltham[5] obtained a similar equation, which we can write

$$\dfrac{dR}{dt} = \dfrac{K}{8R} \ln \dfrac{R}{R_{cr}} \qquad (7)$$

With $K = 8\alpha M \sigma$ the two equations are identical in the neighborhood of $R = R_{cr}$ but they are quite different at the extreme values of R. For large grains our equation (6) predicts that the growth rate approaches a constant value of $\alpha M \sigma / R_{cr}$ which is satisfactory, whereas Feltham's equation (7) predicts unreasonably low values. For small grains our equation (6) predicts that dR^2/dt approaches a constant negative value which is quite correct for a two-dimensional system.[9,10] Feltham's equation (7), on the other hand, predicts indefinitely high negative values. As a consequence, our equation (6) seems to be a much more attractive choice than equation (7).

Comparison of our equation (6) with equation (1) shows a remarkable similarity, thus allowing us to follow the procedure of Lifshitz and Slyozov with very little modification, as will be demonstrated in the next section.

In the case of two dimensions we can use the result of von Neumann[9] and Mullins[10] in order to determine the value of α. They were able to show that equation (3) can be integrated around any grain in a two-dimensional system, yielding a very simple result which we can write in the following way by using the radius R instead of the area,

$$\dfrac{dR}{dt} = \dfrac{M\sigma}{R} \cdot \left(\dfrac{n}{6} - 1\right) \qquad (8)$$

Comparison with equation (6) reveals that our choice of equation (5) implies that the following relation holds between the size R and the average number of neighbors per grain of that size in a two-dimensional system:

$$n = 6 + 6\alpha \left(\dfrac{R}{R_{cr}} - 1\right) \qquad (9)$$

The average number of neighbors per grain in the total system would then be

$$\bar{n} = \dfrac{1}{N} \sum_N 6 \left[1 + \alpha\left(\dfrac{R}{R_{cr}} - 1\right)\right] = 6 + 6\alpha\left(\dfrac{\bar{R}}{R_{cr}} - 1\right) \qquad (10)$$

We can thus conclude that $R_{cr} = \bar{R}$ because there is a topological rule telling that $\bar{n} = 6$ in a two-dimensional array of grains.[4] The same conclusion can be drawn directly from equation (6) because the sum of $R\,dR/dt$ over all the grains must be zero in order to conserve the total size of the two-dimensional system. We can now evaluate α. According to the detailed picture of grain growth given by Smith,[4] a shrinking grain never gets less than three neighbors. With the value $n = 3$ at $R = 0$ equation (9) gives $\alpha = \tfrac{1}{2}$. Another estimate can be carried out for an abnormally large grain, $R \gg R_{cr}$. Its periphery may cut through most of the surrounding grains sufficiently close to their centers to justify the following estimate:

$$2\pi R \cong \sum_n 2R = 2n\bar{R} \qquad (11)$$

Inserting this in equation (9) and remembering that $\bar{R} = R_{cr}$ we find $\alpha = \dfrac{\pi}{6} \cong \tfrac{1}{2}$. It may thus appear that equation (5) can properly describe the geometry of the grains over the whole range of sizes in a two-dimensional system if the value $\alpha = \tfrac{1}{2}$ is chosen.

In three dimensions it is not possible to integrate equation (3) in the general case, obtaining a result like equation (8) for an individual grain. The integration can be carried out in special cases, however, thus allowing us to estimate α. Again we shall examine two cases, $R = 0$ and $R \gg R_{cr}$. A shrinking grain in a three-dimensional array has 4 neighbors just before it disappears, resembling a tetrahedron. Assuming that the four faces of such a grain have spherical shape and meet each other with an angle of 120° along the grain edges a numerical calculation can be carried out yielding $dR/dt = 1.0 \cdot M\sigma/R$ and comparison with equation (6) at $R \ll R_{cr}$ thus gives the value $\alpha = 1.0$ for a three-dimensional system. For a large grain we can compare with the two-dimensional case which gave $\alpha = \frac{1}{2}$ but in view of the fact that $\rho_2 = 0$ in two dimensions but $\rho_2 = \rho_1$ in three, we estimate that α should be about twice as large in three dimensions, $\alpha \cong 1$.

3. CALCULATION OF NORMAL GRAIN GROWTH

Introducing the relative size $u = R/R_{cr}$ we can transform our basic equation (6)

$$\frac{dR^2}{dt} = 2\alpha M\sigma \cdot (u - 1) \qquad (12)$$

and form an expression for the growth of the relative size of a particle

$$\frac{du^2}{dt} = \frac{1}{R_{cr}^2}\left[\frac{dR^2}{dt} - \left(\frac{R}{R_{cr}}\right)^2 \cdot \frac{dR_{cr}^2}{dt}\right] \qquad (13)$$

$$\frac{du^2}{dt} = \frac{1}{R_{cr}^2}\left[2\alpha M\sigma \cdot (u - 1) - u^2 \cdot \frac{dR_{cr}^2}{dt}\right] \qquad (14)$$

After dividing through by dR_{cr}^2/dt we get

$$\frac{du^2}{dR_{cr}^2} = \frac{1}{R_{cr}^2}\left[2\alpha M\sigma \cdot \frac{dt}{dR_{cr}^2} \cdot (u - 1) - u^2\right] \qquad (15)$$

This equation can be written in an extremely simple form by introducing two new variables

$$\frac{du^2}{d\tau} = \gamma \cdot (u - 1) - u^2 \qquad (16)$$

where
$$\tau = \ln R_{cr}^2 \qquad (17)$$

$$\gamma = 2\alpha M\sigma \cdot dt/dR_{cr}^2 \qquad (18)$$

The new variable τ represents the time because R_{cr} increases monotonically with time. The same equation with the exponent 3 instead of 2 was found by Lifshitz and Slyozov. They were able to show that their equation predicts that the coalescence process will asymptotically approach a steady state where the variable γ has a constant value, no matter what the initial size distribution is. The same arguments can be applied to our case but will not here be repeated in detail. In order to find the asymptotic value of γ, we shall discuss Fig. 1 where the function $du^2/d\tau$ has been plotted for three constant values of γ. It is evident

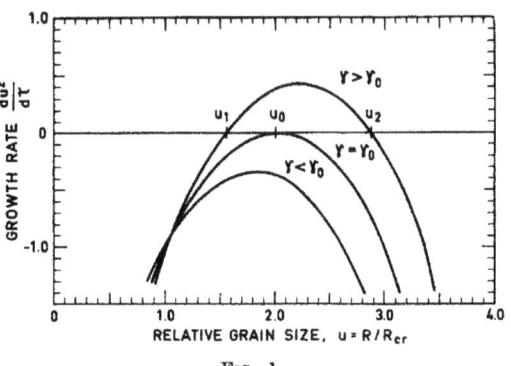

FIG. 1

that a value of $\gamma < \gamma_0$ cannot be maintained during the process because all the grains would then shrink with a nonvanishing rate and after a certain time all the grains would have disappeared. A value $\gamma > \gamma_0$ is also not possible during a steady state because all the grains larger than u_1 would then approach the size u_2 which would mean that they increase their absolute size R indefinitely. On the other hand, all grains smaller than u_1 would disappear within a certain time. As a consequence, the only possible γ value for the steady state is γ_0 which can easily be calculated from equation (16). Let us first calculate u_1 and u_2:

$$u = \tfrac{1}{2}\gamma \pm \sqrt{\tfrac{1}{4}\gamma^2 - \gamma} \qquad (19)$$

For $\gamma = \gamma_0$ the two roots must coincide, thus yielding $\gamma_0 = 4$ and $u_0 = 2$. With the value $\gamma = \gamma_0 = 4$, equation (18) gives the growth rate equation for normal grain growth:

$$\frac{dR_{cr}^2}{dt} = \tfrac{1}{2}\alpha M\sigma \qquad (20)$$

This value is quite close to the value obtained by Feltham, a natural consequence of the similarity between Feltham's growth equation (7) and our growth equation (6).

From the shape of the curve for $\gamma = \gamma_0$ in Fig. 1 we can conclude that all grains larger than u_0 would decrease their relative size toward u_0 but they would never be able to pass through this point. Thus, they would increase their absolute size R indefinitely which is physically impossible. This argument allows the important conclusion to be drawn that there must not be any grains larger than $2R_{cr}$ after the steady state condition of grain growth has been reached. If the initial size distribution is too wide, a fraction of large grains will grow in an abnormal manner until all the other grains have been consumed. When completed, this process has resulted in a more narrow size distribution and, at longer times, the steady state may be

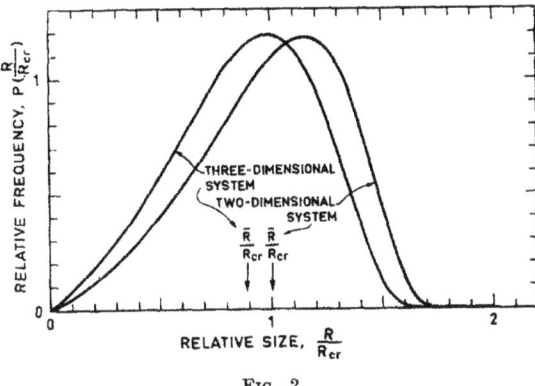

FIG. 2

approached asymptotically. It appears convenient to define this steady state as normal grain growth. Abnormal grain growth may thus sometimes be a necessary initial stage during the development toward normal grain growth.

Following the procedure of Lifshitz and Słyozov, we may calculate the whole distribution of the individual grain sizes. These calculations are presented in the Appendix and they result in the following distribution function:

$$P(u) = (2e)^\beta \cdot \frac{\beta u}{(2-u)^{2+\beta}} \cdot \exp \frac{-2\beta}{2-u} \quad (21)$$

where $\beta = 2$ in two dimensions and $\beta = 3$ in three dimensions.

Figure 2 shows the shape of the function for these two cases. The position of the mean value $\bar{R}$ is indicated. As shown in the Appendix, $\bar{R} = R_{cr}$ in the two-dimensional case, in agreement with the result in Section 2. In a three-dimensional system $\bar{R} = 8/9 R_{cr}$ for this particular grain size distribution.

Experimentally, it is very difficult to determine the true size distribution in a three-dimensional system. It is much easier to make observations on a two-dimensional section and in order to allow comparison with such measurements the size distribution in a section was calculated numerically from the true size distribution assuming that all the grains are spherical. In Fig. 3 the two distributions are compared. The quantity s denotes the apparent radius of a grain as observed in the section. As expected, the distribution curve for a section shows a more pronounced tail on the upper side.

All the distribution curves have a tail reaching up to the theoretical value of 2 but showing negligibly small values long before that limit. For practical purposes one might say that the maximum size is about $1.6 R_{cr}$ in a three-dimensional system and $1.7 R_{cr}$ in a two-dimensional system. If we compare with the average size we find that the maximum size is about $1.8 \bar{R}$ in a three-dimensional system and $1.7 \bar{R}$ in a two-dimensional system. In a section through a three-dimensional system it is about $2\bar{s}$.

4. EFFECT OF SECOND-PHASE PARTICLES

When a grain boundary is moving through a matrix with second-phase particles, there is a tendency for the grain boundary to stick at the particles. As a consequence, the geometry will be considerably more complex and equation (5) will no longer hold. However, Zener[7] has shown that one can take into account the effect of second-phase particles by assuming that the geometry is unaffected and instead introducing a back stress by the particles. For particles of uniform size r and a total volume fraction f, Zener found that the hypothetical back stress would be approximately $S = 3f\sigma/4r$. For the general case we shall write $S = \sigma \cdot z$ where z depends upon the number and sizes of the second-phase particles. In agreement with Zener's calculation we shall assume that z is independent of the grain boundary curvature although this may not be strictly true.

Taking the second-phase particles into account we shall change equation (3):

$$v = M \cdot \Delta P_{\text{true}} = M \cdot (\Delta P \pm S)$$
$$= M\sigma \cdot \left[\left(\frac{1}{\rho_1} + \frac{1}{\rho_2}\right) \pm z\right] \quad (22)$$

where ΔP and $\left(\dfrac{1}{\rho_1} + \dfrac{1}{\rho_2}\right)$ should be estimated without taking the effect of the second-phase particles into account. As a consequence of equation (22), equation (6) will also change:

$$\frac{dR}{dt} = M\sigma \left[\alpha \left(\frac{1}{R_{cr}} - \frac{1}{R}\right) \pm gz\right]$$
$$\simeq \alpha M\sigma \cdot \left(\frac{1}{R_{cr}} - \frac{1}{R} \pm \frac{z}{\alpha}\right) \quad (23)$$

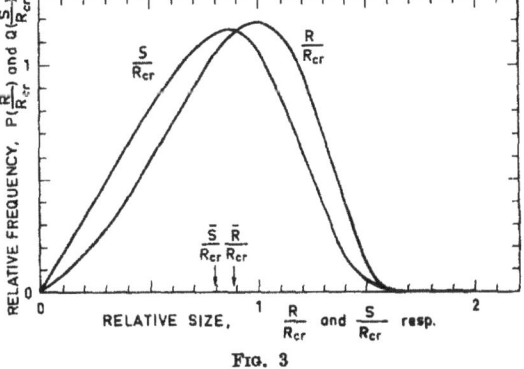

FIG. 3

The sign must be chosen in each case such that the back stress S is acting against the movement of the grain boundary. The negative sign holds when $\frac{1}{R} < \frac{1}{R_{cr}} - \frac{z}{\alpha}$ and the positive sign when $\frac{1}{R} > \frac{1}{R_{cr}} + \frac{z}{\alpha}$. Between these two limits, the back stress would exceed the driving force which is not physically possible. Thus we obtain $dR/dt = 0$ in this range. Considering the fact that the evaluation of z is very approximate, we have neglected the small effect of g being slightly larger than unity due to the non-circular shape of the grains.

Applying the same procedure as in Section 3, we can transform equation (23) into

$$\frac{du^2}{d\tau} = \gamma \cdot \left[u \cdot \left(1 \pm \frac{zR_{cr}}{\alpha}\right) - 1 \right] - u^2 \quad (24)$$

It should immediately be realized that we cannot expect to find any steady state solution in this case, because the second-phase particles grow more and more important as the grain size R_{cr} increases. If we, in spite of this fact, would try to find a steady state solution in the same way as before, we should now obtain

$$\gamma_0 = 4/(1 - zR_{cr}/\alpha)^2 \quad (25)$$

$$u_0 = 2/(1 - zR_{cr}/\alpha) \quad (26)$$

$$dR_{cr}^2/dt = \tfrac{1}{2}\alpha M \sigma \cdot (1 - zR_{cr}/\alpha)^2 \quad (27)$$

It is evident from equation (26) that the size distribution cannot remain constant when the grain size has grown to such values that the Zener effect becomes important. Instead, the maximum size is predicted to increase from twice the average and to approach infinity as the grain size R_{cr} reaches a limiting value of $R_l = \alpha/z$. According to equation (27) the grain growth will cease at this limiting grain size.

It has long been realized that there should be such a limit and Burke[3] has attempted to describe the decrease of the growth rate on approaching the limit, by introducing the factor $(1 - R_{cr}/R_l)$ into the parabolic growth law, obtaining

$$dR_{cr}^2/dt = A\sigma \cdot (1 - R_{cr}/R_l) \quad (28)$$

In view of equation (27) it appears that the square of Burke's factor would have been a better choice. Due to the square, our equation (27) predicts a more gradual retardation of the growth rate.

In view of the fact that equations (25), (26) and (27) do not represent a true steady state, equation (27)

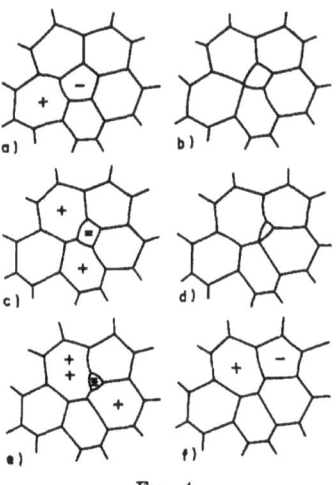

Fig. 4

should only be regarded as a first approximation. It should be noticed, however, that the value of the limit $R_l = \alpha/z$, is not subject to the same uncertainty. It is a direct consequence of equation (23), from which it is evident that a grain cannot grow, however large it is, when the average size is $R_{cr} \geqslant \alpha/z$.

In order for equations (25), (26) and (27) to represent a true steady state, it is necessary that z vary simultaneously with R_{cr} in such a way that the product zR_{cr} remains constant. It is not quite inconceivable that this condition may be satisfied under special experimental conditions. Such possibilities will be discussed in Section 7.

5. DEFECT-MODEL OF GRAIN GROWTH IN TWO-DIMENSIONAL SYSTEMS

In two dimensions we can attack the grain growth problem by another method which, in some respects, is closer to the true mechanism of grain growth.

The topological rule that the average number of neighbors per grain is $\bar{n} = 6$ in a two-dimensional array of grains, has an important consequence. Let us consider an ideal array of hexagonal grains, the so-called bee-hive structure. If we introduce an imperfection by giving a certain grain five neighbors only, then we must give another grain seven neighbors at the same time. This "5–7 pair" may thus be regarded as one defect, introduced into the perfect array. The important properties of such a defect are demonstrated by Fig. 4. In view of equation (8) the 5-grain will shrink and the 7-grain will grow. During the process, the role of being 7-grain is taken over by other neighbors. The area of the shrinking grain will thus be divided among several of its neighbors. The fundamental process of grain growth would thus seem to be the shrinking of a grain with less than 6 neighbors, rather than the growth of a grain with more than 6 neighbors. Another important property is demonstrated by the final arrangement of the grains after the shrinking grain has completely disappeared. The defect is still present. It now affects another grain, causing its number of neighbors to decrease to five. The process will thus be repeated.

The defect will move stepwise through the array of grains and for each step it takes, the number of grains will have decreased by one. As a consequence, the rate of grain growth may be regarded as due to the number of defects per grain, c, and the time a a defect will need to make a grain shrink from normal size to zero.

$$-\frac{dN}{dt} = \frac{cN}{a} \quad (29)$$

The number of grains in the system, N, is related to the size of the grains

$$N \cdot R^2 = \text{constant} \quad (30)$$

where R is the root mean square in the general case. However, assuming a constant distribution function we can choose any mean value for R, by changing the size of the constant in equation (30) appropriately. By choosing R_{cr}, we obtain from equation (30):

$$\frac{2dR_{cr}}{R_{cr}} + \frac{dN}{N} = 0 \quad (31)$$

and by combining with equation (29)

$$\frac{dR_{cr}}{dt} = -\frac{R_{cr}}{2N} \cdot \frac{dN}{dt} = \tfrac{1}{2} R_{cr} \cdot \frac{c}{a} \quad (32)$$

$$\frac{dR_{cr}^2}{dt} = R_{cr}^2 \cdot \frac{c}{a} \quad (33)$$

When there are so many defects that there is a high probability that two or more will cooperate, equation (33) should be slightly modified:

$$\frac{dR_{cr}^2}{dt} = R_{cr}^2 \cdot \sum_p \frac{c_p}{a_p} \quad (34)$$

where c_p is the concentration of grains with 6-p neighbors and a_p is the time it takes such a grain to disappear.

In order to estimate a_p, we shall use the exact equation (8) with $n = 6-p$:

$$\frac{dR}{dt} = -\frac{pM\sigma}{6R} \quad (35)$$

Integration from the normal size R_{cr} to zero gives

$$a_p = -\frac{6}{pM\sigma} \int_{R_{cr}}^0 R \, dR = \frac{3R_{cr}^2}{pM\sigma} \quad (36)$$

The actual length of a may be somewhat less because the shrinkage rate of a grain will increase when it has grown so small that the number of neighbors decreases further. However, this effect is not very important since it seems to affect a comparatively late stage only.

By combination of (36) with (34) we obtain

$$\frac{dR_{cr}^2}{dt} = \tfrac{1}{3} M\sigma \sum_p p \cdot c_p \quad (37)$$

which gives a parabolic growth law as long as the number of defects remains constant. This requirement is analogous to the common assumption of a constant size distribution during normal grain growth. By comparing equation (37) with equation (20) we can evaluate the number of defects during normal grain growth and obtain, remembering that $\alpha = \tfrac{1}{2}$ in a two-dimensional system:

$$\sum_p p \cdot c_p = \tfrac{3}{2}\alpha = \tfrac{3}{4} \quad (38)$$

The quantity $\Sigma\, p \cdot c_p$ may be regarded as the total concentration of defects. The concentration necessary for normal grain growth is very high according to equation (38) and we must expect a considerable number of them to be situated close to each other and thus to be forced to cooperate.

We must now ask how the concentration of defects in a system can approach the value required for normal grain growth. Let us consider a system with very few defects initially. At the beginning of this section it was demonstrated that the defects have a remarkable stability. One may thus be tempted to expect the number of defects to remain fairly constant during grain growth, and thus the concentration of defects to increase. However, there is at least one very important mechanism by which the number of defects will decrease. As a grain has disappeared the active defect must move on and may then encounter another defect in such a way that they will annihilate each other. A qualitative argument seems to suggest that this mechanism to a first approximation should be effective enough to make the concentration of defects remain constant during grain growth. The mechanism by which the concentration is adjusted toward the value required for normal grain growth may thus be of a quite different nature. Even though we have concluded that the fundamental process of grain growth is the shrinking of the grains with less than 6 neighbors and that the role of being 7-grain is passed around among the neighbors, we should now realize that a grain of sufficient size will have a fairly constant number of neighbors and much higher than 6. Such a grain will grow with a fairly constant rate, and it will even be able to increase its relative size if the growth rate of the average size is too low. This occurs if there are too few defects according to equation (33) and will then result in a less uniform size distribution, i.e. in a higher concentration of defects. If this line of argument is correct, the shrinking of an ordinary grain is the fundamental process of normal grain growth only, whereas the growth of a large grain is the fundamental process of abnormal grain growth, the concept of "abnormal grain growth" here taken in a very broad sense.

6. THE EFFECT OF SECOND-PHASE PARTICLES TREATED BY THE DEFECT-MODEL

In Section 4 a mathematical analysis of the effect of second-phase particles on grain growth was carried out using the method of Lifshitz and Slyozov. At the end of the section it was concluded that the analysis could be regarded as a first approximation only, in view of the fact that the mathematical solution did not represent a true steady state. It is thus of considerable interest to examine what predictions the defect-model will lead to in the presence of second-phase particles. Again, we shall consider a two-dimensional system.

Accepting equation (9) we can transform equation (23) into:

$$\frac{dR}{dt} = \frac{M\sigma}{R}\left(\frac{n}{6} - 1 \pm zR\right) \quad (39)$$

which reduces to equation (8) when $z = 0$. With $n = 6 - p$ we obtain

$$\frac{dR}{dt} = -\frac{pM\sigma}{6R}\left(1 - \frac{6z}{p} \cdot R\right) \quad (40)$$

Integration from a normal grain size of R_{cr} to zero yields

$$a_p = \frac{6}{pM\sigma} \cdot \left(\frac{p}{6z}\right)^2 \cdot \left[-\ln\left(1 - \frac{6z}{p} \cdot R_{cr}\right) - \frac{6z}{p} \cdot R_{cr}\right] \quad (41)$$

or, approximately

$$\frac{1}{a_p} = \frac{pM\sigma}{3R_{cr}^2} \cdot \left(1 - \frac{6z}{p} \cdot R_{cr}\right)^{2/3} \quad (42)$$

This particular approximation was chosen because it is good for small R_{cr} and also predicts that a_p approaches infinity at $R_{cr} = p/6z$ in agreement with the basic equation (40). By introducing (42) into (34) we get

$$\frac{dR_{cr}^2}{dt} = \tfrac{1}{3}M\sigma \cdot \sum_p p \cdot \left(1 - \frac{6z}{p} \cdot R_{cr}\right)^{2/3} \cdot c_p \quad (43)$$

where the summation is carried out over all p values yielding positive terms. By a series expansion it can be shown that equation (43) becomes identical to equation (27) for low values of zR_{cr} if the effect of cooperating defects is neglected and the value $c_1 = 3\alpha/2$ is chosen in agreement with equation (38). This result may be taken as an indication that equation (27) is approximately correct. When comparing the two equations at larger values of zR_{cr}, it is realized that equation (43) predicts a series of limiting values for R_{cr}. At the first limit, $R_{l_1} = 1/6z$, most of the single defects are pinned by the second-phase particles because a 5-grain of the average size, $R = R_{cr} = 1/6z$, will not shrink, according to equation (39). Grain growth can continue only by the action of cooperating defects. At the next limit, $R_{l_2} = 1/3z$, most of the groups of two defects are also pinned because a 4-grain of the size $1/3z$ cannot shrink. In view of this effect one should expect the growth rate to decrease gradually as the final limiting grain size is approached, in qualitative agreement with equation (27). One may guess that the final limit corresponds to the largest p value that is physically possible. This occurs for a grain with three neighbors, $p = 6 - n = 3$, and gives $R_{l_3} = 1/2z$. This is in excellent agreement with the grain size limit predicted by equation (27). On the other hand, in view of the defect-model one might expect that the grain growth will stop earlier if the size distribution is fairly uniform because the number of groups with three cooperating defects might then be negligible. As a consequence, the defect-model seems to suggest that normal grain growth should stop already when the groups of two defects are pinned, i.e. when $R_{cr} = 1/3z$. There should thus exist two grain size limits as indicated in Fig. 5. The lower limit is situated at about $1/3z$ and normal grain growth cannot occur above this limit. The upper limit is situated at $1/2z$ and no grain growth at all can take place in a material that has a larger grain size. Between the two limits abnormal grain growth can take place if there already are some grains much larger than the average. According to equation (23) such grains will grow with a rate of

$$\frac{dR}{dt} = \alpha M\sigma \cdot \left(\frac{1}{R_{cr}} - \frac{z}{\alpha}\right) \quad (45)$$

and this expression yields positive values as long as $R_{cr} < 1/2z$ for a two-dimensional system where $\alpha = \tfrac{1}{2}$. If the normal grain growth has stopped close to $1/3z$, an abnormally large grain can still grow with a rate of

$$\frac{dR}{dt} = \frac{M\sigma}{6R_{cr}} \quad (46)$$

According to this estimate, the second-phase particles

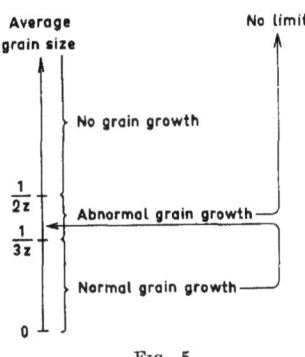

Fig. 5

will decrease the growth rate of an abnormally large grain by a factor of $\frac{1}{3}$ only, when the normal grain growth is completely controlled by them. Of course, the final grain size obtained by abnormal grain growth may be much higher than the upper limit $1/2z$. It is limited only by the number of grains that were large enough to grow and absorb the matrix of grains with normal sizes.

7. THE INITATION OF ABNORMAL GRAIN GROWTH

As demonstrated in Fig. 5 the defect-model seems to predict that abnormal grain growth can develop in a material if three conditions are simultaneously fulfilled.

(1) Normal grain growth cannot take place due to the presence of second-phase particles.

(2) The average grain size has a value below the limit $1/2z$.

(3) There is at least one grain much larger than the average.

Whether these conditions are automatically fulfilled in a material where the normal grain growth has stopped due to the presence of second-phase particles, is a question of considerable practical importance. We shall therefore examine the process in more detail. From Section 3, we know that there are grains as large as $2R_{cr}$ during normal grain growth. From Section 4, we know that this maximum size should increase further as the limit is approached. In fact, equation (26) even predicts that the maximum size approaches infinity as R_{cr} approaches the limit $1/2z$. However, equation (26) is approximately valid and it does not seem justified to use it for predicting that a few grains will actually have time enough to grow to an abnormally large size before all grain growth has stopped. In view of the fact that the Zener effect becomes appreciable only close to the limit, it would seem more probable that the fairly uniform size distribution attained during normal grain growth far below the limit, will not tend to change much until close to the limit. Many individual grains may thus reach a size larger than the limit $1/3z$ and may then be stabilized by the second-phase particles. The smaller grains are not stabilized and may shrink and disappear before grain growth stops completely. As a result, we may tentatively suggest that the final grain size may very well be somewhat higher than the limit $1/3z$, as indicated in Fig. 5, or even higher than the upper limit $1/2z$, but not very much higher. As a consequence, it seems reasonable to expect that normal grain growth will not automatically develop into abnormal growth of a few grains. The lower part of Fig. 6 demonstrates how the normal grain growth proceeds when the effect of the second-phase particles is negligible. All the lines are here calculated from equation (16) with $\gamma = 4$. The upper part of Fig. 6 demonstrates what would happen as the grain growth is affected by second-phase particles. In this part of the figure the lines are tentatively drawn, in accordance with the suggested mechanism.

Next, we should ask whether a very large grain, formed by some extraordinary process, could grow abnormally in a material where the normal grain growth has stopped. The answer depends upon the value of the final grain size reached by the normal grain growth. The defect-model does not give any clear-cut answer to this question because of the uncertainty regarding the exact value of the final grain size. The mathematical analysis, on the other hand, predicts that normal grain growth should proceed up to the limit $1/2z$, thus making abnormal grain growth impossible except for a limited time period when the normal grain growth

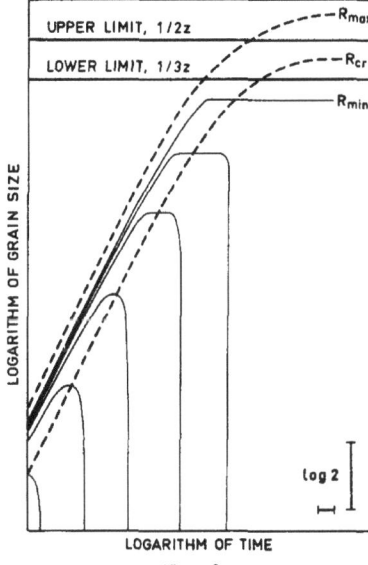

Fig. 6

has started to slow down but has not yet reached the limit.

Finally, we shall consider the possibility of initiating abnormal grain growth by a continuous decrease of the z value. This effect can be accomplished by increasing the particle size r through coalescence or by decreasing the volume fraction f by dissolving the second phase.

During prolonged grain growth experiments one could expect coalescence of the second-phase particles and we could then expect the quantity z to vary as $1/\bar{r}$ and the grain size limits will thus vary as $\bar{r}$, i.e. proportionally to $t^{1/3}$ in view of equation (2). After sufficiently long times the actual grain size will thus grow slowly as

$$R_{cr} = K \cdot t^{1/3} \tag{44}$$

where the value of K mainly depends upon the constant k in equation (2). If there are two "second-phases" in the material, the back stress will be $S = \sigma \cdot z = \sigma \cdot (z_1 + z_2)$ where z_1 and z_2 are evaluated for each phase separately. If z_1 and z_2 are of different orders of magnitude, one can neglect the effect of the least effective phase. On the other hand, if the most effective phase coalesces faster than the other one, the grain size will gradually change over, as demonstrated by Fig. 7. In view of this picture it does not seem surprising that the experimental values of the time exponent are usually appreciably less than the "theoretical" value of $\frac{1}{2}$.

In Fig. 7 we have assumed that the grain growth proceeds in the normal fashion during the coalescence. Figure 8 demonstrates two other possibilities. This figure is intended as a continuation of Fig. 6. First, it shows how the normal grain growth does not start until the lower limit has reached the value of R_{cr}. From then on, R_{cr} is suggested to follow the limit closely. As a consequence, an unusually large grain can now start to grow abnormally even though it might previously have been inhibited because the final grain size, reached by the normal grain growth in Fig. 6, was too high.

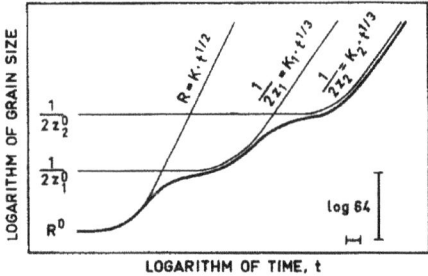

Fig. 7

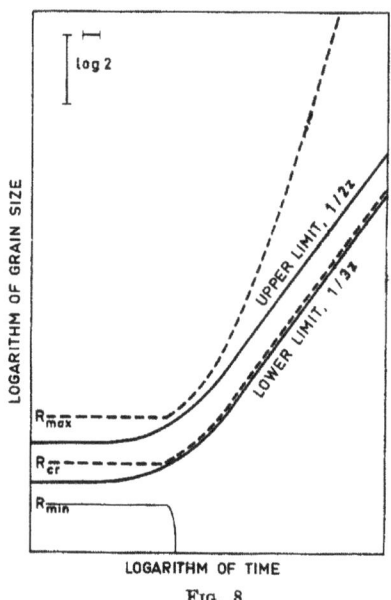

Fig. 8

Secondly, Fig. 8 demonstrates how the grain size distribution grows wider during coalescence by the maximum grain size increasing somewhat faster than the average. This process might develop into abnormal grain growth but it seems impossible at the present to predict how long a time would be needed.

Of course, any process that leads to a slow increase of the grain size limit may initiate the development of abnormal grain growth by the same mechanisms. Most cases of abnormal grain growth, met with in practice, seem to be connected with the dissolving of a second-phase rather than the coalescence. The defect-model does not seem to suggest any immediate explanation of this observation. In terms of the mathematical analysis, on the other hand, there is a slight difference that may prove essential.

As already pointed out, equation (26) seems to suggest that normal grain growth should always develop into abnormal grain growth as the limit $1/2z$ is approached. This conclusion could not be accepted before, in view of the fact that γ_0 is not a constant as the limit is approached. However, if z decreases by coalescence as R_{cr} increases, it seems possible that zR_{cr} could remain constant, thus making the solution described by equations (25), (26) and (27) a true steady state solution. Let us examine the case where R_{cr} follows the limit $1/3z$ closely, thus making $zR_{cr} = \frac{1}{3}$. Equation (26) will then predict that there is a new asymptotic size distribution with a maximum size equal to $6R_{cr}$ instead of $2R_{cr}$. According to the argument by Lifshitz and Slyozov, we should thus expect that a fraction of grains at the upper end of the size distribution curve, established during normal grain growth, would grow rapidly during coalescence.

Even though we have assumed that zR_{cr} remains constant during coalescence, the solution described by equations (25), (26) and (27) still is not a true steady state because equation (27) predicts a parabolic growth of R_{cr} whereas coalescence only permits R_{cr} to increase proportional to $t^{1/3}$. A true steady state can only be obtained if R_{cr} is allowed to grow faster than during coalescence. This might happen in a limited time periode during the dissolution of the second-phase, thus yielding a possible explanation why most of the well-known cases of abnormal grain growth seem to take place when a second-phase is dissolving.

8. ABNORMAL GRAIN GROWTH IN THREE-DIMENSIONAL SYSTEMS

The defect model is strictly applicable only to a two-dimensional array of grains. In a three-dimensional system there is no rule about a constant number of neighbors per grain. In spite of this, it seems reasonable to assume that the general line of argument in Sections 5, 6 and 7 can be applied to a three-dimensional system as well as to a two-dimensional. We may thus suggest that there are always two grain size limits. Normal grain growth can only occur below the lower limit. Between the two limits a large grain would be able to grow abnormally but the risk of developing a grain large enough to grow may be slight under ordinary conditions. The risk of developing a large grain is much higher if the limits slowly move to larger sizes by coalescence or dissolution of the second-phase particles. A safe method of avoiding abnormal grain growth would be to form an average grain size so much larger than the upper limit, that it will remain larger even when the limit moves to larger values during the heat treatment of the material. Such a large size can be obtained by phase transformation or recrystallisation.

Another way to avoid abnormal grain growth would be to decrease the distance between the two size limits. An obvious method to achieve this would be to choose a material with a larger volume fraction of the second phase. Let us consider how the conditions will change if we keep the number of particles constant but increase the volume fraction, f. Both limits will decrease. However, it does not seem possible ever to stop normal grain growth at a grain size smaller than the distance between the particles. After the limit for normal grain growth has reached this value it will stay there but the upper limit will probably decrease further and thus move closer to the lower limit. As a consequence, we may conclude that abnormal grain growth is not very likely to occur in a material where most of the particles are observed to be situated in the grain boundaries.

APPENDIX

Calculation of the Size Distribution During the Steady State

At the steady state $\gamma = \gamma_0 = 4$ and equation (16) can be written

$$\frac{du^2}{d\tau} = \gamma_0(u - 1) - u^2 = -(2 - u)^2 \quad (A1)$$

$$\frac{du}{d\tau} = -\frac{(2-u)^2}{2u} \quad (A2)$$

The number of grains between u and $u + du$ at a certain time τ will be denoted by $\varphi(u, \tau) \cdot du$ and the total number of grains at the same time by $N(\tau)$. The distribution function φ must satisfy the continuity equation in the grain-size space

$$\frac{\partial \varphi}{\partial \tau} + \frac{\partial}{\partial u}\left(\varphi \cdot \frac{du}{d\tau}\right) = 0 \quad (A3)$$

By trying the solution

$$\varphi = \chi(\tau + \psi)\bigg/\frac{du}{d\tau} \quad (A4)$$

where we know that $\dfrac{du}{d\tau}$ is a function of u only, we find

$$\frac{d\psi}{du} = -1\bigg/\frac{du}{d\tau} \quad (A5)$$

and on integration we obtain

$$\psi = \int_0^u \frac{du}{-\dfrac{du}{d\tau}} = \int_0^u \frac{2u\,du}{(2-u)^2}$$

$$= 2\left[\ln(2-u) + \frac{2}{2-u} - \ln 2e\right] \quad (A6)$$

The function χ can now be calculated from the information that the size K of the whole system must be constant:

$$K = \int_0^2 R^\beta \cdot \varphi \cdot du = \int_0^2 u^\beta \cdot R_{cr}^\beta \cdot \varphi \cdot du$$

$$= \int_0^2 \exp \tfrac{1}{2}\beta\tau \cdot u^\beta \cdot \varphi \cdot du$$

$$= \int_0^2 \exp \tfrac{1}{2}\beta\tau \cdot u^\beta \cdot \frac{\chi}{\dfrac{du}{d\tau}} \cdot du \quad (A7)$$

The constant β is 2 in two dimensions and 3 in three dimensions. The variable τ was introduced using equation (17). The integration is only carried out up to the value $u = 2$ because larger grains are not allowed in the steady state. The value of K must be independent of τ which is possible only if $\chi \cdot \exp \tfrac{1}{2}\beta\tau$ is independent of τ. It is thus necessary that χ can be expressed as

$$\chi(\tau + \psi) = B \cdot \exp[-\tfrac{1}{2}\beta \cdot (\tau + \psi)] \quad (A8)$$

The value of the constant B could be calculated by numerical integration, using (A6) and inserting (A8) and (A2) in equation (A7). By inserting (A8) in (A4), we obtain

$$\varphi = \chi \bigg/ \frac{du}{d\tau} = B \cdot \exp -\tfrac{1}{2}\beta\tau \cdot \exp -\tfrac{1}{2}\beta\psi \bigg/ \frac{du}{d\tau} \quad (A9)$$

and we can calculate the total number of grains in the system, using (A5):

$$N(\tau) = \int_0^2 \varphi\, du$$

$$= -B \cdot \exp -\tfrac{1}{2}\beta\tau \cdot \int_0^\infty \exp -\tfrac{1}{2}\beta\psi \cdot d\psi$$

$$= \frac{2}{\beta} \cdot B \cdot \exp -\tfrac{1}{2}\beta\tau \cdot \bigg| \exp -\tfrac{1}{2}\beta\psi \bigg|_0^\infty$$

$$= -\frac{2}{\beta} \cdot B \cdot \exp -\tfrac{1}{2}\beta\tau \quad (A10)$$

Let $P(u) \cdot du$ be the probability that the size of a grain is between u and $u + du$. Then

$$P(u) = \frac{\varphi(u, \tau)}{N(\tau)} = -\tfrac{1}{2}\beta \exp -\tfrac{1}{2}\beta\psi \bigg/ \frac{du}{d\tau} \quad (A11)$$

Inserting (A2) and (A6) yields

$$P(u) = \frac{\beta u}{(2-u)^{2+\beta}} \cdot (2e)^\beta \cdot \exp \frac{-2\beta}{2-u} \quad (A12)$$

This is the steady state distribution during normal grain growth.

From the experimental point of view, the mean value $\bar{R}$ of the individual sizes is of considerable interest. This value can be calculated using the expression for $P(u)$:

$$\bar{u} = \int_0^2 u \cdot P(u) \cdot du =$$

$$= \int_0^2 \frac{\beta \cdot (2e)^\beta \cdot u^2}{(2-u)^{2+\beta}} \cdot \exp \frac{-2\beta}{2-u} \cdot du \quad (A13)$$

Using the new variable $x = 1/(2 - u)$ we obtain

$$\bar{u} = \int_{\frac{1}{2}}^\infty \beta \cdot (2e)^\beta \cdot (4x^\beta - 4x^{\beta-1} + x^{\beta-2}) \cdot$$

$$\exp -2\beta x \cdot dx \quad (A14)$$

With $\beta = 2$ we find $\bar{u} = 1$ and thus $\bar{R} = R_{cr}$. With $\beta = 3$ we find $\bar{u} = \tfrac{8}{9}$ and thus $\bar{R} = \tfrac{8}{9}R_{cr}$. The latter result is formally identical to the case of interface-controlled coalescence which has been treated by Wagner.[11]

ACKNOWLEDGMENT

The author is indebted to many of his students, particularly Ulf Lindborg and Per Hellman for helpful comments at various stages.

He also gratefully acknowledges financial support from the Swedish Technical Research Council.

REFERENCES

1. D. Harker and E. Parker, *Trans. Amer. Soc. Met.* **34**, 156 (1945).
2. P. A. Belk, J. C. Kremer, L. J. Demer and M. L. Holzworth, *Trans. Amer. Inst. Min. (Metall.) Engrs.* **175**, 372 (1948).
3. J. E. Burke, *Trans. Amer. Inst. Min. (Metall.) Engrs.* **180**, 73 (1949).
4. C. S. Smith, "*Metal Interfaces*" p. 65. ASM, Cleveland (1952).
5. P. Feltham, *Acta Met.* **5**, 97 (1957).
6. I. M. Lifshitz and V. V. Slyozov, *Zh. Eksp. Teor. Fiz.* **35**, 479 (1958).
7. C. Zener, Private communication to C. S. Smith *Trans. Amer. Inst. Min. (Metall.) Engrs.* **175**, 15 (1949).
8. G. W. Greenwood, *Acta Met.* **4**, 243 (1956).
9. J. von Neumann, "*Metal Interfaces*" p. 108 (ASM, Cleveland 1952).
10. W. W. Mullins, *J. Appl. Phys.* **27**, 900 (1956).
11. C. Wagner, *Z. Electrochemie* **65**, 581 (1961).

10 Introduction to "Grey and White Solidification of Cast Iron"

published in "The Solidification of Metals", The Iron Steel Institute, London (1968)

Mats Hillert studied the solidification processes of cast irons for a fifteen-year period, beginning in the 1950's. The results of this work were reported in a series of papers published in Swedish and in English. In a paper presented at the 1964 Conference on the Physical Metallurgy of Cast Iron, he outlined the thermodynamics of cast irons, and discussed the main theoretical aspects of their solidification. Two years later, this paper with Subba Rao reported the results of unidirectional solidification experiments on cast irons. A series of well-controlled Bridgman experiments was performed, permitting the evaluation of the coarseness of the structures of white and grey cast irons as a function of growth rate. The experimental results also showed that undercooled graphite grew under conditions different from flake graphite, and it was proposed that the mechanisms of growth of the two structures were different.

Mats Hillert analyzed the results using advanced models of eutectic growth and plate growth of graphite. Similar models are currently used in computer-based simulations of solidification and structure formation in castings. These models are often based on the concepts of temperature/growth transformation diagrams, whose underlying principles were presented here for the first time.

The growth kinetics of white and grey eutectic structures were compared. It was pointed out that white cast iron grew at a much faster rate than that of grey iron with increasing undercooling due to the differences in slopes of the liquidus lines for cementite and graphite. This concept was used to explain the technically important problem called inverse greyness.

Hasse FREDRIKSSON

GREY AND WHITE SOLIDIFICATION OF CAST IRON

M. Hillert and Subba Rao V. V.

Today it is well realized that the ordinary grey and white structures of cast iron are both the direct result of the solidification reaction. However, it has been suggested many times in the past that the grey structure would be the result of a solid-state graphitization process occurring after an initial white solidification reaction. This hypothesis initiated several detailed examinations of the solidification reactions of cast iron around 1950 and very strong evidence was produced to show that the hypothesis was wrong. As a consequence, it is now generally agreed that the ordinary flake graphite structure and the socalled undercooled graphite structure are both formed directly from the melt.

The very intensive study of the solidification of cast iron carried out in several countries has not only resulted in the evidence against the hypothesized 'indirect' formation of graphite; in fact, it yielded new fundamental information which inspired further efforts to increase our understanding of the solidification mechanism of cast iron. In Sweden, the evidence against the hypothesized indirect formation of graphite was produced by Hultgren et al.[1] In the process of their study, Hultgren et al. found that the kinetics of the grey and white solidification reactions were quite different, a fact that was successfully used to prove that graphite forms directly from the melt. The difference in kinetics yields different cooling curves and the cooling curve of a mottled iron reveals that the grey part of the structure in such an iron formed before the white reaction started. This is demonstrated by the cooling curves in Fig.1 which are constructed in accordance with the result of Hultgren et al.

The difference in kinetic behaviour of the two reactions was regarded as very important from the practical as well as theoretical point of view and led to further work at the Swedish Institute for Metal Research and the Royal Institute of Technology in Stockholm. The purpose of the present paper is to review this work and to give a detailed description of the grey and white solidification mechanisms.

SOLIDIFICATION SEQUENCE IN CAST IRON

From practical experience it is well known that a cast iron melt may be made to solidify grey by slow cooling and white by rapid cooling. This observation has led to the conclusion that each cast iron has a certain critical cooling rate below which it solidifies completely grey and a higher critical cooling rate above which it solidifies completely white. Between the two critical rates a mottled structure is formed (see Fig.2a). This concept of 'critical cooling rate' is still widely accepted and considerable effort is being made to determine experimentally the critical cooling rates for various cast iron compositions.[2] It provides a simple explanation of the fact that a casting may have a white case and a grey core. It has also been applied to explain the surprising phenomenon called inverse greyness, where a white casting shows some grey structure close to the surface and especially in thin sections. In order to satisfy the critical cooling rate concept, one has here postulated that the cooling rate actually was lower at the beginning of the solidification process than later on.[3,4]

On the other hand, from Hultgren's result it can be concluded that cast iron sometimes solidifies under such conditions that the following two principles hold:

(i) graphite is more easily nucleated than cementite
(ii) once nucleated, cementite will grow much faster than graphite.

This conclusion indicates a hysteresis as demonstrated in Fig.2b. In a certain range of cooling rate the solidification can be either white or grey. If it has started grey at the upper part of this range it will proceed in that fashion until a cementite nucleus is formed by some mechanism. If and when this happens, the reaction will abruptly change into the white fashion (horizontal arrows pointing to the left in Fig.2b). On the other hand, the grey fashion may be expected to be stable at the lower part of the range. If the reaction by some reason has started in the white fashion, it will change to grey when a graphite nucleus is formed in this range of cooling rate (horizontal arrows pointing to the right in Fig.2b).

In view of this picture, inverse greyness could simply be explained by assuming that the solidification normally starts in the grey fashion irrespective of the cooling rate and due to graphite being more easily nucleated than cementite.[5] The series of pictures in Fig.3 demonstrates the progress of the eutectic solidification in such a casting. It illustrates that the grey structure is to be found where the eutectic reaction started.

It appears that such a hysteresis would be important for the solidification process of cast iron under a wide range of casting conditions. A study was thus started in order to investigate these conditions in detail and to find their explanation.

THERMODYNAMIC INFORMATION

It was realized at an early stage that the explanation of

Experimental information on the white and grey modes of solidification is reviewed and compared with theoretical predictions, mainly for the purpose of explaining how the various modes of solidification compete during the solidification of cast iron. Controlled solidification experiments have yielded information on the coarseness of grey and white structure as function of growth rate. These data do not allow a critical test of the growth mechanism. Measurements of the solidification temperature yield a better test. It is concluded that the white and grey eutectic structures of cast iron grow considerably slower than predicted by the theory of diffusional-controlled cooperative growth. From practical experience of cast iron, it is concluded that the competition between grey and white solidification is not simply a function of cooling rate. A hysteresis is predicted and has been found by controlled-solidification experiments. As expected, it has been found to cover the growth rate where grey and white solidification grow with the same solidification temperature. It is suggested that the structures called undercooled graphite in the Fe–C system and modified structure in the Al–Si system form with a higher degree of cooperation than the usual eutectic structures in these systems.

The authors are at The Royal Institute of Technology, Stockholm 70, Sweden.

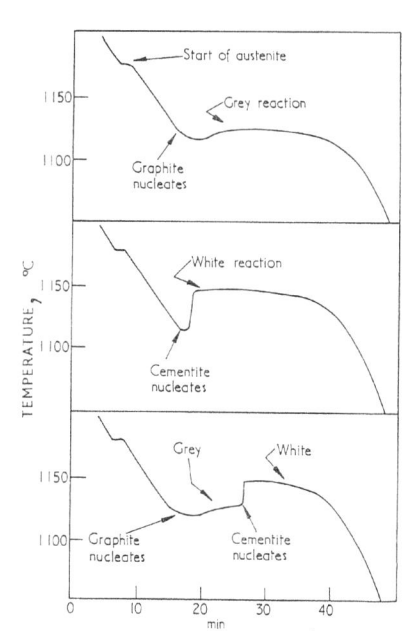

1 Cooling curves for grey, white, and mottled iron

the difference in kinetic behaviour of the grey and white eutectics might be caused by the marked difference in slope of the liquidus lines for graphite and cementite.[6] In order to calculate the slope of the liquidus line for cementite, a thermodynamic evaluation was made of the stable Fe–C phase diagram.[7] Thermodynamic functions were calculated and used to calculate the metastable phase diagram (broken lines in Fig.4). The result confirmed the previous calculations[6,8] but were more detailed and could be used for quantitative calculations of the growth rates of white and grey eutectic. Two models were used, cooperative eutectic growth and separate growth of graphite and cementite, respectively. With both models the white structure gave the highest growth rate at the same degree of undercooling, thus explaining the well known fact that rapid cooling promotes white solidification. The same conclusion was reached by Tiller[9] who carried out a similar study at the same time.

Detailed experimental data from controlled-solidification experiments on cast iron are now available. A more detailed comparison between theory and experiment will thus be presented in this paper. The thermodynamic information obtained in the previous study will again be used.

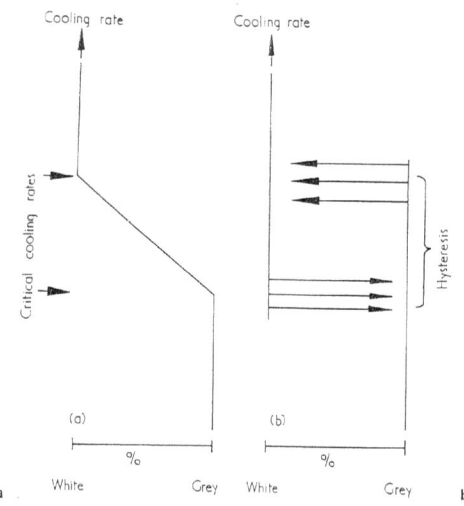

2 *a* concept of critical cooling rates; *b* suggested hysteresis

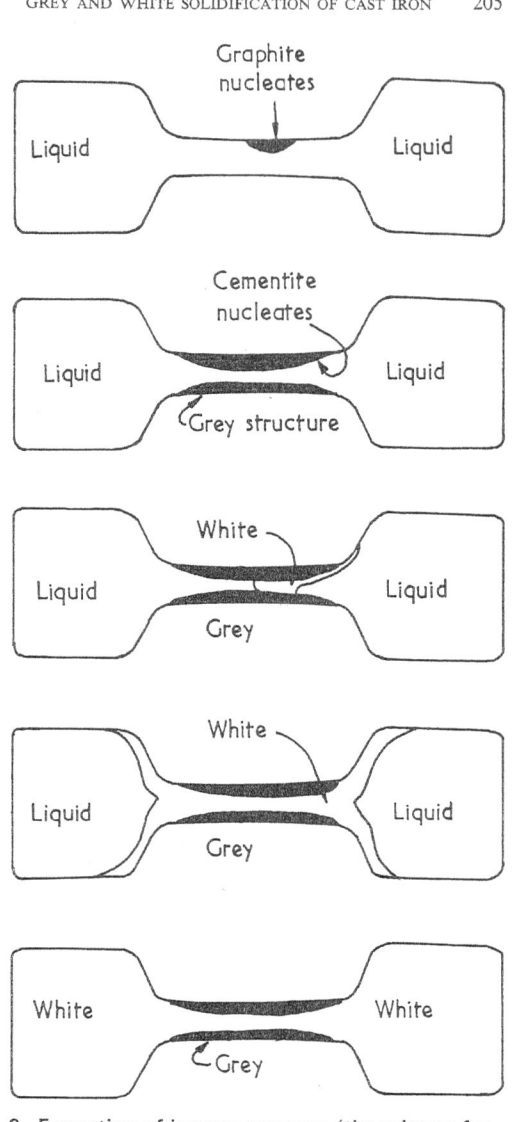

3 Formation of inverse greyness (the primary formation of austenite is not shown in the figure)

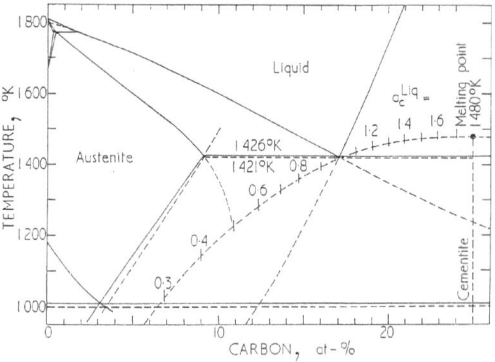

4 Fe–C phase diagram according to reference 7

THEORETICAL CALCULATION OF DIFFUSIONAL-CONTROLLED GROWTH

In eutectic solidification the reaction may proceed in

HILLERT and SUBBA RAO

TABLE 1 Calculation of theoretical growth rates for cooperative growth of grey and white structures in cast iron

	graphite+γ	cementite+γ
T_0, °K	1426	1421
ΔH_m, cal/mol	2073	2135
σ, erg/cm²	800	800
λ^β/λ, (as mole fraction)	0·088	0·5
λ^γ/λ, (as mole fraction)	0·912	0·5
$x_C^{L/\gamma}-x_C^{L/\beta}$	4.6·10⁻⁴ΔT	8.6·10⁻⁴ΔT
$x_C^\beta-x_C^\gamma$	0·909	0·158
D, cm²/s	5·10⁻⁵	5·10⁻⁵
a	0·9	0·54
λ^*, cm	1.8·10⁻⁴/ΔT	1.8·10⁻⁴/ΔT
$v_{max}.^{\beta+\gamma}$	5·10⁻⁴$(\Delta T)^2$	30·10⁻⁴$(\Delta T)^2$
$v.\lambda^2 (=v_{max}.(2\lambda^*)^2)$	70·10⁻¹²	400·10⁻¹²

TABLE 2 Calculation of theoretical growth rates for separate precipitation of graphite and cementite, respectively, in cast iron

	graphite	cementite
T_0, °K	1415	1415
ΔH_m, cal/mol	4633	2439
σ, erg/cm²	500	500
V_m^β, cm³/mol	5·3	7
$x_C^L-x_C^{L/\beta}$	1·10⁻⁴ΔT	5·10⁻⁴ΔT
$x_C^L-x_C^L$	0·83	0·08
r^*, cm	2·10⁻⁵/ΔT	5·10⁻⁵/ΔT
Ω	9·10⁻⁵ΔT	6·10⁻³/ΔT
v_{max}^{Zener}, cm/s	0·3·10⁻⁴$(\Delta T)^2$	8·10⁻⁴$(\Delta T)^2$
$\lambda(=8r^*.\lambda/\lambda^\beta)$, cm	20·10⁻⁴/ΔT	8·10⁻⁴/ΔT
$v.\lambda^2$	120·10⁻¹²	500·10⁻¹²
$v_{max}^{Ivantsov}$, cm/s	2·10⁻⁹$(\Delta T)^3$	4·10⁻⁶$(\Delta T)^3$
$\lambda(=\pi D/4v)$, cm	20000/$(\Delta T)^3$	10/$(\Delta T)^3$
$v.\lambda$	4·10⁻⁵	4·10⁻⁵

many ways, characterized by different degrees of co-operation between the two eutectic phases. There are two extreme cases that have been treated theoretically in some detail; ideally cooperative growth of the two phases and separate growth of one of the eutectic phases. The present paper will be limited to isothermal conditions and it will be assumed that the growth rates of the solid phases into the liquid are completely controlled by diffusion in the liquid phase. Furthermore, Zener's hypothesis will be accepted, stating that the geometry of the growing phase or phases will be such that the maximum growth rate is obtained.[10]

For cooperative growth, the Zener theory predicts an interlamellar spacing, λ, of twice the critical value, λ^*, given by

$$\lambda^* = \frac{2\sigma V_m^L}{(-\Delta G_m)} = \frac{2\sigma V_m^L T_0}{\Delta H_m \Delta T} \quad \ldots\ldots(1)$$

The maximum growth rate is then obtained as

$$v_{max}.^{\beta+\gamma} = \frac{D}{4a\lambda^*} \cdot \frac{1}{f^\beta f^\gamma} \cdot \frac{x_C^{L/\gamma}-x_C^{L/\beta}}{x_C^\beta-x_C^\gamma} \quad \ldots\ldots(2)$$

The quantities f^β and f^γ are the relative amounts of the two phases in the eutectic, measured as mole fractions,[7] and a is a parameter that is slightly larger than 0·5, its exact value depending upon the f values. Detailed calculations[11] have shown that the shape of the solidification front should be rather flat according to this theory when $\lambda=2\lambda^*$. On the other hand, if λ is much larger by some reason, the interface would be more ragged and the growth rate should be lower than predicted by equation (2). The data thus calculated for cooperative growth of white eutectic (austenite+cementite) and of grey eutectic (austenite+graphite) are presented in Table 1.

The consideration of separate growth will be restricted to the edgewise growth of plates, because cementite as well as graphite are known to form as plates or flakes. As pointed out by Zener[10] the effective diffusion distance at the edge of a plate should be shorter, the sharper the edge. However, the sharpness of the edge is restricted by the effect of capillarity and, according to a dimensional argument by Zener, the maximum growth rate is obtained when the edge is twice as thick as the critical value where capillarity will completely stop the growth. The critical value, as represented by the radius of curvature at the edge, r, is given as

$$r^* = \frac{\sigma V_m}{(-\Delta G_m)} = \frac{\sigma V_m T_0}{\Delta H_m \Delta T} \quad \ldots\ldots(3)$$

According to an approximate calculation[11] based on Zener's model, the maximum growth rate is given by the following expression

$$v_{max}. = \frac{D\Omega}{8r^*} \quad \text{at } r=2r^* \quad \ldots\ldots(4)$$

where

$$\Omega = \frac{V_m^\beta(x_C^L-x_C^{L/\beta})}{V_m^L(x_C^\beta-x_C^L)} \quad \ldots\ldots(5)$$

On the other hand, Ivantsov[12] has derived the rigorous solution for the edgewise growth of a plate, neglecting the effect of capillarity. It has been suggested that his solution could be applied to the case of capillarity by simply exchanging the equilibrium composition $x_C^{L/\beta}$ with the value obtained at the edge.[9] For a dilute solution this new value would be $x_C^{L/\beta}.(1-r^*/r)$. As shown in Appendix I, this suggestion leads to the following expression in the case of low supersaturations.

$$v_{max}. = \frac{8D\Omega^2}{27\pi r^*} \quad \text{at } r=3r^* \quad \ldots\ldots(6)$$

The growth rates predicted by these two models are very different, as shown in Table 2. Unfortunately, at present it seems very difficult to tell where the correct value should lie. Trivedi et al.[13] have recently shown that the solution based on Zener's model is approximately correct for large supersaturations only. Trivedi[14] has further improved the solution but his conclusion still seems to be that the model does not give satisfactory results for low supersaturations. On the other hand, Trivedi et al.[13] have also attempted to apply Ivantsov's method to the case of capillarity. At low supersaturations they find a correction factor. However, as shown in Appendix II this factor seems to be very different from unity, thus indicating that Ivantsov's equation cannot be applied to the case of capillarity in the way suggested. Tables 1 and 2 both show that the white structure is predicted to grow considerably faster than grey structure according to all the models.

In controlled-solidification experiments, the coarseness of the eutectic structure is obtained as function of growth rate. In order to compare such results with theory, one needs a theoretical prediction of the coarseness. This is directly obtained as $2\lambda^*$ in the theory of cooperative growth. For separate growth, on the other hand, theory only predicts the shape of the edge of a plate and not the spacing between two neighbouring plates. As a very rough estimate of the spacing according to the Zener model of platelike growth, it may be suggested that the final thickness of a plate will be twice its thickness at the edge. The spacing will then be $\lambda=8r^*\lambda/\lambda^\beta$ where λ/λ^β is obtained from the phase diagram, using the lever rule.

For Ivantsov's model, Trivedi et al.[13] suggested that the spacing should be such that the diffusion fields from the edges of two neighbouring plates will not seriously overlap each other. As shown in Appendix III, this leads to a prediction of $\lambda \simeq \pi D/4v$. The spacings thus calculated are also listed in Tables 1 and 2.

5 Shape of grey eutectic colony growing at slow cooling

GROWTH CHARACTERISTICS OF GREY EUTECTIC

Before proceeding to the result of controlled-solidification experiments, it may be advantageous to review the growth characteristics of the two eutectic structures under discussion.

The growth of a grey eutectic colony* under slow cooling is easily revealed after interruption by rapid quenching. Figure 5 shows a schematic picture drawn in accordance with a microphoto published by Hultgren et al. By repeated sectioning of such a specimen they came to the conclusion that all the graphite flakes in such a colony were interconnected,[1] as was the austenite. Bunin et al. have come to the same conclusion and have published convincing evidence.[15] Consequently, it is now generally agreed that each nucleus of graphite may develop into a eutectic colony of approximately the same extension in all directions. It might even be regarded as a structure of two interwoven crystals, one of each phase, although the crystalline orientation of the graphite lattice varies considerably. In this respect, the grey eutectic resembles a normal eutectic structure formed by a highly cooperative growth process. On the other hand, a normal eutectic ordinarily grows with an approximately flat interface, whereas the ragged interface in Fig.5 is the typical result when the interlamellar spacing λ is larger than characteristic of the temperature of solidification. A normal eutectic will give this shape during the first moment after sudden decrease of the temperature of solidification, and before a smaller λ value has been established by branching. In the case of grey eutectic it could thus be argued that the branching mechanism is not sufficiently effective to yield the proper number of lamellae. On the other hand, it is well known that the coarseness of the graphite eutectic will increase when the undercooling is decreased. One should thus conclude that the coarse spacing demonstrated in Fig.5 is the characteristic spacing of the graphite eutectic. Even without defining the reason for the coarse spacing, it may be justifiable to conclude that the graphite eutectic does not grow with a high degree of cooperation.

A reason for low degree of cooperation may sometimes be that one of the eutectic phases grows faster separately than in cooperation with the other phase It should then form crystals of the same shape as in proeutectic precipitation. It may be tempting to suggest this explanation for the flake-like form of graphite in the graphite eutectic, the characteristic shape for proeutectic graphite being thin plates. However, the theoretical calculations presented in Tables 1 and 2 clearly indicate that ideally cooperative growth should give much higher growth rates than separate growth of graphite. This conclusion does not change if one chooses an even lower value of the surface energy of graphite than 500 erg/cm².

As a consequence, it may be concluded that graphite is not able to cooperate with austenite in such a perfect way as described by the theory of cooperative growth. The growth process does not seem to be completely diffusional controlled. At the present stage it seems impossible to tell whether the growth process should be regarded as separate growth of graphite (i.e. graphite being the leading phase) or as a non-diffusional-controlled cooperative process.

GROWTH CHARACTERISTICS OF WHITE EUTECTIC

A detailed examination has been carried out of the nucleation and growth of the white eutectic in cast iron.[16] The surprising result was that the white solidification can proceed very fast from a single cementite nucleus and spread to all parts of a casting of considerable dimensions. The fast growth takes place by edgewise growth of cementite plates. By some mechanism, probably the breaking-off of fragments, plates of slightly new orientations are formed successively and the reaction spreads out in a fanlike fashion. Figure 6 shows a case where growth has occurred from the bottom upwards. It is quite evident from this microstructure that the high rate of white solidification depends upon the growth characteristics of the cementite plates. The austenite in a hypoeutectic cast iron is usually present as dendrites in the whole casting before cementite nucleates and starts to grow. However, it seems to have no direct effect upon the edgewise growth of the cementite

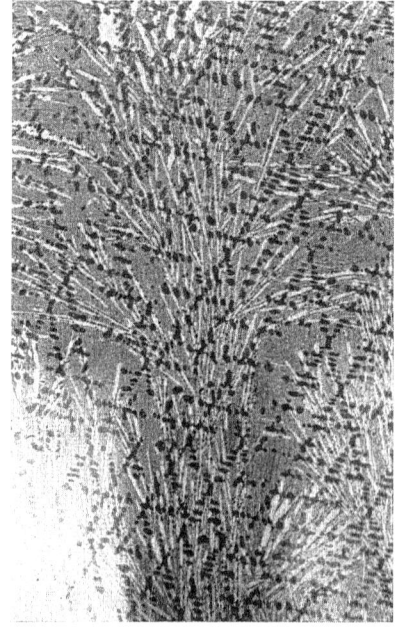

6 Progress of white solidification by fan-like growth of cementite during cooling. The reaction has been interrupted by quenching[16] ×50

---

* In order to avoid confusion with the concept of 'cellular solidification', the term 'eutectic colony' is here being proposed instead of the term 'eutectic cell' usually used in the field of cast iron. According to this proposal, a eutectic colony can split up into several cells during growth, due to cellular solidification.

7 Formation of euctectic structure in white cast iron. White phase is cementite; black phase is austenite

plates. As a second stage of the eutectic reaction, the two phases, cementite and austenite, start to cooperate and the remaining melt between cementite plates and at the sides of cementite plates will now solidify. Figure 7 shows that two different two-phase structures will result. Only the sidewise growth resembles a true cooperative process.

The theoretical data presented in Tables 1 and 2 again indicate that an ideally cooperative growth should yield higher growth rates than separate growth of cementite, although the difference is not as large as for graphite. The reason why edgewise growth is found to dominate may be that the sidewise growth of cementite is not completely diffusional-controlled, a fact that is revealed by the strong tendency of cementite to develop the flat interfaces characteristic of proeutectic cementite. There is also direct evidence that the edgewise growth of cementite is more rapid than the sidewise growth together with austenite.[16] When a steep temperature gradient was applied to a small casting during its solidification, the cementite was found to turn in such a way that the edgewise growth could take place in the direction of the gradient.

CONTROLLED SOLIDIFICATION EXPERIMENTS

During the last decade the technique called 'controlled solidification' or 'directional solidification' has been extensively applied to the study of solidification of metals with low alloy contents and of eutectic alloys. The specimen is placed in a long and thin container which is placed inside a hot furnace and pulled into a colder zone with a constant speed. The solidification will then take place at some position in the temperature gradient and occur with the same speed as the container is pulled. It is thus possible to predetermine the rate and direction of solidification and to study how the rate will affect the shape of the solidification front and the microstructure of the solidified material.

For the particular case of cast iron, the controlled-solidification technique offers another important possibility. In an earlier discussion it was predicted that the solidification of cast iron would exhibit hysteresis. One should then be able to carry out a controlled-solidification experiment and obtain grey structure at a predetermined growth rate, where the white structure should dominate, if it is nucleated. Such a conclusion may be tested if the speed of the container is suddenly increased to a high value where cementite nucleates and then decreased to the original value. If white structure is then observed to form at the speed where grey structure formed before the change, this would confirm the existence of a hysteresis.

In 1962 the authors began a research project on cast iron based on the technique of controlled solidification. Since then several other authors have reported measurements obtained with the same technique. All the data will now be considered in a comparison of theory and experiment. However, it should be realized that such a comparison may be of limited value in view of the fact that there is no satisfactory theory available for predicting the spacing of a eutectic structure, except for Zener's theory for ideally cooperative growth.

COARSENESS OF EUTECTIC STRUCTURES

As expected from the earlier observation[16] that edgewise growth occurs in the direction of a steep temperature gradient, the white structure that forms in a controlled-solidification experiment is completely of the edgewise type (Fig.8). The coarseness as a function of growth rate is shown in Fig.9, together with a theoretical line based on cooperative growth. Except for a slight difference in slope, the agreement may appear satisfactory. In view of the fact that the final structure (Fig.8) resembles the regular lamellar structure typical of cooperative growth, it may be tempting to accept the agreement shown in Fig.9 as an indication of cooperative growth. However, the microscopic evidence of edgewise growth should not be disregarded and, as shown in Tables 1 and 2, the theoretical line for separate growth of cementite according to the Zener model would almost coincide with the line for cooperative growth. On the other hand, the Ivantsov model would yield a much higher line. It should be concluded

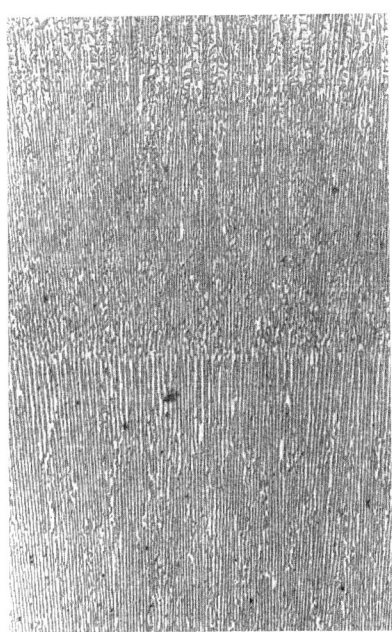

8 Structure of white euctectic when formed during controlled solidification. The rate of growth was increased at the centre of the picture ×60

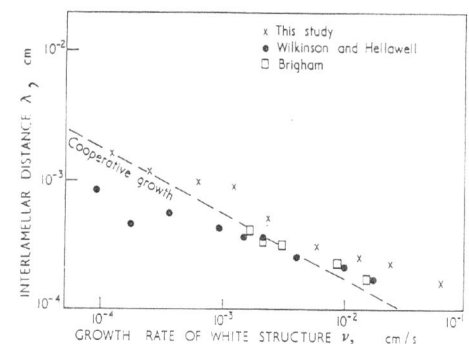

9 Coarseness of white eutectic in cast iron

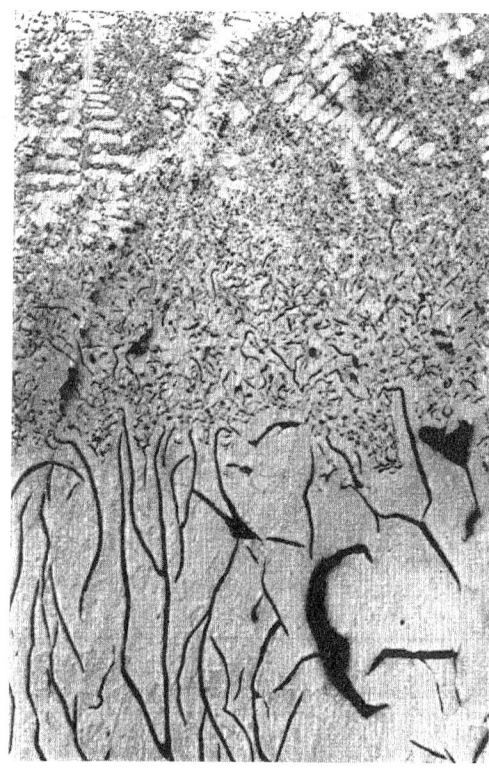

10 Structure of grey eutectic when formed during solidification. The growth rate was increased from 1.2×10^{-4} to 60×10^{-4} cm/s at the centre of the picture ×100

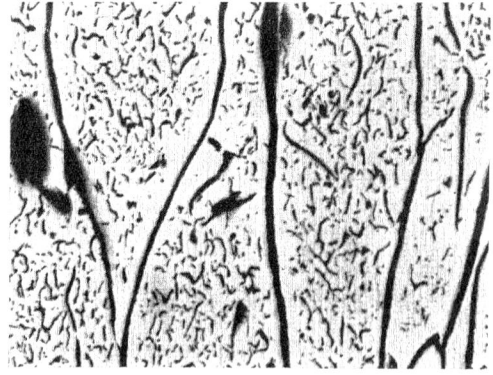

11 Formation of flake graphite and undercooled graphite side by side ×150

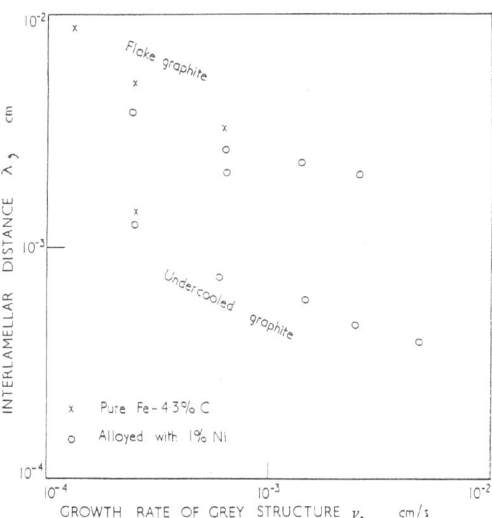

12 Coarseness of grey eutectic structure in cast iron

that the λ–v relationship does not provide a critical test.

The grey solidification results in a less-directional microstructure, in particular at high growth rates where socalled undercooled graphite forms. Figure 10 shows an example where the growth rate was suddenly changed from a low value to a much higher (from the bottom upwards).

In many cases we have obtained coarse and fine grey structure growing side by side for a considerable length (Fig.11). This seems rather puzzling and may be taken as an indication that flake graphite and undercooled graphite are of different nature. Figure 12 shows our results as reported in 1965.[17] This observation has not been reported by other authors, as shown in Fig.13 where all the data published to date are included.

The theoretical line for cooperative growth falls surprisingly close to the experimental data, considering the fact that this model should not be at all applicable to the grey solidification. The theoretical line for separate growth, according to the Zener model, falls slightly above the first line.

UNDERCOOLING AT THE SOLIDIFICATION FRONT

A more powerful method of studying the growth process would be to measure the degree of undercooling at the solidification front as a function of growth rate.

Two attempts have been made to measure directly or indirectly the temperature at the solidification front in controlled solidification experiments.[20,21] Both sets of data are very uncertain and new attempts should be

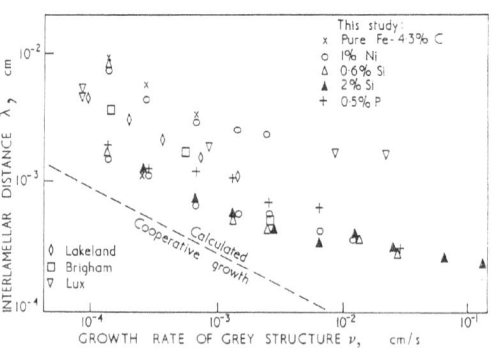

13 Coarseness of grey eutectic structure in cast iron

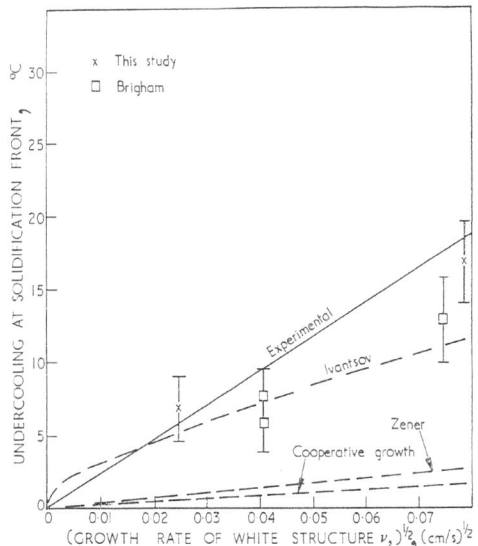

14 Solidification temperature as function of growth rate

made to obtain more reliable data on this important quantity. An attempt has also been made to evaluate the temperature of solidification from the cooling curves of ordinary castings.[28] These data also seem to be rather uncertain, the scatter being appreciable.

The experimental data from the controlled solidification experiments are presented in Figs.14 and 15 together with the theoretical predictions according to Tables 1 and 2. In spite of the large scatter in the experimental data, these diagrams seem to indicate that the theory of cooperative growth is not valid for cast iron. The data for white solidification in Fig.14 seem to be satisfactorily described by the Ivantsov model, but it should be remembered that the spacing of this structure is better described by the Zener model (Fig. 12).

The data for grey solidification in Fig.15 fall considerably below the predictions from the Ivantsov model. In view of these results, no conclusion can be drawn at present regarding the nature of the growth process of white and grey eutectics in cast iron.

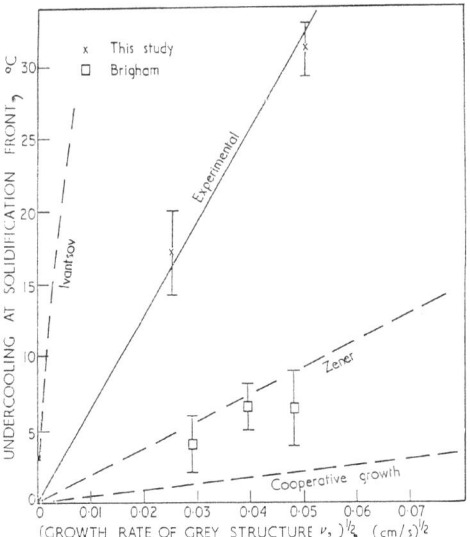

15 Solidification temperature as function of growth rate

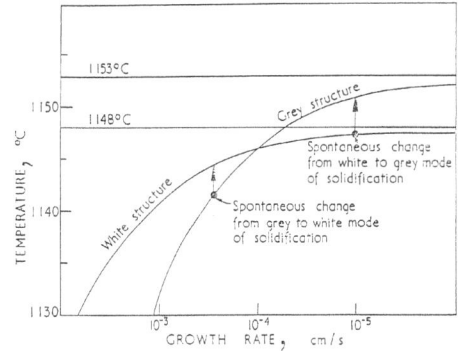

16 Experimental temperature at solidification front as function of growth rate

TRANSFORMATION DIAGRAM FOR CAST IRON

From cooling experiments with cast iron melts, Bunin et al.[24] and Drapal[25] have constructed transformation diagrams resembling the TTT diagrams for steel. Unfortunately, it is difficult to extract basic information from such diagrams in view of the fact that the solidification process in a cast iron melt is very complicated. In order to make a comparison possible with the results of the controlled-solidification technique, these data may be plotted in a diagram with temperature v. the inverse growth rate, $1/v$. Accepting the full, straight lines in Figs.14 and 15 as representing the experimental data, an experimental transformation diagram was constructed (Fig.16).

It is interesting to note that the curves for white and grey solidification intersect at a growth rate of about 10^{-4} cm/s. As a consequence, one should expect grey structure to win at lower growth rates (to the right of the intersection in Fig.16) and white structure to win at higher rates (to the left in Fig.16). In order to test this conclusion, experiments in which the growth rate was changed stepwise were carried out.[20] For a pure Fe–C melt, it was found that the mode of solidification was changed from grey to white when the rate was increased to 0.3×10^{-3} (indicated by an arrow in Fig.16), but the white structure did not change back to grey when the rate was again lowered to half the value. In fact, with the rates available, we were never able to change from white mode of solidification back to the grey mode. For nickel- and silicon-alloyed cast iron, the reverse reaction was observed, but not until the rate was lowered to 1/40 of the value necessary for the change from grey to white. By analogy, it may be suggested that the pure cast iron would change from white to grey at a rate of about 0.01×10^{-3} cm/s (indicated by an arrow in Fig.16).

The hysteresis predicted earlier has thus been confirmed. It should be realized that the size of the hysteresis may be strongly influenced by variations in the experimental technique and the composition of the cast iron. It should be of considerable theoretical interest to make the hysteresis small enough to allow a direct measurement of the critical growth rate and temperature, where the curves for white and grey structure intersect. From the data so far available, it can only be concluded that the intersection as evaluated from the experimental data of Figs.14 and 15 falls within the experimental region of hysteresis.

TRANSFORMATION DIAGRAM FOR GREY CAST IRON

It may now be possible to suggest an explanation for the fact that two grey structures have been observed growing side by side (Fig.11). At the same time the formation of austenite dendrites at the high growth rate can be

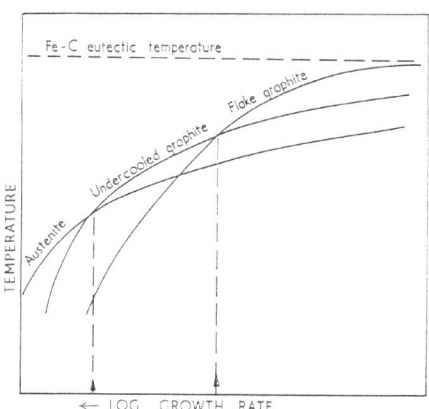

17 Hypothetical transformation diagram for grey cast iron

described as shown at the top of Fig.10. Figure 17 shows a hypothetical transformation diagram that will explain how an increase in growth rate can change the structure. It is important to realize that austenite will never fill all the space. As a consequence, if austenite dendrites start to form when the growth rate is increased, it cannot isolate the undercooled graphite from the melt and stop its growth. Similarly, the ability of flake graphite to branch and give a flat solidification front is quite limited as demonstrated by the shape of the eutectic colony in Fig.5. Consequently, it may also be difficult for the flake-graphite structure to isolate the undercooled graphite from contact with the melt and to stop its growth.

The relative positions of the three curves in Fig.17 may be explained in terms of an increased tendency for austenite to form as the temperature is lowered. At small undercooling graphite is the leading phase, at large undercooling austenite is the leading phase, and in a region of moderate undercooling there is no leading phase. The conclusion would thus be arrived at that the undercooled graphite structure forms with a higher degree of cooperation between graphite and austenite than the flake-graphite structure.

A similar observation has been made of the formation of aluminium dendrites at fast growth rates in an experimental study of Al–Si eutectic using the controlled solidification technique.[26] A very similar transformation diagram as Fig.17 may thus be suggested for Al–Si. The usual, coarse Al–Si eutectic structure would correspond to the flake-graphite structure and the socalled modified Al–Si structure would correspond to the undercooled-graphite structure. In fact the modified Al–Si structure appears to be very similar to a normal, rod-like eutectic. The fact that the silicon phase is actually present as rods and not as isolated globules has been shown by repeated sectioning,[27] by deep etching,[28] and more recently by scanning electron microscopy.[26]

Acknowledgment

This work has been financially supported by the Swedish Council for Applied Research.

References

1 A. HULTGREN et al.: *JISI*, 1954, **176**, 365.
2 K. LÖHBERG and K. RÖHRIG: *Giess. Techn. Zwiss.*, 1966, **18**, 63.
3 CHZHAN CHEN-PU: *Abst. J. Metallurgy*, 1958, Nos. 5–6, 92.
4 G. ÖSTBERG: *JISI*, 1958, **189**, 57.
5 M. HILLERT: *Trans. ASM*, 1961, **53**, 555.
6 M. HILLERT: *Acta Met.*, 1955, **3**, 34.
7 M. HILLERT: ASM Symposium on Cast Iron, Detroit, 1964.
8 A. A. ZHUKOV: *J. Phys. Chem. Akad. Nauk SSSR*, 1962, **36**, 1371.
9 W. A. TILLER: ASM Symposium on Cast Iron, Detroit, 1964.
10 C. ZENER: *Trans. AIME*, 1946, **167**, 550.
11 M. HILLERT: *Jernkon. Ann.*, 1957, **141**, 757.
12 G. P. IVANTSOV: 'Growth of crystals', v.3; 1960, New York, Consultants Bureau.
13 R. TRIVEDI et al.: *J. Appl. Phys.*, (to be published).
14 R. TRIVEDI: Ph.D. Thesis. 1966, Carnegie Inst. of Tech.
15 K. P. BUNIN et al.: *Lit. Proizv.*, 1953, **4**, 21; 25.
16 M. HILLERT and H. STEINHÄUSER: *Jernkon. Ann.*, 1960, **144**, 520.
17 M. HILLERT: Progress Report, Project No.3656 (Swedish Technical Council 1965).
18 M. P. WILKINSON and A. HELLAWELL: *BCIRA J.*, 1963, **11**, 439.
19 K. D. LAKELAND: *ibid.*, 1964, **12**, 634.
20 SUBBA RAO: Thesis 1966, Royal Inst. of Technology.
21 R. BRIGHAM: Thesis 1966, McMaster University, Canada.
22 B. LUX: *Giess. Techn. Wiss.*, 1966, **18**, 219.
23 B. LUX and W. KURZ: *Giessereiforschung*, 1967, **19**, 49.
24 K. P. BUNIN and L. T. KALININA: *Doklady A. N. Ukr. SSR*, 1960, **9**.
25 S. DRAPAL: *Kovové Mat.*, 1966, IV, 104.
26 M. HILLERT and N. LANGE: unpublished research.
27 L. Å. FORSMAN and S-A RAPP: Thesis 1962, Royal Inst. of Techn., Stockholm.
28 J. A. F. BELL and W. C. WINEGARD: *Nature*, 1965, **208**, 177.

APPENDIX I

Ivantsov[12] has derived the following equation for the edgewise growth of a plate

$$\Omega = \sqrt{\pi} \cdot \sqrt{p} \cdot e^p \cdot erfc(\sqrt{p})$$

where $p = \dfrac{vr}{2D}$

The various quantities have the same definition as in paper. It will be assumed that the effect of capillarity can be taken into account by adding the factor $(1 - r^*/r)$ to the left-hand side.

Restricting the calculation to very small p and Ω, we obtain approximately

$$\Omega \cdot (1 - r^*/r) = \sqrt{\pi} \cdot \sqrt{p}$$

Using the definition of p, we find

$$v = \frac{2D\Omega^2}{\pi r}\left(1 - \frac{r^*}{r}\right)^2$$

The maximum value of v is easily calculated.

$$v_{max} = \frac{8D\Omega^2}{27\pi r^*} \quad \text{at } r = 3r^*$$

APPENDIX II

Trivedi et al.[13] have modified the Ivantsov solution to take into account the effect of capillarity on the shape of the plate. They give a modified equation for small values of p and Ω that can easily be changed into the following expression:

$$\Omega \cdot \left(1 - \frac{r^*}{r}\right) = \sqrt{\pi} \cdot \sqrt{p} \cdot e^p \cdot erfc\left(\sqrt{p}\right) \cdot \left[1 - \frac{3\Omega r^*}{2pr}\right]$$

The bracket may be regarded as a correction factor and the term $3\Omega r^*/2pr$ should thus be a small number. However, using the solution of Appendix I as an approximation, the term can be estimated in the following way for the case of $r=3r^*$ that yields the maximum growth rate.

$$\frac{3\Omega r^*}{2pr} = \frac{3\pi\Omega}{2\Omega^2} \cdot \frac{r^*/r}{1-r^*/r} = \frac{9\pi}{8\Omega} \gg 1$$

It may thus be concluded that the Ivantsov solution will be seriously affected by capillarity.

APPENDIX III

According to the Ivantsov solution[12] the concentration c in the matrix in the neighbourhood of the edge of a growing plate is described by the following equation.

$$\frac{c-c^\infty}{c^i-c^\infty} = \frac{erfc(\sqrt{p}.\xi)}{erfc(\sqrt{p})}$$

where $\xi = [(X^2+Y^2)^{1/2} + X]^{1/2}$

c^i is the concentration at the interface and c^∞ is the original concentration of the matrix, X and Y are dimensionless coordinates. Consider two parallel, growing plates, a distance d apart. Their radius of curvature at the edges is r. At the point exactly between the two edges, $Y=d/2r$ and $\xi=\sqrt{d/2r}$. Following a suggestion by Trivedi et al.[13] we shall assume that the two plates are relatively unaffected by each other if their concentration fields do not overlap appreciably. Let us allow each plate to cause a change of $(c-c^\infty)/(c^i-c^\infty)=0.5$ at the point under consideration. For small p we obtain

$$\tfrac{1}{2} = \frac{erfc(\sqrt{p}.\sqrt{d/2r})}{erfc(\sqrt{p})} \simeq \frac{1-2\sqrt{pd/2\pi r}}{1-2\sqrt{p/\pi}} \simeq 1-2\sqrt{pd/2\pi r}$$

The definition $p = vr/2D$ yields $d \simeq \pi D/4v$. It is suggested that this d value can be used as an approximate value of the spacing between plates formed in a controlled solidification experiment.

11 Introduction to two papers by Mats Hillert: "The Role of Interfaces in Phase Transformations"

published in *Institute of Metals Monograph* (1969)

and "Diffusion and Interface Control of Reactions in Alloys"

in *Metall. Trans.* (1975)

In 1960, a symposium on the "*Decomposition of Austenite by Diffusional Processes*" was held at the fall meeting of AIME in Philadelphia with the aim of reviewing the status quo of knowledge about these transitions. The discussions focused on questions about the effect of crystallography on growth, especially of pearlite, the effect of alloying elements on the transformation, and nucleation. Very little attention was drawn on the effect of the interfaces on the mechanism of these phase transformations. Taking this as a starting point, the progress made by Mats Hillert in the 10 years following this stock taking is really impressive, as manifested by the two papers of the years 69 and 75. Both focus on the role played by interfaces in the mechanism and the kinetics of transformations.

The first paper appeared in the Proceedings of a Symposium on the "*Mechanism of Phase Transformations in Crystalline Solids*" in 1968. That may be the reason why it seems to be less frequently cited than the second one. Much attention is given to lamellar structures and their formation such as discontinuous precipitation, pearlite reaction, but also on ferrite precipitation from austenite. The basic message is that local equilibrium concepts, including paraequilibrium, cannot account for the variety of phenomena observed. It is a specialty of Mats Hillert to make extended use of Gibbs energy diagrams in order to explain the adopted approach. These illustrations offer a very useful intuitive interpretation. It is then complemented by a full mathematical treatment and solution of the problem.

The second paper, presented at the occasion of the R.F. Mehl Medal award, focuses on deviations from local equilibrium due to finite interface mobilities and solute drag. The discussion includes massive transformations from γ to α showing that the existence of a plateau in the transformation temperature in iron is due to the presence of Carbon. The competitive treatments of solute drag by Hillert on the one hand and Cahn on the other are critically discussed.

The fundamental concepts presented in these two papers have proven to offer a very powerful approach, since these concepts have been continuously elaborated, particularly with the availability of computer facilities. These concepts were used in numerical solutions which not only permit to reproduce quantitatively existing experimental observations, but more importantly offer a predictive capability. Thus, the two publications are among those forming the basis for the broad field of later developments and every active person in the field should definitely have access to them.

Gerhard INDEN

Mats Hillert at the CALPHAD meeting in 1977 with I. Ansara (left) and G. Inden (middle).

Session VI

The Role of Interfaces in Phase Transformations

M. Hillert

Some effects of interfaces on solid-state phase transformations are examined using a simple model where the interface is regarded as a thin film of boundary "phase" having its own molar free-energy function for binary alloys. The free-energy losses during the movement of such an interface are evaluated from free-energy diagrams. Two mechanisms of free-energy dissipation are found, caused by diffusion ahead of the interface and inside the interface. The force necessary for moving the grain boundaries in a single-phase binary alloy at a given rate is estimated from this model and the result agrees with the impurity-drag effect treated by Cahn. In particular, the force decreases at high rates as a result of an increasing deviation from local equilibrium at the interface. For discontinuous precipitation, the deviation from local equilibrium at the interface results in a force pulling the grain boundary along with the precipitating phase, thus supplying a force necessary for this kind of reaction. It is suggested that discontinuous precipitation can occur only when there is a deviation from local equilibrium. A detailed calculation of the diffusional process during discontinuous precipitation is carried out and compared with equivalent calculations for a eutectoid transformation under volume or grain-boundary diffusion. The dissipation of free energy under "zero-growth-rate" conditions is examined and Zener's optimum-spacing hypothesis is discussed. The conditions for diffusionless transformation are examined in some detail. The effect of alloying elements on the transformation of austenite is discussed considering the behaviour of an alloying element in and close to the moving interface. The simple pile-up model for the effect is accepted at high temperatures. At lower temperatures two new effects occur; first an increasing free-energy loss due to diffusion inside the interface and, secondly, an increasing deviation from local equilibrium at the interface. As an application, the TTT diagram for the ferrite formation in a molybdenum steel is explained, making use of Hultgren's concept of paraequilibrium.

Grain boundaries and phase boundaries are often considered as mathematically sharp boundaries, although they have important physical properties. As an example of such a property, the specific interfacial energy, σ, may give rise to a pressure difference if the surface is curved

$$P = \sigma \left(\frac{1}{\rho_1} + \frac{1}{\rho_2} \right) \quad \ldots (1)$$

The mathematically sharp boundary model was used, e.g. by Becker,[1] in deriving an expression for σ in a binary alloy. On the other hand, it has been shown that important properties of σ can be derived by applying a more realistic model of the boundary where it is treated as a transition region of some width.[2–4]

Another property of considerable importance for phase transformations is the mobility M that relates the velocity of a boundary v to the pulling force,

$$P = v/M \quad \ldots (2)$$

M is sometimes assumed to be a constant; in other cases it varies with v. In particular, the theory of the mobility of grain boundaries in alloys has been based on a model where the boundary is treated as a thin film of some thickness.[5,6]

A third property of great importance for phase transformations is the enhanced diffusion along boundaries. In the treatments of boundary diffusion the predominant model is that the boundary can be regarded as a thin film of grain-boundary material of thickness δ and diffusivity D^B. The grain boundary is thus treated as a film of a separate phase having its own properties. This crude model will be applied in the present paper, and the grain-boundary material will be assumed to have its own molar free-energy function. Free-energy changes in and close to the boundary will be considered and expressions will be derived for the forces acting on the boundary due to such free-energy changes.

Such expressions are based on the fundamental relation

$$P = \Delta G_m / V_m \quad \ldots (3)$$

where V_m is the molar volume of the new phase. Depending on the sign of ΔG_m, the force may act in either direction. For the case of boundary migration in single-phase alloys, equation (3) will be used to derive the retarding force due to the so-called impurity-drag effect. For the case of discontinuous precipitation, it will be shown that equation (3) will yield the force necessary for pulling the grain boundary along as the new phase precipitates. Without this force, which has been overlooked in previous theories of discontinuous precipitation, this process may not be possible.

Free-Energy Diagram of Binary Alloys

The chemical equilibrium between two phases may be displaced by a pressure difference caused by a curved interface. The well-known Thomson–Freundlich equation describes this effect quantitatively for a simple case. Under more complicated conditions, the mathematics of the thermodynamic calculation may be very complex and quite difficult to penetrate. However, the situation can be presented quite

---

Manuscript received 25 April 1968. Professor M. Hillert, Sc.D., is at the Royal Institute for Technology, Stockholm, Sweden.

Session VI: Interface-Controlled Transformations

quantitatively by using a free-energy diagram and the derivation of mathematical expressions may thus be reduced to a matter of arithmetic.

To simplify the discussion in this paper, we shall always assume that the original matrix phase, γ, is under ordinary pressure and any pressure difference according to equation (1) will thus be situated in the growing phase, α or β. Fig. 1 shows an example of the well-known free-energy diagram for a binary system containing two phases. A γ phase with the composition X_1 is supersaturated with respect to the β phase. There is a driving force ΔG^β for the formation of the β phase and β may thus form even under a high pressure P^β, caused by a curved interface. The critical value of this pressure is easily calculated using equation (3). $P^\beta = \Delta G^\beta / V_m$, thus allowing us to evaluate the critical size of a nucleus or the critical curvature of the interface during growth.

To carry out a numerical calculation, the shape of the free-energy curves must be known. Such calculations may be quite complex, in particular if the molar volume V_m of the β phase is not constant but must be represented by the expression

$$V_m = (1 - X) \cdot \bar{V}_A + X \cdot \bar{V}_B \quad \ldots (4)$$

The broken curve in Fig. 1 shows that the free-energy curve of the β phase is distorted when raised a distance $P^\beta V_m$. However, it has been shown[7] that the composition of the β phase in such a case is simply obtained by the point of tangency if a tangent to the β curve is drawn in such a way that the distances between the intercepts at the two sides of the diagram are related as $\bar{V}_A$ to $\bar{V}_B$.

Numerical calculations are greatly simplified if the free-energy curve of the β phase is so narrow that the composition X^β can be treated as a constant. For instance, using the equation

$$\Delta G^\beta = (1 - X^\beta)(\mu_A^1 - \mu_A^e) + X^\beta (\mu_B^1 - \mu_B^e) \ldots (5)$$

where μ represents the chemical potentials, i.e. the level of the intercepts of a tangent on the sides of the diagram (see Fig. 2), and e represents the equilibrium between the two phases, and approximating the γ phase as an ideal or dilute solution, we get

$$\Delta G^\beta = RT \left[(1 - X^\beta) \ln \frac{1 - X_1^\gamma}{1 - X_e^\gamma} + X^\beta \ln \frac{X_1^\gamma}{X_e^\gamma} \right] \quad \ldots (6)$$

If $X_1^\gamma - X_e^\gamma \ll X_e^\gamma$, we get approximately

$$\Delta G^\beta = RT \left[\frac{1 - X^\beta}{1 - X_e^\gamma} - \frac{X^\beta}{X_e^\gamma} \right] (X_e^\gamma - X_1^\gamma) \quad \ldots (7)$$

It is often of interest to calculate the total driving force for the complete reaction. This quantity is represented by ΔG_{total} in Fig. 2 and using the same approximation, equation (6) (with X_1^γ substituted for X^β) will now give us

$$\Delta G_{\text{total}} = RT \frac{(X_e^\gamma - X_1^\gamma)^2}{2 X_1^\gamma} \quad \ldots (8)$$

Free-Energy Sinks

All the free energy available for a transformation cannot always be used to overcome the pressure difference according to equation (1), if the transformation takes place with a measurable speed. Some energy may be used to overcome the friction according to equation (2) and, if diffusion occurs, this process will also require free energy. Consider the formation of β phase from a supersaturated γ matrix with composition X_1 (Fig. 3). Let us assume that a concentration difference of $X_1 - X_i$ is needed in the matrix to make the diffusion sufficiently rapid. X_i will then be the matrix composition close to the growing β phase. Fig. 3 demonstrates that the driving free energy for moving the boundary, ΔG^β, is lowered by the diffusion process. Of the total free energy available for the complete transformation, the quantity ΔG_d is used up by the diffusion process. Its size is easily estimated in the same way as equation (8), yielding

$$\Delta G_d = RT \frac{(X_i^\gamma - X_1^\gamma)^2}{2 X_1^\gamma} \quad \ldots (9)$$

Application of Boundary Model to Grain-Boundary Migration

We shall assume that a grain boundary in a single-phase binary alloy can be regarded as a thin film of a special phase with a free-energy curve such that equilibrium with a γ grain is established when the concentrations are related by

$$X^B = K \cdot X^\gamma \quad \ldots (10)$$

Fig. 4 demonstrates this case with a K value larger than unity. Equilibrium is obtained from any pair of parallel tangents and it is a true, stable equilibrium, since the amount of the grain-boundary phase is constant as the width of the film δ is constant. The vertical position of the free-energy curve for the grain-boundary film was chosen arbitrarily in Fig. 4, this position not being of any importance for the present discussion. On the other hand, it is important if one wants to discuss the value of the interfacial free energy.

The concentration profile is shown at the bottom of Fig. 4. In spite of the high concentration inside the boundary, it should be realized that X_1 will be the average composition of the material flowing down through the boundary, if the boundary is slowly migrating upward with a constant velocity. At the upper side of the boundary, the material entering the boundary is transferred to a higher free-energy state, as illustrated by the arrow ΔG_u, and at the lower side of the boundary the same amount of material is leaving the boundary and thus being transferred to a lower state as illustrated by the arrow ΔG_l. However, no net force is acting on the boundary since

$$P = (\Delta G_u - \Delta G_l)/V_m = 0 \quad \ldots (11)$$

The concentration profile will change somewhat if the boundary is moving with a constant measurable velocity, (Fig. 5). In each phase the profile must be of the exponential form with the value X_1 as the limiting value.

$$X^\gamma = X_1 + (X_u^\gamma - X_1) \exp[-vy/D^\gamma] \quad \ldots (12)$$

$$X^B = X_1 + (X_l^B - X_1) \exp[-v(y + \delta)/D^B] \quad \ldots (13)$$

Assuming chemical equilibrium to be established locally at each side of the boundary, equation (10) will yield

$$X_l^B = K X_l^\gamma = K \cdot X_1 \quad \ldots (14)$$

$$X_u^B = K X_u^\gamma \quad \ldots (15)$$

and the application of equation (13) at $y = 0$ yields

$$X_u^B = X_1 [1 + (K - 1) \exp(-v\delta/D^B)] \ldots (16)$$

The quantity X_u^γ in equation (12) can now be substituted to yield

$$X^\gamma = X_1 \left[1 + \frac{K - 1}{K} (\exp(-v\delta/D^B) - 1) \exp -vy/D^\gamma \right] \quad \ldots (17)$$

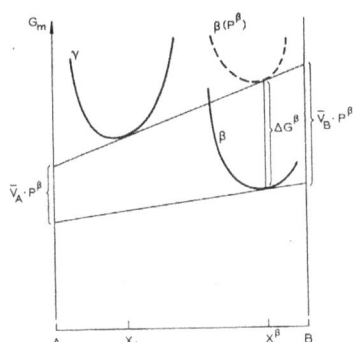

Fig. 1 *Free-energy diagram showing the driving force for precipitation of β from γ.*

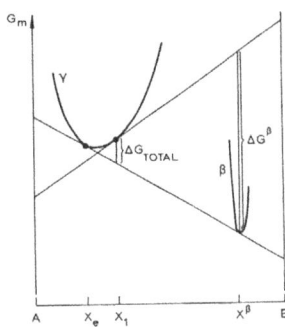

Fig. 2 *Free-energy diagram for precipitation of β.*

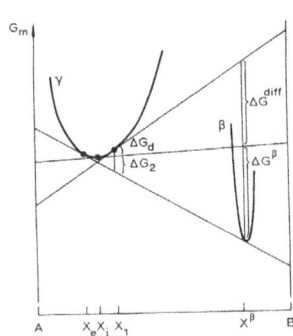

Fig. 3 *Free-energy diagram for diffusion and precipitation of β*

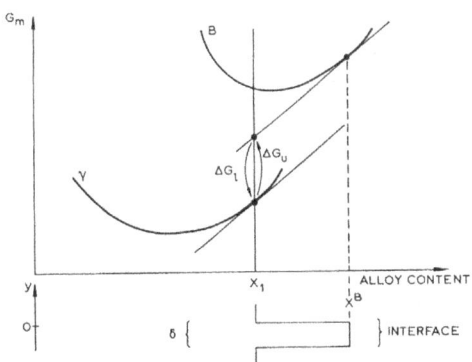

Fig. 4 *Free-energy diagram for interface at rest.*

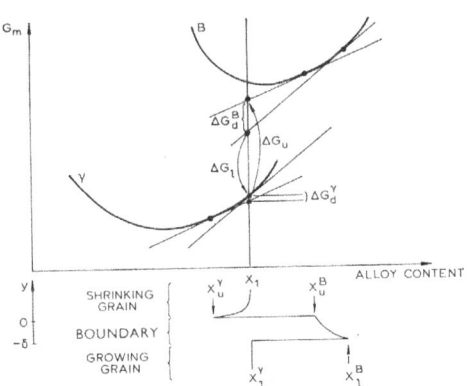

Fig. 5 *Free-energy diagram for slowly moving boundary.*

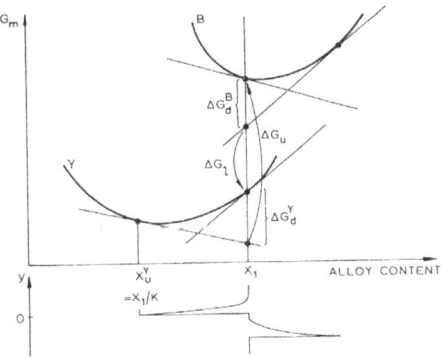

Fig. 6 *Free-energy diagram for moving boundary with maximum friction.*

Session VI: Interface-Controlled Transformations

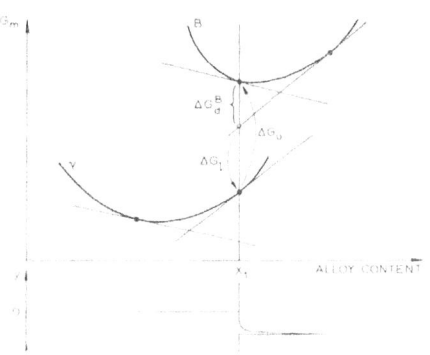

Fig. 7 Free-energy diagram without local equilibrium at the upper side of the boundary.

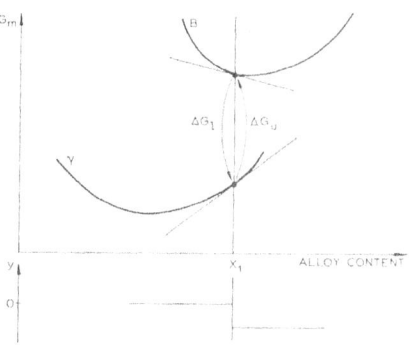

Fig. 8 Free-energy diagram with no local equilibrium at the sides of the boundary.

It is interesting to note that all the compositions at the two sides of the boundary are independent of the diffusivity in the grain, D^γ. The compositions at the upper side X_u^γ and X_u^B are dependent on the diffusivity inside the boundary D^B.

The concentration profiles in Fig. 5 should give rise to diffusion and one should thus expect a loss of free energy in each phase, ΔG_d^γ and ΔG_d^B, i.e. a retarding force on the boundary motion. This fact is also illustrated by the difference in length of the two arrows ΔG_u and ΔG_l in Fig. 5 and, as expected, one finds

$$P = (\Delta G_u - \Delta G_l)/V_m = (\Delta G_d^\gamma + \Delta G_d^B)/V_m \quad \ldots (18)$$

Using equation (9) we find

$$\Delta G_d^\gamma = \frac{RT}{2X_1}(X_1 - X_u^\gamma)^2$$

$$= \frac{RT}{2}\left(\frac{K-1}{K}\right)^2 [1 - \exp(-v\delta/D^B)]^2 \cdot X_1 \quad \ldots (19)$$

$$\Delta G_d^B = \frac{RT}{2X_1}[(X_1 - X_l^B)^2 - (X_1 - X_u^B)^2]$$

$$= \frac{RT}{2}(K-1)^2 [1 - \exp(-2v\delta/D^B)] \cdot X_1 \quad \ldots (20)$$

For low velocities, v, and K not much less than unity, the exponential functions can be expanded in series and inserting equations (19) and (20) in equation (18) we obtain

$$P = \frac{RT\delta(K-1)^2 X_1}{D^B V_m} \cdot v \quad \ldots (21)$$

We have thus derived an expression for the effect of an alloy element on the mobility M of a boundary. Hypothesizing an interaction force between the alloy atoms and the grain boundary, Lücke and Detert[5] were able to derive a similar expression for the same effect, the only difference being that they obtained $(K-1)$ to the first power instead of squared. Their expression thus seemed to predict a negative friction force P for an alloy element with $K < 1$, an unreasonable result as pointed out by Cahn.[6] Cahn carried out a very ambitious treatment of the problem assuming a continuous variation of the attraction energy and diffusivity through the boundary, and was able to resolve the difficulty. For low velocities his result seems to be in agreement with our equation (21).

For high velocities the concentration profiles will change to the situation demonstrated in Fig. 6. The concentrations at the upper side of the boundary have reached their limits, $X_u^B = X_1$ and $X_u^\gamma = X_1/K$ and the friction force has reached its maximum value

$$P = \frac{RT(K-1)^2 X_1}{2V_m} \quad \ldots (22)$$

On the other hand, at high velocities we must consider a new effect; the concentration profiles may become so narrow that they only exist mathematically but not physically. This will first occur in the γ phase because $D^\gamma \ll D^B$ (Fig. 7), and finally in the boundary phase as well (Fig. 8). The difference between the two arrows, ΔG_u and ΔG_l, will decrease and go to zero.

Owing to the atomistic nature of the system, it may never be realistic to treat the extreme value at the tip of a profile as existing physically. Instead we shall choose to use the value existing at a distance of $\delta/4$ from the tip of the profile. This is an arbitrary choice but should be fairly realistic considering the fact that the width of the grain boundary, δ, is probably a few atomic distances. Fig. 9 demonstrates this model and yields the following result:

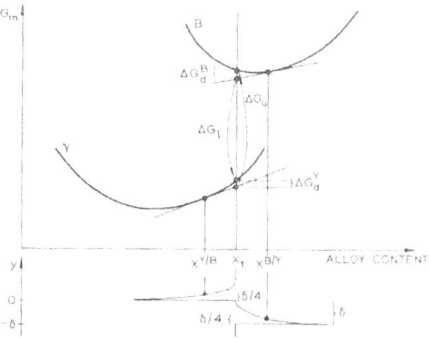

Fig. 9 Realistic free-energy diagram for rapidly moving boundary.

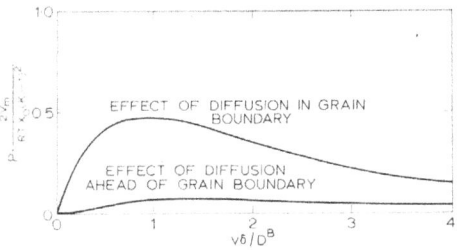

Fig. 10 Impurity-drag effect on the movement of a grain boundary calculated with $D^B = D^\gamma$. The lower curve will be negligible for $D^B \gg D^\gamma$.

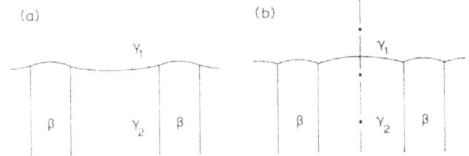

Fig. 11(a) and (b) Possible shapes of the interface of discontinuous precipitation during growth.

$$X^\gamma(y = \delta/4) = X_1 \left[1 - \frac{K-1}{K} [1 - \exp(-v\delta/D^B)] \cdot \exp(-v\delta/4D^\gamma) \right] \quad \ldots (23)$$

$$X^B(y = -3\delta/4) = X_1 [1 + (K-1)\exp(-v\delta/4D^B)] \quad \ldots (24)$$

$$\Delta G_d^\gamma = \frac{RT}{2}\left(\frac{K-1}{K}\right)^2 [1 - \exp(-v\delta/D^B)]^2 \cdot \exp(-v\delta/2D^\gamma) \cdot X_1 \quad \ldots (25)$$

$$\Delta G_d^B = \frac{RT}{2}(K-1)^2 [\exp(-v\delta/2D^B) - \exp(-2v\delta/D^B)] \cdot X_1 \quad \ldots (26)$$

$$P = \frac{RT}{2V_m}(K-1)^2 \left[\exp\left(-\frac{v\delta}{2D^B}\right) - \exp\left(-\frac{2v\delta}{D^B}\right) - \left[1 - \exp\left(-\frac{v\delta}{D^B}\right)\right]^2 \cdot \exp\left(-\frac{v\delta}{2D^\gamma}\right)\frac{1}{K^2} \right] \cdot X_1 \quad \ldots (27)$$

This expression holds for all values of the velocity, low and high. Fig. 10 shows the result of a numerical calculation carried out with $D^\gamma = D^B$ and $K = 2$. The very low curve represents the effect due to diffusion in the γ phase ahead of the boundary. With values of $D^\gamma \ll D^B$ this curve will be even smaller and can be completely neglected for practical purposes. The main effect causing the friction force is thus due to the diffusion process inside the boundary.

In principle, Fig. 10 shows the same result as obtained by Cahn[6] using a more ambitious model. An advantage of the present model may be that it yields simpler mathematics and a final expression in closed form for all values of velocity.

It is interesting to note that the decrease in the friction force occurs as a result of a growing deviation from local equilibrium at the sides of the boundary. In the next section we shall find that such a deviation from local equilibrium may even result in a force pulling the boundary.

Mechanism of Discontinuous Precipitation

Discontinuous precipitation resembles a eutectoid transformation, although the phase diagram indicates that only one new phase should form. However, this phase forms as parallel lamellae growing together with a new grain of the matrix phase

$$\gamma_1 \to \beta + \gamma_2$$

The new γ_2 grain has a much lower alloy content than the original, supersaturated γ_1 grain. The loss of supersaturation thus occurs discontinuously as the γ_1/γ_2 grain boundary advances and it has been suggested that the reaction is controlled by grain-boundary diffusion.[8]

The growth process has been treated theoretically by Turnbull[8] and Cahn.[9] They calculated the rate of grain-boundary diffusion necessary to lower the alloy content in front of the new γ_2 grain but neglected to consider the force actually pulling the grain boundary. As a consequence, their treatments give no indication why discontinuous precipitation occurs. Cahn simply assumed that the grain boundary will move with a rate proportional to the available free energy, without considering by what mechanism part of this energy could be transformed into a force.

To resolve this difficulty, Kirkaldy[10] suggested the existence of a metastable miscibility gap in the γ phase. The reaction could then be treated as a eutectoid transformation where there is a chemical driving force acting on the growth of both the new phases. Shapiro[11] has worked out this theory in detail. On the other hand, it may be argued that discontinuous precipitation occurs in such regions of temperature and composition that it is difficult to imagine that suitable miscibility gaps would exist in all necessary cases.

Another mechanism has been proposed by Sulonen,[12] who suggested that the difference in atomic size would give rise to strain energy in the concentration gradient ahead of the growing γ_2 grain. This effect would result in a force on the boundary because the strains would be released as the new grain advances. This model has not been worked out in detail.

A further possiblity is that the interfacial energies at the three-phase junctions balance each other in such a way that the γ_1/γ_2 grain boundary is actually pulled by the growing β phase, (Fig. 11(a)). This mechanism has been observed occasionally during precipitation of cementite from austenite but may not be a mechanism characteristic of precipitation controlled by grain-boundary diffusion. On the contrary, evidence from electron microscopy[11,13] strongly indicates that the γ_1/γ_2 grain boundary is convex, as illustrated by Fig. 11(b).

To derive an expression for the force acting on the γ_1/γ_2 grain boundary, we should examine the free-energy diagram. The concentration profile at the bottom of Fig. 12 holds along a line perpendicular to the grain boundary (e.g. the line at the centre of the γ_2 lamella in Fig. 11(b)). The concentration across the assumed grain-boundary film is almost constant because of the high value of D^B in relation to the growth rate. On the other hand, there may be a very rapid change in the matrix grain in front of the boundary because

Session VI: Interface-Controlled Transformations

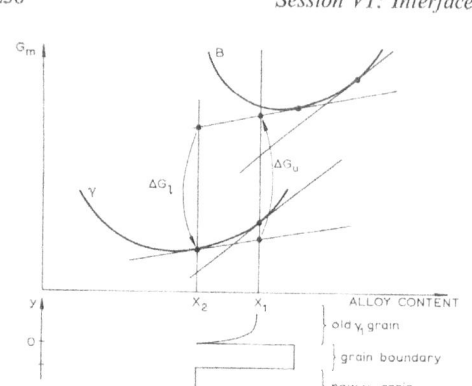

Fig. 12 Free-energy diagram for the grain boundary in discontinuous precipitation assuming local equilibrium.

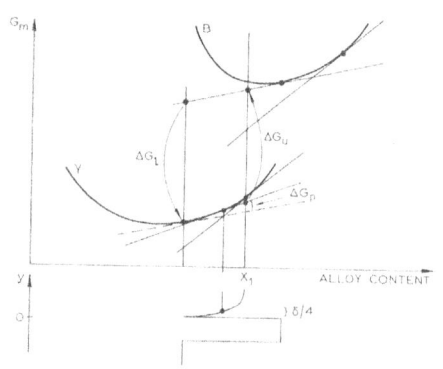

Fig. 13 Free-energy diagram for the grain boundary in discontinuous precipitation with deviation from local equilibrium.

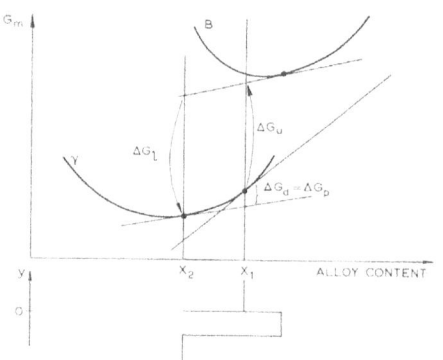

Fig. 14 Free-energy diagram for the grain boundary in discontinuous precipitation with maximum force on the boundary.

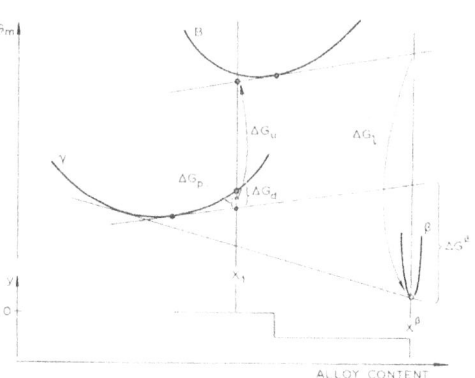

Fig. 15 Free-energy diagram for the γ/β interface in discontinuous precipitation.

D^γ is much smaller. When using the free-energy diagram in this case, it must be realized that the average composition of the material entering the grain boundary is not the same as that of the material leaving the grain boundary to form the new γ_2 grain. The two composition values are X_1 and X_2, respectively. The difference is caused by the sidewise diffusion along the grain boundary necessary for the growth of the β phase. The arrows ΔG_u and ΔG_l at the top of Fig. 12 represent the free-energy changes accompanying the transfer of material. They have the same length and will thus give no net force on the grain boundary.

If the growth rate is high enough in relation to D^γ, the profile may be so sharp that it is not realistic to assume full chemical equilibrium at the upper side of the grain boundary. Fig. 13 shows such a situation using the same approach as in the previous section. The two arrows will now have different lengths, the difference being equal to the free energy that is no longer used on diffusion in the γ phase. In this case we find a force pulling the grain boundary upwards

$$P = (\Delta G_l - \Delta G_u)/V_m = \Delta G_p/V_m \quad \ldots (28)$$

This force will reach its maximum value when the transformation is so rapid that no diffusion occurs in the matrix ahead of the grain boundary (Fig. 14).

$$P_{\max.} = \Delta G_d/V_m \quad \ldots (29)$$

From this model we may predict that discontinuous precipitation would occur only when the grain-boundary diffusion is so much higher than lattice diffusion that the profile will be steep enough. Equation (17) shows that the shape of the profile is determined by the ratio D^γ/v. As an example, Speich[13] has evaluated $D^\gamma = 7 \times 10^{-16}$ cm/sec at 500° C in the Fe–Zn system and the lowest growth rate observed for discontinuous precipitation at that temperature is 1.15×10^{-7} cm/sec. We thus find

$$D^\gamma/v = 7 \times 10^{-16}/1.15 \times 10^{-7} \text{ cm} = 0.6 \times 10^{-8} \text{ cm}.$$

It may thus be safe to conclude that the force is very close to its maximum value in all Speich's experiments.

A detailed theory of the growth of discontinuous precipitation will be worked out in a later section of this paper. It will then be assumed that some fraction, f, of ΔG_d will not be spent on diffusion but give rise to a force

$$P = f \cdot \Delta G_d/V_m \quad \ldots (30)$$

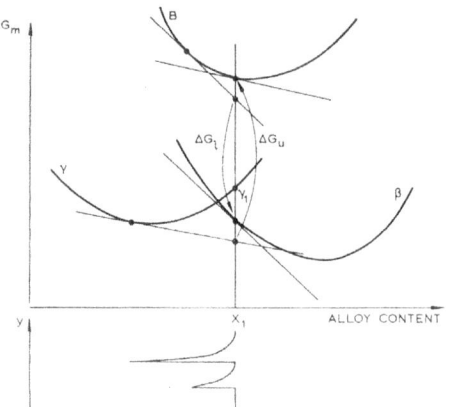
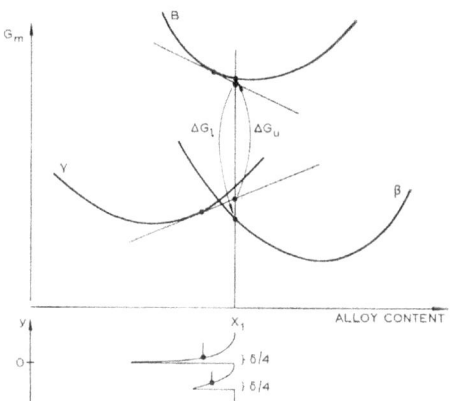

Fig. 16 Diffusionless transformation $\gamma_1 \to \beta$ is not possible with complete local equilibrium.

Fig. 17 Diffusionless transformation $\gamma_1 \to \beta$ is possible with deviation from local equilibrium.

In applying this theory to a specific case, it is necessary to estimate the value of f from the predicted value of v, using the parameter D^γ/v to estimate the steepness of the profile. As in the previous section, this value of f may be estimated from Fig. 13 using the composition on the mathematical shape of the profile at the distance $\delta/4$ from the boundary or the concentration at some other distance, e.g. one atomic distance.

Fig. 15 illustrates the force acting on the movement of the γ_1/β interface

$$P = (\Delta G_l - \Delta G_u)/V_m = (\Delta G^\beta + \Delta G_p)/V_m$$
$$= (\Delta G^\beta + f\Delta G_d)V_m \quad \ldots (31)$$

Diffusionless Transformation in Alloys

If the free-energy curve of the β phase in Fig. 15 was much wider, a diffusionless transformation $\gamma \to \beta$ might be possible. A necessary condition is that the composition of the matrix γ phase lies to the right of the intersection of the two free-energy curves, as illustrated in Fig. 16. However, this condition is not sufficient as shown by the fact that $\Delta G_u > \Delta G_l$. These arrows have been constructed assuming local equilibrium at both sides of the boundary. For the reaction to proceed, we must have $\Delta G_u < \Delta G_l$. This may occur if the rate of reaction is high enough to result in a deviation from local equilibrium at the upper side of the boundary. This situation is illustrated by Fig. 17. We can conclude that a diffusionless transformation may occur if there is some mechanism by which its speed is initially raised to a high value. The critical speed appears to depend upon the free-energy curve of the boundary.

The same free-energy diagram can be used to show that there is no critical speed if the original composition lies to the right of the composition of β phase in stable equilibrium with γ phase. In such an alloy, the precipitation of β phase should automatically develop into a diffusionless growth.

Balance of Forces in Lamellar Structures

We shall now consider the formation of lamellar structures and assume that they are completely regular and contain perfectly parallel lamellae. The radius of curvature at any point on the advancing interface should be such that all the forces will balance each other. Equations (1)–(3) give

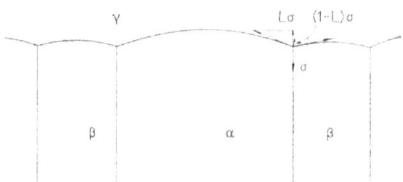

Fig. 18 Balance of forces at three-phase junctions.

$$\frac{\sigma}{\rho} = \frac{\Delta G_m}{V_m} - \frac{v}{M} \quad \ldots (32)$$

The three-phase junctions deserve particular attention. The three surface tensions should here balance each other and thus control the size of the three angles. As demonstrated by Fig. 18, the surface tension of the newly formed α/β interface, σ, is carried partly by the α lamella and partly by the β lamella. The two parts of σ may be denoted as $L\sigma$ and $(1 - L)\sigma$, where L can easily be calculated from information on the three angles or the three surface tensions.

Balancing the forces acting on the whole edge of the α lamella, we find

$$2L\sigma = \int_{-S^\alpha/2}^{S^\alpha/2} (\Delta G_m/V_m - v/M)\, dz \quad \ldots (33\alpha, \beta)$$

By the number (33α, β) we indicate that a similar equation holds for the β lamella. To use this equation, it is necessary to know how ΔG_m varies along the edge of the lamella, i.e. to know the variation in composition along the edge. This kind of information can be obtained from a calculation of the diffusion.

Diffusion during Growth of Lamellar Structures

We shall neglect the possibility of diffusion occurring in the growing phases, thus assuming that they retain the composition obtained at the moment of formation. When treating the diffusion ahead of a growing lamella, it is useful

238 Session VI: Interface-Controlled Transformations

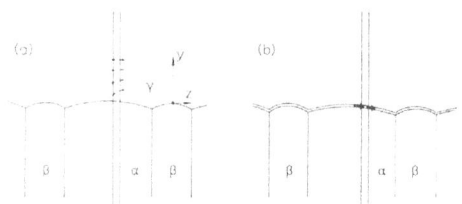

Fig. 19 Growth of lamellar structures by (a) *volume diffusion* and (b) *boundary diffusion*.

to consider a thin, long volume element as shown in Fig. 19(a). As the phase boundary moves upwards within this volume element, its change in composition must be balanced by sidewise diffusion

$$-\int_{-\infty}^{\infty} D \frac{d^2 x}{dz^2} dy = v(X_1^\gamma - X^\alpha) \quad \ldots (34\alpha, \beta)$$

To solve this equation in the general case, we must know the shape of the interface. As a consequence, equation (34α, β) must be solved simultaneously with equation (33α, β). However, in a previous paper[7] it was shown that the procedure is greatly simplified if the edges of the lamellae can be approximated as flat. A representation of the composition in the matrix by a series

$$X^\gamma - X_1^\gamma = \sum_0^\infty A_n \exp(-\lambda_n y) \cdot \cos 2\pi n z/S \quad \ldots (35)$$

will then reduce to a simple Fourier series at the edge, $y = 0$. To satisfy Fick's law

$$\lambda_n = \frac{v}{2D}(1 + \sqrt{1 + 16\pi^2 n^2 D^2/v^2 S^2}) \quad \ldots (36)$$

The application of equation (34α, β) yields the values of the Fourier coefficients

$$A_n = \frac{4\lambda_n v S^2}{D(2\pi n)^3}(X^\beta - X^\alpha) \sin n\pi S^\alpha/S \text{ for } n > 0 \quad \ldots (37)$$

At low temperatures, the volume diffusion may be very slow and boundary diffusion may instead dominate. This situation is described by Fig. 19(b), where the thickness of the boundary has been greatly exaggerated in order to illustrate the effect. Treating the boundary as a thin film of thickness δ and diffusivity D^B, we can easily integrate equation (34α, β), again approximating the interface as flat

$$-\int_{-\infty}^{\infty} D \frac{d^2 x}{dz^2} dy = D^B \frac{d^2 x^B}{dz^2} \cdot \delta = v(X_1^\gamma - X^\alpha) \quad \ldots (38\alpha, \beta)$$

Integrating (38α, β) for the case of constant X^α we obtain

$$\frac{X^B - X_3^B}{X_1^\gamma - X^\alpha} = \frac{v(S^\alpha)^2}{8 D^B \delta}\left[1 - \left(\frac{2z}{S^\alpha}\right)^2\right] \quad \ldots (39)$$

X_3^B is the composition in the boundary at the three-phase junction, i.e., at $z = S^\alpha/2$ and X_1^γ is the original matrix composition.

Assuming a constant distribution coefficient between the boundary phase and the γ phase according to equation (10), we can represent the concentration in the boundary by an equivalent composition that would hold in the γ phase at equilibrium. We shall denote such values by X^b. Equation (39) can thus be transformed,

$$\frac{X^b - X_3^b}{X_1^\gamma - X^\alpha} = \frac{v(S^\alpha)^2}{8KD^B\delta}\left[1 - \left(\frac{2z}{S^\alpha}\right)^2\right] \quad \ldots (40)$$

In discontinuous precipitation, the growing α phase is actually a grain of the same phase as the γ matrix. As long as the α phase is formed in local equilibrium with the grain-boundary material, we have $X^\alpha = X^b$. The composition X^α will then vary along the interface. Cahn[9] has shown that equation (38) will then have the following solution

$$\frac{X^b - X_1^\gamma}{X_3^b - X_1^\gamma} = \frac{\cosh z \sqrt{a}/S^\alpha}{\cosh \sqrt{a}/2} \quad \ldots (41)$$

where

$$a = v(S^\alpha)^2/KD^B\delta \quad \ldots (42)$$

Again, the properties of the boundary are expressed by a single quantity $(KD^B\delta)$.

Eutectoid Transformation at Zero Growth Rate

Some essential features of the eutectoid transformation can be demonstrated by a very simple model where the two growing phases are supposed to be in full equilibrium with the adjacent matrix. Naturally, the growth rate must be zero in this hypothetical case. There will be no sidewise diffusion and the composition of the adjacent matrix can be represented by a single value X^γ. It may be tempting to assume that X^γ is identical to the original matrix composition X_1^γ but this point will require further discussion.

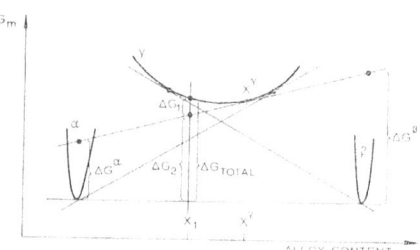

Fig. 20 *Free-energy diagram for eutectoid reaction.*

Fig. 20 shows a free-energy diagram and ΔG_{total} is the available free energy. If all this energy would go into the interfaces between the lamellae in the eutectoid structure, we would have

$$S_{\text{rev.}} = 2\sigma V_m/\Delta G_{\text{total}} \quad \ldots (43)$$

where rev. stands for reversible, indicating that no free energy is lost irreversibly on diffusion or on any other rate process.

We shall now apply equation (33α, β) using the expression of ΔG^β given by equation (7). The integration can easily be carried out since X^γ is constant

$$L\sigma V_m = \int_0^{S^\alpha/2} \Delta G^\alpha dz = RT \left[\frac{1-X^\alpha}{1-X_e^{\gamma/\alpha}} - \frac{X^\alpha}{X_e^{\gamma/\alpha}} \right] \int_0^{S^\alpha/2} (X_e^{\gamma/\alpha} - X^\gamma) dz$$

$$= RT \left[\frac{1-X^\alpha}{1-X_e^{\gamma/\alpha}} - \frac{X^\alpha}{X_e^{\gamma/\alpha}} \right] (X_e^{\gamma/\alpha} - X^\gamma) S^\alpha/2 \quad \ldots (44)$$

$$(1-L)\sigma V_m = RT \left[\frac{X^\beta}{X_e^{\gamma/\beta}} - \frac{1-X^\beta}{1-X_e^{\gamma/\beta}} \right] (X^\gamma - X_e^{\gamma/\beta}) S^\beta/2 \quad \ldots (45)$$

The lever rule relates S^α, S^β, and S. Assuming a constant molar volume for all the phases, we obtain

$$(X_1^\gamma - X^\alpha) S^\alpha = (X^\beta - X_1^\gamma) S^\beta = (X^\beta - X^\alpha) S^\alpha S^\beta/S \quad \ldots (46)$$

Equations (44) and (45) may thus be said to contain two unknown quantities, X^γ and S, and they can be solved easily. We shall denote this particular S value by S_0 because it is characteristic of zero growth rate.

Dividing equations (44) and (45) will give

$$\frac{X_e^{\gamma/\alpha} - X^\gamma}{X^\gamma - X_e^{\gamma/\beta}} = \frac{L}{1-L} \frac{S^\beta}{S^\alpha} \left[\frac{X^\beta}{X_e^{\gamma/\beta}} - \frac{1-X^\beta}{1-X_e^{\gamma/\beta}} \right] \bigg/ \left[\frac{1-X^\alpha}{1-X_e^{\gamma/\alpha}} - \frac{X^\alpha}{X_e^{\gamma/\alpha}} \right] \quad \ldots (47)$$

It is shown that the composition X^γ depends upon the relative values of the three interfacial energies involved (through L) and upon the asymmetry of the phase diagram (through the values of the two brackets). Even for the completely symmetric case, equation (47) predicts $X^\gamma = X_1^\gamma$ only for $X_1^\gamma = 1/2$. In spite of the fact that we set out to treat the case where no sidewise diffusion occurs, we find that lengthwise diffusion cannot be prevented in the general case. We should thus expect some loss of free energy, i.e. we should expect to find $S_0 \neq S_{rev}$.

Adding equations (44) and (45) will give

$$S_0 = \frac{2\sigma V_m}{RT(X_e^{\gamma/\alpha} - X_e^{\gamma/\beta})} \left\{ \frac{LS/S^\alpha}{\left[\frac{1-X^\alpha}{1-X_e^{\gamma/\alpha}} - \frac{X^\alpha}{X_e^{\gamma/\alpha}} \right]} + \frac{(1-L)S/S^\beta}{\left[\frac{X^\beta}{X_e^{\gamma/\beta}} - \frac{1-X^\beta}{1-X_e^{\gamma/\beta}} \right]} \right\} \quad \ldots (48)$$

The relation between S_0 and the composition at the front X^γ is best demonstrated by the free-energy diagram. Direct integration of equation (44) for the case of ΔG^α constant will give

$$L\sigma V_m = \Delta G^\alpha S^\alpha/2 \quad \ldots (49\alpha)$$

and in the same way for the β phase

$$(1-L)\sigma V_m = \Delta G^\beta S^\beta/2 \quad \ldots (49\beta)$$

Adding (49α) and (49β) and using the lever rule,

$$\Delta G^\alpha S^\alpha + \Delta G^\beta S^\beta = \Delta G_2 S \quad \ldots (50)$$

we obtain

$$S_0 = 2\sigma V_m/\Delta G_2 \quad \ldots (51)$$

Fig. 20 demonstrates that part of the total free energy available will be spent on diffusion, ΔG_1, and the rest will go into interfacial energy, ΔG_2. As a consequence, we find that $S_0 > S_{rev}$. However, the difference is not very great, ΔG_1 being only a minor part of ΔG_{total}.

Eutectoid Transformation by Volume Diffusion

This case has been treated by Zener,[14] using a method based on dimensional arguments, and by the present author[7] and by Jackson and Hunt[15] in a more rigorous way, which will now be followed.

We shall apply the same procedure as with zero growth rate in the preceding section. If the friction term v/M in equation (33α, β) is neglected, we again find the same expression as the first part of equation (44) but the integration will be more complicated because X^γ now varies along the edge of the lamella. For volume diffusion we can express X^γ by means of the Fourier coefficients given by equation (37) and after integration of equation (44) we obtain

$$L\sigma V_m = RT \left[\frac{1-X^\alpha}{1-X_e^{\gamma/\alpha}} - \frac{X^\alpha}{X_e^{\gamma/\alpha}} \right] \cdot \left[(X_e^{\gamma/\alpha} - X_1^\gamma - A_0) \frac{S^\alpha}{2} - \frac{vS^2(X^\beta - X^\alpha)}{2D\pi^3} \cdot B \right] \quad \ldots (52)$$

$$(1-L)\sigma V_m = RT \left[\frac{X^\beta}{X_e^{\gamma/\beta}} - \frac{1-X^\beta}{1-X_e^{\gamma/\beta}} \right] \cdot \left[(X_1^\gamma - X_e^{\gamma/\beta} + A_0) \frac{S^\beta}{2} - \frac{vS^2(X^\beta - X^\alpha)}{2D\pi^3} \cdot B \right] \quad \ldots (53)$$

where

$$B = \sum_1^\infty \frac{\lambda_n S}{2\pi n^4} (\sin \pi n S^\alpha/S)^2 \quad \ldots (54)$$

Again equation (46) relates S^α, S^β, and S. Equations (52) and (53) may thus be said to contain three unknown quantities, A_0 (defining the average composition at the front), S, and v. Adding equations (52) and (53) will give

$$\left[X_e^{\gamma/\alpha} - X_e^{\gamma/\beta} - \frac{vS(X^\beta - X^\alpha)S^2}{D\pi^3 S^\alpha S^\beta} \cdot B \right] \cdot S$$
$$= \frac{2\sigma V_m}{RT} \left[\frac{LS/S^\alpha}{\frac{1-X^\alpha}{1-X_e^{\gamma/\alpha}} - \frac{X^\alpha}{X_e^{\gamma/\alpha}}} + \frac{(1-L)S/S^\beta}{\frac{X^\beta}{X_e^{\gamma/\beta}} - \frac{1-X^\beta}{1-X_e^{\gamma/\beta}}} \right] \quad \ldots (55)$$

The right-hand side can be expressed in terms of S_0 according to equation (48) and we obtain

$$v = \frac{D\pi^3 S^\alpha S^\beta (X_e^{\gamma/\alpha} - X_e^{\gamma/\beta})}{BS^2(X^\beta - X^\alpha)} \cdot \frac{1}{S} \left(1 - \frac{S_0}{S} \right) \quad \ldots (56)$$

This relation is equivalent to the expression derived by Zener. The numerical value of B can easily be calculated for any S^α/S and shows good agreement with Zener's estimate.[7]

As pointed out by Zener, a relation such as equation (56) can be satisfied by any value of S larger than S_0. Consequently, it does not directly predict that the lamellar structure should form with a constant spacing S. However, Zener suggested that there should be some self-regulating mechanism by which the spacing would be adjusted close to the value that maximizes the growth rate

$$S_{optimum} = 2S_0.$$

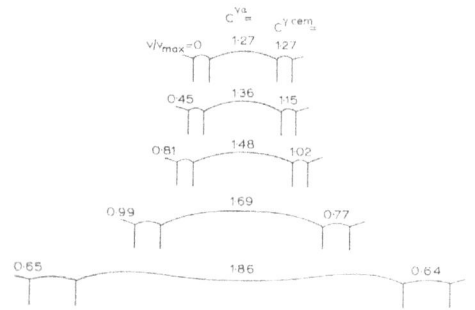

Fig. 21 Calculated shape and growth rate of pearlite with different interlamellar spacings.

Equations (48) and (56) predict

$$S_{optimum} \propto (X_e^{\gamma/\alpha} - X_e^{\gamma/\beta})^{-1}$$

$$v_{maximum} \propto (X_e^{\gamma/\alpha} - X_e^{\gamma/\beta})^2$$

For any value of S one can calculate v as well as A_0 and the concentration is then known in detail along the front. The derivation of our basic equations (44) and (45) was based on the assumption of local chemical equilibrium at all points of the growth front. We can now calculate what shape the front should have in order to be in local equilibrium, using equation (32) but again neglecting the term v/M. The radius of curvature ρ is related to the shape of a lamella by

$$\frac{1}{\rho} = -\frac{d^2y}{dz^2} \bigg/ \left[1 + \left(\frac{dy}{dz}\right)^2\right]^{3/2} \quad \ldots (57)$$

A detailed calculation of the shape of the edges was carried out by Hillert[7] and by Jackson and Hunt.[15] Fig. 21 shows the result for the case $S^\alpha/S = 7/8$ which holds for pearlite in the Fe–C system. It is very interesting to note that the interface is rather flat, thus justifying the assumption used in deriving the concentration profile. It is also interesting to note that the interface will bend back for large S values. The critical S value for this phenomenon can be directly calculated by putting $X^\gamma = X_e^{\gamma/\alpha}$ at the middle of the edge of the α lamella ($z = 0$; $y = 0$). Equation (35) yields

$$X_e^{\gamma/\alpha} - X_1^\gamma = A_0 + \sum_1^\infty A_n = A_0 + \frac{vS(X^\beta - X^\alpha)}{D\pi^2} \cdot C \quad \ldots (58)$$

where

$$C = \sum_1^\infty \frac{\lambda_n S}{2\pi n^3} \cdot \sin n\pi S^\alpha/S \quad \ldots (59)$$

Inserting this in equation (52) and eliminating v by means of equation (56), yields

$$S = S_0 + \frac{2L\sigma V_m S^2/S^\alpha S^\beta}{RT\left[\frac{1-X^\alpha}{1-X_e^{\gamma/\alpha}} - \frac{X^\alpha}{X_e^{\gamma/\alpha}}\right](X_e^{\gamma/\alpha} - X_e^{\gamma/\beta})(\pi C S^\alpha/SB - 1)} \quad \ldots (60\alpha, \beta)$$

For the case of a symmetric phase diagram and symmetric conditions at the three-phase junctions, $L = 1/2$, we obtain approximately

$$S = 3S_0 \quad \ldots (61)$$

For all other cases, the bending back of the edge will occur even sooner either for the α lamellae or for the β lamellae. The fact that this phenomenon is not usually observed experimentally seems to indicate that $S < 3S_0$ and it seems justified to conclude that the spacing is automatically adjusted to a value close to $S = 2S_0$ as suggested by Zener's optimizing principle. On the other hand, experimental data in many cases seem to indicate strongly that S is $\gg$ than S_{rev}. As discussed in the previous section, the free-energy loss due to lengthwise diffusion gives an effect in that direction but it is rather small. It is conceivable that there are other, more important free-energy losses that have not yet been considered. It may, for instance, be pointed out that according to the present approach the two new phases in general form under different pressures at the three-phase junction, in spite of the fact that they have a flat interface which cannot support any pressure difference. This may lead to complications and a free-energy loss of some kind.

Eutectoid Transformation by Boundary Diffusion

This case was treated by Cahn[9] assuming that the composition of the two growing phases would vary with the concentration in the boundary. Equation (41) was thus employed to describe the diffusion. However, we are mainly interested in the same kind of system as in the preceding section, where X^α and X^β are regarded as constants and equation (40) will thus be applied when the transformation is controlled by boundary diffusion rather than volume diffusion. The similar case of monotectoid reaction has been treated by Shapiro[11] in an attempt to explain discontinuous precipitation.

Expressing the composition in front of the edge of a growing lamella by equation (40) we can integrate equation (44), obtaining

$$L\sigma V_m = RT\left[\frac{1-X^\alpha}{1-X_e^{\gamma/\alpha}} - \frac{X^\alpha}{X_e^{\gamma/\alpha}}\right] \cdot \left[X_e^{\gamma/\alpha} - X_3^b - \frac{v(S^\alpha)^2(X_1^\gamma - X^\alpha)}{12KD^B\delta}\right] \cdot \frac{S^\alpha}{2} \quad \ldots (62)$$

$$(1-L)\sigma V_m = RT\left[\frac{X^\beta}{X_e^{\gamma/\beta}} - \frac{1-X^\beta}{1-X_e^{\gamma/\beta}}\right] \cdot \left[X_3^b - X_e^{\gamma/\beta} - \frac{v(S^\beta)^2(X^\beta - X_1^\gamma)}{12KD^B\delta}\right] \cdot \frac{S^\beta}{2} \quad \ldots (63)$$

Again equation (46) relates S^α, S^β, and S and can be used to replace $(X_1^\gamma - X^\alpha)$ and $(X^\beta - X_1^\gamma)$ by $(X^\beta - X^\alpha)$. We shall now simplify the calculations by assuming that the two kinds of boundary have the same value of $(KD^B\delta)$. Adding equations (62) and (63) would then yield

$$\left[X_e^{\gamma/\alpha} - X_e^{\gamma/\beta} - \frac{v(X^\beta - X^\alpha)S^\alpha S^\beta}{12KD^B\delta}\right] \cdot S$$

$$= \frac{2V_m\sigma}{RT}\left[\frac{LS/S^\alpha}{\left(\frac{1-X^\alpha}{1-X_e^{\gamma/\alpha}} - \frac{X^\alpha}{X_e^{\gamma/\alpha}}\right)} + \frac{(1-L)S/S^\beta}{\left(\frac{X^\beta}{X_e^{\gamma/\beta}} - \frac{1-X^\beta}{1-X_e^{\gamma/\beta}}\right)}\right] \quad \ldots (64)$$

and inserting S_0 from equation (48) we obtain

$$v = \frac{12KD^B\delta \; S^2(X_e^{\gamma/\alpha} - X_e^{\gamma/\beta})}{S^\alpha S^\beta(X^\beta - X^\alpha)} \cdot \frac{1}{S^2}\left(1 - \frac{S_0}{S}\right) \quad \ldots (65)$$

Except for the factor S^2, instead of S, this result is very similar to that which was obtained with volume diffusion. The maximum growth rate is now obtained at

$$S_{optimum} = \frac{3}{2} S_0$$

and the dependence of supersaturation becomes

$$S_{optimum} \propto (X_e^{\gamma/\alpha} - X_e^{\gamma/\beta})^{-1}$$

$$v_{maximum} \propto (X_e^{\gamma/\alpha} - X_e^{\gamma/\beta})^3$$

As before, we can calculate the critical spacing where the edge would start to bend back by putting $X^\gamma = X_e^{\gamma/\alpha}$ at $z = 0$, $y = 0$. Equation (40) then yields

$$X_e^{\gamma/\alpha} - X_3^b = (X_1^\gamma - X^\alpha) \frac{v(S^\alpha)^2}{8KD^B\delta} \quad \ldots (66)$$

and inserting this in equation (62) combined with equation (65), we obtain

$$S = S_0 + \frac{4L\sigma V_m(S/S^\alpha)^2}{RT\left[\dfrac{1-X^\alpha}{1-X_e^{\gamma/\alpha}} - \dfrac{X^\alpha}{X_e^{\gamma/\alpha}}\right](X_e^{\gamma/\alpha} - X_e^{\gamma/\beta})} \quad \ldots (67\alpha, \beta)$$

The symmetric case again yields $S = 3S_0$, and the fact that bending back is not normally observed experimentally seems to support the view that Zener's optimizing principle may be fairly correct whether boundary or volume diffusion is the rate-controlling process.

At low temperatures it may be reasonable to expect a eutectoid transformation to take place by means of boundary diffusion with such a high rate that the volume diffusion in the matrix ahead of the interface is too slow to allow local equilibrium between the matrix and the interface. Equation (44) should then be modified by adding $f \cdot \Delta G_l$ to ΔG^α and ΔG^β. However, this will make only a slight change, as demonstrated by the fact that ΔG_l is a small fraction of ΔG_{total}.

NOTE ADDED IN PROOF: Eutectoid transformation by boundary diffusion has recently been treated by Shapiro and Kirkaldy (*Acta Met.*, 1968, **16**, 579) and by Sundquist (personal communication). Their approaches are somewhat different but their results are essentially in agreement with the present result.

Discontinuous Precipitation

This reaction will only be treated assuming rate control by boundary diffusion, since we have not found any driving force for the grain-boundary movement unless there is a deviation from local equilibrium between the γ matrix and the boundary. The new γ grain that grows together with the β phase will be denoted α, to emphasize the similarity with the eutectoid reaction. This case of growth of a lamellar structure does not seem to have been treated adequately before. As pointed out earlier, the treatments by Turnbull[8] and Cahn[9] have a severe limitation.

The concentration in the boundary at the edge of the β lamellae is described by equation (40) as in the eutectoid transformation, but equation (41) must be used for the α lamellae because their composition will vary as X^b. We now get the following result by applying equation (33α), neglecting the term v/M, expressing ΔG_d by equation (9) and integrating

$$L\sigma V_m = \int_0^{S^\alpha/2} f \cdot \Delta G_d\, dz = \frac{fRT}{2X_1^\gamma} \int_0^{S^\alpha/2} (X_1^\gamma - X^b)^2$$

$$= \frac{fRT(X_1^\gamma - X_3^b)^2}{2X_1^\gamma(\cosh\sqrt{a}/2)^2} \int_0^{S^\alpha/2} (\cosh z\sqrt{a}/S^\alpha)^2 dz$$

$$= \frac{fRT(X_1^\gamma - X_3^b)^2}{8X_1^\gamma}(1 + \sqrt{a}/\sinh\sqrt{a}) \cdot \frac{\tanh\sqrt{a}/2}{\sqrt{a}/2} \cdot S^\alpha \quad \ldots (68)$$

For the β lamellae there is also a force due to the deviation from local equilibrium (equation (31)). However, the force due to the chemical change, ΔG^β, may be much larger. The term $f \cdot \Delta G_d$ will therefore be neglected and we then obtain exactly the same result as for the β lamellae in a eutectoid transformation (equation (63)).

$$(1 - L)\sigma V_m = \int_0^{S^\beta/2}(\Delta G^\beta + f\Delta G_d)dz$$

$$\simeq RT\left[\frac{X^\beta}{X_e^{\gamma/\beta}} - \frac{1-X^\beta}{1-X_e^{\gamma/\beta}}\right] \cdot$$

$$\left[X_3^b - X_e^{\gamma/\beta} - \frac{v(S^\beta)^2(X^\beta - X_1^\gamma)}{12KD^B\delta}\right] \cdot \frac{S^\beta}{2} \quad \ldots (69)$$

Again we have two equations and three unknowns, now X_3^b, S, and v. The calculation will be somewhat more complicated than for the eutectoid transformation, mainly because the relation between S^α, S^β, and S depends upon the growth rate, the composition of the α phase not being constant.

The average composition of the α phase can be calculated by means of equation (41) using $X^\alpha = X^b$, i.e. assuming local equilibrium between the boundary and the growing α grain in agreement with the earlier discussion.

$$X_1^\gamma - \bar{X}^\alpha = \frac{2}{S^\alpha}\int_0^{S^\alpha/2}(X_1^\gamma - X^b)dz = \frac{2(X_1^\gamma - X_3^b)}{S^\alpha \cosh\sqrt{a}/2}$$

$$\int_0^{S^\alpha/2} \cosh z\sqrt{a}/S^\alpha dz = (X_1^\gamma - X_3^b) \cdot \frac{\tanh\sqrt{a}/2}{\sqrt{a}/2}$$

$$\ldots (70)$$

The composition of the β phase is constant and the lever rule will thus yield

$$\frac{S^\beta}{S^\alpha} = \frac{X_1^\gamma - \bar{X}^\alpha}{X^\beta - X_1^\gamma} = \frac{X_1^\gamma - X_3^b}{X^\beta - X_1^\gamma} \cdot \frac{\tanh\sqrt{a}/2}{\sqrt{a}/2} \quad \ldots (71)$$

A similar expression was derived by Cahn[9] using the equilibrium value $X_e^{\gamma/\beta}$ instead of X_3^b. This was a serious and unnecessary approximation.

Numerical calculations can now in principle be performed using equations (68), (69), and (71), in combination with the definitions $S = S^\alpha + S^\beta$ and $a = v(S^\alpha)^2/KD^B\delta$ from equation (42). However, for low amounts of β phase, equation (69) can be simplified by neglecting the third term in the bracket. By dividing equation (68) with (69) and using equation (71), we would then obtain

242   Session VI: *Interface-Controlled Transformations*

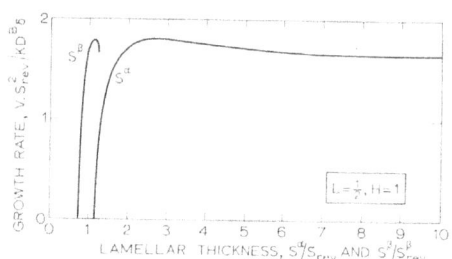

Fig. 22 *Growth rate as a function of lamellar thickness for discontinuous precipitation.*

$$\frac{1-L}{L}$$

$$= \frac{X_3^b - X_e^{\gamma/\beta}}{X_1^\gamma - X_3^b} \cdot \frac{\left[\dfrac{X^\beta}{X_e^{\gamma/\beta}} - \dfrac{1-X^\beta}{1-X_e^{\gamma/\beta}}\right]}{\left[\dfrac{X^\beta}{X_1^\gamma} - 1\right]} \cdot \frac{4}{f(1 + \sqrt{a}/\sinh\sqrt{a})}$$

... (72)

By introducing a new quantity H,

$$H = \frac{1-L}{L} \cdot f \cdot \left[\frac{X^\beta}{X_1^\gamma} - 1\right] / \left[\frac{X^\beta}{X_e^{\gamma/\beta}} - \frac{1-X^\beta}{1-X_e^{\gamma/\beta}}\right]$$

... (73)

we can now write

$$\frac{X_3^b - X_e^{\gamma/\beta}}{X_1^\gamma - X_3^b} = \frac{H}{4}(1 + \sqrt{a}/\sinh\sqrt{a}) \quad \ldots (74)$$

$$\frac{X_1^\gamma - X_3^b}{X_1^\gamma - X_e^{\gamma/\beta}} = 1 / \left[1 + \frac{H}{4}(1 + \sqrt{a}/\sinh\sqrt{a})\right]$$

... (75)

Inserting equation (75) in equation (68), we get a relation between S^α and v (through the parameter a).

$$S^\alpha = \frac{4L\sigma V_m X_1^\gamma}{fRT(X_1^\gamma - X_e^{\gamma/\beta})^2} \cdot \frac{\sqrt{a}}{\tanh\sqrt{a}/2} \cdot \frac{\left[1 + \dfrac{H}{4}(1 + \sqrt{a}/\sinh\sqrt{a})\right]^2}{1 + \sqrt{a}/\sinh\sqrt{a}}$$

... (76)

For any value of a, we can directly calculate S^α and, through the definition of a, we can then obtain the corresponding growth rate v. In expressing the value of S^α it is convenient to use the spacing of the hypothetical structure that would form if all the free energy available went into interfacial energy

$$S_{rev.} = \frac{2\sigma V_m}{\Delta G_{total}} = \frac{4\sigma V_m X_1^\gamma}{RT(X_1^\gamma - X_e^{\gamma/\beta})^2} \quad \ldots (77)$$

It is justifiable to compare S^α directly with this value because the amount of β phase is assumed to be small. The thickness of the β lamellae should be compared with their thickness in the same hypothetical structure,

$$S_{rev.}^\beta = \frac{X_1^\gamma - X_e^{\gamma/\beta}}{X^\beta - X_e^{\gamma/\beta}} \cdot \frac{4\sigma V_m X_1^\gamma}{RT(X_1^\gamma - X_e^{\gamma/\beta})^2}$$

$$= \frac{4\sigma V_m X_1^\gamma}{RT(X_1^\gamma - X_e^{\gamma/\beta})(X^\beta - X_e^{\gamma/\beta})} \quad \ldots (78)$$

Using equation (71) we now obtain

$$\frac{S^\alpha}{S_{rev.}} = \frac{L}{f} \cdot \frac{\sqrt{a}}{\tanh\sqrt{a}/2} \cdot \frac{\left[1 + \dfrac{H}{4}(1 + \sqrt{a}/\sinh\sqrt{a})\right]^2}{1 + \sqrt{a}/\sinh\sqrt{a}}$$

... (79)

$$\frac{S^\beta}{S_{rev.}^\beta} = \frac{2L}{f} \cdot \frac{1 + \dfrac{H}{4}(1 + \sqrt{a}/\sinh\sqrt{a})}{1 + \sqrt{a}/\sinh\sqrt{a}}$$

... (80)

In view of the definition of the quantity a (equation (42)), it is convenient to express calculated values of v in terms of $KD\delta/(S_{rev.})^2$.

Whatever relation we shall find between S^α, S^β, and v, we can immediately see that for any set of values of the parameters L, f, and H the following dependency of the original supersaturation could be expected

$$S_{optimum}^\alpha \propto (X_1^\gamma - X_e^{\gamma/\beta})^{-2}$$

$$S_{optimum}^\beta \propto (X_1^\gamma - X_e^{\gamma/\beta})^{-1}$$

$$v_{maximum} \propto (X_1^\gamma - X_e^{\gamma/\beta})^4$$

Such a strong variation of the growth rate with supersaturation seems to be confirmed by Speich's experimental data.[13] Unfortunately, it is difficult to test the theory quantitatively with those data because the solid solution in the Fe–Zn system is far from ideal and the amount of β phase was as high as 30% in Speich's experiments.

Equations (79) and (80) show that the spacing expected for zero growth rate will in general be different from $S_{rev.}$ and $S_{rev.}^\beta$. At the limit $v \to 0$ we get $a \to 0$ and equations (79) and (80) are simplified to

$$\frac{S_0^\alpha}{S_{rev.}} = \frac{L}{f}\left(1 + \frac{H}{2}\right)^2 \quad \ldots (81)$$

$$\frac{S_0^\beta}{S_{rev.}^\beta} = \frac{L}{f}\left(1 + \frac{H}{2}\right) \quad \ldots (82)$$

For eutectoid transformations we found that, in general, $S_0 > S_{rev.}$ owing to the occurrence of some diffusion in a direction normal to the growth front. Some free energy is then lost irreversibly. For discontinuous precipitation, we have the same situation. The factor $(1 - f)$ represents the amount of ΔG_d in Fig. 14 that is spent on diffusion in the matrix ahead of the boundary and, as expected, equation (81) predicts that S_0^α will grow larger with larger $(1 - f)$, i.e. smaller f. For this transformation it can be a very large effect and in the limit all ΔG_d is spent on diffusion, and the spacing will be infinite. For eutectoid reactions the corresponding effect had minor importance only.

In the other extreme, $f = 1$, no free energy is lost irreversibly at $v = 0$ but, nevertheless, the spacing may become very large depending on the L value. Contrary to the eutectoid transformation, discontinuous precipitation seems to be very sensitive to the interfacial energies that determine the angles at the three-phase junctions and thus the value of L. For $L = 0$, equation (81) in combination with equation (73) predicts an infinitely large spacing and equation (75) reveals that $X_3^b = X_1^\gamma$. No decomposition is taking place and all the free energy available is saved in the "growing α phase". As expected, equation (80) predicts that the fraction of β phase will be negligible in this extreme. S_0^β can increase only as a result of small f values. Keeping f constant, equation

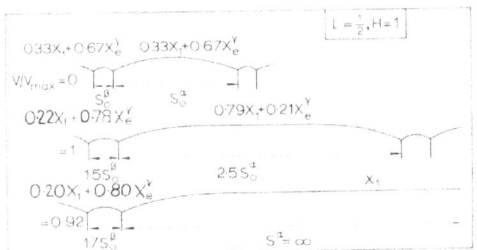

Fig. 23 Shape of the interface of discontinuous precipitation as a function of spacing and growth rate. Numbers give the local concentration at the centre of the edge of each lamella.

TABLE I
Growth Characteristics of Lamellar Structures

	Eutectoid		Discontinuous Precipitation by Boundary Diffusion
	Volume Diffusion	Boundary Diffusion	
v_{max} prop.	$(\Delta T)^2$	$(\Delta T)^3$	$(\Delta T)^4$
$S_{optimum}$ prop.	$(\Delta T)^{-1}$	$(\Delta T)^{-1}$	$(\Delta T)^{-2}$
$S^{\alpha}_{optimum}$ prop.	$(\Delta T)^{-1}$	$(\Delta T)^{-1}$	$(\Delta T)^{-1}$
$S_{optimum}/S_0$	2	1.5	~2
S_0/S_{rev}	close to 1	close to 1	from 1 to ∞
$S_{bending\ back}/S_0$	~3	~3	∞

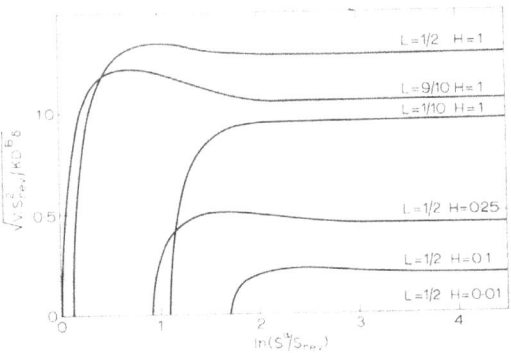

Fig. 24 Growth rate as a function of spacing under various conditions.

(81) predicts that S_0^β will decrease towards a value of $S_{rev.}^\beta/2$ for $L = 0$ as a result of the decreasing degree of decomposition. In fact, this effect will also dominate when we consider the case $v > 0$ and study the variation of spacing with growth rate. Fig. 22 presents the result of a numerical calculation based on equations (79) and (80) and using $f = 1$, $L = 1/2$, $H = 1$. The curve of v versus S^α is similar to the same curve for a eutectoid transformation. The growth rate starts at zero at a critical value S_0^α, it increases rapidly towards a maximum at ~ $2S_0^\alpha$, and decreases slowly as the spacing is further increased. In this transformation, however, the growth rate does not decrease towards zero but towards a value only slightly lower than the maximum value. This difference from a eutectoid transformation is explained by the curve of v versus S^β. Again the growth rate starts at zero at a critical value S_0^β, but as the growth rate increases S^β increases only by a factor of 2. From the viewpoint of a β lamella, the situation does not change much as the spacing S^α increases from the optimum value towards infinity. Fig. 23 illustrates the shape of the growth front and should be compared to Fig. 21 for the eutectoid transformation. In addition to the differences already discussed, Fig. 23 also shows that the edge of the α lamella is not predicted to bend back at large spacings. This result is formally a consequence of the expression for the composition in the boundary (equation (41)), which does not allow the composition to pass through the value of X_1^γ. The composition at the middle of the edges is also given in Fig. 23 as calculated from equations (75) and (41) and demonstrates clearly that the situation for the β lamellae does not change appreciably after the maximum in growth rate.

In view of these results, one could expect almost any spacing S^α to form from the optimum at ~ $2S_0$ to infinity. For Zener's optimizing principle to work, there must be a very effective mechanism for the formation of new β lamellae that is sensitive to a small deviation of the growth rate from its maximum value. It appears difficult to visualize such a mechanism. On the other hand, the experimental information seems to indicate strongly that the spacing is fairly constant, thus suggesting that the curve of v vs. S^α should show a more pronounced maximum. This would happen if the growth rate decreases more at large spacings, e.g. by the effect of the term v/M neglected in our calculation. Such a term may also lead to a drastic change in shape of the growth front at large spacings.

Fig. 24 shows the results calculated for a series of L values and for a series of f values. With the approximations used in the present calculations, the L value will mainly influence the position of S_0^α relative to $S_{rev.}$, whereas the value of f will also affect the growth rate at large spacings.

Some of the results obtained for the various lamellar structures are summarized in Table I. To allow comparison, the different supersaturations have here been represented by the undercooling ΔT, assuming proportionality.

The Effect of Alloying Elements on the Transformation of Austenite

When studying the transformations of austenite in alloyed steels, Hultgren[16] noticed that the kinetic data on the ferrite formation could often be represented by two C-curves in the TTT diagram, instead of one. This indicates that there are two different modes of formation. He also found that the two transformation products sometimes show a difference in appearance under the microscope. Hultgren concluded that the difference was due to the behaviour of the alloying elements during the transformation. At high temperature it is natural to expect that complete chemical equilibrium exists at the phase interfaces and a partitioning of the alloying elements between the matrix and the growing phases could be expected. At low temperatures, the sluggish alloying element may not have sufficient time for diffusion since the growth rate depends on the rate of carbon diffusion. The growing phases may then inherit the alloy content of the matrix. Such transformation products were called paraferrite, paracementite, and parapearlite by Hultgren, who also

244 Session VI: Interface-Controlled Transformations

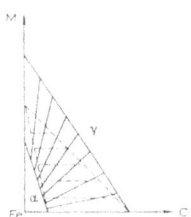

Fig. 25 Isothermal section of ternary phase diagram. Broken lines represent true paraequilibrium.

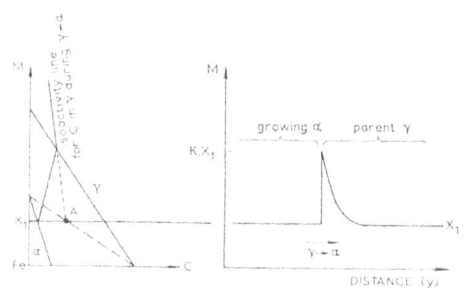

Fig. 26 Pile-up of M ahead of growing α. The original alloy content is X_1. The original composition must lie to the left of A if the growing α is to inherit the alloy content X_1.

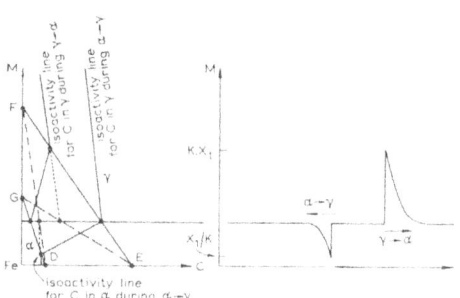

Fig. 27 Difference between the reaction $\gamma \to \alpha$ and the reverse reaction $\alpha \to \gamma$. The difference in carbon activity represents a hysteresis.

discussed the equilibrium at the phase interfaces, introducing the term paraequilibrium to designate the case of partial equilibrium where the two adjoining phases are in equilibrium with respect to carbon only, the iron and the alloying element being too sluggish for redistribution. The case of complete equilibrium at interfaces was called orthoequilibrium. Hillert[17] and Rudberg[18] discussed in detail the position of the paraequilibrium phase boundaries in ternary phase diagrams and Aaronson et al.[19] recently presented detailed calculations of such phase boundaries under the name of no-partition equilibrium curves. Fig. 25 presents a simple case of an isothermal phase diagram for iron-rich alloys in a system Fe–C–M where the alloying element M may stand for Mn and the temperature could be 800° C. The solid lines represent the stable phase diagram and the dashed lines the paraequilibrium. The tie-lines in the paraequilibrium phase diagram are of course directed towards the C corner because they hold under the condition of no partitioning of M between the two phases.

The conditions at the moving interface during transformation of an alloy system were considered in further detail by the present author,[20] who pointed out that the new phase might form with the same alloy composition as the parent phase even if there was complete chemical equilibrium at the interface. This may occur as a result of the formation of a pile-up of the alloying element ahead of the interface, as illustrated in Fig. 26. It was also pointed out that the carbon activity at the interface was controlled by this situation and a method was published by means of which quantitative estimates of the carbon activity could be made.[21] For low alloy contents, it was possible to calculate the change in carbon activity at the two-phase boundary from the equilibrium partition coefficient $K^{\gamma/\alpha}$ of the element between the two phases

$$\ln a_C^1/a_C^0 = -\frac{K^{\gamma/\alpha} \cdot X_{Fe}^\gamma - X_{Fe}^\gamma}{X_C^\gamma - X_C^\alpha} \cdot X_M^\alpha \quad \ldots (83)$$

a_C^0 and a_C^1 are the carbon activity values without and with the alloy content X_M^α, respectively. It was suggested that the main part of the effect of alloying elements on the transformation of austenite to ferrite, cementite, pearlite, and bainite was due to this change in carbon activity which affects the carbon-activity difference available to drive the diffusion of carbon. The same pile-up model has independently been suggested by Popov and Mikhalev,[22] Kirkaldy,[23] and Darken.[24]

The pile-up model predicts that the rate of the reaction $\gamma \to \alpha$ should be controlled by the rate of carbon diffusion away from the moving interface into the interior of the γ grain. This can occur only if the carbon activity is higher at the interface than in the interior of the γ grain. The model may thus work only if the parent phase has an original carbon content to the left of the point A in Fig. 26. The dashed line describes the position of this critical carbon content for various alloy contents. If the original composition lies to the right of this line but inside the equilibrium two-phase field, the reaction can still take place but only by long-range diffusion of the alloy element. The reaction will then be very much slower and the new phase will not inherit the original composition of the parent phase.

The triangle DGE in Fig. 27 may be regarded as the operating two-phase field during the reaction $\gamma \to \alpha$ as long as it is rate-controlled by carbon diffusion. For the reverse reaction $\alpha \to \gamma$ the operating two-phase field is the triangle DFE under the same condition. It is important to note that these are two different triangles. The difference may be described as a kind of hysteresis, considering a repeated process of decarburizing and carburizing, and it is due to the free-energy loss connected with the diffusion of the alloying element in the pile-up. At very low temperatures, or high growth rates, the pile-up may become so thin that it exists only mathematically. No free energy will then be lost for diffusion of the alloying element and the hysteresis has disappeared. The two triangles of Fig. 27 have now moved together and have finally coincided to form the broken-line triangle in Fig. 25, representing the true paraequilibrium. The effect of a given alloy addition on the transformation should thus be expected to decrease con-

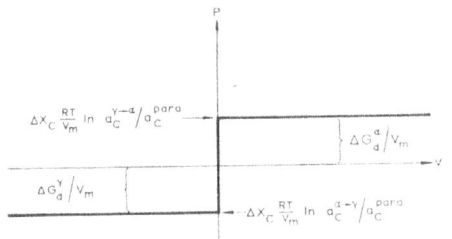

Fig. 28 The force necessary for moving an α/γ interface at a rate v.

Fig. 29 The decrease of the force at high growth rates where a deviation from local equilibrium begins to develop.

siderably when this situation is established at low temperature.

To use the pile-up model in a particular case, where long-range diffusion of the alloying element can be excluded, one simply has to calculate the carbon activity at the α/γ interface, e.g. from the approximate equation (83), and to use this value in the proper kinetic equation. However, there is an alternative method of treating this problem, which uses the true paraequilibrium as a starting point. To introduce this method, we shall consider a hypothetical experiment with a small piece of an Fe–M alloy that is originally in a two-phase state (α + γ) with no partitioning of the alloy element and a constant carbon activity. The temperature is assumed to be low enough to prevent any long-range diffusion of M. The specimen is then subjected to an atmosphere of another carbon activity. The difference in carbon activity gives rise to a force acting on the α/γ phase boundary. A higher carbon activity in the atmosphere would result in a force pulling the boundary in the direction of the α grain, thus causing the reaction α → γ. A lower carbon activity would have the opposite effect. The value of the force supplied by a certain atmosphere can easily be estimated using the basic equation (3)

$$P = -\Delta G_m/V_m = -\Delta X_C \frac{RT}{V_m} \ln a_C^{para}/a_C^{atm} \quad \ldots (84)$$

The quantity ΔX_C is the number of moles of carbon absorbed by the specimen during the transformation of one mole of material. a_C^{para} is the carbon activity of the true paraequilibrium according to Fig. 25. To use this type of equation, it may in general be necessary first to calculate a_C^{para}.

Fig. 28 illustrates the growth rate v as a function of P, neglecting the limiting kinetic factors due to the necessity of carbon transport. (In the following diagrams, P will be directly defined as the force caused by the local carbon activity at the interface, in order to avoid the discussion of such kinetic factors.) As the carbon activity of the atmosphere decreases below a critical level $a_C^{γ→α}$, the growth of α can suddenly start and the growth rate will be limited only by kinetic factors such as the rate of transfer of carbon from the specimen or the rate of carbon diffusion inside the specimen. If the carbon activity of the atmosphere is increased, the reverse transformation does not start until the carbon activity is above another critical value $a_C^{α→γ}$. The two critical carbon-activity values are caused by the free-energy losses due to the diffusion in the pile-up and can easily be calculated by estimating these losses. The relation between these ΔG_d and a_C is given by Fig. 28. For low contents of M and C and for $K^{γ/α}$ close to unity, we obtain approximately from equation (9)

$$\Delta G_d^\gamma = RT\left(\frac{X_1}{K} - X_1\right)^2/2X_1 \simeq RT(1 - K)^2 X_1/2 \quad \ldots (85)$$

$$\Delta G_d^\alpha = RT(X_1 - KX_1)^2/2X_1 \simeq RT(1 - K)^2 X_1/2 \quad \ldots (86)$$

where X_1 is the original alloy content. A more accurate calculation based on an equation such as (6) must be carried out if K is not close to unity.

The size of the hysteresis is obtained by adding equations (85) and (86) and applying the relationships shown in Fig. 28.

$$(X_C^\gamma - X_C^\alpha)RT \ln a_C^{para}/a_C^{γ→α} - (X_C^\gamma - X_C^\alpha) RT \ln a_C^{para}/a_C^{α→γ} = \Delta G_d^\alpha + \Delta G_d^\gamma \quad \ldots (87)$$

$$\ln a_C^{α→γ}/a_C^{γ→α} = (K - 1)^2 X_1/(X_C^\gamma - X_C^\alpha) \ldots (88)$$

The size of the hysteresis can also be estimated from the phase diagram (Fig. 27) using equation (83)

$$\ln a_C^{γ→α}/a_C^0 = -\frac{K - 1}{X_C^\gamma - X_C^\alpha} X_1 \quad \ldots (89)$$

$$\ln a_C^{α→γ}/a_C^0 = -\frac{K - 1}{X_C^\gamma - X_C^\alpha} \cdot X_1/K \quad \ldots (90)$$

Subtracting equation (89) from (90) yields

$$\ln a_C^{α→γ}/a_C^{γ→α} = \frac{K - 1}{X_C^\gamma - X_C^\alpha}\left(1 - \frac{1}{K}\right) X_1 \simeq (K - 1)^2 X_1/(X_C^\gamma - X_C^\alpha) \quad \ldots (91)$$

in agreement with equation (88). This is a demonstration of the fact that the two methods of calculation are equivalent. The following discussion of the effect of alloying elements can thus be based on a consideration of the free-energy loss, ΔG_d, instead of the carbon activity.

If the kinetic factors connected with the transport of carbon are rapid enough in comparison with the rate of diffusion of the alloying element, the growth rate may be so high that the thickness of the pile-up decreases below some measure of the atomic dimensions, d. Exactly as in the case of the impurity-drag effect on the motion of a grain boundary considered earlier, we now find that the free-energy loss decreases and goes to zero. We would then obtain the curve shown in Fig. 29. The effect of the alloying element will now be quite small once the threshold has been overcome.

As in the case of grain-boundary migration, we should consider a further effect: as the growth rate increases a

Session VI: Interface-Controlled Transformations

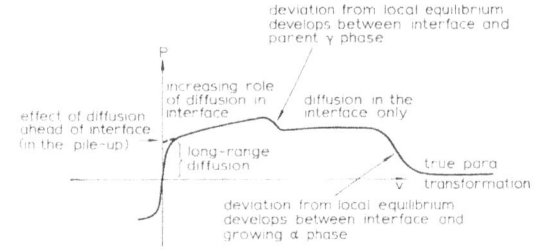

Fig. 30 *Hypothetical diagram demonstrating the variation of the force on the interface as function of the growth rate.*

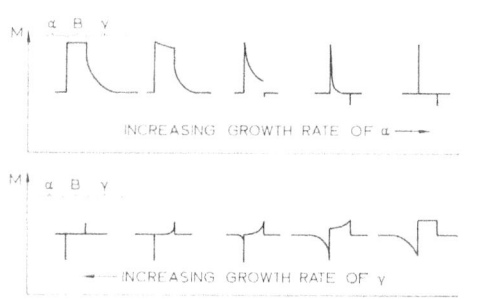

Fig. 31 *Concentration profile inside and ahead of α/γ interface.*

concentration gradient inside the boundary film may form and increase in importance and we then obtain a new loss of free energy. An even higher carbon activity is then necessary to further increase the growth rate. This effect may very well build up considerably before the first effects starts to decline. Fig. 30 shows a possible shape of the curve, taking both the effects into account and also considering the fact that the alloying element can diffuse over large distances if the growth rate is low enough. A detailed calculation of the two effects can be carried out in the same way as for grain-boundary migration. The shape in Fig. 30 is obtained if the alloy content at equilibrium is higher in the boundary than in γ, which in turn is higher than in α. Fig. 31 demonstrates the concentration profiles at a series of growth rates in the two directions. It is evident that the net effect will be rather different for the two reactions γ → α and α → γ.

Depending upon the choice of partition coefficients between α, γ, and the boundary, a great variety of behaviour will be found. The effect due to diffusion inside the boundary may be large or small as compared to the first effect and it may also influence the first effect in such a way that there will be a net decrease of the total effect.

A certain set of conditions will in general result in a definite growth rate where the kinetic factors for the carbon transport are in balance with the carbon activity required at the moving interface. In particular, this is true as long as the curve in Fig. 30 has a positive slope. Cases of instability are theoretically conceivable only if the slope of the curve has negative values, large enough to dominate over the rate of increase of the kinetic resistance with growth rate. This may not be a common case.

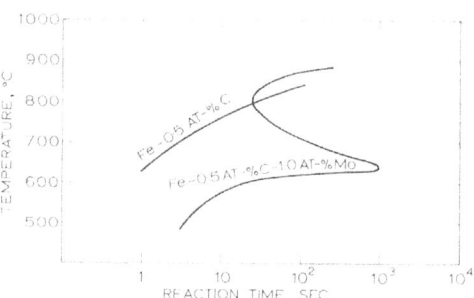

Fig. 32 *Isothermal transformation diagrams for the formation of ferrite (Ref. 25).*

At high temperature and low supersaturations the transformation of austenite occurs close to the origin in Fig. 30 where the rate is determined by long-range diffusion of M. At higher supersaturations, the growth rate may be high enough to prevent long-range diffusion of M. The effect of the alloying element may now be well accounted for by the simple pile-up model, yielding the extrapolated value on the P axis, using the local carbon-activity value at the interface for the calculation of P.

At lower temperatures of transformation, (and at higher growth rates), we move further to the right in Fig. 30 because D^γ and D^B decrease relative to the growth rate. The increase of the curve now becomes more and more important. At even lower temperatures we may reach the descending parts of the curve where the effect of the alloying element decreases rapidly.

By the use of a diagram like Fig. 30 we might understand the effect of molybdenum on the *TTT* diagram, which shows two C-curves for the formation of α.[16] Fig. 32 shows a direct comparison made by Kinsman and Aaronson,[25] who also proved that the effect of molybdenum was not only on the rate of nucleation but also on the growth rate. At high temperature we should expect a higher growth rate with molybdenum because it is a ferrite stabilizer. The retarding effect that starts at $\sim 800°C$ and reaches its maximum at $\sim 630°C$ can be explained as caused by the growing importance of the free-energy loss inside the boundary. To explain the large size of this effect, we must assume that molybdenum has a strong tendency to segregate to the α/γ interface (i.e. a large value of K). In fact, Kinsman and Aaronson, after examining several possibilities without finding the explanation for the existence of the bay in this *TTT* diagram, also concluded that it may be due to an "impurity-drag" effect.

At even lower temperatures, the concentration profiles in front of the boundary and inside the boundary grow steep enough to give an increasing deviation from local equilibrium. The free-energy losses now decrease and the effect of molybdenum grows weaker. If this description is correct, the upper C-curve holds for the formation of ferrite under complete local equilibrium between the moving boundary and the growing α grain. The lower C-curve should hold for ferrite formed under some deviation from local equilibrium and, at a low enough temperature, for ferrite formed under true paraequilibrium conditions. The original suggestion by Hultgren may thus be essentially correct, although he did not develop his hypothesis in sufficient detail.

Acknowledgement

This study is part of a research project supported financially by the Swedish Council for Applied Research.

References

1. R. Becker, *Z. Metallkunde*, 1937, **29**, 245.
2. S. Ono, *Mem. Fac. Eng. Kyushu Univ.*, 1947, **10**, 195.
3. M. Hillert, *Acta Met.*, 1961, **9**, 525.
4. J. W. Cahn and J. E. Hilliard, *J. Chem. Physics*, 1958, **28**, 258.
5. K. Lücke and K. Detert, *Acta Met.*, 1957, **5**, 628.
6. J. W. Cahn, *ibid.*, 1962, **10**, 789.
7. M. Hillert, *Jernkontorets Ann.*, 1957, **141**, 757.
8. D. Turnbull, *Acta Met.*, 1955, **3**, 55.
9. J. W. Cahn, *ibid.*, 1959, **7**, 18.
10. J. S. Kirkaldy, " Decomposition of Austenite by Diffusional Processes ", p. 39. **1962**: New York and London (Interscience Publishers).
11. J. M. Shapiro, Ph.D. Thesis, McMaster Univ., **1966**.
12. M. S. Sulonen, *Acta Met.*, 1964, **12**, 748.
13. G. Speich, " Cellular Precipitation in Fe-Zn Alloys ", to be published.
14. C. Zener, *Trans. Amer. Inst. Min. Met. Eng.*, 1946, **167**, 550.
15. K. A. Jackson and J. D. Hunt, *Trans. Met. Soc. A.I.M.E.*, 1966, **236**, 1129.
16. A. Hultgren, *Jernkontorets Ann.*, 1951, **135**, 403.
17. M. Hillert, *ibid.*, 1952, **136**, 25.
18. E. Rudberg, *ibid.*, 1952, **136**, 91.
19. H. I. Aaronson, H. A. Domian, and G. M. Pound, *Trans. Met. Soc. A.I.M.E.*, 1966, **236**, 768.
20. M. Hillert, *Internal Rep., Swedish Inst. Metal Research*, **1953**.
21. M. Hillert, *Acta Met.*, 1955, **3**, 34.
22. A. A. Popov and M. S. Mikhalev, *Physics Metals Metallography*, 1959, **7**, 36.
23. J. S. Kirkaldy, *Canad. J. Physics*, 1958, **36**, 907.
24. L. S. Darken, *Trans. Met. Soc. A.I.M.E.*, 1961, **221**, 654.
25. K. R. Kinsman and H. I. Aaronson, " Symposium on Transformation and Hardenability in Steel, 1967 ", to be published.

1974 Institute of Metals Lecture
The Metallurgical Society of AIME

Diffusion and Interface Control of Reactions in Alloys

MATS HILLERT
R. F. Mehl Medalist

By applying the local equilibrium concept to moving interfaces, it has been possible to treat the rate of reactions in alloys as controlled by diffusion. Various types of transformations have been examined over the years and satisfactory agreement between theory and experiment have often been found. In recent years, there has been an increasing interest in reactions where local equilibrium is not necessarily established at the moving interface. Such cases are examined and compared. The transition between diffusion control and interface control is discussed.

IN the 1961 Howe Memorial Lecture, Dr. Darken[1] spoke about the role of chemistry in metallurgical research. He covered a wide range of phenomena, but paid particular attention to the kinetics of phase transformations and discussed the model for two extreme cases. In the local equilibrium model[2] one assumes that the compositions of two phases in contact with each other are such that they are in chemical equilib-

---

The Institute of Metals Lecture was established in 1921, at which time the Institute of Metals Division was the only professional Division within the American Institute of Mining and Metallurgical Engineers. It has been given annually since 1922 by distinguished men from this country and abroad. Beginning in 1973 and thereafter, the person selected to deliver the lecture will be known as the "Institute of Metals Division Lecturer and R. F. Mehl Medalist" for that year.

MATS HILLERT has been a Professor of Physical Metallurgy at the Royal Institute of Technology in Stockholm since 1961. Dr. Hillert received his bachelor degree in chemical engineering at Chalmers Technical University in 1947. His graduate studies were conducted at the Massachusetts Institute of Technology, where he received his Doctorate of Science in the field of physical metallurgy in 1956 with a thesis on nucleation and spinodal decomposition.

From 1948 to 1953 he did research work at the Swedish Institute for Metal Research in Stockholm where he studied physical metallurgy. After completing his doctoral requirements, Dr. Hillert returned to the Swedish Institute for Metals Research in 1956 and became a Professor of Physical Metallurgy at the Royal Institute of Technology.

rium with each other. The compositions may vary as one moves from the interface. This assumption allows one to predict the concentration differences within each phase and, in principle, to calculate the rate of diffusion and the rate of reaction as controlled by this diffusion. One calls this a diffusion controlled reaction.

Local equilibrium cannot prevail if the interface is moving during the reaction and the atoms are transferred across the interface by a sluggish mechanism. The rate of reaction will then be less than that predicted by the local equilibrium model. In the extreme case one talks about an interface controlled reaction if the reaction step at the interface is so sluggish that the reaction does not give rise to any concentration differences. Generally, one can define a driving force for a reaction, e.g., the decrease in free energy per volume $\Delta G_m/V_m$, and one may then define the reaction as completely diffusion controlled if practically no part of the driving force is dissipated due to the reaction step at the interface. One may define the reaction as completely interface-controlled if practically no part of the driving force is dissipated by diffusional processes. In principle one should of course expect a mixed control and write

$$\Delta G_m = \Delta G_m^D + \Delta G_m^B \quad [1]$$

where the superscript D denotes diffusion and the superscript B denotes boundary, i.e., the interface. On the other hand, it may be argued that the two free energy sinks should generally be of quite different importance and one should dominate over the other. In the field of solid state transformations in alloys, the model of diffusion control has been the most popular one, partly because it can be used to predict the rate of reaction without any previous knowledge of the reaction itself. The basic information needed is the diffusivity and the phase diagram or, in more general terms, the free energy of the phases involved. Having this information, it is a matter of mathematics to solve the diffusion equation for the proper boundary conditions. This is seldom easy but many simple cases had been solved by 1961, and in particular, the many contributions by Wagner and Zener deserve to be mentioned.

According to the most common version of the model of interface control, the amount of free energy dissipated at the interface is proportional to the migration rate of the interface, v.

$$\Delta G_m^B/V_m = v/M \quad [2]$$

M is a constant of proportionality which is identified as the mobility of the interface. The relationship is usually written in the opposite direction,

$$v = M \cdot \frac{\Delta G_m^B}{V_m} \quad [3]$$

In general, the mobility may vary with the rate and in an attempt to define a constant parameter M one may, for instance, try to apply a relationship of the form

$$v = M \cdot \left(\frac{\Delta G_m^B}{V_m} - F\right)^n \quad [4]$$

The new quantity F would be a constant friction and n is an exponent that may be larger or smaller than unity.

Only a few attempts have been made to relate the mobility M to other quantities that can be determined by independent measurements. Turnbull's model[3] for the mobility of incoherent grain boundaries in a single-phase material deserves mentioning. It assumes that the atoms are transferred individually across the interface by a process related to diffusion and it yields the following type of relationship

$$M = \frac{\delta D^B V_m}{b^2 RT} \quad [5]$$

where δ is the width of the boundary and b is the distance between atoms. D^B is the diffusivity of the atoms in this process and it has been suggested that it is identical to or closely related to the grain boundary diffusivity that controls the rate of diffusion along the grain boundary. However, in general one treats the mobility M as a unique property which can only be obtained by a direct study of the moving interface.

It may be stated that most of the work in this field since 1961 has concerned the aspects of diffusion control. When the experimental rates are found to be smaller than those predicted by pure diffusion control, one sometimes blames this fact onto a sluggish reaction step at the interface, but one rarely attempts to analyze the situation in more detail. As an example of such an attempt the work of Shewmon and collaborators[4,5] may be mentioned. In other cases it has been possible to explain an apparent disagreement by refining the local equilibrium model. This is particularly true in ternary Fe-C alloys and, in particular the work of Kirkaldy and collaborators deserves mentioning.[6,7] By and large, the local equilibrium model has been surprisingly successful, but in recent years it has become increasingly urgent to learn more about its limitations and to start studying the aspects of interface control. The purpose of this lecture is to give a rather personal account of a line of research in this field and to demonstrate how one can perhaps start to treat the interface kinetics in alloys. My hope is to stimulate work in this important but difficult field.

The discussion will mainly deal with incoherent interfaces because coherency adds to the complexity of the problem.

TRANSFORMATIONS IN Fe-C ALLOYS

In view of Eq. [1] it would be advantageous to choose an interstitial alloy system if one wants to study the transition between diffusion control and interface control because interstitial diffusion is rapid. It can yield high growth rates, v, in comparison to the mobility of the interface which might depend upon the movement of the more sluggish lattice atoms and thus make the dissipation of free energy in the interface important. Fe-C is the best known system of this kind and the best kinetic studies concern the transformation of austenite to pearlite. The mathematics of the diffusion controlled growth of a eutectoid like pearlite has been treated in a rather satisfactory manner and it would thus seem that the kinetic data could be used to test if the growth is actually rate controlled by diffusion. The comparison has been made several times over the years. Pearlite growth is predominantly treated as a problem of volume diffusion of carbon in the layer of austenite just ahead of the pearlite interface. Early estimates of the growth rate of pearlite gave lower values than experiments. This situation was not changed when a better mathematical treatment was introduced.[8] However, the discrepancy has gradually decreased during the last two decades and is now so low that some authors conclude that the growth rate of pearlite is actually controlled by volume diffusion.[6] The discrepancy is still such that the actual growth rate is higher than the predicted one. There thus seems to be no room for a sluggish reaction step at the interface, but it has been suggested that diffusion in the interface[9] or in the ferrite lamellae[10] dominates over volume diffusion. In view of Eq. [1] it may be concluded that, in spite of the high diffusivity of carbon, pearlite does not grow fast enough to make the reaction step at the interface important. From an experimental point of view this may be said to be due to the fact that the growth rate of pearlite cannot be much affected by changing the carbon content of the alloy. This is possible in a reaction like $\gamma \to \alpha$ where the diffusion controlled growth rate depends upon the amount of carbon that must diffuse in front of the growing ferrite.

When ferrite forms as a surface layer during decarburization or by precipitation as a film on grain boundaries, the α/γ interface is approximately flat and the growth rate decreases with time by a parabolic law. High growth rates can only be obtained during the very first part of the reaction and are thus difficult to register experimentally. On the other hand, widmanstätten precipitation of α starts at the grain boundaries in this system and the plates seem to grow by a constant rate into the grain interior. This edgewise growth rate is comparatively easy to measure and the theory of diffusion controlled growth in this case was initiated by Zener.[11] One can write his equation in the following way

$$\frac{v\rho}{D} = \frac{1}{a} \cdot \Omega \quad \text{where} \quad \Omega = \frac{x^{\gamma/\alpha} - x^{\gamma 1}}{x^{\gamma/\alpha} - x^{\alpha}} \quad [6]$$

ρ is the radius of curvature at the edge and a is a numerical constant of the order of unity. $x^{\gamma/\alpha}$ is the actual composition of the matrix γ at the interface and $x^{\gamma 1}$ is the initial composition of the γ phase. The equation was later modified by the present author,[8]

$$\frac{v\rho}{D} = \frac{1}{a} \frac{\Omega}{1-\Omega} \quad \text{where} \quad \frac{\Omega}{1-\Omega} = \frac{x^{\gamma/\alpha} - x^{\gamma 1}}{x^{\gamma 1} - x^{\alpha}} \quad [7]$$

and the constant a was estimated as 2. Furthermore, by applying the local equilibrium model and taking into account the pressure difference between α and γ at the plate tip, Zener formulated the following relationship

$$x^{\gamma/\alpha} - x^{\gamma 1} = (^{e}x^{\gamma/\alpha} - x^{\gamma 1})(1 - \rho_c/\rho) \quad [8]$$

where $^{e}x^{\gamma/\alpha}$ is the equilibrium composition of γ at a flat γ/α interface and ρ_c is the critical value of ρ which makes $\Omega = 0$ and $v = 0$. Evidently, the growth rate depends upon the radius and the maximum is found at $\rho = 2\rho_c$. Zener argued that, due to the morphological instability of other shapes, the edge of a growing plate should approach this value and one could expect to observe experimentally a growth rate close to the maximum.

$$\frac{v_{\max}\rho_c}{D} = \frac{1}{8} \frac{\Omega_0}{1-\Omega_0} \quad \text{where} \quad \Omega_0 = \frac{{}^{e}x^{\gamma/\alpha} - x^{\gamma 1}}{{}^{e}x^{\gamma/\alpha} - x^{\alpha}} \quad [9]$$

The value of ρ_c depends upon the alloy composition and by applying Henry's law to the solution of carbon in γ one has, according to the Gibbs-Thomson equation, approximately

$$\rho_c = \frac{\sigma V_m^{\alpha}}{RT({}^{e}x^{\gamma/\alpha} - x^{\gamma 1})} \quad [10]$$

When a comparison was first made between the modified Zener equation and experimental growth rates for alloys of different carbon contents, a considerable discrepancy was found,[12] the experimental values being too low, Fig. 1. It was then suggested that this was due to a constant friction in the interface of the type represented by the quantity F in Eq. [4] with $n = 1$. This should lead to a deviation from local equilibrium at the interface and the concentration difference driving the carbon diffusion in γ should be lower than given by Ω_0. The effect can be represented as a change of Ω_0 and the arrows in Fig. 1 show how the position of the experimental points were thus changed to give a rather good agreement with the theory. The assumption of a constant friction has the consequence that edgewise growth should not at all be possible unless the supersaturation is higher than a critical value. This was evaluated from experiments at a series of temperatures and is plotted as a curve for the start of widmanstätten growth in Fig. 2. This line is denoted by WB_s because the data at the lower temperatures concern edgewise growth of bainite rather than ferrite. The suggestion by Hultgren[13] that the edgewise growth of bainite takes place by the growth of protruding tips of widmanstätten ferrite was supported by these results.

In recent years, the mathematical treatment of the edgewise growth of a plate has been considerably improved. This modern development was started by Ivantsov[14] who obtained the following equation by neglecting the effect of surface tension.

$$\sqrt{\pi p} \, e^p \, \text{erfc}(\sqrt{p}) = \Omega_0 \quad \text{with} \quad p = v\rho/2D \quad [11]$$

The latest development is due to Trivedi[15] and considers the interface kinetics and the surface tension,

$$\frac{1}{\sqrt{\pi p} \, e^p \, \text{erfc}(\sqrt{p})} = \frac{1}{\Omega_0} + \frac{v}{v_c} \cdot S_1 + \frac{\rho_c}{\rho} \cdot S_2 \quad [12]$$

S_1 and S_2 are functions of p and were presented graphically. v_c is a high, critical rate where all the driving force is dissipated at the interface. For low and medium values of p one can approximate S_2 by an analytical expression and, by neglecting the interface kinetics term, one obtains[16]

$$\frac{v_{\max}\rho_c}{D} = \frac{27}{256\pi} \left[\frac{\Omega_0}{1 - \frac{2}{\pi}\Omega_0 - \frac{1}{2\pi}\Omega_0^2} \right]^3 \quad [13]$$

According to Eq. [10], ρ_c is inversely proportional to Ω_0 and $v_{\max}$ would thus be proportional to Ω_0^4 according to Trivedi whereas it would be proportional to Ω_0^2 according to Eq. [9]. The two growth rate equations will thus differ drastically at low supersaturations. This is demonstrated in Fig. 3 and it is evident that the experimental data can be explained fairly well by the new theory. It seems that the previous conclusion regarding the friction in the interface was the result of inadequacies in the theoretical treatment. This is a good demonstration of the danger of studying interface kinetics through the difference between experimental and diffusion controlled growth rates. Direct measure-

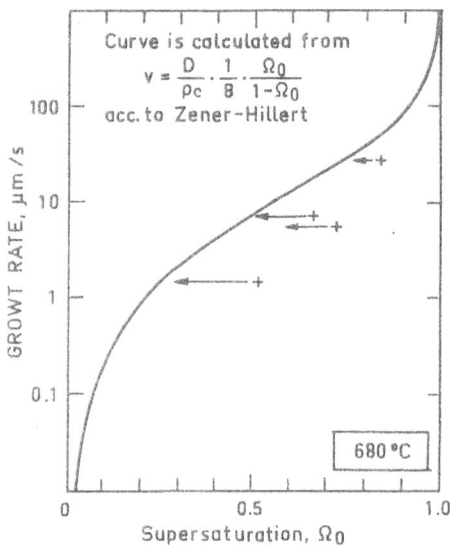

Fig. 1—Edgewise growth rate of widmanstätten ferrite. Experimental points from Ref. 19 are compared with predictions based on rate of diffusion. In order to obtain agreement, a deviation from local equilibrium at the phase interface was postulated. The experimental points were thus displaced as indicated by the arrows.

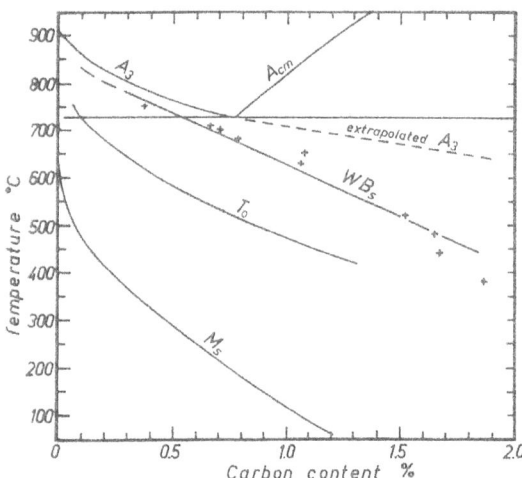

Fig. 2—Comparison between the line where widmanstätten ferrite or bainite can start to form, WB_S, and the corresponding line for martensite, M_S. Ferrite and austenite have equal free energy on the allotropic phase boundary, T_0. (Reproduced from Ref. 12.)

ments of the compositions at the interface would yield much safer results but may often be quite difficult to carry out. Some promising results of this kind have recently been obtained with diffusion couples.[17,18]

The difference between the modified Zener equation and Trivedi's treatment is rather small at large values of Ω_0. This is demonstrated in Fig. 4 which uses logarithmic scales and presents some data from the temperature range of bainite. Again it is seen that Trivedi's treatment provides an adequate description

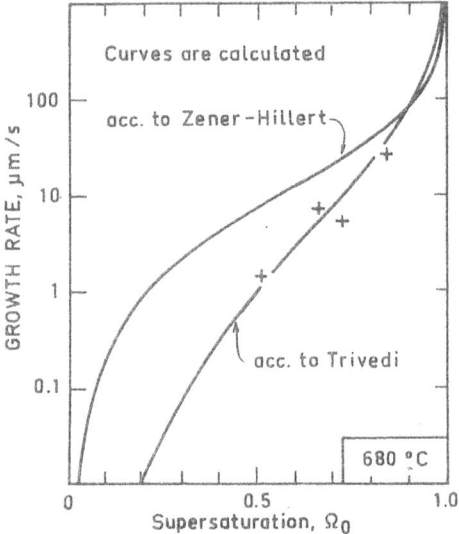

Fig. 3—Edgewise growth rate of widmanstätten ferrite. The experimental points agree well with the improved treatment of diffusion controlled growth according to Trivedi.

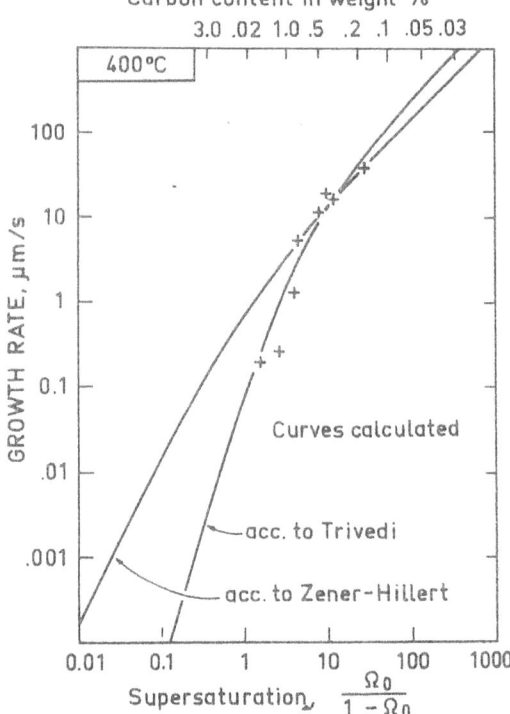

Fig. 4—Edgewise growth rate of bainite. Experimental points from Ref. 12 agree well with theoretical predictions by Trivedi.

of the experimental data and the interface kinetics seems to be of minor importance. This conclusion was also reached by Simonen, Aaronson and Trivedi[19] after a careful examination of the situation. However, they further claim to be able to estimate the small part of the supersaturation which is consumed by the interface kinetics. Their results correspond to M values of $5 \cdot 10^{-12}$ m^4/J, s at 700°C and $8 \cdot 10^{-13}$ m^4/J, s at 450°C. This temperature dependence is surprisingly small and casts some doubt on the validity of the latter part of their analysis. A much safer way to obtain reliable information on the interface kinetics should be to go to even higher growth rates by choosing Ω_0 values close to unity. It is then advantageous to use a new modification of Zener's equation, which describes Trivedi's results very well for medium and high Ω_0,

$$\frac{v_{max}\rho_c}{D} = \frac{1}{4} \cdot \frac{\Omega_0}{1-\Omega_0} \cdot 10^{-2.5(1-\Omega_0)} \quad [14]$$

TRANSFORMATIONS IN PURE Fe

When examining experimental data in which the interface kinetics has a major influence, it would be a great advantage to compare them with theoretical predictions. It is then necessary to know M. As a first attempt in this direction it may be justified to take a value of M evaluated for high angle grain boundaries in pure iron. Their mobility is fairly well known from the rate of grain growth and recrystallization. It has, for instance, been assumed that Eq. [3] holds for grain boundaries. The surface tension of a curved grain boundary gives rise to a pressure difference which enters as the driving force $\Delta G_m^B/V_m$ in Eq. [3]. This assumption leads to a parabolic law for grain growth

$$D^2 = 2M\sigma t \quad [15]$$

where D is an average grain diameter. Strong deviations from this law have been observed experimentally and attempts have recently been made[20,21] to describe such data by means of Eq. [4] with $n > 1$ or $F > 0$. However, at sufficiently small grain size the use of $n > 1$ seems to result in the prediction of unrealistically high rates of grain growth and the inclusion of a value for F would be of little consequence there. An attempt was thus made to evaluate M for zone-refined α-iron by applying Eq. [15] to the rate of grain growth at small grain sizes. Data were taken from studies by Ivanov and Osipov[22] and by Hu[23] and the results are presented in Fig. 5. A value was also extracted from the rate of recrystallization according to Talbot[24] and from the rate of the massive $\alpha \to \gamma$ transformation according to Speich.[25] Comparison is made with a dashed line calculated from the grain boundary diffusivity according to Turnbull's model. It should be noticed that Hu's data do not obey Eq. [15] and presumably one would need information from smaller grain sizes than those used in his study. As a consequence, it was not possible to fix the position of a line representing Eq. [15] in his diagrams except that it should fall to the left of his first measurement. The vertical arrows in Fig. 5 indicate that M should have a value larger than corresponding to this first measurement. It is evident that the dashed line is too low, but it is difficult to decide what should be the most probable line by which to describe M. The solid line was finally chosen. It gives $M = 0.035 \cdot \exp(-17700/T)$ m^4/J, s and will now be used to predict the rate of the $\gamma \to \alpha$ transformation at very low carbon contents. This is a crude attempt, of course,

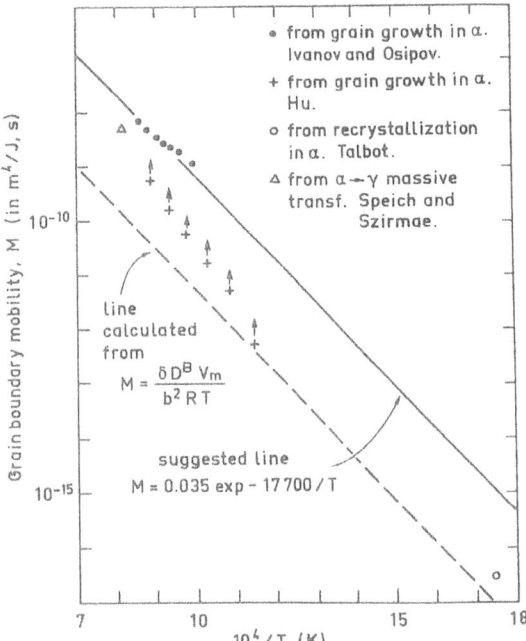

Fig. 5—Comparison between values for the grain boundary mobility of pure iron evaluated from the rate of various transformations.

but some support can be obtained from other transformations. Pearlite is known to grow with a rate of $5 \cdot 10^{-5}$ m/s at 550°C.[26] If an M value is taken from the dashed line, one finds from Eq. [3] that a driving force of more than 2000 J/mole would be required. The total chemical driving force for the transformation of γ to pearlite at this temperature is probably only half this value and, furthermore, it is believed that the major part is needed to drive the diffusion. The M value should thus lie far above the dashed line. The result will be even more dramatic if one examines what driving force is required in order to grow bainite at the experimentally observed rate of $4 \cdot 10^{-5}$ m/s at 400°C.[12] An M value taken from the dashed line would require 30,000 J/mole whereas the total driving force is about 600 J/mole. If a value is taken from the solid line in Fig. 5, one would find that a force of 250 J/mole is required which seems to be rather realistic.

TRANSFORMATIONS IN Fe WITH LOW C CONTENT

It is very difficult to study the growth rate by an isothermal technique if it is higher than 100 μm/s. An exception is the study by Karlyn, Cahn and Cohen[27] of the massive $\beta \to \alpha$ transformation in Cu-Zn where it was possible to freeze the β state by quenching and then to heat to a chosen transformation temperature by pulse-heating. In iron-base alloys a specimen would transform to martensite on quenching. The only experiments available on iron have been carried out by continuous cooling and recording of the arrest temperature caused by the heat of transformation. Fig. 6 gives an example

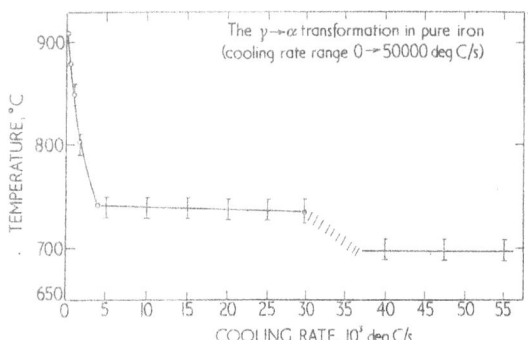

Fig. 6—The $\gamma \rightarrow \alpha$ transformation temperature in pure iron as function of the cooling rate. (Reproduced from Ref. 28.)

taken from a study by Bibby and Parr[28] on zone-refined iron with some carbon. The cooling rates can hopefully be translated into growth rates if it is assumed that the experimental transformation temperature is obtained when the rate of heat evolution balances the heat extraction. One finds roughly

$$v = \frac{Dhc}{L} \qquad [16]$$

Where D is the grain size, h is the cooling rate, c is the specific heat and L is the heat of transformation. With reasonable numbers the following approximate relationship is obtained

$$v\,(\mu m/s) = 2.5 \cdot h\,(K/s) \qquad [17]$$

The use of such a translation is supported by some recent work of Krahe and Grinnaert.[29]

Duwez[30] was the first to carry out this type of experiment on high-purity iron. He found that the transformation temperature decreased gradually from about 900°C at a cooling rate of 10 K/s to about 800°C at 10^4 K/s. Gilbert and Owen[31] obtained similar results with binary iron alloys and pointed out that the micro-structures obtained at high cooling rates resembled those obtained by the so-called massive transformation which had been described in detail by Massalski.[32] The main features of this type of transformation are that it is diffusionless and takes place by the migration of an incoherent interface. Bibby and Parr[28] confirmed the previous results and extended the range of cooling rates to $5.5 \cdot 10^4$ K/s. They reported a constant transformation temperature for the massive transformation from 5 to 30 kK/s and then a sudden drop which indicated that the mode of transformation suddenly changed to the martensite type. Ackert and Parr[33] studied this discontinuity as a function of the carbon content and obtained the results in Fig. 7. The critical cooling rate for the suppression of the massive transformation seems to be rather well established, but the existence of the massive plateau has been recently challenged by Massalski et al.[34]

It is now interesting to compare the experimental results with theoretical predictions for a diffusionless transformation. For pure iron the theoretical calculation is straight forward and makes use of Eq. [3]. Such calculations have previously been carried out for the massive transformation of β to α brass by Karlyn,

Cahn and Cohen.[27] However, in iron there is always some trace of carbon and nitrogen and they may have a decisive influence on the $\gamma \rightarrow \alpha$ transformation because of their high diffusivity and low solubility in α iron. A calculation was carried out assuming local equilibrium for carbon at the α/γ interface. The α phase could still form with the same carbon content as the original γ phase, but there would have to be a carbon-spike just ahead of the advancing interface, as illustrated in Fig. 8. The concept of local equilibrium only for carbon stems from Hultgren's work on transformations in alloyed steels.[13,35] He introduced the name paraequilibrium for the situation where carbon, being a mobile element, has the same chemical potential in two adjoining phases, but iron and substitional alloying elements have not. He only considered ternary and higher systems, but the same concept can be applied to the binary Fe-C system as well[12] and possibly even to other binary systems.[36]

The height of the spike in Fig. 8 should be independent of the growth rate as long as there is local equilibrium for carbon. The free energy dissipation due to diffusion in the spike should also be independent of the growth rate and for reasonable heights it can be estimated[37] by the following equation,

$$\Delta G_m^D = RT\,({}^e x^{\gamma/\alpha} - x^{\gamma 1})^2 / 2x^{\gamma 1} \qquad [18]$$

The shape of the spike is described by

$$x^\gamma - x^{\gamma 1} = ({}^e x^{\gamma/\alpha} - x^{\gamma 1})\exp(-vy/D) \qquad [19]$$

This loss of free energy was subtracted from the total driving force and the growth rate of the massive transformation was then evaluated from Eq. [3]. The result for a series of carbon contents is presented in Fig. 9. The cooling rate at the nose of the curve for pure iron is approximately the correct one according to Bibby and Parr and the value of M chosen from Fig. 5 thus seems to be reasonable. On the other hand, the curve for pure iron does not show the massive plateau reported by Bibby and Parr.

The curves for various carbon contents start at the temperature where the alloy enters the one-phase field for α. This is a natural consequence of the local equi-

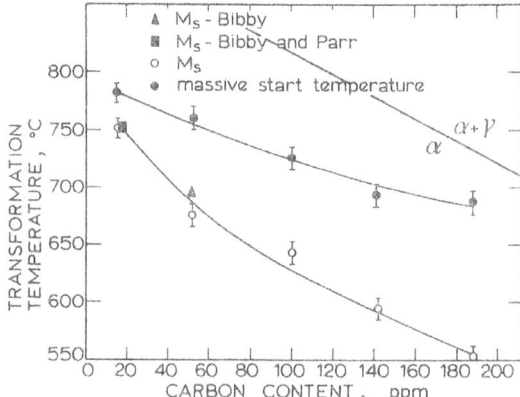

Fig. 7—The massive and martensitic transformation temperatures as functions of the carbon content. (Reproduced from Ref. 33.)

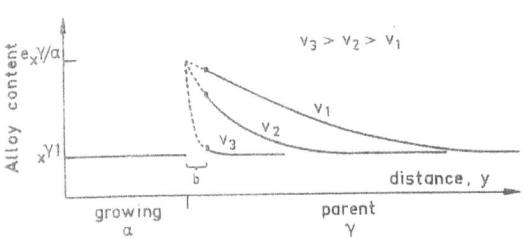

Fig. 8—The formation of a solute spike in the parent phase ahead of an advancing interface.

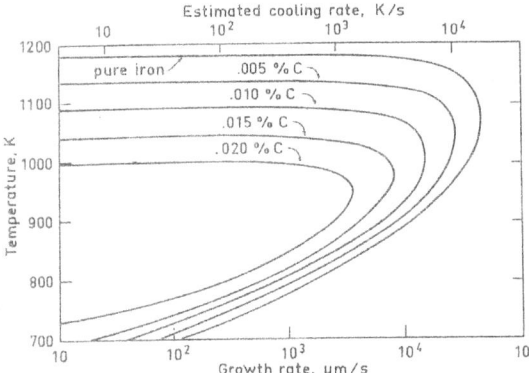

Fig. 9—Theoretical growth rates of the diffusionless $\gamma \rightarrow \alpha$ transformation, considering diffusional loss in a carbon spike.

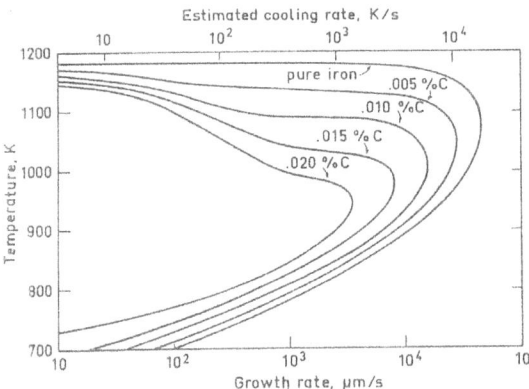

Fig. 10—Diffusional and diffusionless growth of α from γ at various carbon contents. The curves are calculated assuming that a carbon spike exists.

librium model.[27,37] On the other hand widmanstätten growth of low-carbon α can occur in the two-phase field $\alpha + \gamma$ thanks to side-ways diffusion of carbon. If this reaction is also included in the diagram according to Eq. [14] the curves shown in Fig. 10 are obtained. The curves show a gradual transition from diffusion controlled growth to diffusionless growth as the alloy is cooled into the one-phase field of the phase diagram. This is when the supersaturation parameter Ω_0 approaches unity. The transition is predicted to take place at about 300 μm/s. Of course, it is not certain that the crystallography of a widmanstätten plate will

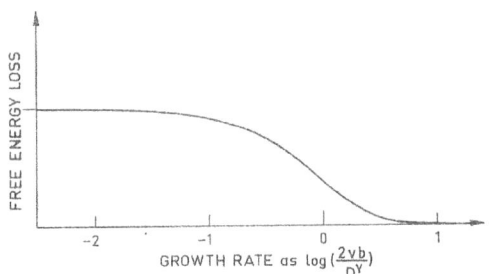

Fig. 11—The free energy loss due to diffusion in a solute spike as a function of the growth rate of the new phase.

actually allow it to develop into the massive mode of growth.

Fig. 10 shows that carbon can result in a plateau and it is interesting that 0.005 pct carbon has already such a strong effect. According to the position of the noses of the curves, the value of the critical cooling rate necessary in order to suppress the massive transformation is predicted to decrease rapidly with the carbon content. That prediction is not supported by the experimental information available, a fact that may be taken as an indication that the local equilibrium model cannot be applied to carbon at these growth rates. It was suggested a long time ago[38] that local equilibrium is no longer established at the interface if the mathematical shape of the spike becomes too thin in comparison with the atomic dimensions and an attempt[37] has more recently been made to treat such cases by simply assuming that the highest value that actually exists in the spike is the value at some small distance from the interface, for instance the atomic distance, b. This suggestion is also illustrated in Fig. 8. The effect on the free energy loss due to diffusion can easily be obtained by exchanging $e_x\gamma/\alpha$ in Eq. [18] for the value of x^γ obtained at $y = b$ according to Eq. [19]. The loss will thus decrease by a factor $\exp(-2vb/D)$. This is illustrated in Fig. 11. As the diffusional loss of free energy decreases, there will be more driving force left to move the interface,

$$v = M[\Delta G_m^{total} - (\Delta G_m^D)_{max} \cdot \exp(-2vb/D)]/V_m \quad [20]$$

It is important to notice that a deviation from local equilibrium at an interface will decrease the free energy dissipation due to diffusion and will thus increase the rate. This is not in contradiction to the fact that a slow reaction step at an interface promotes the deviation from local equilibrium because this is the mechanism by which driving force is transferred to the interface where it is used to drive the slow reaction.

Eq. [20] was applied to a series of carbon contents and Fig. 12 illustrates for two cases how large a driving force must be applied in order to obtain different growth rates. At the low carbon content of 0.0001 pct the curve rises monotonously. At higher carbon contents there is initially a decrease and one may draw the following conclusion. The calculation concerns the growth of α with the same composition as the initial γ. In order to overcome the free energy loss in the spike at low rates, the driving force must be larger than a critical value which is 10^6 N/m^2 in this particular case. If only chemical driving force is available, this amount is not provided until the alloy has been cooled inside the one-phase field in the phase diagram. Other sources of driving force are possible, for instance the surface tension at favorable configurations in the γ grain boundary net-work. Once the reaction gets started, the diffusional loss decreases and the rate increases discontinuously to the other side of the minimum on the curve in Fig. 12. As pointed out before[27,37] there is no thermodynamic reason why a massive transformation could not proceed inside a two-phase field, but some particular type of nucleation event is necessary.

In view of this result we must modify the curves in Fig. 10 by omitting the diffusional loss after the transition to diffusionless growth. The effect of carbon will then be very small due to the low carbon contents involved, and all the curves will approach the curve for pure iron. Fig. 13 shows this result and should now be compared with the experimental information. The critical cooling rate for suppression of the massive transformation is now predicted to be rather independent of the carbon content although the plateau temperature varies with the carbon content. That seems to be in agreement with the experimental information. In this connection it may be noted that the temperature dependency assumed for M is in agreement with the observation by Vyhnal and Radcliffe[39] that the critical cooling rate is decreased to below 10^2 K/s if the transformation temperature is lowered to about 680°C by the application of a hydrostatic pressure of 40 kbar.

It is interesting that Massalski et al have questioned two of the reported observations, namely, the existence of the massive plateau[34] and the suppression of the massive transformation inside a two-phase field.[40,41] For a pure element like iron, the latter problem never arises because the allotropic phase boundary, T_0, coincides with the limit of the one-phase field. The curve for pure iron in Fig. 13 has no plateau. By the same reason, any alloy which could start to transform in the massive manner at the T_0 line would show no massive plateau. However, if the massive transformation is suppressed, there should be a plateau as indicated by the curves for the various carbon contents in Fig. 13. The plateau may be difficult to experimentally establish beyond doubt because of the difficulty of making isothermal experiments. It may thus be better to examine where the transformation temperature falls as compared to the limit of the one-phase field. This phase boundary was thus included in Fig. 7 and the data seem to fall well inside the one-phase field. In such a case there are good reasons to expect a plateau and the model used to calculate the curves in Fig. 13 certainly leads to such a prediction. The observation that the massive transformation in Fe-C alloys does not start until well within the one-phase field may be explained by a recent suggestion of Massalski et al[42] that, in continuous cooling experiments, the arrest temperature is controlled by the rate of nucleation rather than

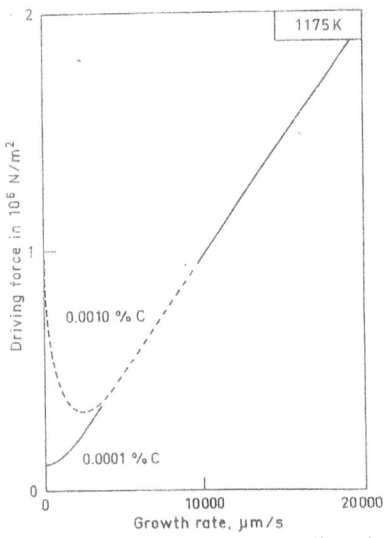

Fig. 12—The theoretical growth rate of the diffusionless $\gamma \rightarrow \alpha$ transformation as a function of the driving force. The curves are calculated assuming that the carbon spike disappears at high growth rates as illustrated in Fig. 11.

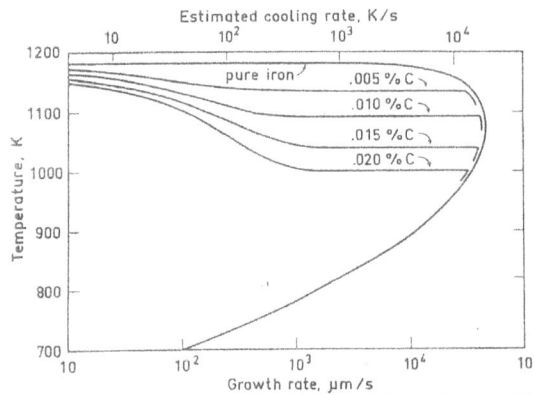

Fig. 13—Diffusional and diffusionless growth of α from γ. The curves are calculated assuming that the carbon spike disappears at high growth rates.

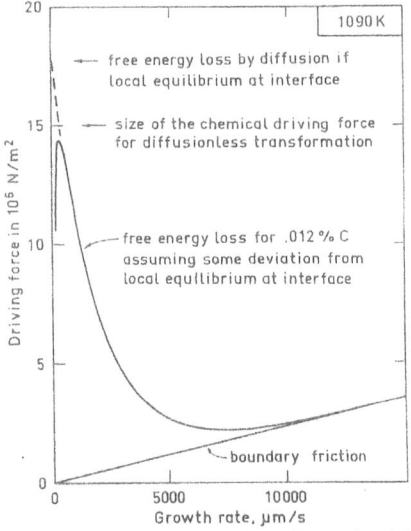

Fig. 14—Theoretical growth rates for diffusional and diffusionless formation of α from γ.

growth. This emphasizes the importance of making isothermal growth measurements.

In connection with the discussion as to whether the massive transformation can occur outside the one-phase field, a further modification of the calculation presented in Fig. 13 should be mentioned. In connection with Fig. 12, it was stated that the diffusionless transformation could not get started at low growth rates unless the driving force is as high as the chemical driving force at the limit of the one-phase field. However, at low rates, there may be side-ways diffusion and α can form with a more favorable carbon content. If that alternative is also included one obtains the curve shown in Fig. 14. The growth rate axis is expanded in order to demonstrate the situation at low growth rates. It is evident that the driving force requirement for the start of the massive transformation would be lowered by about 20 pct in this case if it were possible for widmanstätten ferrite to develop into massive ferrite. This would yield a corresponding expansion of the region for massive transformation outside the limit of the one-phase field. We shall return to this problem in a subsequent section.

OTHER CASES OF DIFFUSIONLESS TRANSFORMATION

It should be interesting to next examine the behavior of an alloy with a substitutional alloying element. Fe-Ni seems to be the system where most experimental information is available. Fig. 15 reproduces the results by Swanson and Parr.[43] It shows that the massive plateau temperature as well as the critical cooling rate for suppression of the massive transformation are both strongly dependent upon the nickel content. Theoretical curves can easily be calculated from the model in this case because the diffusivity of nickel is so low that the diffusion controlled growth rates should fall far to the left of the diagram. For all rates inside the diagram one may thus completely neglect the spike of the alloying element ahead of the advancing α/γ interface. Because of the high alloy contents, the driving force will be affected and the noses of the curves will thus be well separated, Fig. 16. Because no spike was assumed in this calculation, each curve starts at the T_0 temperature of the alloy concerned. On the other hand, if the spike exists during the nucleation event, which is not inconceivable, then the massive transformation will be suppressed to the limit of the one-phase field and one obtains the curves in Fig. 17. Comparison with the experimental data in Fig. 15 reveals again that the massive transformation is suppressed inside the one-phase field. This may be due to nucleation control, as suggested by Massalski et al.[42] It may, in part, be due to the presence of carbon and nitrogen impurities. In this connection it is interesting that the data of Swanson and Parr in Fig. 15 indicate that there is a massive transformation in an alloy of 7 at. pct Ni and Massalski et al[44] have recently reported a massive transformation product in an alloy of 8.7 at. pct Ni. These values are outside the one-phase field.

The question whether a diffusionless transformation can occur outside the one-phase field for the product phase is probably intimately related to the properties of the interface[40] and is thus a very important question

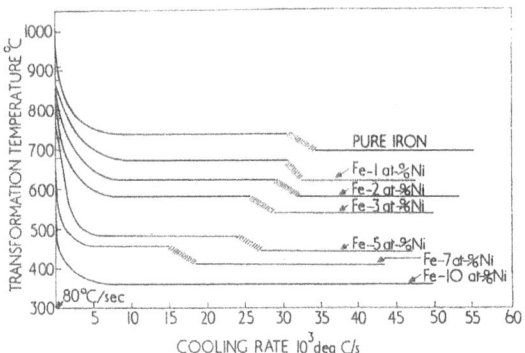

Fig. 15—The $\gamma \rightarrow \alpha$ transformation temperature in Fe-Ni alloys as function of the cooling rate. (Reproduced from Ref. 43.)

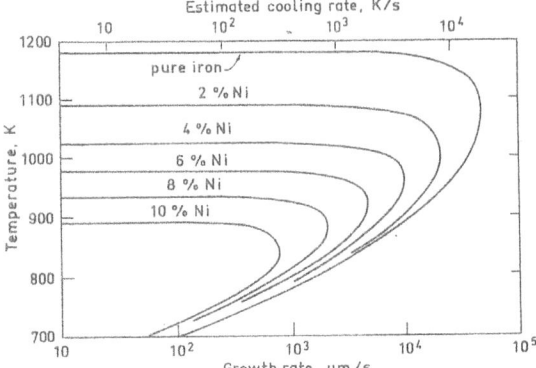

Fig. 16—Theoretical growth rates of the diffusionless $\gamma \rightarrow \alpha$ transformation in Fe-Ni alloys. No free energy loss due to diffusion is considered.

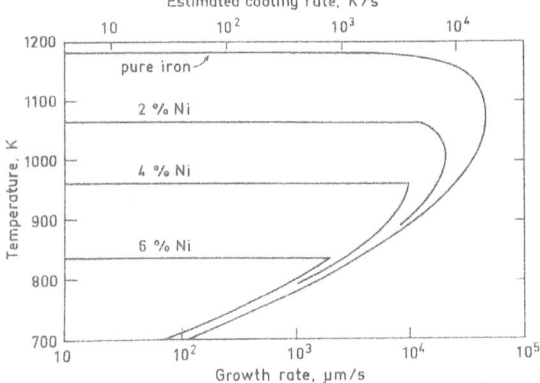

Fig. 17—Theoretical growth rate curves for the diffusionless $\gamma \rightarrow \alpha$ transformation in Fe-Ni alloys, assuming that the reaction cannot start in the $\alpha + \gamma$ two-phase field.

in connection with the discussion of interface control. There are at least three cases where the occurrence of a diffusionless transformation outside the one-phase field is well established. The most well-known case is the formation of martensite in iron with high carbon contents. The concept of the spike in the parent phase should apply to carbon in this case as well as in the case of massive transformation and one may conclude that a slow growing nucleus could not start a diffusionless transformation of either type. Martensite must be nucleated by the help of some additional driving free energy and it may make sense to talk about a dynamic nucleation event.

The second case is sometimes found in the transformation of austenite to ferrite or bainite in alloyed steels. Hultgren[35] found that the ferritic constituent of bainite usually forms with the same alloy content as the parent austenite and it was clearly shown by Rao and Winchell[45] that this can happen in a steel with as much as 10 at. pct Ni. Again one may conclude that the nucleation event must have been so rapid that there was insufficient time for nickel and iron to separate between the α and γ phases. Rao and Winchell accepted that the transformation could be regarded as diffusionless for nickel and iron, but when they evaluated the carbon supersaturation on that basis they found that the growth rates were lower by a factor of 100 than the values predicted from carbon diffusion control. In fact, at their experimental temperature of 400°C, the presence of 10 pct nickel should lower the supersaturation by less than a factor of 2 and we may thus compare their results directly with the growth rate data for bainite in low-alloyed steels, Fig. 4. The comparison is made in Fig. 18. Rao and Winchell discussed several sources of error, but finally admitted that the possibility exists that nickel atoms shuffle at the interface and act as a drag on the interface motion. A similar suggestion was made by Kinsman and Aaronson.[46]

The third case concerns solidification. Baker and Cahn[47] carried out splat quenching experiments on an alloy system with a so-called retrograde phase diagram. Specimens with such a high alloy content that they could never pass into the terminal one-phase region were found to solidify in a diffusionless manner.

In order to begin to understand the role of the interface during transformations in alloys one must examine the processes inside the interface. This is very difficult and may never be accomplished by direct experiments. As a first attempt one may construct and study various theoretical models and compare their predictions with experiment. Such models have been suggested for grain boundaries in single-phase alloys. They will now be examined in some detail and their extension to phase interfaces will then be discussed.

SOLUTE DRAG ON GRAIN BOUNDARIES

Lücke and Detert[48] made the first attempt to treat the interaction between grain boundaries and a solute. Their model is intimately related to Cottrell's theory of the interaction between dislocations and carbon atoms in ferrite. It was further developed by Cahn[49] and by Lücke and Stüwe[50-52] and it predicts a solute-drag effect on the motion of grain boundaries. Another approach to the same problem was developed from the concept of the spike in front of the advancing interface during a phase transformation.[37] Because of its close relation to the preceding discussion it is convenient to describe that model first.

This model assumes that the grain boundary has a finite thickness, δ, and that the material in the interface may be regarded as belonging to a special phase with its own thermodynamic properties. Fig. 19 illustrates the situation at a grain boundary in a single-phase material when there is a tendency for the solute to segregate to the interface. The top diagram is for a stationary boundary. The other diagrams illustrate how spikes appear ahead of the moving boundary as well as inside the boundary layer. The curves are calculated assuming local equilibrium at the contact between the boundary material and each grain. It is only in the top diagram that this assumption leads to local equilibrium between the two grains. It is thus imme-

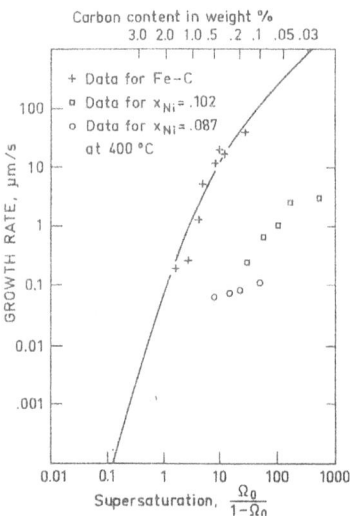

Fig. 18—Comparison between experimental growth rates of bainite in low alloyed steels and high nickel steels.

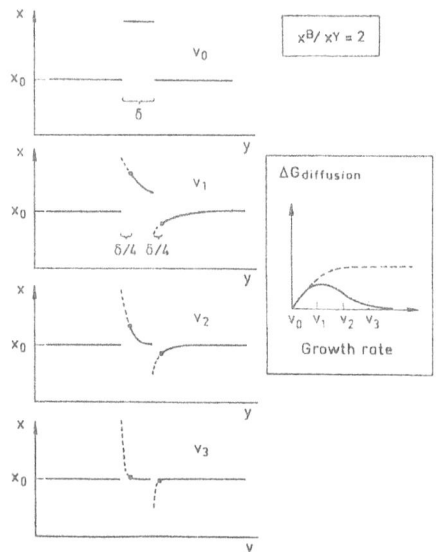

Fig. 19—The development of a solute-spike inside a migrating grain boundary.

diately evident that the model can be used to describe a gradual deviation from local equilibrium as the migration rate of the boundary increases. The dashed curve inside the frame shows how the free energy loss due to diffusion increases monotonously with the rate. The model can be modified in the same way as the model of the spike previously discussed, *i.e.*, by cutting off the top of the spikes. The result is indicated by the full curve inside the frame and predicts that the free energy loss should go through a maximum and then decrease towards zero. Fig. 20 presents the same types of curves calculated with different values for the thickness of the layer of the spike which is cut off. It may be stated that the introduction of diffusion in the boundary layer introduces a deviation from local equilibrium between the two grains which is accompanied by an increase in the diffusional loss of free energy. The removal of the tops of the spikes introduces a deviation from local equilibrium between the boundary layer and each one of the grains and it is accompanied by a decrease in the diffusional loss of free energy. It is evident that the previous statement that a deviation from local equilibrium should result in a higher growth rate must be modified. The decisive factor is the size of the solute drag that appears inside the boundary.

Cahn[49] and Lücke and Stüwe[50] assumed that the properties of the grain boundary layer varies with the distance from the center as illustrated in the box in Fig. 21. The curves in Fig. 21 show the concentration profile calculated for a series of migration rates assuming a constant diffusivity. The drag was then evaluated from the asymmetry of the concentration profiles, using the following equation to describe how the solute atoms are attracted towards the center of the boundary

$$P = \frac{1}{V_m} \int_{-\infty}^{\infty} (x - x_0) \cdot \frac{dE}{dy} \cdot dy \quad [21]$$

x_0 is the alloy composition. It is interesting that this model yields the same type of result as the model with the truncated spike. In order to be able to compare the two models, it may be advantageous to evaluate the con-

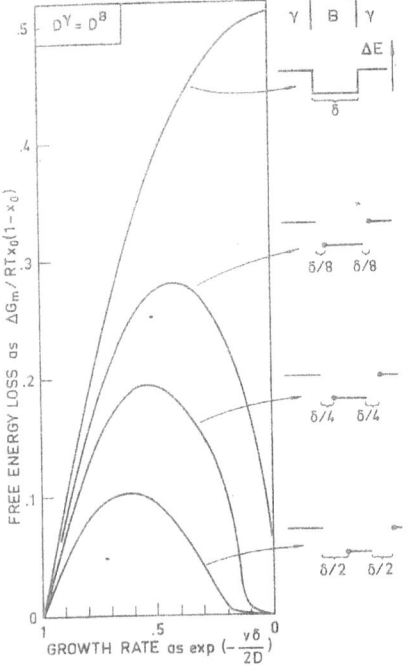

Fig. 20—The free energy loss due to diffusion at a migrating grain boundary according to several versions of the spike model.

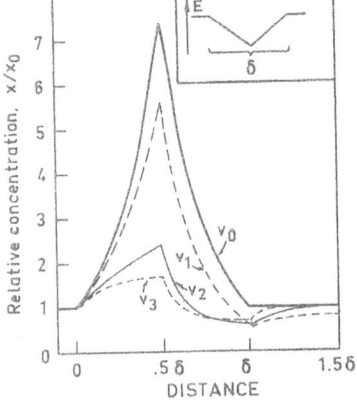

Fig. 21—The concentration profile in a grain boundary at four rates of migration, according to Cahn's model. (Copied from Ref. 49.)

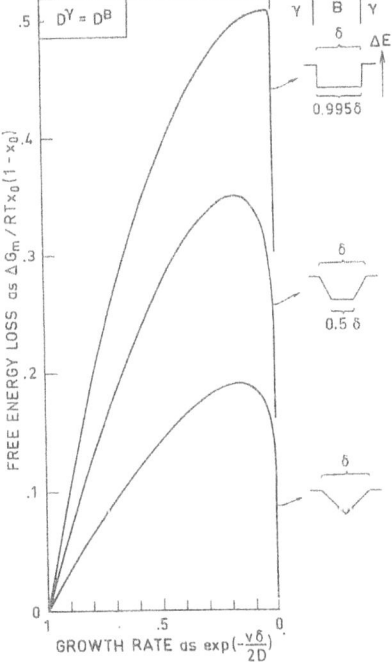

Fig. 22—The free energy loss due to diffusion at a migrating grain boundary, calculated for three grain boundary models. The lower one is identical to Cahn's model and the upper one approaches the spike model.

sequences of the Cahn model by evaluating the free energy loss due to diffusion instead of the attraction between the solute atoms and the boundary. This can be done with the following equation

$$\Delta G_m^D = \int_{-\infty}^{\infty} \frac{vRT}{Dx(1-x)} \cdot (x - x_0)^2 \cdot dy \qquad [22]$$

The result will be identical but in the second type of calculation all the elements will be positive and it is thus much easier to carry out an accurate calculation. This calculation has been programmed for a computer[53] and any case can now be treated with no difficulty. Fig. 22 shows some results. The lowest curve is for the case already treated by Cahn and the upper curve is for the spike model without truncation. It is quite evidict that the important feature of Cahn's model was the assumption of a gradual change in the properties as one goes from the grains into the grain boundary. The rather sharp discontinuity in the upper case has resulted in a movement of the maximum to very high migration rates and in the limit of a sharp discontinuity the maximum will be at an infinite rate. As expected, Cahn's model is then identical to the spike model.

There is no doubt that a realistic model must predict that the free energy loss due to diffusion should decrease above some critical migration rate. Apparently, there are two different ways to accomplish this effect, either by truncating the top of the spike or by levelling out the discontinuities. Both methods may seem a little arbitrary, but they may be justified intuitively. With the help of a computer they are both fairly easy to handle. Some further results[53] of Cahn's model will now be presented in order to demonstrate what kind of predictions one can make.

The model shown in the middle of Fig. 22 was investigated using different values for the ratio of the diffusivity in the grains and in the boundary, D^γ/D^B.

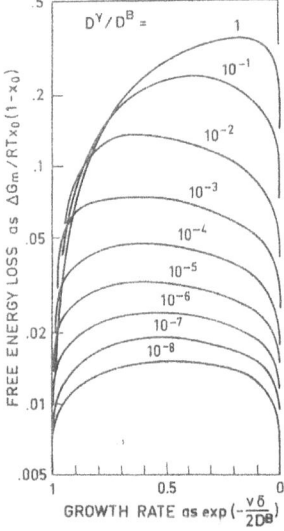

Fig. 23—The free energy loss due to diffusion in a migrating grain boundary, calculated for different ratios between the diffusivities.

Fig. 23 gives the result and demonstrates how important this ratio is. By increasing the grain boundary diffusivity in relation to the volume diffusivity by a factor of 10^8, one can decrease the maximum drag by a factor of 20. It should thus be unrealistic to estimate the drag from any constant value. Very little seems to be known today about the value that should be used for the diffusivity in the interface in a case like this. The most primitive guess would be to take the value for diffusion along the interface but it has been stressed, for instance by Shewmon,[4] that diffusion across the interface may be quite different. This objection may be supported by referring to the fact that the mobility of a boundary according to Fig. 5 is considerably higher than predicted from the ordinary grain boundary diffusivity. On the other hand, it may also be argued that the rate of migration of a boundary does not depend upon the atoms being shuffled as they cross the boundary. The rate of shuffling may thus be lower and may very well be comparable to the rate of diffusion along the interface.

The top and bottom curves in Fig. 23 are presented

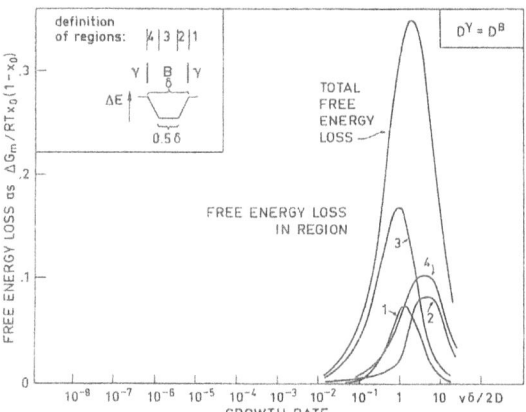

Fig. 24—The free energy loss due to diffusion in different regions of a migrating grain boundary, assuming a constant diffusivity.

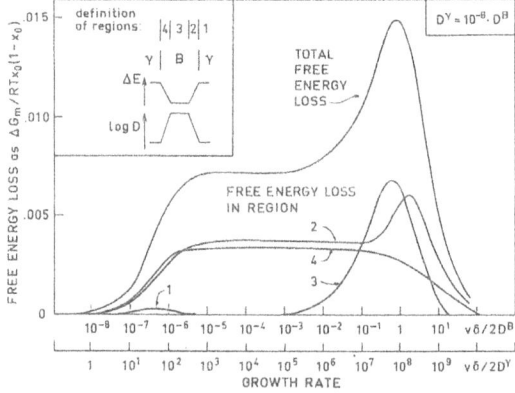

Fig. 25—The free energy loss due to diffusion in different regions of a migrating grain boundary, assuming that the diffusivity is much higher in the boundary than in the grains.

in a slightly different way in Figs. 24 and 25 and it is shown how the free energy loss is divided between four different zones in the boundary.

THE SOLUTE-INTERFACE INTERACTION IN PHASE TRANSFORMATIONS

The new version of Cahn's model can be directly applied to a phase transformation, as could the truncated spike model.[37] A previous attempt in this direction was made by Aaronson, Laird and Kinsman.[54] One example is shown in Fig. 26 and it is there examined what happens when one varies the tendency of the solute to segregate to the boundary. A solute drag effect similar to the one found in a single-phase material is found at positive or negative segregations. However, if there is a tendency that the composition of the boundary should lie between those of the two phases, then the free energy loss starts to decrease as soon as the rate of transformation increases. The result is then of the same type as presented in Fig. 11 which was obtained without considering the properties of the interface.

In a preceding section, the $\gamma \to \alpha$ massive transformation in iron-base alloys was discussed and compared with predictions based upon values of the mobility M taken mainly from results for α/α grain boundaries. The results seem promising and it may be possible to go one step further. One may evaluate the interaction between grain boundaries and various alloy elements in iron from the effect on the grain boundary mobility.

One may then try to apply the same properties to the γ/α boundaries and use the model to predict what happens during a phase transformation. In many cases it may also be possible to make direct measurements of the effect of an alloy element on the rate of an α/γ interface in a transformation. Such studies have been started by Nishizawa and collaborators.[55] It should be possible through a model to compare such results and the result of grain growth experiments.

It should be particularly interesting to study the transition from full local equilibrium to partial local equilibrium for carbon in the transformation of austenite in alloyed steels. This transition was predicted by Hultgren[35] and was intuitively visualized by the present author[38] as illustrated in Fig. 27. The situations 1 to 4 correspond to a series of increasing growth rates. The dashed lines are the phase boundaries at the so-called

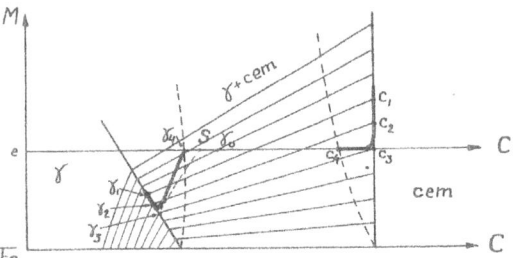

Fig. 27—The transition between full local equilibrium at a cementite/austenite interface and partial local equilibrium for carbon only. The thin lines at an angle to the Fe-C axis are isoactivity lines for carbon in austenite. The more horizontal lines are tie lines in the γ + cementite two-phase field. The dashed lines define the paraequilibrium phase boundaries. (Reproduced from Ref. 38.)

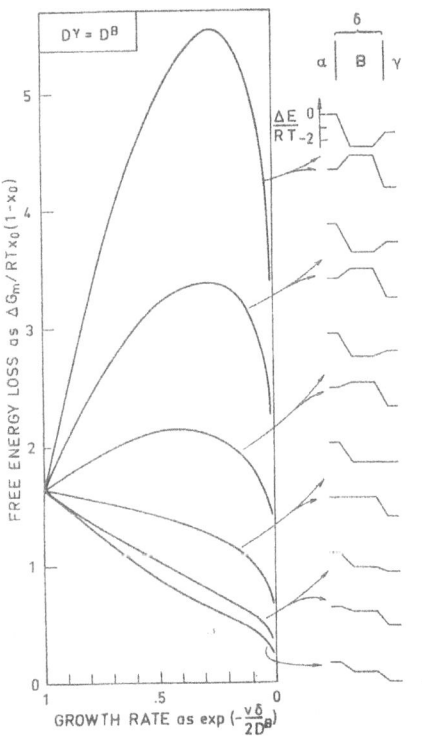

Fig. 26—The free energy loss due to diffusion at an α/γ interface, assuming different segregation tendency of the solute.

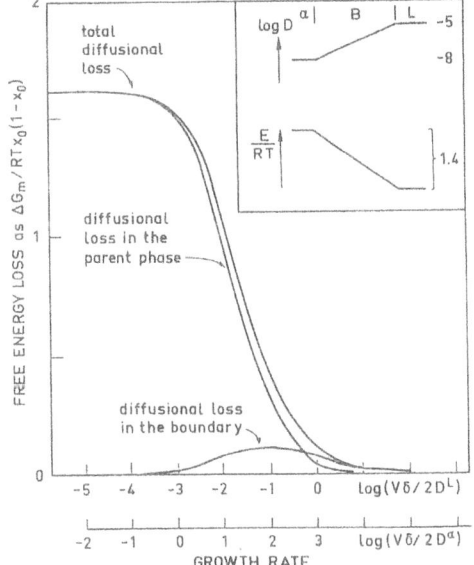

Fig. 28—The free energy loss due to diffusion at a solid/liquid interface during solidification at various rates.

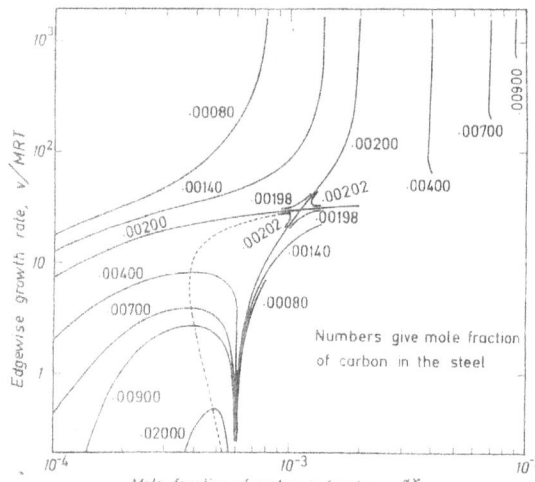

Fig. 29—The relation between the carbon content of growing α and the growth rate. The curves are calculated from a particular model of the properties of the α/γ interface, constructed in order to describe the growth of widmanstätten α as well as martensite. (Reproduced from Ref. 12.)

paraequilibrium. Coates[7,56] has discussed this problem more recently.

There are two cases where one might be able to make a reasonable guess. The first one is the case of solidification at a low solute content. It is there tempting to make the guess that the properties of the solid-liquid interface vary gradually across the interface. Fig. 28 shows the result of a calculation based on that model. It is interesting that the diffusional loss inside the interface is always very small. Furthermore, the diffusional loss in the spike in the liquid phase just ahead of the growing crystal decreases rapidly at a rate of about $0.02 D^L/\delta$ where δ is the width of the interface. This is a comparatively low growth rate and can probably be accomplished during dendritic growth. For instance, already at $\Omega_0 = 0.7$ Eq. [14] yields $v = 0.1 D^L/\rho_c$ and with a rough value of $\rho_c = 5\delta$ one obtains a migration rate where the diffusional loss has already decreased to half its original value. As a consequence, if the model is correct, there should be no difficulties for an alloy to change from dendritic to diffusionless solidification below the T_0 line inside an $\alpha + L$ two-phase field. This may be the explanation of the result of Baker and Cahn.

Baker and Cahn explained their results by reference to a theory by Chernov[57] for trapping surface active solute atoms. The observation that needed an explanation is that solute atoms are transferred from the liquid phase to the solid phase in spite of the fact that their chemical potential is higher in the solid phase. It may be stated that the present model describes how the chemical potential of the solute is gradually increased across the interface and it may be regarded as a model of a trapping mechanism.

A much more evident case of trapping is the martensite transformation in Fe-C alloys. The iron atoms are transferred across the martensitic interface in an orderly fashion and the carbon atoms are caught in their interstitial sites. It is not suggested that the present model be used for this case. A demonstration of what can be done in such a case is given by Fig. 29. It shows the result of a theoretical calculation at 600°C of the transition between widmanstätten ferrite and martensite.[12] It is based upon a growth model that was constructed in order to describe both types of growth. The main feature was the postulated existence of a constant friction for widmanstätten growth which was described in connection with Figs. 1 and 2. It was further assumed that the same friction governs the growth of martensite. Finally, it was assumed that all the carbon atoms of austenite were transferred to the ferrite phase together with the iron atoms, but some of them could subsequently leave the ferrite phase by diffusion across the interface due to the difference in chemical potential. These assumptions may not be very realistic but it is interesting that such a simple approach can be used to produce a number of attractive features in Fig. 29. At high carbon levels, each alloy has a growth curve with a maximum and the result is almost identical to the predictions of the modified Zener equation for widmanstätten ferrite. There is a certain carbon content of ferrite connected with the maximum. See the dashed line. As expected, the carbon content first decreases with the carbon level of the alloy but it then starts to increase and a critical point is reached. Alloys with less than 0.2 at. pct show no maximum in their growth curves. The growth rate can increase gradually from zero to very high values. This leads to the prediction that slow growing widmanstätten ferrite can develop gradually into martensite if the carbon content of the alloy is low enough. The curves at the upper right-hand corner show that martensitic growth can occur at high carbon levels, as well, but must then start off at a high initial rate by means of some dynamic nucleation event.

The models that have been discussed in this lecture are certainly not realistic in all their details and, most probably, even their main features must be modified in the future. However, it is possible that future progress in this field may depend upon the construction of such models and their testing against experiments.

REFERENCES

1. L. S. Darken: *TMS-AIME*, 1961, vol. 221, p. 654.
2. A. A. Noyes and W. R. Whitney: *Z. Phys. Chem.*, 1897, vol. 23, p. 689.
3. D. Turnbull: *Trans. AIME*, 1951, vol. 191, p. 661.
4. P. G. Shewmon: *TMS-AIME*, 1965, vol. 233, p. 736.
5. F. V. Nolfi, P. G. Shewmon, and J. S. Foster: *Met. Trans.*, 1970, vol. 1, p. 2291.
6. M. P. Puls and J. S. Kirkaldy: *Met. Trans.*, 1972, vol. 3, p. 2777.
7. D. E. Coates: *Met. Trans.*, 1973, vol. 4, p. 2313.
8. M. Hillert: *Jernkontorets Ann.*, 1957, vol. 141, p. 757.
9. B. Sundqvist: *Acta Met.*, 1968, vol. 16, p. 1413.
10. J. C. Fisher: *Thermodynamics in Physical Metallurgy*, p. 201, ASM, Cleveland, 1950.
11. C. Zener: *Trans. AIME*, 1946, vol. 167, p. 550.
12. M. Hillert, "The Growth of Ferrite, Bainite and Martensite," Internal Report, Swedish Inst. for Metal Research, 1960.
13. A. Hultgren: *Trans. ASM*, 1947, vol. 39, p. 915.
14. P. G. Ivantsov: *Dokl. Akad. Nauk. SSSR*, 1947, vol. 58, p. 567.
15. R. Trivedi: *Met. Trans.*, 1970, vol. 1, p. 921.
16. W. P. Bosze and R. Trivedi: *Met. Trans.*, 1974, vol. 5, p. 511.
17. J. R. Eifert, D. A. Chatfield, G. W. Powell, and J. W. Spretnak: *TMS-AIME*, 1968, vol. 242, p. 66.
18. T. Nishizawa and A. Chiba: *J. Japan Inst. Metals*, 1969, vol. 33, p. 869 and 1970, vol. 34, p. 627.

19. E. P. Simonen, H. I. Aaronson, and R. Trivedi: *Met. Trans.*, 1973, vol. 4, p. 1239.
20. B. B. Rath and Hsun Hu: *Trans. TMS-AIME*, 1969, vol. 245, p. 1243.
21. E. A. Grey and G. T. Higgins: *Scr. Met.*, 1972, vol. 6, p. 253.
22. V. I. Ivanov and K. A. Osipov: *Sov. Phys. Dokl.*, 1961, vol. 6, p. 425.
23. Hsun Hu: *Canad. Met. Quart.*, 1974, vol. 13, p. 275.
24. J. Talbot: *Recovery and Recrystallization of Metals*, L. Himmel, ed., p. 269, Interscience Publ., New York, 1963.
25. G. R. Speich and A. Szirmae: *Trans. TMS-AIME*, 1969, vol. 245, p. 1063.
26. J. H. Frye, E. E. Stansbury, and D. L. McElroy: *Trans. AIME*, 1953, vol. 197, p. 219.
27. D. Karlyn, J. W. Cahn, and M. Cohen: *TMS-AIME*, 1969, vol. 245, p. 1971.
28. M. J. Bibby and J. Gordon Parr: *J. Iron Steel Inst.*, 1964, vol. 202, p. 100.
29. P. R. Krahe and A. Grinnaert: École des Mines, Paris, private communication.
30. P. Duwez: *Trans. AIME*, 1951, vol. 191, p. 765.
31. A. Gilbert and W. S. Owen: *Acta Met.*, 1962, vol. 10, p. 45.
32. T. B. Massalski: *Acta Met.*, 1958, vol. 6, p. 243.
33. R. J. Ackert and J. Gordon Parr: *J. Iron Steel Inst.*, 1971, vol. 209, p. 912.
34. S. K. Battacharyya, J. H. Perepezko, and T. B. Massalski: *Scr. Met.*, 1973, vol. 7, p. 485.
35. A. Hultgren: *Kungl. Vet. Akad. Handl.*, 1953, vol. 4, bd. 4, no. 3.
36. G. W. Powell and R. Schuhman: *TMS-AIME*, 1969, vol. 245, p. 961.
37. M. Hillert: Monograph and Report Series No. 33, p. 231, Inst. of Metals, 1969.
38. M. Hillert: "Paraequilibrium," Internal Report, Swedish Inst. for Metal Research, 1953.
39. R. F. Vyhnal and S. V. Radcliffe: *Acta Met.*, 1967, vol. 15, p. 1475.
40. J. H. Perepezko and T. B. Massalski: *Scr. Met.*, 1972, vol. 6, p. 743.
41. T. B. Massalski, A. J. Perkins, and J. Jaklovsky: *Met. Trans.*, 1972, vol. 3, p. 687.
42. S. K. Bhattacharyya, J. H. Perepezko, and T. B. Massalski: Carnegie-Mellon Univ., Pittsburgh, unpublished research, 1974.
43. W. D. Swanson and J. Gordon Parr: *J. Iron Steel Inst.*, 1964, vol. 202, p. 104.
44. T. B. Massalski, J. H. Perepezko, and J. Jaklovsky: Carnegie-Mellon Univ., Pittsburgh, unpublished research, 1974.
45. M. M. Rao and P. G. Winchell: *TMS-AIME*, 1967, vol. 239, p. 956.
46. K. R. Kinsman and H. I. Aaronson: "Transformations and Hardenability in Steels," Climax Molybdenum Corp., Ann Arbor, Mich., 1967.
47. J. C. Baker and J. W. Cahn: *Acta Met.*, 1969, vol. 17, p. 575.
48. K. Lücke and K. Detert: *Acta Met.*, 1957, vol. 5, p. 628.
49. J. W. Cahn: *Acta Met.*, 1962, vol. 10, p. 789.
50. K. Lücke and H. Stüwe: *Recovery and Recrystallization of Metals*, p. 131, Interscience, New York, 1963.
51. K. Lücke, R. Rixen, and F. W. Rosenbaum: *The Nature and Behavior of Grain Boundaries*, H. Hu, ed., Plenum Press, New York, 1972.
52. K. Lücke and H. P. Stüwe: *Acta Met.*, 1971, vol. 19, p. 1087
53. B. Sundman: Thesis, Royal Inst. of Tech., Stockholm, 1974.
54. H. I. Aaronson, C. Laird, and K. R. Kinsman: *Phase Transformations*, ASM, Metals Park, Ohio, 1970.
55. F. Togashi and T. Nishizawa: Trans. Japan Inst. Metals, Tohoku Univ., Sendai, unpublished research, 1974.
56. D. E. Coates: *Met. Trans.*, 1972, vol. 3, p. 1203.
57. A. A. Chernov: *Growth of Crystals*, vol. 3, p. 35, Consultants Bureau, New York, 1962.

12 Introduction to "The Regular Solution Model for Stoichiometric Phases and Ionic Melts"

published in *Acta Chemica Scandinavica* (1970)

It is a rare recent issue of *CALPHAD, Journal of Phase Equilibria, Zeitschrift für Metallkunde, Journal of Alloys and Compounds* as well as other journals belonging to the "thermodynamic cluster" that does not contain a paper in which the thermodynamic properties of a solid phase are modeled by using the sublattice model. One can also frequently find papers in which the thermodynamic properties of liquid are described using the ionic two-sublattice model ([1] is an excellent example of how this model can be utilized in practice). An analysis of the literature also reveals that order-disorder transformations can successfully be handled within the framework of the compound energy formalism. Reference [2] is an interesting and sophisticated example of how this formalism can be used for dealing with an interaction between magnetic and chemical ordering.

The compound energy formalism description of thermodynamic properties of phases with several sublattices was originated by Hillert and Staffansson in their widely cited [3]. It was explicitly stated in the abstract that, "The regular solution model is developed for ionic melts and stoichiometric phases... by a formal method". Later, Hillert in [4] re-iterated that, "The compound energy formalism started ... as a purely mathematical method, based on an analytical expression for the Gibbs energy using terms of increasing powers of the mole fraction of atoms within the individual sublattices, so-called 'site fractions'. In addition, random mixing within each sublattice was assumed when constructing the terms for the constitutional entropy. It is thus the natural extension of the regular solution model with higher power terms and it reduces to that model when all the sites in all but one of the sublattices are vacant".

The fruitfulness of the approach suggested by Hillert and Staffansson and its applicability to a wide range of possible situations was quickly recognized. The first enhancement of the formalism, which initially considered "two elements in one sublattice and two in the other" was made a year later by Harvig [5], who still assumed the existence of two sublattices only but allowed them to be filled with an arbitrary number of elements. The next major advance was made a decade later when Sundman and Ågren generalized the formalism to the case of multicomponent phases with an arbitrary number of sublattices [6]. They significantly simplified analytical expressions for the Gibbs energy by introducing an important concept of the component array, and derived formulae for the partial Gibbs energies. It can be said that the publication of [6] catalyzed the wide usage of the sublattice model in computational thermodynamics in general and in thermodynamic optimization in particular. It should be emphasized that, while Sundman and Ågren addressed the mathematical complexities of the Hillert-Staffansson method, they still considered it a purely formal device.

Although nowadays it is difficult to say who first clearly realized that the sublattice model can be used for modeling of ordering phases, [7] is usually cited as a paper in which the applicability of the two-sublattice model to this problem was demonstrated. In spite of certain questionable criticism [8,9] and undisputable limitations [10], the two-sublattice model is now widely used for describing ordered phases. This circumstance highlights an interesting feature of the Hillert-Staffansson approach. It was introduced by its own creators as being somewhat formal, something that was a formalism rather than a model in a true sense of this word. It then turned out that the approach could successfully be used for handling various dissimilar situations. What are the roots of the power and the wide applicability of the method? A hint can be found in [11], which states that, in the sublattice modelling the crystallographic information is mainly used for deciding the number of sublattices to be used and the species present in each of them. Ideally, each sublattice is occupied by only one species (atoms, molecules, ions, vacancies, etc.), but some sublattices may have a mixing of species. The mixing characteristics, in general, are commensurate with the data on the homogeneity range of the phase. In formulating a sublattice model for a phase, due credit must be given to the crystallographic and solubility range data of the phase, not only in the system under consideration, but in all systems (binary as well as higher-ordered) where the same phase is observed. Sometimes it is even necessary to take into account information regarding phases with related structures, as in the case of phases related by higher-order phase transitions. By this approach one can ensure the generality of the model". The Hillert-Staffansson model does allow taking into account crystallography and solubility [12]. This intrinsic characteristic of the model explains why it is capable of tackling so many diverse cases, ranging from oxide systems [13] to liquid solutions with different tendency for ionization [14], from ordering [15] to point defects [16]. We can only wonder whether this feature was apparent to Hillert and Staffansson from the very beginning!

Let me conclude this introduction to the seminal paper "The regular solution model for stoichiometric phases and ionic melts" by Hillert and Staffansson in an unexpected way. Starting from 2004, Master and Ph.D. students of McMaster's Department of Materials Science and Engineering can take the half course "Computational Thermodynamics" as part of their graduate studies. Students enrolled in the course are required to read key publications devoted to thermodynamic modeling. Needless to say, the paper by Hillert and Staffansson falls into the "mandatory reading" category. It was discovered that this article was considered incomprehensible by an overwhelming majority of students. The questions asked by the poor souls helped to reveal the nature of the difficulties encountered. The source of the difficulties lies in Hillert's famous "it is immediately clear that..." which sometimes is formulated as "it is easy to show that..." or as a similar sounding statement, which may keep a reader busy for quite a while. Since a listing of all of the frequently asked questions falls beyond the scope of this introduction, let us present the most commonly asked one: "The authors mentioned that it was convenient to choose the size of the sublattices in $(A,B)_a (C,D)_c$ such that $a + c = 1$". Why is it convenient?

Let us consider a multicomponent system, which was formed by taking n_A moles of the component A, n_B moles of the component B and so on. Let us assume that these components were distributed among the phases $\alpha, \beta, \ldots$ which the system is composed of. If x_i^ϕ is the molar fraction of the component i in the phase ϕ, then the following conditions must be valid:

$$n_i = \sum_\phi n^\phi x_i^\phi, \quad i = A, B, \text{etc.}$$

where n^ϕ is the number of moles of the phase ϕ. Let us calculate the total number of moles of all components in the system:

$$\sum_i n_i = \sum_i \sum_\phi n^\phi x_i^\phi = \sum_\phi n^\phi \sum_i x_i^\phi = \sum_\phi n^\phi$$

One concludes that the total number of moles of all components is equal to the total number of moles of all phases. Keeping this in mind, let us consider the following reaction:

1 mole of A + 1 mole of B → N moles of the compound, in which $x_A = x_B = 1/2$ (1)

It is clear that $N = n_A + n_B = 2$ and that, therefore, the compound should be presented as $A_{1/2}B_{1/2}$ rather than AB. Subsequently, the reaction (1) should be written as $A + B \to 2 A_{1/2}B_{1/2}$ rather than $A + B \to AB$.

In the case of the reaction

$$n_A A + n_B B + n_C C + n_D D \to N \left(A_{1-y'}B_{y'}\right)_a \left(C_{1-y''}D_{y''}\right)_c$$

summation of the conditions of mass balance $n_A = Na(1-y')$, $n_B = Nay'$, $n_C = Nc(1-y'')$ and $n_D = Ncy''$ yields $n_A + n_B + n_C + n_D = N(a+c)$. The total number of moles of components is equal to the number of moles of $\left(A_{1-y'}B_{y'}\right)_a \left(C_{1-y''}D_{y''}\right)_c$ only if $a + c = 1$. Convenient, is it not?

Dmitri V. MALAKHOV

References

[1] O.B. Fabrichnaya, Thermodynamic modelling of melting in the system FeO-MgO-SiO2-O2 at pressure of 1 bar, *CALPHAD* 24 (2000) 113–131.

[2] Ikuo Ohnuma, Osamu Ikeda, Ryosuke Kainuma, Bo Sundman and Kiyohito Ishida, Interaction between magnetic and chemical ordering using the compound energy model, *Zeitschrift für Metallkunde* 89 (1998) 847–854.

[3] M. Hillert and L.-I. Staffansson, The regular solution model for stoichiometric phases and ionic melts, *Acta Chemica Scandinavica* 24 (1970) 3618–3626.

[4] M. Hillert, The compound energy formalism, *Journal of Alloys and Compounds* 320 (2001) 161–176.

[5] H. Harvig, An extended version of the regular solution model for stoichiometric phases and ionic melts, *Acta Chemica Scandinavica* 25 (1971) 3199–3204.

[6] B. Sundman and J. Ågren, A regular solution model for phases with several components and sublattices, suitable for computer application, *The Journal of Physics and Chemistry of Solids* 42 (1981) 297–301.

[7] I. Ansara, B. Sundman and P. Willemin, Thermodynamic modeling of ordered phases in the Ni-Al system, *Acta Metallurgica* 36 (1988) 977–982.

[8] N. Saunders, When is a compound energy not a compound energy? A critique of the 2-sublattice order/disorder model, *CALPHAD* 20 (1996) 491–499.

[9] I. Ansara, N. Dupin, B. Sundman, Reply to the paper: When is a compound energy not a compound energy? A critique of the 2-sublattice order/disorder model, *CALPHAD* 21 (1997) 535–542.

[10] W.A. Oates, F. Zhang, S.-L. Chen and Y.A. Chang, Some problems arising from two sublattice modelling of ordered phases, *CALPHAD* 23 (1999) 181–188.

[11] K.C. Hari Kumar, I. Ansara and P. Wollants, Sublattice modelling of the µ-phas, *CALPHAD* 22 (1998) 323–334.

[12] R. Ferro and G. Cacciamani, Remarks on crystallochemical apects in thermodynamic modeling, *CALPHAD* 26 (2002) 439–458.

[13] M. Hillert, B. Jansson and B. Sundman, Application of the compound-energy model to oxide systems, *Zeitschrift für Metallkunde* 79 (1988) 81–87.

[14] M. Hillert, B. Jansson, B. Sundman and J. Ågren, A two-sublattice model for molten solutions with different tendency for ionization, *Metallurgical Transactions A* 16 (1985) 261–266.

[15] B. Sundman, S.G. Fries and W.A. Oates, A thermodynamic assessment of the Au-Cu system, *CALPHAD* 22 (1998) 335–354.

[16] M. Hillert and M. Selleby, Point defects in B2 compounds, *Journal of Alloys and Compounds* 329 (2001) 208–213.

The Regular Solution Model for Stoichiometric Phases and Ionic Melts

M. HILLERT and L.-I. STAFFANSSON

Institute of Metallurgy, Royal Institute of Technology, S-100 44 Stockholm 70, Sweden

> The regular solution model is developed for ionic melts and stoichiometric phases of the type $(A, B)_a (C, D)_c$ by a formal method. The validity of Flood's equation for exchange reactions in ionic melts is discussed in terms of this model. The model is also applied to phases which contain one substitutional and one interstitial solute. The vacancies in the interstitial sublattice is then regarded as an additional component of the system. Expressions for the integral and partial free energies are derived.

A stoichiometric phase can usually be regarded as composed of two sublattices, the sites of each one being occupied by a certain element. It is often possible to dissolve some amount of a new element by substituting one or the other of the main components. One may thus produce a ternary system composed of two sublattices one of them being filled by a mixture of two elements and the other one being filled by the third element. It has been suggested [1,2] that the thermodynamics of such a system can be described by the classical regular solution model if applied to the sublattice containing two elements. The same procedure should be applicable even if more than two elements are introduced into one sublattice. However, a new problem arises if both sublattices contain more than one element each. The simplest case can be represented by the formula $(A, B)_a (C, D)_c$.

A similar case is encountered in ionic melts containing at least two cations A and B and two anions C and D. A regular solution model for such ionic melts has been derived by Førland [3] as an extension of the quasithermodynamic theory of Flood, Førland and Grjotheim.[4] The theory of conformal ionic solutions by Blander [5] leads to the same model. The same expressions will now be derived by a simple, purely formal method, applicable to ionic melts as well as stoichiometric phases. It will also be shown how the model can be applied to interstitial solutions.

REGULAR SOLUTION MODEL

REPRESENTATION OF COMPOSITION

The coefficients in $(A, B)_a (C, D)_c$ express the number of sites in each sublattice and, for convenience, the size of the sublattices may be chosen such that $a + c = 1$. The number of moles of each kind of atom, n_A etc., are related by the following equation

$$n_e = n_A/a + n_B/a = n_C/c + n_D/c \qquad (1)$$

where n_e represents the size of the system.

In ionic melts containing at least two cations, A and B, and two anions, C and D, the requirement of electroneutrality yields the relation

$$n_e = n_A/a + n_B/b = n_C/c + n_D/d \qquad (2)$$

where, for convenience, the quantities $1/a$ etc. can be identified with the valence of each ion. The size of the system, n_e is then expressed as the number of equivalents.

It is usual to express the concentration of a certain element by its mole fraction,

$$X_A = \frac{n_A}{n_A + n_B + n_C + n_D}; \quad \sum X_A = 1 \qquad (3)$$

For an ionic melt, it is sometimes convenient to define the mole fractions separately for the cations and for the anions.

$$Y_A = \frac{n_A}{n_A + n_B}; \quad Y_A + Y_B = 1 \qquad (4)$$

$$Y_C = \frac{n_C}{n_C + n_D}; \quad Y_C + Y_D = 1 \qquad (5)$$

It is also convenient to define corresponding fractions taking into account the charge of the ions,

$$Z_A = \frac{n_A/a}{n_A/a + n_B/b} = \frac{n_A}{a \, n_e}; \quad Z_A + Z_B = 1 \qquad (6)$$

$$Z_C = \frac{n_C/c}{n_C/c + n_D/d} = \frac{n_C}{c \, n_e}; \quad Z_C + Z_D = 1 \qquad (7)$$

The corresponding quantities for a stoichiometric phase are related in the following way in view of $b = a$ and $d = c$.

$$Z_A = Y_A = X_A \frac{a+c}{a}; \quad Z_B = Y_B = X_B \frac{a+c}{a} \qquad (8)$$

$$Z_C = Y_C = X_C \frac{a+c}{c}; \quad Z_D = Y_D = X_D \frac{a+c}{c} \qquad (9)$$

In the systems under consideration there are four elements but in view of relation (1) or (2) the degrees of freedom in varying the composition is one less than in an ordinary quaternary system. The variation in composition can thus be represented by two parameters and the composition is conveniently plotted on a square where the corners represent the four basic compounds A_aC_c, A_aD_d, B_bC_c, and B_bD_d and the parameters Z_B and Z_D are used to represent the composition of any intermediate point;[6] Fig. 1.

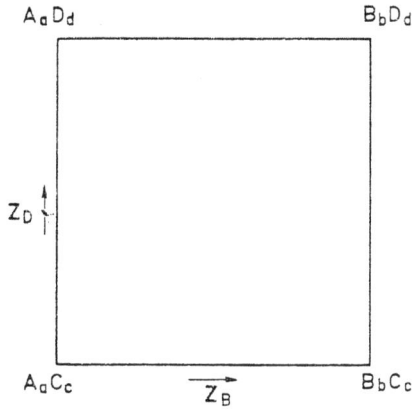
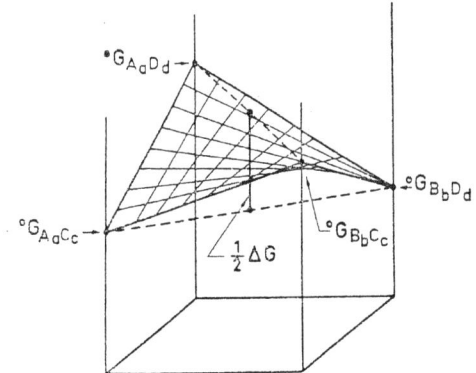

Fig. 1. Representation of composition in a quaternary system where the components mix with each other, two and two.

Fig. 2. Suggested surface of reference for the free energy in a quaternary system where the components mix with each other two and two.

Temkin [7] proposed that the entropy of mixing of an ionic melt can be calculated under the assumption that the anions mix randomly with each other and the cations with each other. This model gives the following expression,

$$-S^{\text{ideal}}/R = n_A \ln Y_A + n_B \ln Y_B + n_C \ln Y_C + n_D \ln Y_D \qquad (10)$$

or, by dividing with the size of the system as defined by n_e from eqn. (2)

$$-S_m^{\text{ideal}}/R = aZ_A \ln Y_A + bZ_B \ln Y_B + cZ_C \ln Y_C + dZ_D \ln Y_D \qquad (11)$$

The same equations with $a=b$ and $c=d$ hold for a stoichiometric phase if the atoms mix randomly within each sublattice.

STATES OF REFERENCE FOR THE FREE ENERGY

In multicomponent systems it is usual to define an excess free energy, $^{\text{E}}G_m$, by the following expression

$$G_m = \sum_i X_i {}^\circ G_i - TS_m^{\text{ideal}} + {}^{\text{E}}G_m \qquad (12)$$

where the quantities $^\circ G_i$ are the free energies of the pure components. They thus define the natural plane of reference for the free energy. As an exemple, a ternary system has three states of reference and they define a plane of reference. In the systems now under consideration there are four states of reference, one each for the compounds A_aC_c, A_aD_d, B_bC_c, and B_bD_d, and it will generally be impossible to construct a plane of reference through all of them. It thus seems to be necessary to choose a non-planar surface of reference. The simplest geometric shape is the one defined by the following choice.

$$G_m = Z_A Z_C {}^\circ G_{A_aC_c} + Z_A Z_D {}^\circ G_{A_aD_d} + Z_B Z_C {}^\circ G_{B_bC_c} + Z_B Z_D {}^\circ G_{B_bD_d} - TS_m^{\text{ideal}} + {}^E G_m \quad (13)$$

where $Z_A Z_C {}^\circ G_{A_aC_c} + Z_A Z_D {}^\circ G_{A_aD_d} + Z_B Z_C {}^\circ G_{B_bC_c} + Z_B Z_D {}^\circ G_{B_bD_d}$ represents the surface of reference. It is illustrated by Fig. 2 and its deviation from a plane shape is defined by the following quantity,

$$\Delta G = {}^\circ G_{A_aD_d} + {}^\circ G_{B_bC_c} - {}^\circ G_{A_aC_c} - {}^\circ G_{B_bD_d} \quad (14)$$

POWER SERIES REPRESENTATION OF EXCESS FREE ENERGY

In an ordinary multicomponent system the excess free energy is often represented by a power series. As can be seen from eqn. (12), however, first power terms are already used in defining the plane of reference. The expression for the excess free energy therefore starts with second power terms which define the regular solution model

$$^E G_m = \sum X_i X_j K_{ij} \quad (15)$$

For the systems now under consideration eqn. (13) shows that four of the second power terms are already used in defining the surface of reference. The excess free energy for a simple case could possibly be described by the remaining two terms of the same power,

$$^E G_m = Z_A Z_B K_{AB} + Z_C Z_D K_{CD} \quad (16)$$

However, it seems rather improbable that the interaction between the A and B atoms should be quite independent of whether the other sublattice is occupied by C or D atoms. It may thus be suggested that a regular solution model for these systems should be defined by the following expression

$$^E G_m = Z_A Z_B Z_C L_{AB}{}^C + Z_A Z_B Z_D L_{AB}{}^D + Z_C Z_D Z_A L_{CD}{}^A + Z_C Z_D Z_B L_{CD}{}^B \quad (17)$$

EXPRESSIONS FOR THE PARTIAL FREE ENERGIES

The partial quantities can be derived from the integral quantities of eqns. (11) and (13) by standard methods, yielding the following rule of calculation with our choice of composition variables

$$\overline{G}_{A_aC_c} = G_m + (1 - Z_A)\frac{\partial G_m}{\partial Z_A} + (1 - Z_C)\frac{\partial G_m}{\partial Z_C} \quad (18)$$

The following expressions are obtained using ΔG from eqn. (14).

$$\overline{G}_{A_aC_c} = {}^\circ G_{A_aC_c} + Z_B Z_D \Delta G + RTa \ln Y_A + RTc \ln Y_C + {}^E G_{A_aC_c} \quad (19)$$

$$\overline{G}_{A_aD_d} = {}^\circ G_{A_aD_d} - Z_B Z_C \Delta G + RTa \ln Y_A + RTd \ln Y_D + {}^E G_{A_aD_d} \quad (20)$$

$$\overline{G}_{B_bC_c} = {}^\circ G_{B_bC_c} - Z_A Z_D \, \Delta G + RTb \ln Y_B + RTc \ln Y_C + {}^E G_{B_bC_c} \quad (21)$$

$$\overline{G}_{B_bD_d} = {}^\circ G_{B_bD_d} + Z_A Z_C \, \Delta G + RTb \ln Y_B + RTd \ln Y_D + {}^E G_{B_bD_d} \quad (22)$$

The well-known fact that these partial quantities can be obtained from a free energy diagram by constructing a plane of tangency and reading its intersections with the component axes, immediately shows that the following relations hold.

$$\overline{G}_{A_aD_d} + \overline{G}_{B_bC_c} - \overline{G}_{A_aC_c} - \overline{G}_{B_bD_d} = 0 \quad (23)$$

$$^E G_{A_aD_d} + {}^E G_{B_bC_c} - {}^E G_{A_aC_c} - {}^E G_{B_bD_d} = 0 \quad (24)$$

It is also of considerable interest to study the change in free energy if component B is exchanged with the corresponding amount of component A, i.e. $dn_A/a = -dn_B/b$ and $dn_e = 0$ in view of eqn. (2). This change may be accomplished in two ways, either by adding A_aC_c and removing B_bC_c or by adding A_aD_d and removing B_bD_d. With the first alternative one obtains

$$\left(\frac{\partial G}{\partial n_A/a}\right)_{n_e, n_C} = \overline{G}_{A_aC_c} - \overline{G}_{B_bC_c} = Z_C({}^\circ G_{A_aC_c} - {}^\circ G_{B_bC_c}) + Z_D({}^\circ G_{A_aD_d} - {}^\circ G_{B_bD_d}) +$$

$$+ RTa \ln Y_A - RTb \ln Y_B + {}^E G_{A_aC_c} - {}^E G_{B_bC_c} \quad (25)$$

It may be pointed out that this partial derivative of the total free energy is identical to $(\partial G_m/\partial Z_A)_{Z_C}$ which in some cases may provide a simpler way of calculation.

An analogous expression is obtained for the exchange of component D with component C,

$$\left(\frac{\partial G}{\partial n_C/c}\right)_{n_e, n_A} = \overline{G}_{A_aC_c} - \overline{G}_{A_aD_d} = Z_A({}^\circ G_{A_aC_c} - {}^\circ G_{A_aD_d}) + Z_B({}^\circ G_{B_bC_c} - {}^\circ G_{B_bD_d})$$

$$+ RTc \ln Y_C - RTd \ln Y_D + {}^E G_{A_aC_c} - {}^E G_{A_aD_d} \quad (26)$$

The expression for the excess free energy defined by eqn. (16) gives the following partial quantities,

$$^E G_{A_aC_c} = Z_B^2 K_{AB} + Z_D^2 K_{CD} \quad (27)$$

$$^E G_{A_aD_d} = Z_B^2 K_{AB} + Z_C^2 K_{CD} \quad (28)$$

$$^E G_{B_bC_c} = Z_A^2 K_{AB} + Z_D^2 K_{CD} \quad (29)$$

$$^E G_{B_bD_d} = Z_A^2 K_{AB} + Z_C^2 K_{CD} \quad (30)$$

$$^E G_{A_aC_c} - {}^E G_{A_aD_d} = {}^E G_{B_bC_c} - {}^E G_{B_bD_d} = (Z_D - Z_C) K_{CD} \quad (31)$$

$$^E G_{A_aC_c} - {}^E G_{B_bC_c} = {}^E G_{A_aD_d} - {}^E G_{B_bD_d} = (Z_B - Z_A) K_{AB} \quad (32)$$

The more complicated expression for the excess free energy defined by eqn. (17) gives the following partial quantities.

$$^E G_{A_aC_c} = Z_B(Z_D Z_A + Z_B Z_C) L_{AB}{}^C + Z_D(Z_D Z_A + Z_B Z_C) L_{CD}{}^A + Z_B Z_D(Z_D - Z_C) L_{CD}{}^B$$

$$+ Z_B Z_D(Z_B - Z_A) L_{AB}{}^D \quad (33)$$

$$^{\mathrm{E}}G_{\mathrm{A_aD_d}} = Z_\mathrm{B}Z_\mathrm{C}(Z_\mathrm{B}-Z_\mathrm{A})L_{\mathrm{AB}}{}^\mathrm{C} + Z_\mathrm{C}(Z_\mathrm{C}Z_\mathrm{A}+Z_\mathrm{B}Z_\mathrm{D})L_{\mathrm{CD}}{}^\mathrm{A} + Z_\mathrm{B}Z_\mathrm{C}(Z_\mathrm{C}-Z_\mathrm{D})L_{\mathrm{CD}}{}^\mathrm{B} +$$
$$+ Z_\mathrm{B}(Z_\mathrm{C}Z_\mathrm{A}+Z_\mathrm{B}Z_\mathrm{D})L_{\mathrm{AB}}{}^\mathrm{D} \tag{34}$$

$$^{\mathrm{E}}G_{\mathrm{B_bC_c}} = Z_\mathrm{A}(Z_\mathrm{D}Z_\mathrm{B}+Z_\mathrm{A}Z_\mathrm{C})L_{\mathrm{AB}}{}^\mathrm{C} + Z_\mathrm{A}Z_\mathrm{D}(Z_\mathrm{D}-Z_\mathrm{C})L_{\mathrm{CD}}{}^\mathrm{A} + Z_\mathrm{D}(Z_\mathrm{D}Z_\mathrm{B}+Z_\mathrm{A}Z_\mathrm{C})L_{\mathrm{CD}}{}^\mathrm{B}$$
$$+ Z_\mathrm{A}Z_\mathrm{D}(Z_\mathrm{A}-Z_\mathrm{B})L_{\mathrm{AB}}{}^\mathrm{D} \tag{35}$$

$$^{\mathrm{E}}G_{\mathrm{B_bD_d}} = Z_\mathrm{A}Z_\mathrm{C}(Z_\mathrm{A}-Z_\mathrm{B})L_{\mathrm{AB}}{}^\mathrm{C} + Z_\mathrm{A}Z_\mathrm{C}(Z_\mathrm{C}-Z_\mathrm{D})L_{\mathrm{CD}}{}^\mathrm{A} + Z_\mathrm{C}(Z_\mathrm{C}Z_\mathrm{B}+Z_\mathrm{A}Z_\mathrm{D})L_{\mathrm{CD}}{}^\mathrm{B}$$
$$+ Z_\mathrm{A}(Z_\mathrm{C}Z_\mathrm{B}+Z_\mathrm{A}Z_\mathrm{D})L_{\mathrm{AB}}{}^\mathrm{D} \tag{36}$$

$$^{\mathrm{E}}G_{\mathrm{A_aC_c}} - {}^{\mathrm{E}}G_{\mathrm{A_aD_d}} = Z_\mathrm{B}Z_\mathrm{A}L_{\mathrm{AB}}{}^\mathrm{C} + Z_\mathrm{B}(Z_\mathrm{D}-Z_\mathrm{C})L_{\mathrm{CD}}{}^\mathrm{B} + Z_\mathrm{A}(Z_\mathrm{D}-Z_\mathrm{C})L_{\mathrm{CD}}{}^\mathrm{A} - Z_\mathrm{A}Z_\mathrm{B}L_{\mathrm{AB}}{}^\mathrm{D} \tag{37}$$

$$^{\mathrm{E}}G_{\mathrm{A_aC_c}} - {}^{\mathrm{E}}G_{\mathrm{B_bC_c}} = Z_\mathrm{C}(Z_\mathrm{B}-Z_\mathrm{A})L_{\mathrm{AB}}{}^\mathrm{C} - Z_\mathrm{C}Z_\mathrm{D}L_{\mathrm{CD}}{}^\mathrm{B} + Z_\mathrm{D}Z_\mathrm{C}L_{\mathrm{CD}}{}^\mathrm{A} + Z_\mathrm{D}(Z_\mathrm{B}-Z_\mathrm{A})L_{\mathrm{AB}}{}^\mathrm{D} \tag{38}$$

APPLICATION TO IONIC SALT MELTS

Flood et al.[4] suggested that the change in free energy when the ion B is exchanged with ion A can be evaluated by the following expression

$$(\partial G_\mathrm{m}/\partial Z_\mathrm{A})_{Z_1,Z_2} = \sum_i Z_i \, (\partial G_\mathrm{m}/\partial Z_\mathrm{A})_{Z_1,Z_2\ldots}{}^{Z_i=1} \tag{39}$$

The summation is carried out over all the ions of the opposite charge, here denoted by 1, 2, 3, *etc.*, and there is no limitation to the number of such ions. The superscript $Z_i = 1$ indicates that the derivative should be evaluated at this value.

For the case of two anions and two cations eqn. (25) yields the following expressions,

$$\left(\frac{\partial G_\mathrm{m}}{\partial Z_\mathrm{A}}\right)_{Z_\mathrm{C}}^{Z_\mathrm{C}=1} = {}^\circ G_{\mathrm{A_aC_c}} - {}^\circ G_{\mathrm{B_bC_c}} + RT a \ln Y_\mathrm{A} - RT b \ln Y_\mathrm{B} + {}^{\mathrm{E}}G_{\mathrm{A_aC_c}}{}^{Z_\mathrm{C}=1} - {}^{\mathrm{E}}G_{\mathrm{B_bC_c}}{}^{Z_\mathrm{C}=1} \tag{40}$$

$$\left(\frac{\partial G_\mathrm{m}}{\partial Z_\mathrm{A}}\right)_{Z_\mathrm{C}}^{Z_\mathrm{D}=1} = {}^\circ G_{\mathrm{A_aD_d}} - {}^\circ G_{\mathrm{B_bD_d}} + RT \, a \ln Y_\mathrm{A} - RT b \ln Y_\mathrm{B} + {}^{\mathrm{E}}G_{\mathrm{A_aC_c}}{}^{Z_\mathrm{D}=1} - {}^{\mathrm{E}}G_{\mathrm{B_bC_c}}{}^{Z_\mathrm{D}=1} \tag{41}$$

and by combination with eqn. (25) one obtains,

$$\left(\frac{\partial G_\mathrm{m}}{\partial Z_\mathrm{A}}\right)_{Z_\mathrm{C}} = Z_\mathrm{C}\left(\frac{\partial G_\mathrm{m}}{\partial Z_\mathrm{A}}\right)_{Z_\mathrm{C}}^{Z_\mathrm{C}=1} + Z_\mathrm{D}\left(\frac{\partial G_\mathrm{m}}{\partial Z_\mathrm{A}}\right)_{Z_\mathrm{C}}^{Z_\mathrm{D}=1} - Z_\mathrm{C}{}^{\mathrm{E}}G_{\mathrm{A_aC_c}}{}^{Z_\mathrm{C}=1} + Z_\mathrm{C}{}^{\mathrm{E}}G_{\mathrm{B_bC_c}}{}^{Z_\mathrm{C}=1}$$
$$- Z_\mathrm{D}{}^{\mathrm{E}}G_{\mathrm{A_aC_c}}{}^{Z_\mathrm{D}=1} + Z_\mathrm{D}{}^{\mathrm{E}}G_{\mathrm{B_bC_c}}{}^{Z_\mathrm{D}=1} + {}^{\mathrm{E}}G_{\mathrm{A_aC_c}} - {}^{\mathrm{E}}G_{\mathrm{B_bC_c}} \tag{42}$$

Flood's expression thus neglects a series of excess free energy terms. However, the sum of these terms will only be $Z_\mathrm{C}Z_\mathrm{D}(L_{\mathrm{CD}}{}^\mathrm{A} - L_{\mathrm{CD}}{}^\mathrm{B})$ according to the regular solution model defined by eqn. (17) and it will be zero according to the model defined by eqn. (16). It thus seems that Flood's expression might be very realistic for many cases.

APPLICATION TO AN INTERSTITIAL AND SUBSTITUTIONAL SOLUTION

An interstitial solution can be regarded as composed of two sublattices. One of them is completely filled by the base element and any substitutionally dissolved element. The other sublattice is only partially filled by the interstitially dissolved element. However, it is possible to regard the vacancies in the interstitial sublattice as a component and the whole system, containing a base element A, a substitutional addition B and an interstitial addition C, can thus be represented as a stoichiometric phase, $(A,B)_a (C, V)_c$ where V stands for the vacancies and $a+c=1$.

The equations, derived for a stoichiometric phase, can thus be applied to this kind of system as well. The mole fractions should then be redefined, taking into account the presence of the vacancies.[8]

$$X_1 = aX_A/(1-X_C) \tag{43}$$

$$X_2 = aX_B/(1-X_C) \tag{44}$$

$$X_3 = aX_C/(1-X_C) \tag{45}$$

$$X_4 = c - aX_C/(1-X_C) \tag{46}$$

The relation of these variables to those defined for a stoichiometric phase is obtained by inserting X_1 instead of X_A in eqns. (8) and (9), yielding

$$Z_A = Y_A = X_1/a = X_A/(1-X_C) \tag{47}$$

$$Z_B = Y_B = X_2/a = X_B/(1-X_C) \tag{48}$$

$$Z_C = Y_C = X_3/c = (a/c)X_C/(1-X_C) \tag{49}$$

$$Z_V = Y_V = X_4/c = 1 - (a/c)X_C/(1-X_C) \tag{50}$$

It may thus be advantageous to present the composition of a ternary system $A-B-C$ on a square like Fig. 1 and using $X_B/(1-X_C)$ and $(a/c)X_C/(1-X_C)$ as the variables.

It should further be noticed that the expressions derived for the molar quantities must be multiplied by $(1-X_C)/a$ in order to hold for one mole of real material.

It may for instance be instructive to discuss how the four terms in eqn. (13), defining the surface of reference, are transformed. We find

$$\sum Z_i Z_j {}^\circ G_{ij}(1-X_C)/a = X_A {}^\circ G_{A_aV_c}/a + X_B {}^\circ G_{B_aV_c}/a + X_C({}^\circ G_{A_aC_c} - {}^\circ G_{A_aV_c})/c +$$
$$+ \frac{X_B X_C}{1-X_C} \Delta G/c \tag{51}$$

The quantity ${}^\circ G_{A_aV_c}/a$ is identical to ${}^\circ G_A$, the free energy of one mole of pure A, because the "compound" A_aV_c is nothing but pure A. By the same reason ${}^\circ G_{B_aV_c}/a$ is identical to ${}^\circ G_B$. However, it is not possible to identify any term in (51) as representing the free energy of one mole of pure C because pure C cannot exist according to our model, C being an interstitially dissolved element. Instead, our model leads to an artificial standard state of C for the solution of C in A and the expression $(°G_{A_aC_c} - °G_{A_aV_c})/c$ represents its molar free energy.

In the same way, the molar free energy of the corresponding standard state for the solution of C in B is equal to $(°G_{B_aC_c} - °G_{B_aV_c})/c$. The definition of ΔG in eqn. (14) shows that the factor $\Delta G/c$ in the last term of (51) represents the difference between the two artificial standard states for C now mentioned.

The two excess free energy terms of eqn. (16) will transform to the following shape

$$(Z_A Z_B K_{AB} + Z_C Z_D K_{CD})(1 - X_C)/a = \frac{X_A X_B}{1 - X_C} K_{AB}/a + X_C \left(1 - \frac{a}{c}\frac{X_C}{1 - X_C}\right) K_{CV}/c \tag{52}$$

As expected the first term describes the interaction between the two elements in the main lattice and the second term describes the interaction within the interstitial sublattice. These interactions may depend upon the elements present in the other sublattice as described in the model defined by eqn. (17), yielding the following more complicated expression.

$$\frac{a}{c}\frac{X_C}{1 - X_C}\frac{X_A X_B}{1 - X_C} L_{AB}{}^C/a + \left(1 - \frac{a}{c}\frac{X_C}{1 - X_C}\right)\frac{X_A X_B}{1 - X_C} L_{AB}{}^V/a +$$
$$+ X_A \frac{X_C}{1 - X_C}\left(1 - \frac{a}{c}\frac{X_C}{1 - X_C}\right) L_{CV}{}^A/c + X_B \frac{X_C}{1 - X_C}\left(1 - \frac{a}{c}\frac{X_C}{1 - X_C}\right) L_{CV}{}^B/c \tag{53}$$

G_m for a system A – B – C according to this regular solution model is described by the sum of (51) and (53) and the ideal entropy.

The partial quantities can be obtained from eqns. (20), (22), and (26) by inserting the expressions for the excess free energy, eqns. (34), (36), and (37).

$$\overline{G}_A = \frac{1}{a}\overline{G}_{A_aV_c} = °G_A + RT \ln Y_A + RT \frac{c}{a} \ln (1 - Y_C) + {}^EG_A \tag{54}$$

$$\overline{G}_B = \frac{1}{a}\overline{G}_{B_aV_c} = °G_B + RT \ln Y_B + RT \frac{c}{a} \ln (1 - Y_C) + {}^EG_B \tag{55}$$

$$\overline{G}_C = \frac{1}{c}\left(\frac{\partial G}{\partial n_c/c}\right)_{n_e, A} = °G_C + RT \ln Y_C/(1 - Y_C) + {}^EG_C \tag{56}$$

where

$$a^E G_A = - Y_B Y_C(\Delta G + L_{AB}{}^C - L_{AB}{}^V + L_{CV}{}^B - L_{CV}{}^A) + Y_B{}^2 L_{AB}{}^V + Y_C{}^2 L_{CV}{}^A +$$
$$+ Y_B{}^2 Y_C 2(L_{AB}{}^C - L_{AB}{}^V) + Y_B Y_C{}^2 2(L_{CV}{}^B - L_{CV}{}^A) \tag{57}$$

$$a^E G_B = Y_A Y_C(\Delta G + L_{AB}{}^V - L_{AB}{}^C + L_{CV}{}^B - L_{CV}{}^A) + Y_A{}^2 L_{AB}{}^V + Y_C{}^2 L_{CV}{}^B +$$
$$+ Y_A{}^2 Y_C 2(L_{AB}{}^C - L_{AB}{}^V) + Y_A Y_C{}^2 2(L_{CV}{}^A - L_{CV}{}^B) \tag{58}$$

$$c^E G_C = Y_B(\Delta G + L_{AB}{}^C - L_{AB}{}^V + L_{CV}{}^B - L_{CV}{}^A) - Y_C 2 L_{CV}{}^A +$$
$$+ Y_B Y_C 2(L_{CV}{}^A - L_{CV}{}^B) + Y_B{}^2(L_{AB}{}^V - L_{AB}{}^C) \tag{59}$$

$$c°G_C = °G_{A_aC_c} - °G_{A_aV_c} + L_{CV}{}^A \tag{60}$$

It is interesting to note that all the parameters used can in principle be determined from experimental information on the binary system AB and information on $\bar{G}_C$ for the ternary system.

Acknowledgement. This work was carried out in connection with a research project on alloying elements in steel supported financially by the *Swedish Board for Technical Development*.

REFERENCES

1. Richardson, F. D. *J.I.S.I.* **175** (1953) 33.
2. Hillert, M., Wada, T. and Wada, H. *J.I.S.I.* **205** (1967) 539.
3. Førland, T. *Norg. Tek. Vetenskapsakad.* Ser. 2, No. 4 (1957).
4. Flood, H., Førland, T. and Grjotheim, K. *Z. anorg. allgem. Chem.* **276** (1954) 289.
5. Blander, M. and Yosim, S. J. *J. Chem. Phys.* **39** (1963) 2610.
6. Ricci, J. E. *The phase rule and heterogeneous equilibrium*, van Nostrand, New York 1951.
7. Temkin, M. *Acta Phys. Chim. USSR* **20** (1945) 411.
8. Hillert, M. *Phase transformations*, A.S.M., Metals Park, Ohio 1970, p. 181. (Symposium on phase transformations, Detroit 1968).

Received April 23, 1970.

13 Introduction to "Diffusion Controlled Growth of Lamellar Eutectics and Eutectoids in Binary and Ternary Systems"

published in *Acta Metallurgica* (1971)

Lamellar structures are beautiful examples of 'self-organized patterns'. The classic 'solid-state' illustrations familiar to the materials scientist include pearlite formation (particularly in steels) and discontinuous precipitation, both of which have kept generations of scientists interested. The regularity of the structures formed has driven theorists to search for a quantitative description of the structural evolution, and emphasis has usually been placed on three aspects: growth velocity, lamellar spacing and phase composition. The earliest treatments, presented in 1945–46, were due to Brant [1], Scheil [2], Zener [3] and Turnbull [4].

Even in the case of idealized binary alloys, the challenges in the theoretical treatment include coupling the thermodynamics and kinetics with interfacial and/or bulk transport along with considerations of interfacial effects, including crystallography. The latter aspect, in particular, is still under study. The treatments all lead to a relationship between the growth velocity and the lamellar repeat distance. The problem is degenerate. An additional criterion must be invoked to identify a unique velocity-spacing pair, and exactly what constitutes an appropriate criterion is a question that is embroiled in controversy.

For ternary alloys, there are new features that need to be explained. These include variations in the volume fractions and compositions of the growing phases, questions of the evolution of the lamellar spacing which can now show non-steady state characteristics and complications arising from the variety of diffusion processes possible for both solute species, either of which can control the overall growth velocity.

Hillert's 1971 article in *Acta Metallurgica* considers almost all of these features of lamellar growth in eutectic and eutectoid systems, for both binary and ternary alloys. The emphasis of the article is on the ternary systems but Hillert begins by recounting the state of affairs for the binary case both for volume diffusion control [5] and interfacial diffusion control [6] with a generalization of the latter to cases where the compositions of the growing phases are coupled to the composition of the matrix.

The treatment for ternary systems is restricted to cases of small ternary additions. Hillert writes a Fourier series for the composition profile(s) in the matrix in the frame of the moving interface, taking into account the effect of the curvature of the interface on the interfacial compositions. By using the level rule, the mass balance across the interface and assuming a relatively flat interface he is able to find expressions for the compositions of the growing phases, their fractions, the shape of the interface and a relationship between the growth velocity and the lamellar spacing under steady state conditions.

For ternary systems with major solute addition (M) and minor addition (N), four different solute transport scenarios are considered:

– Volume diffusion of M and N;

– Volume diffusion of M, Boundary diffusion of N;

– Boundary diffusion of both M and N;

– Boundary diffusion of M, Volume diffusion of N.

Hillert qualifies his relationships by reiterating that care must be taken to use the expressions only under conditions consistent with the assumptions he has made in their derivation. This is particularly important in the case of so-called Fe-C-X steels where the diffusion coefficients of C and X can differ by many orders of magnitude, but pearlite growth might be controlled by the diffusion of either, depending on alloy composition and temperature. As an example of one extreme (where the redistribution of solute N $\to$ 0 and the growing phases inherit the bulk N content) Hillert considers the growth of parapearlite, where the growth rate is controlled by the diffusion of C with the reaction proceeding according to a driving force reduced in comparison to that of the ortho-reaction.

In that sense, this paper opens the path toward a proper understanding of the coupling between interfacial conditions and growth in a ternary system exhibiting pattern formation. The interplay between thermodynamics, kinetics and morphological instabilities in such systems is still an open question for modeling

Christopher R. HUTCHINSON and Yves BRÉCHET

References

[1] W.H. Brandt, *J. Appl. Phys.* 16 (1945) 139.

[2] E. Scheil, *Z. Metallkd.* 37 (1946) 123.

[3] C. Zener, *Trans. AIME* 167 (1946) 550.

[4] D. Turnbull, *Acta metall.* 3 (1955) 55.

[5] M. Hillert, *Jernkont. Ann.* 141 (1957) 757.

[6] M. Hillert, *Mechanism of Phase Transformations in Solids.* The Institute of Metals, Vol. 33 (1968) p. 231.

DIFFUSION CONTROLLED GROWTH OF LAMELLAR EUTECTICS AND EUTECTOIDS IN BINARY AND TERNARY SYSTEMS*

M. HILLERT†

The mathematical treatment of the steady state growth of lamellar eutectics and eutectoids in binary systems, as controlled by volume diffusion in the matrix or boundary diffusion, is extended to include compositional variations in the two growing phases. It is concluded that the effect of such variations on the growth rate is normally rather small.

The effect of a ternary addition is treated for a number of combinations of volume diffusion and boundary diffusion of the alloying elements. Explicit expressions for the growth rate are obtained in all the cases thanks to the assumption that only the content of the ternary addition varies within the two growing phases.

CROISSANCE CONTROLEE PAR DIFFUSION DES EUTECTIQUES LAMELLAIRES ET DES EUTECTOIDES DANS LES SYSTEMES BINAIRES ET TERNAIRES

La théorie mathématique de la croissance en régime permanent des eutectiques et des eutectoïdes dans les systèmes binaires, contrôlée par la diffusion en volume dans la matrice ou par la diffusion aux joints, est étendue aux variations de composition comprises dans les limites des deux phases en cours de croissance. L'auteur conclut que l'influence de telles variations sur la vitesse de croissance est normalement plutôt faible.

L'influence d'une addition ternaire est étudiée pour quelques combinaisons de diffusion en volume et de diffusion aux joints des éléments entrant dans l'alliage. Des expressions précises pour la vitesse de croissance sont obtenues dans tous les cas grâce à l'hypothèse suivant laquelle seule la concentration de l'élément trenaire varie entre des limites situées à l'intérieur des deux phases en cours de croissance.

DIFFUSIONSKONTROLLIERTES WACHSTUM LAMELLARER EUTEKTIKA UND EUTEKTOIDE IN BINÄREN UND TERNÄREN SYSTEMEN

Die mathematische Behandlung des stationären, durch Volumendiffusion in der Matrix ode Grenzflächendiffusion kontrollierten Wachstums lamellarer Eutektika und Eutektoide in binären Systemen wird so erweitert, daß Variationen der Zusammensetzung in den zwei wachsenden Phasen berücksichtigt werden können. Es ergibt sich, daß der Einfluß solcher Variationen auf die Wachstumsgeschwindigkeit meist relativ klein ist.

Der Einfluß ternärer Zulegierungen wird für einige Kombinationen von Volumendiffusion und Grenzflächendiffusion des Legierungselements behandelt. Unter der Annahme, daß nur der Anteil des ternären Bestandteils innerhalb der zwei wachsenden Phasen variiert, werden in allen Fällen explizite Ausdrücke für die Wachstumsgeschwindigkeit gefunden.

INTRODUCTION

There have been many attempts to treat the diffusion-controlled growth of lamellar eutectoid or eutectic structures in binary systems. Brandt[1] and Scheil[2] were the first ones to analyze mathematically the diffusion in the matrix ahead of the growing structure and Zener[3] was the first one to take into account the effect of the interfacial energy between the lamellae in the growing two-phase structure. He was thus able to derive an equation relating the growth rate and the spacing of the lamellar structure.

Hillert[4] considered in detail the shape of the interface and its effect on the composition of the matrix at the interface. Using the simplifying assumption that the interface is approximately flat, he was able to solve the mathematics of diffusion in the matrix and to obtain a more accurate equation relating the growth rate and the spacing of the lamellar structure, and also to calculate the shape of the interface. His treatment was later repeated by Jackson and Hunt[5] who extended it to the case of rod-like structures.

Hillert only considered the case where the composition of the two growing phases can be treated as constant, as is the case for pearlite in the Fe–C system. Donaghey and Tiller[6] have recently shown how the mathematics of diffusion can be solved in the case where the compositions of the two growing phases vary with the composition of the matrix. Bolze and Kirkaldy[7] have combined this solution with the detailed consideration of the shape of the interface to obtain the relation between growth rate and spacing. Apart from the simplifying assumption of a flat interface which is used in calculating the diffusion in the matrix, the treatment of the growth of lamellar structures as controlled by volume diffusion in the matrix seems rather complete.

Turnbull[8] was the first one to give a mathematical treatment of the growth of a lamellar structure as controlled by boundary diffusion. He suggested a simple modification of Zener's crude treatment of volume diffusion. Cahn[9] presented a more rigorous solution for the boundary diffusion when the compositions of the two growing phases vary with the composition of the matrix. He combined this with

---

* Received December 28, 1970.
† Royal Institute of Technology, Stockholm, Sweden.

Zener's crude method of taking into account the effect of the interfacial energy. He also introduced the mobility of the interface but was forced to restrict the treatment to a symmetric case. Shapiro and Kirkaldy,[10] Sundquist[11] and Hillert[12] applied the more detailed consideration of the shape of the interface but restricted their treatment to the case where the two growing phases have constant compositions. The treatment for boundary diffusion is thus less complete than for volume diffusion.

Fisher[13] has modified Zener's crude treatment to the case where diffusion in one of the two growing phases dominates. There does not seem to exist any advanced treatment of this case.

The main purpose of the present work is to examine the effect of ternary additions. In order to do this, it is advantageous first to put the treatments of the binary cases in a convenient form and to extend the treatment of boundary diffusion control to cases where the compositions of the two growing phases vary with the composition of the matrix. An earlier attempt by Bolze and Kirkaldy[7] to treat the ternary effect gave a very complicated result and the only tractable case was very restrictive in character.

It was suggested long ago[14] that small ternary effects in Fe–C alloys could be treated by a simple modification of the treatment for the binary Fe–C system and this approach was recently tested by Hillert[15] considering volume diffusion of carbon and by Sundquist[16] considering boundary diffusion of carbon. Furthermore, Fridberg and Hillert[17] recently presented an attempt to treat large ternary effects by modifying the binary theory for boundary diffusion of the alloying element. The treatment to be presented in the present work will give these cases as limiting cases.

DERIVATION OF GENERAL EQUATIONS

We shall consider the reaction $\gamma \to \alpha + \beta$ in a binary system, B–M, with small amounts of a ternary addition, N. The effect of the shape of the interface γ/α on the local equilibrium can then be obtained by integrating the Gibbs–Duhem equation from the binary side, yielding an expression of the following type if it is assumed that the matrix phase, γ, is under ordinary pressure.

$$\frac{\Delta P^\alpha V_m^\alpha}{RT} = h^\alpha(X_M^{\gamma/\alpha} - {}^eX_M^{\gamma/\alpha}) + k^\alpha X_N^{\gamma/\alpha} \qquad (1)$$

The quantity ${}^eX_M^{\gamma/\alpha}$ represents the ordinary equilibrium value for the binary case. The quantities h^α and k^α are defined as follows and are approximated as constants during the integration,

$$h^\alpha = \frac{X_M^\alpha}{X_M^\gamma} - \frac{X_B^\alpha}{X_B^\gamma} \qquad (2)$$

$$k^\alpha = \frac{X_N^\alpha}{X_N^\gamma} - \frac{X_B^\alpha}{X_B^\gamma} \qquad (3)$$

The pressure difference ΔP^α is related to the interfacial energy of the curved interface and its radii of curvature in the usual way. This type of equation has been used previously in treatments of ternary effects.[7,18]

The corresponding change in composition of the α phase will be represented by

$$X_M^{\alpha/\gamma} - {}^eX_M^\alpha = K_M^\alpha(X_M^{\gamma/\alpha} - {}^eX_M^{\gamma/\alpha}) \qquad (4)$$

$$X_N^{\alpha/\gamma} = K_N^\alpha \cdot X_N^{\gamma/\alpha} \qquad (5)$$

and the quantities K_M^α and K_N^α will be treated as constants. The superscript e on the mole fractions, X, denotes the binary equilibrium values. The choice $K_M^\alpha = 0$ makes the M content in the α phase constant.

The balance of forces at the interface of the growing lamellar structure to the matrix will be treated in the same way as in the previous publication.[12] By integrating the force acting on the edge of an α lamella one obtains

$$\sigma L^\alpha = \int_0^{S^\alpha/2} \Delta P^\alpha \cdot dz \qquad (6)$$

$$\frac{2\sigma L^\alpha V_m^\alpha}{SRT} = f^\alpha h^\alpha(\overline{X_M^{\gamma/\alpha}} - {}^eX_M^{\gamma/\alpha}) + f^\alpha k^\alpha \overline{X_N^{\gamma/\alpha}} \qquad (7)$$

where σ is the interfacial tension between the α and β lamellae and L^α the fraction of this tension which is carried by the α lamellae. The quantities $\overline{X_M^{\gamma/\alpha}}$ and $\overline{X_N^{\gamma/\alpha}}$ are the alloy contents averaged over the whole edge. S is the so-called interlamellar spacing of the lamellar structure and f^α the fraction of the α phase in this structure. Corresponding equations are obtained for the β phase. It should be noticed that $L^\alpha + L^\beta = 1$ and $f^\alpha + f^\beta = 1$.

The treatment will be limited to the stationary growth of the lamellar structure. The average composition of the two-phase structure must then be identical to the initial composition of the γ phase, which will be denoted by $X_M^{\gamma 1}$ and $X_N^{\gamma 1}$. The lever rule will thus give

$$f^\alpha(X_M^{\gamma 1} - \overline{X_M^\alpha}) = f^\beta(\overline{X_M^\beta} - X_M^{\gamma 1})$$
$$= f^\alpha f^\beta(\overline{X_M^\beta} - \overline{X_M^\alpha}) \qquad (8)$$

$$f^\alpha(X_N^{\gamma 1} - \overline{X_N^\alpha}) = f^\beta(\overline{X_N^\beta} - X_N^{\gamma 1})$$
$$= f^\alpha f^\beta(\overline{X_N^\beta} - \overline{X_N^\alpha}) \qquad (9)$$

or, by inserting equations (4) and (5)

$$f^\alpha[X_M^{\gamma 1} - {}^eX_M^\alpha - K_M^\alpha(\overline{X_M^{\gamma/\alpha}} - {}^e\overline{X_M^{\gamma/\alpha}})]$$
$$= f^\beta[{}^eX_M^\beta - X_M^{\gamma 1} + K_M^\beta(\overline{X_M^{\gamma/\beta}} - {}^e\overline{X_M^{\gamma/\beta}})]$$
$$= f^\alpha f^\beta[{}^eX_M^\beta - {}^eX_M^\alpha + K_M^\alpha({}^eX_M^{\gamma/\alpha} - \overline{X_M^{\gamma/\alpha}})$$
$$+ K_M^\beta(\overline{X_M^{\gamma/\beta}} - {}^eX_M^{\gamma/\beta})] \quad (8a)$$

$$f^\alpha(X_N^{\gamma 1} - K_N^\alpha \overline{X_N^{\gamma/\alpha}}) = f^\beta(K_N^\beta \overline{X_N^{\gamma/\beta}} - X_N^{\gamma 1})$$
$$= f^\alpha f^\beta(K_N^\beta \overline{X_N^{\gamma/\beta}} - K_N^\alpha \overline{X_N^{\gamma/\alpha}}) \quad (9a)$$

In general equation (8) will help us to predict the values of f^α and f^β and equation (9) will help us to determine the content of N averaged over the whole interface of the two-phase structure to the matrix. For convenience, the two equations obtained by applying equation (7) to α and β can be combined to give two, more useful equations. The following equation may thus be obtained and is useful for deriving a relation between growth rate and spacing.

$$\frac{2\sigma}{SRT}\left[\frac{L^\alpha V_m^\alpha}{f^\alpha h^\alpha} - \frac{L^\beta V_m^\beta}{f^\beta h^\beta}\right]$$
$$= \overline{X_M^{\gamma/\alpha}} - \overline{X_M^{\gamma/\beta}} - ({}^eX_M^{\gamma/\alpha} - {}^eX_M^{\gamma/\beta})$$
$$+ \frac{k^\alpha}{h^\alpha}\overline{X_N^{\gamma/\alpha}} - \frac{k^\beta}{h^\beta}\overline{X_N^{\gamma/\beta}} \quad (10)$$

In particular, for $v = 0$ we expect that there is no side-wise diffusion at the interface and we can put

$$\overline{X_M^{\gamma/\alpha}} = \overline{X_M^{\gamma/\beta}} = {}^0X_M^\gamma \quad (11)$$

$$\overline{X_N^{\gamma/\alpha}} = \overline{X_N^{\gamma/\beta}} = {}^0X_N^\gamma \quad (12)$$

Equation (10) will thus give

$$S_0 \cdot \frac{RT}{2\sigma} = \frac{\left(\frac{L^\beta V_m^\beta}{f_0^\beta h^\beta} - \frac{L^\alpha V_m^\alpha}{f_0^\alpha h^\alpha}\right)}{\left[{}^eX_M^{\gamma/\alpha} - {}^eX_M^{\gamma/\beta} - {}^0X_N^\gamma\left(\frac{k^\alpha}{h^\alpha} - \frac{k^\beta}{h^\beta}\right)\right]} \quad (13)$$

The fractions of the α and β phases, f^α and f^β, are not necessarily independent of the growth rate. The values at zero growth rate are denoted by f_0^α and f_0^β. The quantity S_0 is the so-called critical spacing which makes the growth rate zero. Introducing this quantity in equation (10), one obtains

$$\overline{X_M^{\gamma/\alpha}} - \overline{X_M^{\gamma/\beta}} + \frac{k^\alpha}{h^\alpha}(\overline{X_N^{\gamma/\alpha}} - {}^0X_N^\gamma) - \frac{k^\beta}{h^\beta}(\overline{X_N^{\gamma/\beta}} - {}^0X_N^\gamma)$$
$$= \left[{}^eX_M^{\gamma/\alpha} - {}^eX_M^{\gamma/\beta} - {}^0X_N^\gamma\left(\frac{k^\alpha}{h^\alpha} - \frac{k^\beta}{h^\beta}\right)\right]$$
$$\times \left[1 - \frac{S_0}{S}\cdot\left(\frac{L^\beta V_m^\beta}{f^\beta h^\beta} - \frac{L^\alpha V_m^\alpha}{f^\alpha h^\alpha}\right)\bigg/\left(\frac{L^\beta V_m^\beta}{f_0^\beta h^\beta} - \frac{L^\alpha V_m^\alpha}{f_0^\alpha h^\alpha}\right)\right]$$
$$(14)$$

This is a convenient form of the equation relating the growth rate to the spacing. The quantity ${}^0X_N^\gamma$ can be immediately calculated from equation (9a) using the relation given by equation 5.

$${}^0X_N^\gamma = X_N^{\gamma 1}/(f_0^\alpha K_N^\alpha + f_0^\beta K_N^\beta) = X_N^{\gamma 1}/p_{0N} \quad (15)$$

The quantity p_{0N} is introduced here for convenience. It will be discussed further in later sections. As the second useful equation, which can be obtained from equation (7), one of the following may be chosen

$$\frac{2\sigma}{SRT}\left[\frac{L^\alpha V_m^\alpha}{h^\alpha} + \frac{L^\beta V_m^\beta}{h^\beta}\right]$$
$$= f^\alpha(\overline{X_M^{\gamma/\alpha}} - {}^eX_M^{\gamma/\alpha}) + f^\beta(\overline{X_M^{\gamma/\beta}} - {}^eX_M^{\gamma/\beta})$$
$$+ \frac{f^\alpha k^\alpha}{h^\alpha}\overline{X_N^{\gamma/\alpha}} + \frac{f^\beta k^\beta}{h^\beta}\overline{X_N^{\gamma/\beta}} \quad (16)$$

$$\frac{2\sigma}{SRT}\left[\frac{L^\alpha V_m^\alpha}{f^\alpha f^\alpha h^\alpha} + \frac{L^\beta V_m^\beta}{f^\beta f^\beta h^\beta}\right]$$
$$= \frac{1}{f^\alpha}(\overline{X_M^{\gamma/\alpha}} - {}^eX_M^{\gamma/\alpha}) + \frac{1}{f^\beta}(\overline{X_M^{\gamma/\beta}} - {}^eX_M^{\gamma/\beta})$$
$$+ \frac{k^\alpha}{f^\alpha h^\alpha}\overline{X_N^{\gamma/\alpha}} + \frac{k^\beta}{f^\beta h^\beta}\overline{X_N^{\gamma/\beta}} \quad (17)$$

Equation (16) will be found to be most convenient when the transportation of M takes place by volume diffusion and equation (17) when boundary diffusion of M is dominating. The M content at zero growth rate, ${}^0X_M^\gamma$, can be obtained from either one by inserting the value of $S = S_0$ and the value ${}^0X_N^\gamma$ from equation (15).

We have thus derived equations for the calculation of all the characteristic quantities describing the stationary growth of a lamellar structure. They can be applied to various cases of diffusion control. What remains to be done in a particular case is to find the mathematical solution to the rate-controlling diffusion mechanism. We shall first examine volume diffusion and boundary diffusion separately, in the binary case and then proceed to various combinations of diffusion mechanisms in the ternary case.

BINARY SYSTEM WITH VOLUME DIFFUSION

As first pointed out by Brandt,[1] the composition of the matrix can be described by a Fourier series of the form

$$X_M^\gamma = X_M^{\gamma 1} + A_0 + \sum_1 A_n \cos b_n z \cdot \exp - \lambda_n y \quad (18)$$

where

$$b_n = 2\pi n/S \quad (19)$$

and

$$\lambda_n = \frac{v}{2D}\left(1 + \sqrt{1 + \frac{4b_n^2 D^2}{v^2}}\right) \quad (20)$$

z is the coordinate normal to the flat sides of the lamellae and y measures the distance from the interface between the matrix and the two phase structure. It is worth noting that λ_n can be transformed in the following way,

$$\frac{1}{\lambda_n} = \frac{1}{b_n^2}\left(\lambda_n - \frac{v}{D}\right) \quad (21)$$

The composition in the matrix at the interface will be directly given by equation (18) with $\exp -\lambda_n y = 1$ if the shape of the interface is assumed to be approximately flat and the y coordinate is chosen in such a way that $y = 0$ at the interface. The average values at the edge of each kind of lamella is thus obtained as follows,

$$\overline{X_M^{\gamma/\alpha}} = \frac{2}{S^\alpha}\int_0^{S^\alpha/2}\left(X_M^{\gamma 1} + A_0 + \sum_1 A_n \cos b_n z\right) dz$$

$$= X_M^{\gamma 1} + A_0 + \frac{2}{S^\alpha}\sum_1 \frac{A_n}{b_n}\sin n\pi f^\alpha \quad (22)$$

$$\overline{X_M^{\gamma/\beta}} = X_M^{\gamma 1} + A_0 - \frac{2}{S^\beta}\sum_1 \frac{A_n}{b_n}\sin n\pi f^\alpha \quad (23)$$

It is interesting that A_0 is given directly by inserting equations (22) and (23) in equation (16) for the binary case.

$$X_M^{\gamma 1} + A_0$$
$$= f^\alpha \overline{X_M^{\gamma/\alpha}} + f^\beta \overline{X_M^{\gamma/\beta}}$$
$$= f^\alpha {}^e X_M^{\gamma/\alpha} + f^\beta {}^e X_M^{\gamma/\beta} + \frac{2\sigma}{SRT}\left(\frac{L^\alpha V_m^\alpha}{h^\alpha} + \frac{L^\beta V_m^\beta}{h^\beta}\right) \quad (24)$$

Following the procedure by Hillert[4] we can evaluate the higher terms, A_n, by considering the amount of material in an infinitely long, thin slice parallel to the lamellae. If we again use the approximation of the interface as a flat surface at $y = 0$, we obtain for a slice going through an α lamella,

$$-\int_0^\infty D\frac{d^2 X_M^\gamma}{dz^2} dy = v(X_M^{\gamma 1} - X_M^\alpha) \quad (25)$$

where v is the steady state growth rate. By inserting equation (18) in the left-hand side and integrating, and inserting equation (18) with $y = 0$ in the right-hand side by means of equation (4) we obtain

$$\frac{D}{v}\sum_1 \frac{A_n b_n^2}{\lambda_n}\cos b_n z$$
$$= X_M^{\gamma 1} - {}^e X_M^\alpha + K_M^{\alpha e} X_M^{\gamma/\alpha}$$
$$- K_M^\alpha\left(X_M^{\gamma 1} + A_0 + \sum_1 A_n \cos b_n z\right) \quad (26)$$

An equivalent equation is obtained for the β phase. They can be combined and written in the following form

$$\frac{D}{v}\sum_1 \frac{A_n b_n^2}{\lambda_n}\cos b_n z$$
$$= a_0 + \sum_1 a_n \cos b_n z + \left(p_0 + \sum_1 p_n \cos b_n z\right)$$
$$\times \left(r_0 + \sum_1 r_n \cos b_n z - X_M^{\gamma 1} - A_0 - \sum_1 A_n \cos b_n z\right)$$
$$\quad (27)$$

where

$$a_0 + \sum_1 a_n \cos b_n z = \begin{cases} X_M^{\gamma 1} - {}^e X_M^\alpha & \text{for } 0 < z < S^\alpha/2 \\ X_M^{\gamma 1} - {}^e X_M^\beta & \text{for } S^\alpha/2 < z < S/2 \end{cases} \quad (28)$$

$$p_0 + \sum_1 p_n \cos b_n z = \begin{cases} K_M^\alpha & \text{for } 0 < z < S^\alpha/2 \\ K_M^\beta & \text{for } S^\alpha/2 < z < S/2 \end{cases} \quad (29)$$

$$r_0 + \sum_1 r_n \cos b_n z = \begin{cases} {}^e X_M^{\gamma/\alpha} & \text{for } 0 < z < S^\alpha/2 \\ {}^e X_M^{\gamma/\beta} & \text{for } S^\alpha/2 < z < S/2 \end{cases} \quad (30)$$

We find by Fourier analysis

$$a_0 = f^\alpha(X_M^{\gamma 1} - {}^e X_M^\alpha) + f^\beta(X_M^{\gamma 1} - {}^e X_M^\beta);$$
$$a_n = ({}^e X_M^\beta - {}^e X_M^\alpha)\frac{2\sin n\pi f^\alpha}{n\pi} \quad (31)$$

$$p_0 = f^\alpha K_M^\alpha + f^\beta K_M^\beta;$$
$$p_n = (K_M^\alpha - K_M^\beta)\frac{2\sin n\pi f^\alpha}{n\pi} \quad (32)$$

$$r_0 = f^\alpha {}^e X_M^{\gamma/\alpha} + f^\beta {}^e X_M^{\gamma/\beta};$$
$$r_n = ({}^e X_M^{\gamma/\alpha} - {}^e X_M^{\gamma/\beta})\frac{2\sin n\pi f^\alpha}{n\pi} \quad (33)$$

Donaghey and Tiller[6] satisfied the boundary condition used in equation (25) by a different method which they apparently regarded as more accurate. However, their method is completely equivalent and their result can be transformed into equation (26) by means of equation (21). As a consequence, there is no real conflict between their result and the result obtained by Hillert[4] for the case of $K_M^\alpha = K_M^\beta = 0$. The coefficients A_n can be evaluated easily from equation (27) for the special case of $K_M^\alpha = K_M^\beta = K$. Using equation (21) we obtain,

$$A_n = \frac{a_n + K r_n}{\frac{D}{v}\lambda_n - 1 + K}$$

$$= \frac{{}^eX_M^\beta - {}^eX_M^\alpha + K({}^eX_M^{\gamma/\alpha} - {}^eX_M^{\gamma/\beta})}{\frac{1}{2}\sqrt{1 + 4b_n^2 D^2/v^2} - \frac{1}{2} + K} \cdot \frac{2 \sin n\pi f^\alpha}{n\pi} \quad (34)$$

Formally, we can also obtain the following

$$X_M^{\gamma 1} + A_0 = f^\alpha\, {}^eX_M^{\gamma/\alpha} + f^\beta\, {}^eX_M^{\gamma/\beta}$$
$$+ \frac{1}{K}\left[f^\alpha (X_M^{\gamma 1} - {}^eX_M^\alpha) + f^\beta (X_M^{\gamma 1} - {}^eX_M^\beta) \right] \quad (35)$$

This expression is in direct conflict with equation (24). It seems that the conflict is caused by the approximation of the interface as a flat surface which entered into the derivation of equation (24) as well as equation (25) which led to equation (35). In the limit $K \to 0$, the coefficient A_0 will drop out of equation (26) and can only be determined from equation (24). It may thus seem reasonable to trust equation (24) rather than equation (35) in the general case as well.

The exact evaluation of the coefficients A_n is very complicated in the general case of $K_M^\alpha \neq K_M^\beta$, because of the cross products of terms in $\cos b_n z$. As discussed by Donaghey and Tiller, this results in an infinite system of equations which must be solved simultaneously in order to yield the individual coefficients A_n. However, they pointed out that approximate values can be obtained when $vS/D \ll 1$, and this is normally the case, as pointed out by Hillert.[4] Using the transformation

$$2 \cos n\pi z/S \cdot \cos m\pi z/S$$
$$= \cos(n-m)\pi z/S + \cos(n+m)\pi z/S \quad (36)$$

and omitting all terms which result in $(vS/D)^2$, we can rewrite equation (27),

$$\frac{D}{v} \sum_1 \frac{A_n b_n^2}{\lambda_n} \cos b_n z$$
$$= a_0 + \sum_1 a_n \cos b_n z + p_0(r_0 - X_M^{\gamma 1} - A_0)$$

$$+ p_0 \sum_1 r_n \cos b_n z + (r_0 - X_M^{\gamma 1} - A_0) \sum_1 p_n \cos b_n z$$
$$+ \tfrac{1}{2} \sum_1 p_n(r_n - A_n) \quad (37)$$

We can thus obtain

$$X_M^{\gamma 1} + A_0 = r_0 + a_0/p_0 + \sum_1 p_n(r_n - A_n)/2p_0 \quad (38)$$

$$A_n = \frac{a_n + p_0 r_n + p_n(r_0 - X_M^{\gamma 1} - A_0)}{\frac{D}{v}\lambda_n - 1 + p_0} \quad (39)$$

It is now difficult to decide what expression for A_0 should be inserted in equation (39). To the present author it seems most logical to trust equation (24) on all occasions and thus to disregard equation (38). By inserting equation (24) in equation (39) and approximating λ_n by $b_n = 2\pi n/S$ and neglecting $1 - p_0$ in comparison with $2\pi D/vS$, we obtain

$$A_n = \frac{vS}{D}\Big[\,{}^eX_M^\beta - {}^eX_M^\alpha + p_0({}^eX_M^{\gamma/\alpha} - {}^eX_M^{\gamma/\beta})$$
$$- (K_M^\beta - K_M^\alpha) \cdot \frac{2\sigma}{SRT}\left(\frac{L^\alpha V_m^\alpha}{h^\alpha} + \frac{L^\beta V_m^\beta}{h^\beta}\right)\Big]$$
$$\times \frac{\sin n\pi f^\alpha}{(n\pi)^2} \quad (40)$$

By inserting equation (40) in equation (18) we can obtain an expression describing the composition of the matrix. Donaghey and Tiller stopped their analysis at this point and were not able to obtain any relation between growth rate v and spacing S. Instead, the combination of variables vS/D appeared as an arbitrary parameter in their expressions.

The expression thus obtained for A_n can be inserted in equations (22) and (23) to give the average compositions and they can in turn be inserted in equation (8a) to yield f^α and equation (14) to give the relation between v and S. However, the equation obtained for the evaluation of f^α contains the growth rate v and vice versa. It seems practically impossible to separate them analytically but numerical calculations can be carried out easily by iteration for any particular case.

If f^α and f^β are approximated by their values at zero growth rate, equation (14) will yield the following expression

$$\frac{vSB}{Df^\alpha f^\beta}\Big[\,{}^eX_M^\beta - {}^eX_M^\alpha + p_0({}^eX_M^{\gamma/\alpha} - {}^eX_M^{\gamma/\beta})$$
$$+ (K_M^\beta - K_M^\alpha) \cdot \frac{2\sigma}{SRT}\left(\frac{L^\alpha V_m^\alpha}{h^\alpha} + \frac{L^\beta V_m^\beta}{h^\beta}\right)\Big]$$
$$= ({}^eX_M^{\gamma/\alpha} - {}^eX_M^{\gamma/\beta})\left(1 - \frac{S_0}{S}\right) \quad (41)$$

where

$$B = \sum_{1} \frac{(\sin n\pi f^\alpha)^2}{(n\pi)^3} \quad (42)$$

In the limit $K_M{}^\alpha \to 0$ and $K_M{}^\beta \to 0$, this equation reduces to the expression derived previously[4] for this case. For the general case, it is interesting to note that equation (41) is not strongly dependent upon the values of $K_M{}^\alpha$ and $K_M{}^\beta$. This seems natural if one considers that the deviation from the ordinary equilibrium, which is caused by the curved edge of an α lamella, can only change the composition of the α phase from the phase diagram value for equilibrium with γ, here denoted by $X_M{}^\alpha$, to the neighbourhood of the phase diagram value for equilibrium with β which may be denoted by $X_M^{\alpha/\beta}$.

Bolze and Kirkaldy[7] have derived an equation similar to equation (41) and observed that it has the following general shape

$$v \text{ prop } \frac{1}{S} \cdot \left(1 - \frac{S_0}{S}\right) \bigg/ \left(1 - \frac{\phi}{S}\right)$$

A questionable point in their treatment is that they in principle accepted both of equations (24) and (35) and instead discarded the lever rule.

BINARY SYSTEM WITH BOUNDARY DIFFUSION

For boundary diffusion, equation (25) is changed to the following shape

$$-D^B \frac{d^2 X_M{}^B}{dz^2} \delta = v(X_M{}^{\gamma 1} - X_M{}^\alpha) \quad (43)$$

Assuming a constant distribution coefficient between the boundary and the matrix

$$K^B = X_M{}^B / X_M^{\gamma/\alpha} \quad (44)$$

and using equation (4) we can transform equation (43),

$$-\frac{K^B D^B \delta}{K_M{}^\alpha \cdot v} \cdot \frac{d^2 X_M^{\gamma/\alpha}}{dz^2} = \frac{X_M{}^{\gamma 1} - {}^e X_M{}^\alpha}{K_M{}^\alpha} - X_M^{\gamma/\alpha} + {}^e X_M^{\gamma/\alpha} \quad (45)$$

The treatment by Cahn[9] as modified by Hillert[12] shows that the solution to this equation takes the following form

$$\frac{X_M^{\gamma/\alpha} - {}^e X_M^{\gamma/\alpha} - (X_M{}^{\gamma 1} - {}^e X_M{}^\alpha)/K_M{}^\alpha}{X_M{}^{\gamma 3} - {}^e X_M^{\gamma/\alpha} - (X_M{}^{\gamma 1} - {}^e X_M{}^\alpha)/K_M{}^\alpha} = \frac{\cosh z\sqrt{a^\alpha}/S^\alpha}{\cosh \sqrt{a^\alpha}/2} \quad (46)$$

where

$$a^\alpha = vS^2(f^\alpha)^2 K_M{}^\alpha / K^B D^B \delta \quad (47)$$

The quantity $X_M{}^{\gamma 3}$ is the composition of the matrix at the three-phase junctions at the interface.

The average value over the edge of an α lamella is easily obtained by integration and becomes

$$\frac{\overline{X_M^{\gamma/\alpha}} - {}^e X_M^{\gamma/\alpha} - (X_M{}^{\gamma 1} - {}^e X_M{}^\alpha)/K_M{}^\alpha}{X_M{}^{\gamma 3} - {}^e X_M^{\gamma/\alpha} - (X_M{}^{\gamma 1} - {}^e X_M{}^\alpha)/K_M{}^\alpha}$$
$$= \frac{\tanh \sqrt{a^\alpha}/2}{\sqrt{a^\alpha}/2} = t^\alpha \quad (48)$$

t^α is a quantity introduced for convenience. For small values of a^α we can use the approximation

$$\frac{\tanh \sqrt{a^\alpha}/2}{\sqrt{a^\alpha}/2} = t^\alpha = 1 - \frac{1}{12}a^\alpha = 1 \bigg/ \left(1 + \frac{1}{12}a^\alpha\right) \quad (49)$$

Equivalent expressions hold for the β phase. In order to eliminate $X_M{}^{\gamma 3}$ from the evaluation of $X_M^{\gamma/\alpha}$ and $X_M^{\gamma/\beta}$, we can proceed as follows. Equation (48) gives

$$X_M{}^{\gamma 3} - {}^e X_M^{\gamma/\alpha} - (X_M{}^{\gamma 1} - {}^e X_M{}^\alpha)/K_M{}^\alpha$$
$$= \frac{-1}{f^\alpha t^\alpha K_M{}^\alpha} \cdot f^\alpha [X_M{}^{\gamma 1} - {}^e X_M{}^\alpha - K_M{}^\alpha (\overline{X_M^{\gamma/\alpha}} - {}^e X_M^{\gamma/\alpha})] \quad (50)$$

$$X_M{}^{\gamma 3} - {}^e X_M^{\gamma/\beta} - (X_M{}^{\gamma 1} - {}^e X_M{}^\beta)/K_M{}^\beta$$
$$= \frac{1}{f^\beta t^\beta K_M{}^\beta} \cdot f^\beta [{}^e X_M{}^\beta - X_M{}^{\gamma 1} + K_M{}^\beta (\overline{X_M^{\gamma/\beta}} - {}^e X_M^{\gamma/\beta})] \quad (51)$$

The last parts of the right-hand sides of these equations are equal according to equation (8a). On taking the difference between the two equations, we thus obtain the following,

$${}^e X_M^{\gamma/\alpha} - {}^e X_M^{\gamma/\beta} + \frac{X_M{}^{\gamma 1} - {}^e X_M{}^\alpha}{K_M{}^\alpha} + \frac{{}^e X_M{}^\beta - X_M{}^{\gamma 1}}{K_M{}^\beta}$$
$$= (1/f^\alpha t^\alpha K_M{}^\alpha + 1/f^\beta t^\beta K_M{}^\beta)$$
$$\times f^\alpha [X_M{}^{\gamma 1} - {}^e X_M{}^\alpha - K_M{}^\alpha (\overline{X_M^{\gamma/\alpha}} - {}^e X_M^{\gamma/\alpha})] \quad (52)$$

$$\overline{X_M^{\gamma/\alpha}} = {}^e X_M^{\gamma/\alpha} + (X_M{}^{\gamma 1} - {}^e X_M^{\alpha/\gamma})/K_M{}^\alpha - \frac{1}{f^\alpha K_M{}^\alpha}$$
$$\times \frac{{}^e X_M^{\gamma/\alpha} - {}^e X_M^{\gamma/\beta} + \dfrac{X_M{}^{\gamma 1} - {}^e X_M{}^\alpha}{K_M{}^\alpha} + \dfrac{{}^e X_M{}^\beta - X_M{}^{\gamma 1}}{K_M{}^\beta}}{1/f^\alpha t^\alpha K_M{}^\alpha + 1/f^\beta t^\beta K_M{}^\beta} \quad (53)$$

and an equivalent equation for the β phase. Their difference gives

$$\overline{X_M^{\gamma/\alpha}} - \overline{X_M^{\gamma/\beta}}$$
$$= \left[{}^eX_M^{\gamma/\alpha} - {}^eX_M^{\gamma/\beta} + \frac{X_M^{\gamma 1} - {}^eX_M^\alpha}{K_M^\alpha} + \frac{{}^eX_M^\beta - X_M^{\gamma 1}}{K_M^\beta} \right]$$
$$\times \left(1 - \frac{1/f^\alpha K_M^\alpha + 1/f^\beta K_M^\beta}{1/f^\alpha t^\alpha K_M^\alpha + 1/f^\beta t^\beta K_M^\beta} \right) \quad (54)$$

For small a^α and a^β we obtain approximately

$$1 - \frac{1/f^\alpha K_M^\alpha + 1/f^\beta K_M^\beta}{1/f^\alpha t^\alpha K_M^\alpha + 1/f^\beta t^\beta K_M^\beta}$$
$$= \frac{\left(\frac{1}{t^\alpha} - 1\right)/f^\alpha K_M^\alpha + \left(\frac{1}{t^\beta} - 1\right)/f^\beta K_M^\beta}{1/f^\alpha t^\alpha K_M^\alpha + 1/f^\beta t^\beta K_M^\beta}$$
$$\simeq \frac{vS^2 f^\alpha f^\beta K_M^\alpha K_M^\beta}{12 K^B D^B \delta (f^\alpha K_M^\alpha + f^\beta K_M^\beta)} \quad (55)$$

Introducing equation (54) with (55) in (14) and and approximating f^α and f^β by their values at zero growth rate, we find the following relation between growth rate and spacing,

$$\frac{vS^2 f^\alpha f^\beta}{12 K^B D^B \delta}$$
$$\times \frac{K_M^\beta(X_M^{\gamma 1} - {}^eX_M^\alpha) + K_M^\alpha({}^eX_M^\beta - X_M^{\gamma 1})}{f^\alpha K_M^\alpha + f^\beta K_M^\beta}$$
$$\qquad + K_M^\alpha K_M^\beta({}^eX_M^{\gamma/\alpha} - {}^eX_M^{\gamma/\beta})$$
$$= ({}^eX_M^{\gamma/\alpha} - {}^eX_M^{\gamma/\beta})\left(1 - \frac{S_0}{S}\right) \quad (56)$$

In the limit $K_M^\alpha \to 0$ and $K_M^\beta \to 0$ we obtain by means of equation (8a)

$$\frac{vS^2 f^\alpha f^\beta}{12 K^B D^B \delta} \cdot ({}^eX_M^\beta - {}^eX_M^\alpha)$$
$$= ({}^eX_M^{\gamma/\alpha} - {}^eX_M^{\gamma/\beta}) \cdot \left(1 - \frac{S_0}{S}\right) \quad (57)$$

This is identical to the expression obtained previously for this limiting case.[10,12]

In the general case $X_M^{\gamma 3}$ can be evaluated by inserting $X_M^{\gamma/\alpha}$ and $X_M^{\gamma/\beta}$ from equation (48) applied to α and β phase into equation (17) and f^α by inserting them into equation (8a). However, each one of the three equations thus obtained for the calculation of v, f^α and $X_M^{\gamma 3}$, after the selection of a desired S value, will contain at least one of the other two quantities to be calculated. It appears practically impossible to separate them analytically but numerical calculations can easily be carried out by iteration for any particular case.

A close examination of equation (56) shows that it is not strongly dependent upon the values of K_M^α and K_M^β. The same conclusion was reached for the case of volume diffusion. It thus seems to be justified to simplify the treatment of the ternary cases by assuming that the content of M does not vary in the two growing phases. The main attention will instead be directed toward the distribution of the ternary addition, N.

TERNARY SYSTEM WITH VOLUME DIFFUSION OF BOTH ALLOYING ELEMENTS

We shall simplify the calculations by assuming that there is no variation of the M content in the two growing phases. The value of f^α can then be calculated directly from the M content of the alloy, $X_M^{\gamma 1}$, using equation (8a). Assuming volume diffusion of M, we can directly take the coefficients A_n from equation (40) with $K_M^\alpha = K_M^\beta = 0$,

$$A_n = \frac{vS}{D_M}(X_M^\beta - X_M^\alpha) \cdot \frac{\sin n\pi f^\alpha}{(n\pi)^2} \quad (58)$$

and the average M content at the edge of an α lamella is obtained from equation (22),

$$X_M^{\gamma/\alpha} = X_M^{\gamma 1} + A_0 + \frac{vSB(X_M^\beta - X_M^\alpha)}{f^\alpha D_M} \quad (59)$$

and an equivalent equation for β is obtained from equation (23). B is still defined by equation (42).

Assuming volume diffusion of the ternary addition N, we can find the Fourier coefficients by the same procedure as used previously. We thus obtain equations (38) and (39). However, this time we cannot discard equation (38) because there is only one equation (16) and it cannot be used to determine the A_0 coefficients for both the alloying elements. In order to make our treatment of the ternary case consistent with the binary case as $X_N^{\gamma 1}$ goes to zero, we shall continue to use equation (16) to calculate A_0 for M and we shall accept equation (38) for N. We thus obtain from equation (38) and (39) if we omit terms in $(vS/D)^2$,

$$A_n = \frac{-p_n \cdot a_0/p_0}{\frac{D}{v}\lambda_n - 1 + p_0}$$
$$\simeq \frac{X_N^{\gamma 1}}{p_0} \cdot \frac{vS}{D_N}(K_N^\beta - K_N^\alpha) \cdot \frac{\sin n\pi f^\alpha}{(n\pi)^2} \quad (60)$$

$$X_N^{\gamma 1} + A_0 = \frac{a_0}{p_0} - \frac{1}{2p_0}\sum_1 p_n A_n$$
$$= \frac{X_N^{\gamma 1}}{p_0}\left[1 + \frac{vSB}{D_N p_0}(K_N^\beta - K_N^\alpha)^2\right]$$
$$\qquad\qquad\qquad\qquad\qquad (61)$$

The quantity p_0 is kept in the last equation for convenience. It is defined by equation (32) when applied to N instead of M. Comparison shows that it is identical to the quantity p_{0N}, defined by equation (15). This symbol will be chosen in order to avoid confusion with p_0 for M. For the average N content at the edge of an α lamella we obtain from equation (22)

$$\overline{X_N^{\gamma/\alpha}} = X_N^{\gamma 1} + A_{0N} + \frac{1}{f^\alpha} \cdot \frac{X_N^{\gamma 1}}{p_{0N}} \cdot \frac{vSB}{D_N}(K_N^\beta - K_N^\alpha) \quad (62)$$

and an equivalent equation for the β phase from equation (23). We now have all the quantities necessary for the evaluation of the relation between growth rate and spacing from equation (14).

$$\frac{vSB(X_M^\beta - X_M^\alpha)}{D_M f^\alpha f^\beta} + \frac{X_N^{\gamma 1}}{p_{0N}} \frac{vSB}{D_N}(K_N^\beta - K_N^\alpha)$$
$$\times \left[\frac{k^\alpha}{f^\alpha h^\alpha} + \frac{k^\beta}{f^\beta h^\beta} + \frac{K_N^\beta - K_N^\alpha}{p_{0N}}\left(\frac{k^\alpha}{h^\alpha} - \frac{k^\beta}{h^\beta}\right)\right]$$
$$= \left[{}^eX_M^{\gamma/\alpha} - {}^eX_M^{\gamma/\beta} - \frac{X_N^{\gamma 1}}{p_{0N}}\left(\frac{k^\alpha}{h^\alpha} - \frac{k^\beta}{h^\beta}\right)\right]\left(1 - \frac{S_0}{S}\right) \quad (63)$$

We can also calculate the average content of M at the whole interface from equation (16).

$$X_M^{\gamma 1} + A_0$$
$$= f^{\alpha\,e}X_M^{\gamma/\alpha} + f^{\beta\,e}X_M^{\gamma/\beta}$$
$$+ \frac{2\sigma}{SRT}\left(\frac{L^\alpha V_m^\alpha}{h^\alpha} + \frac{L^\beta V_m^\beta}{h^\beta}\right) - \frac{X_N^{\gamma 1}}{p_{0N}}$$
$$\times \left\{\frac{f^\alpha k^\alpha}{h^\alpha} + \frac{f^\beta k^\beta}{h^\beta} + \frac{vSB}{D_N}(K_N^\beta - K_N^\alpha)\right.$$
$$\left.\times \left[\frac{k^\alpha}{h^\alpha} - \frac{k^\beta}{h^\beta} + \frac{K_N^\beta - K_N^\alpha}{p_{0N}}\left(\frac{f^\alpha k^\alpha}{h^\alpha} + \frac{f^\beta k^\beta}{h^\beta}\right)\right]\right\} \quad (64)$$

Thanks to the procedure adopted and to the approximation of constant M contents in α and β, we have thus been able to derive analytic expressions for all the interesting quantities, without restricting the treatment in any other way regarding the shape of the ternary phase diagram or the composition of the alloy considered, except that the content of N must be low. The reason why Bolze and Kirkaldy were less successful in their treatment of the ternary case may be partly due to the fact that they were more ambitious and included cross effects in their diffusion equations. In the present approach, the diffusion of M and N were treated as completely independent of each other.

TERNARY SYSTEM WITH VOLUME DIFFUSION OF THE MAJOR ALLOYING ELEMENT AND BOUNDARY DIFFUSION OF THE MINOR ALLOYING ELEMENT

We shall again assume that there is no variation of the M content in α and β. Assuming volume diffusion of M, we can again use equations (58) and (59). Assuming boundary diffusion of N, we can obtain the average N content at the edge of an α lamella from equation (53) by omitting all the concentrations with the superscript e in view of the difference between equations (4) and (5).

$$\overline{X_N^{\gamma/\alpha}}/X_N^{\gamma 1} = \frac{1}{K_N^\alpha} - \frac{1}{f^\alpha K_N^\alpha} \cdot \frac{1/K_N^\alpha - 1/K_N^\beta}{1/f^\alpha t^\alpha K_N^\alpha + 1/f^\beta t^\beta K_N^\beta} \quad (65)$$

This can be transformed into the following shape which is sometimes convenient in view of equation (15).

$$\overline{X_N^{\gamma/\alpha}} = \frac{X_N^{\gamma 1}}{p_{0N}}\left[1 - f^\beta\left(1 - \frac{K_N^\beta}{K_N^\alpha}\right)\right.$$
$$\left.\times \frac{\left(\frac{1}{t^\alpha} - 1\right)\Big/f^\alpha K_N^\alpha + \left(\frac{1}{t^\beta} - 1\right)\Big/f^\beta K_N^\beta}{1/f^\alpha t^\alpha K_N^\alpha + 1/f^\beta t^\beta K_N^\beta}\right] \quad (66)$$

Equivalent expressions can be derived for the β phase. We now have all the quantities necessary for the evaluation of the relation between growth rate and spacing from equation (14).

$$\frac{vSB(X_M^\beta - X_M^\alpha)}{D_M f^\alpha f^\beta}$$
$$+ \frac{X_N^{\gamma 1}}{p_{0N}} \cdot \frac{\left(\frac{1}{t^\alpha}-1\right)\Big/f^\alpha K_N^\alpha + \left(\frac{1}{t^\beta}-1\right)\Big/f^\beta K_N^\beta}{1/f^\alpha t^\alpha K_N^\alpha + 1/f^\beta t^\beta K_N^\beta}$$
$$\times f^\alpha f^\beta (K_N^\beta - K_N^\alpha)\left(\frac{k^\alpha}{h^\alpha f^\alpha K_N^\alpha} + \frac{k^\beta}{h^\beta f^\beta K_N^\beta}\right)$$
$$= \left[{}^eX_M^{\gamma/\alpha} - {}^eX_M^{\gamma/\beta} - \frac{X_N^{\gamma 1}}{p_{0N}}\left(\frac{k^\alpha}{h^\alpha} - \frac{k^\beta}{h^\beta}\right)\right]\left(1 - \frac{S_0}{S}\right) \quad (67)$$

For small a^α and a^β we find approximately

$$\frac{vSB(X_M^\beta - X_M^\alpha)}{D_M f^\alpha f^\beta} + \frac{X_N^{\gamma 1}}{p_{0N}} \cdot \frac{vS^2 f^\alpha f^\beta}{12 K^B D^B \delta}$$
$$\times \frac{f^\alpha f^\beta K_N^\alpha K_N^\beta}{p_{0N}}(K_N^\beta - K_N^\alpha)\left(\frac{k^\alpha}{h^\alpha f^\alpha K_N^\alpha} + \frac{k^\beta}{h^\beta f^\beta K_N^\beta}\right)$$
$$= \left[{}^eX_M^{\gamma/\alpha} - {}^eX_M^{\gamma/\beta} - \frac{X_N^{\gamma 1}}{p_{0N}}\left(\frac{k^\alpha}{h^\alpha} - \frac{k^\beta}{h^\beta}\right)\right]\left(1 - \frac{S_0}{S}\right) \quad (68)$$

We can also calculate the average content of M at the whole interface from equation (16)

$$\begin{aligned}X_M^{\gamma 1} + A_0 &= f^{\alpha\,e}X_M^{\gamma/\alpha} + f^{\beta\,e}X_M^{\gamma/\beta} \\ &+ \frac{2\sigma}{SRT}\left(\frac{L^\alpha V_m^\alpha}{h^\alpha} + \frac{L^\beta V_m^\beta}{h^\beta}\right) - \frac{X_N^{\gamma 1}}{p_{0N}}\left[\frac{f^\alpha k^\alpha}{h^\alpha} + \frac{f^\beta k^\beta}{h^\beta}\right. \\ &+ \frac{\left(\frac{1}{t^\alpha} - 1\right)\Big/ f^\alpha K_N^\alpha + \left(\frac{1}{t^\beta} - 1\right)\Big/ f^\beta K_N^\beta}{1/f^\alpha t^\alpha K_N^\alpha + 1/f^\beta t^\beta K_N^\beta} \\ &\left.\times f^\alpha f^\beta (K_N^\beta - K_N^\alpha)\left(\frac{k^\alpha}{h^\alpha K_N^\alpha} - \frac{k^\beta}{h^\beta K_N^\beta}\right)\right] \quad (69)\end{aligned}$$

The corresponding quantity for N, $X_N^{\gamma 3}$, can be derived from equation (50). After simplification one obtains,

$$X_N^{\gamma 3} = X_N^{\gamma 1} \cdot \frac{f^\alpha t^\alpha + f^\beta t^\beta}{f^\alpha t^\alpha K_N^\alpha + f^\beta t^\beta K_N^\beta} \quad (70)$$

As in the previous case, we have here been able to derive analytic expressions for all the interesting quantities, without restricting the treatment severely.

TERNARY SYSTEM WITH BOUNDARY DIFFUSION OF BOTH ALLOYING ELEMENTS

We shall still assume that there is no variation of the M content in α and β, i.e. $K_M^\alpha = K_M^\beta = 0$. Assuming boundary diffusion of M we can apply equation (48) and in the limit $K_M^\alpha = K_M^\beta \to 0$ we obtain

$$\overline{X_M^{\gamma/\alpha}} = X_M^{\gamma 3} + (X_M^{\gamma 1} - X_M^\alpha) \cdot \frac{vS^2(f^\alpha)^2}{12K^B D^B \delta} \quad (71)$$

and an equivalent equation for the β phase. The lever rule, equation (8), is simplified to the following

$$f^\alpha(X_M^{\gamma 1} - X_M^\alpha) = f^\beta(X_M^\beta - X_M^{\gamma 1})$$
$$= f^\alpha f^\beta(X_M^\beta - X_M^\alpha) \quad (72)$$

Assuming boundary diffusion of the ternary addition, N, we can directly apply equation (66). We thus have all the quantities necessary for the evaluation of the relation between growth rate and spacing from equation (14).

$$\begin{aligned}&\frac{vS^2 f^\alpha f^\beta (X_M^\beta - X_M^\alpha)}{12 K_M^B D_M^B \delta} + \frac{X_N^{\gamma 1}}{p_{0N}} \\ &\times \frac{\left(\frac{1}{t^\alpha}-1\right)\Big/ f^\alpha K_N^\alpha + \left(\frac{1}{t^\beta}-1\right)\Big/ f^\beta K_N^\beta}{1/f^\alpha t^\alpha K_N^\alpha + 1/f^\beta t^\beta K_N^\beta} \\ &\times f^\alpha f^\beta (K_N^\beta - K_N^\alpha)\left(\frac{k^\alpha}{h^\alpha f^\alpha K_N^\alpha} + \frac{k^\beta}{h^\beta f^\beta K_N^\beta}\right) \\ &= \left[{}^eX_M^{\gamma/\alpha} - {}^eX_M^{\gamma/\beta} - \frac{X_N^{\gamma 1}}{p_{0N}}\left(\frac{k^\alpha}{h^\alpha} - \frac{k^\beta}{h^\beta}\right)\right]\left(1 - \frac{S_0}{S}\right) \\ &\quad (73)\end{aligned}$$

For small a^α and a^β we find approximately

$$\begin{aligned}&\frac{vS^2 f^\alpha f^\beta (X_M^\beta - X_M^\alpha)}{12 K_M^B D_M^B \delta} + \frac{X_N^{\gamma 1}}{p_{0N}} \cdot \frac{vS^2 f^\alpha f^\beta}{12 K_N^B D_N^B \delta} \\ &\times \frac{f^\alpha f^\beta K_N^\alpha K_N^\beta}{p_{0N}}(K_N^\beta - K_N^\alpha)\left(\frac{k^\alpha}{h^\alpha f^\alpha K_N^\alpha} + \frac{k^\beta}{h^\beta f^\beta K_N^\beta}\right) \\ &= \left[{}^eX_M^{\gamma/\alpha} - {}^eX_M^{\gamma/\beta} - \frac{X_N^{\gamma 1}}{p_{0N}}\left(\frac{k^\alpha}{h^\alpha} - \frac{k^\beta}{h^\beta}\right)\right]\left(1 - \frac{S_0}{S}\right) \quad (74)\end{aligned}$$

The value of $X_N^{\gamma 3}$ is again given by equation (70) and the value of $X_M^{\gamma 3}$ is obtained from equation (17), using equations (66), (71) and (72).

$$\begin{aligned}&\frac{\overline{X_M^{\gamma/\alpha}}}{f^\alpha} + \frac{\overline{X_M^{\gamma/\beta}}}{f^\beta} = \frac{X_M^{\gamma 3}}{f^\alpha f^\beta} = \frac{{}^eX_M^{\gamma/\alpha}}{f^\alpha} + \frac{{}^eX_M^{\gamma/\beta}}{f^\beta} \\ &+ \frac{2\sigma}{SRT}\left(\frac{L^\alpha V_m^\alpha}{f^\alpha f^\alpha h^\alpha} + \frac{L^\beta V_m^\beta}{f^\beta f^\beta h^\beta}\right) - \frac{X_N^{\gamma 1}}{p_{0N}} \\ &\times \left[\frac{k^\alpha}{f^\alpha h^\alpha} + \frac{k^\beta}{f^\beta h^\beta} + \frac{\frac{(1/t^\alpha - 1)}{f^\alpha K_N^\alpha} + \frac{(1/t^\beta - 1)}{f^\beta K_N^\beta}}{1/f^\alpha t^\alpha K_N^\alpha + 1/f^\beta t^\beta K_N^\beta}\right. \\ &\left.\times f^\alpha f^\beta (K_N^\beta - K_N^\alpha)\left(\frac{k^\alpha}{f^\alpha f^\alpha h^\alpha K_N^\alpha} - \frac{k^\beta}{f^\beta f^\beta h^\beta K_N^\beta}\right)\right] \\ &\quad (75)\end{aligned}$$

TERNARY SYSTEM WITH BOUNDARY DIFFUSION OF THE MAJOR ALLOY ELEMENT AND VOLUME DIFFUSION OF THE MINOR ALLOY ELEMENT

For M we can apply equations (71) and (72) and for N we can apply equations (61) and (62). From equation (14) we obtain the relation between growth rate and spacing,

$$\begin{aligned}&\frac{vS^2 f^\alpha f^\beta (X_M^\beta - X_M^\alpha)}{12 K^B D^B \delta} + \frac{X_N^{\gamma 1}}{p_{0N}} \cdot \frac{vSB}{D}(K_N^\beta - K_N^\alpha) \\ &\times \left[\frac{k^\alpha}{f^\alpha h^\alpha} + \frac{k^\beta}{f^\beta h^\beta} + \frac{K_N^\beta - K_N^\alpha}{p_{0N}}\left(\frac{k^\alpha}{h^\alpha} - \frac{k^\beta}{h^\beta}\right)\right] \\ &= \left[{}^eX_M^{\gamma/\alpha} - {}^eX_M^{\gamma/\beta} - \frac{X_N^{\gamma 1}}{p_{0N}}\left(\frac{k^\alpha}{h^\alpha} - \frac{k^\beta}{h^\beta}\right)\right]\left(1 - \frac{S_0}{S}\right) \\ &\quad (76)\end{aligned}$$

The average N content at the interface is given directly by equation (61) and $X_M^{\gamma 3}$ is obtained from equation (17) using equations (62), (71) and (72)

$$\begin{aligned}\frac{X_M^{\gamma 3}}{f^\alpha f^\beta} &= \frac{{}^eX_M^{\gamma/\alpha}}{f^\alpha} + \frac{{}^eX_M^{\gamma/\beta}}{f^\beta} + \frac{2\sigma}{SRT}\left(\frac{L^\alpha V_m^\alpha}{f^\alpha f^\alpha h^\alpha} + \frac{L^\beta V_m^\beta}{f^\beta f^\beta h^\beta}\right) \\ &- \frac{X_N^{\gamma 1}}{p_{0N}} \cdot \left\{\frac{k^\alpha}{f^\alpha h^\alpha} + \frac{k^\beta}{f^\beta h^\beta} + \frac{vSB}{D}(K_N^\beta - K_N^\alpha)\right. \\ &\left.\times \left[\frac{k^\alpha}{f^\alpha f^\alpha h^\alpha} - \frac{k^\beta}{f^\beta f^\beta h^\beta} + \frac{K_N^\beta - K_N^\alpha}{p_{0N}}\left(\frac{k^\alpha}{f^\alpha h^\alpha} + \frac{k^\beta}{f^\beta h^\beta}\right)\right]\right\} \\ &\quad (77)\end{aligned}$$

APPLICABILITY OF SOLUTIONS FOR TERNARY CASES

It seems reasonable to expect that the solution obtained for the case of volume diffusion of both alloying elements should be capable of describing the effect of a ternary addition to eutectic solidification. However, it should be observed that the solution was derived using approximations which are valid only when $vS/D_N \ll 1$. This may not be the case if $D_N > D_M$ and the growth rate is mainly controlled by M diffusion.

Most eutectoid transformations in solid state seem to be governed by boundary diffusion.[19] A possible exception is the pearlite transformation in iron–carbon alloys where volume diffusion of carbon may be rapid enough to dominate. Whether or not this is true, the rate of diffusion of the alloying element is much slower and one should be careful not to use any equation based upon unjustified approximations. Equation (68) and (74) can only be used if the effect of the ternary addition is so strong that the rate will be mainly controlled by the slow diffusion of this element. However, at a weaker ternary effect the rate will be governed by the rapid carbon diffusion. The quantities a^α and a^β are not small as compared to unity and equations (67) and (73) should be used. In the extreme case, t^α and t^β go towards zero and the ternary term on the left-hand side of equations (67) and (73) goes towards a constant value. In the limit, equation (67) is transformed into the following shape and equation (73) into a very similar one.

$$\frac{vSB(X_M^\beta - X_M^\alpha)}{D_M f^\alpha f^\beta} = \left[{}^eX_M^{\gamma/\alpha} - {}^eX_M^{\gamma/\beta} \right.$$
$$\left. - X_N^{\gamma 1}\left(\frac{k^\alpha}{h^\alpha K_N^\alpha} - \frac{k^\beta}{h^\beta K_N^\beta}\right) \right]\left(1 - \frac{S_0}{S}\right) \quad (78)$$

where

$$S_0 \cdot \frac{RT}{2\sigma} = \frac{\left(\dfrac{L^\alpha V_m^\alpha}{h^\alpha} - \dfrac{L^\beta V_m^\beta}{h^\beta}\right)}{\left[{}^eX_M^{\gamma/\alpha} - {}^eX_M^{\gamma/\beta} - X_N^{\gamma 1}\left(\dfrac{k^\alpha}{h^\alpha K_N^\alpha} - \dfrac{k^\beta}{h^\beta K_N^\beta}\right)\right]} \quad (79)$$

This equation describes the growth of so-called parapearlite where each one of the two growing phases inherits the initial alloy content of the matrix.[20,12] This is less favorable, thermodynamically, and results in a decrease of the driving force for the reaction [the bracket in equation (78)], as compared to the case of ortho-pearlite where there is an equilibrium distribution of the alloying element between the two growing phases [the bracket in equation (67)]. Equations (78) and (79) demonstrate that the growth of para-pearlite can be treated by applying the equations for the binary case but modifying the driving force with respect to the effect of the ternary addition. The justification of the method previously applied to para-equilibrium[14–16] is thus confirmed. The application of equations (67) and (68) to para-pearlite, ortho-pearlite and intermediate cases will be treated in a separate paper.

It is easy to see by comparison that equation (78) is also the paraversion corresponding to equation (63), i.e. to the case of volume diffusion of the ternary addition. In that case it seems to be more difficult to treat intermediate cases between ortho and para.

ACKNOWLEDGEMENT

Dr. Lee Donaghey and Dr. George Bolze have kindly put copies of their theses at the author's disposal and the inspiration obtained therefrom is gratefully acknowledged. This work is part of a larger program supported financially by the Swedish Board for Technical Development.

REFERENCES

1. W. H. Brandt, *J. appl. Phys.* **16**, 139 (1945).
2. E. Scheil, *Z. Metallk.* **37**, 123 (1946).
3. C. Zener, *Trans. Am. Inst. Min. (metall.) Engrs* **167**, 550 (1946).
4. M. Hillert, *Jernkont. Annl.* **141**, 757 (1957).
5. K. A. Jackson and J. D. Hunt, *Trans. metall. Soc. A.I.M.E.* **236**, 1129 (1966).
6. L. F. Donaghey and W. A. Tiller, *Mat. Sci. Engr* **3**, 231 (1968/69).
7. G. M. Bolze, Theory of the Eutectoid Transformation in Binary and Ternary Systems, Thesis, McMaster University (1970).
8. D. Turnbull, *Acta Met.* **3**, 55 (1955).
9. J. W. Cahn, *Acta Met.* **7**, 18 (1959).
10. J. M. Shapiro and J. S. Kirkaldy, *Acta Met.* **16**, 579 (1968).
11. B. E. Sundquist, *Acta Met.* **16**, 1413 (1968).
12. M. Hillert, The Mechanism of Phase Transformations in Crystalline Solids, p. 231, Monograph and Report Series No. 33, Institute of Metals (1969).
13. J. C. Fisher, Thermodynamics in Physical Metallurgy, p. 201. A.S.M. (1950).
14. M. Hillert, Internal Report, Swedish Inst. Metal Research (1953).
15. M. Hillert, *Phase Transformations*, p. 181. A.S.M. (1970).
16. B. E. Sundquist, *Acta Met.* **17**, 967 (1969).
17. J. Fridberg and M. Hillert, *Acta Met.* **18**, 1253 (1970).
18. S. Björklund, L. F. Donaghey and M. Hillert, The Effect of Alloying Elements on the Rate of Ostwald Ripening of Cementite in Steel, to be published.
19. J. W. Cahn and W. C. Hagel, *Decomposition of Austenite by Diffusional Processes* p. 131, edited by V. F. Zackay and H. I. Aaronson. Interscience (1962).
20. A. Hultgren, *K.V.A. Handlingar* **4**, Ser. 4 (1953).

14 Introduction to "On the Theories of Growth During Discontinuous Precipitation"

published in *Metallurgical Transactions* (1972)

As Professor Mats Hillert celebrates his 80th birthday I would like first of all to express my deepest appreciation of his leadership in the science of materials for many years. Although he retired in 1991, his unique way of looking at natural phenomena with tremendous intellectual curiosity and profound respect has remained remarkable for the following generations of scientists. Most of his ideas are still correct today and are presented in a comprehensive way.

Professor Hillerts paper entitled 'On the Theories of Growth During Discontinuous Precipitation' is an excellent example of the critical insight into the existing theories dealing with the problem of discontinuous precipitation, the reaction during which the formation of new phase is heterogeneous and limited to a migrating reaction front.

At the beginning, the Author examined from a free energy point of view various ideas including those of Zener's, Turnbull's, Aaronson and Liu, and Cahn's. This stimulated his very different approach in which:

– the balance of the forces acting on the migrating reaction front of discontinuous precipitation is as follows:

$$\frac{v}{M} = \frac{\Delta F}{V_m} - \frac{\sigma}{r}$$

where v is the velocity for the steady-state growth, M the mobility of the reaction front, V_m molar volume of the phase, ΔF is the change of the local chemical force across the reaction front, σ the specific free energy and r the radius of local curvature of the reaction front;

– each growing phase (α and β) is treated individually;

– and surface balance energy has to be satisfied at the $\alpha_o/\alpha/\beta$ triple point.

Later on, Professor Hillert made some improvements in his model (see *Acta Metall.* 1982, 30 1599) by: (a) removing the limitation to low solute contents; but the final solution was still numerical, (b) approximate treatment of diffusion and providing in this way explicit expressions for all the interesting quantities in terms of the experimental parameters (the growth rate, the spacing, the remaining supersaturation and the composition at the tree-phase junctions). At that time Professor Hillert indicated the necessity of further studies, in particular, the concentration profiles across the α lamellae and shape of the growing front.

One of the most important works showing the impact of the theory developed by Professor Hillert was that made by Solorzano, Purdy and Weatherly (*Acta Metall.* 1984, 32, 1709) in which they presented a new interpretation of the solute concentration profiles obtained using high-resolution chemical analysis across the α lamella. The observed deviation from the equilibrium state of the extrapolated solute content of the α phase lamella at the contact with the β phase lamella was in fact in good agreement with a local equilibrium of the α with the curved β lamella, as suggested by Professor Hillert.

Further experimental data concerning the solute profile across the α lamella were provided by Duly, Cheynet, and Brechet (*Acta Metall. Mater.* 1994, 42, 3844 and 3555) as well as Zieba and Gust (*Inter. Mater. Rev.* 1998, 43, 70 and *Mater. Chem. Phys.* 2000, 62, 183) The idea of balancing forces at each point of the moving reaction front and triple junction was also a central point of the theoretical works concerning the velocity and spacing selection in discontinuous precipitation and kinetics of multilayer homogenization understood as discontinuous dissolution (L. Klinger, Y. Bréchet, G. Purdy, *Acta Mater.* 1997, 45, 4667 and 5005).

On the other hand, thanks to the formula provided by Professor Hillert it was possible to compare the observed and calculated interface shapes of the reaction front of discontinuous precipitates, as was carried out in 1987 by Tashiro and Purdy (*Metall. Trans.* 1989, 20A, 1593) in series of in-situ experiments in the transmission electron microscope and using thin foils of Al-Zn alloy. Taking into account the loss of free energy due to spinodal reaction ahead of the reaction front, as well as different values of the contact angle at $\alpha_o/\alpha/\beta$ junction, Tashiro and Purdy obtained good consistency of the observed shapes of the α_o/α boundary with those calculated. Further insight into this problem, also stimulated by seminal work of Professor Hillert, was done by Zieba (*Def. Diff. Forum* 1997, 143–147, 1557).

The paper 'On the Theories of Growth During Discontinuous Precipitation' by Professor Hillert described also in detail the coherency stresses in the parent grain due to the change in composition close to the grain boundary as a driving force for the movement (an idea reported originally by Sulonen in 1964). Professor Hillert showed that, based on Sulonen's work, it is possible to determine the sign of stresses which confirms that the solute gradient can exist at the reaction front even at a relatively low temperature, at which the volume diffusion is almost negligible. This subject was later studied in detail by Dryden and Purdy (*Acta Metall.* 1990, 38, 1255). They derived formula for virtual forces acting on the reaction from which it is possible to predict both the sign and magnitude of the stresses.

To sum up, by giving highlights of Professor Hillert paper and subsequent examples of its impact for the better understanding of the mechanism and kinetics of discontinuous precipitation, I am deeply convinced that this contribution has been tremendous for the development of materials science.

Pavel ZIEBA

On Theories of Growth During Discontinuous Precipitation

MATS HILLERT

Growth theories for the lamellar eutectoid transformation are reviewed from a free energy point of view in order to form a background for an examination of the various treatments of growth during discontinuous precipitation. The shortcomings of Cahn's theory of discontinuous precipitation are pointed out and the treatment based upon the individual consideration of growth conditions for the two phases is discussed, with particular emphasis on how the total free energy available to drive the reaction is divided into many parts which are used for various purposes. New sets of calculations are presented in order to demonstrate the effect of free energy losses due to boundary friction and volume diffusion. The effect of coherency stresses in the parent grain is shown to decrease the free energy loss due to volume diffusion and may thus provide a driving force for grain boundary movement.

THE purpose of the present paper is to review the mathematically formulated theories of the discontinuous mode of precipitation, sometimes called cellular precipitation. In particular, the various approaches will be examined from a free energy point of view.

The theory of discontinuous precipitation has evolved in close contact with the theory of the lamellar eutectoid transformation which is less complex in some respects. It is thus convenient to start with an examination of the theoretical development in the latter field. In doing so, we shall modify the original treatments in order to emphasize the free energy aspects.

Finally, a new set of numerical calculations will be presented which takes into account the limited mobility of the grain boundary, the effect of free energy losses due to volume diffusion, and the effect of coherency stresses in the layer of the parent grain close to the advancing grain boundary.

In order to simplify the equations it will be assumed throughout this paper that the molar volume, V_m, has the same value in all the phases involved. The products of both the eutectoid transformation and the discontinuous precipitation will be assumed to be lamellar and to have perfect periodicity. Only steady-state growth will be considered. In general, the mathematical treatments will be developed for binary alloys and the mole fraction x will be used to represent the concentration of the minor component; sometimes called the alloying element or the solute.

ZENER'S TREATMENT OF THE EUTECTOID TRANSFORMATION

It is well recognized today that a theoretical treatment of the growth of a eutectoid structure, by the reaction $\gamma \to \alpha + \beta$, must be based upon a consideration of the diffusional processes as well as the effect of the

MATS HILLERT is Professor of Physical Metallurgy, Royal Institute of Technology, Stockholm, Sweden.

This paper is based on a presentation made at a symposium on "The Cellular and the Pearlite Reactions," held at the Detroit Meeting of The Metallurgical Society of AIME, October 20, 1971, under the sponsorship of the IMD Heat Treatment Committee.

surface energies of the interfaces involved. Zener[1] was the first one to take the second factor into account in a mathematical treatment. He pointed out that part of the total driving free energy, ΔG_m^{total} per mole, must be spent on the creation of the new interfaces between the lamellae in the eutectoid two-phase structure,

$$\Delta G_m^{surface} = 2\sigma V_m / S \quad [1]$$

σ is the specific surface energy of the α/β interfaces and S is the interlamellar spacing. For convenience, Zener introduced the critical spacing, where all the free energy available to drive the eutectoid transformation would have to go into the interfaces. We shall denote that spacing by S_{rev} in order to indicate that a reversible change of one type of free energy into another is implied.

$$\Delta G_m^{total} = 2\sigma V_m / S_{rev} \quad [2]$$

The free energy left to drive the diffusion is thus decreased by a factor of $(1 - S_{rev}/S)$.

Assuming that local equilibrium is maintained at the edges of the growing lamellae, one could evaluate the concentration difference which drives the diffusion of the alloying element from the edges of the growing α phase to the edges of the growing β phase. Neglecting the effect of surface energy, one would obtain

$$\Delta x_e = x_e^{\gamma/\alpha} - x_e^{\gamma/\beta} \quad [3]$$

where the subscript e indicates that the values are given by the equilibrium phase diagram. However, Zener pointed out that the actual composition of the parent phase in contact with the edges of the growing lamellae should deviate from the ordinary equilibrium values because of the action of the surface energy on the curved edges. The diffusion is thus driven by a smaller concentration difference,

$$\Delta x = x^{\gamma/\alpha} - x^{\gamma/\beta} \quad [4]$$

Zener realized that the total free energy driving the eutectoid transformation is approximately proportional to Δx_e and suggested that the free energy remaining to drive the diffusion would be proportional to Δx. He could thus evaluate Δx from the relation

$$\Delta x/\Delta x_e = (\Delta G_m^{total} - \Delta G_m^{surface})/\Delta G_m^{total}$$

$$= 1 - S_{rev}/S \qquad [5]$$

The remaining part of Zener's treatment concerns the growth rate as it is controlled by the rate of diffusion through the parent phase. The total flow away from the edge of an α lamella was estimated from Ficks law in the following way,

$$\frac{dm}{dt} = \frac{A^\alpha D}{V_m} \frac{dx}{dy} = \frac{A^\alpha D \Delta x}{V_m l_{eff}} = \frac{bS^\alpha D \Delta x}{V_m S^\alpha/2} = \frac{2bD\Delta x}{V_m} \qquad [6]$$

The quantity b is an arbitrary distance perpendicular to the growth direction and parallel to the planar faces of the lamellae. It is interesting to note that the result is independent of the width of the α lamellae and that the same value would have been obtained if one had instead estimated the total flow arriving at the edge of a β lamella. In order to obtain this result Zener was forced to assume that the effective diffusion distance, l_{eff}, is proportional to the width of the lamella under consideration. Zener suggested that the proportionality constant is equal to $\frac{1}{2}$.

The mass flow causes both phases to grow and their growth rates must be equal. In fact, this result is obtained by considering the material balance at the edges of α or β lamellae, because the lever rule relates the lamellar thicknesses in such a way that the following equalities hold

$$dm/dt = vbS^\alpha(x_1 - x^\alpha)/V_m = vbS^\beta(x^\beta - x_1)/V_m$$
$$= vSbf^\alpha f^\beta(x^\beta - x^\alpha)/V_m \qquad [7]$$

The growth rate is here denoted by v and the initial composition of the γ phase is denoted by x_1. The quantities f^α and f^β are the relative amounts of the two phases. It should be noted that the amount must be measured in moles in order to make the equation correct for a case where the molar volumes are different. The relation $f^\alpha = S^\alpha/S$ will be used here. By combining Eqs. [6] and [7], Zener obtained his final growth rate equation,

$$v = \frac{2D\Delta x}{f^\alpha f^\beta(x^\beta - x^\alpha)} \cdot \frac{1}{S} = \frac{2D\Delta x_e}{f^\alpha f^\beta(x^\beta - x^\alpha)}$$
$$\cdot \frac{1}{S}(1 - \frac{S_{rev}}{S}) \qquad [8]$$

It is possible to present Zener's derivation in another way. The combination of Eqs. [6] and [7] may be regarded as a method of evaluating what concentration difference Δx is required in order to allow a growth rate v,

$$\Delta x = f^\alpha f^\beta(x^\beta - x^\alpha) \cdot vS/2D \qquad [9]$$

From Δx one can calculate how much of the total free energy is used to drive the diffusion

$$\Delta G_m^{diff}/\Delta G_m^{total} = \Delta x/\Delta x_e \qquad [10]$$

A growth rate equation is finally obtained from the assumption that the remaining part of the free energy goes into surface energy

$$\Delta G_m^{total} - \Delta G_m^{diff} = \Delta G_m^{surface} = 2\sigma V_m/S \qquad [11]$$

The result already presented in Eq. [8] is obtained by inserting Eqs. [9] and [10] into [11].

APPROXIMATE TREATMENT FOR BOUNDARY DIFFUSION

When discussing the growth rate of discontinuous precipitation, Turnbull[2] pointed out that Zener's treatment of the eutectoid transformation could be modified to the case where the growth rate is boundary diffusion controlled by making appropriate changes in the derivation. Actually, it is only necessary to change Eq. [6]. The cross section through which the diffusion takes place is no longer the area of the edge itself, $A^\alpha = bS^\alpha$, but rather the cross section of the boundary, $2b\delta$, where δ is the thickness of the boundary layer which contributes to the enhanced diffusion represented by the boundary diffusion coefficient D^B. The factor of 2 is due to the fact that diffusion away from an α lamella takes place to both sides. The effective diffusion distance must now be taken as proportional to S rather than S^α in order to make the result independent of whether one considers an α or a β lamella. In order to conform most closely with Zener's treatment the diffusion distance should be approximated by $S/4$ because in the symmetric case this is equal to $S^\alpha/2$. Finally, if Δx is still used to represent the composition of the parent phase close to the boundary, a coefficient K must be introduced, where K is defined as the distribution coefficient for the alloying element between the boundary material and the parent phase. The following expression is thus obtained

$$\frac{dm}{dt} = \frac{8b\delta D^B K \Delta x}{V_m S} \qquad [12]$$

and by combination with Eqs. [5] and [7] the following rate equation is now obtained

$$v = \frac{8KD^B \delta \Delta x_e}{f^\alpha f^\beta(x^\beta - x^\alpha)} \cdot \frac{1}{S^2}(1 - \frac{S_{rev}}{S}) \qquad [13]$$

It is interesting to note that this result differs only by a factor of 1.5 from that given by a more ambitious treatment based upon Fick's second law and a detailed consideration of the balance of surface tensions at the growth front.[3]

INDIVIDUAL TREATMENT FOR EACH GROWING PHASE IN A EUTECTOID

It is important to note that Zener's treatment involves the magnitude of the concentration difference Δx but says nothing about the individual values of $x^{\gamma/\alpha}$ and $x^{\gamma/\beta}$. Fisher[4] later pointed out that the average concentration in the parent phase may not necessarily be the same close to a growing eutectoid colony and further away. Such a difference should lead to free energy losses due to some diffusion in the growth direction. Furthermore, it should be realized that the growing eutectoid is not completely described by its growth rate and spacing. The shape of the interface between the parent phase and the eutectoid may also be of interest and it is intimately related to the individual values of $x^{\gamma/\alpha}$ and $x^{\gamma/\beta}$ through the Gibbs-Thomson relation.

In essence, for a more complete description it is necessary to determine one additional quantity which defines the concentration level at the interface. In order to determine the additional quantity, an additional

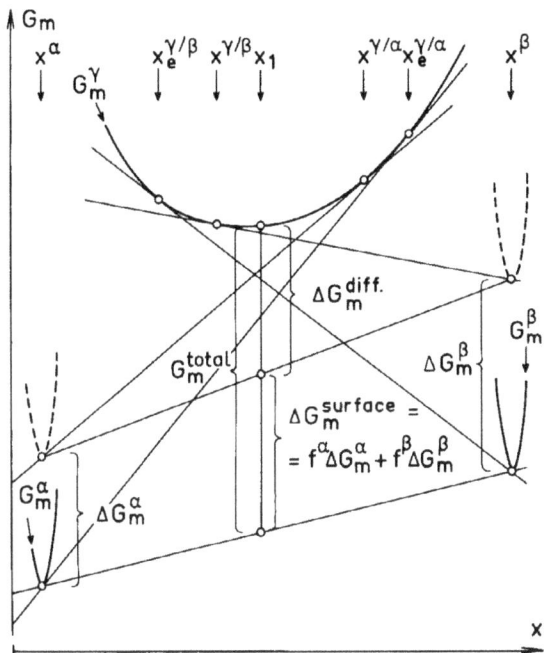

Fig. 1—Free energy diagram for a eutectoid transformation. x_1 is the initial composition of the parent phase. ΔG_m^{diff} is the free energy loss due to diffusion.

It is thus a matter of dividing that part of the free energy which goes into surface energy into two parts. The method is best illustrated by a free energy diagram, Fig. 1. It can be shown that a reasonable shape for the free energy curve of the parent γ phase will yield the following expressions[3]

$$\Delta G_m^\alpha = RT(x_e^{\gamma/\alpha} - x^{\gamma/\alpha})(x_1 - x^\alpha)/x_1(1-x_1) \quad [17]$$

$$\Delta G_m^\beta = RT(x^{\gamma/\beta} - x_e^{\gamma/\beta})(x^\beta - x_1)/x_1(1-x_1) \quad [18]$$

They can be transformed by means of the lever rule,

$$f^\alpha(x_1 - x^\alpha) = f^\beta(x^\beta - x_1) = f^\alpha f^\beta(x^\beta - x^\alpha) \quad [19]$$

The following result is thus obtained by combining Eqs. [14] and [15] with Eqs. [17] and [18],

$$f^\alpha \Delta G_m^\alpha = L^\alpha \cdot 2\sigma V_m/S = (x_e^{\gamma/\alpha} - x^{\gamma/\alpha})$$
$$\cdot RTf^\alpha f^\beta(x^\beta - x^\alpha)/x_1(1-x_1) \quad [20]$$

$$f^\beta \Delta G_m^\beta = L^\beta \cdot 2\sigma V_m/S = (x^{\gamma/\beta} - x_e^{\gamma/\beta})$$
$$\cdot RTf^\alpha f^\beta(x^\beta - x^\alpha)/x_1(1-x_1) \quad [21]$$

Assuming rate control by volume diffusion in the parent phase, we have from Eq. [9],

$$x^{\gamma/\alpha} - x^{\gamma/\beta} = f^\alpha f^\beta(x^\beta - x^\alpha)vS/2D \quad [22]$$

We have thus obtained three relations among the four unknown quantities, $x^{\gamma/\alpha}$, $x^{\gamma/\beta}$, v, and S and can in principle obtain the relation between v and S, corresponding to Eq. [8]. In view of the similarity between Eqs. [16] and [11], we should expect to obtain a very similar result. By adding Eqs. [20] and [21] and

relationship is required. This problem was first solved in 1957[5] and was later[3] discussed in more detail using free energy diagrams. A simplified treatment will now be presented using the approximation that the concentration of the parent phase has a constant value $x^{\gamma/\alpha}$ along the whole edge of an α lamella and another constant value $x^{\gamma/\beta}$ along the whole edge of a β lamella.

The additional relationship is obtained by analyzing how the total free energy available to drive the eutectoid transformation is divided into several parts, two of them being used to drive the growth of the α and β lamella, respectively. These parts will be denoted by ΔG_m^α and ΔG_m^β. Assuming that they are used to overcome the effect of the surface tension of the curved edges of the lamellae, we obtain the following two relations instead of the single relation given previously by Eq. [11].

$$\Delta G_m^\alpha = L^\alpha \cdot 2\sigma V_m/S^\alpha \quad [14]$$

$$\Delta G_m^\beta = L^\beta \cdot 2\sigma V_m/S^\beta \quad [15]$$

L^α and L^β represent the fractions of the surface tension of the α/β interfaces which are carried by each phase. Their values depend upon the balance of the surface tensions among the three types of interfaces involved, α/β, γ/α, and γ/β. Their sum is always equal to unity and the close relation between Eqs. [14] and [15] on one hand and Eq. [11] on the other may thus be illustrated by adding Eqs. [14] and [15] after multiplying each one by the fraction of the phase to which they apply.

$$f^\alpha \Delta G_m^\alpha + f^\beta \Delta G_m^\beta = L^\alpha \cdot 2\sigma V_m/S + L^\beta \cdot 2\sigma V_m/S$$
$$= 2\sigma V_m/S \quad [16]$$

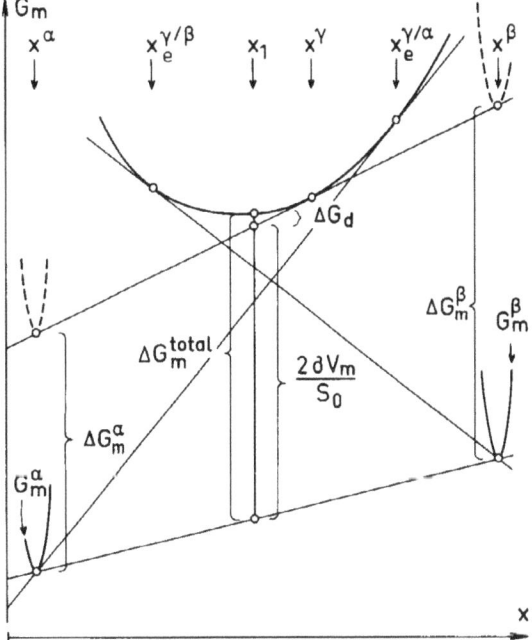

Fig. 2—Free energy diagram for a eutectoid transformation at the critical spacing S_0, where the growth rate is zero. x_1 is the initial composition of the parent phase and x^γ is its composition close to the eutectoid colony. ΔG_d is the free energy loss due to diffusion in the growth direction.

substituting the concentration difference given by Eq. [22], we obtain

$$v = \frac{2D\Delta x_e}{f^\alpha f^\beta (x^\beta - x^\alpha)} \cdot \frac{1}{S}\left(1 - \frac{S_0}{S}\right) \qquad [23]$$

where the new quantity S_0 is defined by the following relation

$$2\sigma V_m/S_0 = \Delta x_e \cdot RT f^\alpha f^\beta (x^\beta - x^\alpha)/x_1(1-x_1) \qquad [24]$$

When comparing Eq. [23] with the previous result, Eq. [8], it should be noticed that S_0 is not necessarily identical to S_{rev} because the right hand side of Eq. [24] is not necessarily identical to ΔG_m^{total}. This can be seen in the following way. According to Eq. [23], S_0 can be regarded as the critical spacing when the growth rate goes to zero, i.e. when $x^{\gamma/\alpha} = x^{\gamma/\beta}$. This composition, x^γ, can be obtained if we divide Eq. [20] by Eq. [21].

$$x^\gamma = L^\beta x_e^{\gamma/\alpha} + L^\alpha x_e^{\gamma/\beta} \qquad [25]$$

It is thus seen that the composition of the parent phase close to the growing eutectoid colony can be equal to the original composition, x_1, only by coincidence. This result confirms the suggestion by Fisher and the free energy diagram in Fig. 2 illustrates what part of the total free energy is spent on diffusion in the growth direction in the absence of sidewise diffusion at zero growth rate. This free energy loss will be denoted by ΔG_d. In the eutectoid case, it is generally very small as shown by Fig. 2 but it plays a very important role in discontinuous precipitation.

TURNBULL'S TREATMENT OF DISCONTINUOUS PRECIPITATION

We shall consider the precipitation of β from a supersaturated α phase. Following a suggestion by Smith,[6] Turnbull[2] developed mathematically a growth model for discontinuous precipitation based upon boundary diffusion. He modified Zener's equations for the eutectoid transformation, Eqs. [6] and [7]. The quantity A^α in Eq. [6] was estimated as $b\delta$, δ being the thickness of the boundary layer permitting enhanced diffusion, and D was identified with the diffusivity in this layer, D^B. The effective diffusion distance was estimated as S^α. By insertion into Eq. [7] we thus obtain

$$D^B \delta \Delta x / S^\alpha = v S^\alpha (x_1 - x^\alpha) \qquad [26]$$

or

$$v = \frac{D^B \delta}{(S^\alpha)^2} \cdot \frac{\Delta x}{x_1 - x^\alpha} \qquad [27]$$

The main difficulty is now to estimate the concentration ratio. In approximating the concentration difference, Δx, which drives diffusion, Turnbull neglected the effect of surface energy and used $\Delta x = x_1 - x_e^{\alpha/\beta}$, where x_1 denotes the initial composition of the supersaturated α solution. On the other hand, in the denominator Turnbull neglected x^α in comparison with x_1 and was thus able to obtain the following expression

$$v = \frac{D^B \delta}{(S^\alpha)^2} \cdot \frac{x_1 - x_e^{\alpha/\beta}}{x_1} \qquad [28]$$

Turnbull's equation was later modified by Aaronson and Liu[7] who, in essence, pointed out that the quantity x^α in the denominator cannot be neglected. By setting x^α equal to $x_e^{\alpha/\beta}$ in the denominator, equating $A^\alpha = 2b\delta$ because material is leaving the edge of an α lamella in both directions, and keeping Zener's approximation of $l_{eff} = S^\alpha/2$, we can obtain their growth equation,

$$v = \frac{4 D^B \delta}{(S^\alpha)^2} \qquad [29]$$

CAHN'S TREATMENT OF DISCONTINUOUS PRECIPITATION

Turnbull's own experiments[8] had shown that the growing α grain does not necessarily have the equilibrium concentration but rather is appreciably supersaturated. Cahn[9] argued that the remaining supersaturation should be an important parameter in the theoretical treatment of growth and concluded that an additional equation is required in order to calculate this unknown quantity. He apparently failed to realize that the situation is analogous to the situation in eutectoid transformations when one wants to evaluate the composition at the interface. As a consequence, he did not make use of the two relations offered by Eqs. [14] and [15] but considered the effect of surface energy in terms of Zener's derivation, Eq. [11]. In an attempt to find another relation, he drew attention to the possible effect of limited boundary mobility which should require that part of the available free energy is lost as friction in order to make the boundary move with a certain rate v,

$$v = M \cdot \Delta G_m^{friction}/V_m \qquad [30]$$

It is worth noting that Cahn used a slightly different definition of the mobility M by leaving the molar volume, V_m, out of the equation. Inserting the frictional loss into Eq. [11] and also including the amount of free energy retained by the remaining supersaturation of the growing α grain, $\Delta G_m^{retained}$, we obtain

$$\frac{2\sigma V_m}{S} = \Delta G_m^{surface} = \Delta G_m^{total} - \Delta G_m^{retained} - \Delta G_m^{diff} - \Delta G_m^{friction} \qquad [31]$$

The last term is in principle known from Eq. [30] and its addition has not changed the nature of Eq. [31] although it has added to the mathematical complexity. One relation is still missing in order to evaluate the remaining terms $\Delta G_m^{retained}$ and ΔG_m^{diff} on the right-hand side of Eq. [31]. The equivalent evaluation for the eutectoid case was performed by finding the solution to the diffusion problem, Eq. [9]. ΔG_m^{diff} was then calculated by inserting Eq. [9] into Eq. [10] and the growth rate was finally obtained by inserting ΔG_m^{diff} in Eq. [11]. Cahn followed the same procedure for discontinuous precipitation. His solution to the diffusion problem had the following form.

$$\frac{x_1 - x}{x_1 - x_3} = \frac{\cosh z\sqrt{a}/S^\alpha}{\cosh \sqrt{a}/2} \qquad [32]$$

z is the coordinate perpendicular to the growth direction and the planar faces of the lamellae and a is a parameter defined as

$$a = v(S^\alpha)^2/KD^B \delta \qquad [33]$$

This is how the growth rate v, enters into the derivation. Eq. [32] thus corresponds to Eq. [9]. However, Eq. [32] contains an additional parameter, x_3, which is defined as the composition of the parent grain at the points where the three phases meet. It plays an important role in controlling what part of the supersaturation will remain in the growing α phase. In essence, an additional equation is required in order to evaluate x_3. Cahn proceeded by simply putting $x_3 = x_e^{\alpha/\beta}$ and could thus evaluate $\Delta G_m^{\text{retained}}$ from Eq. [32]. Inserting that value into Eq. [31] and, for some reason, neglecting the term ΔG_m^{diff} he obtained a relation between growth rate and spacing, containing the boundary mobility, M, as a parameter. The validity of his results is doubtful in view of his approximation of x_3 by $x_e^{\alpha/\beta}$ and his neglect of the term ΔG_m^{diff}.

THE AARONSON AND LIU TREATMENT

An interesting approach was taken by Aaronson and Liu[7] in an attempt to consider the effect of the curving edge of the β lamellae on the solubility according to the Gibbs-Thomson equation. For low alloy contents in the initial α phase, the β lamellae will be much thinner than the α lamellae and it is reasonable to consider the edge of the β lamellae as circular. For a radius of curvature, r, the Gibbs-Thomson equation would yield

$$RT\ln x_r/x_e^{\alpha/\beta} = \sigma V_m/r(x^\beta - x^\alpha) \quad [34]$$

The radius r can be related to the spacing by $r = S^\beta/2L^\beta = f^\beta S/2L^\beta$ and for dilute solutions Eq. [34] is actually identical to Eq. [21]. However, the possibility of relating the composition at the three-phase corners, x_3, to the spacing by identifying it with x_r was missed by Aaronson and Liu who instead used the erroneous identification of x_3 with $x_e^{\alpha/\beta}$, adopted by Cahn. As a consequence, the shortcomings of Cahn's theory were only partly corrected.

INDIVIDUAL TREATMENT FOR EACH GROWING PHASE

The treatment of discontinuous precipitation was finally[3] carried out using the method previously applied to the eutectoid transformation.[5] For the β phase, Eq. [15] yields the same expression as in the eutectoid case, Eq. [18], but for the growing α grain the conditions are quite different because it is of the same phase as the parent grain. The conditions were considered in detail in the previous paper[3] assuming that the grain boundary material can be represented by its own free energy curve. In order to simplify the present discussion it will here be assumed that this free energy curve is identical to the ordinary free energy curve for the parent phase. That assumption does not affect the result except for a constant K which can easily be inserted.

Fig. 3 shows a free energy diagram for the situation at the edge of a β lamella. x_1 is the initial composition of the parent grain and $x^{\alpha/\beta}$ is the composition in the boundary. Assuming that the boundary is in local equilibrium with the parent grain, the value of $x^{\alpha/\beta}$ would also hold in this grain close to the boundary. There will thus be a concentration difference, $x_1 - x^{\alpha/\beta}$, inside the parent grain. All the material has to pass through the corresponding gradient before it can take part in the reaction. The free energy loss due to volume diffusion in the parent grain is represented by ΔG_d^β in the diagram and the energy driving the growth of the β phase, ΔG_m^β, is found in the usual way. The dashed curve represents the free energy of the growing β phase if all the driving force, ΔG_m^β, is used to overcome the pressure effect caused by the surface tension of the curved edge of the β lamella.

Fig. 4 shows the corresponding free energy diagram for the situation at the edge of an α lamella. x_2 represents the composition in the boundary and, as pointed out by Cahn,[9] the growing α grain has the same composition. Furthermore, the same value should also hold in the parent grain close to the boundary if local equilibrium is assumed. There will then be a free energy loss due to volume diffusion in the growth direction in the parent phase, represented by ΔG_d^α in the diagram. It is now important to notice that there will be no free energy driving the growth of the new α grain under the assumption of local equilibrium between the boundary and the parent phase. On the other hand, if local equilibrium is not established, the composition of the parent grain will not have to move from x_1 to x_2 and the free energy loss will be less. As shown in the previous paper,[3] the corresponding free energy will then act as a driving force for the growth of the new α grain. The dashed curve shows the free energy curve for the growing α grain assuming that $f \cdot \Delta G_d^\alpha$ is the fraction of ΔG_d^α which is not lost to vol-

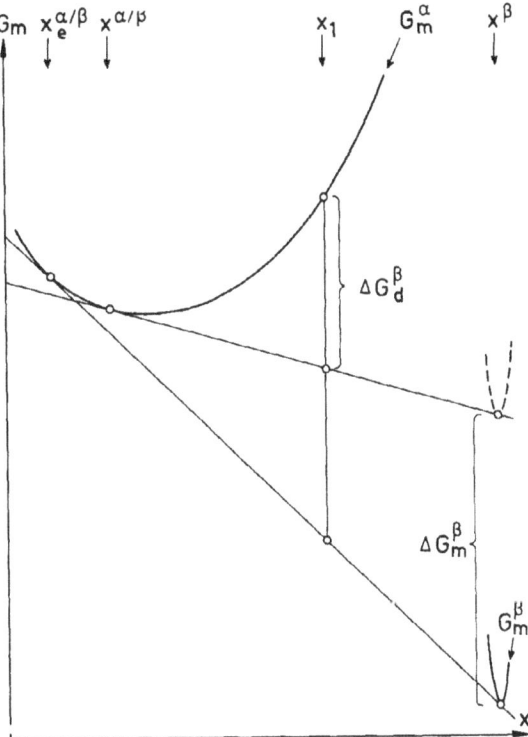

Fig. 3—Free energy diagram for the edge of a growing β lamella in discontinuous precipitation. ΔG_d^β is the free energy loss due to volume diffusion in the parent grain ahead of the advancing edge. ΔG_m^β is the driving force for the growth of β phase. x_1 is the initial composition of the parent phase and $x^{\alpha/\beta}$ is its composition close to the edge.

ume diffusion and also assuming that this fraction is used to overcome the pressure caused by the surface tension of the curved edge of the α lamella.

A similar effect should occur near the β phase. Fig. 5 gives the complete picture of the reaction and is based upon Figs. 3 and 4. The division of the total free energy, available to drive the reaction, is shown along the vertical line at the initial composition x_1. The part spent on boundary diffusion, $\Delta G_m^{B \cdot \text{diff}}$, is found between the point representing the initial parent phase and the intersection with the line $\alpha 1$-$\beta 1$ obtained under the assumption that there is no loss due to volume diffusion. The distance down to the next intersection, given by the line $\alpha 2$-$\beta 2$, represents the actual loss due to volume diffusion, $\Delta G_m^{V \cdot \text{diff}}$. The next line, $\alpha 3$-$\beta 3$, is drawn assuming that part of the driving force is used to overcome the low mobility of the α/α grain boundary and the α/β phase boundary. The distance to the previous intersection represents the free energy loss thus spent on friction in the boundaries. The intersection obtained with the line $\alpha 4$-$\beta 5$ gives the free energy of the final two-phase structure, assuming that there is no effect of surface energy, whereas the previous intersection took surface energy into account. The distance thus represents the energy stored in surface energy. As pointed out by Cahn, the situation represented by the intersection of the line $\alpha 4$-$\beta 5$ does not represent the final stable equilibrium and the diagram shows that part of the free energy, denoted by $\Delta G_m^{\text{retained}}$, is left in the growing grain due to the remaining supersaturation.

Assuming that the initial parent phase is a dilute solution and has a low degree of supersaturation, the following expressions can be derived from the free energy diagram.

$$\Delta G_m^{\text{total}} = \frac{RT}{2x_1}(x_1 - x_e^{\alpha/\beta})^2 \quad [35]$$

$$\Delta G_m^{\text{retained}} = f^\alpha \cdot \frac{RT}{2x_1}(x_2 - x_e^{\alpha/\beta})^2$$

$$\simeq \frac{RT}{2x_1}(x_2 - x_e^{\alpha/\beta})^2 \quad [36]$$

$$\Delta G_m^{B \cdot \text{diff}} = \frac{RT}{2x_1}(x_1 - x^{\alpha/\beta})^2 - f^\alpha \cdot \frac{RT}{2x_1}(x_2 - x^{\alpha/\beta})^2$$

$$- f^\alpha \cdot \frac{RT}{2x_1}(x_1 - x_2)^2 - f^\beta \frac{RT}{2x_1}(x_1 - x^{\alpha/\beta})^2$$

$$\simeq \frac{RT}{x_1}(x_1 - x_2)(x_2 - x^{\alpha/\beta}) \quad [37]$$

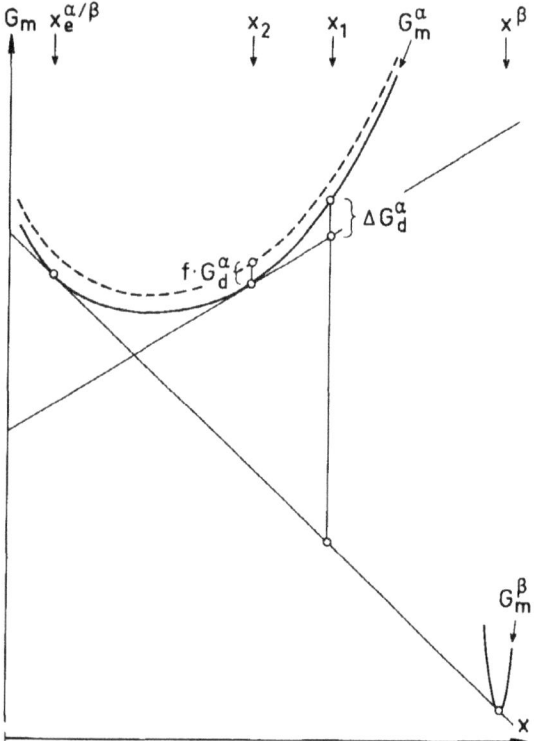

Fig. 4—Free energy diagram for the edge of a growing lamella of a new α grain in discontinuous precipitation. x_1 is the initial composition of the parent α grain and x_2 is the composition of the boundary material and of the growing α grain. G_m^α represents the free energy of the unstressed α phase and of the boundary. ΔG_d^α is the free energy loss due to volume diffusion in the parent α grain ahead of the advancing edge and $f \cdot \Delta G_d^\alpha$ is the fraction which can be used as a driving force for the growth of the new α grain if, by some reason, the composition of the parent grain at the boundary is displaced from x_2 in the direction towards x_1. The dashed curve represents the free energy of the growing α grain.

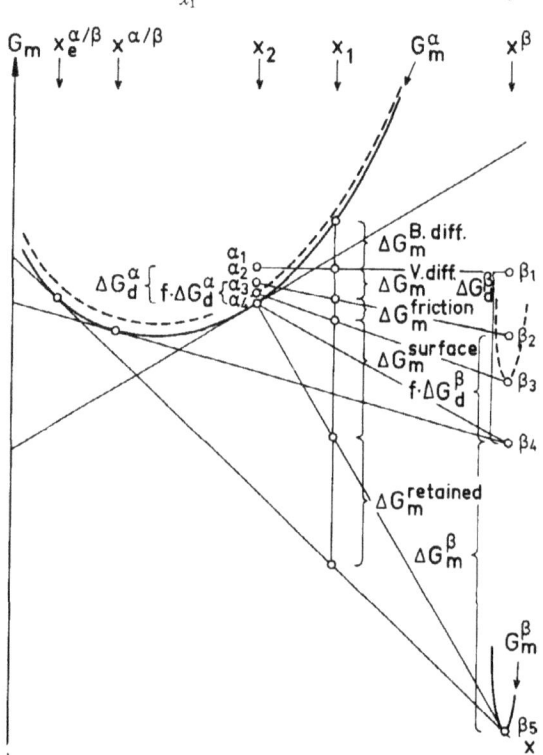

Fig. 5—Complete free energy diagram for discontinuous precipitation showing the various sinks for the total driving free energy. The dashed curves represent the free energy of the growing α and β phases. x_1 is the initial composition of the parent α grain.

The sum of $\Delta G_m^{\text{surface}}$ and $\Delta G_m^{\text{friction}}$ is conveniently divided into the parts which act on each phase. Introducing the surface energy contributions according to Eqs. [14] and [15] and the friction loss according to Eq. [30], we obtain,

$$L^\alpha \cdot 2\sigma V_m/S^\alpha + vV_m/M = f\Delta G_d^\alpha = f \cdot \frac{RT}{2x_1}(x_1 - x_2)^2 \quad [38]$$

$$L^\beta \cdot 2\sigma V_m/S^\beta + vV_m/M^\beta = RT(x^{\alpha/\beta} - x_e^{\alpha/\beta})$$
$$\times (x^\beta - x_1)/x_1(1 - x_1) + f \cdot \frac{RT}{2x_1}(x_1 - x^{\alpha/\beta})^2 \quad [39]$$

The treatment could now be completed by applying an approximate solution of the diffusion equation similar to Zener's solution for the eutectoid case. However, the more exact treatment based upon the appropriate solution to the diffusion equation, given by Eq. [32], is not much more complicated when the initial parent phase has a low supersaturation. That case has already been treated in the previous publication[3] but the friction term vV_m/M, was omitted. The treatment involves the integration of Eqs. [38] and [39] along the edges of the lamellae. Repeating the calculations with the friction terms left in the equations, the following result is now obtained.

$$\frac{S^\alpha}{S_{\text{rev}}} = \frac{L^\alpha}{g} \cdot \frac{\sqrt{a}}{\tanh\sqrt{a}/2} \cdot \frac{[1 + gI(1 + \sqrt{a}/\sinh\sqrt{a})]^2}{1 + \sqrt{a}/\sinh\sqrt{a}} \quad [40]$$

$$\frac{S^\beta}{S_{\text{rev}}^\beta} = \frac{2L^\alpha}{g} \cdot \frac{1 + gI(1 + \sqrt{a}/\sinh\sqrt{a})}{1 + \sqrt{a}/\sinh\sqrt{a}} \quad [41]$$

where

$$S_{\text{rev}} = \frac{2\sigma V_m}{\Delta G_m^{\text{total}}} = \frac{4\sigma V_m x_1}{RT(x_1 - x_e^{\alpha/\beta})^2} \quad [42]$$

$$S_{\text{rev}}^\beta = \frac{4\sigma V_m x_1}{RT(x_1 - x_e^{\alpha/\beta})(x^\beta - x_e^{\alpha/\beta})} \quad [43]$$

$$g = f/(1 + S_{\text{rev}}/2L^\alpha S^\alpha m) \quad [44]$$

$$m = S_{\text{rev}} M\sigma/KD^B\delta \quad [45]$$

$$I = \frac{x_e^{\alpha/\beta} L^\beta}{4x_1 L^\alpha}\left(1 + \frac{f^\beta a S_{\text{rev}}}{2L^\beta S^\alpha m^\beta}\right) \simeq \frac{x_e^{\alpha/\beta} L^\beta}{4x_1 L^\alpha} \quad [46]$$

$$m^\beta = S_{\text{rev}} M^\beta \sigma/KD^B\delta \quad [47]$$

$$Q \simeq \frac{S^\beta/S^\alpha}{S_{\text{rev}}^\beta/S_{\text{rev}}} = \frac{\tanh\sqrt{a}/2}{\sqrt{a}/2}/[1 + gI(1 + \sqrt{a}/\sinh\sqrt{a})] \quad [48]$$

$$\Delta G_m^{\text{retained}}/\Delta G_m^{\text{total}} = 1 - \frac{1}{\sqrt{a}} \tanh\sqrt{a}/2$$
$$\cdot \frac{3 + 4gI(1 + \sqrt{a}/\sinh\sqrt{a})}{[1 + gI(1 + \sqrt{a}/\sinh\sqrt{a})]^2}$$
$$+ \frac{1}{2}\text{sech}^2\sqrt{a}/2/[1 + gI(1 + \sqrt{a}/\sinh\sqrt{a})]^2 \quad [49]$$

$$\Delta G_m^{B \cdot \text{diff}}/\Delta G_m^{\text{total}} = \left[\frac{2}{\sqrt{a}}\tanh\sqrt{a}/2 - \text{sech}^2\sqrt{a}/2\right]/$$
$$[1 + gI(1 + \sqrt{a}/\sinh\sqrt{a})]^2 \quad [50]$$

The quantities S_{rev} and S_{rev}^β are introduced in order to make the representation of S^α and S^β dimensionless. They are defined as the spacing and the thickness of the β lamellae, respectively, obtained if all the driving free energy could be transformed into surface energy,

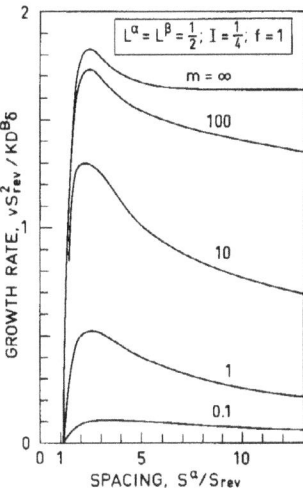

Fig. 6—Growth rate of discontinuous precipitation as function of the spacing for various values of the mobility, m.

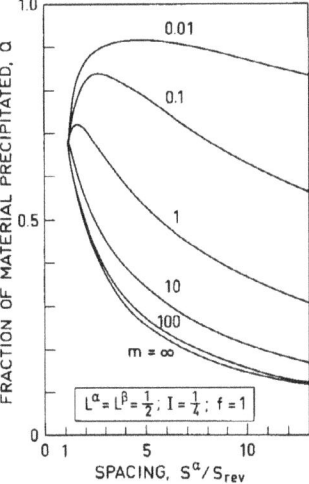

Fig. 7—The fraction of material precipitated as function of the spacing for various mobilities.

which is not physically possible. Because a low initial supersaturation has been assumed, the α lamellae will be much thicker than the β lamellae and S^α can be approximated by S. The parameters m and m^β are dimensionless representations of the mobility of the grain boundary, M, and the phase boundary, M^β. The parameter Q was introduced by Cahn[9] and represents the fraction of material precipitated.

NUMERICAL CALCULATIONS FOR VARIOUS MOBILITIES

The significance of the equations derived in the preceding section will now be illustrated by some numerical calculations. However, it should be noticed that the theory contains many quantities which may vary from system to system and which are usually not known, the most important ones being L^α, L^β, M, and f. As a consequence, it is difficult to illustrate all aspects with

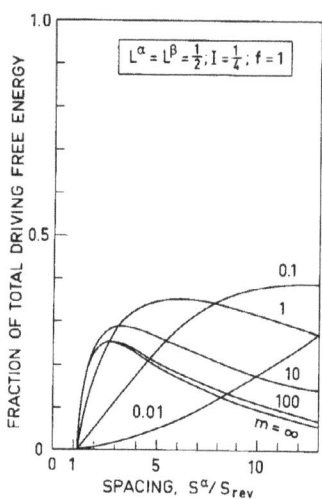

Fig. 8—The free energy loss due to grain boundary diffusion during discontinuous precipitation.

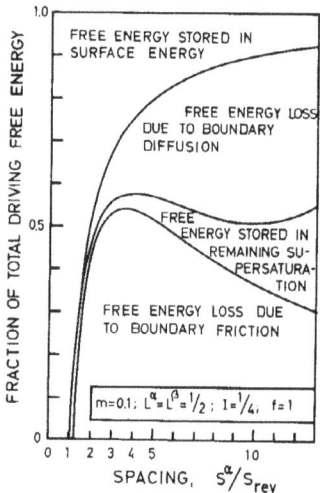

Fig. 9—The sizes of the various free energy sinks during discontinuous precipitation, as functions of the spacing.

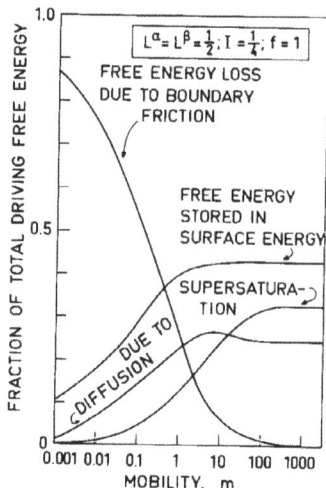

Fig. 10—The sizes of the various free energy sinks during discontinuous precipitation as functions of the boundary mobility.

a limited number of calculations. The effect of the grain boundary mobility M will first be examined using the values $L^\alpha = L^\beta = \frac{1}{2}$, $I = \frac{1}{4}$, and $f = 1$. The latter value implies that no free energy will be lost to volume diffusion and that the total of ΔG_d^α is available to drive the growth of the new α grain. In order to make the numerical calculations valid for various alloy compositions, the growth rate v and spacing S will be represented by the dimensionless parameters $vS_{rev}^2/KD^B\delta$ and S^α/S_{rev}, where S_{rev} contains the alloy composition through Eq. [42]. For convenience, the mobility M will also be represented in a dimensionless manner using the parameter m, defined by Eq. [45], but it should be noticed that m depends upon the alloy composition through S_{rev}.

Eq. [40] gives the growth rate vs spacing through the parameter a which was defined by Eq. [33]. The results are presented in Fig. 6 for various mobilities.

The curve for $m = \infty$ is reproduced from the previous publication[3] and indicates that the growth rate does not tend towards zero at large spacings if there is no grain boundary friction. The explanation is that the edge of the α lamellae will be almost flat except in the vicinity of the β lamellae. The new α grain can thus grow with no driving force and attain the initial composition except in the vicinity of the β lamellae. The side-wise diffusion in the grain boundary thus takes place only in regions close to the β lamellae and the "effective diffusion distance" does not approach infinity when the spacing is increased. As a consequence, the maximum in growth rate at some optimum spacing is not very pronounced and it was argued[3] that it is difficult to visualize a mechanism by which the system could choose a spacing close to the optimum under such conditions. It was suggested that the maximum would be more pronounced if the mobility term was not neglected and that prediction is now confirmed by the other curves in Fig. 6. The explanation is that a limited mobility will require that there is some driving free energy even if the grain boundary is flat. As a consequence, the α grain can no longer grow with the initial composition at any point and the effective diffusion distance will now go towards infinity when the spacing is increased.

Fig. 6 demonstrates that the curves are depressed to low growth rate values when the mobility is decreased through the region of $m = 10$ to 1, the reason being that the free energy loss due to boundary friction is then becoming significant.

Fig. 7 illustrates the variation of Q, the fraction of material precipitated. For high mobility values, Q decreases with increasing spacing because an increasing fraction of each α lamella can grow with little or no change in composition from the initial value. On the other hand, the low mobility values show that there is initially a drastic increase in Q, the reason being that the growth rate is increasing towards its maximum value and the low mobility requires that an increasing part of the total free energy be liberated and spent on friction.

It is natural that all the curves in Fig. 7 start from the same value at a small spacing where the growth rate is zero. It should be noticed that this value is determined by the values of L^α and L^β. For $a \to 0$, Eq. [48] reduces to

$$Q = 1/(1 + f \cdot x_e^{\alpha/\beta} L^\beta / 2x_1 L^\alpha) \qquad [51]$$

or, for $f = 1$ and low supersaturations where $x_1 \cong x_e^{\alpha/\beta}$,

$$Q = 1/(1 + L^\beta/2L^\alpha) = 2L^\alpha/(1 + L^\alpha) \qquad [52]$$

These aspects were not demonstrated by Cahn's treatment because his substitution of $x_e^{\alpha/\beta}$ for x_3 made Q independent of L^α and L^β and therefore a unique function of a. In particular, at zero growth rate ($a = 0$) it gave the value $Q = 1$. People have attempted to use experimental Q values for estimation of the a parameter, by applying Cahn's expression for Q.[7] However, as demonstrated by Eq. [48], such an estimation can be done only if L^α and L^β are known.

Fig. 8 gives the fraction of the total free energy available to drive the reaction, which is spent on boundary diffusion. For each value of the mobility, the curve shows increasing values towards a maximum which occurs at a considerably larger spacing than the optimum spacing which gives the highest growth rate. The reason is, partly, that less free energy is absorbed by surface energy as the spacing increases and, partly, that less free energy is lost by friction as the growth rate decreases above the optimum spacing. The decrease of the curves at even larger spacings is related to the lowering of the Q value which leads to an increasing part of the total free energy being retained in the remaining supersaturation of the new α grain.

The various free energy sinks are compared in Fig. 9 for the case of $m = 0.1$. In order to allow a similar comparison for various mobilities, the conditions obtained at the optimum spacing were selected partly because all Cahn's numerical calculations were carried out at the optimum spacing and partly because there may be some reason to expect the experimental spacing to lie close to that value. Fig. 10 presents the results and, in particular, it shows how the effect of low mobility starts to become important below $m = 10$ and dominates below $m = 0.01$. It would thus be very important to know the correct m value. It may be guessed intuitively that the mobility M is closely related to the boundary diffusivity and it is interesting to note that the dimensionless parameter m depends upon their ratio, Eq. [45]. It would thus be reasonable to expect that the m parameter should not vary much with temperature or composition or even from system to system. A rough estimate based upon the comparison with boundary diffusion would yield

$$M \simeq V_m D^B / RT\delta \qquad [53]$$

$$m \simeq S_{rev} V_m \sigma / RT\delta^2 \simeq 10^4 \qquad [54]$$

This estimate seems to indicate that one is actually far to the right of the diagram in Fig. 10 and the boundary friction should not be very important. In this connection, it is interesting to discuss the result obtained by Speich[10] in an attempt to test Cahn's theory by evaluating the free energy loss due to boundary friction and comparing it with the growth rate. Cahn's theory as well as the present treatment would predict a linear dependency but Speich found a cubic relationship. How-

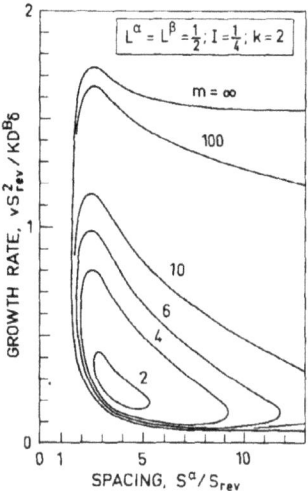

Fig. 11—The effect of volume diffusion on the growth rate of discontinuous precipitation. The parameter value, $k = 2$, was chosen to give a negligible effect of volume diffusion at high growth rates but a large effect at low rates. The upper branch of each curve represents steady-state growth conditions and the lower branch represents a barrier for growth.

ever, an examination reveals that the expression given by Cahn for the calculation of the fraction of the free energy, which is used to drive the boundary, also includes the fraction used to drive boundary diffusion. As a consequence, Speich tested the relation between the sum of these two terms and the growth rate. It is now interesting to note from Fig. 10 that the loss due to friction is negligible and the loss due to diffusion is almost a constant fraction of the total free energy at large m values. If the mobility lies in that region, it may be concluded that Speich actually tested the relationship between the total free energy and the growth rate. The theory would then predict a constant a value, provided that the optimum spacing is obtained, and that result would in turn imply that the growth rate is inversely proportional to S_{rev}^2 or directly proportional to the square of the total free energy. This is not so far from Speich's results. It is possible that the remaining descrepancy is due to the usage of the dilute-solution approximations in the present calculations.

NUMERICAL CALCULATIONS FOR VARIOUS f VALUES

It was argued in the previous publication[3] that the whole of ΔG_d would not be lost to volume diffusion when volume diffusion is slow in comparison with the growth rate. A fraction, f, would then be saved and could be used to drive the movement of the boundary. This fraction was estimated as

$$f = 1 - \exp(-v\delta/2D) \qquad [55]$$

where D is the volume diffusion coefficient. An estimate based on Speich's experimental data gave $D/v = 0.6 \cdot 10^{-8}$ cm and f would then be of the order of 0.9. Consequently, it seemed reasonable to carry out calculations with $f = 1$, $i.e.$ assuming that no free energy is lost by volume diffusion. All the previous calculations

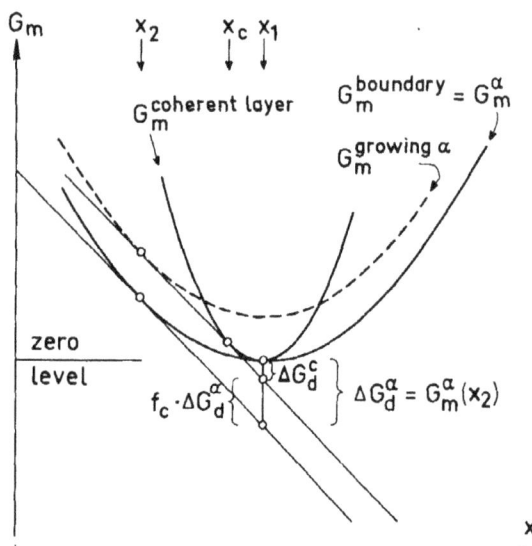

Fig. 12—Free energy diagram illustrating the effect of coherency stresses in the parent grain on the driving force for boundary movement. G_m^α represents the free energy of unstressed α phase and of the boundary. G_m^c represents the free energy of the stressed layer of the parent grain. The equilibrium is obtained from the parallel tangents and the dashed curve represents the free energy of the growing α grain.

were based on that choice. On the other hand, it is evident from Eq. [55] that f should rapidly decrease with decreasing v values and it is that effect which will now be examined. Eq. [55] will be rewritten, using a new dimensionless parameter k,

$$f = 1 - \exp[-kaS_{\text{rev}}^2/(S^\alpha)^2] \qquad [56]$$

where

$$k = KD^B \delta^2 / 2DS_{\text{rev}}^2 \qquad [57]$$

Numerical calculations have been carried out with $k = 2$ which roughly corresponds to the conditions of Speich's experiments on FeZn.[10] As before, the following values were used: $L^\alpha = L^\beta = \frac{1}{2}, I = \frac{1}{4}$. The results are presented in Fig. 11. For large growth rate values the results are not much different from those presented in Fig. 6 but there is a drastic difference at low growth rates. In particular, the curves do not approach a small critical spacing as the growth rate goes to zero. Instead, the curves bend back towards larger spacings. The reason is that the loss of free energy due to volume diffusion is increasing at low growth rates and there will be less free energy available to go into surface energy. As a consequence of this behavior, each curve will show two possible growth rates for each spacing. An examination of the magnitude of the various parts of the free energy as a function of growth rate at constant spacing reveals that the lower branch of each curve in Fig. 11 represents unstable situations such that an infinitesimal increase of the growth rate would lead to an acceleration and the upper branch would be approached. An infinitesimal decrease of the growth rate would lead to a further retardation of the growth and the reaction would stop completely.

It may be concluded from Fig. 11 that discontinuous precipitation, as described in the present treatment, cannot start unless the barrier, represented by the lower branch of each curve, is overcome by some additional mechanism. It may be stated that this is an example of a nucleation barrier where the critical nucleus must be a dynamic one and have a supercritical growth rate. A similar situation may occur in the massive mode of transformation.[3] It is also interesting to note from Fig. 11 that discontinuous precipitation is predicted to be quite impossible if the mobility is too low. The value of $k = 2$ has led to the complete disappearance of the curves for $m = 1$ and less.

THE EFFECT OF COHERENCY STRESSES

It has been suggested by Sulonen[11] that the driving force for the movement of the grain boundary stems from the coherency stresses in the parent grain which arise due to the change in composition close to the grain boundary. This mechanism has the advantage that the size of the driving force would not go to zero at zero growth rate. It should thus be an important mechanism during the early stages of discontinuous precipitation and it is possible that its importance later decreases as the growth rate approaches the high values where the role of volume diffusion decreases.[12]

In order to include Sulonen's mechanism in the present treatment, it is necessary first to examine how coherency stresses will affect the free energy diagram. The numerical calculations in the present paper have been carried out under the assumption that the free energy curve for the unstressed phase has a shape such that the quantity ΔG_d^α can be represented by $RT(x_1 - x_2)^2/2x_1$. This implies that the difference in free energy between the unstressed α phase and the tangent through the point at x_1 is a parabola which can be represented by the same expression. Fig. 12 was constructed using this tangent as the zero level and the free energy curve for the α phase in this diagram is thus given by the following expression,

$$G_m^\alpha(x) = RT(x_1 - x)^2/2x_1 \qquad [58]$$

Let us now consider a thin layer of the parent grain with a composition x_c which is fully coherent with the bulk of the parent grain of composition x_1 and is thus stressed to the same lattice parameter. In the unstressed condition, the layer would have a lattice parameter $d + \Delta d$ where

$$\Delta d = \frac{dd}{dx}(x - x_1) \qquad [59]$$

and dd/dx is the variation of the lattice parameter d with the alloy content, x. In order to bring the dimensions in the plane of the layer back to the dimensions of the parent grain, stresses of the following size are needed in the plane

$$\sigma_1 = \sigma_2 = \frac{E}{1-\nu}\frac{d\ln d}{dx}(x_1 - x) \qquad [60]$$

The elastic energy of the layer is thus obtained as follows,

$$w = \frac{E}{1-\nu}\left(\frac{d\ln d}{dx}\right)^2(x_1 - x)^2 \qquad [61]$$

and the free energy curve for the coherent layer is

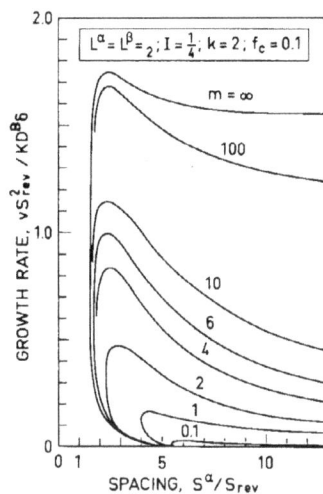

Fig. 13—The effect of coherency stresses on the growth rate of discontinuous precipitation. The parameter value, $f_c = 0.1$, corresponds to the critical value for obtaining discontinuous precipitation in copper alloys and it eliminates the growth barrier down to $S^\alpha = 5.5\, S_\text{rev}$.

obtained by superimposing Eqs. [58] and [61],

$$G_m^c(x) = \left[\frac{RT}{2x_1} + \frac{E}{1-\nu}\left(\frac{d\ln d}{dx}\right)^2\right](x_1 - x)^2 \quad [62]$$

The free energy diagram in Fig. 12 illustrates the conditions for local equilibrium between the parent grain and the boundary, assuming that the free energy curve for the boundary material can be represented by the normal curve for the α phase. It was shown in the previous publication[3] that the local equilibrium between a grain and a boundary is obtained by a construction using parallel tangents rather than a common tangent. The actual composition of the boundary is still denoted by x_2. The composition of the coherent layer, x_c, is thus obtained as follows, assuming local equilibrium,

$$-2\cdot\frac{RT}{2x_1}(x_1 - x_2) = -2\left[\frac{RT}{2x_1} + \frac{E}{1-\nu}\left(\frac{d\ln d}{dx}\right)^2\right](x_1 - x_c) \quad [63]$$

and the free energy loss due to volume diffusion is easily obtained, using the fact that the free energy curve is a parabola,

$$\Delta G_d^c = G_m^c(x_c) = \frac{RT}{2x_1}(x_1 - x_2)^2 \Big/ \left[1 + \frac{2Ex_1}{(1-\nu)RT}\left(\frac{d\ln d}{dx}\right)^2\right] \quad [64]$$

Fig. 12 demonstrates that the presence of coherency stresses has decreased the free energy loss due to volume diffusion and the difference will act as a driving force for the movement of the grain boundary. The dashed curve shows the free energy curve for the growing α grain assuming that all the driving force goes into surface energy. The driving force due to the coherency stresses may be regarded as a fraction of the free energy ΔG_d^α which would be lost due to volume diffusion if there were no coherency stresses. This fraction is obtained as follows,

$$f_c = 1 - \Delta G_d^c / \Delta G_d^\alpha = 1 \Big/ \left[1 + \frac{(1-\nu)RT}{2Ex_1}\left(\frac{dx}{d\ln d}\right)^2\right] \quad [64a]$$

It is interesting to note from Fig. 12 that the driving free energy due to coherency stresses is not directly derived from the elastic energy but is obtained indirectly through the effect on the composition of the coherent layer. The coherency stresses will decrease the concentration differences within the parent grain and thus decrease the loss due to volume diffusion.

Böhm[13] examined the occurrence of discontinuous precipitation in binary copper alloys and concluded that the difference between the atomic radii of copper and the alloying element must exceed 11 pct. Using the following approximation

$$\frac{d\ln d}{dx} = \frac{d_B - d_{Cu}}{d_{Cu}} = 0.11 \quad [65]$$

and choosing 400°C as a reasonable temperature for the reaction, we thus obtain $f_c = 0.09$ as the critical value for the occurrence of discontinuous precipitation. In order to test theoretically the effect of such a value, a set of calculations were carried out using $f_c = 0.1$ and assuming that the remaining part of the free energy loss due to diffusion may also be transformed into a driving force for the boundary movement when the growth rate is high enough to prevent volume diffusion. The total magnitude of the driving force is thus given by

$$f = f_c + (1 - f_c) \cdot [1 - \exp(-v\delta/2D)] \quad [66]$$

Except for the use of Eq. [66] instead of Eq. [55], the same parametric values were chosen as in Fig. 11. The new results are presented in Fig. 13. As expected, there are no differences at high growth rates but there is a drastic change at low growth rates. In particular, the lower branches of the curves do not extend to large spacings but end at a certain critical spacing which can be easily calculated from Eq. [40] using $a = 0$, $v = 0$. As a consequence, the barrier formed by the lower branches has vanished for larger spacings and there is no longer a critical value for the mobility.

This calculation has demonstrated that coherency stresses may be very important for the initiation of discontinuous precipitation but also that it is possible that the effect grows less important as the reaction gets started. However, it must be emphasized that the latter result depends upon the choice of k values as large as 2. For lower k values, the coherency stresses may be important during the whole reaction. Sulonen has argued that this is the case. He has studied the effect of an externally applied force on the steady-state growth rate in various alloys and the results seem to support his suggestion. In particular, with large solute atoms, the growth rate decreased in the direction of a tension force and increased at right angles. With small solute atoms the reverse effect was found.

In order to develop the theory mathematically for the case of an externally applied force, it is necessary to know how the force will affect the free energy curve of the grain boundary and the growing α grain. If a planar case is considered, it is self-evident that the grain boundary material as well as the material in the new α grain will be stressed by the external force. On the other hand, it is not known how the atoms support the external force during their transfer from the parent to the growing grain through a boundary. It seems reasonable to expect that they can support the

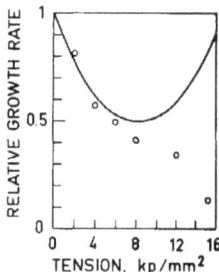

Fig. 14—The effect of an externally applied tension on the growth rate of discontinuous precipitation in Cu-Cd according to experiments by Sulonen (circles) and calculations (curve).

external force better the more coherent the boundary is. Two extreme cases will now be considered. It will first be assumed that the free energy driving the movement of the boundary can be evaluated under the assumption that the externally applied force only acts on the parent grain and it is thus implied that the work done by the external force, when stressing the material which is being incorporated in the boundary and in the new α grain, respectively, has no effect on the reaction.

Consider an externally applied stress σ_3 in the growth direction. It will modify the previous equations in the following way

$$w = \frac{E}{1-\nu}\left(\frac{d\ln d}{dx}\right)^2 (x_1 - x)^2 - \frac{2\nu\sigma_3}{1-\nu}\frac{d\ln d}{dx}(x_1 - x)$$
$$+ \frac{\sigma_3^2}{2E} \qquad [67]$$

$$-2 \cdot \frac{RT}{2x_1}(x_1 - x_2) = -2\left[\frac{RT}{2x_1} + \frac{E}{1-\nu}\left(\frac{d\ln d}{dx}\right)^2\right](x_1 - x_c)$$
$$+ \frac{2\nu\sigma_3}{1-\nu} \cdot \frac{d\ln d}{dx} \qquad [68]$$

$$f_c = 1 - \left[1 + \frac{2\nu\sigma_3 x_1}{(1-\nu)RT(x_1-x_2)} \cdot \frac{d\ln d}{dx}\right]^2 /$$
$$\left[1 + \frac{2Ex_1}{(1-\nu)RT}\left(\frac{d\ln d}{dx}\right)^2\right] + \frac{x_1\sigma_3^2}{ERT(x_1-x_2)^2} \quad [69]$$

Numerical calculations for the alloy Cu-3.81 wt pct Cd, which was studied in detail by Sulonen,[14] give the following expression at 400°C, using the rather arbitrary choice of $x_2 = 0.75 x_1$,

$$f_c = 0.20 - 1.5 \cdot 10^{-2}\sigma_3 + 9 \cdot 10^{-4}\sigma_3^2 \qquad [70]$$

where σ_3 is measured in kp/mm^2. This expression was inserted in Eq. [40] with $L^\alpha = \frac{1}{2}$, $l = \frac{1}{4}$ and $m = \infty$ and the maximum growth rates were evaluated for various values of σ_3. The results were compared with the growth rate at $\sigma_3 = 0$ and the relative growth rates, thus obtained are compared with Sulonen's experimental results in Fig. 14. Except for the region of high tension, the diagram shows a remarkable agreement which in part is due to the value of $0.75 x_1$ which was chosen for x_2. The disagreement at high tension may depend upon the σ_3^2 term in Eq. [70] and stem from the corresponding term in Eq. [67]. It is thus possible that the result can be taken as an indication that it was not correct to treat atoms as unstressed when they are being incorporated in the boundary or in the new α grain. On the other hand, if the square term in Eq.

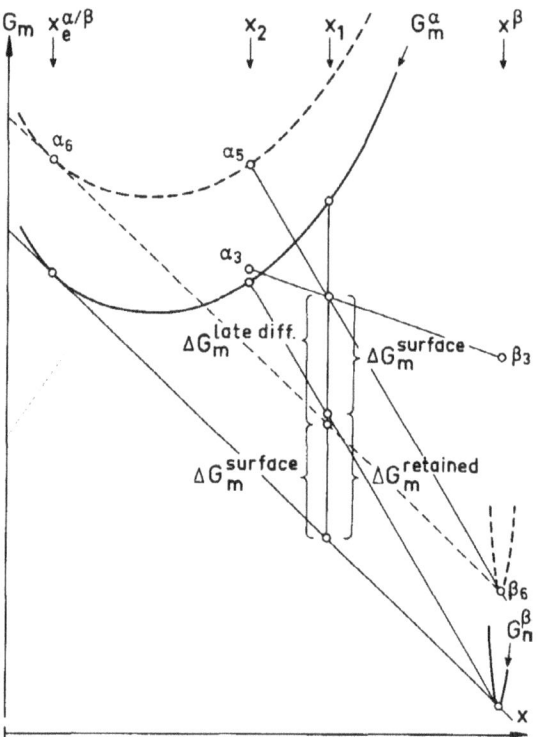

Fig. 15—Free energy diagram illustrating the pressure changes in the lamellae after their formation. The dashed curves represent the free energy of the α and β phases after the pressure has equilibrated.

[67] is completely omitted, which would represent the other extreme, the theoretical growth rate will go to zero at a tension of 14 kp/mm^2. The fact that the experimental data fall between these two extremes may be significant but it is too early to draw conclusions in any particular direction.

THE EQUILIBRIUM CONDITION BETWEEN THE LAMELLAE

The difficulty encountered in the detailed treatments of the eutectoid transformation as well as discontinuous precipitation is the assumption of planar interfaces between the growing lamellae all the way up to the contact with the parent grain. It may make it tempting to suggest that the α and β lamellae are growing under the same pressure but, in general, the curvatures of the edges require that they grow under different pressures.[3] This paradox may be resolved by assuming that the plane of the interface between the lamellae represents a pronounced energy cusp in Wulff's surface energy plot. This assumption may also explain why the β lamellae are not observed to grow sidewise close to the parent grain as they could otherwise be expected to do by means of enhanced diffusion down the interface between the lamellae. This latter aspect has been pointed out by Aaronson and Liu.[7]

Regardless of the nature of the interface between the lamellae, it should be expected that there will soon be a redistribution of the internal stresses and a uniform pressure will develop in the lamellar structure. This redistribution is illustrated by Fig. 15 and it is demonstrated that it takes place without changing the amount of free energy stored in surface energy. The line $\alpha 5 - \beta 6$, which represents the new situation, gives the same intersection with the vertical line at the alloy composition x_1 as did the previous line $\alpha 3 - \beta 3$. The dashed curves represent the free energy of the two phases under the new pressure and the dashed tangent, $\alpha 6 - \beta 6$, demonstrates that local equilibrium between the lamellae now requires that the new α grain change its composition to the equilibrium value $x_e^{\alpha/\beta}$. This is the reason why Cahn[9] and Aaronson and Liu[7] believed that this value should hold at the point where the interface makes contact with the boundary to the parent grain and thus put x_3 equal to $x_e^{\alpha/\beta}$. However, even though it is not known by what mechanisms or how strongly the final equilibrium composition of the α lamellae at their planar faces affects the conditions at the boundary with the parent grain, it appears that a more detailed theory should start from the individual treatment of each growing phase assuming that the balance of the surface energies determines how the pressure is initially divided between the two growing phases.

SUMMARY AND CONCLUSIONS

The various treatments of the lamellar eutectoid transformation and discontinuous precipitation have been reviewed and discussed from a free energy point of view. It is emphasized that Zener's treatment of the eutectoid transformation involves an estimate of the free energy loss due to volume diffusion. A related treatment is offered for the case of boundary diffusion.

The free energy diagram for a eutectoid transformation is discussed with particular emphasis on the free energy loss due to volume diffusion. The driving free energy for the growth of each phase is identified and the mathematical treatment based upon the individual consideration of each phase is reviewed.

Various treatments of discontinuous precipitation, as controlled by boundary diffusion, are reviewed and Cahn's theory is examined in detail. Some approximations are pointed out, which may limit the validity of his results.

The free energy diagram for discontinuous precipitation is discussed with particular emphasis on the various free energy sinks. The driving free energy for the growth of each phase is identified and the mathematical treatment based upon the individual consideration of each phase is developed with attention to the mobility of the boundary. A new set of calculations is presented which examines the effect of different mobilities. It is suggested that the mobility may normally be so high that the free energy loss due to boundary friction can be neglected. It is demonstrated that there is no simple way of relating experimental information on the fraction of material precipitated to the mobility.

The driving free energy for the growth of the new α grain is identified with some fraction of the free energy which would be lost due to volume diffusion if certain mechanisms did not interfere. Two mechanisms are considered. One depends upon the sluggishness of volume diffusion and becomes more important the higher the growth rate. Numerical calculations based upon this mechanism predict the existence of a barrier to the initiation of discontinuous precipitation.

The other mechanism depends upon coherency stresses in the parent grain, as suggested by Sulonen. It is shown (using a free energy diagram) that coherency stresses will decrease the free energy loss due to volume diffusion and thus provide a driving force for the growth of the α grain. Numerical calculations show that this mechanism may eliminate the barrier and thus be important for the initiation of discontinuous precipitation. A comparison between experimental data and theoretical calculations of the effect of an externally applied force indicates that the mechanism may also be important during steady-state growth.

Finally, the equilibrium conditions between the lamellae are discussed and some unsolved problems are pointed out.

REFERENCES

1. C. Zener: *AIME Trans.*, 1946, vol. 167, p. 550.
2. D. Turnbull: *Acta Met.*, 1955, vol. 3, p. 55.
3. M. Hillert: *Inst. Met. Proc. Int. Symp. on The Mechanism of Phase Transformations in Crystalline Solids*, Manchester, July 1968, 1969, p. 231.
4. J. C. Fisher: *Thermodynamics in Physical Metallurgy*, p. 201, Amer. Soc. Metals, Cleveland, 1950.
5. M. Hillert: *Jernkont. Ann.*, 1957, vol. 141, p. 757.
6. C. S. Smith: *Trans. ASM*, 1953, vol. 45, p. 562.
7. H. I. Aaronson and Y. C. Liu: *Scripta Met.*, 1968, vol. 2, p. 1.
8. D. Turnbull and H. N. Treaftis: *Acta Met.*, 1955, vol. 3, p. 43.
9. J. W. Cahn: *Acta Met.*, 1959, vol. 7, p. 18.
10. G. R. Speich: *Trans. TMS-AIME*, 1968, vol. 242, p. 1359.
11. M. S. Sulonen: *Acta Polytech. Scand.*, Chem. Incl. Met. Series No. 28, Helsinki, 1964.
12. M. Hillert and R. Lagneborg: *J. Mater. Sci.*, 1971, vol. 6, p. 208.
13. H. Böhm: *Z. Metallk.*, 1961, vol. 52, p. 564.
14. M. S. Sulonen: *Acta Met.*, 1964, vol. 12, p. 749.

15 Introduction to "The Effect of Alloying Elements on the Rate of Ostwald Ripening of Cementite in Steel"

published in *Acta Metallurgica* (1972)

In the 1990s a series of articles appeared in the metallurgical literature on Ostwald ripening of ternary and multicomponent systems [1–3]. These articles dealt with coarsening on a global scale and were primarily concerned with the kinetics of the critical radius during the later stages when the radius is proportional to the cube-root of time. None of these cited this gem written by Mats Hillert and co-authors in 1972. In fact the ISI Web of Science Citation index failed to find a single reference to the paper in their database of over 1100 papers written on the subject of Ostwald ripening.

A reason for the oversight may be that the paper appears from the title to be limited to the narrow topic of cementite coarsening. However, it is more general for it treats coarsening in ternary systems when one solute element diffuses orders of magnitude faster than the other (*e.g.* carbon and manganese in steel). Key assumptions made are "local equilibrium" at the interface and linearized phase boundary and activity functions. From these assumptions a surprisingly detailed picture of concentration profiles adjacent to the ferrite/cementite interface is derived. Here Mats Hillert and his group show their ability to analyze a problem that is so complex that no one else had attempted it before their contribution or tried to improve on it after. The profile is shown to depend on both cementite particle size and the stage of coarsening. When the coarsening time is in the later stages, a rate constant for ternaries is derived (equation 22). That equation is comparable to equations derived in the 1990's and is another example of work by Mats Hillert being well ahead of its time.

John E. MORRAL

References

[1] H.M. Lee, S.M. Allen and M. Grujicic, Coarsening resistance of M2C carbides in secondary hardening steels. 1. Theoretical model for multicomponent coarsening kinetics, *Metallurgical Trans. A* 22 (1991) 2863.

[2] J.E. Morral and G.R. Purdy, Particle coarsening in binary and multicomponent alloys, *Scripta Metall. Mater.* 30 (1994) 905.

[3] C.J. Kuehmann and P.W. Voorhees, Ostwald ripening in ternary alloys, *Metall. Mater. Trans. A* 27 (1996) 937.

THE EFFECT OF ALLOYING ELEMENTS ON THE RATE OF OSTWALD RIPENING OF CEMENTITE IN STEEL*

S. BJÖRKLUND,† L. F. DONAGHEY†‡ and M. HILLERT†

The thermodynamic basis for the effect of ternary additions on the rate of coarsening of cementite in steel is considered in detail and a rate equation for the change in size of individual particles is derived using a crude model to estimate the rate of diffusion. The rate equation can be interpreted analytically for long times yielding a cube law for the rate of coarsening. The rate constant is predicted to be inversely proportional to the alloy content and to $(1 - K)^2$ where K is the distribution coefficient for the element between cementite and ferrite.

The theoretical predictions are compared with experimental information and fair agreement is found regarding the rate of coarsening. The agreement is somewhat less satisfactory regarding the rate by which the alloying element is distributed between the two phases.

INFLUENCE DES ELEMENTS DE L'ALLIAGE SUR LA VITESSE DE MATURATION D'OSTWALD DE LA CEMENTITE DANS L'ACIER

Les bases thermodynamiques relatives à l'influence d'additions ternaires sur la vitesse de grossissement de la cémentite dans l'acier sont étudiées en détails, et une équation de vitesse pour la variation de la taille des particules individuelles est obtenue à l'aide d'un modèle sommaire d'évaluation de la vitesse de diffusion. L'équation de vitesse peut être interprétée analytiquement pour les longues durées, ce qui donne une loi cubique pour la vitesse de grossissement. La constante de vitesse est prévue comme étant inversement proportionnelle à la concentration de l'alliage et à $(1 - K)^2$ où K est le coefficient de distribution pour l'élément entre la cémentite et la ferrite.

Les prévisions théoriques sont comparées aux résultats expérimentaux et les auteurs trouvent un bon accord pour la vitesse de grossissement. Quant à la vitesse de distribution de l'élément entre les deux phases, l'accord est un peu moins satisfaisant.

DER EINFLUSS DER LEGIERUNGSELEMENTE AUF DIE GESCHWINDIGKEIT DER OSTWALD-REIFUNG VON ZEMENTIT IN STAHL

Die thermodynamischen Grundlagen für den Einfluß ternärer Zulegierungen auf die Vergröberungsgeschwindigkeit von Zementit in Stahl werden ausführlich diskutiert; eine Ratengleichung für die Größenänderung individueller Teilchen wird abgeleitet anhand eines groben Modells zur Abschätzung der Diffusionsgeschwindigkeit. Die Rategleichung kann für lange Zeiten analytisch gelöst werden und liefert ein kubisches Gesetz für die Vergröberungsgeschwindigkeit. Es wird vorhergesagt, daß die Geschwindigkeitskonstante umgekehrt proportional zum Legierungsgehalt und zu $(1 - K)^2$ ist, wobei K die Verteilung des Elements auf das Zementit und Ferrit beschreibt.

Die theoretischen Vorhersagen wurden mit experimentellen Ergebnissen verglichen und für die Vergröberungsgeschwindigkeit wurde befriedigende Übereinstimmung gefunden. In Bezug auf die Verteilung des Legierungselements auf die beiden Phasen ist die Übereinstimmung weniger befriedigend.

1. INTRODUCTION

Martensite, formed on quenching carbon steels, is highly supersaturated with carbon. The driving force for the precipitation of cementite is thus very large and it is impossible to heat martensite without starting its decomposition. A very fine precipitate is formed rapidly and the average particle size is then increased gradually by so-called Ostwald ripening. The smaller particles shrink and disappear and the larger particles grow at their expense. The theory for this process has been developed for binary alloys by Greenwood,[1] Lifshitz and Slyozov[2] and Wagner.[3]

The coarsening of the average particle size results in a gradual decrease of hardness during prolonged tempering of martensite. It is a common practical experience that alloying additions may have a strong retarding effect on the rate of tempering even if the addition does not lead to the formation of special carbides and it may thus be concluded that alloying elements can have a strong effect on the rate of Ostwald ripening of cementite. The theoretical basis for this effect will be outlined in the present paper and a mathematical treatment will be carried out for a simple case. The complete picture of the effect is very complicated and cannot be treated analytically. An attempt to treat more complicated cases by a computer simulation method will be reported in a subsequent paper.

The model will be based upon the assumption of local equilibrium at the cementite/ferrite interfaces. This assumption has been successfully applied to explain the effect of alloying elements on several reactions occurring during various heat treatments of steel.[4] A valuable result from this previous work is the experience that the rate of a moving interface in most cases is either controlled by the diffusion of carbon or by the diffusion of the alloying element. The difference in carbon activity or alloy content, driving the diffusion, can be evaluated from the ternary phase diagram. An important feature of the present case is the fact that the solubility of cementite

---

* Received December 27, 1971.
† The Royal Institute of Technology, Stockholm 70, Sweden.
‡ Now at: University of California, Berkeley, California.

in a ferrite matrix depends upon the particle size as well as the alloy content. This dependence forms the thermodynamic basis for the model and will be discussed before the kinetics is considered.

2. THERMODYNAMIC BASIS

In ternary iron alloys containing carbon as an interstitial alloying element and a substitutional alloying element M, it is advantageous to define new concentration parameters from the ordinary mole fractions, x, as shown by Hillert et al.[5]

$$\left.\begin{array}{l} u_{Fe} = x_{Fe}/(1 - x_C) = x_{Fe}(1 + u_C) \\ u_M = x_M/(1 - x_C) = x_M(1 + u_C) \\ u_C = x_C/(1 - x_C) = x_C(1 + u_C) \end{array}\right\} \quad (1)$$

The molar concentration of the alloying element per unit volume then becomes

$$c_M = x_M/V_m = u_M/V_m(1 + u_C) \quad (2)$$

It may often be a reasonable approximation to treat $V_m(1 + u_C)$ as independent of composition and Fick's law can then be written as follows,

$$J_M = -D_M \frac{dc_M}{dy} = -\frac{D_M}{V_m(1 + u_C)} \cdot \frac{du_M}{dy} \quad (3)$$

Furthermore, the rate equation for a moving interface between α (ferrite) and β (cementite) based upon alloy diffusion takes the following simple form if $V_m(1 + u_C)$ has the same value in both phases.

$$v(u_M^\beta - u_M^\alpha) = D_M^\alpha du_M^\alpha/dy - D_M^\beta du_M^\beta/dy \quad (4)$$

The surface tension of the curved interface between a spherical β particle and an α matrix gives rise to a pressure difference, $P^\beta = 2\sigma/r$. The equilibrium between the two phases is affected by this pressure difference as well as by the presence of an alloying element. For each phase one can apply the Gibbs-Duhem equation,

$$\left.\begin{array}{l} u_{Fe}^\alpha d\ln a_{Fe} + u_M^\alpha d\ln a_M + u_C^\alpha d\ln a_C = 0 \\ u_{Fe}^\beta d\ln a_{Fe} + u_M^\beta d\ln a_M + u_C^\beta d\ln a_C \\ \qquad = (1 + u_C^\beta)V_m^\beta dP^\beta/RT \end{array}\right\} \quad (5)$$

For each component, the activity has the same value in the two phases at the interface if local equilibrium is assumed. By elimination of a_{Fe} one then obtains,

$$d\ln a_C = \frac{u_M^\alpha u_{Fe}^\beta - u_M^\beta u_{Fe}^\alpha}{u_C^\beta u_{Fe}^\alpha - u_C^\alpha u_{Fe}^\beta} \cdot d\ln a_M$$
$$+ \frac{u_{Fe}^\alpha(1 + u_C^\beta)V_m^\beta/RT}{u_C^\beta u_{Fe}^\alpha - u_C^\alpha u_{Fe}^\beta} \cdot dP^\beta \quad (6)$$

For low alloy contents we can assume that a_M is proportional to u_M which yields $d\ln a_M = du_M^\alpha/u_M^\alpha$. By introducing a distribution coefficient $K = u_M^\beta u_{Fe}^\alpha/u_M^\alpha u_{Fe}^\beta$ we obtain from equation (6),

$$d\ln a_C = \frac{1 - K}{u_C^\beta u_{Fe}^\alpha/u_{Fe}^\beta - u_C^\alpha} \cdot du_M^\alpha$$
$$+ \frac{(1 + u_C^\beta)V_m^\beta/RT}{u_C^\beta - u_C^\alpha u_{Fe}^\beta/u_{Fe}^\alpha} \quad (7)$$

For ferrite u_C^α is very small and can be neglected in equation (7). For cementite $u_C^\beta = \frac{1}{3}$ and the molar volume is usually defined for Fe$_3$C, i.e. $V_m^{cem} = 4V_m^\beta$. For low alloy contents $u_{Fe}^\alpha = u_{Fe}^\beta = 1$ and $K = u_M^\beta/u_M^\alpha$ and equation (7) is simplified to the following form,

$$d\ln a_C = 3(1 - K)du_M^\alpha + V_m^{cem} dP^{cem}/RT \quad (8)$$

The coefficients in this equation can be treated as constants for reasonably small variations of u_M^α and P^{cem}. By integration from $u_M^\alpha = 0$ and $P^{cem} = 0$ to an alloy content u_M^α and a pressure $P^{cem} = 2\sigma/r$, we obtain

$$\ln a_C/a_C^0 = 3(1 - K)u_M^\alpha + 2\sigma V_m^{cem}/RTr \quad (9)$$

In the subsequent calculations we shall use the values $V_m^{cem} = 24$ cm^3/mole at 700°C[6] and $\sigma = 700$ ergs/cm^2. We shall treat K as a constant, independent of particle size, which may be a reasonable approximation. For Mn we shall use the value $K = 12$ at 700°C[7] and for Si the value $K = 0.03$.[8]

3. KINETIC MODEL

At any moment during Ostwald ripening, there is a critical size, r_c, such that smaller particles shrink and larger particles grow. The carbon activity at the interface of a particle of this size will be denoted by a_C^c and can be calculated from equation (9) if one knows the value of r_c and the alloy content in the matrix at its interface, u_M^c. Particles of almost the same size will grow or shrink very slowly. It is reasonable to assume that their rate is controlled by alloy diffusion and it may be concluded that they must have almost the same carbon activity as a particle of the critical size. Otherwise, they should exchange carbon rapidly and change their size. The conditions in the matrix at the interface of such particles are thus described by points very close to the isoactivity line for the value a_C^c. This is demonstrated schematically in the phase diagram in Fig. 1 for an alloying element with $K = 0.5$ and quantitative information can be obtained by applying equation (9)

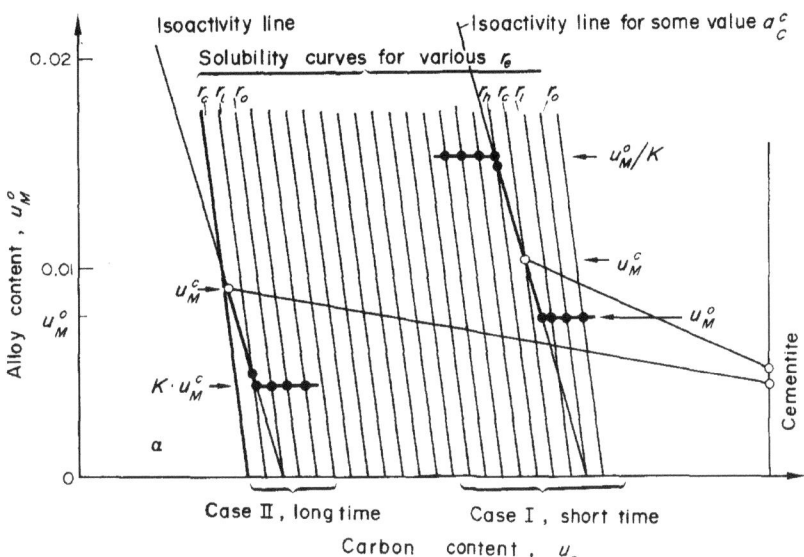

FIG. 1. Ternary phase diagram with $K = 0.5$ showing the effect of particle size on the solubility of cementite in ferrite. The conditions at the interface of all the particles present at the same time are represented by points which fall on a broken line, as demonstrated by thick lines for two cases.

to r_c and an arbitrary value, r, and equilibrating the carbon activities,

$$\frac{1}{r} - \frac{1}{r_c} = \frac{3(1-K)RT \cdot (u_M^c - u_M)}{2\sigma V_m^{\text{cem}}} \quad (10)$$

The distance between the solubility lines for various particle sizes in Fig. 1 is greatly exaggerated. Case I in the diagram shows that particles of a size smaller than a limiting value r_0 cannot have the same carbon activity because there is no intersection between their solubility lines and the isoactivity line for a_C^c. Such particles must have a higher carbon activity. They will thus lose carbon rapidly and shrink with a rate controlled by carbon diffusion rather than alloy diffusion. The value of r_0 can be calculated from equation (10),

$$\frac{1}{r_0} - \frac{1}{r_c} = \frac{3(1-K)RTu_M^c}{2\sigma V_m^{\text{cem}}} \quad (11)$$

The same diagram demonstrates that there is another limiting value at large particle sizes because the intersection between the isoactivity line for a_C^c and the solubility line for larger particles would yield a tie-line α + cementite which indicates that cementite will obtain a higher alloy content than the original content in the matrix, u_M^0. This is unrealistic for an element with $K < 1$. Instead, such particles will grow rapidly with a rate controlled by carbon diffusion and they will inherit the alloy content of the original matrix. The alloy content in the matrix will have a much higher value just at the interface, u_M^0/K, but the increased alloy content will exist in a very thin spike,

only, as demonstrated to the right in Fig. 2. The value of this limiting size at the high side can also be calculated from equation (10).

$$\frac{1}{r_h} - \frac{1}{r_c} = \frac{3(1-K)RT(u_M^c - u_M^0/K)}{2\sigma V_m^{\text{cem}}} \quad (12)$$

The left hand side of Fig. 2 demonstrates that the same reasoning can be applied at the low side and indicates that particles of a size somewhat larger than r_0 can also shrink by a rate controlled by carbon diffusion. Assuming that the shrinking particles have an alloy content of u_M^0, which may be reasonable at an early stage, we conclude that the growing matrix could inherit this content without any long-range diffusion of the alloying element. The conditions in the matrix at the interface of particles smaller than a limiting size of r_l are described by the intersection of the solubility lines with a horizontal line at u_M^0 in Fig. 1 and the value of r_l is again obtained from equation (10).

$$\frac{1}{r_l} - \frac{1}{r_c} = \frac{3(1-K)RT(u_M^c - u_M^0)}{2\sigma V_m^{\text{cem}}} \quad (13)$$

Figure 1 has yielded the conclusion that Ostwald ripening in the presence of an alloying element is controlled partly by carbon diffusion and partly by alloy diffusion and the various regions are illustrated in Fig. 2.

The diffusion constant of a substitutional alloying element is several orders of magnitude smaller than for carbon. It may thus be suggested that the over-all rate of coarsening will be controlled by alloy diffusion unless the region of alloy control between r_h and r_l is

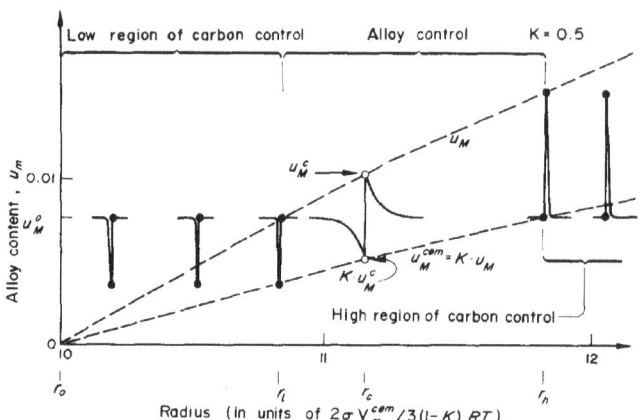

Fig. 2. Concentration profiles for particles of various size present at some early stage of coarsening (case I from Fig. 1).

negligibly small. For example, the condition $r_h - r_l < 0.1 r_c$ would yield the following criterion for an over-all rate control by carbon diffusion,

$$r_c < \frac{0.2\sigma V_m^{\text{cem}} K}{3(1-K)^2 RT u_M^0} \qquad (14)$$

Figure 3 shows a corresponding phase diagram for an element with $K = 2$. The main difference is that r_0 falls on the other side of r_c. Equations (11)–(14) are still valid.

4. APPROXIMATE TREATMENT OF ALLOY DIFFUSION

The shapes of the concentration profiles are very complicated when the rate is controlled by alloy diffusion because the size of an individual particle changes in a complicated manner which depends upon the size and concentration profiles of the other particles in the system. In order to make calculations possible, we shall introduce several simplifying assumptions or approximations.

The profile suggested for a particle of critical size, r_c, in the middle of Fig. 2 can be used as a demonstration of the complexity of the problem. It should fulfil the following two criteria. First, the amount of the alloying element in the spike to the right of the interface must be equal to the amount missing in the depleted region to the left and, second, the diffusional flux reaching the interface from the left must be equal to the flux leaving the interface to the right, because this particular interface is not moving in the present moment. However, it has previously been moving from the left-hand side and the depleted region should thus extend much further to the left than indicated by the random-walk distance for diffusion from the present position of the interface. Furthermore, the value of the diffusion constant in cementite is not

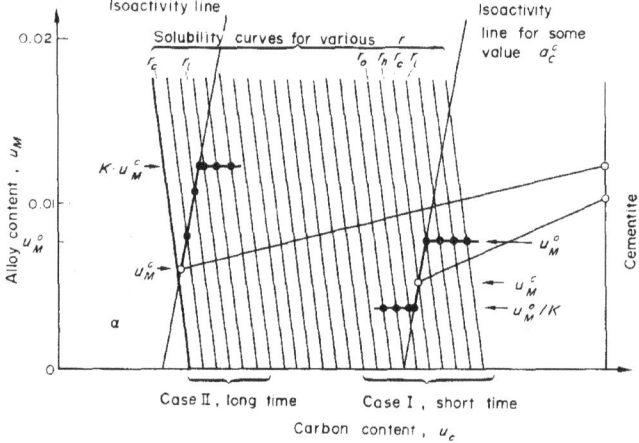

Fig. 3. Ternary phase diagram with $K = 2$, showing the effect of particle size on the solubility of cementite in ferrite. The conditions at the interface of all the particles present at the same time are represented by points which fall on a broken line, as demonstrated by thick lines for two cases.

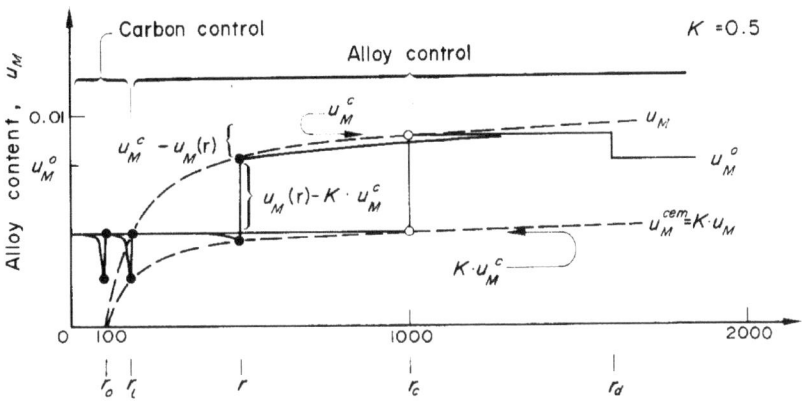

Fig. 4. Concentration profiles for particles of various size present at a later stage of coarsening than in Fig. 2 (case II from Fig. 1.)

known with any certainty. As a consequence, we shall not attempt to calculate this profile and cannot predict the exact position of r_c between the two limits, r_l and r_h.

After sufficiently long times it may be a reasonable approximation to assume that the depleted zone covers the whole volume of the particle and the conditions at the interface of a particle of critical size may then be estimated from the simplified profile shown in the middle of Fig. 4. The enriched zone outside the particle is here described by a region of constant composition and it seems reasonable to estimate its width from the random-walk distance,

$$r_d - r_c = \sqrt{2Dt} \qquad (15)$$

A material balance would then give the following result for this simplified profile.

$$r_c^3(u_M^c - Ku_M^c) = r_d^3(u_M^c - u_M^0) \qquad (16)$$
$$u_M^c = u_M^0/[1 - (1-K)(r_c/r_d)^3] =$$
$$u_M^0/[1 - (1-K)/(1+\sqrt{2Dt}/r_c)^3] \qquad (17)$$

Provided that the critical particle can be used to represent the average particle, this equation can be used to predict the gradual change of the alloy content of the particles, Ku_M^c, from the initial value u_M^0. However, good agreement with experiments for short times should not be expected in view of the limitations of the model. Furthermore, after very long times, the enriched zones from different particles should impinge and an equilibrium value should be approached. This value can be calculated directly from the volume fraction of the precipitate phase, f,

$$u_M^{eq} = u_M^0/[1 - (1-K)f] \qquad (18)$$

The deviation from equation (17) and the approach of the value given by equation (18) should take place in the neighborhood of $(r_c/r_d)^3 = f$. The interface compositions of particles with different sizes after long times is obtained from the phase diagram as illustrated by case II in the left-hand side of Fig. 1. The corresponding variation of interface composition with size is plotted as dashed lines in Fig. 4. It is interesting to note that the high region of carbon control has now disappeared. A concentration profile for a particle of the size $0.5\ r_c$ is indicated in Fig. 4, making use of the dashed lines to define the interface compositions and assuming that the profile in both phases should approach the constant levels defined for the critical size. The justification for comparing with the critical size profile comes from the fact revealed by the binary coarsening theory that all the particles still present, recently were of approximately the same size. The shrinking rate of the particle can now be evaluated from an estimate of the gradients at the interface. In the matrix it will be estimated as $-(u_M^c - u_M)/r$ by comparing with Zener's well known steady state approximation for spherical particles. Lacking information on the diffusion constant in cementite, the concentration spike in this phase will be neglected which should be a good approximation if the diffusion constant is considerably smaller in cementite than in the ferritic matrix. We can thus obtain the following rate equation from equation (4).

$$\frac{dr}{dt} = -\frac{D}{r} \cdot \frac{u_M^c - u_M}{u_M - Ku_M^c} \qquad (19)$$

We shall use this equation for the whole region of alloy control although its justification for particles larger than the critical size is not so strong.

At the late stages where equation (19) is supposed to apply, the high limiting value r_h has increased to infinity and does not exist any more. Our model predicts that the low limiting value r_l is still given by equation (13) if the alloy content of cementite in a critical size particle, $Ku_M{}^c$, is inserted instead of the original alloy content $u_M{}^0$.

5. PROPERTIES OF THE RATE EQUATION

The rate equation given by equation (19) correctly predicts an infinite rate at the size r_l where u_M approaches $Ku_M{}^c$. The rate at and below this size will be limited by the rate of carbon diffusion and it may often be a good approximation to assume that a particle will disappear immediately when it reaches the size r_l. It should then be possible in principle to calculate the particle size distribution and the rate of coarsening from equation (19) by combination with equation (10). This procedure has been followed in the coarsening theory for binary alloys[2,3] where the rate equation takes the following form

$$\frac{dr}{dt} = \frac{2\sigma V_m^{\text{cem}} D_C u_C{}^0}{RT} \cdot \frac{1}{r}\left(\frac{1}{r_c} - \frac{1}{r}\right) \quad (19a)$$

where $u_C{}^0$ is the ordinary solubility of cementite in ferrite. The coarsening rate after a steady state size distribution has been established is then given by the following equation,

$$r_c{}^3 = \frac{8\sigma V_m^{\text{cem}} D_C \cdot u_C{}^0}{9RT} \cdot t \quad (20)$$

Our rate equation for ternary systems can be reduced to the same general form by approximating $u_M - Ku_M{}^c$ by its value at $r = r_c$ which is $u_M{}^c - Ku_M{}^c$. This is not a good approximation for short times, as indicated by the shape of the dashed lines in Fig. 2, but it may be a reasonable approximation for the majority of the particles after long times as indicated by the shape of the dashed lines in Fig. 4. In particular, it should be noticed that the limiting size r_l for long times decreases to very small values compared with r_c and can then be approximated by zero. Using this approximation, the ternary rate equation takes the following form,

$$\frac{dr}{dt} = \frac{2\sigma V_m^{\text{cem}} D_M}{3RT(1-K)^2 u_M{}^c} \cdot \frac{1}{r}\left(\frac{1}{r_c} - \frac{1}{r}\right) \quad (21)$$

and the rate of coarsening after long times is given by

$$r_c{}^3 = \frac{8\sigma V_m^{\text{cem}} D_M}{27RT(1-K)^2 u_M{}^c} \cdot t \quad (22)$$

The ratio of the rate constants for the binary and ternary cases will thus be

$$\frac{k(\text{ternary})}{k(\text{binary})} = \frac{D_M}{3(1-K)^2 D_C u_C{}^0 u_M{}^c} \quad (23)$$

The value of $u_M{}^c$ may be taken from equation (17) or (18).

6. COMPARISON WITH EXPERIMENTAL RESULTS

Detailed experimental information on the effect of alloying elements on the process of coarsening in steel is very scarce. The gradual change of the alloy content in cementite during coarsening has been studied already by Kuo and Hultgren[7] but they did not report on the coarsening itself. Koch, Dittman and Keller[9] have recently reported on a similar study and they present some micrographs, containing important information on the coarsening and also hardness values which can be related to the particle size. The first serious attempt to investigate the effect of alloying elements on coarsening seems to be a study by Björklund[10] who examined the effect of the alloying elements in a low-alloyed steel studied by Bannyh et al.[11] by comparing with the rate in a pure Fe–C alloy. His results indicated that the alloying elements had decreased the rate constant by a factor of 40 or more. Vedula and Heckel[12] have recently obtained a similar result with pure Fe–C alloys. Some information on the effect of Mn, Cr and Ni is available in two recent papers.[13,14] In addition, the effect of Mn and Si has been studied by one of the present authors.[15] The information thus available will now be used to test the theory on some critical points.

(a) Effect on the absolute rate of coarsening

In order to test the applicability of equation (22), it is convenient to evaluate the quantity $r^3(1-K)^2 \times u_M{}^c/D_M t$ from experimental data and to compare the result with a theoretical value calculated from equation (22), as $8\sigma V_m/27RT$. For the present case, the theoretical value is obtained as $6 \cdot 10^{-8}$ cm at 700°C. The evaluation based upon the experimental data is presented in Table 1 where diffusion data from Ref. 16 were used. In cases where there were more than one alloying element of major importance, the values of $(1-K)^2 u_M{}^c/D_M$ for each of them were added. The numbers presented in the last column should be compared with the theoretical value of 6. Two of the experimental values are considerably higher but they may be in error because they relate to steels with very low allow contents where a considerable fraction may not actually be dissolved in the

TABLE 1. Effect of alloying elements on the coarsening rate at 700°C

	Alloy composition (wt. %)					$(1-K)^2 u_M{}^c/D_M \cdot 10^{-10}$ (sec/cm²)				Sum	Exp $r^3/t \cdot 10^{20}$ (cm³/sec)	Sum · $r^3/t \cdot 10^8$ (cm)
	C	Cr	Mn	Ni	Si	Cr	Mn	Ni	Si			
Bannyh et al.	0.75	0.03	0.19		0.23	30	70		1	101	35	35*
Airey et al.	0.15	0.002				6				6	100	6
	0.15	0.01		1.04		25		10		35	7	2.5
Mukherjee et al.	0.21	0.02	0.07			42	47			89	2	1.8
	0.21	0.87				1900				1900	0.06	1.1
Vedula et al.	0.79	0.004		0.075		4		0.7		5	2000	100†
Björklund‡	0.4			1.0				3.3		3.3	180	6
				0.1				0.3		0.3	1000	3
			1.0				500			500	10	50
			0.01				5			5	3300	170†

* The alloy effect is probably less than calculated because Cr, Mn and Si may be bound to impurities like O and S in this commercial steel.

† At these low alloy contents, it is uncertain how much is actually dissolved in the metal. The alloy effect may be much less than calculated.

‡ The experimental values of r^3/t from Björklund are preliminary estimates.

metal. The other values are reasonably close to the theoretical value of 6 and it should be noticed that they are obtained from experimental rate constants over more than 5 orders of magnitude. It may thus be concluded that the theory is remarkably successful in explaining the effect of alloying elements on the coarsening rate of cementite in ferrite.

For binary FeC alloys, it has been concluded in a number of recent papers[12-14] that the rate of coarsening of cementite in ferrite is too slow to be explained by the ordinary rate of carbon diffusion. However, it should be noticed that the present theory predicts that already a manganese content of 0.001 wt. % should decrease the growth rate by a factor of 10 if the rate is controlled by carbon diffusion and it is thus necessary to use extremely pure materials in order to study the rate characteristic of the binary systems. As a consequence, the nature of the rate-controlling mechanism in the binary system still appears to be uncertain. However, this uncertainty does not necessarily affect the discussion of the rate of coarsening in ternary alloys where the rate constant is appreciably lower. It thus seems reasonable to accept the result presented by Table 1 as an indication that the coarsening rate in ternary alloys is controlled by alloy diffusion after the particle size has passed a certain level of coarseness which might be defined approximately by equation (14).

(b) *Relative effect of different alloy contents*

Koch et al. investigated two steels with 0.36 and 1.60% Mn, respectively. The relative rate of coarsening in the two steels can be estimated from the hardness measurements. For instance, the low-alloyed steel tempered for 2 hr gave the same hardness as the high-alloyed steel tempered for 10 hr, indicating a ratio of 5 between their rate constants. This value is nicely explained by equation (22) which contains $u_M{}^c$ as a measure of the alloy content and predicts that the rate constant is inversely proportional to this quantity. A crude estimate of the effect of alloy content on the rate can be made by simply comparing the average contents in the two steels. This ration is 4.5. A more accurate estimate can be based on the experimental values of the alloy content of cementite, presented in the paper. This estimate yields a ration of 5.8. Both estimates are in excellent agreement with the value of 5.

From the above comparisons between theory and experiment, it seems justified to conclude that equation (22) is essentially correct for late stages at 700°C which is the temperature where most of the experiments have been carried out.

(c) *Gradual change of alloy content in cementite*

The model developed for long times in Section 4 was based on the assumption that the amount of the alloying element in a particle would depend upon the size of the affected zone in the matrix around the particle. The width of this zone was estimated from the random-walk distance and a constant composition was assumed in the particle. The validity of the model would be tested by examining how the distribution of the alloying element between cementite and ferrite varies with the ratio between the random-walk distance and the distance between the particles. According to the experiments by Kuo and Hultgren, the cementite particles increase their Mn content during tempering, in general agreement with equation (17). At 700°C this process has proceeded half-way after 260 sec in a steel with 0.7% Mn and 0.5% C. The one-dimensional random-walk distance for this time is 0.08 μm. The particle size, as estimated from equation (25) is $r_c = 0.02$ μm and the average distance between neighbouring particles may thus be estimated as 0.05 μm. This is smaller than the random-walk distance and it is thus somewhat surprising that the distribution of Mn between cementite and ferrite is not closer to the equilibrium value than half way. On the other hand, the order of magnitude seems to be correct. A similar result is obtained from the data by Koch et al. They report that the equilibrium distribution is approached after 18,000 sec at 700°C in the steel with 0.36% Mn. The random-walk distance for this time is 0.7 μm and the particle distance can be estimated to 0.20 μm.

A comparison between the two steels examined by Koch et al. shows that the change of alloy content in the cementite is slower in the low-alloy steel by a factor of 5–10. This effect is qualitatively explained by the fact that the higher coarsening rate in the low-alloy steel results in larger diffusion distances. By combination of equations (17) and (22) we find for large times that the two steels would show the same distribution of the alloying element when they have the same value for $t \cdot (u_M{}^c)^2$. The times would thus be inversely proportional to the square of the alloy content of the steels and a ration of about 20 is thus predicted for the two steels examined by Koch et al. This is somewhat larger than the experimental result.

It seems safe to conclude that the distribution of Mn between cementite and ferrite does not take place quite as fast as predicted by our model. This result may be taken as an indication that the diffusion in the cementite phase is so slow that a constant composition inside the particles is not approached until very late. The result thus casts some doubt also on the validity of equation (22), in particular for alloying elements like Mn which tend to segregate very strongly to the cementite phase. However, it is also conceivable that the experimental results are misleading, for instance due to the presence of non-spherical particles at grain boundaries. Further experimental work is needed to clarify this.

7. SUMMARY AND CONCLUSIONS

The theoretical growth rate equation for individual particles, (19), was derived using a rather accurate treatment of the thermodynamic basis but a crude treatment of the rate of diffusion. The derivation of an equation for the growth of the average size, equation (22), was then performed with a method which becomes exact as time approaches infinity. The comparison with experimental coarsening data seems to support equation (22) but the examination regarding the details of the model appears to cast some doubt on its validity.

An important question which remains to be answered is how rapidly the change takes place from an overall carbon control according to equation (20) at the average particle size given by equation (14) and to an overall alloy control according to equation (22) at large average particle size. This problem will be attacked in a subsequent paper using a computer simulation method. An improvement is also needed regarding the treatment of the rate of diffusion. Some improvements may be possible, in particular if computer methods are employed. However, a detailed comparison of theory and experiment in the important region between the ranges where equations (20) and (22) are applicable, also requires better information on the diffusivity of alloying elements in cementite.

ACKNOWLEDGEMENT

Financial support by the Swedish Board for Technical Development is gratefully acknowledged.

REFERENCES

1. G. W. GREENWOOD, *Acta Met.* **4**, 243 (1956).
2. I. M. LIFSHITZ and V. V. SLEZOV, *Soviet Phys. JETP* **35**, 331 (1959).
3. C. WAGNER, *Z. Elektrochem.* **65**, 581 (1961).
4. M. HILLERT, The Effect of Alloying Elements on Diffusional Transformations in Steel, *Proceedings of the Int. Conf. on the Science and Techn. of Iron and Steel*, Tokyo (1970).
5. M. HILLERT, K. NILSSON and L. TÖRNDAHL, *J. Iron Steel Inst.* **20**, 949 (1971).
6. H. STUART and N. RIDLEY, *J. Iron Steel Inst.* **204**, 711 (1966).
7. K. KUO and A. HULTGREN, *K. V. A. Handlingar* **4**, 22 (1953).
8. T. SATO and T. NISHIZAWA, *Technol. Rep. Tôhoku Univ.* **15**, 65 (1960).
9. W. KOCH, J. DITTMAN and H. KELLER, *Arch. Eisenhutt-Wes.* **39**, 457 (1968).
10. S. BJÖRKLUND, Thesis, Royal Inst. of Tech., Stockholm (1962).
11. O. BANNYH, H. MODIN and S. MODIN, *Jernkont. Annlr* **146**, 774 (1962).
12. K. M. VEDULA and R. W. HECKEL, *Metall. Trans.* **1**, 9 (1970).
13. G. P. AIREY, T. A. HUGHES and R. F. MEHL, *Trans. metall. Soc. A.I.M.E.* **242**, 1853 (1968).
14. T. MUKHERJEE, W. E. STUMPF, C. M. SELLARS and W. J. McG TEGART, *J. Iron Steel Inst.* **207**, 621 (1969).
15. S. BJÖRKLUND, to be published.
16. J. FRIDBERG, L. TÖRNDAHL and M. HILLERT, *Jernkont. Annlr* **153**, 263 (1969).
17. R. P. SMITH, *Trans. Am. Inst. Min. Engrs* **224**, 105 (1962).

16 Introduction to "The Uses of Gibbs Free Energy-Composition Diagrams"

published in *Lectures on the Theory of Phase Transformations* (1975)

During the summer of 1972, when I began preparing lecture notes for the graduate course in phase transformations that I was to teach that fall as a newly hired professor in the Department of Metallurgical Engineering at Michigan Technological University, the need for a good overview of free energy-composition diagrams became apparent. However, no such paper was found in the literature. This experience led me to organize a symposium "On the Teaching of a Graduate Course in Phase Transformations" [1]. This took place during the May 1974 Spring Meeting of TMS-AIME held at the University of Pittsburgh. Naturally, Mats Hillert [2] was invited to give the lead-off talk on "The Uses of Gibbs Free Energy-Composition Diagrams" (G-x Diagrams). Other talks dealt with capillarity [3], nucleation [4] and growth [5]. Papers based on these talks were published the following year as a TMS conference volume entitled "Lectures on the Theory of Phase Transformations" [6].

Helped by the fact that the price set for this book by TMS Executive Director Alex Scott was low enough for students, two printings of the book were completely sold out. Xerox machines operating in many countries provided an unknown number of additional copies. A second edition was published in 1999 [7], with a new paper added on elastic stress effects [8]. Mats Hillert's paper [9] was reproduced almost unchanged from the 1st edition. Developments during the intervening 24 years in the other areas required that papers dealing with them undergo various amounts of rewriting and enlargement. G-x diagrams, on the other hand, had become an integral part of courses on thermodynamics per se; hence a job well done the first time did not need to be further treated in the 2nd Edition. However, these diagrams did provide much of the core in Hillert's monumental treatise on "Phase Equilibria, Phase Diagrams and Phase Transformations" [10].

Mats Hillert's G-x Diagrams paper was precisely what I needed for my graduate course. In order to communicate to this book's presumably predominantly academic audience the pedagogical value of his paper, a brief summary of this paper is now offered. He began by developing the general properties of molar diagrams. He defined extensive properties of a system as those that are the sum of the individual parts of the system. Examples emphasized included G, the Gibbs Free Energy and V, the Volume. Both are conveniently defined on a molar basis. The molar composition of an A-B alloy is described as the ratio of the number of moles of B divided by the total number of moles in the alloy, *i.e.*, as the mole fraction of B, x_B. It is then demonstrated mathematically that in the $\alpha + \beta$ region the points representing x_B^α and x_B^β must be connected by a straight line. On this approach, graphical determination of partial molar quantities is presented (particularly in the context of computer-based calculation of phase diagrams that had become so important in Hillert's research). The Gibbs-Duhem equation is then derived on this basis. After examining criteria for internal stability, the driving forces for diffusion and for a phase transformation as a whole are evaluated. The influence of surface tension upon the driving force for precipitation leads directly to considerations of nucleation theory. A similar approach leads to derivation of the Gibbs adsorption isotherm. Phase transformations such as the cellular or discontinuous precipitation reaction, in which equilibrium compositions cannot be reached during growth, are described as raising problems with dissipation of the Gibbs energy thereby stored. The fine curvature of the precipitate lamellae requires such storage in reversible fashion. Interfacial reaction processes (*e.g.*, growth by means of the ledge mechanism) cause equilibrium to be approached irreversibly. Evaluation of the amount of Gibbs energy reversibly stored is demonstrated by means of G-x Diagrams.

My research interest in G-x Diagrams has been almost entirely in their application to evaluating the driving force for diffusional phase transformations. Examples will now be given of cases in which the driving force alone does, and does not, reproduce experimental observations.

Effects of X on ferrite formation in Fe-C-X alloys

In the presence of a substitutional alloying element, X, both the nucleation [11] and the growth [12] kinetics of ferrite allotriomorphs at austenite grain faces in Fe-C-X alloys can be markedly altered. As now long recognized [13–15], alloying elements that raise the applicable $\gamma/(\alpha+\gamma)$ equilibrium-type temperature should increase the driving force for nucleation and growth at a given isothermal reaction temperature, and *vice versa*. Figure 1 shows that when the steady state nucleation rate of grain boundary ferrite allotriomorphs formed at austenite grain faces in various Fe-0.5 at.% C-ca.3 at.% X alloys (see the caption of this figure for the Fe-C-Mo alloy composition) as a function of isothermal reaction temperature is plotted as a function of ΔF_v, (the Helmholtz free energy change driving nucleation [16]), the individual effects of the various alloying elements are largely suppressed at driving forces less than that corresponding to the calculated (and measured) maximum nucleation rate [11]. In the Fe-C-Ni, and especially in the Fe-C-Mn alloy, however, the nucleation kinetics are orders of magnitude slower predicted by this correlation. Experimental determinations of X adsorption at former austenite grain boundaries [17] provided the basis for evaluating the effects of X upon the average austenite grain boundary energy. The deduction was made that the reductions in grain boundary energy in the Fe-C-Ni and Fe-C-Mn alloys were less than the counterpart reductions in γ:α boundary energy. For a cylindrical pillbox-shaped critical nucleus one of whose faces is coplanar with an austenite grain face, the steady state nucleation rate is proportional to $\exp\{-[4\pi(\sigma_{\alpha\gamma}^e)^2\varepsilon]/[(\Delta F_v+W)^2 kT]\}$ [18]. In this relationship, W = elastic transformation strain energy, k = Boltzmann's constant, T = absolute temperature, $\sigma_{\alpha\gamma}^e$ = edge energy of a pillbox-shaped critical nucleus and $\varepsilon = \sigma_{\alpha\gamma}^c + \sigma_{\alpha\gamma}^{cb} + \sigma_{\gamma\gamma}$, where $\sigma_{\alpha\gamma}^c$ = energy of pillbox broad face lying within one austenite grain, $\sigma_{\alpha\gamma}^{cb}$ = energy of pillbox broad face coplanar with the austenite grain boundary and $\sigma_{\gamma\gamma}$ = austenite grain boundary energy. The interfacial energy terms can thus overwhelm the driving force, ΔF_v, when X exerts different effects upon the interfacial energies involved than when X = Co, Si and Mo [11].

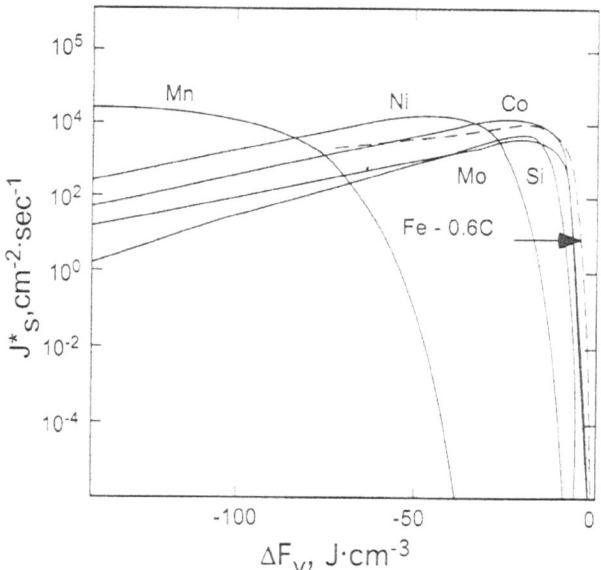

FIG. 1 – Steady state nucleation rate, J_s^*, of ferrite allotriomorphs at austenite grain faces in Fe-ca.0.5 at.% C-ca.3 at.% X (except when X = Mo: 0.84.at% C-2.5 at.% Mo) [11].

Sequences of precipitate nucleation

When more than one product phase can form diffusionally at a given temperature and alloy composition, the factor determining the first phase to appear is the relative nucleation kinetics of all such phases at the available nucleation sites. Although the various interfacial energies in the foregoing equation have just been noted usually to play the predominant role in determining J_s^*, these energies are as yet largely inaccessible to either accurate calculation or measurement. Hence there is a strong tendency to estimate relative nucleation rates upon the basis of ΔF_v; thanks to THERMO-CALC [19, 20], this quantity can now be readily and accurately calculated. In a paper on "Sequences of Precipitate Nucleation" [21], a particularly egregious case where this practices yields incorrect answers was pointed out and discussed. This is the phase sequence in

Al-rich Al-Cu alloys (*e.g.*, 4% Cu). At sufficiently large undercoolings below the $\alpha/(\alpha+\theta)$ equilibrium solvus, in addition to the stable equilibrium θ phase, the transition phases θ′, θ, and GP zones can also form. As illustrated in the schematic G-x Diagram in Figure 2, the driving forces operative upon these phases predict that at a sufficiently large undercooling the θ phase would form first, followed by the θ′, θ, and GP zones transition phases. Experimentally, however, just the opposite sequence obtains [22, 23]. Mainly qualitative considerations of the relative values of the operative interfacial energies suggest that GP zones should have the lowest interfacial energy with the α matrix and that θ″, θ′ and θ ought to have successively such higher energies, now in accordance with experimental observations of the precipitation sequence.

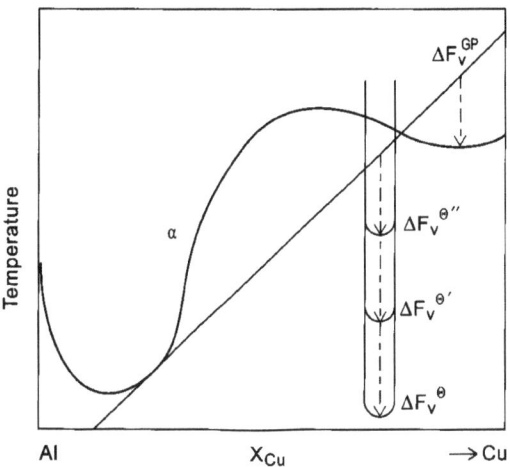

FIG. 2 – Schematic G-x Diagram for Al-Cu alloys, based on transition solvi evaluated by [22, 23].

Hubert I. AARONSON

References

[1] D. Turnbull, J.S. Kirkaldy, J.E. Hilliard and H.I. Aaronson, *Journal of Metals* 27 (Sept. 1975) 24.

[2] M. Hillert, *Lectures on the Theory of Phase Transformations* (TMS-AIME, Warrendale, PA, 1975) p. 1.

[3] R. Trivedi, *ibid*, p. 51.

[4] H.I. Aaronson and J.K. Lee, *ibid*, p. 83.

[5] R.F. Sekerka, C.L. Jeanfils and R.W. Heckel, *ibid*, p. 117.

[6] *Lectures on the Theory of Phase Transformations*, 1st ed. (TMS-AIME, Warrendale, PA, 1975).

[7] *Lectures on the Theory of Phase Transformations*, 2nd ed. (TMS-AIME, Warrendale, PA, 1999).

[8] W.C. Johnson, *ibid*, p. 35.

[9] M. Hillert, *ibid*, p. 1.

[10] M. Hillert, *Phase Equilibria, Phase Diagrams and Phase Transformations* (Cambridge Univ. Press, Cambridge, UK, 1998).

[11] M. Enomoto and H.I. Aaronson, *Metall. Trans. A* 17 (1986) 1385.

[12] J.R. Bradley and H.I. Aaronson, *Metall. Trans. A* 12 (1981) 1729.

[13] J.S. Kirkaldy, *Can. J. Phys.* 35 (1958) 435.

[14] J.S. Kirkaldy, *Can. J. Phys.* 36 (1958) 907.

[15] M. Hillert, *The Mechanism of Phase Transformations in Crystalline Solids* (The Institute of Metals, London, 1969) p. 231.

[16] J.W. Cahn and J.E. Hilliard, *Jnl. Chem. Phys.* 31 (1959) 539.

[17] M. Enomoto, C.L. White and H.I. Aaronson, *Metall. Trans.* A 19 (1988) 1807.

[18] W.F. Lange III, M. Enomoto and H.I. Aaronson, *Metall. Trans.* A 19 (1988) 427.

[19] M. Hillert, *Phase Transformations* (ASM, Metals Park, OH, 1970) p. 181.

[20] M. Hillert, B. Sundman and J. Agren, *Computerized Metallurgical Databases*, ed. by J.R. Cuthill, N.A. Gokcen and J.E. Morral (TMS-AIME, Warrendale, PA, 1988).

[21] K.C. Russell and H.I. Aaronson, *Jnl. Materials Science* 10 (1975) 1991.

[22] E. Hornbogen, *Aluminum* 43 (1967) 9.

[23] V. Gerold, *Zeit. Metallkünde* 45 (1954) 593; 599.

Mats Hillert with John Cahn (left) and Hub Aaronson (middle) at PTM'05.

THE USES OF GIBBS FREE ENERGY-COMPOSITION DIAGRAMS

Mats Hillert
Royal Institute of Technology
Stockholm, Sweden

Introduction

The discussion of equilibria and of the driving force for reactions in alloys is usually based upon Gibbs free energy, G. The purpose of this paper is to present the basis for the use of molar Gibbs free energy diagrams for such purposes. However, molar diagrams for all extensive quantities have many properties in common. This will be emphasized by using the molar volume V_m as the illustrative example when such properties are discussed in the first part of the paper.

Under conditions of constant temperature and pressure the Gibbs free energy, G, is particularly interesting because equilibrium is then characterized by a minimum in this quantity and the decrease in Gibbs free energy during a spontaneous reaction can be regarded as the driving force. These facts give the molar diagram of the Gibbs free energy special properties and the main part of the paper will thus be concerned with G_m diagrams. It should be emphasized, however, that the same properties hold for the Helmholtz energy, F, at given temperature and volume, for the enthalpy, H, at given entropy and pressure and for the energy, U, at given entropy and volume.

The treatment is based upon the fundamentals of thermodynamics as given in textbooks like E.A. Guggenheim, Thermodynamics, An Advanced Treatment for Chemists and Physicists, Amsterdam 1949.

Notation

The two components in a binary system will be denoted by A and B. Following international recommendations, the partial quantities will be denoted as V_B instead of the more common notation $\bar{V}_B$. The phase to which the quantity refers is given as a superscript to the right of the symbol, as V_B^α. For mole fractions, the symbol $x_B^{\alpha/\beta}$ denotes the B content of the α phase in contact with the β phase. The superscript position to the left will be used to denote a standard state of the pure component, as $^o V_B^\alpha$. For mole fractions this position will be used to characterize particular compositions like $^1 x_B^\alpha$ or $^2 x_B^\alpha$ or the equilibrium composition like $^e x_B^{\alpha/\beta}$. The composition of a binary α phase will usually be identified by giving x_B^α but x_A^α can immediately be obtained as $1 - x_B^\alpha$.

General Properties of Molar Diagrams

Any property whose value for the whole system is equal to the sum of its values for the separate parts of the system is called an extensive property. Examples of such are the volume V and the Gibbs free energy G, but also the amount of total substance n or the amount of a particular substance n_B. It is often convenient to divide by the amount of substance, n. The amount n is usually expressed in moles and the quantity obtained by division with n is called a molar quantity and is denoted by a subscript m, for example the molar volume, $V_m = V/n$. In the special case of composition one uses a completely new symbol, $x_B = n_B/n$, and the other molar quantities are often plotted as functions of x_B. This paper is concerned with the properties of such diagrams and it is important to emphasize that, in principle, one is permitted to express n in a number of ways, for instance by the weight or the number of atoms.

One is even permitted to define the molar quantities by dividing by n_A, for example. The important point to remember is that the composition must be defined in the same way. The diagram then has a simple, but important property which will now be derived.

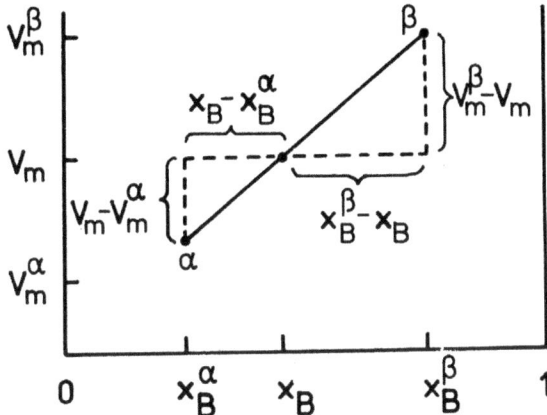

Fig. 1. Definition of straight line between points representing two different phases in a molar diagram.

For a single phase α one can write $V_m^\alpha = V^\alpha/n^\alpha$ and $x_B^\alpha = n_B^\alpha/n^\alpha$. Consider the position of two such phases in a diagram of V_m versus x_B, Fig. 1. Any point (x_B, V_m) on the straight line between α and β is subject to a simple condition

$$\frac{V_m^\beta - V_m}{V_m - V_m^\alpha} = \frac{x_B^\beta - x_B}{x_B - x_B^\alpha} \qquad (1)$$

We shall now show that any point representing a two-phase alloy, $\alpha+\beta$, fulfils this condition and thus lies on the straight line between the two points representing α and β. Since V and n are both extensive quantities we have $n = n^\alpha + n^\beta$ and $n_B = n_B^\alpha + n_B^\beta$. Dividing V and n_B for the alloy by n one obtains

$$V_m^{alloy} = \frac{V^{alloy}}{n^{alloy}} = \frac{V^\alpha + V^\beta}{n^\alpha + n^\beta} = \frac{n^\alpha V_m^\alpha + n^\beta V_m^\beta}{n^\alpha + n^\beta} = \frac{n^\alpha}{n^\alpha + n^\beta} \cdot V_m^\alpha + \frac{n^\beta}{n^\alpha + n^\beta} \cdot V_m^\beta \quad (2)$$

$$x_B^{alloy} = \frac{n_B^{alloy}}{n^{alloy}} = \frac{n_B^\alpha + n_B^\beta}{n^\alpha + n^\beta} = \frac{n^\alpha x_B^\alpha + n^\beta x_B^\beta}{n^\alpha + n^\beta} = \frac{n^\alpha}{n^\alpha + n^\beta} \cdot x_B^\alpha + \frac{n^\alpha}{n^\alpha + n^\beta} \cdot x_B^\beta \quad (3)$$

n^α and n^β can be eliminated in the following way,

$$\frac{V_m^\beta - V_m^{alloy}}{V_m^{alloy} - V_m^\alpha} = \frac{n^\alpha}{n^\beta} = \frac{x_B^\beta - x_B^{alloy}}{x_B^{alloy} - x_B^\alpha} \quad (4)$$

This relationship is identical to eq. 1 with $V_m = V_m^{alloy}$ and $x_B = x_B^{alloy}$ and the point $(x_B^{alloy}, V_m^{alloy})$ thus lies on the straight line between α and β. This conclusion holds independently of how the amount of material, n, is defined as long as the same definition is used in deriving the molar quantity and the composition. This simple property will now be used to derive a number of relationships.

First, consider a two-phase system containing a very large quantity of α phase and a very small amount of β phase. Suppose, for the sake of argument, that the molar volume diagram is such that the point representing the β phase lies on the tangent to the curve for the α phase. According to the above mentioned property of such diagrams, the point representing the two-phase mixture also lies on the tangent, as is demonstrated in Fig. 2. When the β phase is actually dissolved in the α phase, the molar volume of α is changed along the V_m^α curve, i.e. along the tangent, as long as the amount of β is very small. As a consequence, the point representing the alloy will not move and the total volume of the system will not change as the β phase is being dissolved.

In the general case, the β phase does not lie on the tangent and the volume of the system will thus change as β is dissolved. If

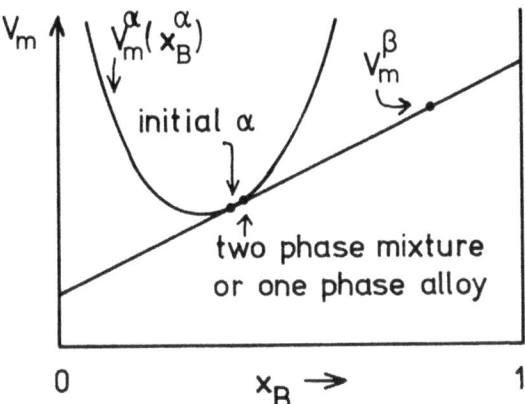

Fig. 2. Demonstration of the property of a curve in a molar diagram.

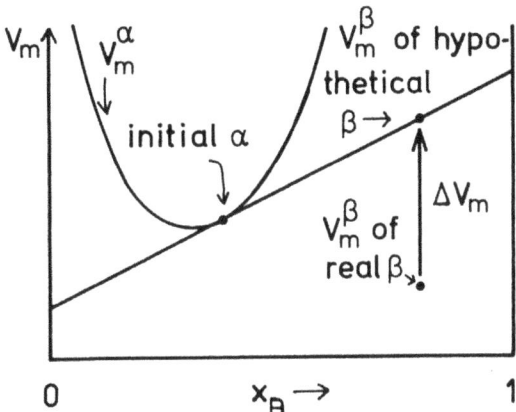

Fig. 3. Change in volume accompanying the dissolution of β in α.

the β phase was first transformed into a hypothetical state, situated on the tangent, the subsequent reaction would not change the volume. The total change would be the change connected with the preliminary hypothetical transformation of β, Fig. 3. Remembering that we are concerned with a molar diagram, we understand that the length of the vertical arrow represents the increase in volume per mole of β phase being dissolved in the α phase. It should again be emphasized that this construction is valid only as long as the compositional change of α is small.

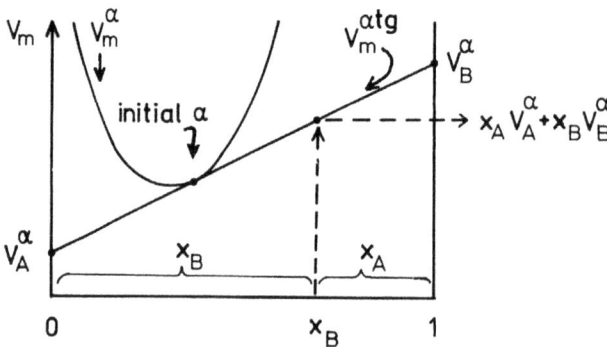

Fig. 4. Derivation of an equation for a tangent in a molar diagram.

The magnitude of this change can easily be expressed mathematically if the intersections of the α tangent with the component axes, V_A^α and V_B^α, are known. Fig. 4 illustrates that any point on the α tangent is given by

$$V_m^{\alpha tg} = x_A V_A^\alpha + x_B V_B^\alpha \qquad (5)$$

Fig. 5 will thus yield the following expression

$$\Delta V_m = x_A^\beta V_A^\alpha + x_B^\beta V_B^\alpha - V_m^\beta \qquad (6)$$

As a special case we shall now assume that the β phase is pure B and its molar volume is situated on the α tangent, Fig. 6. There will now be no change in total volume when some B is dissolving. On the other hand, the α phase will now increase by the volume occupied by the B atoms before dissolution and it is thus evident that

$$(\partial V^\alpha / \partial n_B)_{n_A} = V_m^\beta \qquad (7)$$

This is the thermodynamic definition of the partial molar volume of B in the α phase, V_B^α. This quantity may thus be obtained for any composition of interest by constructing the tangent to the α phase at that composition. Fig. 6 immediately yields the well-known relation

$$V_B^\alpha = V_m^\alpha + x_A^\alpha \cdot dV_m^\alpha / dx_B \qquad (8)$$

The intersection of the same tangent on the left-hand side of the diagram defines the equivalent quantity for the other component in the same α phase, V_A^α, Fig. 7. Two relations are obtained directly from this diagram.

$$V_m^\alpha = x_A^\alpha V_A^\alpha + x_B^\alpha V_B^\alpha \qquad (9)$$

$$dV_m^\alpha / dx_B = V_B^\alpha - V_A^\alpha \qquad (10)$$

Of course, all the quantities in these equations must be evaluated for the alloy composition x_B^α.

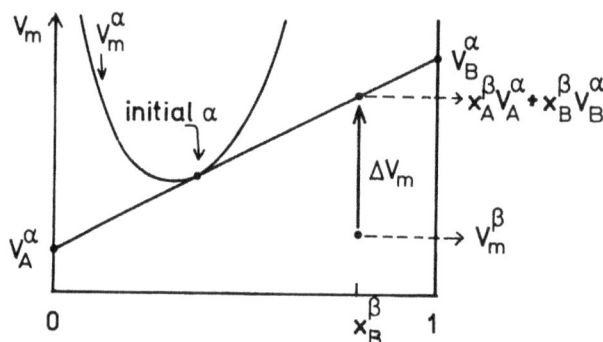

Fig. 5. Evaluation of the change in volume accompanying the dissolution of β in α.

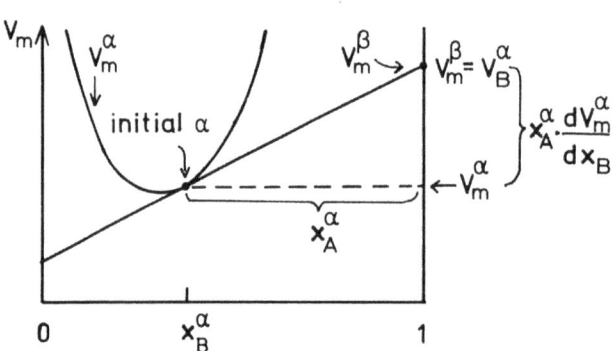

Fig. 6. Relation between partial molar volume, V_B, and molar volume, V_m.

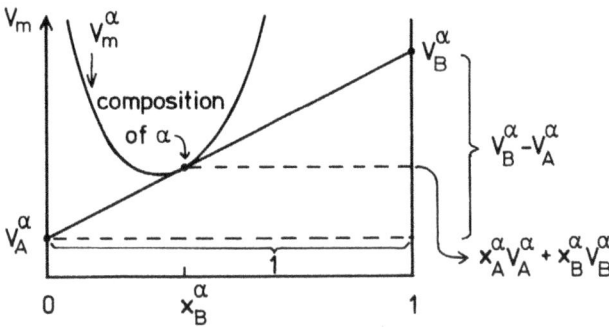

Fig. 7. Evaluation of molar volume, V_m, from the two partial molar volumes, V_A and V_B.

Having identified the intersections of the tangent with the two component axes as the partial molar volumes of the components, we may now interpret eq. 5 in the following way. Reference is also made to Fig. 4. A sufficiently large amount of an α phase may be regarded as a reservior of both A and B atoms. It decreases its volume by V_A^α per mole of A atoms removed and by V_B^α per mole of B atoms. If one simultaneously removes x_A of A atoms and x_B mole of B atoms, one has removed a total of one mole. The volume has then decreased by $x_A V_A^\alpha + x_B V_B^\alpha$ which is the value that can be read on the tangent at the composition of the material being removed.

So far, we have regarded the α phase as a reservoir with which one can exchange a small quantity of A or B atoms without appreciably changing its composition. In order to treat the changes due to a larger variation in composition one is interested in the rate of change of V_A^α and V_B^α when the composition of the phase is changed. The following relations can be obtained directly from the molar diagram in Fig. 8 by comparing triangles:

$$-dV_A^\alpha / x_B^\alpha = dV_B^\alpha / x_A^\alpha = d(V_B^\alpha - V_A^\alpha)/1 \tag{11}$$

As already shown by eq. 10, we can substitute $V_B^\alpha - V_A^\alpha$ with dV_m^α/dx_B. Multiplying eq. 11 by $x_A^\alpha x_B^\alpha$ we obtain

$$-x_A^\alpha dV_A^\alpha = x_B^\alpha dV_B^\alpha = x_A^\alpha x_B^\alpha d(V_B^\alpha - V_A^\alpha) = x_A^\alpha x_B^\alpha \cdot d^2 V_m^\alpha/dx_B^2 \cdot dx_B \qquad (12)$$

The left-hand equality on this line is identical to the Gibbs-Duhem relation.

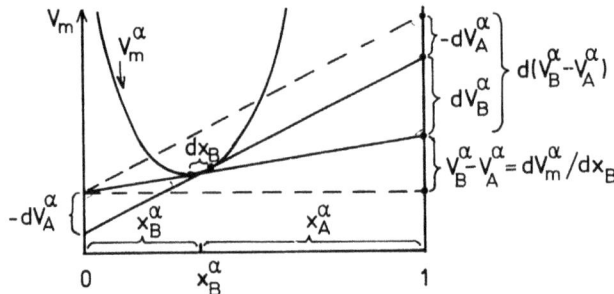

Fig. 8. Evaluation of the change in partial molar volume due to a change in composition.

Internal Stability

A state of equilibrium at given temperature and pressure is characterized by a minimum in Gibbs free energy. This criterion will now be discussed in relation to the molar diagram for Gibbs free energy.

In order for a single phase to have internal stability, every separation of the two components must result in an increase of G_m for the system. Let us first consider a splitting of the phase in two parts of equal size, one slightly richer in A and the other slightly richer in B. Since the molar Gibbs free energy for such a two-phase state is obtained on the straight line between the points representing the two parts, the change in G_m is obtained

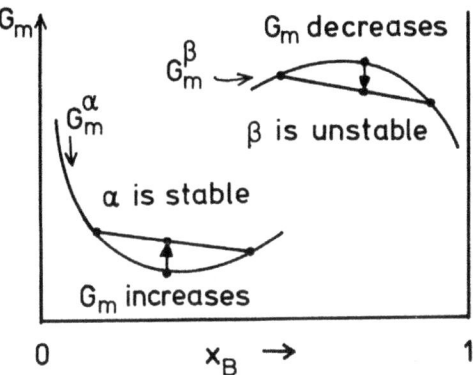

Fig. 9. Test of the internal stability of phases.

as the length of the arrows in Fig. 9. It is evident that the criterion for stability is

$$d^2 G_m^\alpha / dx_B^2 > 0. \tag{13}$$

In order to make the V_m diagrams in the preceding paragraph directly applicable to considerations of Gibbs free energy, the V_m curves were always drawn with a positive second derivative although there is no such requirement for V_m.

In view of the importance of the second derivative of the G_m^α curve, one must keep at least three terms if one wants to represent the G_m^α curve in the neighborhood of a certain composition $^1x_B^\alpha$ by a Taylor series expansion

$$G_m^\alpha = a + b \cdot \Delta x_B + c \cdot (\Delta x_B)^2 \tag{14}$$

where $\Delta x_B = x_B^\alpha - {}^1x_B^\alpha$ and $c = \frac{1}{2} d^2 G_m^\alpha / dx_B^2$. In general, such a parabolic representation is satisfactory only for small changes but it is sometimes illustrative to use the parabolic shape for a phase with a narrow range of existence. For instance, a very steep parabola can be used to describe a stoichiometric phase with a well defined composition. The G_m diagram in Fig. 10 demonstrates that there is no unique way of placing the tangent to such a phase. The individual values of G_A^α and G_B^α are not defined for such a phase, but a strict relation exists between them and is defined by the value of G_m^α at the minimum.

$$G_m^\alpha = x_A^\alpha G_A^\alpha + x_B^\alpha G_B^\alpha \tag{15}$$

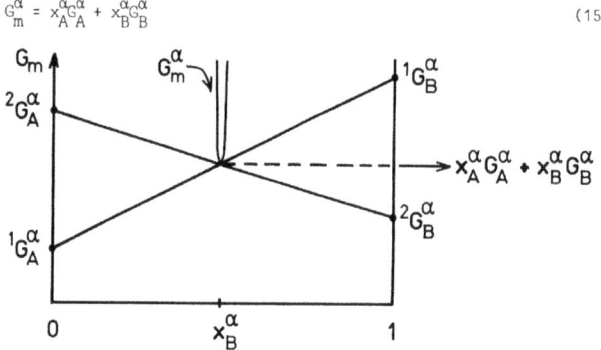

Fig. 10. Relation between the two partial molar Gibbs free energies, G_A and G_B, in a stoichiometric phase.

The fact that one cannot represent the whole of a G_m^α curve by a parabola becomes particularly evident as one approaches the side of the diagram. It may seem self-evident that the chemical potential of B, G_B^α, cannot have a definite value in a phase that contains no B atoms. However, a definite value is predicted by a parabola in the limit $x_B^\alpha \to 0$, as illustrated in the left-hand part of Fig. 11. Instead, the shape of the G_m^α curve must be such that the tangent becomes vertical at the limit. This property is adequately described by the term containing the ideal entropy of

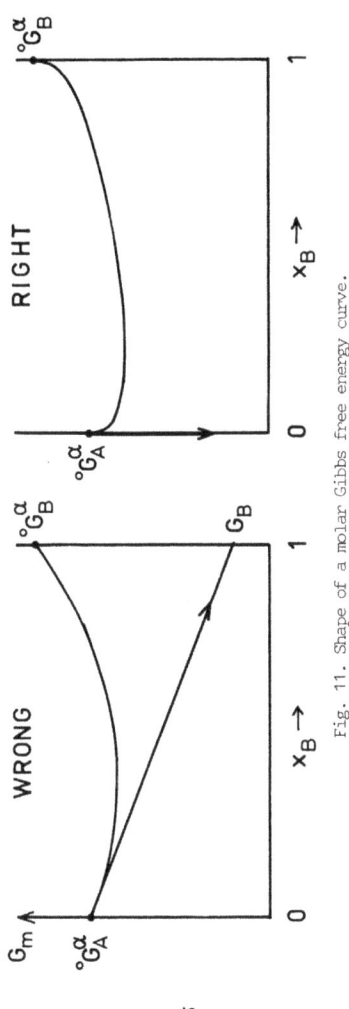

Fig. 11. Shape of a molar Gibbs free energy curve.

mixing derived by statistical thermodynamics,

$$-TS^{ideal} = RT(x_A \ln x_A + x_B \ln x_B) \qquad (16)$$

The remaining part of G_m^α can sometimes be adequately described by a power series. It should be noted that, in order to obtain the correct values at the sides of the diagram one must represent the power series by the following expression, $x_A^\alpha \cdot {}^oG_A^\alpha + x_B^\alpha \cdot {}^oG_B^\alpha + x_A^\alpha x_B^\alpha \cdot L^\alpha$. The quantities ${}^oG_A^\alpha$ and ${}^oG_B^\alpha$ are the molar Gibbs free energy for pure A and B, respectively, and L^α is a constant in the regular solution model. The following expressions are obtained from this approximation.

$$G_m^\alpha = x_A^\alpha \cdot {}^oG_A^\alpha + x_B^\alpha \cdot {}^oG_B^\alpha + RT(x_A^\alpha \ln x_A^\alpha + x_B^\alpha \ln x_B^\alpha) + x_A^\alpha x_B^\alpha \cdot L^\alpha \qquad (17)$$

$$G_A^\alpha = {}^oG_A^\alpha + RT \ln x_A^\alpha + (x_B^\alpha)^2 \cdot L^\alpha \qquad (18)$$

$$G_B^\alpha = {}^oG_B^\alpha + RT \ln x_B^\alpha + (x_A^\alpha)^2 \cdot L^\alpha \qquad (19)$$

$$c = \tfrac{1}{2} d^2 G_m^\alpha / dx_B^2 = \tfrac{1}{2} RT / x_A^\alpha x_B^\alpha - L^\alpha \qquad (20)$$

Let us now calculate the increase in Gibbs free energy in an α phase of composition ${}^1x_B^\alpha$ due to a compositional fluctuation by which a small region has changed its composition to x_B. According to the properties of a molar diagram, there would be no change in Gibbs free energy if the point representing that region were situated on the tangent. However, if the region can still be regarded as α phase, it is situated on the G_m^α curve and the distance to the tangent represents the increase in Gibbs free energy per mole of atoms in the region, Fig. 12. If the G_m^α curve is represented by eq. 14, the tangent is given by

$$G_m^{\alpha tg} = a + b \cdot \Delta x_B \qquad (21)$$

and the distance is obtained from the difference between eqs. 14 and 21,

$$\Delta G_m = c \cdot (\Delta x_B)^2 = \tfrac{1}{2} d^2 G_m^\alpha / dx_B^2 \cdot (\Delta x_B)^2 \tag{22}$$

With the regular solution model one obtains

$$\Delta G_m = (\tfrac{1}{2} RT / x_A^\alpha x_B^\alpha - L^\alpha) \cdot (\Delta x_B)^2 \tag{23}$$

The calculation is valid only for small Δx_B. It should be noticed

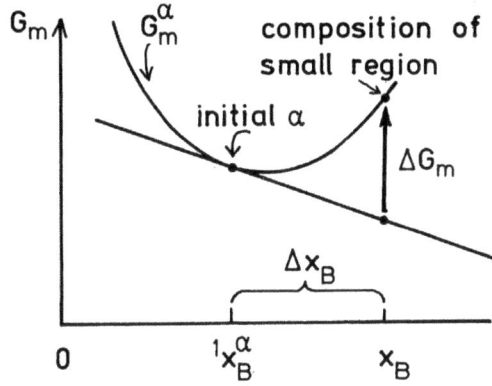

Fig. 12. Increase in Gibbs free energy due to a compositional fluctuation.

that the size of Δx_B should be judged by comparison with the composition of the fluctuation and not of the alloy. The criterion is thus $\Delta x_B \ll x_B$ and $\Delta x_B \ll x_A$. This limits the use of the calculation when the fluctuation approaches a pure component, as demonstrated by Fig. 13. The exact value of ΔG_m can, of course, be obtained as follows. The point on the α curve is $x_A \cdot G_A^\alpha(x_B) + x_B \cdot G_B^\alpha(x_B)$ and the point on the tangent is
$x_A \cdot G_A^\alpha(^1 x_B^\alpha) + x_B \cdot G_B^\alpha(^1 x_B^\alpha)$. The distance is thus

$$\Delta G_m = x_A [G_A^\alpha(x_B) - G_A^\alpha(^1 x_B^\alpha)] + x_B [G_B^\alpha(x_B) - G_B^\alpha(^1 x_B^\alpha)] \tag{24}$$

and, by the application of the regular solution model,

$$\Delta G_m = RT[x_A \ln(x_A/{}^1x_A^\alpha) + x_B \ln(x_B/{}^1x_B^\alpha)] - L^\alpha \cdot (x_B - {}^1x_B^\alpha)^2 \quad (25)$$

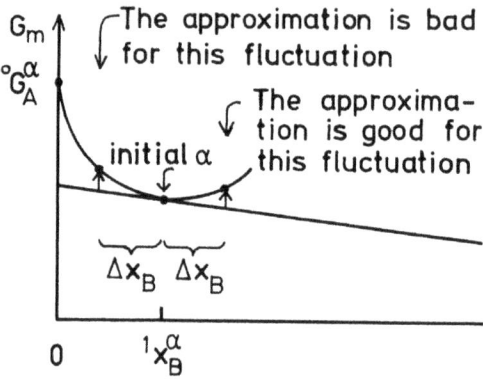

Fig. 13. Validity of approximate method of evaluating a change in Gibbs free energy.

The limit of stability is reached when $d^2 G_m^\alpha/dx_B^2 = 0$ and in the regular solution model this occurs when $RT = 2x_A^\alpha x_B^\alpha \cdot L^\alpha$, according to eq. 20. This defines the spinodal. Inside the spinodal, the G_m^α curve has a negative curvature. The molar Gibbs free energy diagram can be used to demonstrate that, in an alloy outside the spinodal, a compositional fluctuation can become stable and grow spontaneously after passing a critical composition, which is situated inside the spinodal. The critical composition is easily found for each alloy composition ${}^1x_B^\alpha$ by constructing a second tangent parallel to the usual one, Fig. 14. The distance between the two parallel tangents defines the activation energy necessary in order to form a region of the critical composition. It is easy to see that the two tangents move closer to each other if the alloy com-

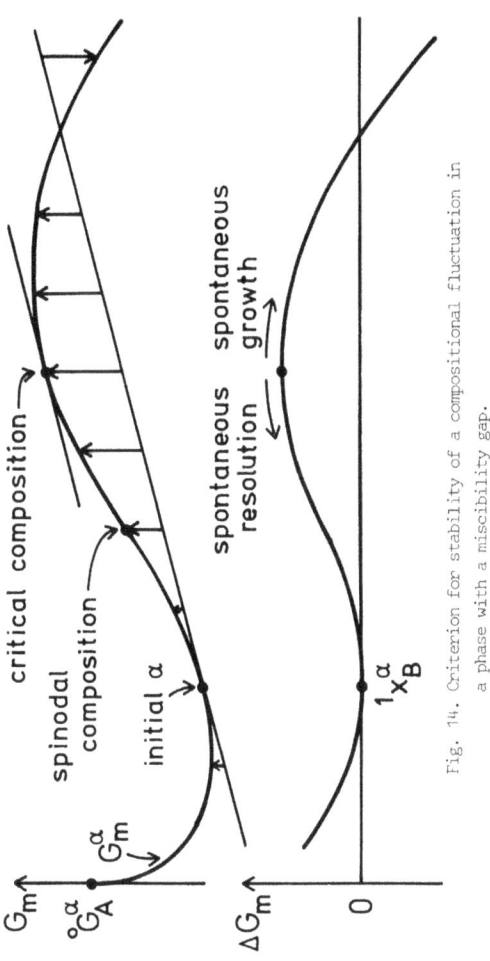

Fig. 14. Criterion for stability of a compositional fluctuation in a phase with a miscibility gap.

position is moved closer to the spinodal composition. They coincide at the spinodal composition and the barrier to decomposition will thus vanish at the spinodal.

Driving Force for Diffusion

It is natural that the positive curvature of a G_m curve which makes the phase stable against fluctuations in composition, also provides the driving force for the elimination of differences in composition within the phase if such are present, i.e. the driving force for diffusion. We shall now discuss the force which makes the individual B atoms diffuse from a region with a high B concentration to a region with a low B concentration. Each region may be regarded as a reservoir of B with its own value of G_B and the difference in G_B is identical to the decrease in Gibbs free energy when one mole of B is transferred, Fig. 15. Assuming that

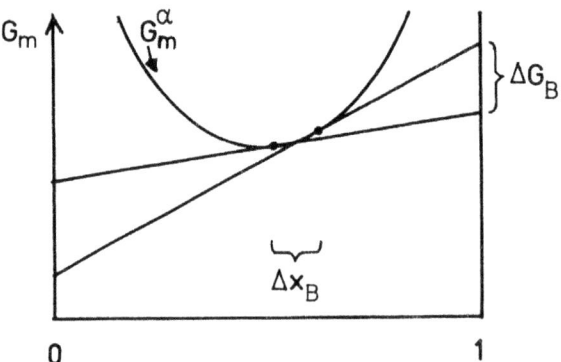

Fig. 15. Evaluation of the driving force for diffusion of B atoms.

the rate of transfer is proportional to the decrease in Gibbs free energy and the number of B atoms per volume, x_B/V_m, and inversely proportional to the transport distance, Δy, one obtains the following expression for the flux of B atoms

$$J_B = - M_B \cdot x_B/V_m \cdot \Delta G_B/\Delta y = - M_B/V_m \cdot x_B \cdot dG_B/dx_B \cdot \Delta x_B/\Delta y \qquad (26)$$

The constant of proportionality, M_B, may be regarded as the mobility of the B atoms. The curvature of the G_m curve can be introduced by the use of an equation equivalent to eq. 12

$$J_B = - M_B/V_m \cdot x_A x_B \cdot d^2 G_m/dx_B^2 \cdot \Delta x_B/\Delta y \qquad (27)$$

This is recognized as Fick's law and the diffusion constant for B is obtained as

$$D_B = M_B \cdot x_A x_B \cdot d^2 G_m/dx_B^2 \qquad (28)$$

The mobility is thus multiplied by a thermodynamic factor. By introducing the activity or the activity coefficient for B this factor can be transformed to the shape which was introduced by Darken

$$x_A x_B \cdot d^2 G_m/dx_B^2 = x_B \cdot dG_B/dx_B = dG_B/d\ln x_B$$

$$= RTd\ln a_B/d\ln x_B = RT(1+d\ln f_B/d\ln x_B) \qquad (29)$$

It is evident that a similar derivation can be carried out for the other component and that the same factor is obtained:

$$D_A = M_A \cdot x_A x_B \cdot d^2 G_m/dx_B^2 \qquad (30)$$

Calculation of Two-Phase Equilibria

The requirement that the Gibbs free energy is a minimum in a state of equilibrium, can be used in calculations of the compositions of two phases in equilibrium with each other. It is becoming increasingly more common to solve such problems by computerized calculations and it may be instructive to base the discussion on two different methods for such calculations. Consider a system containing one mole of material with the average composition x_B^{av}. Let it first be in a two-phase state α1+β1. Its average molar Gibbs free energy, G_m^{av}, is given at x_B^{av} on the line α1-β1, Fig. 16. The compositions of α and β are then varied in such a way that G_m^{av} decreases gradually. The diagram demonstrates that the minimum of G_m^{av} will finally be reached when α and β are situated on the points of tangency on the common tangent. By this consideration we have thus shown that the equilibrium compositions, $^e x_B^{\alpha/\beta}$ and $^e x_B^{\beta/\alpha}$, can be found by constructing the common tangent.

If the average composition was outside the two-phase region, e.g. at α3 in the diagram, the calculation would stop when α3 is reached because the amount of β phase would then be zero. That will be the stable state for such an alloy.

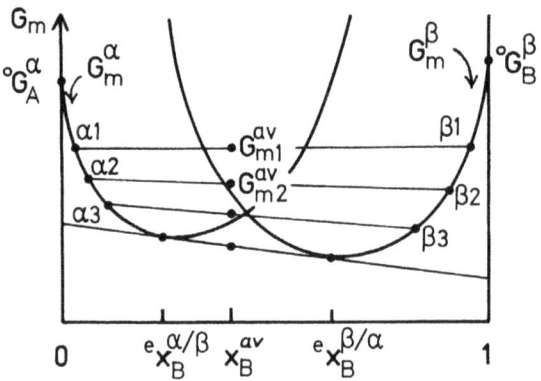

Fig. 16. Calculation of a two-phase equilibrium by finding the minimum in Gibbs free energy.

An alternative method to proceed from the initial α1+β1 state to more stable states is to compare the values of G_B and

G_A, respectively, in the two phases and to transfer B to the phase with the lowest G_B value and A to the phase with the lowest G_A value. Fig. 17 shows that $\Delta^1 G_B$ would drive the transfer of B from α1 to β1 and $\Delta^1 G_A$ would drive the transfer of A from α1 to β1. In this particular case, the two phases would thus move closer to each other and the driving forces would gradually decrease, as illustrated by $\Delta^2 G_B$ and $\Delta^2 G_A$ when the phases are at α2 and β2. Both driving forces will finally vanish when the two tangents coincide. We have thus again shown that the equilibrium compositions, $^e x_B^{\alpha/\beta}$ and $^e x_B^{\beta/\alpha}$, can be found by the common-tangent construction. This procedure simulates the physical process of attaining equilibrium and sometimes it may even be advantageous to adjust the rate of transfer of A and B atoms to the size of ΔG_A and ΔG_B.

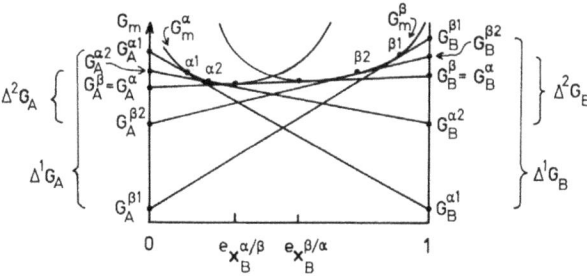

Fig. 17. Calculation of a two-phase equilibrium by moving atoms in the direction of decreasing partial molar Gibbs free energy.

In principle, the two-phase equilibrium can be calculated directly by solving the two equilibrium conditions given by $G_A^\alpha = G_A^\beta$ and $G_B^\alpha = G_B^\beta$, which of course define the common tangent. However, this can seldom be done analytically. If one chooses to carry out the numerical calculations with a Newton-Raphson method one will actually follow a procedure related to the method just described. Alternatively, an iteration procedure can be used, as illustrated in Fig. 18. Start by guessing β1. Calculate $G_B^{β1}$. Calculate α1 from $G_B^{α1} = G_B^{β1}$. Calculate $G_A^{α1}$. Calculate β2 from $G_A^{β2} = G_A^{α1}$. Etc.

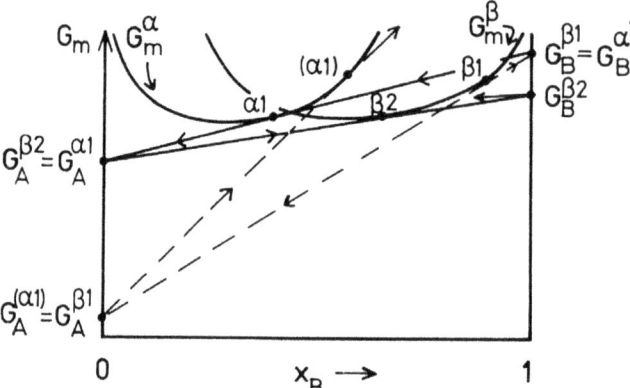

Fig. 18. Calculation of a two-phase equilibrium by an iteration method.

According to the diagram, this iteration seems to converge to the common tangent, but it would have diverged if α was calculated from $G_A^α = G_A^β$ and β from $G_B^β = G_B^α$. This is illustrated by the dashed lines. The iteration procedure, like any other numerical method, must be constructed carefully in order to converge.

The calculation is greatly simplified if, by some reason, the composition of one of the phases, e.g. β, is known, Fig. 19. One can then calculate the common tangent from the fact that the β phase is situated on the tangent to the α phase. Compare eq. 5.

$$x_A^β \cdot G_A^α(x_B^α) + x_B^β \cdot G_B^α(x_B^α) = G_m^β(x_B^β) \qquad (31)$$

The composition of the α phase, $x_B^α$, is the only unknown quantity and only one equation is thus required. This will be the case when one of the phases is a so-called stoichiometric phase which has a fixed composition.

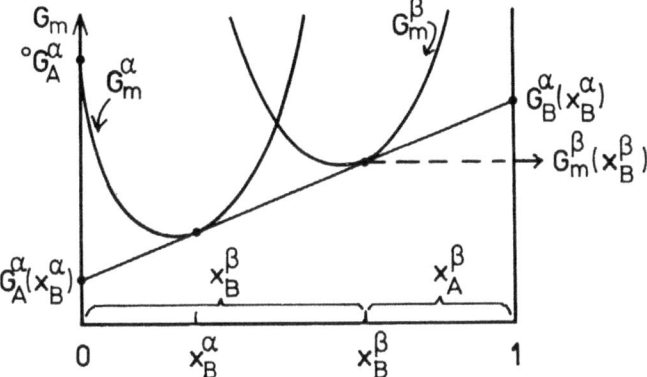

Fig. 19. Calculation of the composition of an α phase in equilibrium with a β phase of known composition.

There will of course be no unknown equilibrium composition to calculate if both phases are stoichiometric. However, it may still be interesting to calculate the values of G_A and G_B established by such an equilibrium. The equilibrium is defined by the common tangent. By comparing triangles in Fig. 20 one obtains

$$\frac{G_B - G_m^\beta}{G_B - G_m^\alpha} = \frac{x_A^\beta}{x_A^\alpha} \tag{32}$$

and by rearranging this equation one finds

$$G_B = \frac{x_A^\alpha G_m^\beta - x_A^\beta G_m^\alpha}{x_A^\alpha - x_A^\beta} \tag{33}$$

23

In the same way one obtains

$$G_A = \frac{x_B^\alpha G_m^\beta - x_B^\beta G_m^\alpha}{x_B^\alpha - x_B^\beta} \tag{34}$$

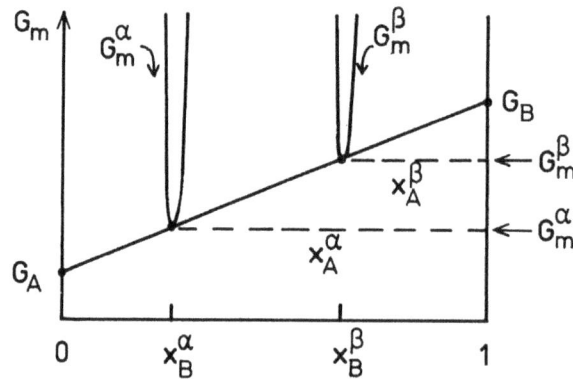

Fig. 20. Evaluation of the partial molar Gibbs free energies for an equilibrium between two stoichiometric phases.

Driving Force for Precipitation

Consider a one-phase alloy, α1, which is inside the two-phase region, α+β. The Gibbs free energy would decrease by precipitation of β and the total driving force for the complete reaction in one mole of the alloy is given by the arrow in Fig. 21. According to eq. 22, one has approximately

$$\Delta G_m = \frac{1}{2} \cdot d^2 G_m^\alpha / d x_B^2 \cdot (\Delta x_B^\alpha)^2 \tag{35}$$

On the other hand, the driving force for the formation of a very small quantity of β from a large quantity of α1 is obtained from the tangent representing the supersaturated α1 matrix. This construction is shown in Fig. 22 and the magnitude of this driving force is obtained as follows

$$\Delta G_m = x_A^\beta G_A^{\alpha 1} + x_B^\beta G_B^{\alpha 1} - G_m^\beta \tag{36}$$

$$G_m^\beta = x_A^\beta G_A^{\alpha e} + x_B^\beta G_B^{\alpha e} \tag{37}$$

$$\Delta G_m = x_A^\beta \cdot (G_A^{\alpha 1} - G_A^{\alpha e}) + x_B^\beta \cdot (G_B^{\alpha 1} - G_B^{\alpha e}) \tag{38}$$

This is the driving force during the early stages of precipitation, in particular the nucleation stage.

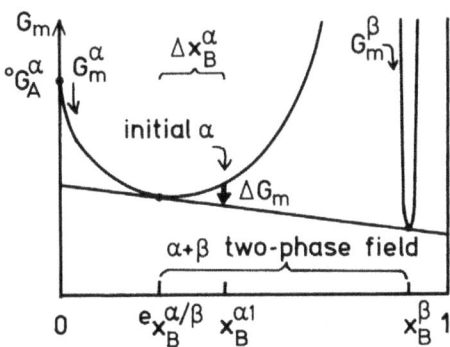

Fig. 21. Demonstration of the driving force for a complete precipitation reaction.

For low supersaturations, i.e. small Δx_B^α, one can introduce the curvature of the G_m^α curve. By comparing triangles in Fig. 23 one obtains

$$\frac{\Delta G_m}{x_B^\beta - x_B^\alpha} = \frac{\Delta(G_B - G_A)}{1} \tag{39}$$

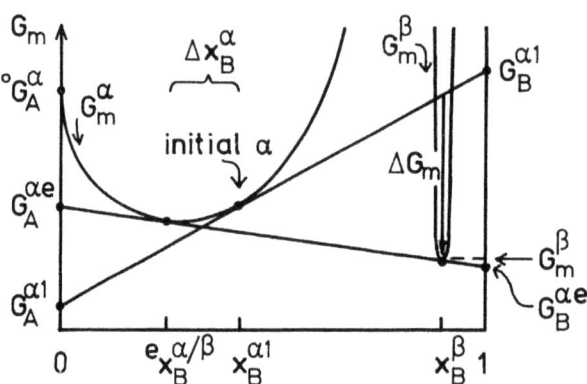

Fig. 22. Demonstration of the driving force at the start of a precipitation reaction.

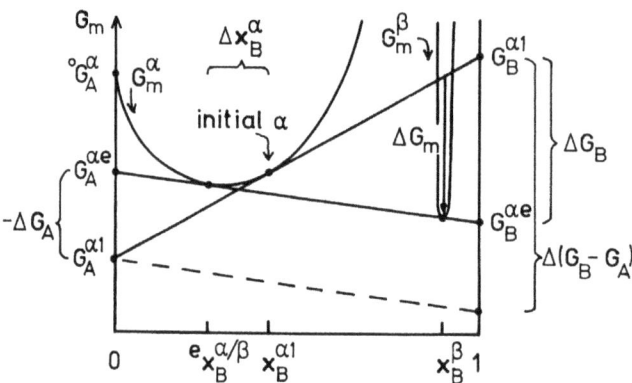

Fig. 23. Construction used in calculating the driving force at the start of a precipitation reaction.

Application of eq. 10 yields

$$\Delta G_m = (x_B^\beta - x_B^\alpha) \cdot \Delta(G_B - G_A) = (x_B^\beta - x_B^\alpha) \cdot \Delta(dG_m^\alpha/dx_B) \quad (40)$$

$$\Delta G_m = d^2 G_m^\alpha/dx_B^2 \cdot (x_B^\beta - x_B^\alpha) \cdot \Delta x_B^\alpha \quad (41)$$

The diagram illustrates that x_B^α in the factor $(x_B^\beta - x_B^\alpha)$ should be taken somewhere between $x_B^{\alpha 1}$ and $^e x^{\alpha/\beta}$. However, its exact position is not critical when Δx_B^α is small. Eq. 41 resembles the expression for the driving force of the complete reaction, eq. 35, but it is important to notice that the previous expression contained the square of the supersaturation. It is also interesting to notice that the last expression contains a factor $(x_B^\beta - x_B^\alpha)$ which is the difference in composition between the two phases. At the start of precipitation, a new phase may thus be favoured if it differs much in composition even if it cannot be in stable equilibrium with the matrix phase. Such a case is illustrated in Fig. 24. It is thus conceivable that the nucleation of a stable precipitate phase can be very difficult because it has a similar composition to that of the matrix, but a metastable phase with a quite different composition can form and assist in nucleating the stable phase.

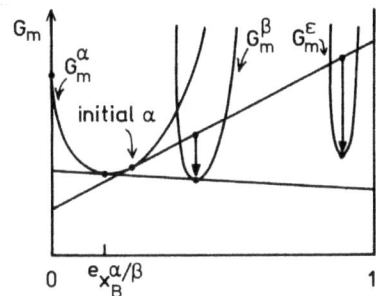

Fig. 24. Demonstration of the effect of phase composition on the driving force at the start of precipitation.

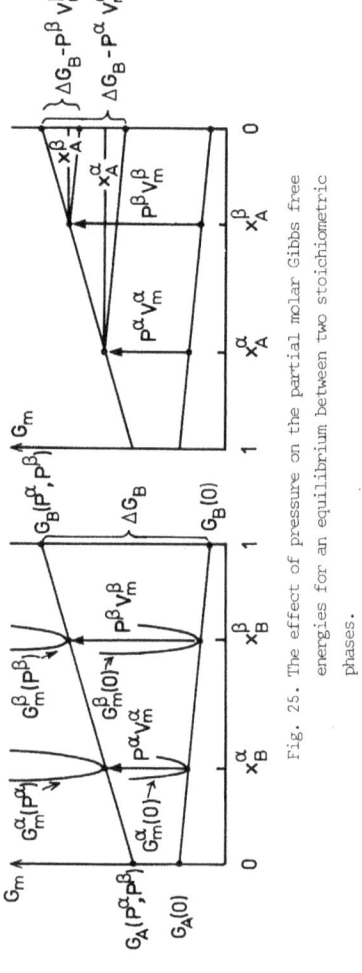

Fig. 25. The effect of pressure on the partial molar Gibbs free energies for an equilibrium between two stoichiometric phases.

Effect of Surface Tension on Two-Phase Equilibria

According to the definiton, the Gibbs free energy depends upon the pressure according to $G_m(P) = G_m(0) + PV_m$ where 0 denotes atmospheric pressure. For condensed phases the PV_m term can often be neglected if the pressure is not very high. However, it plays an important role for the equilibrium between two phases when they are under different pressures. This occurs when the interface is curved and is caused by the surface tension of the curved interface. For instance, a spherical α/β interface gives $P^\beta - P^\alpha = 2\sigma/r$ where σ is the surface tension and r is the radius of curvature.

Let us consider how the equilibrium between two phases is changed by the introduction of increased pressures. The two G_m curves are displaced upwards by the amounts of $P^\alpha V_m^\alpha$ and $P^\beta V_m^\beta$, respectively. The left-hand part of Fig. 25 illustrates that the values of G_A and G_B will change even if the compositions of the phases are so well defined that they do not change. The right-hand side illustrates that the change of G_B can be evaluated by the method used in deriving eq. 33 from Fig. 20. By comparing triangles one obtains

$$\frac{\Delta G_B - P^\beta V_m^\beta}{\Delta G_B - P^\alpha V_m^\alpha} = \frac{x_A^\beta}{x_A^\alpha} \qquad (42)$$

where ΔG_B stands for $G_B(P^\alpha, P^\beta) - G_B(0)$. Eq. 42 can be rearranged and yields

$$G_B(P^\alpha, P^\beta) - G_B(0) = \frac{x_A^\alpha \cdot P^\beta V_m^\beta - x_A^\beta \cdot P^\alpha V_m^\alpha}{x_A^\alpha - x_A^\beta} \qquad (43)$$

In the same way one obtains

$$G_A(P^\alpha, P^\beta) - G_A(0) = \frac{x_B^\alpha \cdot P^\beta V_m^\beta - x_B^\beta \cdot P^\alpha V_m^\alpha}{x_A^\alpha - x_A^\beta} \qquad (44)$$

When considering a spherical β particle in an α matrix, it is usually assumed that the α matrix is under atmospheric pressure and one puts $P^\alpha=0$. One then obtains a diagram, Fig. 26, which is similar to the one used for estimating the driving force at the start of a precipitation, Fig. 22. The term $P^\beta V_m^\beta$ is thus given by the expressions for ΔG_m given by eqs. 38 and 41. For small changes, Δx_B^α, eq. 41 yields

$$\Delta x_B^\alpha = \frac{P^\beta V_m^\beta}{d^2 G_m^\alpha/dx_B^{\alpha 2} \cdot (x_B^\beta - x_B^\alpha)} \tag{45}$$

and by inserting $P^\beta = 2\sigma/r$ and applying eq. 20 according to the regular solution model one obtains

$$\Delta x_B^\alpha = \frac{2\sigma V_m^\beta x_A^\alpha x_B^\alpha}{r \cdot (RT - 2L^\alpha x_A^\alpha x_B^\alpha) \cdot (x_B^\beta - x_B^\alpha)} \tag{46}$$

This is an extended form of the Gibbs-Thomson equation. For large Δx_B^α the calculation must involve an integration which becomes particularly easy when the α phase is a dilute solution of B in A. In this case x_A^α can be approximated as unity and $2L^\alpha x_A^\alpha x_B^\alpha$ can be neglected in comparison with RT.

$$dx_B^\alpha = \frac{dP^\beta \cdot V_m^\beta x_B^\alpha}{RT \cdot (x_B^\beta - x_B^\alpha)} \tag{47}$$

The following result is obtained by integration from atmospheric pressure if the difference in composition between the two phases, $x_B^\beta - x_B^\alpha$, is not affected markedly by pressure:

$$\ln\left[1 + \frac{\Delta x_B^\alpha}{{}^e x_B^{\alpha/\beta}}\right] = \frac{2\sigma V_m^\beta}{rRT \cdot (x_B^\beta - x_B^\alpha)} \tag{48}$$

${}^e x_B^{\alpha/\beta}$ is the equilibrium composition of the α phase when both phases are under atmospheric pressure.

The calculation becomes much more involved if both phases can vary in composition. Fig. 27 demonstrates that the composition of

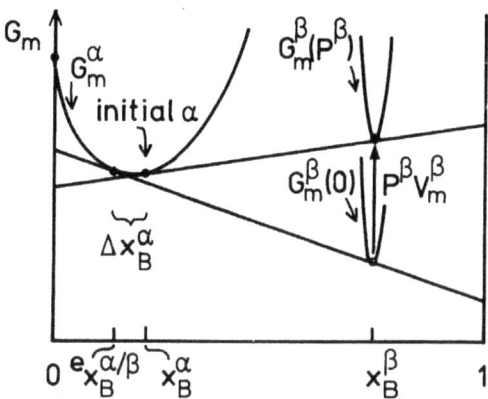

Fig. 26. Change in composition of a phase in equilibrium with a second phase on which a pressure is applied.

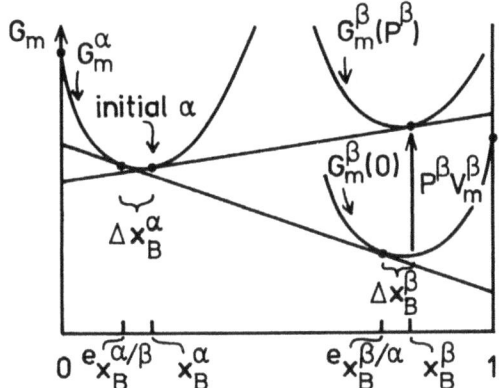

Fig. 27. Change in compositions when a pressure is applied to one of the phases in a two-phase equilibrium.

the β phase will also then depend upon the pressure P^β. It would take too much space to present the calculation for this case but, for small changes the calculation of Δx_B^α is still valid and for the β phase one can find an expression similar to eq. 45.

$$\Delta x_B^\beta = \frac{P^\beta \cdot (x_A^\alpha V_A^\beta + x_B^\alpha V_B^\beta)}{d^2 G_m^\beta / dx_B^{\beta\,2} \cdot (x_B^\beta - x_B^\alpha)} \tag{49}$$

It may further be mentioned that the composition of the β phase can be obtained graphically by drawing a tangent to β such that the distance between the intersections on the sides of the diagram are in the ratio V_A^β / V_B^β. This is illustrated in Fig. 28.

If the β phase is a dilute solution of B in A, eq. 49 can be rewritten as follows.

$$dx_B^\beta = \frac{dP^\beta \cdot V_m^\beta x_B^\beta}{RT \cdot (x_B^\beta - x_B^\alpha)} \tag{50}$$

This resembles eq. 47 and the following relationship holds if both phases are dilute solutions of B in A,

$$\frac{dx_B^\alpha}{x_B^\alpha} = \frac{dx_B^\beta}{x_B^\beta} \tag{51}$$

Integration from atmospheric pressure yields

$$\frac{x_B^\alpha}{e_{x_B^\alpha}^{\alpha/\beta}} = \frac{x_B^\beta}{e_{x_B^\beta}^{\beta/\alpha}} \tag{52}$$

As a consequence, the factors $x_B^\alpha/(x_B^\beta-x_B^\alpha)$ and $x_B^\beta/(x_B^\beta-x_B^\alpha)$ in eqs. 47 and 50 are constant and both equations can be integrated easily,

$$\Delta x_B^\alpha = \frac{2\sigma V_m^\beta \cdot {}^e x_B^{\alpha/\beta}}{rRT \cdot ({}^e x_B^{\beta/\alpha} - {}^e x_B^{\alpha/\beta})} \tag{53}$$

$$\Delta x_B^\beta = \frac{2\sigma V_m^\beta \cdot {}^e x_B^{\beta/\alpha}}{rRT \cdot ({}^e x_B^{\beta/\alpha} - {}^e x_B^{\alpha/\beta})} \tag{54}$$

These equations hold even if Δx_B^α is not smaller than ${}^e x_B^{\alpha/\beta}$ and Δx_B^β is not smaller than ${}^e x_B^{\beta/\alpha}$. It should be emphasized that they apply only to cases where both phases are dilute solutions of B in A, whereas eq. 48 applies when the composition of the β phase is so well defined that $x_B^\beta - x_B^\alpha$ can be treated as a constant. That case is found if β is a stoichiometric phase or a dilute solution of A in B.

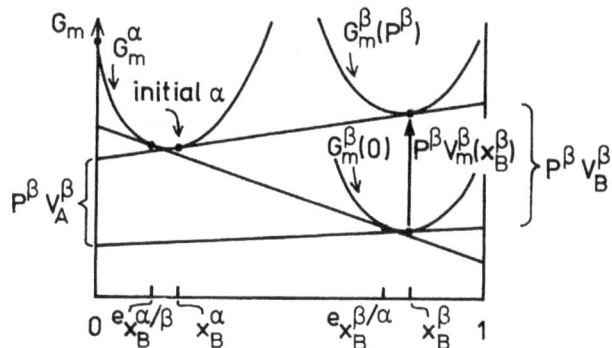

Fig. 28. Construction for finding the new compositions when a pressure is applied to one of the phases in a two-phase equilibrium.

Nucleation in a Binary System

A critical β nucleus may be regarded as being in equilibrium with the α matrix although the α matrix is supersaturated with respect to β. The nucleus would shrink spontaneously if it were a little smaller and grow if it were a little larger. The treatment presented in the preceding section can thus be applied directly to calculate the size and composition of a critical nucleus. A diagram identical to the one shown in Fig. 27 can be used to illustrate the unstable equilibrium between a critical nucleus and the matrix phase, Fig. 29. For instance, at small supersaturations, Δx_B^α, one obtains from eqs. 45 and 49

$$\frac{2\sigma V_m^\beta}{r^*} = P^* V_m^\beta = d^2 G_m^\alpha / dx_B^2 \cdot (x_B^\beta - x_B^\alpha) \cdot \Delta x_B^\alpha \tag{55}$$

$$\Delta x_B^{\beta*} = \frac{x_A^\alpha V_A^\beta + x_B^\alpha V_B^\beta}{V_m^\beta} \cdot \frac{d^2 G_m^\alpha / dx_B^2}{d^2 G_m^\beta / dx_B^2} \cdot \Delta x_B^\alpha \tag{56}$$

It should be emphasized that complications arise if the surface tension varies with the composition of α and β. For instance, the composition of the nucleus would move closer to the composition of the matrix if the surface tension could thus decrease. This might be the case for coherent precipitation in a miscibility gap and it results in a transition from a nucleation controlled reaction to spinodal decomposition at the spinodal curve in the phase diagram.

The discussion has not yet considered the activation energy for nucleation. At the first glance it may even seem puzzling that the diagram in Fig. 29 shows that the Gibbs free energy of a system will not change if one takes a small quantity of A and B atoms with an average composition of $x_B^{\beta*}$ from the α phase of composition x_B^α and transforms it into β phase of a constant pressure P^*. The point is that a new, spherical nucleus must form from zero size and grow gradually. Its pressure varies with its size and is thus higher than P^* until it reaches the critical size $r^*=2\sigma/P^*$. In order to add A and B atoms to the nucleus during its formation, one must thus use higher values for G_A and G_B than are available from the α matrix. If we evaluate the Gibbs free energy of the nucleus by integrating over its growth from zero size and all the time add balanced amounts of A and B atoms to keep the composition constant by taking $dn_A = x_A^\beta \cdot dn$ and $dn_B = x_B^\beta \cdot dn$, we would obtain the following for a spherical nucleus which contains $n = \frac{4}{3}\pi r^3/V_m^\beta$ mole of atoms.

$$G^{\beta nucleus} = \int [G_A^\beta(P^\beta)dn_A + G_B^\beta(P^\beta)dn_B] = \int_0^{n^*}[x_A^\beta G_A^\beta(P^\beta) + x_B^\beta G_B^\beta(P^\beta)] \cdot dn =$$
$$\int_0^{n^*} G_m^\beta(P^\beta) \cdot dn = \int_0^{n^*} [G_m^\beta(0) + P^\beta V_m^\beta] \cdot dn = G_m^\beta(0) \cdot n^* + \int_0^{r^*} \frac{2\sigma}{r} \cdot V_m^\beta \cdot \frac{4\pi r^2 dr}{V_m^\beta} =$$
$$= G_m^\beta(0) \cdot n^* + 4\pi(r^*)^2 \sigma \qquad (57)$$

By combination with $n^* = \frac{4}{3}\pi(r^*)^3/V_m^\beta$ and $r^* = 2\sigma/P^*$ one obtains

$$G^{\beta nucleus} = [G_m^\beta(0) + \frac{3}{2}P^*V_m^\beta] \cdot n^* \qquad (58)$$

The Gibbs free energy per mole of the nucleus, $G^{\beta nucleus}/n^*$, is thus higher than the ordinary value for the β phase by the amount $\frac{3}{2}P^*V_m^\beta$. Fig. 30 illustrates that the driving force for precipitation provides 2/3 of this quantity and the reamining part, $\frac{1}{2}P^*V_m^\beta$ must be provided by some process of activation. The activation energy for a nucleus of size n^*, is thus $\frac{1}{2}P^*V_m^\beta n^*$ and by inserting the expressions for n^* and r^* one obtains $\frac{16\pi}{3}\sigma^3/(P^*)^2$ which is identical to the more common expression for the activation energy, $\frac{16\pi}{3}\sigma^3/(\Delta G_m/V_m^\beta)^2$.

In connection with Fig. 30, it should be pointed out that a tangent to the curve for $G_m^{\beta nucleus}$ does not give the values of G_A and G_B. If a small quantity of B atoms is added to the nucleus without changing its A content in view of eq. 7, it will move it to the right on the curve but the nucleus will also be larger and the curve will move downwards. It is shown in Fig. 31 that the net movement will be in the direction towards the end point of the tangent representing the α matrix, in agreement with our starting point that the critical nucleus is in equilibrium with the matrix.

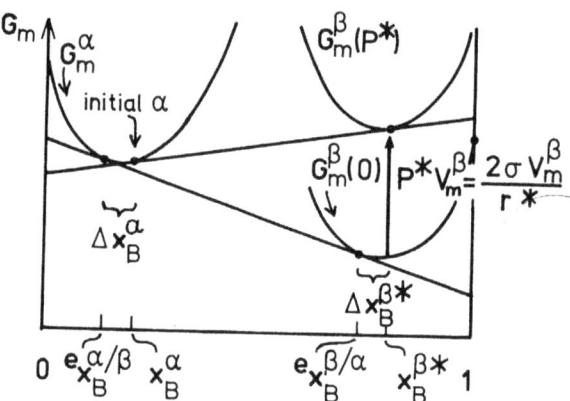

Fig. 29. Evaluation of size and composition of a critical nucleus.

Segregation to Grain Boundaries

Simple physical models of a grain boundary in metals have been used many times in discussions of segregation to grain boundaries and the effect of this segregation on surface tension. The properties of such a model in a binary system will now be discussed with the help of the molar diagram.

The model that will be discussed simply states that the grain boundary has a constant thickness and that the material in the boundary can be regarded as belonging to a separate phase which will here be denoted by b. A G_m function can thus be assigned to the boundary material and by applying the regular solution model one would, for instance, have

$$G_m^b = x_A^b \cdot {}^oG_A + x_B^b \cdot {}^oG_B + RT[x_A^b \ln x_A^b + x_B^b \ln x_B^b] + L^b x_A^b x_B^b \qquad (59)$$

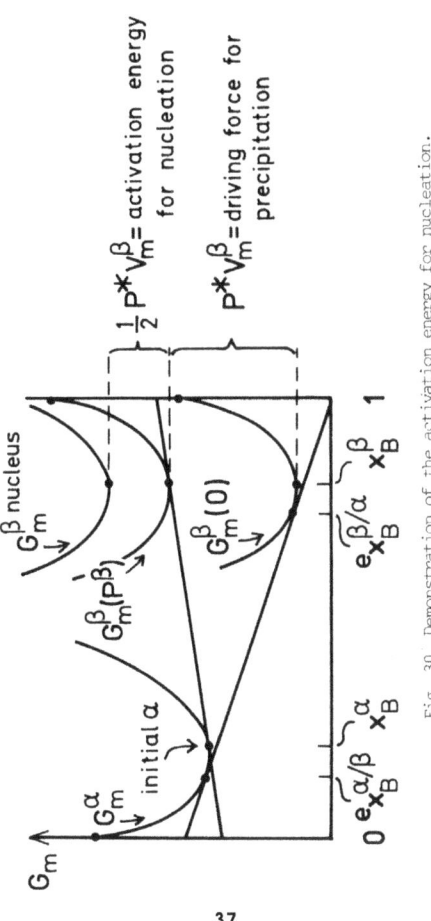

Fig. 30. Demonstration of the activation energy for nucleation.

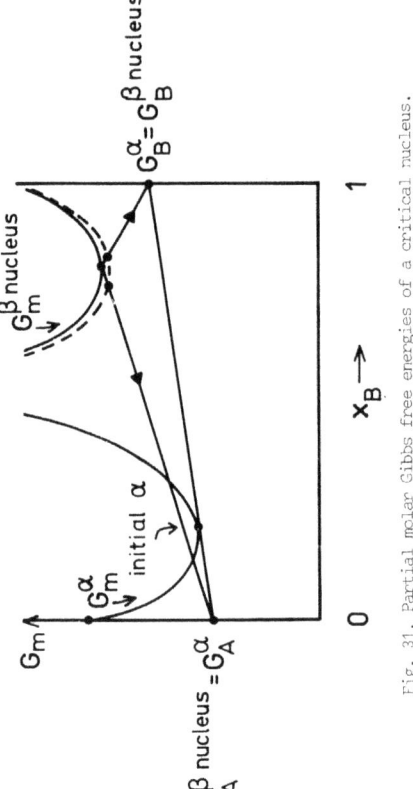

Fig. 31. Partial molar Gibbs free energies of a critical nucleus.

The value for a boundary in pure A, $^{o}G_A^b$, is higher than the value for the matrix in pure A, $^{o}G_A^{\alpha}$, by the amount given by the surface tension of grain boundaries in pure A, σ_A. By transforming to energy per mole we obtain

$$^{o}G_A^b - {}^{o}G_A^{\alpha} = \sigma_A V_m^b / t \tag{60}$$

where t is the thickness of the boundary.

For an alloy one must consider the equilibrium between the boundary phase and the α matrix. However, it must be remembered that the model assumes that the number of atoms in the boundary is constant. The usual construction of the common tangent for finding the equilibrium is thus too restrictive. It is sufficient to require that the Gibbs free energy should not change when an A atom is moved from the boundary and placed in the matrix and a B atom is moved in the opposite direction at the same time. The equilibrium condition is thus,

$$G_A^{\alpha} - G_A^b = G_B^{\alpha} - G_B^b \quad \text{or} \quad G_B^b - G_A^b = G_B^{\alpha} - G_A^{\alpha} \tag{61}$$

In view of Fig. 7 and eq. 10 one obtains

$$dG_m^b/dx_B = dG_m^{\alpha}/dx_B \tag{62}$$

The equilibrium concentration in the boundary is thus found by a parallel-tangent construction based on the composition of the matrix, Fig. 32. With the regular solution model, eq. 17, the equilibrium condition yields

$$RT\ln \frac{x_B^b x_A^{\alpha}}{x_A^b x_B^{\alpha}} = {}^{o}G_A^b - {}^{o}G_A^{\alpha} - {}^{o}G_B^b + {}^{o}G_B^{\alpha} + L^{\alpha}(1-2x_B) - L^b(1-2x_B^b) \tag{63}$$

The distance between $^{o}G_A^b$ and $^{o}G_A^{\alpha}$ on the left-hand axis in Fig. 32

represents the increase in Gibbs free energy if one mole of new boundary is created in a system of pure A in the state of α. It is given by eq. 60 and an equivalent expression holds for pure B. By inserting these expressions in eq. 63 we obtain for low B contents,

$$\ln x_B^b/x_B^\alpha \cong (\sigma_A - \sigma_B)V_m^b/RTt + (L^\alpha - L^b)/RT \tag{64}$$

For ordinary metals σ might be of the order of 1 J/m^2 and with V_m=7 cm^3/mole, t=10^{-7} cm and T=1000 K the first term on the right-hand side of eq. 64 is less than 1. It is evident that strong segregations (large x_B^b/x_B^α) must be due to the L term and in particular to a large negative value of L^b, i.e. a strong tendency for A and B atoms to mix in the boundary.

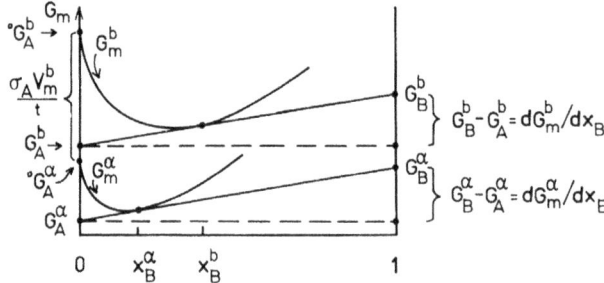

Fig. 32. Evaluation of the segregation to a boundary.

So far, we have not discussed the significance of the distance between the two tangents. It can be obtained from the length of the arrow in Fig. 33. By comparison with Fig. 3 it is evident that the arrow represents the increase in Gibbs free energy if one mole of new boundary with a composition x_B^b is created in a large system consisting of α phase with the composition x_B^α. The distance is thus $\sigma V_m^b/t$ where σ is the surface tension in a material where the matrix has the composition x_B^α.

40

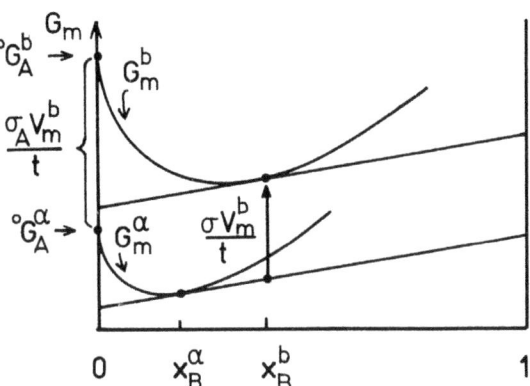

Fig. 33. Evaluation of the surface tension of a boundary in a binary alloy.

The shapes of the curves in Fig. 33 were chosen such that the parallel-tangent construction predicts a segregation of the alloying element B to the boundary. The diagram demonstrates that σ of the alloy is lower than the value σ_A, which holds for pure A, i.e. the surface tension is decreased by the addition of an element that segregates to the boundary. This fact can be described in mathematical terms. Consider a small increase of the matrix composition and make the parallel-tangent construction in Fig. 34 before and after the change. The change increases the value of G_B^α by dG_B^α and decreases the distance between the parallel tangents by $-d\sigma \cdot V_m/t$. The relation between these two changes is directly given by comparing triangles.

$$\frac{-d\sigma \cdot V_m^b/t}{x_A^\alpha - x_A^b} = \frac{dG_B^\alpha}{x_A^\alpha} \tag{65}$$

41

By rearranging and inserting x_B instead of x_A we find

$$-d\sigma = \frac{x_B^b - x_B^\alpha}{1 - x_B^\alpha} \cdot \frac{t}{V_m^b} \cdot dG_B^\alpha \tag{66}$$

This is the version of Gibbs adsorption equation for the present model. That equation is usually written $-d\sigma = \Gamma_{B(A)} \cdot d\mu_B$ where $d\mu_B$ is equal to dG_B^α and $\Gamma_{B(A)}$ stands for the following combination of parameters: $\Gamma_B - \Gamma_A x_B^\alpha / (1 - x_B^\alpha)$. The quantities Γ_B and Γ_A are the excess amounts of B and A per unit area. For the present model, where the total number of atoms in the boundary is constant, one obtains $-\Gamma_A = \Gamma_B = (x_B^b - x_B^\alpha) t / V_m^b$ and thus

$$\Gamma_{B(A)} = \frac{x_B^b - x_B}{1 - x_B^\alpha} \cdot \frac{t}{V_m^b} \tag{67}$$

in complete agreement with eq. 66.

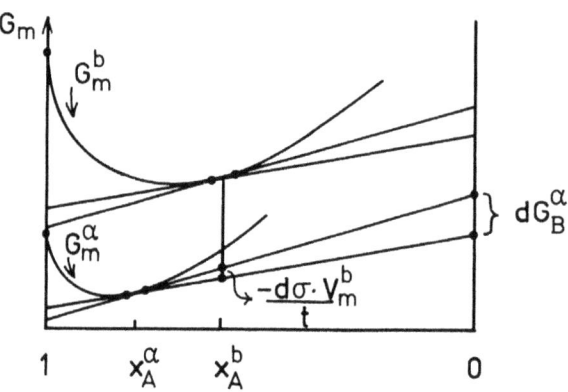

Fig. 34. Relation between the changes in surface tension and partial molar Gibbs free energy.

16 – "The Uses of Gibbs Free Energy-Composition Diagrams"

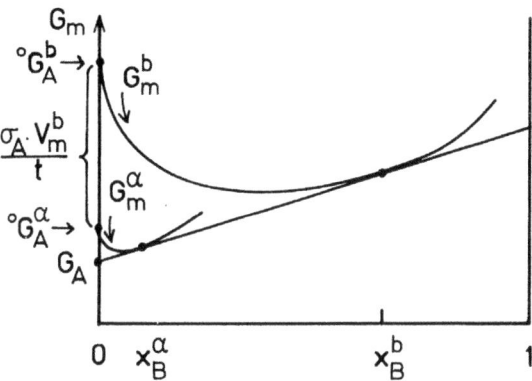

Fig. 35. Hypothetic case of vanishing surface tension.

The change of σ due to an appreciable addition of B can be calculated by integrating the Gibbs adsorption equation, but can also be obtained directly from the molar diagram. For example, suppose that one is interested to estimate the maximum segregation that can possibly occur to a grain boundary. It is certainly less than the segregation needed in order to make the surface tension vanish because that case has not yet been observed experimentally. An upper limit may thus be obtained by considering such a hypothetic case. It would occur if the distance between the two parallel tangents goes to zero, i.e. if the two parallel tangents coincide. Fig. 35 illustrates this situation. Let us limit the discussion to cases where the B content in the matrix is low. The equilibrium condition can then be written as follows if one applies eq. 18 according to the regular solution model.

$$°G_A^\alpha \cong G_A = °G_A^b + RT\ln(1-x_B^b) + L^b(x_B^b)^2 \tag{68}$$

It was concluded that L^b should have a large negative value if B segregates strongly. By omitting the L^b term and inserting $\sigma_A V_m^b/t$ instead of ${}^oG_A^b - {}^oG_A^\alpha$ in view of eq. 60 we thus obtain

$$-\ln(1-x_B^b) < \sigma_A V_m^b/RTt \tag{69}$$

Let us assume for the sake of argument that the boundary is very thick (t large). The right-hand side of eq. 69 is then small and the left-hand side can be represented by its first term in a series expansion,

$$x_B^b < \sigma_A V_m^b/RTt \tag{70}$$

This is equivalent to a layer of pure B of a thickness of $x_B^b \cdot t < \sigma_A V_m^b/RT$. With $\sigma_A = 1$ J/m², $V_m^b = 7$ cm³/mole and T=1000 K we find that the equivalent layer of pure B is less than 10Å thick. One may thus conclude that, no matter how thick a boundary may be, in order to increase segregation over that equivalent to a few monolayers, a much larger σ_A or a much lower temperature is required.

Dissipation of Gibbs Free Energy

Let us again consider the total driving force for precipitation of β from a supersaturated α phase, ΔG_m in Fig. 21. It would be completely dissipated during the reaction if the composition of the α matrix actually decreased to the equilibrium value ${}^ex_B^{\alpha/\beta}$, i.e. if all the supersaturation vanished during the reaction. Assuming that the composition of the α matrix at the β interface is always ${}^ex_B^{\alpha/\beta}$ during the reaction, one can conclude that all the extra B atoms diffuse down-hill in the α matrix and reach a region of the composition ${}^ex_B^{\alpha/\beta}$ before being incorporated in the growing β phase. All the driving force is then consumed by driving that diffusion. Of course, the dissipation of Gibbs free energy due to diffusion in α can only depend upon the changes within the α phase itself.

If, by some reason, the extra B atoms leave the α phase from a region of a different composition $x_B^{\alpha/\beta}$, the dissipation of Gibbs free energy due to diffusion in α can be evaluated from a similar construction based upon $x_B^{\alpha/\beta}$, Fig. 36. Evidently, the remaining part of ΔG_m must then be used for some other purpose. Such cases will now be discussed.

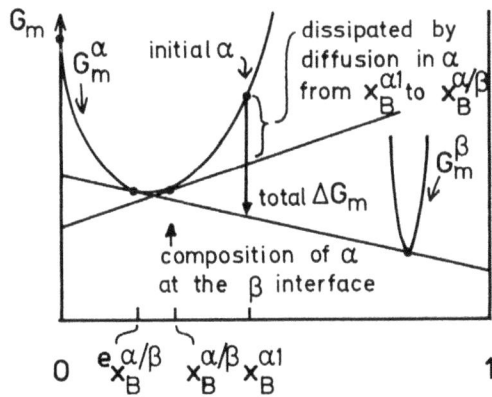

Fig. 36. Evaluation of Gibbs free energy dissipation due to diffusion.

Discontinuous precipitation takes place by a steady-state process. Alternating lamellae of the precipitate and of the depleted matrix grow side by side. The growth occurs at the edges of the lamellae and the shape stays constant. Knowing the curvature of the edges of the β lamellae, r^β, one knows the pressure under which the β phase is growing, $P^\beta = \sigma/r^\beta$. One can then evaluate the term $P^\beta V_m^\beta$ by which the G_m curve for β is raised, Fig. 37. The common-tangent construction shows that all the supersaturation cannot disappear by such a reaction. The composition of the α matrix in contact with the growing β phase, $x_B^{\alpha/\beta}$, differs from the equilibrium

45

composition $^e x_B^{\alpha/\beta}$. The part of the total driving force that is dissipated by diffusion is denoted by 1 in Fig. 34. In the depleted α matrix there still remains some supersaturation, $x_B^{\alpha/\beta} - ^e x_B^{\alpha/\beta}$, and a driving force for further precipitation of β which is denoted by 4 in Fig. 37. If evaluated per mole of the original alloy instead of per mole of the depleted α phase, this remaining part of the driving force corresponds to the distance denoted by 3 in Fig. 37. The part of the total driving force which is denoted by 2 remains to be discussed. It corresponds to the extra work done during the growth of the β phase under the pressure P^β and it is exactly equal to $P^\beta V_m^\beta$ if evaluated per mole of β that is formed instead of per mole of the original alloy. It may all be transformed into interfacial energy of the boundaries between the α and β lamellae or it

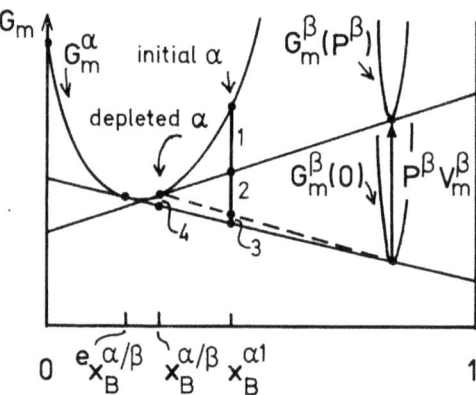

Fig. 37. Identification of different sinks for Gibbs free energy when a β phase is growing under a high pressure.

may partly be dissipated by some irreversible reaction.

In the next case we shall assume that the β phase can grow under ordinary pressure and we shall discuss what happens if the transfer of atoms across the α/β interface requires a driving force. Different cases can be imagined. If the A and B atoms move across the interface separately, it is necessary to have some driving force for the transfer of each element if the growing β phase should contain both. It then seems natural to apply a two-tangent construction such that the driving forces ΔG_A and ΔG_B yield balanced amounts of A and B in the growing β phase, considering the individual mobilities of A and B in the interface. Fig. 38 illustrates such a case and it is evident that the α phase at the interface cannot have the equilibrium composition. Some supersaturation remains and all the driving force available in the initial α phase cannot be used for diffusion. It is again divided into three parts. As in the previous case, the distance 1 represents the loss due to diffusion and the distance 3 represents the driving force remaining in the depleted matrix. The distance 2 now represents the loss due to the interface reaction. The relative sizes of part 1 and part 2 must be such that the diffusional flow of atoms in α down to the interface keeps pace with the interface reaction.

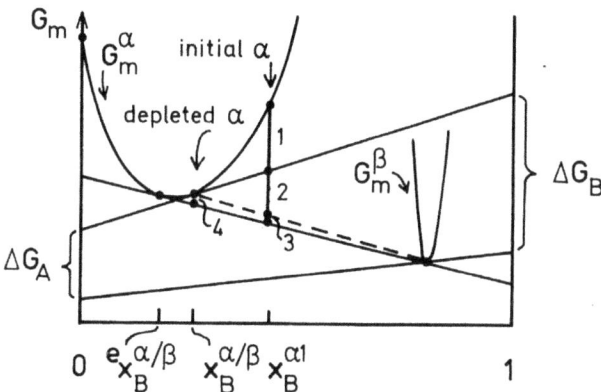

Fig. 38. Identification of different sinks for Gibbs free energy during interface-controlled growth of a β phase.

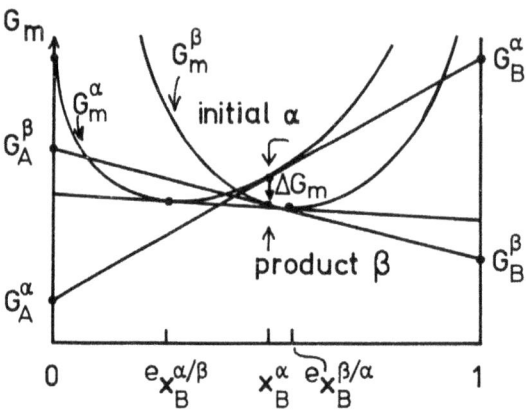

Fig. 39. Evaluation of the driving force for a diffusionless transformation of α to β.

In many systems the α phase may have a higher Gibbs free energy than β phase of the same composition, provided that the supersaturation of α is high enough. This condition occurs below the line of equal Gibbs free energy in the phase diagram, sometimes called the T_o-line or the allotropic phase boundary. At such a composition there is a driving force for a diffusionless transformation α→β and it is interesting to ask what factor could hold such a reaction back. Fig. 39 illustrates such a case and it demonstrates that G_A is lower in the initial α phase than in the product β phase. Under the conditions described by this diagram, β cannot grow if its growth depends upon the individual transfer of A and B atoms across the interface. However, the β phase could grow if the two kinds of atoms cross the interface by some cooperative process. This is probably what happens in a martensitic transformation and it may be due to the special type of interface that a martensitic transformation makes use of. When examining, from the thermodynamic point of view, whether a martensitic transformation could proceed, it is sufficient to evaluate the magnitude of ΔG_m and to examine if it is large enough to overcome various factors such as mechanical stresses.

There is a massive type of transformation which makes use of an incoherent interface and allows an initial α phase to transform into a β phase of the same composition. Such a transformation has often been observed inside the one-phase region of the β phase. Fig. 40 illustrates such a case and it is shown that it can be explained as a reaction by which the A and B atoms cross the interface individually but, in order to obtain a positive value of ΔG_A as well as of ΔG_B, it is then necessary to assume that some diffusion takes place in the α matrix just ahead of the migrating interface. The composition at the interface is thus changed to a point, $x_B^{\alpha/\beta}$, such that the α tangent is now completely above the β tangent. It is easy to see that this construction is possible only if the β tangent does not intersect the G_m^α curve, i.e. if the alloy composition falls within the one-phase region of β. The diagram also illustrates what part of ΔG_m is dissipated by diffusion in α, part 1, and how much is left to drive the atoms across the interface and thus make the interface migrate, part 2. The relative sizes of the two parts must be such that the two reactions keep pace with each other.

Fig. 40 thus illustrates a possible mechanism for massive growth by individual transfer of atoms across the interface. However, it should be realized that the construction may not apply to a rapidly moving interface. It is quite possible that less free energy will then dissipate by diffusion in the α phase and the massive transformation may occur inside the two-phase region. In order to understand such cases it seems necessary to have detailed knowledge about the reactions inside the incoherent interface.

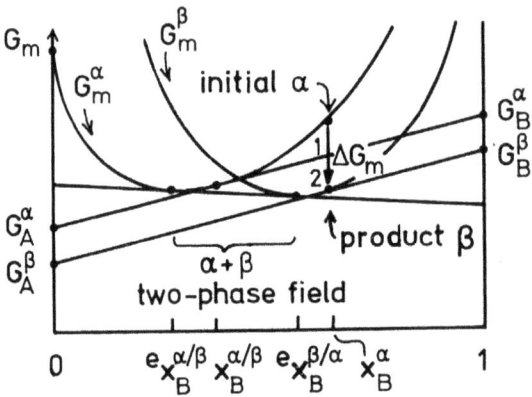

Fig. 40. Demonstration of the conditions for a massive transformation of α to β when the two kinds of atoms are transferred individually.

17 Introduction to "A Treatment of the Solute Drag on Moving Grain Boundaries and Phase Interfaces in Binary Alloys"

published in *Acta Metallurgica* (1976)

Hillert and Sundman [1] proposed a general model for solute drag on moving grain and interphase boundaries that, along with their subsequent solidification model [2], set the stage for work that followed and gave us a unified picture of the solute drag phenomenon. It followed theoretical treatments of dilute solute drag on grain boundary migration [3, 4], and developed in parallel with theoretical treatments of solute drag focused specifically on diffuse, coherent interfaces [5, 6].

Hillert and Sundman presented a general expression, applicable to a diffuse interface in a binary alloy comprised of solvent ('A') and solute ('B'), for the force P_B that the solute exerts upon a unit area of interface for an arbitrary profile of solute adsorption free energy. In one of its equivalent forms they wrote it as:

$$P_B = -\frac{1}{v} \int_{-\infty}^{\infty} J_B(y) \cdot \frac{d}{dy}(G_B(y) - G_A(y))dy \tag{12a}$$

where v is the interface velocity, J_B is the interdiffusion flux of B relative to A, y is the spatial coordinate in the direction normal to the interface, and G_B and G_A are the chemical potentials of B and A, respectively.

They presented a model for the profile across the interface of the standard free energy of adsorption, $^0G_B - {}^0G_A$, that can be applied broadly to a wide variety of intergrain and interphase boundaries. They advanced the correspondence between force per unit area of interface and the dissipation of free energy per mole of material transformed, ΔG_m. They showed how the free energy dissipation rate varies from zone to zone across an interface, depending on y-dependent profiles of $^0G_B - {}^0G_A$ and of the interdiffusivity. And they unified the previous pictures of solute drag under this single framework.

Hillert and Sundman proposed an interdiffusivity profile $D(y)$ that varies exponentially with position, *i.e.* log D linear in y. I believe this profile to be the key to understanding when solute drag effects are substantial or negligible, as discussed below.

The Hillert-Sundman model inspired, in works reviewed by Hillert [7], the generalization of the solute drag treatment to include the force P_{TOT} due to both solute and solvent on the interface. This force can be written [8]:

$$P_{TOT} = -\frac{\Delta G_m}{V_m} + \frac{1}{v} \int_0^\delta J_B(y) \cdot \frac{d}{dy}(G_B(y) - G_A(y))dy \tag{31}$$

where the limits of integration are the boundaries of the interface region. This "dissipation integral" represents the free energy dissipated by interdiffusion within the interface.

The question has arisen, "how is the interface velocity determined by these forces?" Our experiments [9] in rapid solidification, in which we measured an interface velocity-undercooling function in Si-As alloys with virtually the same slope as in pure Si, are consistent with $v \propto \Delta G_m$. Although there is disagreement on whether $v \propto \Delta G_m$ is permissible based on theoretical considerations [7, 9, 10], I think everyone agrees that $v \propto P_{TOT}$ with the dissipation integral on the r.h.s of (31) being negligible for solidification is a plausible explanation of the solidification experiments.

This is currently the only plausible qualitative explanation of which I am aware of the observations that solute drag effects are significant in the migration of grain [11] and anti-phase [6] boundaries, but negligible in solidification. Let us see how this explanation [9] arises naturally out of the Hillert-Sundman interface model of an exponentially varying interdiffusivity within an interphase boundary between two phases with vastly different interdiffusivities. In steady-state solidification the first factor in the dissipation integral in (31), J_B, is the interdiffusion flux; the second factor, $\frac{d}{dy}(G_B - G_A)$, is the thermodynamic driving force for interdiffusion. When either factor approaches zero, dissipation is insignificant. A steep interdiffusivity gradient implies a narrow dissipative zone within a moving interface. In the region of the interface between the dissipative zone and the solid, the mobility is so low that the interdiffusion flux is negligible and the first factor kills the integrand. On the other side, between the dissipative zone and the liquid, the mobility is so high that chemical potential gradients are negligible and the second factor kills the integrand. The steeper the gradient in log D, the narrower the dissipative zone, and the smaller is the contribution of the dissipation integral to the r.h.s. of (31). These effects have been worked out quantitatively for the Hillert-Sundman model for solidification of Si-As [9].

For grain and anti-phase boundaries, two effects within the Hillert-Sundman model seem to be responsible for the observation of significant solute drag effects. First, the mobility for interdiffusion may not vary across the interface as markedly as for the crystal/melt interface, thereby leading to a wider dissipative zone than for the crystal/melt interface. This is likely to be the case for anti-phase boundaries. Second, the driving free energy for the transformation, ΔG_m, may be so small that the dissipation integral, though small itself, is not negligible by comparison. This is likely to be the case for both grain and anti-phase boundaries.

I am grateful to Professor Hillert for the pleasant and stimulating discussions that we have had on the subject of solidification and for his profound influence on my thinking about phase transformations, and I commend him on this celebratory occasion.

Michael J. AZIZ

References

[1] M. Hillert and B. Sundman, *Acta Metall.* 24 (1976) 731.

[2] M. Hillert and B. Sundman, *Acta Metall.* 25 (1977) 11.

[3] J.W. Cahn, *Acta Metall.* 10 (1962) 789.

[4] K. Lucke and H. Stuwe, *Acta Metall.* 19 (1971) 1087.

[5] J.S. Langer and R.F. Sekerka, *Acta Metall.* 23 (1975) 1225.

[6] J.E. Krzanowski and S.M. Allen, *Acta Metall.* 31 (1983) 213.

[7] M. Hillert, *Acta Materialia* 47 (1999) 4481.

[8] M.J. Aziz and T. Kaplan, *Acta Metall.* 36 (1988) 2335.

[9] J.A. Kittl, P.G. Sanders, M.J. Aziz, D.P. Brunco and M.O. Thompson, *Acta Mater.* 48 (2000) 4797.

[10] T. Kaplan, M.J. Aziz, and L.J. Gray, *J. Chem. Phys.* 99 (1993) 8031.

[11] J.P. Drolet and A. Galibois, *Acta Metall.* 16 (1968) 1387.

A TREATMENT OF THE SOLUTE DRAG ON MOVING GRAIN BOUNDARIES AND PHASE INTERFACES IN BINARY ALLOYS

MATS HILLERT and BO SUNDMAN

Division of Physical Metallurgy, Royal Institute of Technology, Stockholm 70, Sweden

(Received 25 September 1975)

Abstract—A previous treatment of the solute drag effect on the movement of grain boundaries and phase interfaces, which was based upon the evaluation of the free-energy dissipation, is developed further. The new treatment becomes identical to the treatments by Cahn and Lücke and Stüwe for grain boundaries in dilute solutions but is not limited to low solute contents, to ideal solutions or to single-phase systems.

Some numerical calculations for grain boundaries are presented and the dissipation of free energy in different zones of the interface is discussed. The maximum solute drag, calculated with a variable diffusivity, is always lower than the value found by using any average value of the diffusivity. The simple expression for the estimation of the solute drag, previously given, is approximately correct at constant diffusivity, only. A number of calcultions for diffusionless phase transformations are presented and allow some conclusions regarding the spontaneous start of such a transformation.

Résumé—On continue de développer un traitement antérieur de l'effet du freinage par le soluté du mouvement des joints de grains et des interfaces entre phases, traitement basé sur l'évaluation de l'énergie libre dissipée. Ce nouveau traitement est identique à ceux de Cahn, et de Lücke et Stüwe pour les joints de grains dans les solutions diluées, mais il n'est limité ni aux faibles teneurs en soluté, ni aux solutions idéales, ni aux systèmes monophasés.

On présente des calculs numériques pour les joints de grains, et on discute la dissipation de l'énergie libre dans différentes zones de l'interface. Le freinage maximal par le soluté, calculé avec une diffusivité variable, est toujours inférieur à la valeur que l'on trouve en utilisant n'importe quelle valeur moyenne de la diffusivité. L'expression simple que l'on avait donnée antérieurement pour estimer le freinage n'est à peu près correcte que dans le cas d'une diffusivité constante. On présente pour les transformations de phases sans diffusion, des calculs qui permettent de tirer quelques conclusions concernant le début spontané.

Zusammenfassung—Es wird eine frühere Behandlung des Lösungseinflusses auf die Bewegung von Korngrenzen und Phasengrenzflächen, die auf der Bestimmung der Dissipation freier Energie aufbaute, weiter entwickelt. Die neue Behandlung wird identisch mit denjenigen von Cahn und Lücke und Stüwe für Korngrenzen in verdünnten Lösungen, ist aber nicht auf niedrige Lösungsgehalte, auf ideale Lösungen oder auf einphasige Systeme beschränkt.

Einige numerische Rechnungen für Korngrenzen werden vorgelegt; die Dissipation der freien Energie in verschiedenen Zonen der Grenzfläche wird diskutiert. Die maximale lösungsinduzierte Behinderung, berechnet mit einer variablen Diffusivität, ist immer niedriger als der Wert, der mit irgendeinem mittleren Wert für die Diffusivität erhalten wird. Der einfache, früher angegebene Ausdruck zur Abschätzung des Lösungseinflusses ist annähernd richtig nur bei konstanter Diffusivität. Vorgelegt wird eine Anzahl von Rechnungen über diffusionslose Phasentransformationen, welche einige Schlüsse im Hinblick auf den spontanen Einsatz einer solchen Tranformation gestatten.

INTRODUCTION

In their studies of the effect of alloy additions on the recrystallization of aluminium, Lücke *et al.* [1, 2] found that the rate of recrystallization could decrease by many orders of magnitude without exceeding the solubility limit for the alloy addition. They concluded that the effect was caused by the presence of the foreign atoms in solid solution and was due to a direct interaction between moving grain boundaries and the foreign atoms in solution. They suggested that the effect was related to a tendency of the foreign atoms to segregate to the grain boundaires and they attempted to calculate this tendency, assuming that a foreign atom is attracted to a grain boundary because it will there cause less elastic stresses than in the interior of a crystal. Lücke and Detert [3] developed the model in more detail. They supposed that foreign atoms, segregated to a boundary, will be left behind if the boundary starts to move. The boundary will then exert an attraction force on them which makes them diffuse in the same direction as the boundary moves. The average drift velocity of atoms can be obtained from

$$v = \frac{D}{kT} \cdot f, \quad (1)$$

where f is the average value of the attraction force. The movement of the boundary will be held back by the same attraction force. The solute drag on the boundary, P, can thus be evaluated as the sum of the attraction force on all the foreign atoms lagging behind the migrating boundary.

$$P = n \cdot f \qquad (2)$$

n is the number of segregated, foreign atoms per area of the boundary. Lücke and Detert did not estimate how the number of foreign atoms per area depends upon the velocity of the boundary but pointed out that it should fall drastically if the velocity is larger than the drift velocity of atoms obtained from equation (1) when the largest possible value of f is inserted. For lower velocities they eliminated f by combining the two equations and thus they obtained a relation between the solute drag and the velocity for the case where the velocity is low enough to permit the majority of the segregated foreign atoms to diffuse with the boundary,

$$P = \frac{nkTv}{D}. \qquad (3)$$

Cahn [4] and Lücke and Stüwe [5, 6] developed the model further by considering how the interaction energy between the foreign atoms and the boundary, E, varies with the distance y of an atom from the middle of the boundary. The interaction force is $-dE/dy$ and the total drag is obtained by integrating over all the atoms. Lücke and Stüwe thus obtained the following equation

$$P = -\int_{-\infty}^{\infty} \frac{x}{V_m} \cdot \frac{dE}{dy} \cdot dy \qquad (4)$$

V_m is the molar volume and x is the mole fraction of the solute.

Cahn chose to add a term and wrote

$$P = -\int_{-\infty}^{\infty} \frac{x - x^0}{V_m} \cdot \frac{dE}{dy} \cdot dy. \qquad (5)$$

This is permitted for a single-phase material where

$$\int_{-\infty}^{\infty} \frac{dE}{dy} dy = \int_{-\infty}^{\infty} dE = E(\infty) - E(-\infty) = 0, \qquad (6)$$

x^0 is the constant composition of the material sufficiently far away from the boundary.

The direction of the force depends upon the sign of dE/dy. Forces acting in opposite directions will exactly cancel in a stationary boundary. For a moving boundary, the distribution of the solute will be changed and there will be a net force. In order to evaluate the net force, one must first calculate how the solute mole fraction, x, varies across the boundary. For any rate of migration, this can in principle be done from information on how E varies across the boundary. Using a diffusion equation, related to equation (1), Cahn [4] and Lücke and Stüwe [5, 6] carried out this calculation for low solute contents and Cahn was able to obtain analytical expressions for the solute drag at very low and very high velocities. He also carried out some numerical calculations for intermediate velocities. A characteristic feature of the results was that the solute drag reached a maximum at an intermediate velocity and decreased towards zero at high velocities.

A different approach to the solute drag problem was taken by Hillert [7] who argued that the work put into the movement of a boundary in order to overcome the solute drag, must dissipate by the diffusion of the solute taking place as a result of the boundary movement. It should thus be possible to evaluate P as $\Delta G_m/V_m$ where ΔG_m is the dissipation of free energy due to diffusion when the boundary passes through a volume containing one mole of atoms.

In principle, both treatments could apply to any type of function describing the variation of E through the boundary but, quite naturally, the authors chose functions which would simplify their own calculations. As a consequence, Cahn [4] and Lücke and Stüwe [5] chose a wedge-shaped function which gave a constant value of dE/dy which changed sign in the middle of the boundary. On the other hand, Hillert [7] chose a square well function yielding constant properties in the boundary and abrupt changes at the two sides of the boundary. It is interesting that Hillert did not find a maximum of the solute drag but a monotonous increase towards an asymptotic value at increasing velocities. He then proposed that one should take into account the atomistic nature of the system and showed that one could thus obtain the same type of variation with velocity for the square well function that Cahn and Lücke and Stüwe had found for the wedge-shaped function without considering the atomic nature.

Recently, Lücke and Stüwe [6] have developed an atomistic model of grain boundary mobility in a binary alloy, which yields results similar to their previous treatment.

The purpose of the present work was to develop Hillert's treatment further. It will be demonstrated that it reduces to the treatment of Cahn [4] and Lücke and Stüwe [5] for dilute solutions. It will be applied to various cases of general interest including phase transformations. Although there have been several suggestions of a solute drag effect in phase transformations [7–10] there have previously been no serious attempts to carry out detailed calculations.

FORMULATION OF NEW MODEL

The treatment will be limited to binary systems where the two components have the same mobility. The driving force for interdiffusion is $d(G_B - G_A)/dy$ and the fluxes can be obtained from the absolute reaction rate theory,

$$-J_A = J_B = -\frac{D}{RTV_m} x_A x_B \frac{d(G_B - G_A)}{dy}. \qquad (7)$$

It is well known that the rate of the free energy dissipation due to diffusion is given by the product of the flux and the driving force

$$\frac{dG}{dt} = -\int_V J_B \cdot \frac{d(G_B - G_A)}{dy} \cdot dV. \tag{8}$$

This equation will be applied to an infinitely long volume element with a cross section A, which is parallel to a boundary, migrating with a constant velocity v. Let y be the distance of any point from the boundary and let us calculate the free energy dissipation in the volume element during a time period Δt when the boundary passes 1 mole of material. We shall denote this free energy by ΔG_m.

$$\Delta t = V_m/Av \tag{9}$$

$$dV = A\,dy \tag{10}$$

$$\Delta G_m = (dG/dt)\cdot \Delta t = -\frac{V_m}{v}\int_{-\infty}^{+\infty} J_B \cdot \frac{d(G_B - G_A)}{dy}\cdot dy. \tag{11}$$

According to Hillert's suggestion, the solute drag is evaluated by dividing this ΔG_m with V_m. By the use of equation (7) the result can be expressed in three different ways.

$$P = -\frac{1}{v}\int_{-\infty}^{\infty} J_B \cdot \frac{d(G_B - G_A)}{dy}\cdot dy \tag{12a}$$

$$P = \frac{1}{vRTV_m}\int_{-\infty}^{\infty} Dx_A x_B \left[\frac{d(G_B - G_A)}{dy}\right]^2 \cdot dy \tag{12b}$$

$$P = \frac{RTV_m}{v}\int_{-\infty}^{\infty} \frac{J_B^2}{Dx_A x_B}\cdot dy. \tag{12c}$$

It should be noticed that Hillert's suggestion can hold only in a stationary state. If there is a change of the concentration profile during a transient period, the corresponding increase of the free energy of the system must also be taken into account.

For a stationary state, one can relate the local value of the flux to the local composition because of the following relation

$$\frac{\partial x_B}{\partial t} = -v\frac{\partial x_B}{\partial y}. \tag{13}$$

If there is no side-ways diffusion, Ficks' second law yields for the interdiffusion of A and B,

$$-\frac{\partial x_A}{\partial t} = \frac{\partial x_B}{\partial t} = -V_m \frac{\partial J_B}{\partial y} = \frac{\partial x_B}{\partial y}. \tag{14}$$

We thus have

$$-\frac{dJ_A}{dx_A} = \frac{dJ_B}{dx_B} = \frac{v}{V_m} \tag{15}$$

or by integration

$$-J_A = J_B = \frac{v}{V_m}(x_B - x_B^0) = -\frac{v}{V_m}(x_A - x_A^0). \tag{16}$$

x_B^0 is an integration constant which can be identified with the alloy composition far away from the boundary because the diffusion flux must be zero there. By using equation (16), two additional ways of expressing the solute drag are obtained,

$$P = -\int_{-\infty}^{\infty} \frac{x_B - x_B^0}{V_m}\cdot \frac{d(G_B - G_A)}{dy}\cdot dy \tag{12d}$$

$$P = \frac{RTv}{V_m}\int_{-\infty}^{\infty} \frac{(x_B - x_B^0)^2}{Dx_A x_B}\cdot dy. \tag{12e}$$

It is evident, particularly from equations (12b, c and e), that the integrand is everywhere positive which is natural since it represents the dissipation of free energy by diffusion.

In order to carry out numerical calculations one must choose a thermodynamic model for the system. For instance, Cahn [4] considered a dilute solution for which he wrote, according to Henry's law

$$G_B = E(y) + RT\ln x_B. \tag{17}$$

The dependency on position and composition were thus supposed to come through separate terms. We should prefer to denote $E(y)$ by $^0G_B(y)$ because it is a free energy rather than an internal energy. For the other component one obtains from Raoult's law,

$$G_A = {}^0G_A(y) + RT\ln x_A. \tag{18}$$

For a dilute or ideal solution one thus obtains

$$\frac{d(G_B - G_A)}{dy} = \frac{d(^0G_B - {}^0G_A)}{dy} + \frac{RT}{x_A x_B}\cdot \frac{dx_B}{dy}. \tag{19}$$

Let us denote $^0G_B - {}^0G_A$ by $\Delta^0 G$. By inserting equation (19) into (12d) the following expression is obtained for the solute drag,

$$P = -\int_{-\infty}^{\infty} \frac{x_B - x_B^0}{V_m}\cdot \frac{d\Delta^0 G}{dy}\cdot dy \tag{20}$$

because the contribution from the second term in equation (19),

$$-\int_{-\infty}^{\infty} \frac{RT}{V_m}\cdot \frac{x_B - x_B^0}{x_A x_B}\cdot \frac{dx_B}{dy}\cdot dy \tag{21}$$

vanishes when the composition has the same value in the two limits of integration and this is true in the case under consideration. It should be noticed that this term vanishes not only for a dilute or ideal solution but for all cases where the dependency of $G_B - G_A$ on position and composition come through separate terms.

Although the term defined by equation (21) vanishes, its integrand is generally not zero but has a positive or negative value depending upon the sign of $x_B - x_B^0$. As a consequence, the remaining integrand in equation (20) is not necessarily positive everywhere in spite of the great similarity with equation (12d). It is thus evident that the integrand in equation (20) does not represent the dissipation of free energy. Instead it seems to represent the local contributions to the drag because it is identical to equation (5) which was postulated by Cahn [4], except that it contains the difference $\Delta^0 G = {}^0G_B - {}^0G_A$ instead of E.

As a further demonstration of the fact that the new treatment becomes identical with the treatment by Cahn [4] and Lücke and Stüwe [5] for dilute solutions, one may compare the expressions obtained by Cahn for the limiting case of very high and very low velocities with the predictions obtained for the same cases from the present treatment. At high velocities it is convenient to apply equation (12b) and one may neglect the movement of the atoms when evaluating $d(G_B - G_A)/dy$. Since the composition is then everywhere close to x_B^0 one finds $dx_B/dy \cong 0$ and any thermodynamic model yields

$$\frac{d(G_B - G_A)}{dy} = \frac{d\Delta^0 G}{dy}. \quad (22)$$

Equation (12b) can thus be approximated by the following equation for high velocities

$$P = \frac{x_A^0 x_B^0}{vRTV_m} \int_{-x}^{+x} D\left[\frac{d\Delta^0 G}{dy}\right]^2 dy. \quad (23)$$

This is identical to Cahn's equation for the same case but it is now evident that it is not limited to dilute solutions but has a much wider application.

For low velocities it is convenient to apply equation (12c). x_B can be approximated by the equilibrium distribution which is governed by $d(G_B - G_A)/dy = 0$ in view of equation (7). The thermodynamic model defined by equations (17 and 18) yields

$$\frac{x_B}{x_A} = \frac{x_B^0}{x_A^0} \cdot \exp -\Delta^0 G/RT. \quad (24)$$

By noting that $x_B - x_B^0$ is equal to $x_B x_A^0 - x_A x_B^0$ one can insert equation (24) into (12c). We thus obtain for very low velocities

$$P = \frac{RTv}{V_m} x_A^0 x_B^0 \int_{-x}^{x} \frac{1}{D}$$
$$\cdot [\exp \Delta^0 G/2RT - \exp -\Delta^0 G/2RT]^2 \, dy. \quad (25)$$

This is identical to Cahn's equation for the same case. Cahn pointed out that equation (25) does not depend upon the sign of $\Delta^0 G$ and it thus predicts the same drag for solutes which avoid the boundary as for solutes which segregate to the boundary.

The present treatment has several advantages. (1) It can easily be applied to the whole range of compositions in a binary system and it can handle any model for the thermodynamic properties of the system. (2) It can be applied to phase transformations as well as to the migration of a grain boundary in a single-phase material. (3) It allows comparatively accurate numerical calculations of the total magnitude of the solute drag because the integrand is everywhere positive. (4) It provides information on where the free energy dissipates in the boundary.

METHOD OF CALCULATION

For the present purposes it is sufficient to apply the regular solution model,

$$G_m = x_A^0 G_A + x_B^0 G_B + RT[x_A \ln x_A + x_B \ln x_B]$$
$$+ x_A x_B L \quad (26)$$

and ${}^0G_A$, ${}^0G_B$ and L will be assumed to vary with the position relative to the boundary, y. It yields the following expression for the driving force for interdiffusion

$$\frac{d(G_B - G_A)}{dy} = \frac{d\Delta^0 G}{dy} + (x_A - x_B)\frac{dL}{dy}$$
$$+ \left(\frac{RT}{x_A x_B} - 2L\right)\frac{dx_B}{dy}. \quad (27)$$

By combination with equations (7 and 16) one obtains the following expression which can be used for calculations of the concentration profile

$$\frac{dx_B}{dy} = \left[-\frac{d\Delta^0 G}{dy} - (x_A - x_B)\frac{dL}{dy} - \frac{RTv}{D} \cdot \frac{x_B - x_B^0}{x_A x_B}\right] /$$
$$\left[\frac{RT}{x_A x_B} - 2L\right]. \quad (28)$$

The solute drag can then be calculated most easily from equation (12e). In the general case, this calculation can only be solved numerically. The numerical calculation must start at the point behind the migrating boundary where G_m is supposed to start to vary with y and the composition must be x_B^0 up to that point. The change in composition is then calculated step by step in the y-direction using equation (28) and the integrand in equation (12e) is evaluated at the same time. This procedure was programmed for computer calculations and Fig. 1 gives a typical example. The L parameter was here assumed to be zero everywhere and the diffusion coefficient D was assumed to be a constant. The top picture of Fig. 1 shows the $\Delta^0 G$ profile through the boundary. It has a constant value in the two grains, $\Delta^0 G^\circ$, and a lower constant value in the central zone 2 of the boundary, $\Delta^0 G^i$. It varies linearly in the two side zones of the boundary, zones 1 and 3. The middle picture shows that the solute distribution at a low rate of migration, curve 1, is close to the equilibrium segregation. As the rate is increased, the total amount of segregation is gradually lowered, curves 2 and 3, and a negative spike is developed in the grain in front of the boundary, zone 4. This spike goes through a maximum at some rate, curve 2, and then decreases, curve 3. The bottom picture shows the integrand in the calculation of the drag, dP/dy, as evaluated from $(x_B - x_B^0)^2/x_A x_B$ according to equation (12e) at constant D. In order to compare the different curves, one should multiply them with their migration rates, v. As a consequence, the low-velocity curve, curve 1, represents a very small drag and it is evident that the drag originates in the central zone of the boundary. As the velocity increases, the contributions from the two side zones of the boundary and from the spike in the grain ahead of the boundary grow in importance, curve 2. At even higher velocities, the contribution from the spike decreases and only the two side zones are important. These results are equivalent to those reported by Cahn.

In the following discussions diagrams will be presented, showing the variation of the drag with the rate of migration or the alloy composition. From the practical point of view, it is then impossible to present curves for dP/dy. Instead, the diagrams will show the total drag obtained by integration of dP/dy over the whole boundary and in some cases the integrated contributions from each one of the four zones defined at the top of Fig. 1. It should be noticed that the contribution from the region lying in front of the boundary, zone 4, can be obtained by direct calculation of $x_A^0[G_A(\infty) - G_A(\delta/2)] + x_B^0[G_B(\infty) - G_B(\delta/2)]$ as soon as the composition at $y = \delta/2$ has been found. With the regular solution model one obtains the following expression for this contribution

$$\frac{V_m}{RT}\int_{\delta/2}^{\infty}\frac{dP}{dy}dy = x_A^0 \ln\frac{x_A^0}{^3x_A} + x_B^0 \ln\frac{x_B^0}{^3x_B} - \frac{L}{RT}(^3x_B - x_B^0)^2, \quad (29)$$

where ^3x_A, ^3x_B denotes the composition at $y = \delta/2$, i.e. at the point of discontinuity on the right hand side of the region concerned. The same procedure could be used for any part of the boundary where the properties stay constant, for instance in the central zone of the boundary described in Fig. 1. This is how Hillert [7] calculated the dissipation of free energy and why he chose to consider a boundary with uniform properties.

The calculation can be carried out analytically under some other conditions as well. A comparatively simple case is found when D and $d\Delta^0 G/dy$ are constant and $L = 0$. Equation (28) can then be rewritten in the following form by introducing a dimensionless coordinate $z = y/\delta$ where δ is the total width of the boundary.

$$\frac{dx_B}{dz} = -\frac{1}{RT}\frac{d\Delta^0 G}{dz}x_A x_B - \frac{v\delta}{D}(x_B - x_B^0). \quad (30)$$

This equation has the solution

$$x_B = a + b[1 + c\exp(m(z - z_0))]^{-1}, \quad (31)$$

where

$$a = \frac{RT}{2d\Delta^0 G/dz}\left(\frac{v\delta}{D} + \frac{1}{RT}\frac{d\Delta^0 G}{dz} - m\right) \quad (32a)$$

$$b = \frac{RTm}{d\Delta^0 G/dz} \quad (32b)$$

$$m = \left[\left(\frac{v\delta}{D} + \frac{1}{RT}\frac{d\Delta^0 G}{dz}\right)^2 - \frac{4v\delta}{DRT}\frac{d\Delta^0 G}{dz}x_B^0\right]^{1/2} \quad (32c)$$

c is the integration constant. This expression of x_B can be inserted into equation (12e) yielding the solute drag. The integration of equation (12e) must be carried out in several steps where each step is limited to a region where D and $d\Delta^0 G/dz$ have constant values. The model defined at the top of Fig. 1 has four such regions. The expression thus obtained for the solute drag is

$$\frac{PV_m}{RT} = -\frac{v\delta RT}{D|d\Delta^0 G/dz|}\left[\frac{(x_B^0)^2}{a(a+b)}\ln\frac{^1x_B}{x_B^0}\right.$$
$$-\frac{(x_A^0)^2}{(1-a)(1-a-b)}\ln\frac{^1x_A}{x_A^0}$$
$$-\frac{(a-x_B^0)^2}{ab(1-a)}\ln\left|\frac{^1x_B - a}{x_B^0 - a}\right|$$
$$\left.+\frac{(a+b-x_B^0)^2}{b(a+b)(1-a-b)}\ln\left|\frac{^1x_B - a - b}{x_B^0 - a - b}\right|\right]$$
$$+\left[x_A^0\ln\frac{^2x_A}{^1x_A} + x_B^0\ln\frac{^2x_B}{^1x_B}\right]$$
$$+\frac{v\delta RT}{D|d\Delta^0 G/dz|}\left[\frac{(x_B^0)^2}{a'(a'+b')}\ln\frac{^3x_B}{^2x_B}\right.$$
$$-\frac{(x_A^0)^2}{(1-a')(1-a'-b')}\ln\frac{^3x_A}{^2x_A}$$
$$-\frac{(a'-x_B^0)^2}{a'b'(1-a')}\ln\left|\frac{^3x_B - a'}{^2x_B - a'}\right|$$
$$\left.+\frac{(a'+b'-x_B^0)^2}{b'(a'+b')(1-a'-b')}\ln\left|\frac{^3x_B - a' - b'}{^2x_B - a' - b'}\right|\right]$$
$$+\left[x_A^0\ln\frac{x_A^0}{^3x_A} + x_B^0\ln\frac{x_B^0}{^3x_B}\right]. \quad (33)$$

The four terms on the right-hand side originate from the four regions. The first and third terms originate from the two sides where $d\Delta^0 G/dz$ is not zero. The quantities a, a', b and b' are all defined by equations (32a and b), respectively, and have different values because the values of $d\Delta^0 G/dz$ have different signs. ^1x_B, ^2x_B and ^3x_B are the concentrations of solute at the points of discontinuity between regions 1 and 2, regions 2 and 3 and regions 3 and 4, respectively. Their values are

$$^1x_B = a + b\left[\left(1 + \left(\frac{b}{x_B^0 - a} - 1\right)\exp\left(\frac{m}{4}\right)\right)\right]^{-1} \quad (34a)$$

$$^2x_B = x_B^0 + (^1x_B - x_B^0)\exp\left(-\frac{v\delta}{2D}\right) \quad (34b)$$

$$^3x_B = a' + b'\left[1 + \left(\frac{b'}{^2x_B - a'} - 1\right)\exp\left(\frac{m'}{4}\right)\right]^{-1}. \quad (34c)$$

736   HILLERT and SUNDMAN: A TREATMENT OF THE SOLUTE DRAG

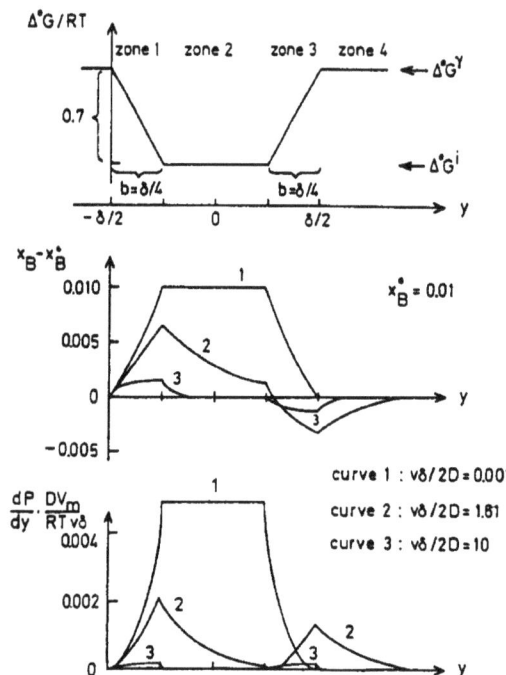

Fig. 1. Top diagram: A grain boundary model with four different zones. Middle diagram: Concentration profiles calculated for three migration rates. Bottom diagram: the integrand in the calculation of solute drag from the concentration profiles.

The second and fourth terms are simpler because they originate from regions where $d\Delta^0 G/dz = 0$. They can be compared with equation (29).

A similar solution can be obtained with the regular solution model if L is so small that $(1 - 2Lx_A x_B/RT)^{-1}$ can be approximated by $1 + 2Lx_A x_B/RT$.

DISCUSSION OF RESULTS FOR GRAIN BOUNDARIES

(a) *Comparison between previous treatments of grain boundaries*

We shall first use the new treatment for a comparison between the previous treatments. This can be done by representing the properties of the grain boundary with the model described at the top of Fig. 1 and using various b values. b is the width of the discontinuities and Cahn [4] and Lücke and Stüwe [5] treated the case $b = \delta/2$ where there is no homogeneous region in the middle of the boundary. Hillert [7] treated the case $b = 0$ where the discontinuities are sharp. Figure 2 shows the results obtained with $b = \delta/2$, $\delta/4$ and $\delta/10000$ and with the regular solution parameter $L = 0$. The calculations were carried out for a dilute solution, low x_B^0, and the results are then proportional to $x_A^0 \cdot x_B^0$. The results were plotted vs the velocity expressed as $\exp(-v\delta/2D)$ in order to cover the whole v range from zero to infinity and to make the curves at low velocities approximate the curves obtained in a plot vs v. Figure 2 demonstrates

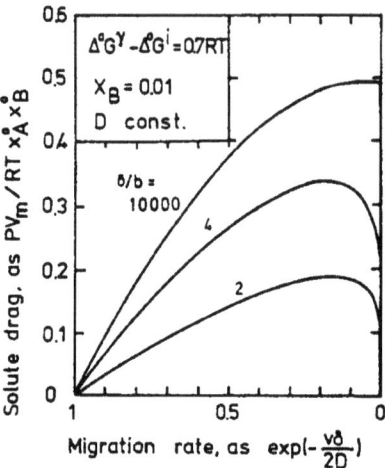

Fig. 2. The solute drag as function of the migration rate, v, calculated for three grain boundary models. The quantities b and δ are defined in Fig. 1.

that the solute drag has a maximum at a velocity which moves to the infinity if one makes the discontinuity sharp by setting $b = 0$. It is thus demonstrated that the absence of a maximum in Hillert's first calculation was due to his choice of sharp discontinuities.

Hillert obtained a maximum by introducing the atomistic nature of the system. It may be stated that the model chosen by Cahn and Lücke and Stüwe reflects the atomistic nature of the system by the width of the discontinuity, b. It thus appears essential never to give b a value which is smaller than the atomic dimensions. As a consequence, b will be chosen as

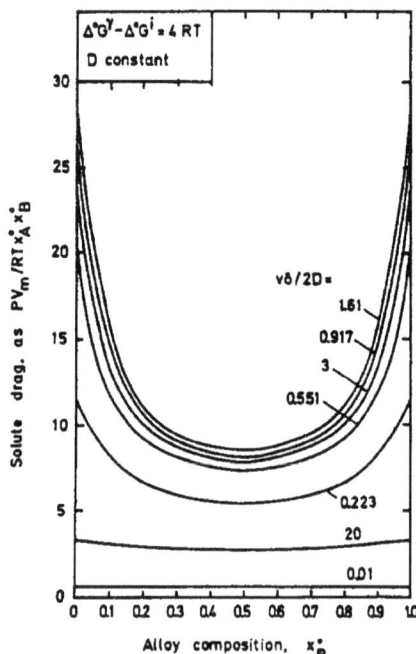

Fig. 3. The solute drag as function of the alloy composition, calculated for different migration rates, v. All the curves are symmetric.

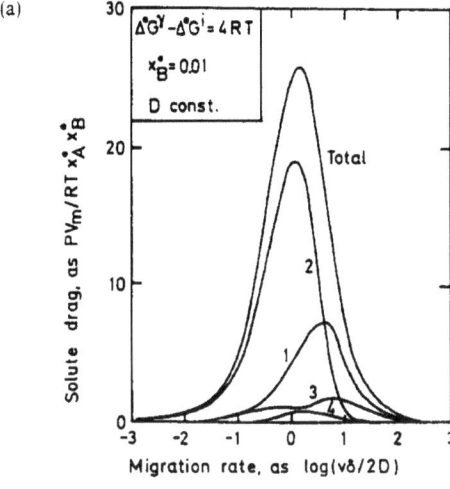

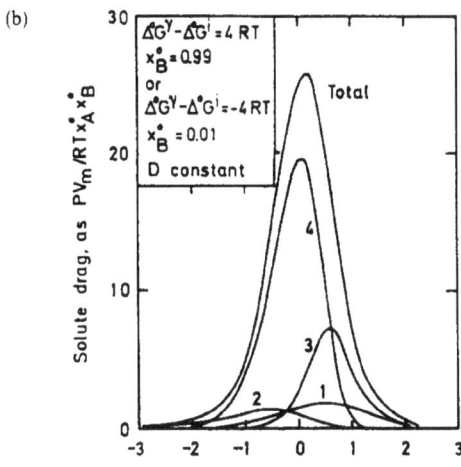

Fig. 4. The solute drag contribution from different zones in the grain boundary. The zones are defined in Fig. 1. The relative importance of the four zones varies with the migration rate. (a) is for a case with a positive segregation to the boundary. (b) is for a case with negative segregation. The total solute drag is the same in the two cases but the roles of the four zones are different.

$\delta/4$ in most of the following calculations since the width of a boundary, δ, is usually assumed to be a few atomic distances.

(b) *Examination of the solute drag over the complete range of composition*

Figure 3 shows the size of the solute drag over the complete range of composition, each curve representing a constant velocity. The only difference in the calculation for solutions of B in A and of A in B is the difference in sign between $({}^0G_B - {}^0G_A)$ and $({}^0G_A - {}^0G_B)$ and it is evident from equations (23 and 25) that the sign of $\Delta^0 G$ does not affect the total magnitude of the solute drag at high and low velocities. One should thus expect to find symmetric curves for high and low velocities. In fact, the diagram seems to indicate that all the curves are symmetric and it

thus appears that the total drag is independent of the sign of $\Delta^0 G$ at all velocities. One can thus obtain the same result with positive segregation of the minor element to the boundary and with negative segregation. It should be noticed that the diagram shows the solute drag divided by $x_A^0 x_B^0$. For high and low velocities, the solute drag is proportional to $x_A^0 x_B^0$, according to equations (23 and 25). The shapes of the curves for intermediate velocities show that the solute drag then increases less rapidly than $x_A^0 x_B^0$ but the maximum drag is still obtained at the central composition.

Figures 4(a and b) show some results from the same calculation, but now plotted for constant composition and versus the velocity. One composition was chosen at each side of the alloy system and the curves for the total solute drag are identical. However, curves representing the dissipation of free energy in the four regions defined in Fig. 1 show large differences. The free energy dissipation in the central zone 2 of the boundary is most important when the solute is attracted to the boundary, Fig. 4(a) but the dissipation in front of the boundary, zone 4, is most important if the solute is trying to avoid the boundary, (Fig. 4b). Equation (23) indicates that the side regions where $\Delta^0 G$ varies, zone 1 and 3, should dominate at higher velocities and, since the sign of $\Delta^0 G$ is here of no importance, the contributions from zone 1 and 3 should here be equal. This is confirmed by both the diagrams.

All the calculations presented so far were obtained under the assumption of a constant value of the diffusivity, D. Figure 5 shows the effect of a variation of

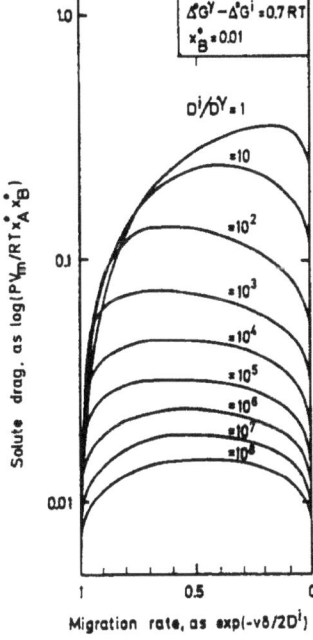

Fig. 5. The solute drag as function of the migration rate, calculated for a variable diffusivity. D^i holds in the center of the boundary and D^γ holds in the undisturbed crystals.

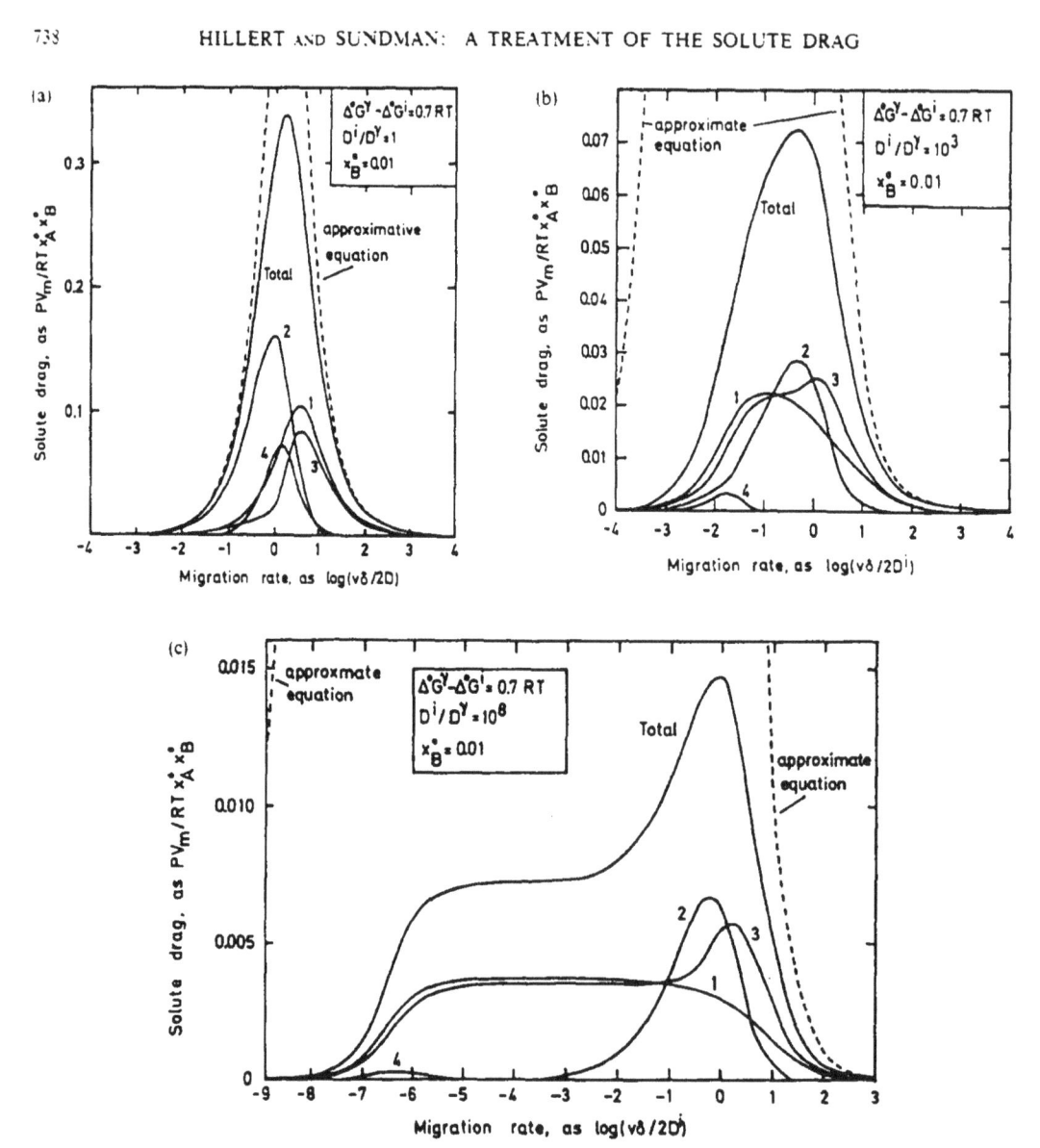

Fig. 6. (a–c) The solute drag contributions from different zones in the grain boundary, evaluated for three of the curves in Fig. 5. The dashed curves are calculated from an approximate equation (Refs. 4 and 5).

D. It was assumed to have different values in the crystals, $D^{..}$, and in the central region of the boundary, D^i, and it was further assumed that log D varies linearly in the intermediate regions. These calculations were only carried out for a dilute solution.

The most interesting feature of Fig. 5 is that the maximum solute drag decreases with increasing values of $D^i/D^{..}$. Previous suggestions [3, 6, 11] that one could employ some average value of D are not justified. The position of the maximum is also interesting. At $D^i = D^{..}$ it occurs at about $v = 4D^i/\delta = 4D^{..}/\delta$. At increasing values of $D^i/^{..}$ the maximum quite naturally moves to values somewhat smaller than $4D^i/\delta$ but increasingly larger than $4D^{..}/\delta$.

In order to examine the reasons for these changes, it is advantageous to plot the solute drag and its four components versus the velocity on a logarithmic scale. Figures 6(a–c) show the results for $D^i/D^{..} = 1$, 10^3 and 10^8. It is evident that the maximum solute drag occurs at a velocity which is closely related to D^i but it is important to note that a considerable drag is obtained much earlier for the case $D^i/D^{..} = 10^8$ and there is an almost constant plateau for 4 orders of magnitude of the velocity.

Cahn as well as Lücke and Stüwe constructed a simple interpolation formula by combining the asymtotic expressions for low and high velocities, equations (23 and 25). The predictions obtained from this

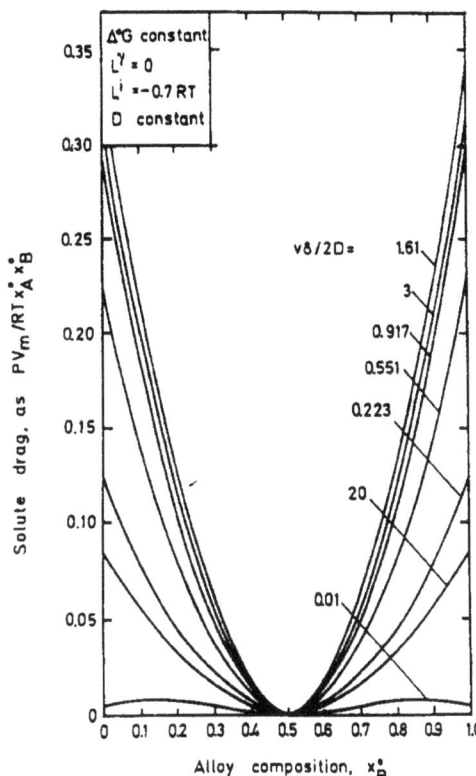

Fig. 7. The solute drag as function of alloy composition calculated for different migration rates using a regular-solution description of boundary segregation.

because there is no tendency for segregation at that composition. This fact is described by the term $(x_A - x_B)dL/dy$ in equation (28).

DISCUSSION OF RESULTS FOR PHASE TRANSFORMATIONS

(a) *The effect of the tendency of segregation to the interface*

The mathematical treatment in the present paper is limited to cases where there is no side-ways diffusion. As a consequence, the equations can be rigorously applied to phase transformations only if the growing phase β has the same composition x_B^0 as the parent phase α had initially. In its present form, the theory can thus be used to predict when such a diffusionless transformation is thermodynamically possible. We shall first examine the simple case where D is a constant and the solute content is low. The distribution constant for the solute between the two phases was chosen to be $K_B^{\alpha/\beta} = 4$ which corresponds to $\Delta^0G^\alpha - \Delta^0G^\beta = 1.4\,RT$. The tendency of the solute to segregate to the central zone of the phase interface was varied from $K_B^{i/\beta} = 1/12$ to 3, i.e. $K_B^{\alpha/i} = 1/3-12$. This corresponds to values of $\Delta^0G^i - \Delta^0G^\beta$ from $-2.48\,RT$ to $+1.1\,RT$ and a linear variation was

approximate equation are presented as dashed lines in Figs. 6(a–c). Cahn tested the equation for constant D and this test is repreated in Fig. 6(a). It shows a very good agreement. except that the maximum is somewhat too high. Lücke and Stüwe [5] tested the equation for very high values of $\Delta^0G^\alpha - \Delta^0G^i$ and found that it is less satisfactory there. Figs. 6b and c show it cannot at all be used to describe cases where the diffusivity varies across the boundary.

(c) *The effect of an interaction parameter*

Most physical models of grain boundary segregation predict that a solute should segregate to the boundary because the internal stresses will be less if the dissimilar atoms are mixed in a region of disordered atomic arrangement. Such a situation cannot be adequately described by allowing $({}^0G_B - {}^0G_A)$ to vary but, at least to a first approximation, it should be possible to describe the situation by giving the L parameter negative values in the boundary. Figure 7 shows the result obtained with $L^\gamma = 0$ in the two grains, $L^i = -0.7\,RT$ in the central zone of the boundary and a linear variation in the two side zones. A constant value of Δ^0G was employed and equation (28) demonstrates that the effect of A in B and of B in A should then be the same. As a consequence, all the curves in Fig. 7 are symmetric. They all go to zero at the central composition where $x_A^0 = x_B^0$

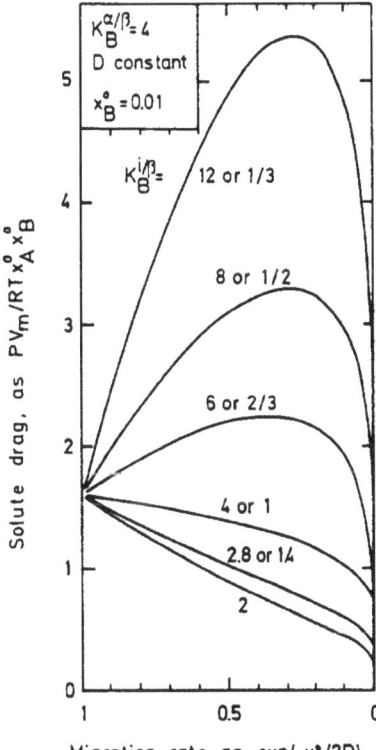

Fig. 8. The free energy dissipation due to diffusion in a so-called diffusionless phase transformation, expressed as a solute drag. The K values represent the equilibrium distribution of the solute. Some combinations of K values yield a monotonously decreasing drag, other combinations yield a maximum at some migration rate.

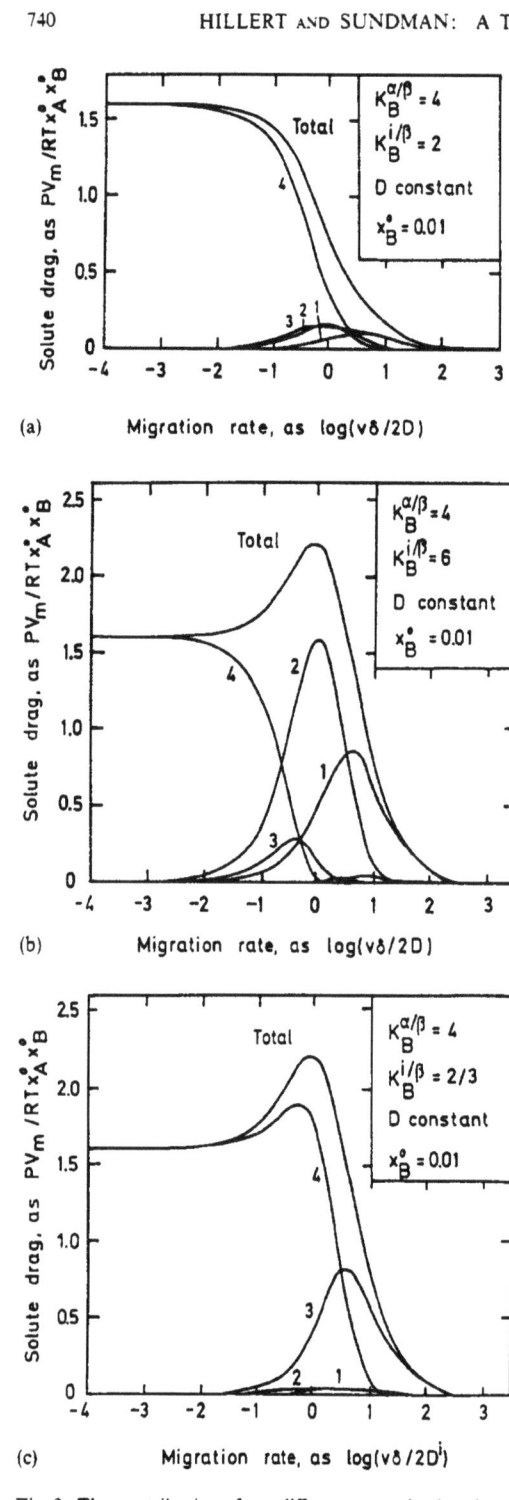

Fig. 9. The contributions from different zones in the phase interface, evaluated for three of the curves in Fig. 8. (a) represents a case with monotonously decreasing drag. (b) and (c) show two cases with a maximum. The total drag has the same shape but the maximum is due to different zones.

assumed at the two sides of the interface. Compare the top diagram in Fig. 1. The results are shown in Fig. 8. All the curves start from a common point at a positive value, which can be easily calculated from the value of $K_B^{\alpha/\beta}$ because, at a low enough velocity, equilibrium should be established between the two phases. In order for the growing β phase to obtain the initial solute content x_B^0, the α phase must then have a solute content of $K_B^{\alpha/\beta} \cdot x_B^0$ in front of the interface. There will thus be a pile-up of solute atoms which are being pushed forward in front of the migrating interface. The height of the starting point for all the curves corresponds to the free energy loss due to that diffusion, which can be directly calculated from

$$x_A^0[G_A(x_B^0) - G_A(K_B^{\alpha/\beta} \cdot x_B^0)]$$
$$+ x_B^0[G_B(x_B^0) - G_B(K_B^{\alpha/\beta} \cdot x_B^0)].$$

Figure 8 shows that the solute drag may increase or decrease with increasing velocity. The largest decrease is obtained with $K_B^{i/\beta} = \sqrt{K_B^{\alpha/\beta}}$, i.e. when the value of $\Delta^0 G$ in the center of the interface lies at the average between the values of the two phases. Figure 9(a) gives the four components of the solute drag for this case and it is evident that the decrease of the total drag is due to a decrease of the height of the pile-up ahead of the migrating interface, zone 4.

Figure 8 shows that the solute drag is affected in exactly the same way if the value of $\Delta^0 G$ in the interface is decreased or increased by a certain amount from the average value. As mentioned in a previous section, the same result was obtained for grain boundaries where the two grains are of the same phase.

Furthermore, Fig. 8 shows that when $\Delta^0 G$ in the interface differs too much from the average, there will be an initial increase of the solute drag and a maximum will be reached at some critical velocity. Figures 9(b and c) show the four components for the cases of $K_B^{i/\beta} = 6$ and $2/3$. The diagrams can be compared with the corresponding diagrams for grain boundaries (Figs. 4a and b). For $K_B^{i/\beta} = 6$ Figure 9(b) shows how a strong drag originates in the central zone 2 of the boundary, as the drag due to the pile-up ahead of the boundary, zone 4, decreases. The same total drag is obtained with $K_B^{i/\beta} = 2/3$ and Fig. 9(c) shows that the maximum is here due to an initial increase of the drag due to the pile-up.

It could be argued that the free energy dissipation in the pile-up of solute atoms in the parent phase ahead of the advancing interface is not caused by the presence of the interface itself and should thus be regarded as separate from the solute drag which by definition is caused by the interface. However, there is no practical way of keeping such a distinction as the velocity is increased and the pile-up is affected by what goes on inside the interface.

From the thermodynamic point of view, the question whether the diffusionless transformation can occur or not is a question of whether the driving force for the diffusionless transformation is large enough. This driving force can be calculated from

$$x_A^0[G_A^\alpha(x_B^0) - G_A^\beta(x_B^0)] + x_B^0[G_B^\alpha(x_B^0) - G_B^\beta(x_B^0)].$$

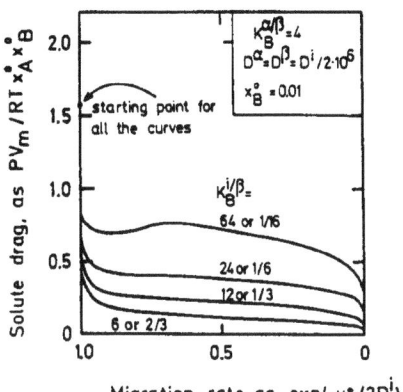

Fig. 10. The free energy dissipation in a phase transformation, calculated for a case with high diffusivity in the boundary.

phase. Figure 8 now indicates that a diffusionless transformation is then possible only if $K_B^{i/\beta}$ lies between the values of approx. 1 and $K_B^{\alpha/\beta}$. The solute drag will then decrease at increasing growth rates and the growth rate may thus attain high values where a considerable part of the driving force becomes available to overcome the physical friction in the interface. The resulting growth rate cannot be calculated without information on the mobility of the interface.

For $K_B^{i/\beta}$ values lower than about 1 or higher than about $K_B^{\alpha/\beta}$, a higher driving force is necessary and the initial solute content must lie within the single-phase region for the new phase in order for a diffusionless transformation to occur. A growth rate limited by the solute drag is then obtained if the driving force is not larger than the maximum on the solute drag curve.

It should further be mentioned that it is in principle possible for growth to occur below and to the right of the maximum point once such a situation has been established but some special mechanism may be required in order to pass the maximum. The same problem has been discussed for grain boundaries by Cahn [4].

It has been shown [12, 7] that the driving force for the diffusionless transformation corresponds exactly to the height of the starting point in Fig. 8 if the initial solute content is exactly equal to the content of the new phase at equilibrium with the parent

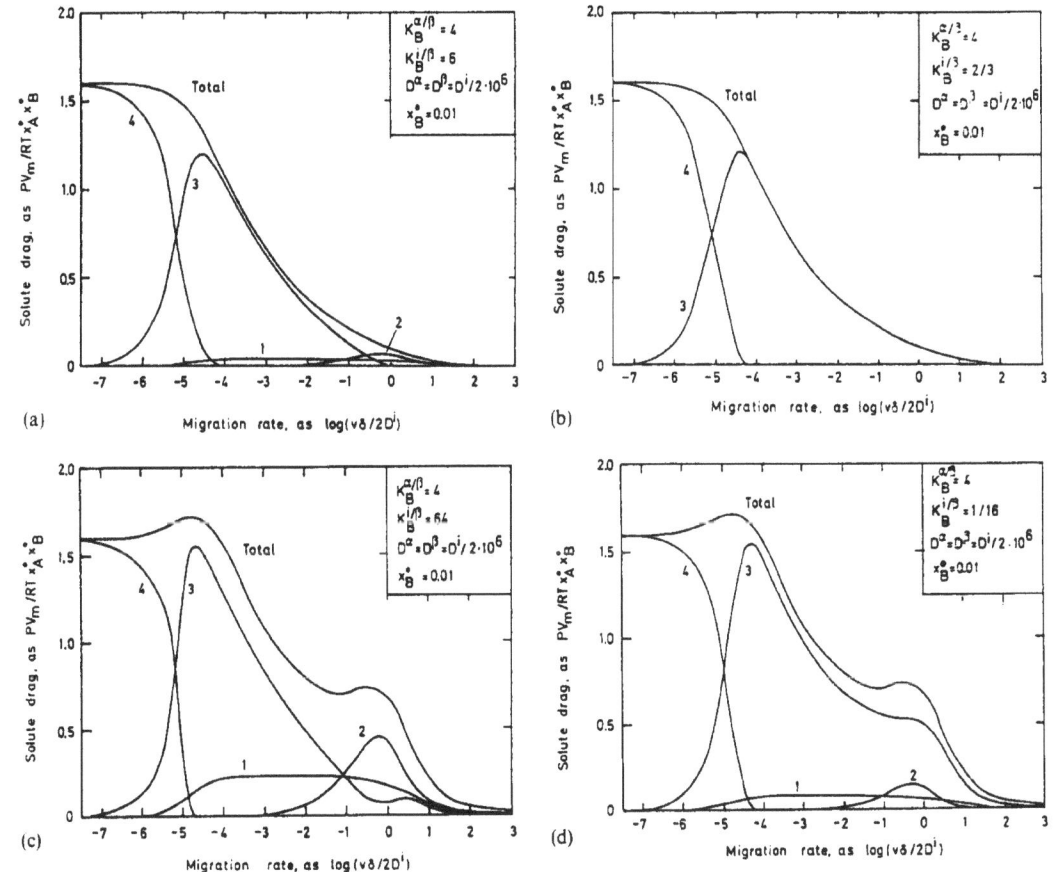

Fig. 11. The contributions from different zones in the phase interface, evaluated for two of the curves in Fig. 10. (a) and (b) show the same total drag, obtained with two different sets of K values. (c) and (d) show another curve which has two maxima.

(b) *The effect of a high diffusivity in the interface*

The calculations of the preceding section were carried out with a constant D value. However, it is often supposed that the diffusivity is considerably higher in an interface than in the crystals. The calculations were thus repeated with $D^\alpha = D^\beta = D^i/2 \cdot 10^6$ and with a linear variation of log D at the two sides of the interface. The results are given in Fig. 10. Two features are obvious. There is a drastic decrease of the drag at a low velocity. The rest of the drag remains almost constant until a velocity, related to D^i/δ, is reached. Figures 11(a and b) show how the four components change with the velocity for the lowest curve in Fig. 10. In both cases the contribution due to the pile-up in front of the boundary, zone 4, decreases strongly above a velocity of about $4D^\alpha/\delta$. Instead a contribution develops in zone 3 and the total drag does not decrease strongly until this contribution starts to decrease. The contributions from the other zones are not very large and in Fig. 11(b) they are even too small to be shown.

Figures 11(c and d) show how the four contributions change with the velocity for the upper curve in Fig. 10. This curve shows a maximum but Figs. 11(c and d) reveal that it actually has two maxima. One occurs at a velocity of about $80D^\alpha/\delta$ and is due to the development of a very strong contribution in zone 3 and some in zone 1 as the role of the pile-up in zone 4 decreases. The other maximum occurs at a velocity of about $0.8\ D^i/\delta$ and is due to the development of a contribution in the central zone 2 in the boundary.

Figure 11(d) shows that the contribution from zone 3 decreases very slowly at the same velocities where zone 2 reaches its maximum. Not until the solute concentration in zone 2 comes close to its equilibrium value in the matrix, does the contribution from zone 3 decrease strongly again.

(c) *The effect of a high diffusivity in the initial phase*

In solidification, the diffusivity is much larger in the initial, liquid phase than in the growing, solid phase and it seems reasonable to assume that the properties vary monotonously through the interface. A calculation was carried out for such a case and a linear variation was assumed for $\Delta^0 G$ as well as log D through the interface. A value of $b = \delta/2$ was thus chosen and there is no central zone 2 with constant properties. The results are shown in Fig. 12 and it is particularly interesting to note that the contributions from the zones inside such a boundary are very small. The contribution from the pile-up ahead of the migrating interface, zone 4, dominates and the total drag decreases when the pile-up starts to decrease at a velocity of about $D^{i'}/\delta$. On the other hand, the diffusion-controlled growth of a β dendrite into the liquid phase is governed by D^L which is much larger than $D^{i'}$. It is thus quite possible that the growth rate of a dendrite becomes so large that the drag decreases

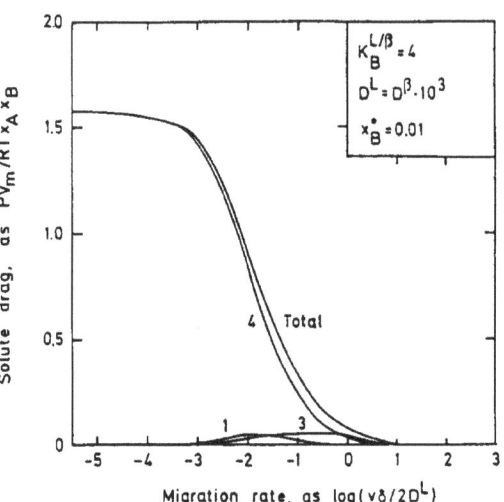

Fig. 12. The solute drag evaluated from a model for solidification. The drag decreases strongly already at a migration rate, which is very low in comparison with the diffusivity in the liquid.

appreciably below its value at low rates. A change into a diffusionless growth may then take place spontaneously even if the total driving force for the diffusionless transformation is less than the total drag at low rates. This may provide the explanation why a diffusionless solidification can occur in an alloy inside the solid + liquid two-phase field [13]. As already mentioned, the driving force for the diffusionless transformation is lower than the solute drag for all alloys inside the two-phase field and it is thus natural to expect that a solid–solid diffusionless transformation does not start until the alloy has been brought inside the one-phase field for the new phase [12, 7].

SUMMARY

The further development of Hillert's treatment of the solute drag in binary alloys becomes identical to the previous treatment by Cahn and Lücke and Stüwe in the limit of dilute solutions. It has been programmed for computerized calculations and a number of calculations have been carried out.

For an ideal solution it is found that the total solute drag is independent of whether the solute is attracted or repelled by the boundary. This result was obtained for all compositions and migration rates. On the other hand, the free energy dissipation takes place in different zones of the boundary.

The effect of a variable diffusivity can be very large and a similar result cannot be produced by using an average value for the diffusivity. A higher diffusivity inside the interface will decrease the maximum value of the drag and make it rather constant over many orders of magnitude of velocity. By combining the asymptotic behavior at very low and very high velocities, Cahn and Lücke and Stüwe constructed a rather simple formula covering the whole range of velocity.

This approximate equation is satisfactory at constant diffusivity, only.

The new treatment can use a regular-solution model in order to describe the case where the solute atoms tend to segregate to the boundary because a mixture of dissimilar atoms will cause less strain energy there.

For diffusionless phase transformations it is found that the free energy dissipation, which starts from an appreciable value at low rates of transformation, in some cases decrease monotonously as the rate is increased. In other cases the free energy dissipation goes through a maximum at an intermediate rate of transformation. This can happen if a strong solute drag develops inside the interface, when the pile-up of solute atoms in front of the advancing interface decreases at high rates of transformation. It can also happen by an increase of the pile-up of solute atoms in front of the interface.

A model, used to describe a liquid–solid interface indicates that the free energy dissipation during diffusionless solidification decreases strongly already at a growth rate thich is related to the low diffusivity in the solid phase and not to the diffusivity in the liquid. It is possible that such growth rates can be obtained by diffusion-controlled growth of dendrites and this result may thus explain why it is easy to obtain a diffusionless solidification in an alloy inside the liquid + solid two-phase field.

Note added in proof—Lücke et al. [14, 15] have recently made further calculations on the atomistic model, using discrete values of the properties at the atomic planes in the boundary and without any restriction regarding the rate of change from one plane to the next.

Acknowledgements—The authors wish to thank professor John Cahn for an interesting discussion and valuable comments.

This research was supported by the Swedish Board for Technical Development.

REFERENCES

1. Lücke K., Masing G. and Nölting P., Z. Metallk. 64 (1956).
2. Detert K. and Lücke K., The influence of defined small amounts of impurities on the recrystallization of aluminium. Brown Univ. Report. AFOSR-TN-56-103; AD-82016-March 1956.
3. Lücke K. and Detert K., Acta Met. 5, 628 (1957).
4. Cahn J. W., Acta Met. 10, 789 (1962).
5. Lücke K. and Stüwe H., Recovery and Recrystallization of Metals, p. 131. Interscience, New York, (1963).
6. Lücke K. and Stüwe H., Acta Met. 19, 1087 (1971).
7. Hillert M., Monograp and Report Series No. 33, p. 231. Inst. of Metals, (1969).
8. Rao M. M. and Winchell P. G., Trans. AIME 239, 956 (1967).
9. Kinsman K. R. and Aaronson H. I., Transformations and hardenability in steels. Climax Molybdenum Corp., Ann Arbor, MI (1967).
10. Aaronson H. I., Laird C. and Kinsman K. R., Phase Transformations. ASM, Metals Park, OH (1970).
11. Gordon P. and Vandermeer R. A., Trans. AIME 224, 917 (1962).
12. Karlyn D., Cahn J. W. and Cohen M., Trans. AIME 245, 194 (1969).
13. Baker J. C. and Cahn J. W., Acta Met. 17, 575 (1969).
14. Rosenbaum F. W., Doctor-Thesis, T. H. Aachen (1971).
15. Lücke K., Rixen R. and Rosenbaum F. W., in Nature and Behavior of Grain Boundaries (Ed. Hsun Hu). Plenum, New York (1972).

18 Introduction to "Chemically Induced Grain Boundary Migration"

published in *Acta Metallurgica* (1978)

Although connections between the movement of grain boundaries and chemical changes in a specimen had been recognized and studied before, especially in the context of discontinuous precipitation, this paper introduced to the scientific world in a direct and unambiguous way the role that such motion has in effecting chemical changes. In particular, it demonstrated and elucidated a process by which boundary migration, together with rapid diffusion of the solute along the grain boundary, can result in a very efficient chemical equilibration in the daughter grain. The driving force for the motion is of course associated somehow with the chemical disequilibrium in the parent grain.

The experiments described in the paper involve supplying, from outside the specimen, solute atoms which are destined to be incorporated into the daughter crystal. This introduction of solute species into the growing grain is accomplished by the diffusion of those atoms from the outside along the grain boundary, together with the migration of the boundary itself. The migration allows the growing grain to capture (or release, in the case of de-alloying) the diffusing atoms, bringing it closer to equilibrium. This novel process has provoked a great deal of imagination and curiosity over the years.

The paper reported on experiments confirming this mechanism in the case of zincification and dezincification of iron. They are revealing and suggestive. Features brought out include the high diffusivity of solute material along the boundary, the appearance of a bed of dislocations accompanying the abrupt change of chemical composition when the boundary begins to move, the bidirectional, sometimes double-seam behaviour of the grain boundary as seen on the surface of a thin film, the geometry of its motion as seen in a section of a thick film, and the resulting solute concentration profile along a boundary which extends through a thin film.

Besides providing experimental results, this paper discussed a theoretical basis by which this action might occur. An extension of earlier ideas due to Cahn, Hillert, and Sundquist, which were focused on discontinuous precipitation, were used. The basis consisted of a force balance equation involving the increment in chemical Gibbs energy across the boundary (postulated to be a driving force), and a solute concentration balance equation. The authors demonstrated very good qualitative agreement between theoretical predictions and experimental results.

Later, this paper inspired a great many experimental and theoretical treatments of this and similar physical scenarios. The phenomenon of chemically (or diffusion) induced grain boundary migration (CIGM or DIGM) was found to occur in many alloys. The authors themselves, together with collaborators, continued exploring this and related subjects.

Besides the mere existence of the CIGM process, a recurring task on the theoretical side has been to better explain the driving mechanism for the motion: how does the presence of an external source of solute material actually induce the migration? Experiments by Rhee and Yoon have pointed to elastic stress occasioned by a nonuniform concentration distribution as being the principal player. In work by Cahn, Elliott, Fife, Penrose and others, it was found that phase field modeling incorporating this effect leads to a dynamical law for the motion of grain boundaries in CIGM compatible with the experiments presented in the Hillert-Purdy paper and elsewhere. In addition, those authors provided properties of steady solutions of the resulting interface dynamical equation with a rigorous mathematical treatment.

Paul C. Fife

CHEMICALLY INDUCED GRAIN BOUNDARY MIGRATION

MATS HILLERT and GARY R. PURDY*

Division of Physical Metallurgy, Royal Institute of Technology, S-100 44 Stockholm 70, Sweden

(Received 21 June 1977; in revised form 10 August 1977)

Abstract—Experimental evidence is presented, showing that grain boundary motion can be induced by changing the composition by means of grain boundary diffusion. The experiments were carried out by treating specimens of pure iron in an atmosphere of zinc or iron–zinc specimens in an atmosphere with a lower zinc potential. The reaction resembles the grain boundary migration in discontinuous precipitation and provides an experimental method of isolating grain boundary migration for study.

The experimental results can be used to evaluate rather directly the grain boundary diffusivity and mobility. The diffusivities thus obtained are several orders of magnitude larger than the values reported for stationary boundaries. It is concluded that the boundary diffusivity is greatly enhanced by grain boundary motion.

Résumé—On présente une preuve expérimentale du fait que l'on peut faire migrer un joint grains en changeant la composition par diffusion intergranulaire. Dans ces expériences, on a traité des échantillons de fer pur dans une atmosphère de zinc et des échantillons de fer et de zinc dans une atmosphère dont le potentiel de zinc était plus faible. La réaction ressemble à la migration des joints de grains dans la précipitation discontinue et elle fournit une méthode expérimentale pour isoler la migration intergranulaire en vue de son étude.

On peut utiliser les résultats expérimentaux pour évaluer assez directement la diffusivité et la mobilité des joints de grains. Les valeurs de la diffusivité que l'on obtient ainsi sont supérieures de plusieurs ordres de grandeur à celles qu'on a publiées pour les joints stationnaires. On en déduit que la diffusivité d'un joint est fortement augmentée par son déplacement.

Zusammenfassung—Es werden experimentelle Hinweise vorgelegt dafür, daß eine Korngrenzbewegung durch die Änderung der Zusammensetzung über Korngrenzendiffusion eingeleitet werden kann. Zur Durchführung der Experimente wurden Proben aus reinem Eisen in einer Zink-Atmosphäre und Proben aus Zink-Eisen in einer Atmosphäre mit kleinerem Zink-Potential behandelt. Die Reaktion ähnelt der Korngrenzbewegung unter diskontinuierlicher Ausscheidung und stellt eine experimentelle Methode dar, die Korngrenzbewegung für die Untersuchung zu isolieren.

Mit den experimentellen Ergebnissen lassen sich Korngrenzendiffusivität und beweglichkeit ziemlich direkt auswerten. Die so erhaltenen Diffusivitäten sind um mehrere Größenordnungen größer als diejenigen, die für stationäre Korngrenzen angegeben werden. Aus den Ergebnissen wird gefolgert, daß die Korngrenzdiffusivität durch Korngrenzbewegung beträchtlich verstärkt wird.

INTRODUCTION

The structure and the structure-sensitive properties of moving solid–solid interfaces need not be the same as for their static counterparts. Indeed, a treatment due to Cahn [1] suggests that a variation of migration mechanism with driving force is generally to be expected.

Previous experiments on moving grain boundaries have focussed on such phenomena as the rate of absorption of run-in dislocations [2] or of nucleation at moving boundaries [3]. Although these observations are of interest, they do not lend themselves to simple quantitative analysis. In the present contribution, we show that it is possible to induce grain boundary motion in response to grain boundary diffusion. For certain specimen geometries, an analysis of the solute concentration profile left in the wake of the moving boundary yields rather directly the grain boundary diffusivity and mobility. The diffusivity obtained in this way can be compared with that measured for static boundaries. We will show here that the two diffusivities differ by about four orders of magnitude. This, in turn, implies that major structural changes occur when grain boundaries are set in motion.

Since the conditions of the present experiments simulate accurately a major component of the discontinuous precipitation reaction in iron–zinc, it should be interesting to compare the newly measured diffusivity and mobility with the values obtained for the same quantities from Speich's kinetic measurements on discontinuous precipitation. This comparison will be based upon a re-evaluation of his data.

---

* Permanent address: Department of Metallurgy and Materials Science, McMaster University, Hamilton, Ontario, Canada.

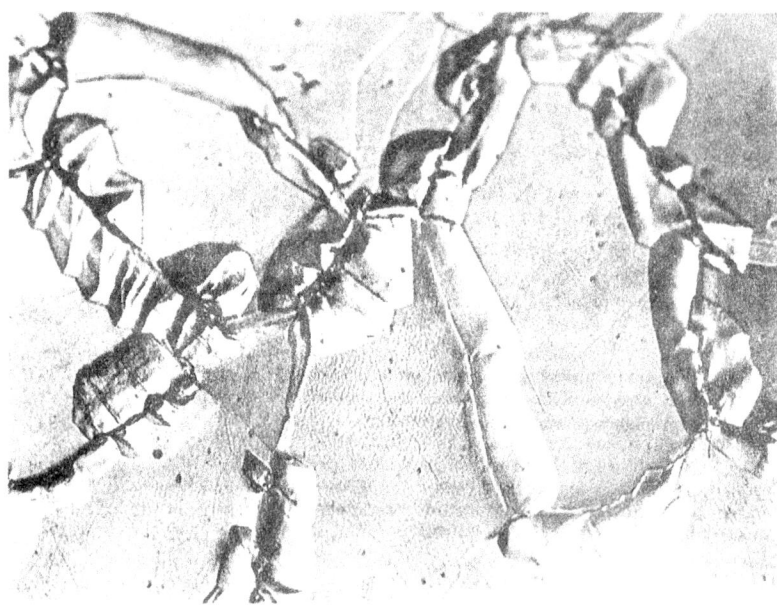

Fig. 1. An optical micrograph illustrating the surface relief that accompanies the zincification of an iron foil, heated in Zn vapour for 4 h at 600°C. Interference contrast. ×200.

EXPERIMENTAL

Most of the present results concern the zincification of pure iron. Thin discs, 3 mm in diameter, were electropolished to perforation, to yield wedge specimens with a maximum thickness of about 100 μm. These specimens were sealed in evacuated silica capsules, and recrystallized at 900°C. The specimens were then resealed in silica capsules, along with several grains of turnings of Fe 11.3 wt.% Zn alloy, heat treated at temperatures ranging from 545 to 600°C, and water-quenched. In a few cases, discs of uniform (50 μm) thickness were treated as described above and subsequently electropolished for examination in a JEOL 1000D electron microscope. In one case, the dezincification of the Fe 11.3 wt.% Zn alloy was studied by heating the recrystallized sheet material in a long evacuated silica capsule, one end of which protruded from the furnace. In addition, the dezincification of the Fe 11.3 wt.% zinc alloy was observed in the heating stage of the electron microscope.

The zincified specimens were sectioned and examined unetched in an ARL SEMQ electron microprobe at 15 kV, 10 nA. Iron and zinc were detected by opposed spectrometers, and the specimens were oriented such that the X-rays were taken-off approximately parallel to the plane of the specimen surfaces.

RESULTS

1. *Metallography*

Figure 1 is a micrograph showing the surface relief that occurs when the grain boundaries move in the thin pure-iron specimen exposed to zinc vapour at

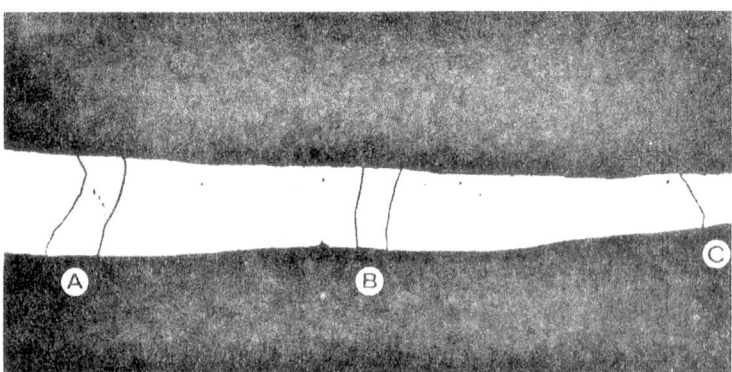

Fig. 2. A section through an iron wedge, exposed to zinc vapour for 3 h at 580°C. Etched in 2% nital. The bands marked A, B and C are enriched in zinc to a level of about 8% by weight. ×360.

Fig. 3. A zinc enriched band in a specimen exposed to zinc vapour for 1 h at 600°C, and subsequently thinned. × 10000.

a temperature of about 600°C. The zinc potential was such that a solid solution of zinc in ferritic iron would result from complete equilibration of the iron specimen with the zinc source. In order to investigate what reaction had actually taken place, wedge specimens were sectioned and examined. As shown in Fig. 2, the areas swept by grain boundaries, A, B and C, have etched more deeply and it was established by electron probe microanalysis that they represent zinc-enriched regions. The remaining areas are essentially pure iron, an observation that is consistent with the very small extrapolated rates of volume diffusion at these temperatures [4]. Evidently, the transport of zinc into the specimens was accomplished almost entirely by boundary diffusion coupled with grain boundary sweeping.

A similar specimen was electrolytically thinned and examined in the high-voltage electron microscope subsequent to the zincification treatment. The grain boundary can now be seen at the upper left corner of Fig. 3. It showed no internal structure and is probably an incoherent high-angle grain boundary. The original position of the grain boundary near the bottom right of Fig. 3, is marked by a remnant wall of dislocations. Similar behaviour was observed in all cases. Figure 4 shows such a wall at a higher magnification. It was imaged in a number of two-beam diffraction conditions and on analysis it proved to be composed of misfit compensating dislocations, primarily of edge character with Burgers vectors lying within 30° of the interface plane. The particular wall in Fig. 4 was found to lie on a (211) plane and the dislocations to lie roughly parallel to the $[0\bar{1}1]$, $[\bar{1}20]$ and $[10\bar{2}]$ directions. The density of dislocations corresponds roughly to the misfit one can expect due to the effect of zinc on the lattice parameter of ferrite, if there is a discontinuous change of the zinc content from zero to about 10%. This observation indicates that the effect of volume diffusion must have been negligible, at least in the early stage when the grain boundary started to move.

Dislocations were often generated in the growing grain. Figure 3 indicates that dislocations actually form at the migrating boundary. It is not clear whether they are extensions from the dislocation wall.

Figure 5 shows a case with a much higher dislocation density. It was taken from the thinnest region of a wedge foil after exposure to zinc vapour in vacuo at 580°C with no further preparation. This specimen contained many small precipitate particles, presumably oxides at or near the surfaces. It is clear that the dislocations trail the boundary as it is forced through the array of particles. The particles may be responsible for the high dislocation density in this case.

It was often observed that a boundary moved faster close to the surface than in the interior of the specimen. The effect became more pronounced as the

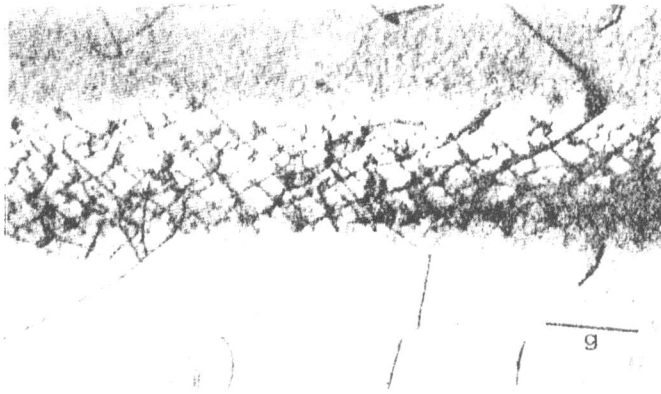

Fig. 4. A wall of dislocations left at the original position of a grain boundary, $g = 0\bar{1}1$. × 35000.

Fig. 5. Generation of dislocations at a grain boundary as it is forced through an array of particles in a thin foil specimen dezincified for 1 h at 580°C. ×15000.

Fig. 7. A section taken parallel to the surface of an iron specimen exposed to zinc vapour for 2 h at 580°C. Note that the present boundary position is marked by a row of etch pits. ×800.

thickness of the specimen was increased and for bulk specimens the boundary migration was confined to the surface region. An example is given in Fig. 6. Similar behaviour was observed for the zincification of thick specimens of an Fe 0.5 wt.% Mo solid solution at 600°C.

It was generally observed that different parts of a grain boundary moved in different directions thus making each one of the two adjoining grains grow into the other at different places. As a result, all the original volume will eventually be swept by grain boundaries and thus transform into a recrystallized high-zinc material. The reaction is demonstrated in Fig. 7 where the present grain boundaries have been attacked by an etch-pitting etchant. The grain boundary in the center of the micrograph is now S-shaped and the middle of it is situated at the original position but rotated 180°. The application of oblique illumination reveals the sharp change in zinc content at the original position of the grain boundary on both sides of the S.

Dezincification was observed to yield the same behaviour and the morphology was very similar. As an example, Fig. 8 shows evidence of extensive boundary motion near the surface of a partially dezincified Fe 11.3 wt.% Zn specimen.

Dezincification could also be studied directly in the hot stage of the high-voltage electron microscope either due to surface oxidation or evaporation of zinc. An observation of grain boundary motion in the Fe 17 wt.% Zn alloy is reproduced in Fig. 9. Here the initial boundary position A A was straight, and the nonplanar morphology developed as migration proceeded at 480°C. In other cases the migrating boundaries remained nearly planar.

2. *Microanalytical results*

A number of electron microprobe step-scans were made of the zinc-enriched regions left by the moving boundaries in the zincification experiments carried out at about 600°C. No significant variations in concentration were found from scans taken parallel to the specimen surfaces. Also, for the thinner regions of the foils, up to about 30 μm, the concentration profiles were found to be flat from one surface to the other, for regions like those marked A, B and C on Fig. 2. Thicker regions of these foils showed some concentration variation in the direction normal to the specimen surface.

Figure 10 shows a concentration profile obtained for a boundary which has moved about 5 μm in 2 h

Fig. 6. Section of a thick specimen, exposed to zinc vapour for 80 min at 600°C. The grain boundary was initially perpendicular to the surface. It has moved to the left. The metal particles at the surface are electrolytic copper introduced into the mounting material to provide electrical conductivity. ×1300.

Fig. 8. A section taken parallel to the surface of an Fe 11.2 wt.% Zn specimen dezincified for 2 h at 580°C. ×1300.

Fig. 9. High voltage electron micrograph of an Fe–17 wt.% Zn foil taken during dezincification in the hot stage at 480°C. The original position of the grain boundary was such that the boundary vertically bisected the micrograph area (A–A). × 10000.

at 580°C. Initially it went straight through the specimen which was 45 μm thick. It was still fairly straight after the movement. We consider this a typical case.

Figure 11 shows two other cases observed in the same specimen. They show a more uniform concentration, for a somewhat thinner section of the foil.

Although we have not managed to make the quantitative experiments completely reproducible the present results are good enough to justify some conclusions and evaluations.

THEORETICAL TREATMENT

The migration of grain boundaries in the present experiments has great similarities with the migration of the grain boundary in discontinuous precipitation, in particular at early stages where the precipitating phase has not yet aligned itself [5]. That case was first given a satisfactory treatment by Cahn [6]. Assuming that the reaction is governed by boundary diffusion and approximating the shape of the boundary as planar, he found the following solution

$$\frac{x - x_1}{x_0 - x_1} = \frac{\cosh z \sqrt{a}/S}{\cosh \sqrt{a}/2} \quad (1)$$

where $a = vS^2/KD^b\delta$, x_1 is the solute content of the parent grain and x_0 is the solute content of the growing lamella at its sides. z is the coordinate in the direction of diffusion in the boundary. S is the width of the lamella, v is the velocity of the boundary. K is the equilibrium distribution coefficient of the solute between the boundary and the growing grain, D^b is the diffusivity and δ the width of the boundary. Several attempts have been made to modify Cahn's treatment in order to take into account the nonplanar shape of the boundary and the effect of capillarity. The most detailed treatment has been presented by Sundquist [7] and is based upon two equations. Using l as a measure of the distance along the curved boundary Sundquist formulated the following equation for the boundary diffusion.

$$KD^b\delta \cdot dx/dl = \int_l^L v(x_1 - x) \, dl. \quad (2)$$

L is the value of l in the middle of the specimen and v is the migration rate of the boundary in the direction of the normal. The balance of forces at the boundary yields the second equation.

$$\Delta G_m = \frac{vV_m}{M} - \sigma V_m \frac{d\theta}{dl}. \quad (3)$$

where θ is the angle between the macroscopic growth direction and the local normal of the boundary. The last term is due to capillarity and it has a decisive influence in the case of discontinuous precipitation. However, in the present case the width of the specimen is very much larger than the width of the lamellae in discontinuous precipitation. As a consequence, the values of $d\theta/dl$ will normally be very much smaller. For the present purpose, this term will be neglected. M is the mobility of the boundary. It contains the intrinsic mobility of the boundary and in alloys it may also be strongly affected by solute-drag effects. It has often been suggested that the mobility varies with the size of the driving force but it will now be treated as a constant. ΔG_m is the driving force acting on the boundary. In the present case it comes from the chemical Gibbs energy driving the reaction

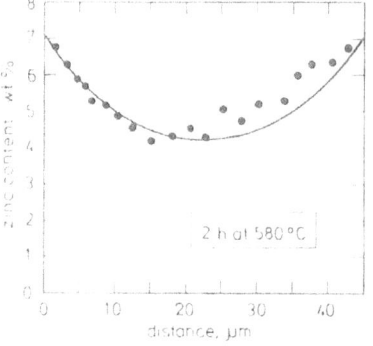

Fig. 10. Experimental concentration profile of zinc from one side of a thin specimen to the other in a zincified region like A, B and C shown in Fig. 2.

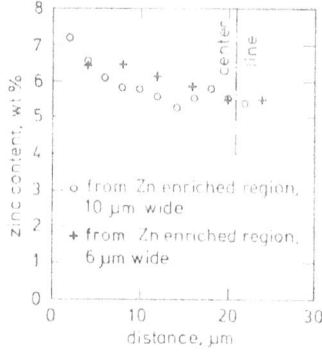

Fig. 11. Two other examples of concentration profiles in zincified specimens.

which can be evaluated from the following equation if the phase under consideration can be described as a sub-regular solution.

$$\Delta G_m^{chem} = (1 - x_1)\left[RT \ln \frac{1 - x_1}{1 - x} + (A + 3B)\right.$$
$$\times (x_1^2 - x^2) - 4B(x_1^3 - x^3)]$$
$$+ x_1 \left\{ RT \ln \frac{x_1}{x} + (A - 3B)[(1 - x_1)^2 \right.$$
$$- (1 - x)^2] + 4B[(1 - x_1)^3$$
$$\left.\left. - (1 - x)^3 \right] \right\}. \quad (4)$$

There are several sinks for this chemical driving force [7, 8] but they will not be considered in the present discussion.

Equations (2) and (3) were used in a series of calculations in order to simulate the present experiments in the Fe–Zn system where the α phase can be described with the following parameters $A = 4384.8 + 6.2781\ T$ J/mol, $B = 4439.2$ J/mol [9]. For the planar case equation (2) reduces to Cahn's solution, equation (1). The boundary was reasonably flat in the case illustrated in Fig. 10 and the concentration profile was calculated from Cahn's solution using $a = 4.4$, $x_1 = 0$ and $x_0 = 0.06$. The experimental data seem to be well represented by this curve, thus indicating that the reaction is governed by boundary diffusion. In combination with the experimental rate of migration, $v = 7 \times 10^{-8}$ cm/s, the a value yields $KD^b\delta = 3.2 \times 10^{-13}$ cm^3/s at the experimental temperature which was 580°C. This value is much larger than values usually observed for boundary diffusion in iron and iron alloys, which are around 10^{-17} cm^3/s [10]. It should be emphasized that the concentration profiles presented in Fig. 11 would yield even larger values.

It is easy to apply equation (3) to the case presented in Fig. 1 since that boundary was fairly planar during its migration. The last term can thus be neglected and the rate of migration can be represented by an average which was 7×10^{-8} cm/s. According to equation (4), the driving force varies from 367 J/mol at the surface to 259 J/mol at the center of the specimen. The average is 292 J/mol and inserted in equation (3) it yields a value of $M = 2 \times 10^{-17}$ m^4/J,s. This is much lower than the mobilities observed in recrystallization and grain growth which have yielded values of 10^{-12} m^4/J,s or higher at 580°C [11].

Accepting the values of $KD^b\delta$ and M obtained for the case shown in Fig. 1, a series of calculations were carried out from equations (2) and (3) for different widths of the specimen and without the simplication of a constant rate of migration for the whole boundary. Instead, the local rate was evaluated from equation (3) using the local value of the driving force. The calculation was carried out under the simplifying assumption that the rate is independent of time, which should be true for short times when the boundary is still reasonably planar. The resulting concentration profiles are presented in Fig. 12. For thin specimens the concentration is thus predicted to be fairly constant but for thick specimens the concentration will be much lower in the middle. The shapes of the boundaries were calculated simultaneously but are not plotted because they gave almost identical curves to those shown in Fig. 12, the reason being that the migration distance is assumed to be proportional to the rate, the rate is proportional to the driving force in view of equation (3) and the driving force is approximately proportional to the Zn content, x, in view of equation (4), if the initial specimen is free of Zn. The shape of the migrating boundary, thus predicted for the case shown in Fig. 10, $S/2 = 22.5\ \mu$m, is not quite as planar as the shape actually observed. However, the agreement may still be regarded as qualitatively correct. The lowest curve in Fig. 12 was obtained for an infinite thickness and it approaches zero asymptotically. For longer times, the boundary becomes increasingly more non-planar. The diffusion distance along the boundary down to a certain depth under the surface will thus increase and the concentration as well as the migration rate will decrease with time at any given depth. One may thus conclude that there seems to be a qualitative agreement with the shape for a bulk specimen presented in Fig. 6.

In spite of the scatter in the experimental information obtained for boundary migration in the present study there is a qualitative agreement on a sufficient number of points to justify the conclusion that the rate of reaction is governed by boundary diffusion, that the driving force is supplied by the chemical

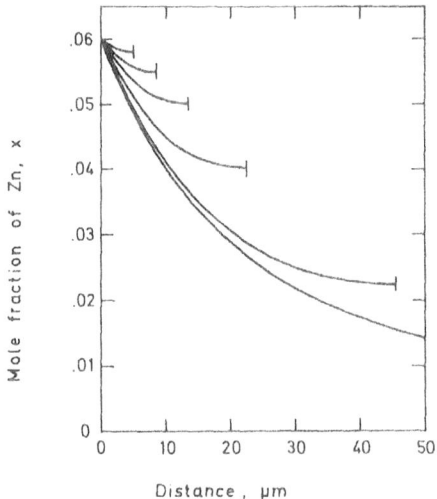

Fig. 12. Concentration profiles calculated for a series of specimen thickness. The specimen centers are marked by a small vertical line. The calculations were carried out with $M = 2 \times 10^{-17}$ m^4/J, s and $KD^b\delta = 3.2 \times 10^{-13}$ cm^3/s.

Gibbs energy and that the mobility is fairly independent of the size of the driving force. On the other hand, the values of the boundary diffusivity and mobility do not agree with values obtained for pure iron. This disagreement may be due to the presence of Zn in the present experiments and it should thus be interesting to compare with values evaluated from information on discontinuous precipitation in the Fe-Zn system.

COMPARISON WITH DISCONTINUOUS PRECIPITATION

Speich [4] has made a detailed experimental study of discontinuous precipitation in the Fe-Zn system. He treated the data theoretically and evaluated the diffusivity and the relation between the driving force and the rate of migration of the grain boundary. His diffusivity data yielded a value of $KD^b\delta = 4 \times 10^{-15}$ cm^3/s which is two orders of magnitude larger than values previously reported for iron and iron alloys but it is two orders of magnitude lower than the values obtained from the present experiments.

Speich found a cubic relation between the migration rate and the driving force. If applied to the experiment presented in Fig. 1, such a law would predict a migration rate in the middle of the specimen which is only a quarter of the migration rate close to the surface but the shape of the migrating boundary indicated an almost constant rate. In order to test if the cubic law is actually the best description of Speich's experimental data, they were now reevaluated. Speich's evaluation of the chemical driving force has previously been criticized [12] and it was now evaluated using equation (4). A difficulty was that Speich only measured the average composition of the growing α grain. He then assumed that the composition at the sides of the α lamellae, x_0, was the equilibrium composition according to the phase diagram. There is no justification for such an assumption because the edges of the adjoining γ lamellae are expected to be curved. Lacking experimental information on the curvature one might guess on any x_0 value between the equilibrium value and the experimental average value reported by Speich. As a consequence, Speich's evaluation of the diffusivity should be regarded as an evaluation of a minimum value only. The closer x_0 lies to the average x, the higher will the resulting diffusivity be. It is thus possible to obtain agreement between the very high diffusivity evaluated from the present experiments and Speich's experimental data by choosing an x_0 value close to his experimental average of x.

The mobility was now evaluated from Speich's experiments using both of the extreme x values. The results are presented in Table 1. ΔG_S is the Gibbs energy absorbed by surface energy in the interfaces between the lamellae. It was subtracted from the driving force before the mobility was evaluated. In passing, it is interesting to note that the driving force seems to be only a few times larger than the surface energy, a result which seems to be in a rather good agreement with Zener's maximum growth rate hypothesis. An accurate comparison is impossible since the value of the specific surface energy between the lamellae is unknown. Speich used a value of 0.7 J/m^2 and the same value was used here. The surprising result that the value of ΔG_m evaluated for the Zn rich alloy is not larger than the estimated surface energy may be removed by choosing a lower specific surface energy.

The mobility values obtained with the two extreme x_0 values do not deviate much from each other. In both cases there is a strong variation of the mobility with the alloy content but, contrary to Speich's evaluation, this variation now does not seem to be related to a variation in the driving force but rather to the variation in alloy content. The mobility decreases with decreasing Zn content and an attempt to extrapolate these data to the conditions of the present experiments yielded a value of M between 10^{-15} and

Table 1. Evaluation of experimental data by Speich on discontinuous precipitation in the Fe-Zn system

x_1	T (°C)	x^{eq}	$\bar{x}$	$v \times 10^{10}$ (cm/s)	ΔG_S (J/mol)	ΔG_m (J/mol)	$x_0 = x^{eq}$ $\Delta G_m - \Delta G_s$ (J/mol)	$M \times 10^{16}$ m^4/J.s	ΔG_m (J/mol)	$x_0 = \bar{x}$ $\Delta G_m - \Delta G_s$ (J/mol)	$M \times 10^{16}$ m^4/J.s
0.097	400	0.020	0.05	2.8	2.9	21	18	1	14	11	2
	450	0.034	0.065	5.2	2.4	12	10	4	6	4	9
	500	0.053	0.07	12	1.7	6	4	20	5	3	30
0.152	400	0.020	0.074	9.6	6.4	34	28	2	19	12	6
	450	0.034	0.084	29	6.9	26	28	2	19	12	6
	500	0.053	0.095	75	4.5	16	11	50	10	5	100
	550	0.081	0.11	170	2.5	8	5	200	5	2	600
0.235	400	0.020	0.107	48	12	46	34	10	19	7	50
	450	0.034	0.12	380	11	46	34	10	19	7	50
	500	0.053	0.125	1100	9.7	31	21	400	17	7	1000
	550	0.081	0.139	1600	9.0	20	11	1000	14	5	2000
	600	0.131	0.144	1100	4.8	13	8	1000	14	9	900
0.305	400	0.020	0.124	190	22	59	37	40	19	-	-
	450	0.034	0.139	1300	20	48	28	300	18	-	-
	500	0.053	0.15	2600	19	40	21	900	19	-	-
	550	0.081	0.17	5400	14	39	25	1500	15	-	-
	600	0.131	0.165	5600	15	26	11	4000	22	7	6000

10^{-14} m^4/J.s to be compared with the value 2×10^{-17} m^4/J.s obtained from the present experiment and 10^{-12} m^4/J.s obtained for pure iron.

CONCLUSIONS

As compared to previous information for iron or iron-base alloys, the present experiments have yielded surprising values for the boundary diffusivity. It is four orders of magnitude larger than the values obtained from experiments with stationary grain boundaries. The discrepancy cannot be explained by errors in the experiments or evaluations. The explanation may be that the properties of a grain boundary is quite different when its is stationary and when it is moving.

It is not certain whether the diffusivity is different in the present experiments and in discontinuous precipitation in the same alloy system because the driving concentration gradient has not been measured in the latter case. On the other hand, there is a recent study of the concentration gradient in another case of discontinuous precipitation [13] which indicates that the concentration gradient may be rather high. If that result can also be applied to the Fe–Zn system, the growth rates for discontinuous precipitation would yield a diffusivity which is two orders of magnitude lower than the present result but it is still two orders of magnitude higher than the values obtained for stationary grain boundaries. It is thus difficult to avoid the conclusion that grain boundary diffusion is greatly enhanced by the movement of the boundary although the degree of enhancement seems to differ from one case to another.

The low values obtained for the mobility as compared to pure iron are not surprising and the difference is most probably due to solute-drag effects. The fact that the present results are even lower than the values obtained from discontinuous precipitation is surprising and no explanation will be offered here. The fact that dislocations were sometimes observed to trail the migrating boundary does not seem to provide any explanation because the dislocation energy is negligible compared to the chemical force driving the boundary motion in these experiments.

It has been suggested that the grain boundary migrates in discontinuous precipitation because of a driving force that arises when the composition is changed by the action of grain boundary diffusion [8]. The present experiments confirm this mechanism and provides a technique for the direct study of it without the complications that occur in the complex process of discontinuous precipitation. This new experimental technique may be described as discontinuous zincification or dezincification and may yield valuable information on the mechanism of discontinuous precipitation. Other applications suggest themselves, however, and we will briefly list some of these:

As a probe for the structure of moving grain boundaries, the method appears to offer promise as a quantitative, if indirect, tool. We consider that the present results constitute the clearest evidence for the need to distinguish between the static and dynamic structures of grain boundaries, and one can imagine a series of similar experiments on boundaries of known misorientation. The remnant wall of epitaxial dislocations is also of interest; it is possible to prepare for study a planar epitaxial array on any prescribed crystallographic plane. The stability of these walls under conditions when volume diffusion is not negligible also invites investigation. It is possible to construct experiments for the study of dynamic boundary interactions with dispersed particles, or with dissolved solutes (although, in the latter case, one can expect complications due to the presence of zinc) and it may prove possible to extend this type of study to include the dynamics of interphase boundaries.

We note that this process of discontinuous zincification or dezincification may be of practical significance, as it provides a rapid method of changing the zinc concentration in iron. We expect that these same methods will be effective in other systems as well, provided that the rate of boundary diffusion is much greater than that of volume diffusion, and that the mobile solute is capable of being supplied or removed via a gaseous or fluid medium.

Acknowledgements—This work was financially supported by the Swedish Board for Technical Development, and the National Research Council of Canada.

REFERENCES

1. J. W. Cahn, *Acta Met.* **8**, 554 (1960).
2. P. H. Pumphrey and H. Gleiter, *Phil. Mag.* **32**, 881 (1975).
3. H. Westengen, Doctoral Thesis, Institut for Fysikalisk Metallurgi, N.T.H., Trondheim (1977).
4. G. R. Speich, *Trans. A.I.M.E.* **242**, 1359 (1968).
5. M. Hillert and R. Lagneborg, *Mater. Sci.* **6**, 208 (1971).
6. J. W. Cahn, *Acta Met.* **7**, 18 (1959).
7. B. E. Sundquist, *Metall. Trans.* **4**, 1919 (1973).
8. M. Hillert, *Inst. Met. Proc. Int. Symp. on The Mechanism of Phase Transformations in Crystalline Solids*, Manchester, July 1968, p. 231 (1969).
9. G. Kirchner, H. Harvig, K.-R. Moqvist and M. Hillert, *Arch. Eisenhüttenw.* **44**, 227 (1973).
10. J. Fridberg L.-E. Törndahl and M. Hillert, *Jernkont. Ann.* **153**, 263 (1969).
11. M. Hillert, *Metall. Trans.* (A) **6**, 5 (1975).
12. M. Hillert, *Metall. Trans.* **3**, 2729 (1972).
13. D. Porter, Personal communication.

19 Introduction to "An Analysis of the Effect of Alloying Elements on the Pearlite Reaction"

published in *Solid-Solid Phase Transformations*, Eds. H.I. Aaronson, D.E. Laughlin, R.F. Sekerka and C.M. Wayman, *AIME* (1982)

Mats Hillerts' contributions have been central to the development of our current understanding of the pearlite transformation. This is especially true for the case of ferrous pearlite where Hillert has contributed both pioneering theoretical treatments and novel experimental studies. A significant portion of these contributions have concerned pearlite formation in so-called Fe-C-X systems and the contribution introduced here, prepared for the 'International Conference on Phase Transformations' held in Pittsburg, Pennsylvania in 1981, is one of Hillert's seminal contributions on pearlite formation in alloyed steels.

The central theme of the article is the question: What are the appropriate interfacial conditions during the growth of pearlite in alloyed steels?

Hillert considers the experimental evidence available at that time in the Fe-C-Ni, Fe-C-Mn, Fe-C-Si and Fe-C-Cr systems and conducts a test of the chemical conditions prevailing at the migrating pearlite/austenite interface through a comparison of model predictions and the available experimental data. The comparisons with experiment take on several forms but attention is mostly given to comparisons on the basis of the compositional and temperature limits of the different modes of pearlite formation.

The beauty of this article lies in the simplicity of the framework Hillert uses to describe the large range of different pearlite growth modes possible in Fe-C-X steels. To illustrate this richness in possible behavior consider the now well studied Fe-C-Mn system [1–5] where under isothermal conditions, depending on the treatment temperature and alloy composition, it is possible to form:

- Pearlite which grows under steady state conditions with a constant velocity and interlamellar spacing but without partitioning of the X element.

- Pearlite which grows under steady state conditions with a constant velocity and interlamellar spacing behind either a zone *depleted* in C (giving an increase in lamellar spacing upon impingement of pearlite colonies) or a zone *enriched* in C (presumably giving a decrease in lamellar spacing upon impingement of pearlite colonies) and with partitioning of the X element.

- Pearlite growth under non-steady state conditions with a growth velocity that decreases and an interlamellar spacing that increases with time with partitioning of the X element. This is the so-called 'Divergent Pearlite' first observed by Cahn and Hagel [2] in the Fe-C-Mn system and Mannerkoski [6] in the Fe-C-Cr system.

- The closely related 'Divergent Pearlite' formation above the Upper $Ae1$ [4,8]. This is sometimes referred to as 'Porous Cementite' in the Fe-C-Si system [7,8] and involves the formation of a metastable ferrite + cementite aggregate within the austenite + cementite two phase field of the phase diagram.

Hillert successfully rationalizes all these observations with a framework based on isoactivity lines, phase diagrams and assuming local equilibrium conditions at the migrating pearlite/austenite interface. It is worth noting that the seeds of this framework were already sown in his now famous 1953 Internal Report on 'ParaEquilibrium' [9], prepared three years before Hillert received his PhD from MIT.

In pure Hillert style, this article also contains guidance on promising areas for further research. Hillert particularly emphasizes the need to consider not only the different types of pearlite growth but also the transitions between the different modes. This is a question that has attracted much recent attention on for the growth of ferrite in Fe-C-X steels [10–12] but has not yet been extended to the case of pearlite growth. In the case of ferrite growth in Fe-C-X systems we are interested in transitions where growth control varies from C volume diffusion to X volume diffusion. These diffusivities are very different and the transition is relatively abrupt. Hillert points out that the interesting transitions in pearlite growth may occur from C volume diffusion control to X interface diffusion control. These diffusivities are not very different and the transitions in growth may be much more extended in the case of pearlite than for ferrite, and may offer a promising avenue for experimental studies of transitions in behavior.

I would like to take this opportunity to thank Mats Hillert for the profound influence he has had on the development of my view of phase transformations, and also to wish him the very best on the occasion of his 80th birthday.

Christopher. R. HUTCHINSON

References

[1] M.L. Picklesimer, D.L. McElroy, T.M. Kegley Jr., E.E. Stansbury and J.H. Frye, *Trans. TMS-AIME* 218 (1960) 473.

[2] J.W. Cahn and W.C. Hagel, *Acta metall.* 11 (1963) 561.

[3] J. Fridberg, Thesis, Royal Institute of Technology, Stockholm, Sweden, 1968.

[4] C.R. Hutchinson and G.J. Shiflet, *Scripta Mater.* 50 (2004) 1.

[5] C.R. Hutchinson, R.E. Hackenberg and G.J. Shiflet, *Acta Mater.* 52 (2004) 3565.

[6] M. Mannerkoski, *Acta Polytech. Scand.*, 1964, Ser A., Chap. 26.

[7] A. Hultgren and O. Edström, *Jernkont. Ann.* 121 (1937) 163.

[8] J. Fridberg and M. Hillert, *Acta Metall.* 18 (1970) 1253.

[9] M. Hillert, ParaEquilibrium, *Internal Report*, Swedish Inst. Metals Res., 1953[1].

[10] M. Hillert, *Scipta Mater.*, 46 (2002) 447.

[11] Odqvist, M. Hillert and J. Agren, *Acta Mater.* 50 (2002) 3211.

[12] C.R. Hutchinson, A. Fuchsmann and Y. Brechet, *Mater. Metall. Trans.* 35A (2004) 1211.

---

[1] This volume, chapter 2.

AN ANALYSIS OF THE EFFECT OF ALLOYING ELEMENTS ON THE PEARLITE REACTION

Mats Hillert

Division of Physical Metallurgy
Royal Institute of Technology
S-100 44 Stockholm, Sweden

The growth characteristics of pearlite in alloyed steels are predicted under the assumption of full local equilibrium (ortho-equilibrium). Comparison with experimental information for nickel, manganese, silicon and chromium yields a satisfactory agreement. This is surprising since the spike of an alloy element, in front of the advancing phase interface, can often be calculated to be much thinner than the atomic dimensions.

The understanding of the effect of alloying elements on the transformation behaviour of austenite in steels must be based upon an understanding of how the alloying elements partition between the various phases during the transformation. This phenomenon was first studied by Bowman in 1945 (1,2) who reported that proeutectoid ferrite inherits the molybdenum content of the parent austenite whereas pearlite grows under partitioning of molybdenum between ferrite and cementite. Hultgren (3) made a more detailed study of manganese and silicon in steels and found that these alloying elements also partition between ferrite and cementite in pearlite but not in bainite. Without hypothesizing why this was so, he proposed that it was due to some major difference between pearlite and bainite. Many years earlier (4) he had suggested that local equilibrium is established for carbon at the phase interfaces during the formation of pearlite in Fe-C alloys. Now he extended this suggestion to include the alloying elements for pearlite but not for bainite. For bainite he suggested that the local equilibrium at the phase interfaces only holds for carbon. Such a restricted local equilibrium he called paraequilibrium. Full local equilibrium he called orthoequilibrium.

In a subsequent study (5) Hultgren confirmed that bainite grows without partitioning of the alloying elements but he now found that the same holds for pearlite in most cases. Partitioning only seemed to occur in a few cases of pearlite. Hultgren was thus forced to conclude that there are two types of pearlite, which he called parapearlite and orthopearlite because he believed that they must have formed under the two different types of local equilibrium.

As a student of professor Hultgren the present author was confronted with the idea of paraequilibrium. It seemed easy to understand that paraequilibrium should be favoured by high growth rate and low temperature since it should occur when there is not sufficient time for iron and the substitutional alloying elements to diffuse. In trying to understand exactly when the transition from ortho- to paraequilibrium should occur, he realized that the criterion of full local equilibrium usually requires that there is a wave (negative or positive) of the alloying elements in front of the advancing interface (6). The quantity D_M/v (D_M being the lattice diffusivity of the alloying element M and v being the growth rate) is a measure of the width of the wave (now usually called spike) and paraequilibrium may thus be expected when D_M/v is less than the atomic dimensions, d. He suggested that the transition between ortho- and paraequilibrium should occur between 10d and 0.4d. Another important finding was that the growing phase could inherit the alloying content of the parent phase even if the spike exists, i.e. even under orthoequilibrium conditions at the interface. Fig. 1 demonstrates this for a simple case of ferrite forming from austenite with an alloying element like Mn or Ni. This growth can only occur if there is a driving force for the diffusion of carbon away from the interface and into the austenite. The diagram shows that the carbon activity at the interface is represented by an isoactivity line which goes through the austenite marked with a cross. A steel with the alloy content under consideration must thus fall to the left of a critical point marked with a filled circle. The open circle is an example and the growth rate of ferrite in such a steel will be controlled by carbon diffusion driven by a carbon activity difference to be evaluated between the two isoactivity lines, Δa_c. It can with good accuracy be inserted in the ordinary diffusion equation for a binary system after replacing the concentration gradient with the activity gradient.

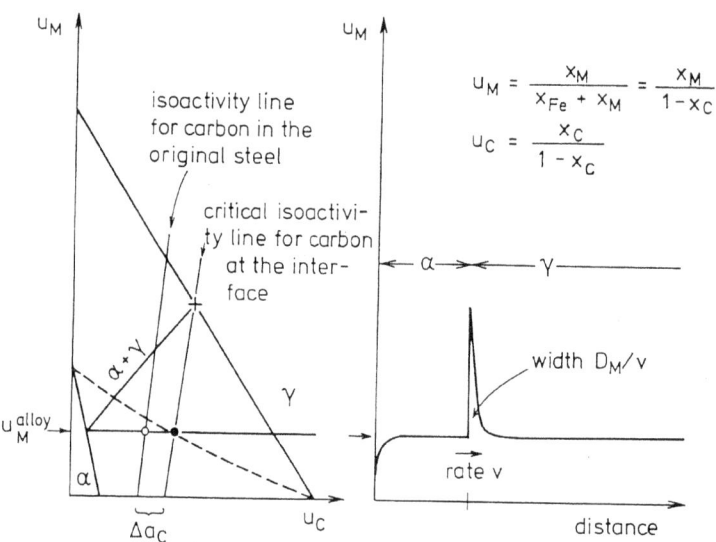

Fig. 1 Construction of the critical limit for the growth of ferrite with the same alloy content as the parent austenite. The open circle represents the composition of a steel and the filled circle is the critical limit at that level of alloy content.

The critical point varies with the alloy content and a critical line can thus be drawn. See the dashed line in Fig. 1. Any alloy above this line can form ferrite only by means of diffusion of the alloy element away from the region to be transformed. The growth will thus be very slow and it may even be a reasonable approximation to assume that there will be a uniform carbon activity in the whole system during such a reaction. Choosing the same isoactivity line we now get the construction shown in Fig. 2. This time the alloy composition (open circle) lies above the critical line. The intersection between the critical line and the isoactivity line through the alloy point is marked with a filled circle. Their vertical distance gives the difference in alloy content driving the diffusion of the alloy element. This difference can also be inserted in the ordinary diffusion equation for a binary system with good accuracy.

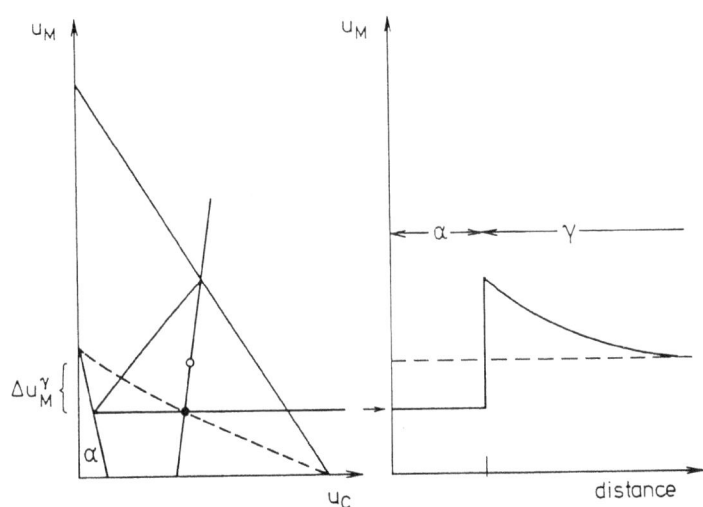

Fig. 2 Construction of the conditions at the ferrite/austenite interface for slow growth of ferrite, controlled by volume diffusion of the alloying element.

The same ideas were independently developed by Kirkaldy and his coworkers (7-10) who used a more ambitious treatment of diffusion in ternary systems rather than working with the difference in carbon activity. The two treatments are equivalent for all practical purposes. The discussion in the present report will be based upon the use of the isoactivity lines.

In a very thorough study of the pearlitic transformation in some manganese steels, Picklesimer et al. (11) were able to show that there is a gradual decrease of the partitioning as the transformation temperature is lowered. Orthopearlite formed at small undercoolings and, below a wide transition region, paraequilibrium appeared. The existence of this gradual transition may be taken as an indication that parapearlite, as well as orthopearlite, forms under full local equilibrium. A transition to paraequilibrium should have been more abrupt. The situation has been further clarified in more recent years because of improved experimental techniques. By a high-resolution electron microprobe Ridley et al. (12-15) were able to make in-situ measurements of the partitioning of alloying elements between ferrite and cementite just behind the growth front and were thus able to decrease the experimental uncertainty due to the possible partitioning by volume diffusion inside the pearlite after its formation. They could thus prove that partitioning can actually occur as a part of the growth process. Using a field-ion microscope Miller and Smith (16) have even been able to measure concentration profiles across the ferrite and cementite lamellae. Many other aspects of the pearlite reaction have also been studied over the last two decades. The aim of the present report is to examine the information now available in order to test what kind of local equilibrium conditions applies at the pearlite/austenite interface. The test will be carried out by comparing various types of experimental information with predictions obtained by applying the full local equilibrium conditions, i.e. orthoequilibrium. The test will be based upon a thermodynamic evaluation by Uhrenius (17) of the

792 M. Hillert

ternary Fe-C-M systems involved. It will be assumed that the dominating mechanism of partitioning is by boundary diffusion in the pearlite/austenite interface. Unfortunately, there is no experimental information on this diffusion constant. However, a compilation of grain boundary diffusion in pure austenite and pure ferrite specimens of various binary Fe-M alloys showed almost the same results for both phases and for different alloying elements as well as radioactive iron (18). It thus seemed reasonable to use the same value in the present case.

Agreement between calculated and experimental values will be taken as an indication that the local equilibrium conditions are close to orthoequilibrium because predictions based upon paraequilibrium will usually be quite different. This is demonstrated by Fig. 3 where the dotted line shows the critical limit for the formation of ferrite under paraequilibrium conditions. It starts from the so-called T_o line in the binary Fe-M phase diagram. It is evident that the negative effect of this alloying element on the formation of ferrite would be much less if paraequilibrium applies instead of orthoequilibrium.

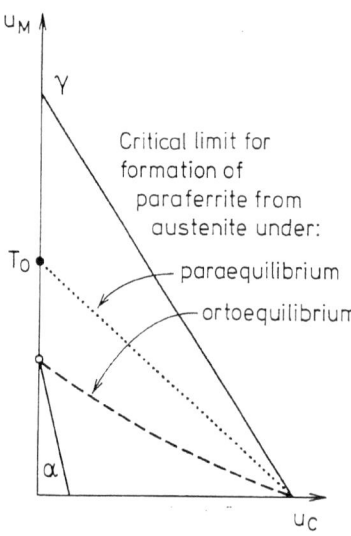

Fig. 3 Comparison of the critical limit for the formation of paraferrite under para- and orthoequilibrium. T_o is the point of equal Gibbs energy in the binary Fe-M system.

Comparison with the Massive Transformation in Binary Alloys

The question whether T_o or the α phase boundary represents the natural limit for the growth of ferrite in a binary system, when there is no change in composition, has been discussed in connection with the massive transformation (19). The conclusion was that the phase boundary seems to be the natural limit when the spike is wider than atomic dimensions and it was suggested to be the limit even at higher growth rates if the phase interface has a high atomic mobility which may be the case if it has a disordered structure.

An exception was found for Fe-Ni and Fe-Mn alloys at fairly low temperatures and it was proposed that this was due to the interface becoming more coherent. When testing the applicability of the full (ortho) equilibrium hypothesis in Fe-C-M alloys one should thus start with a case where the interface is most probably incoherent. This might be the case for pearlite (20).

ALLOYING ELEMENTS AND THE PEARLITE REACTION 793

Parapearlite in Fe-C-Ni

In order for parapearlite to form it is necessary that each one of ferrite and cementite can form with the parent alloy content. We must thus be below two critical lines like the dashed line in Fig. 1. This is demonstrated by Fig. 4, where the triangular region marked A is the "theoretical" region for parapearlite, assuming full local equilibrium. This construction is originally due to Puls and Kirkaldy (9).

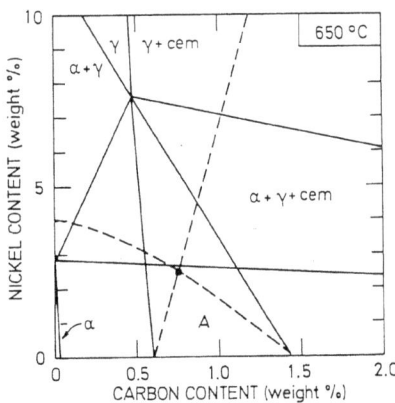

Fig. 4 Construction of the region A where pearlite can grow without partitioning of nickel.

Before proceeding we should emphasize two facts. The first one is that the ordinary equilibrium phase diagram holds for flat phase interfaces only. The displacements of phase boundaries caused by the Gibbs-Thomson effect at curved phase interfaces will be neglected in the remainder of this report. The second fact is that pearlite must usually compete with proeutectoid ferrite or cementite and there is no sharp boundary between regions where one or the other will win. The region A should thus be taken as a very rough outline of where parapearlite may form. In any case, the alloy content of the top of the triangle A is of highest importance. Below that critical level, the reaction may start by the formation of paraferrite or paracementite but after some time parapearlite should form. Above that level parapearlite should never form. The variation with temperature of this critical alloy content has been plotted in Fig. 5 together with experimental information on the critical temperature above which pearlite cannot form in nickel steels (5,21). The agreement is excellent and one may conclude that pearlite in nickel steels always forms as parapearlite under full local equilibrium. This is not the case for bainite (22) which, by some reason, seems to almost follow ferrite in binary Fe-Ni alloys (23). Bainite may be regarded as a eutectoid microconstituent but the effect of nickel on its growth is definitely quite different from the effect on pearlite. We must conclude that it does not form under full equilibrium.

In order for pearlite to grow at an alloy level above the top of the A region in Fig. 5, it is necessary that nickel diffuses away from both ferrite and cementite, i.e. into the austenite. Such a reaction is common in eutectic solidification and leads to segregation and the formation of eutectic cells. However, in that case the parent phase has a high diffusivity. For the pearlite reaction, the diffusion of substantial amounts of an alloying element into the parent austenite would take a very long time and this kind of pearlite (once called metapearlite (24)) can be completely neglected.

794 *M. Hillert*

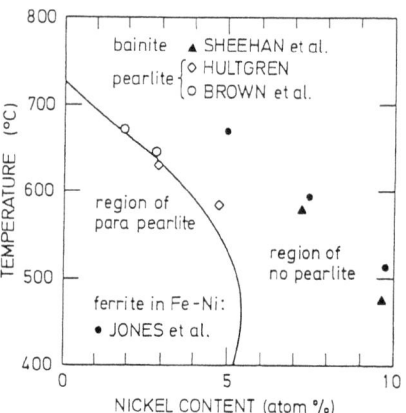

Fig. 5 Comparison between experimental results and predictions of the region where pearlite can form in nickel steels. Comparison is also made with the limits for the formation of bainite and ferrite.

Parapearlite in Fe-C-Mn

A rather abrupt stop of the pearlite reaction in nickel steels as the temperature is increased above the line in Fig. 5 may also be demonstrated by curves for the growth rate. In Fig. 6 such curves are compared for nickel and manganese (21,12). It is evident that the decrease of the growth rate is much more gradual for manganese than for nickel. This is an indication that

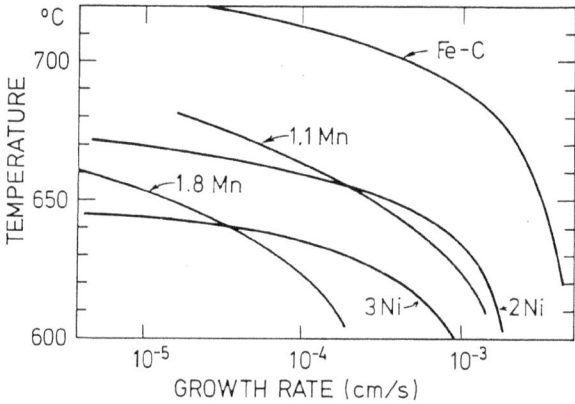

Fig. 6 Comparison of growth rate curves for pearlite in nickel and manganese steels.

there is a mechanism by which pearlite can grow above the top of the A triangle in manganese steels. This is by the partitioning of manganese between ferrite and cementite, which has a strong affinity for manganese. Before discussing the formation of such orthopearlite we should take a more detailed look at the growth conditions for parapearlite. See the schematic diagram in Fig. 7. The open circle represents an alloy and the two crosses represent the austenite composition at the interface, provided that both ferrite and cementite inherit the parent alloy content (as demonstrated by the horizontal line). The partition of carbon between ferrite and cementite is thus driven by the difference between the two isoactivity lines. By inserting this difference in the theory for pearlite growth in pure Fe-C we can easily evaluate the effect of the alloying element on the growth rate of parapearlite. This test of the existence of full local equilibrium was

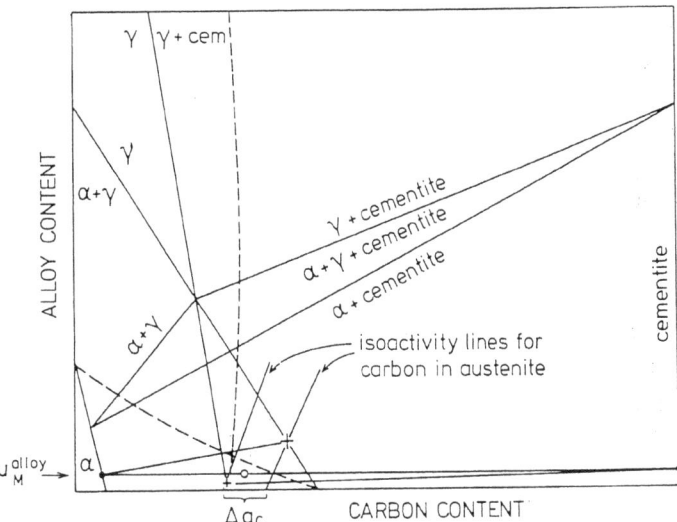

Fig. 7 Schematic phase diagram showing the carbon activity difference driving the growth of parapearlite. The open circle represents the composition of the steel. The two crosses show the austenite compositions at the ferrite/austenite and cementite/austenite interfaces.

done more than a decade ago, assuming rate control by volume diffusion of carbon in austenite (25) as well as boundary diffusion (26). Good agreement was found in both cases. The test was repeated more recently with essentially the same result (12).

Orthopearlite in Fe-C-Mn

It is evident that ferrite and cementite should be able to help each other grow because at the interface manganese should like to diffuse away from ferrite but to cementite. However, this may happen only if the growth rate is slow enough for the diffusion of manganese to occur. As a consequence, it should be a reasonable approximation to assume that the growth rate is so slow that there is sufficient time for carbon to establish a uniform carbon activity along the pearlite/austenite interface. We may thus examine the conditions at the interface by choosing some carbon activity value and plotting the corresponding isoactivity line. See the schematic phase diagram in Fig. 8. We can thus find the points (crosses) representing the composition of the austenite at the ferrite and cementite interfaces. From the corresponding tie-lines we find the points (filled circles) representing the growing ferrite and cementite. It is evident that the growing pearlite must fall on the straight line joining these two points. Let us now suppose that the initial alloy composition falls on the isoactivity line chosen. See the open circle. It must then transform to the orthopearlite represented by the square because the average alloy content cannot change. This orthopearlite has a much higher carbon content than the alloy. The growing orthopearlite will thus draw carbon from the parent austenite and a zone depleted of carbon will form. The growth will then continue according to isoactivity lines further to the left. This means that the difference in alloy content, Δu_M^γ, which drives the partitioning of the alloy element, will decrease. As a consequence, the interlamellar spacing should increase gradually and the growth rate should decrease gradually. This phenomenon was first observed by Kuo (27) in the so-called δ-eutectoid reaction in some high-alloyed steels. It was first observed on pearlite by Cahn and Hagel (28)

796 M. Hillert

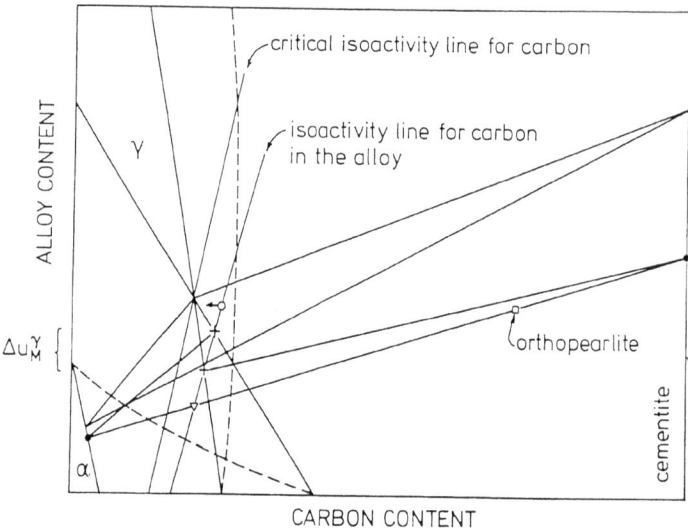

Fig. 8 Schematic phase diagram showing how the composition of the growing orthopearlite (open square) can be found from the composition of the steel (open circle). The two crosses show the austenite compositions at the ferrite/austenite and cementite/austenite interfaces. The filled circles represent the compositions of the two growing phases. The triangle represents a steel for which the orthopearlite inherits the carbon content as well as the alloy content.

who coined the name divergent pearlite. As the isoactivity line going through the corner of the stable γ phase field is approached, Δu_M^γ goes to zero and the growth will cease before complete transformation. Actually, orthopearlite should be able to form in the whole triangle formed by the two critical (dashed) lines and this critical isoactivity line. This is shown in Fig. 9 which is the Fe-C-Mn diagram calculated for 700°C (17).

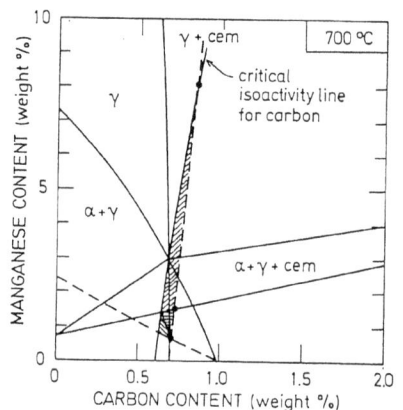

Fig. 9 The Fe-C-Mn phase diagram at 700°C showing the region for orthopearlite (shaded area). The three filled circles represent the upper limit for the formation of parapearlite, constant orthopearlite and divergent orthopearlite.

Constant Orthopearlite

It is often believed that orthopearlite is always divergent. However, if the alloy composition falls at the point marked with a triangle in Fig. 8, the orthopearlite will get the same composition. It should thus be possible to form orthopearlite with a constant interlamellar spacing in such an alloy and this reaction should not stop before completion. We shall call this "constant orthopearlite" as different from "divergent orthopearlite".

If the alloy composition is slightly to the right of the triangular point, there will be a depleted zone but there is no reason why not its thickness will get stabilized and constant growth conditions will be established after a transient period. By the same token, if the alloy composition is to the left, there will be a zone enriched in carbon and constant growth conditions will be established after some time. The only case when constant growth conditions can never be established is when the alloy composition falls above the α+cementite side of the three-phase triangle. Thus the whole region of orthopearlite should be divided into three parts. See B, C and D in Fig. 10 which shows the Fe-C-Mn diagram calculated for 650°C.

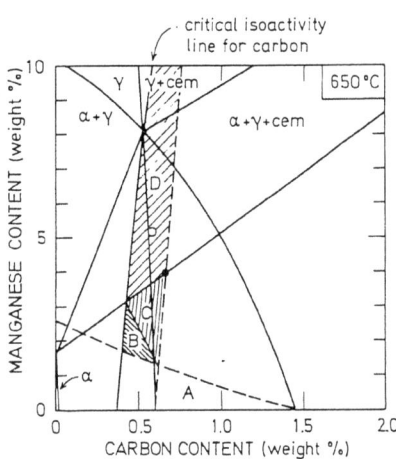

Fig. 10 The Fe-C-Mn phase diagram at 650°C showing the regions for parapearlite (A), constant orthopearlite growing behind a carbon-enriched zone (B), constant orthopearlite growing behind a carbon-depleted zone (C) and divergent orthopearlite (D). The open circle represents a steel examined by Cahn and Hagel.

The alloy content at the top of each one of the regions A, C and D represents the highest level where each one of the three types of pearlite can form. They have thus been plotted as functions of temperature in Fig. 11 where a comparison is made with various types of experimental information. We now get the explanation of the conflicting results obtained by Hultgren (3,5). In his first study he examined a steel with 3.3% Mn which falls well inside the region for constant orthopearlite but in his second study the steel had only 0.7% Mn and fell well inside the parapearlite region.

When Picklesimer (11) made his observation of the gradual variation of partitioning with temperature, it was difficult to decide how much of the partitioning had occurred after the formation of the pearlite. It was even possible that all the partitioning in some specimens had occurred after the formation. Using reasonable judgement one finds that the points for para- and orthopearlite bracket the calculated limit in Fig. 11. In a more recent study, Razik et al. (12) tried to define the exact temperature where perfect paracomposition was obtained by the pearlite transformation itself. Their

798 M. Hillert

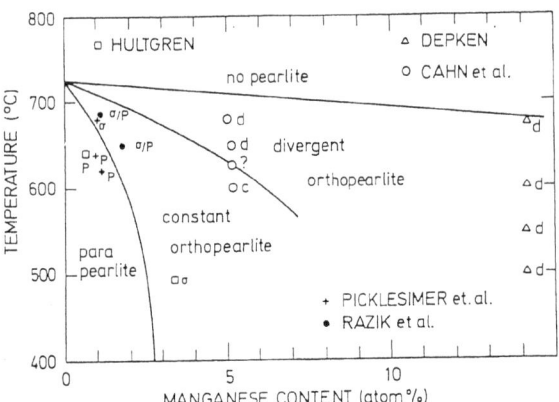

Fig. 11 Comparison between experimental results and predictions of the regions for various types of pearlite in manganese steels.

points fall well above the calculated line. On the other hand, some new information obtained with an improved experimental technique (16) indicates that some partitioning actually occurs at lower temperatures. More exact information is required. Rather than discussing the theoretical upper limit for parapearlite, one should discuss the whole transition from para- to orthopearlite. This will be done for silicon in a subsequent section.

The upper limit for constant orthopearlite is even less well established. In fact, Cahn and Hagel (29) reported divergence for their 5.2% Mn steel at all temperatures examined. On the other hand, their data for 600°C show a constant growth rate up to about 1000 minutes and the divergence observed towards the end of the reaction could very well be due to impingement of the carbon depleted zones originating from pearlite nucleated at opposite boundaries in a grain. Depken (30) found divergence at all his experimental temperatures for two steels with 14% Mn and 0.38 and 0.87% C, respectively. So far, there does not seem to be any experimental information which is in disagreement with the calculated limit for constant orthopearlite.

The upper limit for divergent pearlite is rather well established by Depken's experiment at 675°C. The result was orthopearlite with approximately 90% cementite.

Growth Rate and Coarseness of Orthopearlite

The growth of constant para- and orthopearlite in ternary alloys has been treated theoretically on a sound thermodynamic basis but it was necessary to simplify the thermodynamic properties of the system by linearization (31). This treatment does not seem to have been applied which is not surprising in view of the fact that constant orthopearlite has not been widely recognized. Fridberg (32) made a similar calculation by identifying the carbon activity at the interface by trial and error. By neglecting the Gibbs-Thomson effect he obtained a result which can be illustrated by Fig. 12. The vertical distance between the two crosses gave an alloy difference Δu_M^γ which drives the partitioning and could be inserted in the binary theory for pearlite growth governed by boundary diffusion. The carbon activity difference, Δa_c, can be used for an estimate of the width of the carbon depleted zone but it does not enter into the calculation of the growth characteristics of the constant orthopearlite. For a steel with 2.9% Mn Fridberg calculated growth rates that were 1/4 of the experimental value at 600°C. For Cahn and Hagel's steel with 5.2% Mn he obtained $0.5 \cdot 10^{-6}$ cm/s at 600°C as compared to an experimental value twice as high. The agreement is thus surprisingly good. In

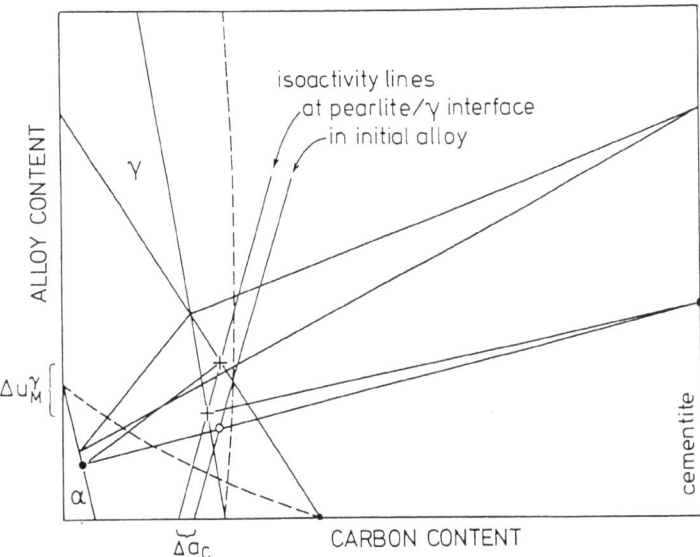

Fig. 12 Schematic phase diagram showing the growth conditions for constant orthopearlite. The open circle represents the composition of a steel and the two crosses represent the austenite compositions at the ferrite/austenite and cementite/austenite interfaces. The two filled circles represent the compositions of the two growing phases.

this connection it may be worth noting that the increase from 2.9 to 5.2% Mn has a surprisingly small effect on the growth rate (by a factor of about 5) and the calculations seem to confirm this experimental result.

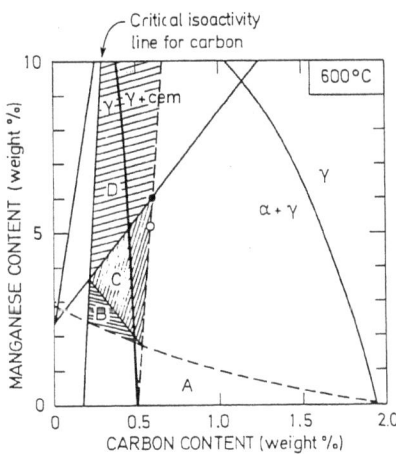

Fig. 13 The Fe-C-Mn phase diagram at 600°C showing the regions for various types of pearlite. The open circle represents a steel examined by Cahn and Hagel and the filled circle represents the critical alloy content for constant orthopearlite.

The phase diagram Fe-C-Mn at 600°C shows that Cahn and Hagel's alloy is here very close to the limit for constant orthopearlite. See the open circle in Fig. 13. At 650 one must certainly consider divergent pearlite. For that case Fridberg carried out a step by step calculation where the growth rate is mainly controlled by the diffusion of carbon through the depleted zone and the boundary diffusion of the alloying element controls the spacing.

800 M. Hillert

Some of his results are presented in Figs. 14 and 15 for the steels with 5.2 and 14.0% Mn. The calculated curves are roughly parabolic due to the rate control by carbon diffusion through the depleted zone. The agreement is satisfactory. He also obtained a reasonable agreement for the spacing.

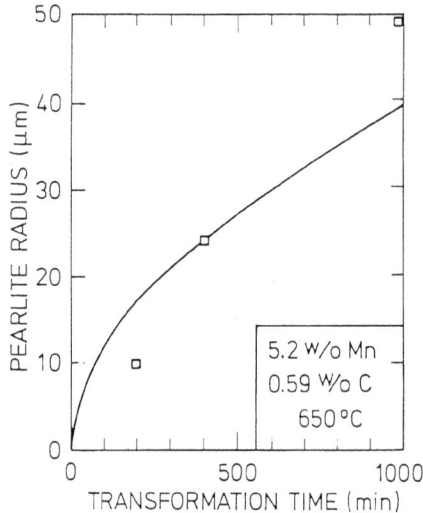

Fig. 14 Comparison of experimental growth data for divergent orthopearlite according to Cahn and Hagel (open squares) and predictions according to Fridberg (curve).

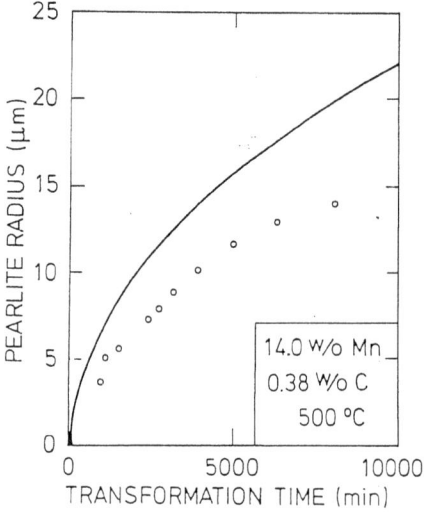

Fig. 15 Comparison of experimental growth data for divergent orthopearlite according to Depken (open circles) and predictions according to Fridberg (curve).

Pearlite in Fe-C-Si

Divergent pearlite was described in the Fe-C-Si system long ago under the name of porous cementite (33). It forms well above the eutectoid temperature which is not more surprising than the divergent orthopearlite region in the Fe-C-Mn system which extends to alloy contents well above the eutectoid level. Fig. 16 shows the various regions for orthopearlite at 750°C. At that temperature there is no parapearlite region.

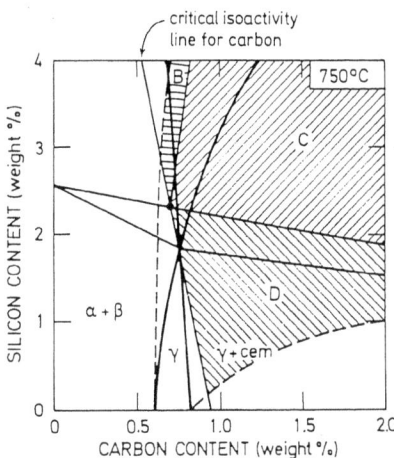

Fig. 16 The Fe-C-Si phase diagram at 750°C showing the regions for constant orthopearlite growing behind a carbon-enriched zone (B), constant orthopearlite growing behind a carbon-depleted zone (C) and divergent orthopearlite (D). No parapearlite can form at this temperature.

In the lower part of the D region the growth rate is so low compared to the rate of carbon diffusion that it is a good approximation to assume that the carbon activity is uniform through the whole of the austenite grains during the pearlite reaction. This was done by Fridberg and Hillert (34) and satisfactory agreement was again obtained. In this case, the growth rate is controlled by the rate of boundary diffusion of silicon.

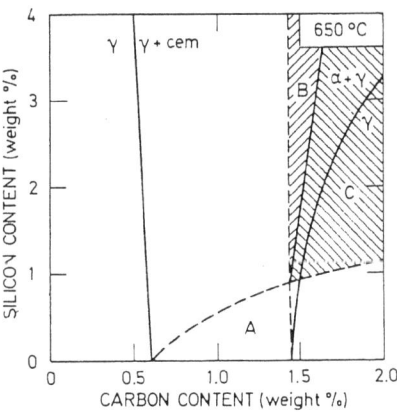

Fig. 17 The Fe-C-Si phase diagram at 650°C showing the regions for parapearlite (A), constant orthopearlite growing behind a carbon-enriched zone (B) and constant orthopearlite growing behind a carbon-depleted zone (C). Divergent orthopearlite will not form at this temperature except for a transient period and a final period with impinging carbon zones.

The region for parapearlite appears below the eutectoid temperature as shown by Fig. 17. It is thus possible to evaluate the critical alloy content for parapearlite (the top of the triangular A region) but the results are very uncertain because they depend strongly on the value of the distribution coefficient of silicon between cementite and ferrite, a quantity which is not well known. Recent measurements by Al-Salman et al. (15) indicate a value of 0.11 but at lower temperatures one should normally expect increasing values for this kind of quantity. The value of $K_{Si}^{cem/\alpha} = 0.11$ was used for constructing the boundary between the parapearlite and constant orthopearlite regions in Fig. 18. The divergent orthopearlite is predicted to grow above 727°C only. The limit shown in Fig. 18 was calculated as the locus of points illustrated by the filled circle in Fig. 16. It represents the lower limit for constant orthopearlite rather than the upper limit for divergent orthopearlite.

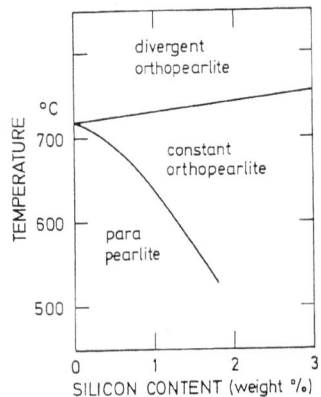

Fig. 18 Predicted regions for various types of pearlite in silicon steels.

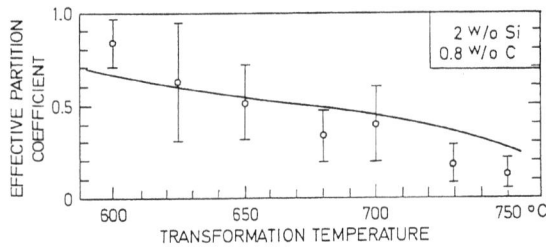

Fig. 19 The effective partition coefficient for silicon between cementite and ferrite in pearlite as function of temperature in a 2% Si steel. The data points with error bars are experimental. The curve is evaluated from the calculated silicon profiles in Fig. 20.

The experimental information on the partitioning of silicon is quite detailed and shows a very gradual change with temperature (15). See Fig. 19. From Fig. 18 we should of course not expect parapearlite in a steel with 2% Si until below 500°C (if the K value of 0.11 is correct). Knowing the experimental growth rate and spacing it is a relatively simple matter to calculate the distribution of silicon at various temperatures, using a part of the theory for constant pearlite in ternary systems (31). The detailed results of a series of such calculations are presented in Fig. 20 and the "experimental" K values, evaluated from the average alloy content of each phase, are presented as a curve in Fig. 19. The agreement with experiments is remarkably good and could be improved further by allowing the equilibrium K value to increase above 0.11 as the temperature is lowered.

By the use of the improved techniques for microanalysis it is today possible to study experimentally profiles like those in Fig. 20. Some results

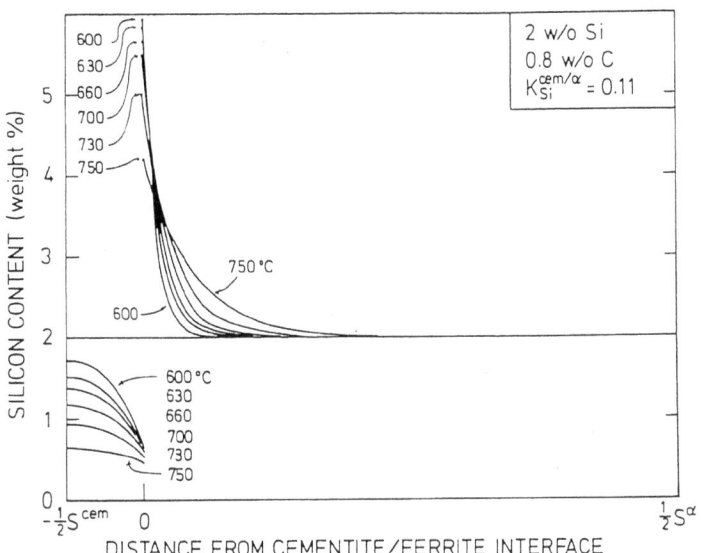

Fig. 20 The concentration profiles for silicon across the cementite and ferrite lamellae in pearlite, calculated theoretically, using the experimental growth rate and lamellar spacing.

of this type have already been obtained (16) and the general form of the profiles has been confirmed. It is particularly interesting to note that the grain boundary diffusion is fast enough to affect the interior of the cementite lamellae when the interior of the ferrite lamellae is still unaffected. It should be realized that the growth of pearlite in a silicon steel depends more critically upon the lowering of the silicon content in the cementite than the increase in the ferrite. This is because cementite cannot grow with too much silicon. In order to satisfy this practical requirement it is thus sufficient to diffuse silicon over distances comparable with the thickness of the cementite lamellae rather than the ferrite lamellae which are about 7 times as large. This is why silicon does not increase the hardenability as much as the low K value seems to suggest.

It should be mentioned that Al-Salman et al. (15) made a similar calculation of grain boundary diffusion and reached similar conclusions although their method of calculation was more schematic.

In this connection it may be emphasized that the transition from para- to orthoferrite is much more abrupt than from para- to orthopearlite. In the former case the rate control is changed from volume diffusion of carbon to volume diffusion of the alloying element. These diffusion coefficients differ by many orders of magnitude. In the latter case the rate control is changed from volume diffusion of carbon to boundary diffusion of the alloying element. These diffusion coefficients are not so different and the transition may thus be rather gradual.

Pearlite in Fe-C-Cr

It is rather difficult to analyze the experimental information in the Fe-C-Cr system because of the uncertainty of how to extrapolate the thermodynamic properties from higher temperatures. It may for instance be mentioned that Coates (10) reached the conclusion that the full local equilibrium hypothesis was not enough for an explanation of the experimental information and suggested some "solute drag" effect. It now seems that his phase diagram was in serious error. A new evaluation by Sharma et al. (35) indicates that the full local equilibrium is indeed sufficient for explaining the ex-

The Existence of the Alloy Spike

We have seen that a large amount of information on the effect of alloying elements on pearlite can be accounted for by the use of the full equilibrium assumption. This is a very surprising result because the width of the alloy spike, when evaluated as D_M/v, is usually much smaller than the atomic dimensions. In order to demonstrate this fact, a curve corresponding to $D_M/v = 1 \text{Å}$ is compared in Fig. 21 with the growth rate curves for all the manganese steels discussed. Practically all the measurements are carried out under such conditions that the spike can be estimated to orders of magnitude smaller than the atomic dimensions. In view of this result one should of course have expected true paraequilibrium.

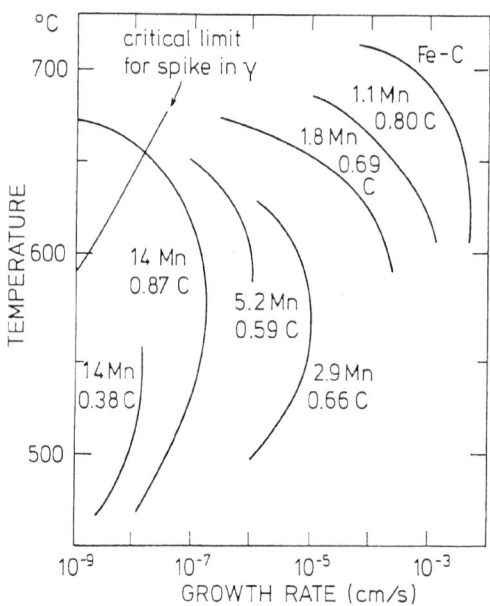

Fig. 21 The critical limit for the manganese spike in the austenite in front of the pearlite (defined by the width of 1 Å), compared with the positions of the experimental observations.

Already when the width of the spike was first discussed (6) it was emphasized "that it is not certain that the diffusion constant D_M has the same value immediately in front of the growing interface as in the undisturbed austenite". By following this idea we should now look into the interface itself and examine what goes on there. When making that suggestion more than 10 years ago (36), I had no real hope that this should ever be possible but the new improvements of the microanalytical methods now seem to have made such studies possible (37). This is a very important field for future research.

The result, that there seems to be full local equilibrium at an interface even if the calculated width of the spike is less than the atomic dimensions, was already reached when the massive transformation was reviewed recently (19). The suggestion was then made that a condition close to full equilibrium may be established between two phases if the phase interface has a sufficiently well disordered structure to give a high atomic mobility in the interface layer separating the two phases. The present results may be taken as support for that suggestion because they indicate that pearlite with its incoherent interface grows under full local equilibrium while bainite with a much more coherent interface does not grow under full local equilibrium. It may seem that Hultgren was right when suggesting that the difference in alloy partitioning is due to a major difference between pearlite and bainite. This difference may be the properties of the phase interfaces which are probably responsible for the development of the two different structures (20).

In this connection it may be emphasized that true paraequilibrium can only be expected if the phase interface does not allow individual movements of the metallic atoms and, when this condition is fulfilled, the growth rate should be as high as the diffusion of the carbon atoms would permit. For the austenite-to-ferrite reaction this may require a martensitic interface. Since the bainite reaction in nickel steels can occur somewhat inside the $\alpha+\gamma$ two-phase field but not very close to the paraequilibrium line and since it can be a very slow reaction, it seems that the bainitic interface does not move by a martensitic mechanism.

Another astonishing result of the present evaluation is that the normal grain boundary diffusion coefficient, determined on stationary boundaries, is enough for explaining the rate of partitioning of the alloying elements. This is not in agreement with the new results that the grain boundary diffusivity may increase by several orders of magnitude when the grain boundary is moving. The same effect could have been expected for diffusion in phase boundaries.

Conclusions

Although based upon uncertain thermodynamic extrapolations and uncertain values for the boundary diffusion coefficients, a comparison between calculations and experiments has yielded strong indications that pearlite in alloyed steels is growing under conditions that approach full local equilibrium, called orthoequilibrium by Hultgren. The growth conditions for bainite deviate from orthoequilibrium but do not seem to approach true paraequilibrium very closely.

Hultgren originally suggested that the difference in the partitioning of the alloying elements observed for pearlite and bainite is due to a major difference between them. The present results support his view. The basic difference between the two microstructures is probably the structure of the austenite/ferrite interface which might be incoherent for pearlite but partially coherent for bainite.

References

1. F.E. Bowman, Trans. ASM 37, 112 (1945).
2. F.E. Bowman, Trans. ASM 38, 61 (1946).
3. A. Hultgren, Trans. ASM 39, 915 (1947).
4. A. Hultgren, Metallographic Study on Tungsten Steels, John Wiley and Sons, New York (1920).
5. A. Hultgren, Jernkont. Ann. 135, 403 (1951).
6. M. Hillert, Internal Report, Swedish Inst. for Metal Research 1953.
7. J.S. Kirkaldy, Can. J. Phys. 36, 907 (1958).
8. G.R. Purdy, D.H. Weichert and J.S. Kirkaldy, Trans. TMS-AIME 230, 1025 (1964).
9. M.P. Puls and J.S. Kirkaldy, Met. Trans. 3, 2777 (1972).
10. D.E coates, Met. Trans. 4, 2313 (1973).
11. M.L. Picklesimer, D.L. McElroy, T.M. Kegley Jr., E.E. Stansbury and J.H. Frye, Trans. TMS-AIME 218, 473 (1960).
12. N.A. Razik, G.W. Lorimer and N. Ridley, Acta Met. 22, 1249 (1974).
13. N.A. Razik, G.W. Lorimer and N. Ridley, Met. Trans. 7A, 209 (1976).
14. S. Al-Salman, G.W. Lorimer and N. Ridley, Met. Trans. 10A, 1703 (1979).
15. S. Al-Salman, G.W. Lorimer and N. Ridley, Acta Met. 27, 1391 (1979).

806 M. Hillert

16. M.K. Miller and G.D.W. Smith, Metal Sci. 11, 249 (1977).
17. B. Uhrenius, Internal Report, K.T.H. 1970.
18. J. Fridberg, L.-E. Törndahl and M. Hillert, Jernkont. Ann. 153, 263 (1969).
19. M. Hillert, Met. Trans. (in press).
20. M. Hillert, Decomposition of Austenite by Diffusional Processes, p. 241, John Wiley (1962).
21. D. Brown and N. Ridley, J. Iron Steel Inst., 207, 1232 (1969).
22. J.P. Sheehan, C.A. Julien and A.R. Troiano, Trans. ASM 41, 1165 (1949).
23. F.W. Jones and W.I. Pumphrey, J. Iron Steel Inst. 163, 121 (1949).
24. M. Hillert, Darken Conference on Physical Chemistry of Steel Making, U.S. Steel (1977).
25. M. Hillert, Phase Transformations, p. 181, ASM (1970)
26. B.E. Sundqvist, Acta Met. 16, 1413 (1968).
27. Kehsin Kuo, J. Iron Steel Inst. 176, 433 (1955).
28. J.W. Cahn and W.C. Hagel, Decomposition of Austenite by Diffusional Processes, p. 131. John Wiley (1962).
29. J.W. Cahn and W.C. Hagel, Acta Met. 11, 561 (1963).
30. H. Depken, Thesis, K.T.H. Stockholm 1971.
31. M. Hillert, Acta Met. 19, 769 (1971).
32. J. Fridberg, Thesis, K.T.H. Stockholm 1971.
33. A. Hultgren and O. Edström, Jernkont. Ann. 121, 163 (1937).
34. J. Fridberg and M. Hillert, Acta Met. 18, 1253 (1970).
35. R.C. Sharma, G.R. Purdy and J.S. Kirkaldy, Met. Trans. 10A, 1129 (1979).
36. M. Hillert, The Mechanism of Phase Transformations in Crystalline Solids, p. 231, Monograph and report series No. 33, Inst. Metals (1969).
37. P.W. Bach, A FIM-Atom Probe Investigation of the Bainite Transformation in a Cr-Mo Steel. Thesis, Technische Hogeschool Twente, 1981.

20 Introduction to "Thermodynamics of the Massive Transformations"

published in *Metallurgical Transactions A* (1984)

and "Massive Transformations in the Fe-Ni system"

in *Acta Materialia* (2000)

It is a pleasure to acknowledge Mats Hillert's contributions to the study and understanding of massive transformations. The two papers on which I was invited to comment, both refer to the intriguing question of whether or not the massive transformation can occur in a two-phase field, and if it does, how close to the T_0 line can it occur?

Using his extensive knowledge of phase transformations, precipitation phenomena and thermodynamics, Prof. Hillert addressed the above question in a number of papers both by himself and with associates. During the 1980's he was more inclined to conclude that what he called "the critical limit" for the occurrence of the massive transformation was more likely to lie very close to the single phase field representing the massive phase (*i.e.* α in Cu-Zn, and α in Fe-Ni), but in his later papers he concluded that the occurrence in the two-phase field was a demonstrated fact, and he concentrated on the factors that may contribute to the extension of the critical limit towards the T_0 line. The discussions of these topics are well recorded in the literature, particularly following the Symposium on Massive Transformations in 2002, in St. Louis (*Met. Trans.* 2002, 33A, 2277–2445).

The basic contribution from Mats ideas is the need to consider under what conditions the expected short-range local equilibrium at the moving transformation interface can become sufficiently unimportant to allow critical extensions towards the T_0. Here, ultimately, he concluded that "one could add the argument that it may not even be correct to locate the critical limit by extrapolation to zero velocity", as some earlier workers tried to do for the Cu-Zn alloys. By this statement Prof. Hillert demonstrated his quest for a truly scientific approach to a controversy, which allows one to modify one's ideas as knowledge increases. In their paper on the Fe-Ni system, Borgenstam and Hillert demonstrated that the "critical limit" may also depend on the temperature at which the massive transformation occurs. At lower temperatures it may move inside the two-phase field, but the authors still argue that the "limiting composition never approaches the T_0 line". This is still a proposition that may require future research.

I would like to congratulate Mats on his 80th birthday celebration, and on his numerous and lasting contributions to further advance our knowledge of solid state transformations.

Thaddeus B. MASSALSKI

Thermodynamics of the Massive Transformation

MATS HILLERT

The thermodynamic nature of the massive transformation may be revealed by finding the limiting conditions for the massive mode of growth. A review of the experimental information available indicates strongly that the solvus line rather than the T_0 line is the natural limit. Some observations of massive transformation inside the two-phase field may be explained by the effect of coherency strains in the composition spike formed in front of the interface. This spike has a significant thickness in many cases and can explain the role of the solvus. It is exceedingly thin in iron-base alloys, and the role of the solvus may there be explained by higher diffusivity inside the interface. A comparison is made with the transformation of γ in Fe-M-C alloys where the situation is very similar.

I. INTRODUCTION

THE rate of growth of a new phase in an alloy is often controlled by the rate of diffusion. In order to calculate the rate of diffusion it is necessary to know the boundary conditions at the phase interface. It is then customary to assume that there is local equilibrium at the phase interface. This local equilibrium concept is illustrated for a $\beta \rightarrow \alpha$ transformation in Figure 1, and it is evident that it predicts that the new phase should grow with a composition on the solvus. The composition may be somewhat displaced into the α single-phase region by the action of the surface tension if the new phase develops protruding tips.

Even though the local equilibrium concept has been very successful in a great number of investigations, it may be expected to fail at high growth rates where an appreciable part of the driving force may be needed for driving the interface. Very high growth rates may be obtained during solidification, in particular when the splat cooling technique is used. This possibility of testing the limitations of the local equilibrium concept was used by Baker and Cahn[1,2] who studied Zn-Cd alloys. This system has a retrograde solidus (Figure 2), and the applicability of the local equilibrium concept can easily be tested by determining whether an alloy with a higher solute content than the maximum solidus composition can solidify to a homogeneous solid phase. They found that this is possible up to about twice the maximum. Composition-invariant solidification may thus be obtained well inside the two-phase field. The conclusion was that the local equilibrium concept does not apply to very rapid solidification. It may thus seem reasonable to expect that the natural limit for composition invariant solidification is not the solidus but the T_0 line where the two phases have equal Gibbs energy values. How close to the T_0 line it can occur would depend upon other properties of the system.

II. COMPOSITION INVARIANT TRANSFORMATION IN SOLID STATE

In solid alloys there are two different kinds of composition invariant phase transformation with high growth rates. They are usually called martensitic and massive and are known from allotropic transformations of pure metals. The martensitic transformation presumably takes place by a cooperative movement of the atoms across the phase interface and results in a shape change and a surface tilt. A martensitic unit grows preferentially in some crystalline direction and cannot cross a grain boundary in the parent phase. The phase interface is probably a highly coherent boundary which can move quickly by the motion of glissile dislocations. A massive transformation usually proceeds with approximately the same rate in all directions and can easily cross the grain boundaries in the parent phase. It does not produce any surface tilt. The phase interface is probably incoherent and, presumably, it moves by thermally activated jumps of individual atoms.

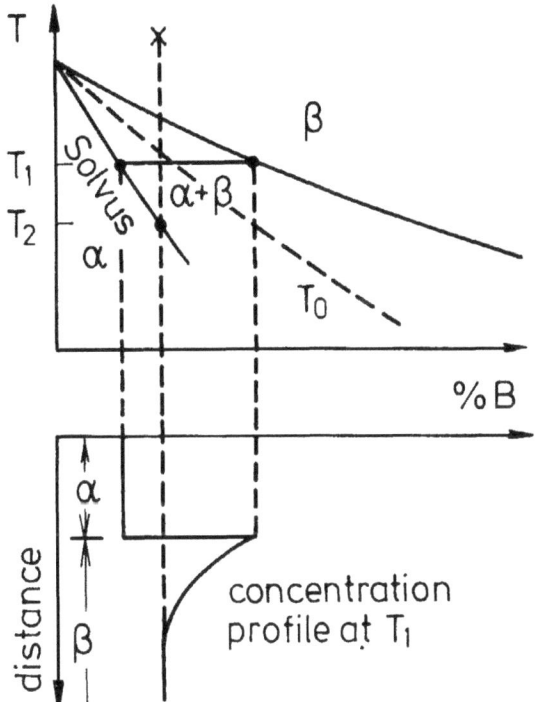

Fig. 1—Growth of α from β under local equilibrium at T_1. The alloy composition is marked with an x. The growing α is on the solvus curve. Composition invariant transformation of this alloy could occur at T_2.

MATS HILLERT is Professor of Physical Metallurgy, Royal Institute of Technology, S-10044 Stockholm 70, Sweden.
This paper is based upon a presentation made at a symposium on The Massive Transformation, held at the Pittsburgh meeting of The Metallurgical Society of AIME and the Materials Science Division of ASM, October 9, 1980, under the sponsorship of the MSD Phase Transformations Committee.

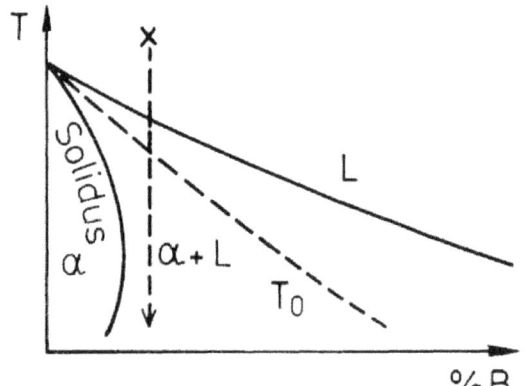

Fig. 2 — Solidification in a system with a retrograde solidus. The indicated alloy cannot undergo a composition invariant solidification at any temperature if local equilibrium is required.

It is well known that the martensitic transformation can occur inside a two-phase field and it is usually believed to be intimately related to the T_0 line which is the theoretical limit for a composition invariant transformation, from a simple thermodynamic point of view. However, due to the mechanism of interface migration and the shape change, a considerable driving force is needed for a martensitic transformation. As a consequence, the martensitic transformation does not start until sometimes appreciable supercooling is provided. The so-called M_s line can thus be regarded as a displaced T_0 line.[5]

Much less is known about the massive transformation, but recently there have been many efforts to establish under what phase diagram conditions it can occur. The results of this work will now be examined. The interesting question is whether the maximum temperature of the massive transformation is the solvus line or the T_0 line.

III. EXPERIMENTAL INFORMATION ON NONFERROUS ALLOYS

Karlyn, Cahn, and Cohen (KCC)[3] chose to study the $\beta \rightarrow \alpha_m$ transformation in the Cu-Zn system which has a retrograde solvus (Figure 3). In contrast to the result of the solidification experiments which were actually made subsequently, they did not find the massive transformation in alloys with a higher zinc content than the maximum of the α solvus line. With an alloy of 38 at. pct Zn they obtained the massive transformation between T_1 and T_2 but not in the two-phase fields below T_1 and above T_2. On the other hand, when the massive transformation was allowed to start between T_1 and T_2, it was found to be able to continue when the temperature was raised just above T_2. They concluded that the nucleation of the massive reaction is confined to the α single-phase region, but the natural limit for growth is the T_0 line. Their experimental results have been checked in several investigations by Massalski and his collaborators,[5-8] and it now seems well established that the massive $\beta \rightarrow \alpha_m$ transformation in the Cu-Zn system can occur inside the $\alpha + \beta$ two-phase field up to a distance of 0.45 at. pct Zn from the solvus. They extended the work to related systems. In the Cu-Al system the region for the massive transformation was found to extend 1.2 at. pct into the two-phase

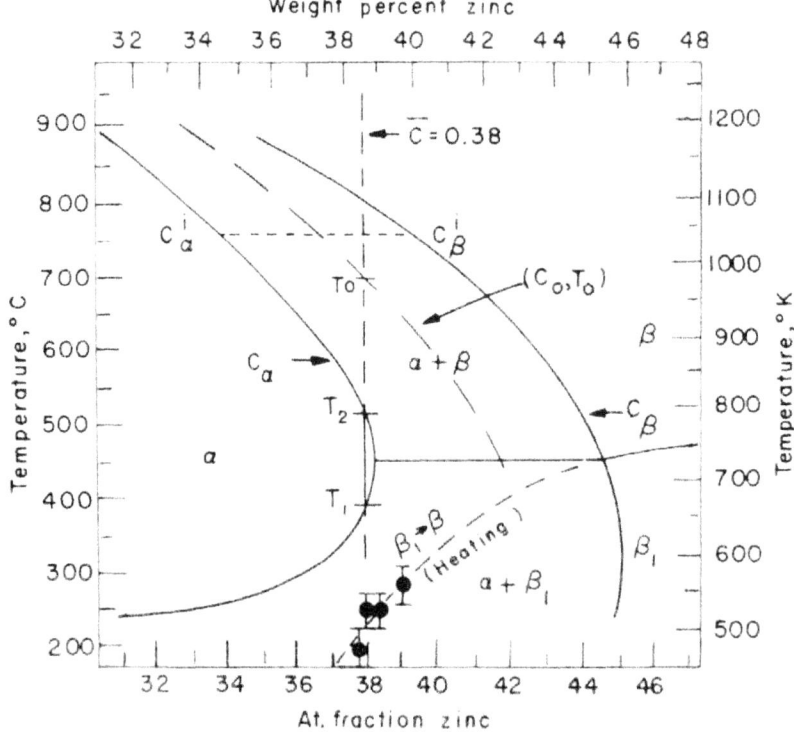

Fig. 3 — The Cu-Zn phase diagram showing the conditions of the experiment by KCC. From Ref. 3.

field and in the Ag-Cd system 1.5 at. pct. On the other hand, no extension was found in the Ag-Zn system. An uncertainty in these investigations is the exact position of the solvus line, and the conclusions may thus be modified to some degree. For example, an examination of the experimental determination of the α solvus line in the Ag-Cd system indicates that this line should be drawn at a slightly higher Cd content. This is shown by the full α solvus line in Figure 4. The dashed line is taken from Reference 5. The experimental data might thus be only 0.6 at. pct inside the solvus. In any case, it seems to be well established that the massive transformation can extend somewhat into the two-phase field. Even though this extension is small, Massalski et al.[7] seem to maintain that the T_0 line is the natural limit for the massive transformation and that the distance between the T_0 line and the line for the start of the massive transformation represents the undercooling needed to obtain the massive transformation in competition with other transformation modes.

Another way to find the natural limit for the massive transformation would be to study its growth rate and to examine where it goes to zero. Such a study has been made by KCC, and the dashed curve in Figure 5 shows their interpretation which was based upon a simple growth model where the undercooling below T_0 gives the driving force. Unfortunately, the growth rate measurements were not very accurate and the solid line indicates that they could just as well be extrapolated to zero at the temperature limit for the observation of the massive transformation (525 °C), according to Massalski et al.[7] This curve was calculated with the same type of equation used by KCC and with the same activation energy of 61 kJ/mol.

Ayers and Massalski made a study of the overall reaction kinetics of the transformation in the Cu-Zn system. A logarithmic plot of the time for half reaction in a Cu-38 at. pct Zn alloy vs the reciprocal temperature gives the diagram presented in Figure 6. 525 °C is the experimental limit of the massive transformation in this alloy, and it is again found that the data may be extrapolated to the value zero at this limit. The curve was calculated with an expression closely related to the one used by KCC,

$$1/t = K(T_2 - T)(T - T_1) \exp(-Q/RT) \quad [1]$$

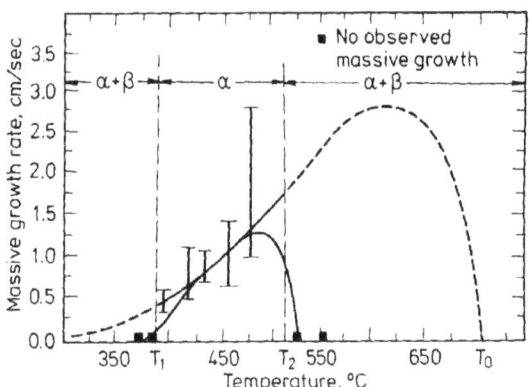

Fig. 5 — Growth rate measurements on the massive $\beta \rightarrow \alpha$ transformation in a Cu-38 at. pct Zn alloy by KCC. The dashed line is their theoretical interpretation. The full line which goes to zero at 377 and 525 °C represents an alternative interpretation.

and the values $T_1 = 650$ K, $T_2 = 798$ K, and $Q = 120$ kJ/mol were used. It was not possible to get a reasonable representation using the Q value of 61 kJ/mol selected by KCC. Some support for an abrupt extrapolation toward low rates, shown by the curve in Figure 6, is lent by the observation of simultaneous growth of massive α and Widmanstätten α in the Cu-Zn system, reported by Ayers.[6]

It is evident from Figures 5 and 6 that higher precision data are required in order to define an experimental point of zero growth rate. In particular, it is important to make measurements close to the experimental limit for the occurrence of the massive transformation. Such data have recently been published for the $\beta \rightarrow \zeta$ massive transformation in the Ag-Al system by Perepezko and Massalski.[10] Unfortunately, they studied alloys where the solvus line and the T_0 line are very close together, and their data cannot be used to test whether the growth rate should go to zero at one or the other.

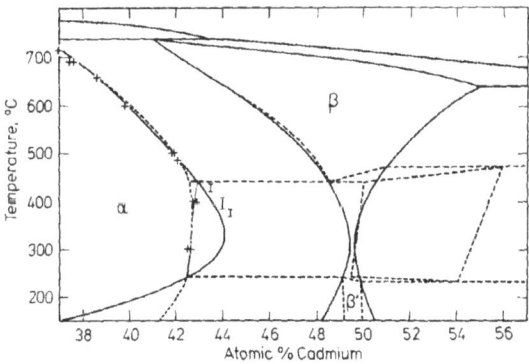

Fig. 4 — The Ag-Cd phase diagram with the starting temperature of the massive $\beta \rightarrow \alpha$ transformation according to Ayers. The phase diagram has been modified according to equilibrium measurements by Owen and Roberts, Phil. Mag., 1939, vol. 27, p. 295. The ζ phase has been excluded in order to show the extension of the α solvus.

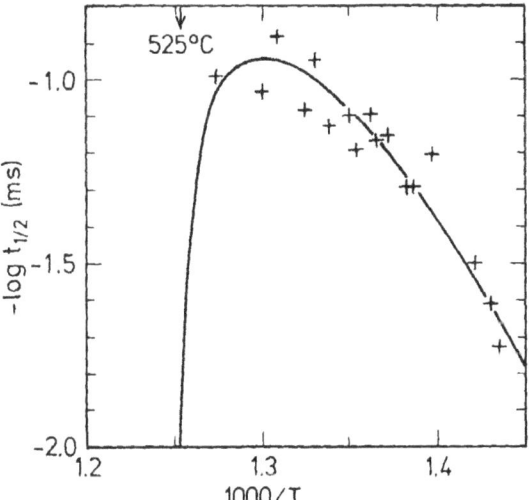

Fig. 6 — Kinetic data for the $\beta \rightarrow \alpha$ massive transformation in a Cu-38 at. pct Zn alloy according to Ayers and Massalski. The curve represents a possible interpretation.

On the other hand, their results show that the massive transformation does not need a special driving force in order to occur when there are no competing reactions. In this connection, it may be mentioned that Kittl, Serebrinsky, and Gomez[11] were also able to measure a very slow rate of transformation at the experimental limit of observation of the $\zeta \rightarrow \beta'$ massive transformation in the Ag-Cd system.

The massive transformation has also been studied in binary alloys of an element with an allotropic transformation. Two recent studies by Plichta, Williams, and Aaronson[12] and Plichta, Aaronson, and Perepezko[13] have failed to produce any evidence that the massive $\beta \rightarrow \alpha_m$ transformation in the Ti-Ag, Ti-Au, and Ti-Si systems can occur in the two-phase field.

IV. EXPERIMENTAL INFORMATION ON FERROUS ALLOYS

Finally, we should consider the experimental information obtained from binary iron alloys. Bibby and Parr[14] studied the allotropic transformation in pure iron over a wide range of cooling rates. They observed two plateaux, one connected with the massive mode of transformation, the other connected with the martensitic mode. Swanson and Parr[15] extended this study to binary Fe-Ni alloys and could follow the two plateaux into the binary system. For the massive plateau they obtained a very smooth curve up to 5 at. pct Ni, which is inside the α single-phase field, and then an additional point at 7 at. pct Ni which does not seem to fall in line with the rest (Figure 7). One may only speculate whether this is connected with the fact that the latter point falls within the $\alpha + \gamma$ two-phase field.

The nature of the massive plateau is unknown and its existence has even been questioned.[16] At present one should not base any conclusion on it. For our discussion it seems more important to examine the starting temperature of the massive transformation. In their classical work, reported in 1949, Jones and Pumphrey[17] studied the starting temperature of the $\alpha \rightarrow \gamma$ and $\gamma \rightarrow \alpha$ transformations in Fe-Ni alloys. Their results are compared with the equilibrium lines of the phase diagram in Figure 7. At high temperatures there is a remarkably good agreement between the data and the equilibrium lines. The data obtained on cooling start to deviate at about 5 at. pct Ni and 900 K. The data obtained on heating start to deviate at about 25 at. pct Ni and 800 K. In order to interpret this information it is necessary to know in what range it represents the massive mode of transformation. The authors did not concern themselves with that problem, and the more recent literature is rather conflicting. For instance, many authors distinguish between equiaxed α and massive α. Equiaxed α in this system is certainly identical to massive α according to our definition. One may conclude that the information from low nickel contents concerns the massive mode of transformation. It seems that the transformation product changes with the nickel content and possibly also with the cooling rate. The dislocation density increases and the phase interface becomes more jagged. As an example, Owen, Wilson, and Bell[18] observed a high dislocation density in a 6 at. pct Ni alloy after quenching at 600 K per second. Speich and Swann[19] found a cell structure and an increased yield strength at and above 6.7 pct Ni. It seems possible that these observations are related to the fact

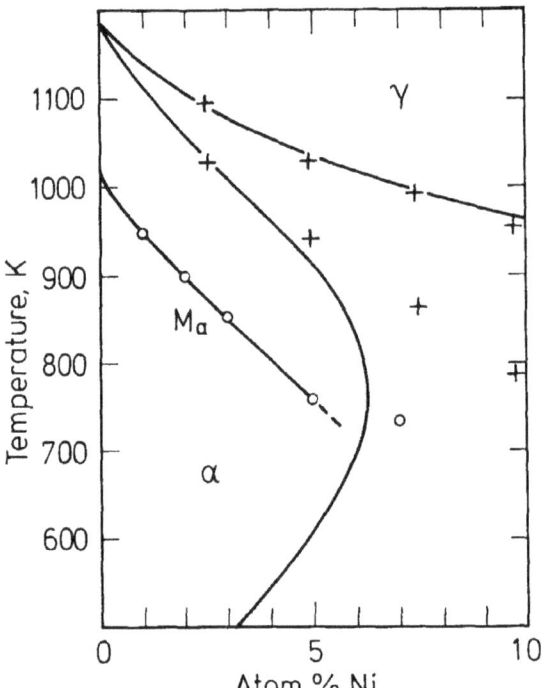

Fig. 7—The $\gamma \rightarrow \alpha$ massive plateau (M_a) in the Fe-Ni system according to Swanson and Parr and the start of the $\alpha \rightarrow \gamma$ and $\gamma \rightarrow \alpha$ transformations according to Jones and Pumphrey. The phase diagram lines are from Hillert, Wada, Wada, *J. Iron Steel Inst.*, 1967, vol. 205, p. 539.

that the massive transformation from γ to α starts to enter the $\alpha + \gamma$ field at about 5 at. pct Ni. It is very uncertain how far the massive $\gamma \rightarrow \alpha$ transformation extends into this phase field. Massalski, Perepezko, and Jaklovsky[20] claim to have observed massive α at 8.7 at. pct Ni.

The results of Jones and Pumphrey on Fe-Mn alloys are very similar (Figure 8). On cooling, their data start to move into the two-phase field at about 3 at. pct Mn and 900 K.

V. THE ROLE OF COHERENCY STRAINS IN THE SPIKE

The general impression is thus that there is remarkably close agreement between the starting temperature for the massive transformation and the solvus temperature. It does not seem reasonable to explain this agreement as a coincidence caused by the need for some undercooling below the T_0 line. It may be more profitable to accept that the solvus is the natural limit for the massive transformation and to look for a special explanation when a deviation is found experimentally. Such an explanation for the deviations found in the Ag-Cd, Cu-Al, and Cu-Zn systems may be found if we accept the following description of the role of the solvus.[21]

Let us modify the diagram in Figure 1 by decreasing the transformation temperature in such a way that the initial alloy falls on the solvus (Figure 9). According to the local equilibrium concept the α phase can now grow with

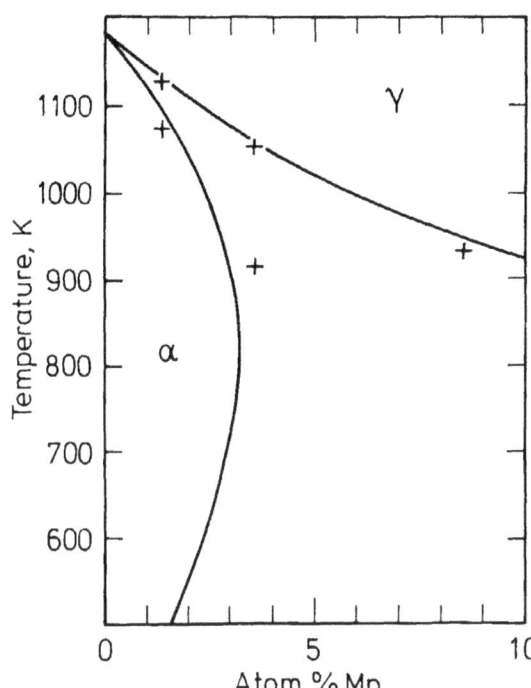

Fig. 8—The start of the $\alpha \rightarrow \gamma$ and $\gamma \rightarrow \alpha$ transformations in the Fe-Mn system according to Jones and Pumphrey. The phase diagram lines are from Hillert, Wada, Wada, *J. Iron Steel Inst.*, 1967, vol. 205, p. 539.

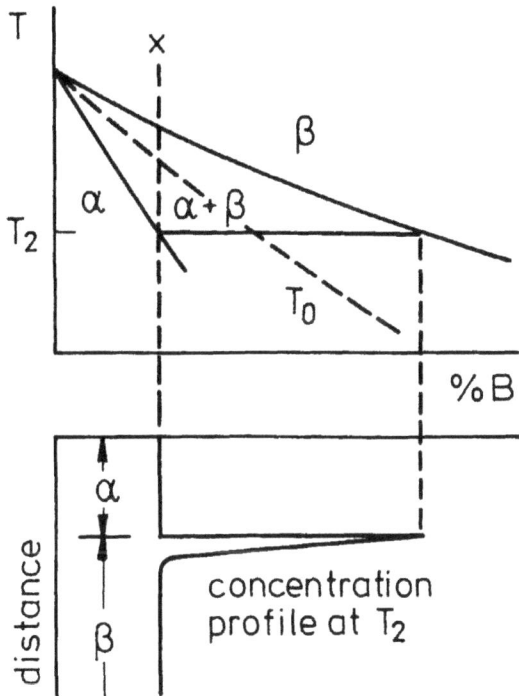

Fig. 9—Composition invariant $\beta \rightarrow \alpha$ transformation under local equilibrium at T_2.

the same composition as the initial parent phase, and a composition invariant transformation thus seems possible. However, there will be a local enrichment in front of the advancing interface. The thickness of this composition spike will be determined by the rate of diffusion D in comparison with the rate of growth v, and it may be estimated as D/v. Its height will be independent of the growth rate as long as the local equilibrium concept applies, and it can be shown that all the driving force available for the composition-preserving transformation will be absorbed by the process of diffusion at this particular temperature. The rate of the composition-invariant reaction would thus be negligible. On the other hand, the driving force will increase further if the parent phase is moved into the α one-phase field by a further lowering of the temperature. Some net driving force will then be available for driving the composition-invariant transformation itself, and it may then take place with a rate controlled by the net driving force.

A factor which has not yet been considered is the coherency strains in the spike caused by the difference in atomic sizes of the two components. A very similar situation was recently considered in connection with the discontinuous type of precipitation.[20] Provided that the spike is very thin and completely coherent with the rest of the parent phase, one may derive the following expression for the strain energy per mole of a layer of composition x^β,

$$G_{el}^\beta = \frac{E}{1-\nu}\left(\frac{d \ln a}{dx}\right)^2 (x_1^\beta - x^\beta)^2 \qquad [2]$$

x_1^β is the alloy composition, a is the lattice parameter of the parent β, E is the modulus of elasticity, and ν the Poisson ratio. The material has been assumed to be isotropic. By including this energy in the Gibbs energy of the layer, one obtains the following elastic contributions to the chemical potentials,

$$G_{Ael}^\beta = K[(x_1^\beta)^2 - (x^\beta)^2] \qquad [3]$$

$$G_{Bel}^\beta = K[(1 - x_1^\beta)^2 - (1 - x^\beta)^2] \qquad [4]$$

where

$$K = \frac{E}{1-\nu}\left(\frac{d \ln a}{dx}\right)^2 \qquad [5]$$

The effect on the two-phase equilibrium is demonstrated in Figure 10. For simplicity we shall now assume that the two phases are ideal solutions. The following equilibrium conditions are then obtained,

$$°G_A^\alpha + RT \cdot \ln(1 - x^\alpha) = °G_A^\beta + RT \cdot \ln(1 - x^\beta)$$
$$+ K[(x_1^\beta)^2 - (x^\beta)^2] \qquad [6]$$

$$°G_B^\alpha + RT \cdot \ln x^\alpha = °G_B^\beta + RT \cdot \ln x^\beta$$
$$+ K[(1 - x_1^\beta)^2 - (1 - x^\beta)^2] \qquad [7]$$

The equilibrium compositions at the interface when the growing α phase has the same composition as the parent β phase, $x^\alpha = x_1^\beta$, may thus be calculated from the following two equations,

$$1 - x^\beta = (1 - x^\alpha) \exp \frac{1}{RT}$$
$$\cdot \{°G_A^\alpha - °G_A^\beta - K[(x^\alpha)^2 - (x^\beta)^2]\} \qquad [8]$$

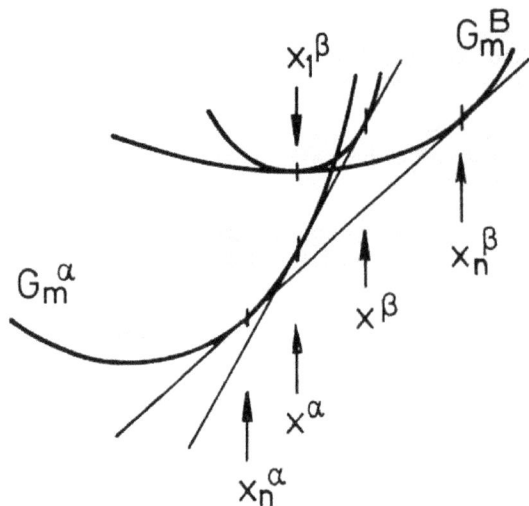

Fig. 10 — Gibbs energy diagram for the $\alpha + \beta$ two-phase equilibrium. x_n^α and x_n^β are the normal equilibrium compositions. x^α and x^β are the equilibrium compositions modified by the strain energy in the spike if the bulk of the β phase x_1^β is equal to x^o.

$$x^\alpha = x^\beta \exp \frac{1}{RT} \{{}^oG_B^\beta - {}^oG_B^\alpha + K[(1 - x^\alpha)^2 - (1 - x^\beta)^2]\} \quad [9]$$

By introducing the normal equilibrium compositions, here identified with a subscript n, we find

$$1 - x_\beta = (1 - x_\alpha)\left(\frac{1 - x^\beta}{1 - x^\alpha}\right)_n \exp \frac{K}{RT}[(x^\beta)^2 - (x^\alpha)^2]$$

[10]

$$x^\alpha = x^\beta \left(\frac{x^\alpha}{x^\beta}\right)_n \exp \frac{K}{RT}[(1 - x^\alpha)^2 - (1 - x^\beta)^2] \quad [11]$$

This system of equations can easily be solved numerically by iteration for any case. This was done for some systems, and the results are presented in Table I. For simplicity, the variation of the lattice parameter was taken from the terminal solutions. The temperatures and compositions were chosen close to the regions where there is experimental information on the massive transformation. The compositions are given in at. pct. The last column of the table lists the values thus predicted for the displacement of the $\alpha/\alpha + \beta$ solvus, due to coherency strains. The magnitude found for Cu-Zn and Ag-Cd is in excellent agreement with the experimental difference between the start of the massive transformation and the normal solvus. One may thus conclude that the solvus line is the natural limit for the massive transformation in these systems, and this indicates that the condition of local equilibrium may be established very well during the massive transformation in these systems.

VI. THE WIDTH OF THE SPIKE

We have thus been led to conclude that local equilibrium may hold rather well at the phase interface during a massive transformation. Indeed, this should be a natural expectation if the width of the spike is large enough. It is thus of interest to estimate this width from D/v. For the Cu-Zn system[23] we find a value of $D = 6 \cdot 10^{-8}$ cm^2 per second. With the growth rate $v = 1$ cm per second we obtain a width of 6 Å. This is only a few times the atomic dimensions, and one may wonder how such a thin spike should look. On the other hand, a spike of such thickness cannot be ruled out completely, and it cannot be justified to expect T_0 to be the natural limit for the massive transformation under such conditions. The situation may be similar in the other systems based upon Cu and Ag. Even without access to any diffusion data it may be justified to conclude that the spike has a considerable thickness in all those cases where the Widmanstätten precipitation is a severe competitor to the massive transformation, because the Widmanstätten precipitation depends upon volume diffusion. In all such cases we should thus expect the solvus line to be the natural limit for the massive transformation.

In the iron-base alloys of the type Fe-Ni and Fe-Mn one must employ exceedingly low cooling rates in order to get Widmanstätten precipitation, and this is related to the fact that the estimated width of the spike in the massive transformation might be 10^{-5} Å or less. From a simple thermodynamic consideration it may thus be tempting to expect the massive transformation to start very close to the T_0 line which is situated approximately in the middle of the $\alpha + \gamma$ field; it is thus very surprising that the start of the massive transformation in the Fe-Ni and Fe-Mn systems falls so close to the phase boundaries at low alloy contents. Before discussing this problem further it may be instructive to consider a similar situation in ternary iron alloys containing carbon.

VII. ALLOY INVARIANT TRANSFORMATIONS IN Fe-M-C ALLOYS

It was observed by Hultgren[24,25] that pearlite in alloyed steels usually forms with partitioning of the alloying elements between the growing ferrite and cementite, whereas bainite always grows without any such partitioning, i.e., under an invariant alloy content. With respect to the alloying

Table I. Estimation of the Effect of Coherency Strains in the Spike on the Local Equilibrium at the Interface of a Growing Massive Phase. The Mole Fractions Are Given in Pct.

A	B	$\Delta a/a$	T °C	K/RT	x_n^α	x_n^β	x_{coh}^α	x_{coh}^β	Δx^α
Ag	Al	0.029	600	0.14					≤0.1
Ag	Cd	0.044	400	0.42	43.6	49.0	44.1	48.5	0.5
Ag	Zn	0.047	400	0.48	39.0	44.0	39.5	43.5	0.5
Cu	Al	0.067	565	0.81	19.6	24.0	20.0	23.5	0.4
Cu	Zn	0.056	500	0.62	38.0	44.6	38.7	43.8	0.7
Fe	Mn	0.028	550	0.27	3.2	19.0	3.3	18.4	0.1

elements this is identical to the massive transformation, the only difference being that, in addition, there is a mobile element present, carbon. Hultgren was familiar with applying the local equilibrium concept to the Fe-C system,[26] and he now suggested that bainite grows under a partial equilibrium at the interface which concerns carbon but not the alloying elements. This he called paraequilibrium. He suggested that the paraequilibrium lines in the phase diagram fall inside the normal two-phase field. This was later proved by the present author.[27] In fact, the paraequilibrium can be related to the T_0 line in the Fe-M system where M stands for the alloying element. Figure 11 illustrates this relation in an isothermal section for an alloying element that stabilizes austenite.[21] However, the present author also pointed out that it is possible to preserve the alloying composition even if there is complete local equilibrium at the interface.[27] What is required is the same type of spike which was demonstrated in Figure 9. This growth condition is demonstrated in Figure 12 for a $\gamma \rightarrow \alpha$ precipitation. The critical alloy composition for this growth condition is found at the intersection of the horizontal line representing the alloying content and an isoactivity line for carbon (dashed line in Figure 12). The locus of all such points obtained for different levels of the alloying content is given in Figure 13 and compared with the critical line for obtaining α precipitation from γ under paraequilibrium. This diagram illustrates that we here have exactly the same problem as for the massive transformation. Should we expect the alloy invariant reaction below the paraequilibrium line starting at the T_0 point or not until below the "local-equilibrium" line starting at the solvus point? In fact, the massive $\gamma \rightarrow \alpha$ transformation in the Fe-M system may be regarded as a limiting case of this more complicated phenomenon. The answer proposed by the present author[27] was that the lower line should hold whenever the spike had a sufficient width in

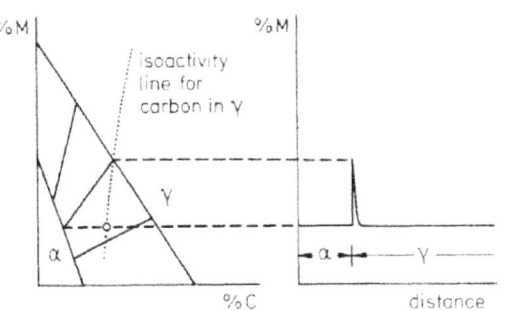

Fig. 12 — Composition invariant growth conditions for α from γ in a ternary Fe-M-C alloy assuming local equilibrium. The circle represents the critical carbon content at the selected level of alloying content.

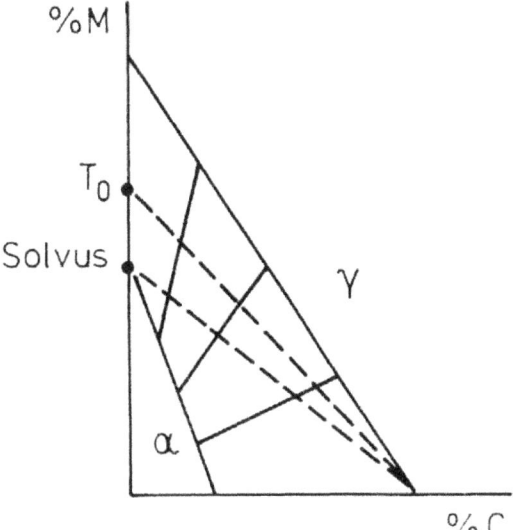

Fig. 13 — Comparison of the critical lines for alloy invariant growth of α from γ according to paraequilibrium (dashed line starting at T_0) and according to full local equilibrium (dashed line starting at the solvus point).

comparison with the atomic dimensions. Evidently, this suggestion is identical to the suggestion made later for the massive transformation in binary alloys.

When estimating the width of the spike in Fe-M-C alloys one should realize that the growth rate cannot be faster than allowed by the diffusion of carbon. The rates are thus much lower than in the massive transformations in binary systems. In spite of this, the width is very seldom found to be of atomic dimensions or larger. As a consequence, one has the same reason to expect the line starting at the T_0 point to be the natural limit for the alloy invariant reactions in Fe-M-C alloys as for the massive transformation in Fe-M alloys.

The fact that the alloying elements usually partition between ferrite and cementite during the growth of pearlite is easily explained by the rapid boundary diffusion. However, one may wonder how to evaluate the driving force for this diffusion if the volume diffusion is so slow that the parent phase cannot modify its alloy content at the phase interface. The chemical potential of an alloying element should then

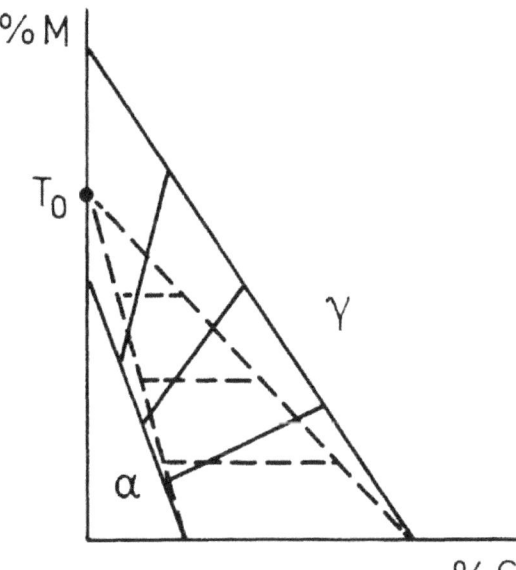

Fig. 11 — The Hultgren $\alpha + \gamma$ paraequilibrium lines (dashed) in comparison to the ordinary equilibrium lines in a ternary Fe-M-C system at a constant temperature.

be different in the two phases which are separated by the phase interface and, in the limit, one should apply Hultgren's paraequilibrium. However, there are strong indications that one obtains reasonable growth rates for pearlite if one bases the calculation on the assumption of full local equilibrium, *i.e.*, on the assumption that the parent phase has the same chemical potentials at the interface that it would have if there was a spike of significant width. The strongest indication may be that this assumption leads to a realistic prediction of the limit where pearlite should form without partitioning of the alloying element. The agreement with experiments for pearlite in the Fe-Mn-C system[28,29] is demonstrated in Figure 14 where a comparison is made with the Fe-Mn phase diagram. This result may appear very surprising because the width of the spike in the γ matrix may be calculated to about 10^{-2} Å. On the other hand, it appears rather reasonable if one considers the rapid rate of diffusion inside the interface. It is true that the boundary diffusion, by which partitioning occurs, takes place along the interface whereas the two phases have contact with each other by atoms jumping across the interface. However, it would be surprising if the rate of the two diffusion processes were not of the same order of magnitude. We may thus guess that there is a good chance for some kind of local equilibrium between the growing phase and the parent phase as long as the rate of boundary diffusion is appreciable and, from the agreement illustrated in Figure 14, we may conclude that this local equilibrium is close to full local equilibrium. In view of the experimental evidence for the role of the solvus line in the massive transformation, it is tempting to generalize this conclusion to all massive transformations whether or not they depend upon boundary diffusion. It may even seem reasonable to expect this conclusion to hold as long as the interface can move by individual jumps of the atoms, *i.e.*, in all cases where the transformation takes place by the movement of an incoherent interface, *i.e.*, for all the massive transformations according to our definition.

VIII. CONCLUSIONS

From the above considerations it appears natural to expect all the massive transformations in binary alloys and the alloy invariant reactions in Fe-M-C alloys to obey the line starting at the solvus point rather than the line starting at the T_0 point. Of course, the exact nature of the local equilibrium at any combination of temperature and interface velocity depends upon the properties of the interface. This is something we can only speculate about at present. One can construct models and simulate the reaction on a computer. As an example, one may mention a simulation of solidification[30] which demonstrated that it is easy to construct a model which can explain that composition invariant solidification in a binary alloy can occur close to the T_0 line. Similarly constructed models for solid/solid interfaces[31] will probably predict that one comes close to the solvus in a binary Fe-M alloy or close to the line starting at the solvus in an Fe-M-C alloy. On the other hand, the highly coherent, martensitic interface has quite different properties, and it is very easy to understand that the martensitic transformation can take place far inside the two-phase field if the conditions are such that the width of the spike is less than the atomic dimensions. Furthermore, the mere fact that the martensitic transformation can take place much further inside a two-phase field than a massive transformation can is a strong indication that the processes occurring inside the incoherent, massive interface are decisive for the nature of the massive transformation. Finally, one may wonder if the cases of massive transformation, which seem to have been observed somewhat inside the $\alpha + \gamma$ two-phase field in the Fe-Ni and Fe-Mn systems, may not be due to the interface gradually becoming partially coherent.

REFERENCES

1. J.C. Baker and J.W. Cahn: *Acta Metall.*, 1969, vol. 17, p. 575.
2. J.C. Baker and J.W. Cahn: in *Solidification*, ASM, Metals Park, OH, 1971, p. 23.
3. C. Zener: *Trans. AIME*, 1946, vol. 167, p. 550.
4. D.A. Karlyn, J.W. Cahn, and M. Cohen: *Trans. TMS-AIME*, 1969, vol. 245, p. 197.
5. E.B. Hawbolt and T.B. Massalski: *Metall. Trans.*, 1970, vol. 1, p. 2315.
6. J.D. Ayers: Ph.D. Thesis, Carnegie-Mellon Univ., Pittsburgh, PA, 1970.
7. T.B. Massalski, A.J. Perkins, and J. Jaklovsky: *Metall. Trans.*, 1972, vol. 3, p. 687.
8. J.D. Ayers: *Metall. Trans.*, 1974, vol. 5, p. 2389.
9. J.D. Ayers and T.B. Massalski: *Metall. Trans.*, 1972, vol. 3, p. 3185.
10. J.H. Perepezko and T.B. Massalski: *Acta Metall.*, 1975, vol. 23, p. 621.
11. J.E. Kittl, H. Serebrinsky, and M.P. Gomez: *Acta Metall.*, 1967, vol. 15, p. 1703.
12. M.R. Plichta, J.C. Williams, and H.I. Aaronson: *Metall. Trans. A*, 1977, vol. 8A, p. 1885.
13. M.R. Plichta, H.I. Aaronson, and J.H. Perepezko: *Acta Metall.*, 1978, vol. 26, p. 1293.

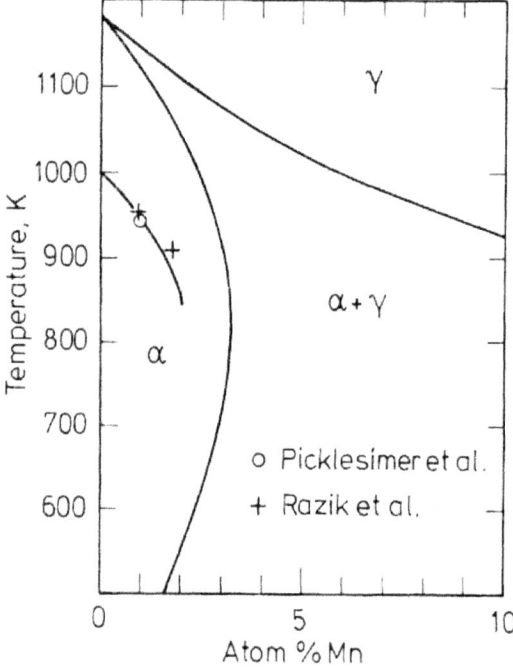

Fig. 14—The critical limit for the growth of pearlite without partitioning of Mn according to experiments and a calculation based upon the local equilibrium concept.

14. M. J. Bibby and J. G. Parr: *J.I.S.I.*, 1964, vol. 202, p. 100.
15. W. D. Swanson and J. G. Parr: *J.I.S.I.*, 1964, vol. 202, p. 104.
16. S. K. Bhattacharyya, J. H. Perepezko, and T. B. Massalski: *Scripta Met.*, 1973, vol. 7, p. 485.
17. F. W. Jones and W. I. Pumphrey: *J.I.S.I.*, 1949, vol. 163, p. 121.
18. W. S. Owen, E. A. Wilson, and T. Bell: in *High-Strength Materials*, V. F. Zackay, ed., John Wiley and Sons, New York, NY, 1965, pp. 167-208.
19. G. R. Speich and P. R. Swann: *J.I.S.I.*, 1965, vol. 203, p. 480.
20. T. B. Massalski, J. H. Perepezko, and J. Jaklovsky: *Materi. Sci. Eng.*, 1975, vol. 18, p. 193.
21. M. Hillert: Monograph and Report Series no. 33, Inst. of Metals, 1969, p. 231.
22. M. Hillert: *Metall. Trans.*, 1972, vol. 3, p. 2729.
23. U. Landegren, C. E. Birchenall, and R. F. Mehl: *Trans. AIME*, 1956, vol. 206, p. 73.
24. A. Hultgren: *Trans. ASM*, 1947, vol. 39, p. 915.
25. A. Hultgren: *Kungl. Vet. Akad. Handl.*, 1953, vol. 4, bd 4, no. 3.
26. A. Hultgren: *A Metallographic Study on Tungsten Steels*, John Wiley and Sons, New York, NY, 1920, p. 30.
27. M. Hillert: "Paraequilibrium," Internal Report, Swedish Inst. for Metal Research, 1953.
28. M. L. Picklesimer, D. L. McElroy, T. M. Kegley, E. E. Stansbury, and J. H. Frye: *Trans. TMS-AIME*, 1960, vol. 218, p. 473.
29. N. A. Razik, G. W. Lorimer, and N. Ridley: *Acta Metall.*, 1974, vol. 22, p. 1249.
30. M. Hillert and B. Sundman: *Acta Metall.*, 1977, vol. 25, p. 11.
31. M. Hillert and B. Sundman: *Acta Metall.*, 1977, vol. 24, p. 731.

Acta mater. 48 (2000) 2765–2775

www.elsevier.com/locate/actamat

MASSIVE TRANSFORMATION IN THE Fe–Ni SYSTEM

A. BORGENSTAM and M. HILLERT†

Department of Materials Science and Engineering, KTH, SE-10044 Stockholm, Sweden

(Received 9 March 2000; accepted 12 April 2000)

Abstract—The critical limit for the massive $\gamma \rightarrow \alpha$ transformation in the Fe–Ni system has been measured by isothermal heat treatment of diffusion couples. The position of the α/γ interface at the end of the treatment could be identified but some growth occurred during the quench. Growth is probably hindered at the beginning of the quench by a redistribution of Fe and Ni at the interface during the isothermal treatment. During the quench it often happens that the α/γ interfaces develop jagged shapes and even very fine plates although there should usually be no orientation relationship. The critical limit agrees fairly well with classical results for the formation of equiaxed ferrite from continuous cooling of homogeneous specimens. At 1023 K it coincides with the $\alpha/\alpha + \gamma$ phase boundary. At lower temperatures it moves inside the $\alpha + \gamma$ two-phase field. The driving force for the massive transformation increases at decreasing temperature and the limiting composition never approaches the T_0 line. © 2000 Acta Metallurgica Inc. Published by Elsevier Science Ltd. All rights reserved.

Keywords: Massive transformation; Microstructure; Thermodynamics; Iron alloys; Interface

1. INTRODUCTION

Alloys may transform without a change in composition from the parent phase to a new one, so-called partitionless transformation. This usually requires high cooling rates because slower cooling would, at least at higher temperatures, give time for long-range diffusion and changes of composition in the transformed region. For the $\gamma \rightarrow \alpha$ transformation in iron with a substitutional alloying element, which is the subject of the present paper, the temperature is about half the melting point or less and long-range diffusion may be neglected under many circumstances. There are two well-known types of partitionless transformation in solids, the martensitic and massive types. However, in iron alloys without interstitials there may also be a partitionless, acicular type, probably related to the bainitic transformation in Fe–M–C alloys, which occurs without partitioning of the substitutional elements [1]. The massive type will be considered in the present work. It is usually easy to distinguish in the microscope because the massive type got its name from the fact that the growing grains are blocky or massive and the whole of the parent phase transforms as the reaction proceeds.

The early work on the massive transformation was reviewed by Massalski in 1968 [2]. At that time it had been established that there is no obvious orientation relationship between parent and product phases, the interface is usually smoothly curved and it may cross a grain boundary in the parent phase. It was thus concluded that the massive transformation proceeds by the migration of highly disordered, incoherent interfaces. However, long planar boundary segments and ledge-like growth may sometimes appear occasionally and then disappear again according to a cinematographic study by Kittl and Massalski [3]. Aaronson *et al.* [4] connected the observations of planar segments with the theoretical expectation of some orientation relationship between a nucleus, formed at a grain boundary, and both parent grains. Thus they envisaged a growing massive crystal as being bounded in some areas by disordered interfaces and in the remaining areas by ledged, partially coherent interfaces. This aspect was further elaborated by Plichta *et al.* [5]. On the other hand, Perepezko and Massalski [6] showed that a polycrystalline specimen of the parent phase can transform into a single crystal of the product phase by a massive transformation, which is clear evidence that the growth process in the massive transformation is not dependent on any orientation relationship even if it can proceed if there is some.

Karlyn *et al.* [7] studied the massive $\beta \rightarrow \alpha$ transformation in Cu–Zn alloys by first retaining the high temperature β phase by quenching and then studying the rate of reaction on up-quenching to

---

† To whom all correspondence should be addressed.

various temperatures. They observed massive transformation to α within the α one-phase field, only. They proposed that this was not due to slow massive growth inside the α + β phase field but due to the difficulty of the reaction to get started there. They proposed that the massive transformation in their alloys initiates at small pre-existing α particles which have rejected Zn into the surrounding β matrix. There would be no driving force for partitionless growth from such Zn enriched regions inside the α + β phase field. They predicted that massive growth might take place in the two-phase field if it could only be initiated. They reported that this prediction had been confirmed in a double-pulsing experiment [8] where massive growth was first allowed to start inside the α one-phase field. It was then observed to continue when the temperature was increased just inside the α + β phase field.

Hawbolt and Massalski [9] investigated the same alloy system by studying thermal arrests in cooling curves and reported that the massive transformation can occur some tenths of a per cent Zn inside the α + β phase field, which is in close agreement with the results from the double-pulsing experiment. In a following study, Massalski et al. [10] detected massive transformation up to approximately 38.75 at.% Zn which was supposed to be 0.45 at.% inside the α + β phase field. They proposed that the critical limit for the massive transformation is not related to the α/α + β phase boundary but to the T_0 line below which the driving force for the partitionless transformation turns positive. They estimated that some 150 K of undercooling are required to bring about a massive transformation in this system.

On the other hand, Plichta et al. [11] studied three binary Ti systems and reported that the massive reaction took place solely in the (metastable) α region and not in the (extended) α + β range.

When examining the experimental information from the Cu–Zn system, Hillert [12] found that the critical limit for the massive transformation, reported by Hawbolt and Massalski [9] and Massalski et al. [10] could coincide with the coherent α/α + β phase boundary. He also showed that the growth rates reported by Karlyn et al. [7] do not necessarily extrapolate to positive values inside the coherent α + β phase field. Thus, it seemed possible that the critical limit for the massive transformation is closely related to the α/α + β phase boundary and not to the T_0 line. It should be mentioned that Hillert [13] had previously pointed out that the α/α + β phase boundary would be the limit for partitionless growth if local equilibrium between parent and product phases prevails at the migrating interface.

It may seem that the Cu–Zn system is not very well suited for studying the position of the critical limit for the massive transformation relative to the α/α + β phase boundary or the T_0 line because the distance between those two lines is only about 3 at.% in a wide range of temperature. The γ→α transformation in some binary Fe systems may be better suited because there the distance between the two lines increases towards lower temperatures. Furthermore, it is easier to prevent diffusional transformations because the parent phase is not b.c.c., as in Cu–Zn, but f.c.c. with a lower diffusivity, and the homologous temperature is lower. Thus, the partitionless transformations have been extensively studied in the Fe–Ni system, which will also be used in the present study. However, previous studies have employed the rapid quenching technique and the present work will introduce an isothermal, composition gradient technique.

The rapid quenching technique is a very efficient method of studying partitionless transformations in Fe alloys. One identifies the temperatures for arrests in the cooling curves and studies how they vary with the cooling rate. Usually the arrest temperature for a certain transformation first decreases with increasing cooling rate but then levels out to a plateau. Several plateaus have been found and each one has been related to a particular mode of the γ→α transformation, all of them partitionless due to the low temperature and short time. Figure 1 is taken from a compilation of data for the Fe–Ni system [1] and it shows that there is some confusion among the data [14–18]. However, four modes are identified and they were denoted I, II, III and IV by Mirzayev et al. [17]. Plateaus III and IV were related to lath and plate martensite, respectively. Plateau I is usually related to equiaxed ferrite and plateau II to acicular ferrite. Swanson and Parr [15] only observed two plateaus and they are roughly represented by the two dashed curves in Fig. 1. They related the upper one of them to massive ferrite, as did Wilson [18] who in addition found an

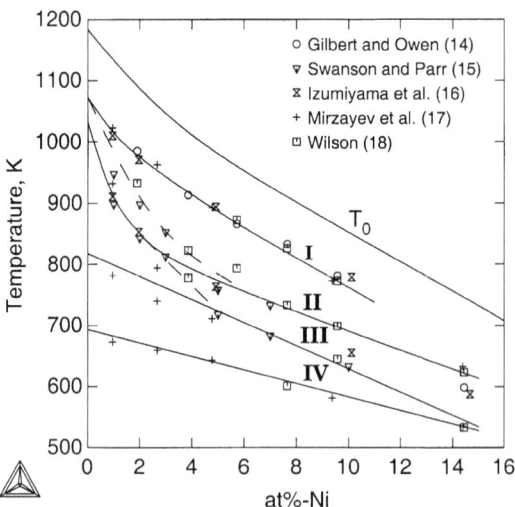

Fig. 1. Plateau temperatures for various partitionless transformations in the Fe–Ni system, determined from rapid cooling experiments [14–18].

equiaxed ferrite plateau in close agreement with plateau I.

An attempt will now be made to evaluate the critical limit for massive growth under isothermal conditions by using specimens with a composition gradient. The transformation can thus start well inside the one-phase field where the initiation is easy. It will be allowed to proceed isothermally into the gradient and come to a natural stop at some composition. The critical question is whether that position can be observed after a final quench to room temperature. The present authors have recently used this technique in a study of the limit for growth of martensite in the Fe–C system [19] and in that case the martensite formed at the experimental temperature could be recognized because it had been tempered before quenching. Of course, that will not happen in the present case. Instead it was expected that some redistribution of Fe and Ni would occur at the interface when it has come to a stop and that should prevent further growth for some time during the quench, in accordance with the proposal by Karlyn et al. [7].

An additional advantage with the isothermal, composition gradient technique is that the growing crystals reach new parent grains with which they can hardly share a specific orientation relationship. One may thus expect to obtain information on the critical limit for the migration of incoherent interfaces.

2. EXPERIMENTAL

Diffusion couples were prepared by pressing together two rods of pure Fe and an Fe–32 mass% Ni alloy, each with a 2×2 mm^2 cross section, and annealing the couple for 400 h at 1500 K. After cooling the couple was sectioned into 6 mm long specimens. They were austenitized in flowing argon at 1323 K and dropped into a lead pot of a temperature in the range of 700–1050 K and then quenched in iced brine after various times.

The specimens were sectioned perpendicularly to the original interface, prepared metallographically and etched in 4% nital. After the microstructure had been examined, the composition profile perpendicular to the original interface was recorded with an EDS instrument and the Ni contents were calibrated by comparing with a profile measured by a microprobe equipped with a wavelength dispersive spectrometer, which was regarded as a more accurate instrument. The correction was very slight, though.

Evidently, in the lead pot the massive transformation must have started in the high Fe part and in specimens treated isothermally at higher temperatures the microstructure revealed very well how far it had proceeded into regions of higher Ni before the quench. In specimens treated at lower temperatures it was less evident. The Ni content was evaluated at the position where the massive transformation had stopped before the quench, when that could be seen. Otherwise, the measurements were taken where the α/γ interface was as planar as possible. Those values are presented in Table 1 together with information on the Ni gradient at the point of measurement.

3. MICROSTRUCTURAL OBSERVATIONS

Typical microstructures are presented in Figs 2–11. Figure 7 was taken at low magnification and shows that the massive transformation $\gamma \rightarrow \alpha$ started in the high Fe part of the specimens (below the picture, not shown) and proceeded roughly perpendicularly into the Ni gradient until the growing α grains came to a halt at a certain Ni content, higher the lower the isothermal temperature was. Above the front of the α phase there is a thick layer of martensite with variable microstructure extending further into the Ni gradient. This specimen was heavily etched in order to bring out the martensite. The α phase at the bottom of the picture is thus fairly dark but closer to the front it has not been attacked as much due to the higher Ni content. The first martensite is again dark, probably due to the high dislocation density. The many black dots are probably pores, some of which were present in the original Ni alloy which was produced in the laboratory. In the composition gradient they have grown considerably, probably due to a Kirkendall effect during the diffusion treatment.

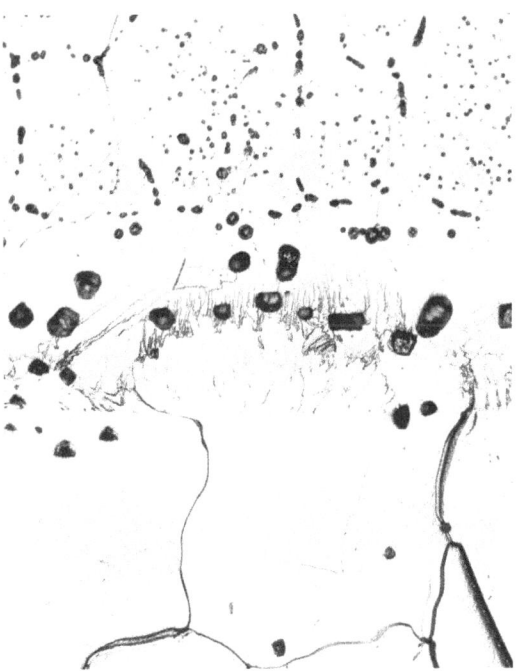

Fig. 2. Specimen from 5 s at 1023 K, showing extensive growth during the quench, on top of the uniform massive layer formed at 1023 K (lower part of picture). Magnification 200×.

2768 BORGENSTAM and HILLERT: MASSIVE TRANSFORMATION IN THE Fe–Ni SYSTEM

Table 1. Maximum nickel content in massive α and nickel gradient in same position

Temperature (K)	Time	at.% Ni						at.% Ni/μm
		1	2	3	4	5	Aver.	
777	2 min	8.78	9.04	9.14			8.99	0.083
832	2 min	8.00	8.49				8.25	
873	2 min	6.08	6.30				6.19	0.083
	2 h	6.10	6.35	6.61	6.86	7.27	6.64	0.096
	7 h	5.99	6.07				6.03	0.081
878	30 s	5.91	6.16	6.40			6.16	0.075
	15 min	6.26	6.55	6.56			6.46	0.085
	2 h	6.43					6.43	0.089
	7 h	6.34	6.75				6.55	0.116
894	2 min	5.46	5.84	5.92			5.74	0.092
923	2 min	4.77	5.07				4.92	0.100
953	2 min	3.04	3.24	3.25	3.31		3.21	0.051
971	5 s	3.26	3.46				3.36	0.083
	2 h	4.30	4.51				4.41	0.084
973	2 h	3.46	3.76				3.61	0.057
1023	5 s	2.12	2.21				2.16	0.086
	2 min	1.97	2.16				2.06	0.075
	7 h	2.09	1.91				2.00	0.076

At the highest temperatures it was easy to see where the α phase stopped during the isothermal treatment because there was a distinct difference from the structure formed during the final quench. As an example, Fig. 3 from 2 min at 1023 K shows a thin layer of a very fine structure which evidently represents a further growth of the α grains. Figures 4(a)–(c) show further examples from the same specimen. The specimen treated for 7 h at the same temperature showed the same structures but the specimen treated for only 5 s showed much more extensive growth during the quench, Fig. 2. Here it is more difficult to see exactly the final position of the α front during the isothermal treatment but there is no doubt about its approximate position. Specimens from 971 and 973 K showed the same two types of structure after 5 s (Fig. 5) and 2 h

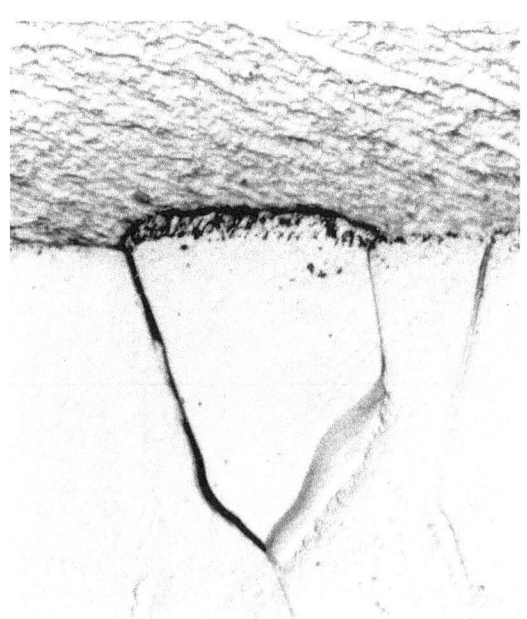

Fig. 3. Specimen from 2 min at 1023 K, showing a thin layer formed by resumed growth (upwards) during the quench. Magnification 1250 ×.

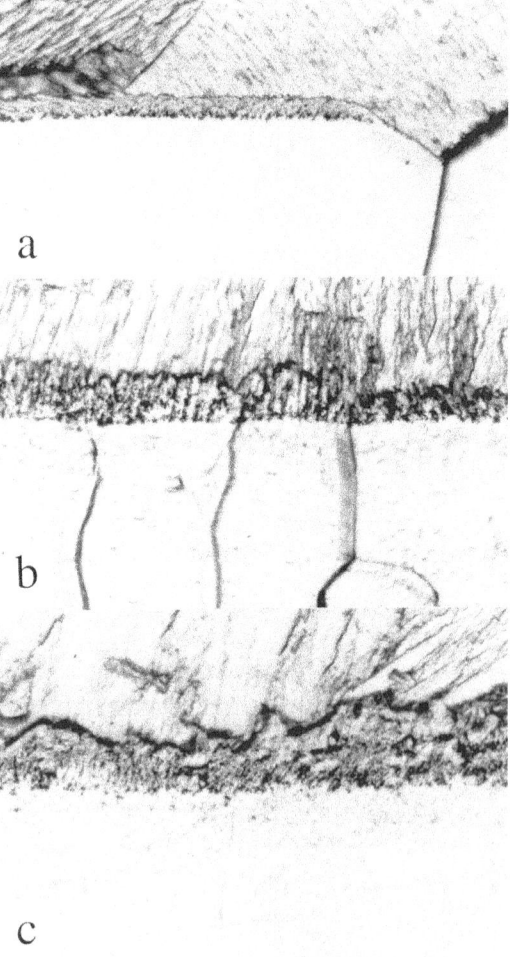

Fig. 4. Three more observations of the thin layer formed in the specimen shown in Fig. 3.

[Figs 6(a)–(d)], respectively. Figure 5 shows the same extensive growth during the quench as Fig. 2 but the position before the quench is better defined in this particular region. The structures in Fig. 6 resemble those in Fig. 4 but it is more difficult to see the internal structure that was so evident in Fig. 4. Figure 7 was taken at a much lower magnification on the same specimen to show the general appearance which was the same in all the specimens.

For some reason, the specimen from 2 min at 953 K had rather few planar segments and Fig. 8 shows a drastic example of a non-planar segment. Similar shapes but less pronounced were sometimes observed in the other specimens, as well. The specimen from 923 K again showed many planar segments but occasionally there were exceptions. Figure 9 shows a step in the planar front and that was also noticed at the higher temperatures. These two cases will be further discussed in Section 7.

The thin α layers, representing further growth during the quench in Fig. 6, could be seen down to 873 K but at lower temperatures it was not possible to distinguish the α layer formed during the quench and it was not even certain that there was any such layer. Jagged shapes were often observed and, due to the suspicion that they could sometimes originate from the quench, measurements were taken on the most planar segments of the front.

In a specimen from 2 min at 832 K the front was generally rather jagged and, even though it was difficult to decide whether that was due to further growth during the quench or not, it was concluded that it originated from the isothermal growth whenever it was connected to the formation of subgrain boundaries in the α phase as in Fig. 10. In a specimen from 2 min at 777 K some segments were even very acicular, Fig. 11, and that tendency was predominant in a specimen treated for 2 min at 696 K.

That specimen looked almost as the front of the outgrowth shown in Fig. 2, which had formed during the quench. It should again be stressed that the data in Table 1 were taken from the most planar segments found in each specimen. No data are included from 696 K because of the difficulty in identifying the position of the front before the quench and the large scatter obtained when an attempt to measure was made.

4. COMPARISON WITH RESULTS FROM CONTINUOUS COOLING

The results in Table 1 are compared with the phase diagram in Fig. 12. The experimental points represent the maximum Ni content in the regions transformed to massive α. There is some experimental scatter. However, it is evident that the massive growth at 1023 K does not extend very far into the α + γ two-phase field, if at all. At lower temperatures it extends deeper and deeper into the two-phase field. However, in the examined temperature range it does not seem to approach the T_0 line, i.e. the line where α and γ

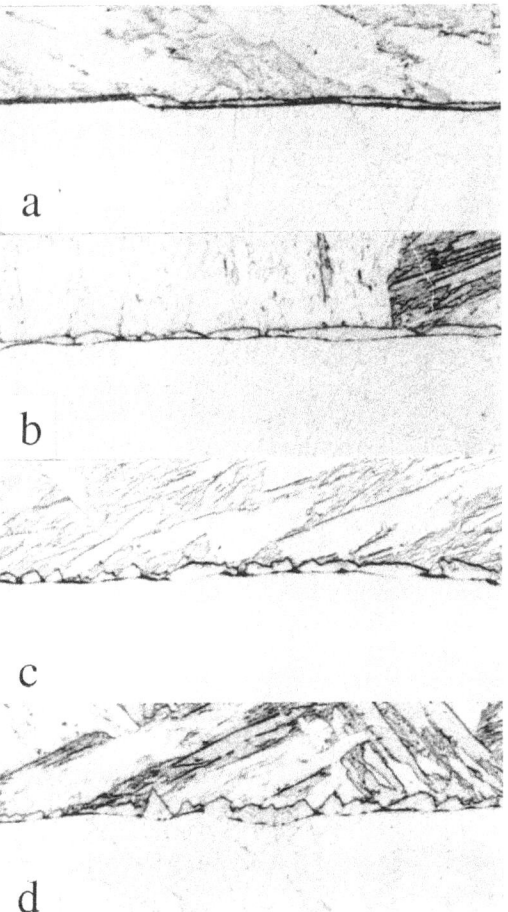

Fig. 6. Specimen from 2 h at 973 K, showing several examples of thin layers formed by resumed growth during the quench. Magnification 1250×.

Fig. 5. Specimen from 5 s at 971 K, showing extensive growth (upwards) during the quench, starting from a fairly flat α/γ interface established at 971 K. Magnification 500×.

2770   BORGENSTAM and HILLERT: MASSIVE TRANSFORMATION IN THE Fe–Ni SYSTEM

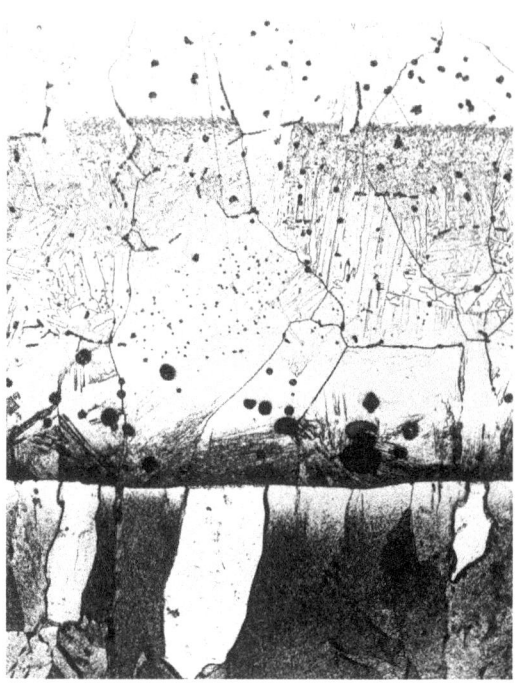

Fig. 7. Typical structure of the specimens shown at low magnification (100×). This specimen was from 2 h at 973 K. The Ni content increases from very low at the bottom to about 30% at the top. The lower part has transformed to massive α and the middle part to martensite. The top part is still γ.

have the same Gibbs energy. Even though there is an experimental uncertainty in the values, and the calculation of the T_0 line is also subject to some uncertainty, this conclusion seems safe. It should be mentioned that the phase diagram and the T_0 line were calculated from a CALPHAD assessment of the Fe–Ni system [20].

Figure 13 compares the present results (dashed line) with the line for plateau I from rapid quenching in Fig. 1 (now represented by a dash-dotted line). That line has been related to the formation of equiaxed ferrite, which should be identical to the massive α studied in the present work. The fact that the present results fall at higher temperatures (or higher Ni contents if a constant temperature is considered) is expected for at least two reasons. First, in rapid quenching the arrest occurs at a temperature where the rate of formation is high enough to balance the rate of heat extraction. Evidently, the upper limit for massive growth should be at a somewhat higher temperature. Secondly, the arrest temperature decreases with increasing cooling rate and seems to level out to a plateau in a range of high cooling rates. All the data in Fig. 1 represent such plateaus and somewhat higher arrest temperatures have been observed at lower cooling rates except for the martensitic transformations which are not time dependent.

Figure 13 also presents arrest temperatures obtained by Jones and Pumphrey [21] using a con-

Fig. 8. Specimen from 2 min at 953 K, showing a very jagged segment of the α/γ front. Magnification 500 × .

stant and rather low cooling rate. Those data are similar to values one can read for low cooling rates in diagrams from the rapid quenching studies by Izumiyama et al. [16] and Mirzayev et al. [17]. It is thus evident that they are all related to the formation of equiaxed ferrite. The fair agreement between the present results and those by Jones and Pumphrey supports the conclusion that all this information is due to the same mode of transformation, i.e. equiaxed ferrite and massive α are identical. If that is correct, the results of the continuous cooling experiments have not been much affected by a slow nucleation process. On the other hand, the nature of the plateau related to massive ferrite by Wilson [18] and roughly represented by the upper one of the two dashed curves in Fig. 1, and probably identical to the transformation

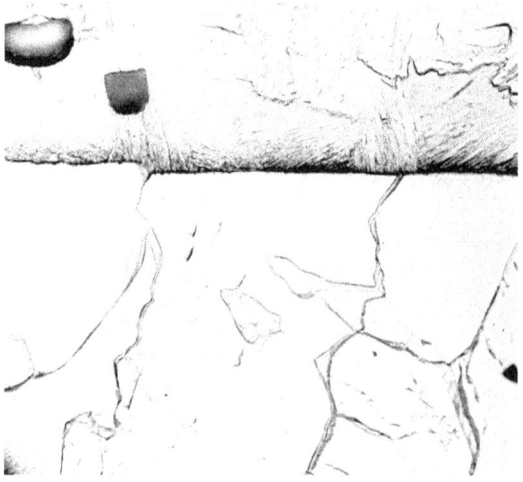

Fig. 9. Specimen from 2 min at 923 K, showing planar segments at different positions in the composition gradient. Magnification 500 × .

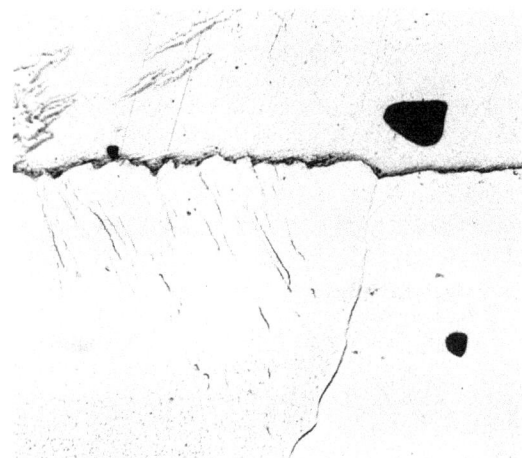

Fig. 10. Specimen from 2 min at 832 K, showing a jagged segment connected with the formation of subgrain boundaries in the α phase (lower part of picture). Magnification 500 ×.

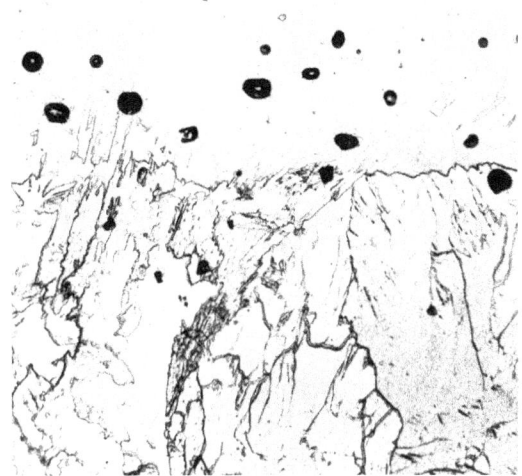

Fig. 11. Specimen from 2 min at 777 K, showing a very acicular segment. The α phase (lower part of picture) shows very strong substructure. Magnification 200 ×.

described as massive by Swanson and Parr [15], is uncertain.

The maximum Ni content for massive growth, established in the present work, coincides well with the value used by Massalski et al. [22], 8.7% Ni, when applying continuous cooling to establish that the massive transformation can occur outside the one-phase field for α which is retrograde and has a maximum of 4.6% Ni.

5. NATURE OF THE MASSIVE TRANSFORMATION

In the Fe–Ni system the volume diffusivity is so low that it is easy to prevent diffusion. Various types of partitionless transformations can thus be observed depending on the orientation relationship between grains of the parent and product phase. The result obtained without any special orientation relationship, and thus with an incoherent interface, may be defined as the ideal massive type. This is probably the one studied with the isothermal, composition gradient technique in the present work. The very strict orientation relationship, which would be expected to give rise to Widmanstätten precipitate if there were time for long-range diffusion, may be responsible for the acicular type that causes plateau II in Fig. 1. The orientation of a crystal nucleated at a grain boundary would probably be related to both parent grains in most cases. The orientation relationship to each one would then be less strict and one can expect to find a wide spectrum of partially coherent interfaces. The resulting types of partitionless transformation should simply be regarded as intermediate cases. In view of the results by Plichta et al. [5] partially coherent interfaces may be very common if the gradient technique is not used. However, the close agreement, demonstrated in Fig. 13, indicates that the result generally does not differ much from what has here been called the ideal massive type.

The fact that the massive transformation in the Cu–Zn system does not seem to extend much into the two-phase field was explained by Karlyn et al. [7] with reference to the inability of a nucleus to break through the pile-up of alloy element surrounding it. Hillert [12] gave an alternative explanation by calculating the steady-state width of the spike of the alloy element in front of the interface, if it would migrate, and finding that it is not negligible as compared with atomic distances. The two phases could then be reasonably close to local equilibrium with each other during migration and, even if initiated, partitionless growth would not proceed if the initial parent phase is substantially inside the

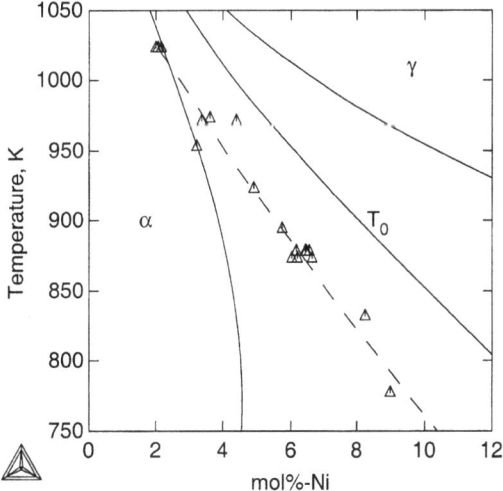

Fig. 12. Limit of massive growth of ferrite in the Fe–Ni system according to the present results (triangles and dashed line), compared with calculated phase boundaries and the equal Gibbs energy curve, the so-called T_0 line.

two-phase field. On the other hand, in the Fe–Ni system the width of the spike would be orders of magnitude smaller than atomic distances and the spike should not exist. The partitionless transformation could then proceed and closely approach the T_0 line unless some other energy consuming process interferes. That could be diffusion of Ni relative to Fe inside the migrating interface. That phenomenon would be related to the solute drag effect on grain growth in single-phase materials. The position of the critical limit for massive growth relative to the phase boundary and the T_0 line gives valuable information on the size of such an effect. In the present case it is interesting to note that the critical limit falls close to the α phase boundary at 1023 K. The dissipation of Gibbs energy is thus roughly the same as in a well-developed spike in front of a migrating interface. At lower temperatures the critical limit moves further inside the two-phase field and eventually falls closer to the T_0 line than to the α phase boundary. From the position of the critical limit one may evaluate the dissipation of Gibbs energy inside the interface, assuming there are no other sinks for driving force. In Fig. 14 it is plotted vs temperature and is found to increase at decreasing temperature in spite of the fact that the critical limit moves further into the two-phase field. There is no tendency that it would ever decrease, which should be necessary if it would ever approach the T_0 line. On the other hand, it would decrease relative to the dissipation in a well-developed spike, which is plotted as a second curve in Fig. 14. That curve increases even faster with decreasing temperature. The dissipation of Gibbs energy by diffusion inside the interface and, hence, the position of the critical limit probably depend on how the chemical properties of the interface material and the transverse diffusivity in the interface vary with temperature. These factors are not known at all and the present kind of data may become a valuable source of information. Modelling of this phenomenon has only begun [23–26].

6. RESUMED MIGRATION DURING QUENCHING

Even though it was usually possible to identify the position where the massive transformation had stopped during the isothermal treatment, signs of resumed migration during the following quench were often found. The most evident cases were found after isothermal treatments for only 5 s (Figs 2 and 5) which may have been too short for an efficient redistribution of Fe and Ni at the interface after it had stopped. It is even possible that 5 s was not even enough for the growth to come to a complete stop. Thus, the interface is able to continue its migration soon after the temperature starts to decrease by the quench. In Fig. 2 the additional growth has proceeded further into the gradient for almost 8 at.% and in Fig. 5 for almost 4 at.%. Related effects but much less pronounced were observed in many specimens treated for a longer time. Figures 3 and 4 from the specimen treated for 2 min at 1023 K show that the resumed migration has there resulted in a thin layer of a very fine structure, often suggesting the formation of a long series of very thin parallel plates, Figs 4(a) and (b). The etching effect in the thin layers is probably due to some defect structure left in the α crystal where the fine plates have met side by side. It is proposed that these thin layers have formed when the mi-

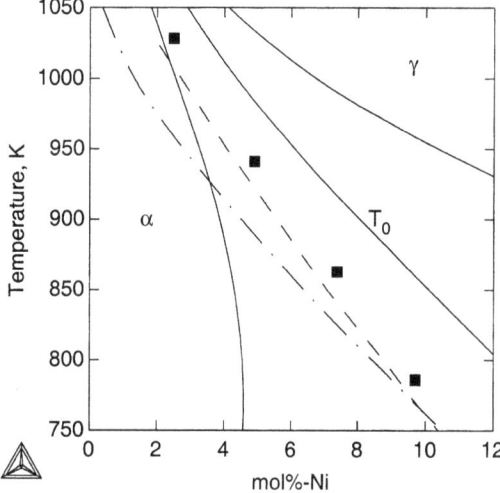

Fig. 13. Present results for the limit of massive growth of ferrite in the Fe–Ni system (represented by the dashed line), compared with the plateau temperature for equiaxed ferrite, plateau I in Fig. 1 (represented by dash-dotted line) and with the beginning of the $\gamma \rightarrow \alpha$ transformation at relatively slow cooling, according to Jones and Pumphrey (squares).

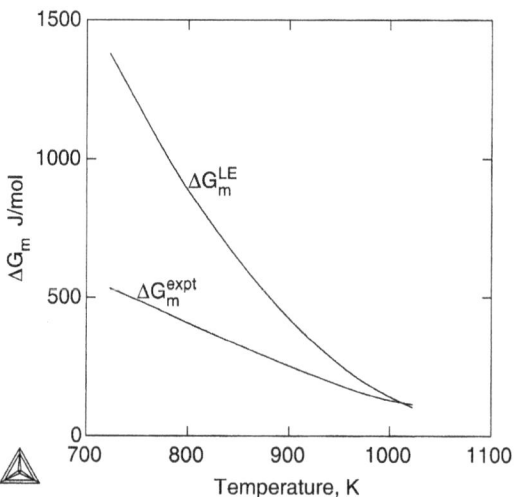

Fig. 14. The driving force for massive growth of ferrite, evaluated at the critical limit according to the present experiments, ΔG_m^{expt}, and evaluated under the assumption of local equilibrium at the α/γ interface, ΔG_m^{LE}.

gration was resumed at some low temperature during the quench, when the driving force had increased to a high enough value to overcome the resistance due to the redistribution of Fe and Ni at the interface when it stood still. The tendency to form such layers in the specimen from 2 h at 973 K was weaker, probably because the supersaturation achieved on quenching is less due to the higher Ni content at the front. Figure 6 shows that the layers are even thinner here and it was very difficult to see the internal structure. Both effects may have been caused by the lower supersaturation. In the specimens treated below 873 K, the supersaturation seems to have been too low to give such α layers.

7. CRYSTALLOGRAPHIC FEATURES

In the specimen from 2 min at 1023 K there were thin layers consisting of very fine plates on almost all the massive grains and they can hardly be due to any orientation relationship to the parent grains. On the other hand, the fact that they often seem to consist of parallel plates, whose direction varies from one α grain to the next, indicates that they are crystallographic features. This can be seen in Fig. 4(b) although the magnification is not quite sufficient. The boundary to the right is an α/α grain boundary. Those in the middle and to the left are subboundaries and they seem to have had only a small effect on the direction of the fine plates. It may be concluded that, even with a random orientation relationship between parent and product grains, the interface can find a direction that is more favourable than others, but only if the driving force is large enough. A planar interface may then develop a more jagged shape and even turn into plate-like growth. Figure 4(c) shows that a very jagged shape formed during the quench although the same α grain had a planar interface before the quench. The series of micrographs in Fig. 6 illustrates that the further growth during the quench sometimes develops a layer of even thickness, sometimes small segments of even thickness and sometimes more triangular outgrowths, probably related to the jagged shapes discussed before.

It is difficult to decide if the acicular structure to the left in Fig. 11 was formed isothermally or during the quench. However, the similar structure in Fig. 2 was certainly formed during the quench and that is additional evidence that a planar interface can even form well-developed plates if the driving force is increased sufficiently by quenching.

It should be mentioned that there were usually a few grains in each specimen that certainly had a very jagged shape already before the quench. Figure 8 shows the most dramatic case. This is most probably due to an orientation relationship between the α and γ grains which could occur occasionally even though a gradient technique was used. It should also be mentioned that neighbouring grains, each one with a planar interface, sometimes have grown differently far into the Ni gradient, which is revealed by steps in the planar front. Figure 9 shows a typical example. This is an indication that even incoherent interfaces may have different properties and give rise to different amounts of dissipation of Gibbs energy by interface diffusion.

The fact that α/α grain boundaries and even subgrain boundaries give rise to re-entrant angles is not surprising. Figure 4(a) shows an example. It is more surprising that this is not always the case, as illustrated in Fig. 4(b).

8. COMPARISON WITH TERNARY ALLOYS CONTAINING CARBON

The $\gamma \rightarrow \alpha$ transformation without partitioning of the substitutional elements has been extensively studied in Fe alloys in the presence of carbon. It has long been known that Mn and Ni have a strong retarding effect on the formation of α in spite of the fact that the theoretical thickness of a spike formed in front of the migrating α interface is well below atomic dimensions. Under such growth conditions there are two possibilities. Either, the reaction is completely diffusionless with respect to Mn and Ni and so-called "paraequilibrium", as defined by Hultgren [27, 28], would prevail, i.e. local equilibrium for carbon across the interface but no movements of Mn or Ni relative to Fe. In that case the retarding effect of Mn or Ni could be directly calculated from the thermodynamic properties and should be rather low. However, experimental information seems to indicate that the retarding effect is much stronger. As a second alternative, it was thus proposed, e.g. by one of the present authors [13], that the spike does not disappear but "moves inside the interface" if the volume diffusivity in the γ phase is too low. There it could have a similar retarding effect as already explained for the binary Fe–Ni case in Section 5. It could be argued that this requirement should always be satisfied if the atomic mobility inside the interface is high enough to allow the interface to migrate. However, it is now realized that there may be a strong coupling between the fluxes of substitutional elements across phase interfaces [26]. This implies that much of the atomic mobility that makes an interface migrate may be of a co-operative nature even though the interface has an incoherent structure. An indication is that the mobility of interfaces in Fe and binary Fe alloys, predicted from the interfacial diffusivity [29] with Turnbull's model [30], is only about a hundredth of the experimental one. This implies that the interface migrates with a process much faster than allowed by the atomic process of diffusion. Thus, one should now be open to the possibility that the dissipation of Gibbs energy due to a spike inside the interface may be considerably lower than in a fully developed spike. It is an interesting question how close the growth conditions can come to those for paraequilibrium. This was recently investigated by an attempt to model the massive tranformation in an Fe–M alloy presented by Hillert and Schalin [25] and of the formation of ferrite in Fe–M–C alloys by Liu and Ågren [31] and Purdy and Brechet [32].

When studying the lengthening of Widmanstätten plates or bainite in high Ni steels, Goodenow et al. [33] as well as Rao and Winchell [34] found appreciable growth rates although the Ni content was well above the solubility of Ni in α-Fe. Thus it is quite clear that there could not have been a well-developed spike in the parent γ. On the other hand, Rao and Winchell compared with growth rates predicted without a spike (based on a calculation similar to one based on paraequilibrium [35]), finding that the growth rates were much too low. After proposing several possible factors to explain their results they wrote "Finally the possibility that the shuffling of nickel atoms slows down the interface must be admitted". Evidently, they had an effect like the solute drag in mind. It should also be mentioned that Kinsman and Aaronson [36] proposed that the effect of some alloying elements on the growth rate of ferrite in Fe–M–C systems is due to an "impurity drag effect", later developed into a "solute drag-like effect" [37]. Their proposal was not developed in detail but it is evidently closely related to the picture given above. A particularly interesting feature is that they explained the strong effect of some carbide forming elements by a strong segregation to the α/γ interface at locally high C contents.

Over the years there have been several attempts to locate the critical limit for the $\gamma \rightarrow \alpha$ transformation in ternary Fe–C–M alloys or commercial steels and relate it to the lines for paraequilibrium and local equilibrium with a spike. It is generally agreed that the critical limit falls between those two calculated lines, just as the present results for the Fe–Ni system falls between the T_0 line and the $\alpha/\alpha + \gamma$ phase boundary. As mentioned by Hillert [26], the calculation of the dissipation of Gibbs energy in front of and inside the migrating interface in a binary Fe–M system could easily be extended to ternary Fe–M–C systems by imposing the condition of constant carbon activity. Similar results should thus be expected. It should be very interesting to try to connect the limiting point for the massive transformation in an Fe–M system with the limiting line for the $\gamma \rightarrow \alpha$ transformation in the Fe–M–C system at the same temperature. This is what Oi et al. [38] recently tried to do in a careful study of the growth of α in Fe–Mn–C and Fe–Ni–C alloys with various carbon contents. The values for the binary Fe–Mn and Fe–Ni systems they took from the decarburized surface zones of the specimens. The value for Fe–Ni is in fair agreement with the present results and, as expected, fall close to the line extrapolated from the ternary information.

It should be emphasized that in this kind of test one should compare information obtained with the same kind of interface because the dissipation of Gibbs energy could very well be quite different for incoherent and partially coherent interfaces. It is thus satisfactory that Oi et al. studied the formation of grain boundary allotriomorphs which may be compared with the massive transformation in carbon-free alloys. There is much more information on the lengthening rate of Widmanstätten plates or bainite but, unfortunately, it is difficult or even impossible to study the corresponding acicular growth of α in carbon free alloys (which forms at plateau II in rapid cooling) due to the competition from the massive transformation.

9. SUMMARY

The critical limit for massive growth of ferrite in the Fe–Ni system can be studied isothermally using diffusion couples. By that technique one can make sure that most of the interfaces studied are incoherent. The final position of the interface can be detected after quenching even though some further growth can occur during the quench.

The critical limit for the massive growth seems to be related to the curve for the first plateau temperature studied by rapid quenching and presumably due to the formation of equiaxed ferrite. It falls even closer to the temperature for the beginning of the transformation observed with slower cooling rates. It is thus confirmed that equiaxed ferrite is identical to the massive ferrite formed by the migration of incoherent interfaces.

The critical limit coincides with the α phase boundary at 1023 K but moves into the $\alpha + \gamma$ two-phase field at lower temperatures. However, it does not approach the T_0 line. The dissipation of Gibbs energy at the critical limit increases at lower temperatures in spite of the fact that it moves further and further into the two-phase field.

The microstructures formed during the quench give strong indications that an incoherent and planar interface can develop into a jagged shape and even form a series of parallel plates when the driving force increases during the quench. There is also some indication that the interaction between a migrating α/γ interface and an alloying element is not of the same strength for all planar, incoherent phase interfaces.

Acknowledgements—Thanks are due to Martin Schwind for valuable experimental help.

REFERENCES

1. Borgenstam, A. and Hillert, M., *Metall. Mater. Trans.*, 1996, **27A**, 1501.
2. Massalski, T. B., in *Phase Transformations, Seminar*

of the Am. Soc. Metals 1968. ASM, Metals Park, OH, 1970, pp. 433–486.
3. Kittl, J. E. and Massalski, T. B., *Acta metall.*, 1967, **15**, 161.
4. Aaronson, H. I., Laird, C. and Kinsman, K. R., *Scripta metall.*, 1968, **2**, 259.
5. Plichta, M. R., Clark, W. A. T. and Aaronson, H. I., *Metall. Trans.*, 1984, **15A**, 427.
6. Perepezko, J. H. and Massalski, T. B., *J. Mater. Sci.*, 1974, **9**, 899.
7. Karlyn, D. A., Cahn, J. W. and Cohen, M., *Trans. metall. Soc. A.I.M.E.*, 1969, **245**, 197.
8. Emmer-Szerbesko, C. D. M., S.B. thesis, MIT, Cambridge, MA, 1967.
9. Hawbolt, E. B. and Massalski, T. B., *Metall. Trans.*, 1970, **1**, 2315.
10. Massalski, T. B., Perkins, A. J. and Jaklovsky, J., *Metall. Trans.*, 1972, **3**, 687.
11. Plichta, M. R., Aaronson, H. I. and Perepezko, J. H., *Acta metall.*, 1978, **26**, 1293.
12. Hillert, M., *Metall. Trans.*, 1984, **15A**, 411.
13. Hillert, M., in *The Mechanism of Phase Transformations in Crystalline Solids*, , Inst. Metals Monographs No. 33. Institute of Metals, London, 1969, pp. 231–247.
14. Gilbert, A. and Owen, W. S., *Acta metall.*, 1962, **10**, 45.
15. Swanson, W. D. and Parr, J. G., *J. Iron Steel Inst.*, 1964, **204**, 104.
16. Izumiyama, M., Tsuchiya, M. and Imai, Y., *Sci. Rep. Res. Inst. Tohoku Univ.*, 1970, **22A**, 93.
17. Mirzayev, D. A., Morozov, O. P. and Shteynberg, M. M., *Physics Metals Metallogr.*, 1973, **36**, 96.
18. Wilson, E. A., *Metals Sci.*, 1984, **18**, 471.
19. Borgenstam, A., Hillert, M. and Ågren, J., *Acta metall. mater.*, 1995, **43**, 945.
20. Zhong, S. X., Gohil, D. D., Dinsdale, A. T. and Chart, T. G., Natn. Phys. Lab., DMA(a)103, London, 1985.
21. Jones, F. W. and Pumphrey, W. I., *J. Iron Steel Inst.*, 1949, **163**, 121.
22. Massalski, T. B., Perepezko, J. H. and Jaklovsky, J., *Mater. Sci. Engng*, 1975, **18**, 193.
23. Hillert, M. and Sundman, B., *Acta metall.*, 1976, **24**, 731.
24. Liu, Z.-K., Ågren, J. and Suehiro, M., *Mater. Sci. Engng*, 1998, **A247**, 222.
25. Hillert, M. and Schalin, M., *Acta mater.*, 2000, **48**, 461.
26. Hillert, M., *Acta mater.*, 1999, **47**, 4481.
27. Hultgren, A., *Trans. Am. Soc. Metals*, 1947, **39**, 915.
28. Hultgren, A., *Kungl. Vet. Akad. Handl.*, 1953, **4**(3), 1.
29. Hillert, M., *Metall. Trans.*, 1975, **6A**, 553.
30. Turnbull, D., *Trans. Am. Inst. Min. Engrs*, 1951, **191**, 661.
31. Liu, Z.-K. and Ågren, J., *Acta metall.*, 1989, **37**, 3157.
32. Purdy, G. R. and Brechet, Y. J. M., *Acta mater.*, 1993, **43**, 3763.
33. Goodenow, R. H., Matas, S. J. and Heheman, R. F., *Trans. Am. Inst. Min. Engrs*, 1963, **227**, 651.
34. Rao, M. M. and Winchell, P. G., *Trans. Am. Inst. Min. Engrs*, 1967, **239**, 960.
35. Rao, M. M., Russell, R. J. and Winchell, P. G., *Trans. Am. Inst. Min. Engrs*, 1967, **239**, 634.
36. Kinsman, K. R. and Aaronson, H. I., in *Transformation and Hardenability in Steels, 1967.* Climax Molybdenum Co, Ann Arbor, MI, 1967, pp. 39–55.
37. Bradley, J. R. and Aaronson, H. I., *Metall. Trans.*, 1981, **12A**, 1729.
38. Oi, K., Lux, C. and Purdy, G. R., *Acta mater.*, 2000, **48**, 2147.

21 Introduction to "On the Nature of the Bainite Transformation in Steel"

published in *Acta Metallurgica* (1984)

The variety of austenite decomposition products in steel remains a fascinating subject, and a challenge to the sagacity of physical metallurgists. It also allows for the apparently endless variety of thermal treatments of steels. At the very beginning of physical metallurgy a lot of work was devoted to a description of these decomposition products, which received names derived from the pioneers in these studies: Roberts-Austen, Widmanstätten, Martens, Bain, and not to be forgotten, Sorby and Troost. Metallurgists were striving for accurate characterizations of these products, and also to establish a clear typology, thanks to a classification, that any student can learn. But the definition of classes involves clear cut distinctions. According to the authors of this paper these distinctions are too sharp, and the borders between the various products are rather diffuse.

Putting aside pearlite, the authors emphasize the continuity between Widmanstätten ferrite, upper and lower bainite and lath martensite, on the basis of morphological, structural and kinetic factors. According to a classic "Hillert style" experiment, if the transformation is first performed in the bainitic range, then continued at higher temperatures, Widmanstätten ferrite simply takes over, without any need for re-nucleation (Figure 1). The reverse situation, where allotriomorphic ferrite is first generated, allows for the formation of bainite at lower temperatures without any nucleation step. The continuity between the different products in the $\gamma \rightarrow \alpha$ phase transformation of the 'diffusive' type is put in parallel with similarities, from the crystallographic and morphologic viewpoint, between bainite and lath martensite: orientation relationship, dislocation generation... Apart from this essential claim, other valuable points in this paper deserve our attention. The growth kinetics derives from a lengthening and/or a thickening of the crystallites of the transformation product. The choice between these two modes is based on the crystallographic nature of the interface, the deformation strain (dilatation and/or shear) and solute drag in the case of alloyed steels (this last problem is discussed in other papers in this volume). The growth can also result from the simple growth of a crystallite, or at lower temperatures by the formation of repeated sub-units as a result of some autocatalytic effect (bainitic "sheaves"). The shape and the growth rate depend on a delicate balance between interface anisotropy and minimization of the elastic energy generated by the transformation stresses. Hence the "lively discussions" on the limiting factors for bainite transformation: is it to be seen as a *diffusive* process limited by Carbon diffusion out of the ferritic product or as a repeated *displacive* nucleation followed by Carbon diffusion? The first interpretation points towards modeling the diffusive growth of needle shaped particles, the second towards a nucleation limited process. The difficulty met in discriminating between these two types of approaches arises from: i) the poor reliability of thermodynamic data at low temperatures, and ii) the unknown conditions at the ferrite needle/austenite interface (local equilibrium without partitioning, paraequilibrium, correction because of the Gibbs-Thomson effect, role of crystallographic faceting on a possible discontinuity of chemical potentials, etc.). Another suggestion of the authors is that the difference between Widmanstätten ferrite and upper bainite could be related to a 'Cahn's transition' in interface migration, from lateral (edgewise) to normal. The growth of subunits is seen as diffusion controlled, whereas their repeated nucleation is autocatalytic because of the elastic stresses generated by the transformation.

The paper offers a good basis for further research. A critical test of the authors model would be the chemical analysis of the ferrite crystals, which requires a sub-micron method and in-situ measurements (in order to avoid post transformation by Carbon diffusion). New high resolution microprobes could perhaps allow this determination.

The line of thought presented in this paper was pursued by Hillert and his school in various aspects both for the analysis of the kinetics and for the thermodynamic prediction of the B_S temperature. However, the current state of modeling remains unsatisfactory, since the coupling between crystallographic constraints and thermodynamic conditions is still absent and the question of the interfacial conditions is still to be solved in a self-consistent manner. The question of Carbon diffusion as a controlling factor remains, at least in the opinion of the authors of this introduction, pretty much settled. Moreover the possible displacive nature of the Iron atoms rearrangement is still open.

One question the authors do not mention is the nature of platelet martensite. Their paper is mainly devoted to bainite as heralded by their title, to which they "annex" lath martensite. One can wonder about the nature and kinetics of plate martensite, characterized by a high driving force and a "needle" like morphology, where the stresses generated by the martensite platelets play a double role; preventing further transformation (the so called "stabilisation" phenomenon), and also initiating the nucleation of new crystallites as soon as the temperature is decreased, because of the need of a stronger driving force. Is there a real "frontier", a transition of catastrophic type or does continuity still prevail?

It is worth noting that the very fundamental issues raised by this paper came back on the scene with the recent renewed interest in TRIP steels. Industrial developments in the processing route leading to metastable retained austenite rely on an understanding of Carbon and alloying element partitioning during intercritical annealing. The work of Hillert and his school on these questions appears in retrospect as being a perfect example of fundamental research, which later directly feeds on industrial applications.

Yves BRÉCHET and Jean PHILIBERT

OVERVIEW NO. 38

ON THE NATURE OF THE BAINITE TRANSFORMATION IN STEELS

G. R. PURDY[1] and M. HILLERT[2]

[1]McMaster University, Hamilton, Ontario, Canada L85 4L8 and
[2]Materials Center, Royal Institute of Technology, S-100 44 Stockholm, Sweden

(Received 30 November 1983)

Abstract—An overview is given of the similarities and differences among the members of the series of products, Widmanstätten ferrite, upper bainite, lower bainite, lath martensite. It is argued that this should be regarded as a continuous series even though there are significant distinctions between neighbouring products. It is proposed that important changes occur in the properties of the α/γ interface with decreasing temperature and result in changes from one product to the next.

Résumé—Nous présentons une revue des ressemblances et des différents produits de la série ferrite de Widmanstätten, bainite supérieure, bainite inférieure et martensite en lattes. Nous montrons qu'on devrait les considérer comme une série continue même s'il y a des différences notables entre des produits voisins. Nous pensons que des changements importants se produisent dans les propriétés de l'interface α/γ lorsqu'on abaisse la température, ce qui fait passer d'un produits à l'autre.

Zusammenfassung—In einem Überblick werden Ähnlichkeiten und Unterschiede zwischen den Teilen der Serie von Produkten Widmannstätten-Ferrit, oberer, unterer Bainit und Lattenmartensit behandelt. Es wird dargelegt, daß diese als eine kontinuierliche Serie angesehen werden sollte, auch wenn benachbarte Produkte sich deutlich unterscheiden. Es wird vorgeschlagen, daß wesentliche Veränderungen in den Eigenschaften der α/γ-Grenzfläche mit abnehmender Temperatur auftreten und zum Übergang von einem Produkt zum nächsten führen.

1. INTRODUCTION

Over the years there has been a lively discussion regarding the nature of the bainitic transformation of austenite in carbon containing steels [1]. There seem to be two main schools of thought, consisting of those who believe that bainite forms by a mechanism which is essentially martensitic in nature, and those who believe that bainite is a diffusional transformation product more closely related to Widmanstätten ferrite. A recent discussion [2] gives the impression that there remains a strong polarization of opinion, dividing the proponents of the two standpoints into opposed schools of thought. In contrast, there have been some efforts to stress the similarities between the various products of austenite decomposition. Widmanstätten ferrite, upper bainite, lower bainite and lath martensite have been regarded as a continuous series of related reaction products [3]. In view of recent discussions on the nature of the bainitic transformation [2], it is instructive to review experimental observations of morphological, structural and kinetic factors which are supportive of the idea that these transformation products are distinct, yet closely related. It is our hope that the overview developed here will serve as a useful focus for further discussion, and ultimately as a guide for further critical experiment in this fascinating and elusive topic.

2. AN OVERVIEW OF THE EXPERIMENTAL OBSERVATIONS

2.1. Structure

On comparing Widmanstätten ferrite which originates from austenite grain boundaries with its upper bainitic counterpart, one notes immediately the microstructural continuity of the two products. If the same low carbon steel is subjected to isothermal transformation at progressively lower temperatures, it is found that the spacing between Widmanstätten figures becomes smaller, and their aspect ratio (length/width) increases in a continuous fashion. A typical series of microstructures, due to Aaronson, is shown in Shewmon's textbook [4].

As the reaction temperature is decreased, it is often observed that the dislocation content of the ferrite phase becomes significantly greater than that of the parent austenite, and that carbides appear within the ferrite [5]. Both of these changes occur without a significant or abrupt change in gross ferrite morphology. Further features of the lower temperature transformation include a real or apparent tendency for the formation of intergranular (as opposed to grain-boundary-initiated) ferrite units, and for the formation of bainitic "sheaves", consisting of repeated discrete ferrite subunits, of the type studied by Oblak and Hehemann [5] and more recently by Bhadesia and Edmonds [6].

If bainite is formed at one temperature, and the specimen is then up-quenched, it is found that bainitic ferrite continues to grow as Widmanstätten ferrite (Fig. 1) [7].

At still lower temperatures, and in an athermal or near-athermal fashion much of the remaining austenite may transform to martensite, a product which bears a strong morphological resemblance to the bainite units formed at low reaction temperatures. The similarity is especially marked for low carbon "lath" martensite and low carbon bainite, although it is not uncommon to find rather similar habits and morphologies for martensites and bainites in higher carbon steels as well. It has even been suggested that lath martensite forms initially with a reduced carbon content, and is therefore not strictly the product of a martensite transformation [8].

It is frequently deduced or demonstrated that the crystallographic habits of bainitic ferrite and martensite units are similar, that their orientation relationships with the parent austenite are close to Kurdjumov–Sachs, and that both products give rise to similar relief effects at a free plane surface. Widmanstätten ferrite is also K–S oriented with respect to austenite, and its growth has also been observed to yield shear-type surface relief effects [9]. Figure 2 shows how Widmanstätten ferrite and lath martensite can have essentially the same morphology, orientation and habit; the implication is that the ferrite can somehow continue to grow as lath martensite as the temperature is lowered.

In this connection it may be emphasized that side-wise growth of lath martensite seems to be kinetically limited in distinction to plate martensite where most of the volume of a plate may be the result of side-wise growth. This may be the reason for the morphological similarity between lath martensite and the Widmanstätten ferrite and bainite.

Fig. 1. Formation of bainite during 10 s at 445°C and its continued growth as Widmanstätten ferrite during 20 s and 693°C in a 0.6% C steel. Magnification ×1800. From Ref. [7].

2.2. Kinetics

The kinetics of Widmanstätten ferrite growth is well documented, and can be understood in a relatively quantitative manner. The edgewise growth of Widmanstätten plates or laths into uniformly supersaturated austenite proceeds at a constant velocity; the thickening of Widmanstätten figures is evidently discontinuous on a microscopic scale, and dependent on the rate of supply of growth ledges for lateral growth. The usual interpretation of these observations [4] is that lengthening is determined by the volume diffusion of carbon in austenite, while precipitate thickening is slower than that expected for full diffusion control, on account of the lower mobility of the interfaces bounding the sides of the precipitates. The morphological stability of the sides of Widmanstätten precipitates has been attributed to this low mobility, and to the anisotropy of interfacial energy for the f.c.c./b.c.c. bicrystal [10]. Figure 3 shows clearly the faceted character of the Widmanstätten ferrite/austenite interface.

Application of theoretical models for volume diffusion controlled lengthening of such precipitates gives quite good agreement with experiment, both in simple steels [11], and in nonferrous alloy analogues [12]. This suggests that the assumption of diffusion controlled growth of Widmanstätten ferrite is essentially correct. There remain uncertainties in phase equilibria and volume diffusion coefficients, and in interfacial free energies that render the closure between theory and experiment less than complete. It is quite likely that a small departure from local equilibrium at the tip of a growing Widmanstätten precipitate is hidden in the experimental error.

Constant lengthening rates, in approximate agreement with the predictions of models for diffusional precipitate lengthening, are again observed for upper bainite, and this has been interpreted as evidence for volume diffusion controlled growth [13]. However, the uncertainties in extrapolated thermodynamic and kinetic quantities are greater at the lower transformation temperatures, and it is perhaps equally appropriate to argue for the diffusional lengthening of bainitic ferrite on the basis of continuity with the higher temperature transformation: the transition from Widmanstätten ferrite to upper bainite appears to possess kinetic as well as microstructural continuity.

Nemoto [14] studied the growth of a single bainite lath using hot stage high voltage electron microscopy. He determined that the lath consisted of a single crystal of ferrite, which lengthened at a constant rate.

The lengthening kinetics of bainite sheaves, consisting of discrete ferritic subunits, are more difficult to quantify. Clearly, the overall growth process is determined by the rates of nucleation of the subunits, as well as by their rates of growth. One suggestion [5], currently supported by Bhadesia and his colleagues [15], is that the growth to final size of individual ferrite subunits is essentially instantaneous, and that

PURDY and HILLERT: THE BAINITE TRANSFORMATION IN STEELS

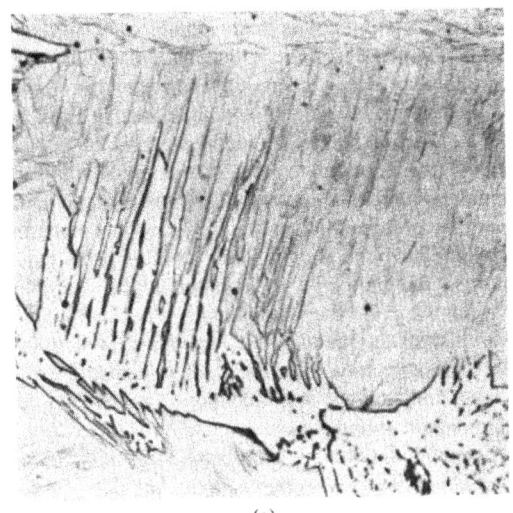

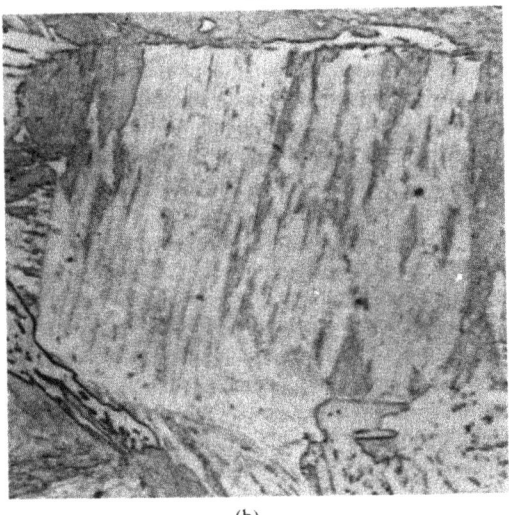

Fig. 2. (a) Grain boundary initiated plates of Widmanstätten ferrite formed in 0.5 s at 630°C. Etched in picral. Magnification ×1200. (b) Same area etched in nital. The Widmanstätten plates happened to have such a crystalline orientation that the nital attack is slow. The micrograph indicates that much of the adjoining martensite, formed on quenching has the same orientation. From Ref. [3].

the macroscopic rate of lengthening is determined by the rate of formation of new subunits.

Further insight into the problem of bainite growth kinetics may be obtained from consideration of the effects of alloying element additions to simple steels. If the rates of lengthening of upper bainite in low alloy nickel steels are compared with those in plain carbon alloys at the same temperature, it is found that the relative effects of the alloy addition can be accounted for on the basis of a simple change in supersaturation, applied to a volume diffusion control model [8]. In this way, one can obtain a further indication that volume diffusion of carbon in austenite plays a role in the growth of bainite. It is in the higher nickel alloys, however, that one finds the more interesting kinetic behaviour. Rao and Winchell [16]

Fig. 3. A Widmanstätten ferrite plate, formed *in situ* in a 0.4 wt% molybdenum 0.36 wt% carbon steel foil, isothermally transformed at 540°C. The dark band is a crystal of ferrite in a matrix of austenite. Hot stage, high voltage electron micrograph. Magnification ×10,000.

studied the lengthening of upper bainite in a 9% Ni steel at 673 K, and found rates that were many orders of magnitude too high for analysis on the basis of a local equilibrium model. At this concentration of nickel, local equilibrium would require that nickel be partitioned to the austenite during growth. At the same time, the rates are orders of magnitude too low for analysis on the basis of a strict paraequilibrium (no-partitioning) model which should result in rate-control by diffusion of carbon [17, 18].

3. DISCUSSION

3.1. Structure

The main microstructural and crystallographic similarities and distinctions among the products of austenite decomposition have been described above. On reflection, it seems that the repeated subunit morphology of bainite sheaves has been the focus of much past and current controversy, and some further discussion of its occurrence may therefore be pertinent here. The repeated subunit structure is thought to result from some kind of autocatalysis, a term often associated with martensite growth although the new units of plate martensite have new orientations whereas the subunits of bainite have the same orientation. It may be important to emphasize that these repeated subunit structures are also observed in diffusional Widmanstätten precipitation e.g. θ' in aluminium–copper alloys [19] and cementite in steels. It seems that the formation of chains or sheets of precipitates is the result of elastic interaction between existing precipitates and nucleating ones. The precipitate stacks found in Al–Cu can be understood on the basis of the interaction of elastic tetragonal formation strains for existing and nucleating particles.

It has been shown that the addition of a shear component to the transformation strain need not change this picture appreciably [20]. An expansion or contraction normal to the broad face of the plates is still required in order to produce a noncoplanar energy minimum position for the new particle. Thus, we can take the observation of repeated subunit morphologies as an indication that elastic strains are becoming important, perhaps in the cessation of growth of the existing subunit, and more surely in the formation of the new subunit.

The occurrence of dislocation substructures in bainitic ferrite is also worthy of discussion. The source of the dislocations has been a matter for conjecture, and they may be attributed to mechanical strain attending the austenite–ferrite transformation or ascribed in some way to the lattice–invariant strain required by the crystallographic theory of martensite growth. A third alternative, which has received little previous attention will be put forward here: we submit that it is possible that the dislocations are generated by the motion of K–S oriented ferrite/austenite interfaces when these interfaces are subjected to sufficiently large driving forces. This hypothesis requires, for example, that an f.c.c./b.c.c. interface which moves by lateral ledge migration at low driving force undergoes a transition in migration mode, from lateral to normal, as the driving force is increased in the manner described by Cahn [21]. The normal migration mode would then be connected with high defect densities in the product phase, perhaps as a result of nonconservative motion of boundary structural units. Some evidence for a transition of this type is found in *in situ* transmission electron microscope studies [14, 22], which suggest that low temperature α/γ interface migration gives rise to dislocated ferrite, for a wide range of precipitate morphologies including the bainitic lath morphology in thin foils. Malcolm and Purdy [23] found evidence for a similar transition in interface migration mode in their studies of bulk specimens of α/γ brass: Widmanstätten α precipitates formed at high supersaturation contained high defect densities, while, on reheating, the α/β interfaces moved by ledge motion, and generated defect-free α layers.

Thus we propose that the planar Widmanstätten ferrite/austenite interface, which is thought to be essentially immobile under low driving forces, responds to higher driving forces (for instance when temperature is lowered into the bainitic range), by migrating normal to itself, and that this transition corresponds to departures from the strictly planar habit, and to the development of the bainitic character, i.e. to the formation of dislocated lath ferrite. However, we expect that the interface mobility in this normal growth mode is highly anisotropic, and that the growth form of the resulting laths reflects this anisotropy.

3.2. Kinetics

As noted above, the analysis of the growth kinetics of Widmanstätten ferrite plates indicates that they lengthen at a rate controlled by the volume diffusion of carbon, and thicken at a rate influenced by the low mobility of the faceted α/γ interface. Even for the lengthening process, a small departure from local equilibrium is possible. We now propose that there is such a departure and that it increases with decreasing temperature as we enter the bainitic range. For upper bainite, we would argue that the lengthening is still determined approximately by carbon volume diffusion, provided that the bainite laths grow continuously as single crystals, as observed, for example by Nemoto [14]. For bainite lath thickening, our prior discussion leads us to propose that a large departure from local equilibrium is sustained during growth, and that the interface migration mode is, in consequence, a normal one.

As noted previously, the lengthening and thickening of bainite sheaves consisting of ferritic subunits is related only indirectly to the rates of growth of individual units. Since the subunits are in other respects similar to the single crystalline laths, we see no reason or need to invoke a new mechanism for their growth. We believe that the subunits are an autocatalytically nucleated variant of the bainite lath product, which grows with some departure from local equilibrium, and which nevertheless requires carbon diffusion during growth.

Finally, on the basis of microstructural similarity, as illustrated by reference to Fig. 2, we put forward the proposition that lath martensite is related to the series of reaction products, Widmanstätten ferrite, upper bainite, and lower bainite by a continuum of transformation states. In this view, lath martensite growth represents a kinetically unstable branch of the bainite reaction, one which is accompanied by essentially complete solute trapping. The rate of martensite formation would then be determined by the mobilities of the transformation interfaces under the high driving forces attending martensite growth. In this connection, we present the results of an earlier modelling study of the bainite transformation, which permits interfacial discontinuities in both the iron and carbon chemical potentials [3]. This analysis, whose results are summarized in Fig. 4, predicts such a kinetic instability, leading to very high rates of interface-controlled growth at the highest supersaturation.

3. CONCLUDING REMARKS

We have argued that the series of transformation products in steel: Widmanstätten ferrite, upper bainite, lower bainite, lath martensite can be interpreted as a continuous series, and we have chosen to emphasize similarities, rather than differences among them. There are of course, rather important distinctions between neighbouring products in the series. Upper bainite is distinguished from Widmanstätten ferrite, in our view, mainly by its morphology and dislocation content. (The latter characteristic is emphasized in the present discussion. We believe that

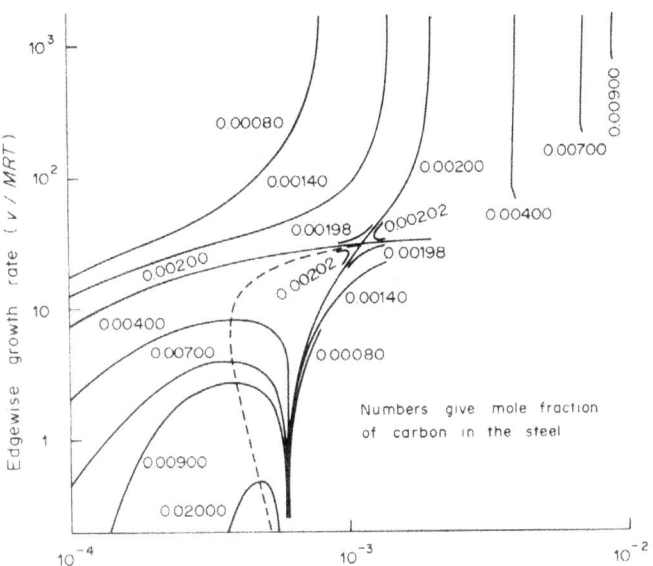

Fig. 4. Model-calculated growth rate for edgewise growth of ferrite as function of the carbon content of the growing ferrite. The initial carbon contents of the steels are given by numbers in the diagram. The maximum growth rate hypothesis predicts diffusion-controlled growth according to points on the dashed line. A point of instability is reached at a carbon content of 0.002. At lower carbon content of the steel, the growth is not diffusion-controlled and high growth rates may be obtained under little or no diffusion of carbon. From Ref. [3].

it may be related to a change in interfacial migration mode.) Upper and lower bainite are both thought to form with significant departures from local equilibrium, but with the requirement that some carbon is rejected by diffusion into austenite during growth. The main distinction between the two products lies in the presence of internal carbides in lower bainite. If attention is focused only on the behaviour of substitutional atoms, then lower bainite is thought to be closely similar to lath martensite, the main source of their distinction lying in the absence of carbon diffusion in the martensite transformation. A further possible difference may be postulated by reference to the analysis of growth kinetics of bainite in Fe–C–9% Ni alloys [16], which indicates that growth (lengthening and thickening) is too slow to be determined solely by carbon diffusion. This observation suggests that the substitutional solute atoms can interact with the bainite interfaces in a way that is not consistent with our concept of the behaviour of the martensite/austenite interface. It would therefore be of particular interest to learn more about the detailed structure of the bainite/austenite interfaces, in these alloys.

Given that there are undeniable differences between the products of austenite decomposition in low carbon steels, and that these differences are important enough to warrant further scrutiny using the full range of available experimental methods, we wish to conclude by restating our emphasis on the family resemblances shared by the members of the series. It is our hope that the viewpoint adopted here will generate constructive debate, and perhaps serve as a basis for the pursuit of new theoretical arguments and for the design of critical experiments.

Acknowledgement—This work was supported by the Natural Sciences and Engineering Research Council of Canada through an International Collaboration Grant.

REFERENCES

1. R. F. Hehemann, K. R. Kinsman and H. I. Aaronson, *Metall. Trans.* **3**, 1077 (1972).
2. G. J. Shiflet and H. I. Aaronson, *Proc. Int. Conf. on Solid–Solid Phase Transformations* (edited by Aaronson, Laughlin, Sekerka and Wayman), p. 1581. Am. Inst. Min. Engrs, New York (1982).
3. M. Hillert, Internal Report, Swedish Institute for Metal Research (1960).
4. P. G. Shewmon, *Transformations in Metals*. McGraw-Hill, New York (1969).
5. J. M. Oblak and R. F. Hehemann, *Transformation and Hardenability in Steels*, p. 15. Climax Molybdenum Co. (1967).
6. H. K. D. H. Bhadesia and D. V. Edmonds, *Acta metall.* **28**, 1265 (1980).
7. M. Hillert, *Värmländ. Bergsmann. Ann.* pp. 1–34 (1958).
8. G. Thomas and M. Sorikaya, *Proc. Int. Conf. on Solid–Solid Phase Transformations* (edited by Aaronson, Laughlin, Sekerka and Wayman), p. 999. Am. Inst. Min. Engrs, New York (1982).
9. K. R. Kinsman and H. I. Aaronson, *Transformation and Hardenability in Steels*, p. 39. Climax Molybdenum Co. (1967).
10. P. G. Shewmon, *Trans. metall. Soc. A.I.M.E.* **233**, 936 (1965).
11. H. I. Aaronson, *Decomposition of Austenite by Diffusional Processes* (edited by Zackay and Aaronson), p. 387. Interscience, New York (1960).

12. G. R. Purdy, *Metals Sci. J.* **5**, 81 (1971).
13. L. Kaufman, S. V. Radcliffe and M. Cohen, *Decomposition of Austenite by Diffusional Processes* (edited by Zackay and Aaronson), p. 313. Interscience, New York (1960).
14. M. Nemoto, *High Voltage Electron Microscopy* (edited by P. R. Swann, C. J. Humphreys and M. J. Goringe), p. 230. Academic Press, New York (1974).
15. H. K. D. H. Bhadesia and A. R. Waugh, *Proc. Int. Conf. on Solid–Solid Phase Transformations* (edited by Aaronson, Laughlin, Sekerka and Wayman), p. 1581. Am. Inst. Min. Engrs, New York (1982).
16. M. M. Rao and P. G. Winchell, *T.M.S.-A.I.M.E.* **239**, 956 (1967).
17. M. Hillert, *Metall. Trans.* **6A**, 5 (1975).
18. G. R. Purdy, *Metallography* **3**, 131 (1975).
19. V. Perovic, G. R. Purdy and L. M. Brown, *Acta metall.* **29**, 889 (1981).
20. V. Perovic, G. R. Purdy and L. M. Brown, *Scripta metall.* **15**, 217 (1981).
21. J. W. Cahn, *Acta metall.* **8**, 556, (1960).
22. G. R. Purdy, *Acta metall.* **26**, 481 (1978).
23. J. A. Malcolm and G. R. Purdy, *Trans. metall. Soc. A.I.M.E.* **239**, 1391 (1967).

22 Introduction to "Solute Drag, Solute Trapping and Diffusional Dissipation of Gibbs Energy"

published in *Acta Materialia* (1999)

Except perhaps for the Calphad period (ca. 1975-1995), Mats Hillert has been deeply concerned with the local state of migrating phase interfaces during phase transformations. This is evident already in the very early work, for example in his analysis of Paraequilibrium from 1953 [1]. This concern is easy to understand: If a phase transformation involves a change in composition, the transformation rate may be controlled by the rate of diffusion and may thus be calculated by solving a diffusion problem. In the conventional approach, introduced by Stefan [2] more than a century ago, boundary conditions are needed at the moving interface and the simplest assumption is that of local equilibrium, *i.e.* the composition on each side of the interface is obtained as an equilibrium tieline in the appropriate phase diagram. Paraequilibrium may then be regarded as a thermodynamic extreme that allows the calculation of a tieline under the constraint that the transformation involves no redistribution of substitutional atoms whatsoever. That calculation is straightforward and requires thermodynamic data only, *i.e.* no kinetic information is needed.

But what happens if neither local equilibrium nor paraequilibrium are good approximations for the situation at the phase interface? In that case one needs to consider the details of the interfacial reactions. This was the topic of another early paper by Hillert [3] where he considered the transformation of austenite (FCC) to various forms of BCC in Fe-C alloys. He assumed that the interface is atomistically sharp and that the transformation involves two physical processes occurring at the interface; diffusive jumps of carbon atoms across the interface and change of the crystalline lattice from FCC to BCC. A very interesting feature of such a model is that it predicts a spontaneous transition to a transformation mode that involves no carbon diffusion at all, *i.e.* a massive or a martensitic transformation, at sufficiently high supersaturations. This result is qualitatively in agreement with the experimental observations and could never have been obtained within the conventional local equilibrium paradigm.

Over the coming years Hillert extended his analysis by allowing the phase interface to have a finite width as well as its own properties. In two papers, co-authored by Sundman, [4, 5] this analysis was developed further and was shown to apply not only to phase transformations but to grain growth as well. Moreover, in the latter case the analysis was demonstrated to give the same predictions as Cahn's [6] treatment from 1962. This result was intriguing because the Hillert-Sundman analysis is based on the dissipation of Gibbs energy by diffusional processes both inside and outside the interface, whereas Cahn's analysis is based on a force balance that only involves the properties of the grain boundary itself.

After the work by Hillert and Sundman several contributions appeared. These were based on the sharp-interface approach, *e.g.* a series of papers by Aziz [7] and Ågren [8] or the finite-width approach, *e.g.* Bréchet and Purdy [9]. Moreover, in the early 90's the phase-field method [see *e.g.* 10,11], based on the diffuse interface concept [12,13], took off. It was shown by Ahmad *et al.* [14] that the phase-field formulation is capable of representing qualitatively the same behavior as the sharp- and finite-width models, *i.e.* a transition to a transformation mode without any composition change at high supersaturations. In writing, it is interesting to note that one of the first computer simulations based on a diffuse interface was actually presented by Hillert [15] as early as 1961, in his analysis of spinodal decomposition.

Despite their differences, the mentioned approaches have much in common and the topic of Hillert's overview "Solute Drag, Solute Trapping and Diffusional Dissipation of Gibbs Energy", in *Acta Mater.* 1999, 57, 4481, is to analyze and compare the various approaches. In addition to giving an in-depth overview Hillert further emphasizes some major concepts: In general a phase transformation in a binary alloy proceeds by two physical processes occurring at the migrating interface, 1) change in crystalline structure and 2) interdiffusion, which both need a driving force to proceed. Hillert presents in detail the irreversible thermodynamic analysis of these two processes and in particular he discusses how the result changes if a different set of processes is considered, for example transfer of the individual atoms across the interface. If the interdiffusion and change in crystalline lattice are independent processes his analysis shows that the transfer of the different kind of atoms across the interface cannot be independent but there must be a coupling. He also demonstrates that the Onsager reciprocal relations are obeyed for the coupling coefficients and why Baker and Cahn [16] concluded that the reciprocal relations were not obeyed in their analysis.

Hillert then turns to the representation of the interface properties for the case where the interface has a finite width and compares the Cahn's solute-drag approach, using the force balance, and Hillert-Sundman's approach, using dissipation of Gibbs energy by diffusion. However, the latter discussion is rather a clarification of the earlier conclusion from Hillert and Sundman than an explanation why the two so different methods give the same predictions for the case of grain-boundary migration. In fact, that explanation came 5 years later, and was presented by Hillert only very recently [17].

It may finally be added that the paper "Solute Drag, Solute Trapping and Diffusional Dissipation of Gibbs Energy" is based on Hillert's Hume-Rothery Lecture in San Diego 1999.

John ÅGREN

References

[1] M. Hillert, Paraequilibrium, in *Internal report*, Swedish institute for Metal Research, Stockholm, 1953[1].

[2] J. Stefan, S.B. Wield, *Akad. Mat. Natur* 98 (1889) 473.

[3] M. Hillert, The growth of Ferrite, Bainite and Martensite, in *Internal report*, Swedish institute for Metal Research, Stockholm, 1960[2].

[4] M. Hillert and B. Sundman, *Acta Metall.* 24 (1976) 731[3].

[5] M. Hillert and B. Sundman, *Acta Metall.* 25 (1977) 11.

[6] J.W. Cahn, *Acta Metall.* 10 (1962) 789.

[7] See for example: M.J. Aziz, *J. Appl. Phys. Lett.* 43 (1983) 552.

[8] J. Ågren, *Acta Metall.* 37 (1989) 181.

[9] G.R. Purdy and Y.J.M. Bréchet, *Acta Metall. Mater.* 43 (1995) 3763.

[10] See for example: R. Kobayashi, *Physica D* 63 (1993) 410.

[11] A.A. Wheler, W.J. Boettinger and G.B. McFadden, *Phys. Rev. A* 45 (1992) 7424.

[12] J.D. van der Waals 1893: See for example English translation in: J.S. Rowlinson, *Journal of statistical physics* 20 (1979) 197.

[13] J.W. Cahn and J.E. Hilliard, *Journal of Chemical Physics* 28 (1958) 258.

[14] N.A. Ahmad, A.A. Wheeler, W.J. Wheeler and G.B. McFadden, *Phys. Rev. E* 58 (1998) 3436.

[15] M. Hillert, *Acta Metall.* 9 (1961) 525.

[16] J.C. Baker and J.W. Cahn, in *Solidification*, ed. by T.J. Hughel and G.F. Bolling (ASM, Metals Park OH, 1971) p. 23.

[17] M. Hillert, *Acta Mater.* 52 (2004) 5289.

---

[1] This volume, chapter 2.
[2] This volume, chapter 6.
[3] This volume, chapter 17.

PERGAMON

PII: S1359-6454(99)00336-5

OVERVIEW NO. 135

SOLUTE DRAG, SOLUTE TRAPPING AND DIFFUSIONAL DISSIPATION OF GIBBS ENERGY†

MATS HILLERT

Department of Materials Science, KTH (Royal Institute Technology), SE-10044 Stockholm, Sweden

(Received 15 June 1999; accepted 31 August 1999)

Abstract—There are several phenomena that depend on the interaction between solute atoms and migrating grain boundaries or phase interfaces. In order to explain them several models have been proposed and these are now reviewed. There are two approaches to this problem: the solute drag approach and the Gibbs energy dissipation approach. The latter has often been applied to a sharp interface model resulting in a treatment that is in accord with expectations from irreversible thermodynamics. When applied to the homogeneous interface model there are problems at high velocities. When applied to a wedge-shaped description of the properties of the interface, a very flexible treatment is obtained but it is difficult to decide how to choose the model parameters. The application of the solute drag approach to phase transformations is discussed. It is postulated that it should give the same result as the dissipation approach and this can be proved in simple cases. Applications to massive transformation, the effect of alloying elements on the formation of ferrite from austenite and DIGM are discussed. © *1999 Acta Metallurgica Inc. Published by Elsevier Science Ltd. All rights reserved.*

Keywords: Alloys; DIGM; Diffusion; Interface; Solute drag

1. INTRODUCTION

It has long been realized that grain boundaries and phase interfaces in metallic materials may have some width and in trying to understand various phenomena, interfaces as thin regions of special properties have been modelled. In its simplest form this model type assumes that the interfacial region has a fixed width and homogeneous properties within that width. It can thus be treated as a special phase with its own Gibbs energy as a function of composition. It is schematically represented in Fig. 1(b). However, when Gibbs developed his treatment of "Equilibrium of Heterogeneous Substances", he was able to show how the energy of interfaces could be treated by considering a sharp, imaginary interface [1]. It is illustrated in Fig. 1(a). This concept has in modern times been applied to migrating interfaces although it does not have a firm theoretical basis under non-equilibrium conditions. It may still be a useful model but important aspects may be missed. In attempts to cover such aspects, thermodynamic models have been developed that take the width into account.

In recent years such phenomena as solute drag and solute trapping have attracted much attention and in order to describe them more sophisticated models have been developed where the properties of the interface vary continuously from those of one phase to those of the other. Those continuous models can be developed to various degrees of sophistication as illustrated in Figs 1(c)–(e). Today there is strong interest even in using a diffuse interface model where the width is not fixed *a priori* and is not even well defined [Fig. 1(f)]. The present work will review the basic aspects of the development of interface modelling.

2. CONVENTIONAL PICTURE OF PHASE TRANSFORMATIONS IN ALLOYS

Since long ago, the conventional picture of the migration of an interface is that it is a result of individual atoms leaving the parent grain (phase), diffusing through an interface and depositing on the side of the growing grain (phase). As an example, Turnbull in 1951 [2] derived an expression for the mobility of a grain boundary in a pure metal from this picture obtaining

$$M = \frac{\delta D^\mathrm{I} V_\mathrm{m}}{b^2 RT} \quad (1)$$

δ is the thickness of the interface, D^I is the diffusion coefficient in the interface, V_m is the molar volume and b is Burgers' vector. M is the mobility and is

---

†This paper is based on the Hume-Rothery Lecture presented at the 128th TMS Annual Meeting, 1 March 1999, San Diego, U.S.A.

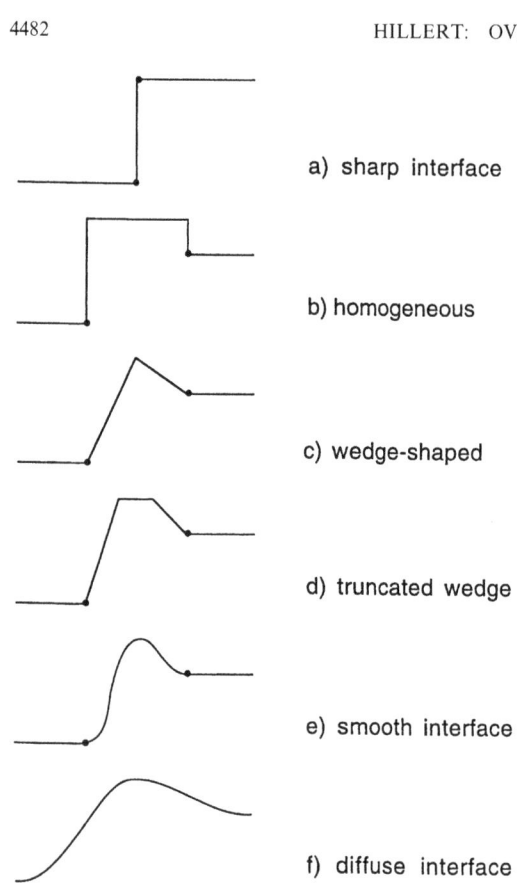

Fig. 1. Schematic characterization of interface models.

a) sharp interface
b) homogeneous
c) wedge-shaped
d) truncated wedge
e) smooth interface
f) diffuse interface

defined from a kinetic equation for the interface migration,

$$D^m = \frac{v}{M} \qquad (2)$$

v is the migration velocity, $1/M$ may be regarded as the friction coefficient and D^m is the driving force making the interface migrate. It will often come from the chemical driving force available for the phase transformation. It may thus be expressed in J/mol, but can be transformed to a pressure by dividing with the molar volume, $P = D^m/V_m$, and to a real force by further multiplying with the area. The molar volume, V_m, will be approximated as a constant all through the present work.

When applying the same model to the migration of an interface in an alloy, one considers the diffusion of each component as a separate process with a driving force $\mu_i^\gamma - \mu_i^\alpha$, where γ is the shrinking phase and α is the growing phase. If there is no coupling between the diffusion processes, it is necessary that all $\mu_i^\gamma - \mu_i^\alpha$ are positive. This conventional picture is illustrated in a molar Gibbs energy diagram in Fig. 2 for a simple binary case. Thus, the conventional picture predicts that the growing phase will never fall inside the two-phase field. The relative magnitudes of the decrease in chemical potential of A or B must be balanced with the atomic

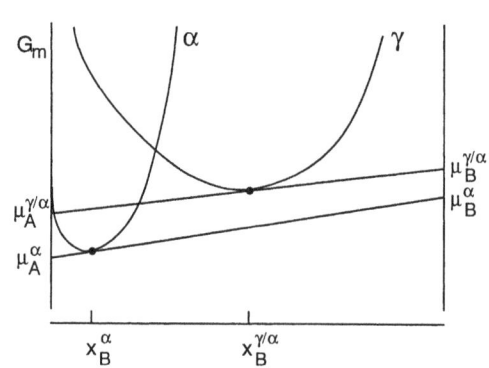

Fig. 2. Molar Gibbs energy diagram for a $\gamma \to \alpha$ phase transformation. According to the conventional picture, the chemical potentials of both components must decrease as shown here.

mobilities in the interface in order to give the correct composition of the growing phase. This model was discussed, for instance, by Jackson [3].

In view of this conventional picture it is not expected that the chemical potential of a solute would increase as a result of the phase transformation. When that phenomenon occurs, it is often called "solute trapping" because the solute atoms seem to be caught by the growing phase although there is a driving force for them to escape. Evidently, this phenomenon would need a more refined model for its explanation.

The driving force of a process will result in a production of entropy and, under constant temperature and pressure, there would be a corresponding dissipation of Gibbs energy. The present review will make use of molar Gibbs energy diagrams and it will thus be convenient to introduce the Gibbs energy dissipation from the beginning. For a process, p, it is given by

$$dG^p = D^p d\xi^p \qquad (3)$$

where D^p is its driving force in J/mol and ξ^p is the extent of the process expressed in moles. (See Ref. [4], Sections 1.8 and 1.11, for instance). For a steady-state process D^p is constant and one can integrate to get

$$\Delta G^p = D^p \Delta \xi^p \qquad (4)$$

The driving forces will always be denoted by D with a superscript to identify the process. Unfortunately, diffusion coefficients are also denoted by D but in the present work only D^I for interface and boundary diffusion and D^γ for diffusion in the γ phase will be used and there should be no risk of confusion.

3. THE SHARP INTERFACE MODEL

In contrast to the conventional picture, it has long been known that the new phase in a martensitic transformation can fall inside a two-phase field and in 1960 Hillert [5] constructed a model describing the possibility that the $\gamma \rightarrow \alpha$ transformation in the Fe–C system could yield Widmanstätten α with its composition falling inside the $\alpha + \gamma$ two-phase field and eventually result in a spontaneous change to a diffusionless, martensitic transformation. The model assumed that there are two independent processes involved. The first one transforms the lattice from γ to α and would by itself be diffusionless. It may be regarded as a co-operative process by which the interface migrates and D_{Fe}^m is the driving force per mole of Fe necessary for overcoming the friction of the interface. When α grows by 1 mole of Fe atoms, then $\Delta\xi^m = 1$ and under steady-state conditions $\Delta G_{Fe}^m = D_{Fe}^m$. At the same time, the other process would allow some carbon to diffuse back into the γ phase by trans-interface diffusion. The combination of processes is illustrated by the molar Gibbs energy diagram in Fig. 3, which shows a case of solute trapping because $\mu_C^\alpha > \mu_C^{\gamma/\alpha}$. Here the axes are chosen as G/N_{Fe} and z_C, defined as N_C/N_{Fe}, which are more convenient than the usual ones, $G_m = G/N$ and $x_C = N_C/N$, when the solute dissolves interstitially and the diffusion of the solvent atoms is negligible (Ref. [4], pp. 82–83). This advantage was not realized originally. N_{Fe} and N_C are the numbers of Fe and C atoms and N is the total number of atoms, all these numbers being expressed in mole.

In order for α to grow with the C content z_C^α, it is necessary that $z_C^{\gamma/\alpha} - z_C^\alpha$ mole of C, per mole of Fe, diffuse back into the γ phase by trans-interface diffusion and thus $\Delta\xi^t = z_C^{\gamma/\alpha} - z_C^\alpha$. The driving force, D^t, is $\Delta\mu_C = \mu_C^\alpha - \mu_C^{\gamma/\alpha}$ per mole of C and the dissipation of Gibbs energy by trans-interface diffusion, counted per mole of Fe, will be

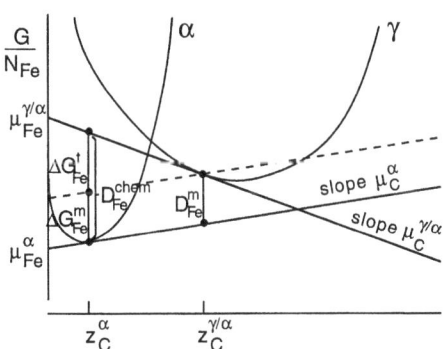

Fig. 3. Molar Gibbs energy diagram for the $\gamma \rightarrow \alpha$ transformation in the Fe–C system under steady-state conditions, illustrating so-called trapping because the chemical potential of carbon increases. ΔG_{Fe}^t is the dissipation of Gibbs energy by trans-interface diffusion, necessary for the change in composition, D_{Fe}^{chem} is the chemical driving force over the interface and ΔG_{Fe}^m is what remains to be dissipated by the migration of the interface. It is equal to the driving force for that process, D_{Fe}^m. All quantities are expressed per mole of Fe atoms.

$$\Delta G_{Fe}^t = D^t \Delta\xi^t = \left(\mu_C^\alpha - \mu_C^{\gamma/\alpha}\right)\left(z_C^{\gamma/\alpha} - z_C^\alpha\right) \quad (5)$$

It is illustrated with a molar Gibbs energy diagram in Fig. 3 using the special choice of axes. The chemical potential of the solute is given by the slope of the tangent in this special case. Equation (5) could thus be obtained graphically from this diagram. From the diagram one can also obtain the dissipation of Gibbs energy by the whole process of formation of α with a composition different from that of the γ phase. It is usually regarded as the chemical driving force for the whole process,

$$D_{Fe}^{chem} = \left(\mu_{Fe}^{\gamma/\alpha} - \mu_{Fe}^\alpha\right) - z_C^\alpha\left(\mu_C^\alpha - \mu_C^{\gamma/\alpha}\right) \quad (6)$$

where all the terms are counted per mole of Fe atoms entering into the growing α phase. All this driving force will be dissipated and by subtracting the dissipation of Gibbs energy by diffusion one finds how much will be dissipated by friction in the migration process,

$$\Delta G_{Fe}^m = D_{Fe}^{chem} - \Delta G_{Fe}^t$$
$$= \left(\mu_{Fe}^{\gamma/\alpha} - \mu_{Fe}^\alpha\right) - z_C^{\gamma/\alpha}\left(\mu_C^\alpha - \mu_C^{\gamma/\alpha}\right) \quad (7)$$

This is also equal to the driving force for the migration, D_{Fe}^m, because $\Delta\xi^m = 1$ (see Fig. 3). When considering the pressure difference occurring when α is growing under a pressure increase due to the interface being curved, one should also subtract a term $V_{Fe}\Delta P$ where V_{Fe} is the partial molar volume for Fe.

The material going into the α phase comes from the γ phase at the interface and it has a different composition. Figure 3 is constructed for steady-state conditions and in order for it to apply there must be diffusion inside the γ phase to compensate for the difference. The composition of the initial γ phase, γ_0, would normally fall somewhere between those of the α phase and of γ at the interface γ/α. This is illustrated in Fig. 4, which was drawn with ordinary axes and would thus hold for a substitutional solute, which is the more common case. In the following we shall retain the notation γ for the parent phase and α for the growing phase. It will be assumed that there are steady-state conditions and no diffusion inside the α phase. Solute atoms will pile up in γ in front of the advancing interface and form a spike of a height $x_B^{\gamma/\alpha} - x_B^{\gamma_0}$. Diffusion will occur down that spike but in order to get a steady-state process it is also necessary to transport a quantity of solute, $x_B^{\gamma_0} - x_B^\alpha$, sidewise out of the way for the advancing α phase. That could take place either by lateral diffusion in the interface or by sidewise diffusion inside γ, which would happen in front of the edge of a Widmanstätten plate or a tip of a dendrite. In the present work, the lateral interface diffusion will be neglected except for the case of DIGM, to be discussed in Section 21. It is

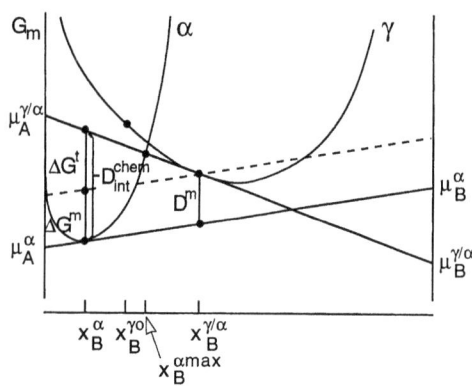

Fig. 4. Molar Gibbs energy diagram for the $\gamma \to \alpha$ transformation in a binary with a substitutional alloy element. ΔG^t, ΔG^m, D_{int}^{chem} and D^m are now expressed per mole of atoms.

not possible to illustrate the Gibbs energy dissipation by two- or three-dimensional diffusion inside the γ phase in Fig. 4 because the exchange of atoms by sidewise diffusion is not included. This problem will be further discussed in Section 18.

In the substitutional case we should consider interdiffusion across the interface and its driving force would be $D^t = \Delta(\mu_A - \mu_B)$. In order to accomplish the change in composition the extent of this trans-interface diffusion will be $\Delta \xi^t = (x_B^{\gamma/\alpha} - x_B^\alpha)$ mole of B atoms per mole of atoms added to the growing α phase. The Gibbs energy dissipation is obtained as

$$\Delta G^t = D^t \Delta \xi^t = \Delta(\mu_A - \mu_B)\left(x_B^{\gamma/\alpha} - x_B^\alpha\right)$$

$$= \left(x_B^{\gamma/\alpha} - x_B^\alpha\right)\left[\left(\mu_A^{\gamma/\alpha} - \mu_A^\alpha\right) - \left(\mu_B^{\gamma/\alpha} - \mu_B^\alpha\right)\right] \quad (8)$$

This relation can be obtained graphically from Fig. 4, which also yields the chemical driving force acting over the interface, counted per mole of α formed,

$$D_{int}^{chem} = x_A^\alpha \left(\mu_A^{\gamma/\alpha} - \mu_A^\alpha\right) + x_B^\alpha \left(\mu_B^{\gamma/\alpha} - \mu_B^\alpha\right) \quad (9)$$

The subscript "int" is used to emphasize that this is the driving force for the processes inside the interface. If no part of D_{int}^{chem} is "wasted" on side-reactions, the dissipation due to friction in the migration process will be

$$\Delta G^m = D_{int}^{chem} - \Delta G^t$$

$$= x_A^{\gamma/\alpha}\left(\mu_A^{\gamma/\alpha} - \mu_A^\alpha\right) + x_B^{\gamma/\alpha}\left(\mu_B^{\gamma/\alpha} - \mu_B^\alpha\right) \quad (10)$$

and this will also be the driving force for the migration counted per mole of atoms, D^m, because $\Delta \xi^m = 1$. This is the quantity to be inserted in equation (2). Of course, it is the net flux of atoms across the interface that causes the interface to migrate. That flux will thus be denoted by J^m and from equation (2) we obtain

$$J^m V_m = v = M D^m = \frac{M \Delta G^m}{\Delta \xi^m} = M \Delta G^m \quad (11)$$

Equation (11) represents the response function for the migration process. Actually, it was defined already by accepting equation (2). It is interesting to note that equation (2) assumes that the velocity of the migrating interface is proportional to its driving force and independent of the driving force for any other simultaneous process.

Equations (8)–(10) correspond to equations (5)–(7) for the interstitial case. It is important to notice that the parallel tangents construction in Fig. 3 can also be applied with the axes used in Fig. 4 because it is now the difference in $\mu_A - \mu_B$ that drives diffusion and all driving forces and dissipations of Gibbs energy are expressed per mole of atoms.

Equations (8) and (10) can be used as a definition of the so-called sharp interface model, which is based on Fig. 1(a). In order to apply the sharp interface model, one must select kinetic equations relating the rate for each process to its driving force. The model can thus be developed in different ways depending on what response functions, i.e. kinetic equations, are chosen but also on how the Gibbs energy of the two phases are described as functions of composition. The sidewise diffusion in the parent phase can also be treated in many ways. There may thus be a great variety of treatments based on the sharp interface model, but the relation between driving forces and dissipation of Gibbs energy, illustrated in Fig. 4, should apply to all those treatments.

As an example we may discuss the kinetic equation for the process causing the change in composition across the interface. The driving force for transportation of A in one direction across the interface in exchange for the same amount of B going in the opposite direction is $D^t = \Delta(\mu_A - \mu_B)$. It may be assumed that this process is independent of the driving force for the migration process because through equation (2) it was assumed that the migration is independent of the driving force for any other process, i.e. also of the process now under consideration. The flux of this process may thus be expressed as

$$J^t = \left(\frac{1}{V_m}\right) L \Delta(\mu_A - \mu_B) \quad (12)$$

where L is the kinetic coefficient. The superscript t was introduced in equation (5) to denote trans-interface diffusion in the interstitial case. The notation is kept here and it now represents trans-interface transportation by interdiffusion, whether or not diffusion is also part of the flux causing the interface to migrate, J^m.

The flux J^t will give α a lower B content than γ by an amount depending on the velocity v,

$$\left(x_B^{\gamma/\alpha} - x_B^\alpha\right)v = J^t V_m = L\Delta(\mu_A - \mu_B) \quad (13)$$

Inserting $\Delta(\mu_A - \mu_B)$ in equation (8) we get

$$v = \frac{L\Delta G^t}{\left(x_B^{\gamma/\alpha} - x_B^\alpha\right)^2} \quad (14)$$

The intersection of the γ/α tangent with the α curve is marked as $x_B^{\alpha_{max}}$ in Fig. 4 because it is often regarded as the upper limit of the B content of α since that is where the driving force, D_{int}^{chem}, goes to zero. However, the situation is more complicated and will be discussed further in Section 5. For the migration process we get the response function by accepting equation (2).

4. SPONTANEOUS TRANSITION OF DIFFUSIONAL TO DIFFUSIONLESS TRANSFORMATION

When applying the sharp interface model to acicular growth of α-Fe in a matrix of γ-Fe, Hillert treated the diffusion of carbon in the γ phase with a modification [6] of a growth model for plates, originally proposed by Zener [7], that takes into account the sidewise diffusion and the effect of interfacial tension of the curved edge of a plate. For the trans-interface diffusion he assumed that the flux of carbon atoms across the interface, J_C^t, is proportional to the chemical potential difference acting over the interface, $\mu_C^\alpha - \mu_C^{\gamma/\alpha}$. Using the z variable we now express this with a kinetic equation

$$\left(z_C^{\gamma/\alpha} - z_C^\alpha\right)v = J_C^t V_{Fe} = V_{Fe} L_{CC} \left(\mu_C^\alpha - \mu_C^{\gamma/\alpha}\right) \quad (15)$$

where L_{CC} is a kinetic coefficient. This is the interstitial correspondence to equation (12). For the interface migration the friction expressed by equation (2) was regarded as less important than a threshold which must be climbed before growth can get started,

$$D^m = K \quad (16)$$

The height of the threshold K was estimated as a function of temperature from kinetic information on Widmanstätten α, bainite and martensite. For martensite it is evident that K contains the resistance due to elastic and plastic strains. Of course, the term v/M in equation (2) could have been kept beside K, but M was regarded as very large for martensitic types of interfaces and the term was thus regarded as less important for describing the transition from growth of a Widmanstätten plate to the diffusionless martensitic growth.

From the kinetic equations for these three processes Hillert calculated $z_C^{\gamma/\alpha}$, z_C^α and v for given values of the alloy composition, $z_C^{\gamma_0}$, and the curvature of the edge. Some results from 600°C are reproduced in Fig. 5. Each curve was calculated for a constant value of the carbon content of the alloy but the curvature of the edge varied as a result of the calculation. At high carbon contents (e.g. $x_C = 0.02$) the curve is rather parabolic and according to Zener [7] the curvature of the edge should adjust itself to give the maximum velocity. At the same time, the carbon content of α would be adjusted. At a lower carbon content in the alloy (e.g. $x_C = 0.009$) a second branch appears at very high velocities (at the upper right corner), but a nucleus starting from a low velocity cannot spon-

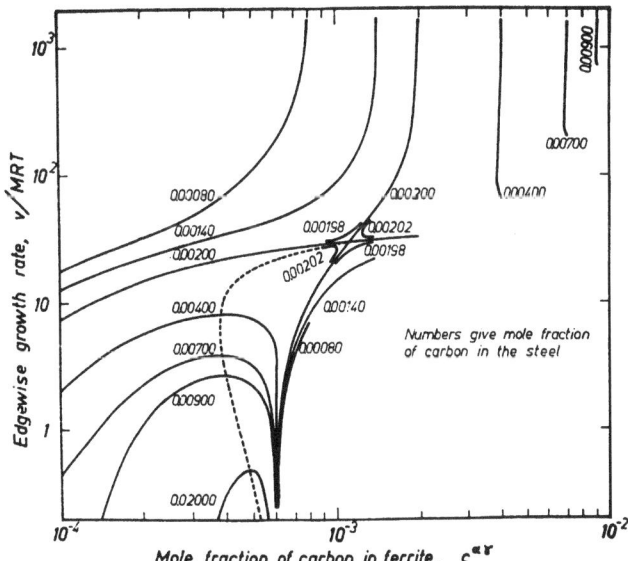

Fig. 5. Diagram from Ref. [5] predicting that spontaneous transition from diffusional to diffusionless $\gamma \to \alpha$ transformation in Fe–C alloys can occur below a critical solute content ($x_C = 0.002$).

taneously reach that branch. It would probably be limited to the lower branch and only reach its point of maximum. However, below a critical carbon content ($x_C = 0.002$) the two branches are connected and a nucleus, initially with a low carbon content, could continuously move up along the curve and reach high velocities, and the carbon content of the growing α would then approach the initial carbon content of the alloy. The model thus predicts a spontaneous transition into a diffusionless transformation if the carbon content is low enough, presumably the martensitic transformation. Even for a slightly higher carbon content it is conceivable that a fluctuation can move a growing edge from a point on the lower branch to a point on the higher branch if they are not too far apart. Hillert also discussed the possibility that a small nucleus can form with a carbon content close to the initial one and quickly develop a high velocity. The critical nucleus of martensite would then be dynamic rather than static. The result shown in Fig. 5 is typical of the sharp interface model and similar results have later been reported several times from various treatments based on the sharp interface model [8–13].

5. FACTORS GOVERNING THE COMPOSITIONS AT THE INTERFACE

In Figs 3 and 4 the composition of the growing α phase was chosen arbitrarily. Figure 6 illustrates the range within which it must fall. For $x_B^\alpha = x_{B1}^\alpha$ there will be no dissipation of Gibbs energy due to diffusion and, thus, there can be no diffusion. For $x_B^\alpha = x_{B3}^\alpha$ there will be no Gibbs energy available for dissipation by interface migration and the interface cannot move. It should be realized that x_{B3}^α is found by making the two parallel lines coincide and thus making their distance, ΔG^m, equal to zero. Evidently, x_B^α must fall in between and x_{B2}^α is a reasonable possibility. The dashed curve in Fig. 6 illustrates how the chemical driving force, $D_{\text{int}}^{\text{chem}}$ from

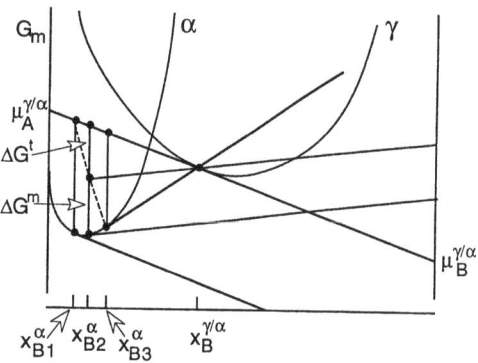

Fig. 6. The sharp interface model. The possible range of x_B^α for the given $x_B^{\gamma/\alpha}$ is x_{B1}^α to x_{B3}^α. The dashed curve illustrates the division of the chemical driving force between dissipation of Gibbs energy by trans-interface diffusion, ΔG^t, and by the migration of the interface, ΔG^m.

Fig. 4, would be divided between ΔG^t and ΔG^m for various values of x_B^α.

The exact position of x_B^α is determined by the balance between the two parts through their kinetic equations since they must be related to the same velocity. However, there is another important factor. The upper limit of the range for x_B^α, x_{B3}^α, depends directly on $x_B^{\gamma/\alpha}$ as shown in Fig. 6 but $x_B^{\gamma/\alpha}$ depends on the initial composition $x_B^{\gamma 0}$, because the difference $x_B^{\gamma/\alpha} - x_B^{\gamma 0}$ drives the diffusion in the γ phase and makes the difference $x_B^{\gamma 0} - x_B^\alpha$ possible by sidewise diffusion. Thus, x_{B3}^α is limited by the fact that $x_B^{\gamma/\alpha}$ cannot come too close to $x_B^{\gamma 0}$. To emphasize the importance of $x_B^{\gamma 0}$, a possible position of that composition was included in Fig. 4.

If the initial composition of the alloy, $x_B^{\gamma 0}$, is chosen to the left of the intersection of the two G_m curves, the so-called T_0 point, then the situation can change drastically by both x_B^α and $x_B^{\gamma/\alpha}$ approaching $x_B^{\gamma 0}$. That limiting situation would describe the diffusionless transformation and the exact value of the alloy composition, $x_B^{\gamma 0}$, where this would first occur, cannot be decided without a numerical calculation as the one presented in Fig. 5. The result would undoubtedly depend on the details of the model chosen for the interface and the values chosen for the model parameters. This question will be further discussed in Section 17, which concerns the massive transformation.

6. OTHER TREATMENTS BASED ON THE SHARP INTERFACE MODEL

Aziz et al. [14,15] have also used the sharp interface model but with two differences from Hillert's treatment. They did not use equation (16), because they considered solidification, and instead they used a modification of equation (2) by applying an expression equivalent to

$$D^m = -RT\ln\left(1 - \frac{v}{v_0}\right) \quad (17)$$

where v_0 is a constant. At low rates it reduces to equation (2) with $M = v_0/RT$. Furthermore, they did not use equation (12) directly but proposed the following equation,

$$\left(x_B^{L/s} - x_B^s\right)v = J^t V_m$$
$$= v_D\left(x_B^s x_A^{L/s} - \kappa_e x_B^{L/s} x_A^s\right) \quad (18)$$

where v_D is a constant they called maximum diffusive speed. For dilute solutions they transformed equation (18) into

$$\frac{x_B^s}{x_B^L} = \left(\frac{v}{v_D} + \kappa_e\right)\Big/\left(\frac{v}{v_D} + 1\right) \quad (19)$$

which was previously derived by Brice [16]. Aziz has made extensive use of Brice's equation because he has been mainly interested in the variation of the partition coefficient, $\kappa = x_B^s/x_B^L$, with the rate v.

However, equation (18) can be related to equation (13) because the parameter κ_e was defined as

$$\kappa_e = \left(\frac{x_A^{L/s} x_B^s}{x_A^s x_B^{L/s}}\right) \exp\left[\frac{\Delta(\mu_B - \mu_A)}{RT}\right] \quad (20)$$

It is thus possible to transform equation (18),

$$\left(x_B^{L/s} - x_B^s\right)v = v_D \kappa_e x_A^s x_B^{L/s}\left[\exp\frac{\Delta(\mu_A - \mu_B)}{RT} - 1\right]$$

$$\approx \left(\frac{v_D \kappa_e x_A^s x_B^{L/s}}{RT}\right) \Delta(\mu_A - \mu_B) \quad (21)$$

for small $\Delta(\mu_A - \mu_B)/RT$. This is identical to equation (12) if v_D is expressed in the other parameters. Except for the approximations, the use of equation (19) is thus equivalent to using $\Delta(\mu_A - \mu_B)$ as the driving force for the trans-interface diffusion and, by combination with equation (8), one obtains the response function equation (13). Furthermore, Aziz inserted ΔG^m from equation (10) as D^m in equation (17). Except for the fact that Aziz did not consider the diffusion inside the parent phase, his treatment is almost identical to Hillert's. However, Aziz has also proposed a second version of his treatment which he calls "without solute drag". This will be discussed in Sections 7 and 8.

It should be emphasized that Aziz considered a planar interface and no sidewise diffusion in the liquid. The experimental parameter was the velocity as controlled by heat conduction. Thus, he did not need anything like Zener's maximum velocity hypothesis.

Jackson et al. [17] were also mainly interested in the variation of the partition coefficient with velocity. Following Jackson [3] they started with J_A proportional to $[x_A^L - x_A^s \exp(-\Delta^\circ G_A/RT)]$ and similarly for J_B. For low driving forces this is equivalent to equations (22) and (23) without cross-terms, to be discussed in the next section, and cannot account for trapping. Thus they modified Jackson's expressions by adding $fx_B^s(\Delta^\circ G_A - \Delta^\circ G_B)$ to $-\Delta^\circ G_A$ in J_A and $fx_A^s(\Delta^\circ G_B - \Delta^\circ G_A)$ to $-\Delta^\circ G_B$ in J_B where f was a function of the velocity, going from 0 at $v = 0$ to 1 at $v = \infty$. In the latter limit the result will be $J_A/x_A^L = J_B/x_B^L$, which implies that the transformation is partitionless. They were thus able to describe the transition from local equilibrium at the interface for low velocities to completely partitionless transformation at very high velocities. This may be regarded as a convenient but crude way of introducing the cross-terms in the diffusion equation (see Section 7).

Another variant of the sharp interface model was presented by Olson et al. [18]. They compared with Hillert's treatment and listed some differences. They did not regard the curvature of the edge as a free variable but estimated a fixed value. Thus they could subtract the interfacial energy effect directly from the available driving force which is equivalent to including it in the K parameter in equation (16). They also subtracted specifically the effect of elastic strains, which is equivalent to including it in the K parameter, also. Furthermore, they used Ivantsov's solution for a parabolic cylinder [19] to calculate the rate of diffusion in the γ phase, instead of the modified Zener equation. So far, their treatment has not been very different from Hillert's.

Finally, Olson et al. [18] introduced a kinetic treatment for the migration of a martensitic interface, which is equivalent to adding a term $a[(RT/Q_0)\ln(v_0/v) - 1]^2$ to the K parameter in equation (16). They showed how to evaluate the constants a, Q_0 and v_0. However, the most important difference is that they failed to consider the rate of trans-interface diffusion of carbon across the interface and seem to have used the whole driving force over the interface to balance the frictional loss in the interface, D^m, which is logical since they neglected to consider the need for trans-interface diffusion. Since they eliminated one variable, the curvature, and one equation, the one for trans-interface diffusion, they still had a degree of freedom and again used Zener's maximum velocity hypothesis. In a second paper [20] they added a calculation of the rate of trans-interface diffusion of carbon, using the same equation as Aziz, namely equation (19), but did not subtract the corresponding dissipation of Gibbs energy, ΔG^t, from the driving force they used for the interface migration in the first paper. In this respect their treatment is similar to that of Aziz "without solute drag". Because they had now added an equation to their previous treatment, they had removed the degree of freedom and no longer needed Zener's hypothesis. In a third paper [21] it seems that ΔG^t still was not subtracted.

The treatment by Salwén [22] should also be mentioned. He considered planar growth from an initial situation where two homogeneous phases are not in equilibrium with each other. The highest velocity thus occurred at the beginning of the process and he studied the change of growth conditions as the rate decreased. It should be mentioned that he used the difference in μ_B as the driving force for trans-interface diffusion, which is correct for an interstitial solute but not for a substitutional solute. With the notations in Fig. 4, he used D_{int}^{chem} instead of $D_{int}^{chem} - G^t$ as the driving force for the interface migration. Again, this is similar to Aziz's treatment "without solute drag".

Finally, a similar treatment by Krielaart [23] should be mentioned. He considered the $\gamma \to \alpha$ transformation in the Fe–C system. As the driving force for the interface migration he used $\mu_{Fe}^{\gamma/\alpha} - \mu_{Fe}^\alpha$

and he thus neglected the second term in equation (7). Figure 3 clearly demonstrates the difference between D_{Fe}^m and $\mu_{Fe}^{\gamma/\alpha} - \mu_{Fe}^{\alpha}$.

7. APPLICATION OF IRREVERSIBLE THERMODYNAMICS

Baker and Cahn in 1971 [24] examined the sharp interface model using irreversible thermodynamics. They started by considering two interface processes in a binary alloy, the transfers of the two components across the interface, and applied the phenomenological relations

$$J_A = L_{AA}\Delta\mu_A + L_{BB}\Delta\mu_B \quad (22)$$

$$J_B = L_{BA}\Delta\mu_A + L_{BB}\Delta\mu_B \quad (23)$$

where $\Delta\mu_j = \mu_j^\gamma - \mu_j^\alpha$ and J_A and J_B are counted as positive when the fluxes go from γ to α. Compared to the conventional approach, discussed in Section 2, they thus assumed that the two diffusional processes could be coupled through the cross-terms.

As discussed by Baker and Cahn, one can change to a more convenient set of fluxes. This will now be demonstrated using the procedure later presented by Caroli et al. [25]. Assuming that there is no diffusion inside the growing phase, the composition of the new phase is completely controlled by the diffusion across the interface in an interface-fixed frame,

$$\frac{J_A}{x_A^\alpha} = \frac{J_B}{x_B^\alpha} = J_A + J_B \quad (24)$$

Baker and Cahn continued by defining the two processes used in the sharp interface model by considering the migration of the interface in a number-fixed frame by the net flux of atoms in the interface-fixed frame

$$J^m = J_A + J_B \quad (25)$$

and the process J^t by which the composition changes from $x_B^{\gamma/\alpha}$ to x_B^α across the migrating interface,

$$J^t = J^m\left(x_B^{\gamma/\alpha} - x_B^\alpha\right) \quad (26)$$

It may be noted that J^m is identical to v/V_m and equation (26) is thus identical to equation (13). For the process causing the change in composition across the interface, the flux must be equal to the difference in interdiffusional flux in the two adjoining phases. Under steady-state conditions $J^\alpha = 0$ because the composition of α is constant. Using the well-known expression for interdiffusion when J_A and J_B are positive in the same direction, we then get

$$J^t = J^\gamma - J^\alpha = J^\gamma = x_B^{\gamma/\alpha}J_A - x_A^{\gamma/\alpha}J_B \quad (27)$$

where J^γ and J^α are the interdiffusion fluxes in γ and α close to the interface. By inserting equations (24) and (25) into equation (27), we can reproduce equation (26).

The driving force for the migration is

$$D^m = x_A^{\gamma/\alpha}\Delta\mu_A + x_B^{\gamma/\alpha}\Delta\mu_B \quad (28)$$

in agreement with equation (10), because $\Delta\xi^m = 1$, and the driving force for the trans-interface exchange of A and B atoms, J^t, is,

$$D^t = \Delta(\mu_A - \mu_B) = \Delta\mu_A - \Delta\mu_B \quad (29)$$

The dissipation of Gibbs energy for the two processes in the sharp interface model is now obtained as

$$\begin{aligned}J^m D^m + J^t D^t &= \left(x_A^{\gamma/\alpha}\Delta\mu_A + x_B^{\gamma/\alpha}\Delta\mu_B\right)(J_A + J_B) \\ &+ \left(x_B^{\gamma/\alpha}J_A - x_A^{\gamma/\alpha}J_B\right)(\Delta\mu_A - \Delta\mu_B) \\ &= J_A\left[\left(x_A^{\gamma/\alpha} + x_B^{\gamma/\alpha}\right)\Delta\mu_A \right.\\ &\left. + \left(x_B^{\gamma/\alpha} - x_B^{\gamma/\alpha}\right)\Delta\mu_B\right] \\ &+ J_B\left[\left(x_A^{\gamma/\alpha} - x_A^{\gamma/\alpha}\right)\Delta\mu_A \right.\\ &\left. + \left(x_B^{\gamma/\alpha} + x_A^{\gamma/\alpha}\right)\Delta\mu_B\right] \\ &= J_A\Delta\mu_A + J_B\Delta\mu_B \quad (30)\end{aligned}$$

It is thus equal to the Gibbs energy dissipation of the two processes from which Baker and Cahn started. Finally, we can solve equations (28) and (29) for $\Delta\mu_A$ and $\Delta\mu_B$ and calculate coefficients for J^m and J^t from the L coefficients in equations (22) and (23) by the use of equations (25) and (27).

$$\Delta\mu_A = D^m + x_B^{\gamma/\alpha}D^t \quad (31)$$

$$\Delta\mu_B = D^m + x_A^{\gamma/\alpha}D^t \quad (32)$$

$$\begin{aligned}J^m &= J_A + J_B = (L_{AA} + L_{BA})\Delta\mu_A \\ &+ (L_{AB} + L_{BA})\Delta\mu_B \\ &= (L_{AA} + L_{BA})\left(D^m + x_B^{\gamma/\alpha}D^t\right) \\ &+ (L_{AB} + L_{BB})\left(D^m - x_A^{\gamma/\alpha}D^t\right) \\ &= (L_{AA} + L_{BA} + L_{AB} + L_{BB})D^m \\ &+ \left[x_B^{\gamma/\alpha}(L_{AA} + L_{BB}) - x_A^{\gamma/\alpha}(L_{AB} + L_{BB})\right]D^t\end{aligned}$$

$$(33)$$

$$J^t = x_B^{\gamma/\alpha} J_A - x_A^{\gamma/\alpha} J_B$$

$$= \left(x_B^{\gamma/\alpha} L_{AA} - x_A^{\gamma/\alpha} L_{BA}\right)\left(D^m + x_B^{\gamma/\alpha} D^t\right)$$

$$+ \left(x_B^{\gamma/\alpha} L_{AB} - x_A^{\gamma/\alpha} L_{BB}\right)\left(D^m - x_A^{\gamma/\alpha} D^t\right)$$

$$= \left[x_B^{\gamma/\alpha}(L_{AA} + L_{AB}) - x_A^{\gamma/\alpha}(L_{BA} + L_{BB})\right]D^m$$

$$+ \left[\left(x_B^{\gamma/\alpha}\right)^2 L_{AA} - x_A^{\gamma/\alpha} x_B^{\gamma/\alpha}(L_{BA} + L_{AB})\right.$$

$$\left. + \left(x_A^{\gamma/\alpha}\right)^2 L_{BB}\right]D^t \qquad (34)$$

Caroli *et al.* were thus able to show that the cross-coefficients are equal for the new set of fluxes and thus satisfy Onsager's reciprocal relation [26], if they are equal for the first set of fluxes, i.e. if $L_{AB} = L_{BA}$. Baker and Cahn failed to show this because they did not have equation (27) but derived J^t from equation (26). Thus, their expression for J^t contained x_B^α as well as $x_B^{\gamma/\alpha}$. However, equations (33) and (34) show that J^m and J^t are both functions of D^m, D^t and $x_B^{\gamma/\alpha}$. By inserting them in equation (26) one could thus express x_B^α as a function of D^m, D^t and $x_B^{\gamma/\alpha}$. After eliminating x_B^α from Baker and Cahn's equation for J^t by inserting this expression for x_B^α, one would obtain equation (34), showing that Onsager's relation is satisfied. One might say that the cross-term in Baker and Cahn's expression for J^t is not a true cross-term because the expression contains x_B^α, which is a function of D^m, D^t and $x_B^{\gamma/\alpha}$.

8. CHOICE OF INDEPENDENT INTERFACIAL PROCESSES IN THE SHARP INTERFACE MODEL

It is evident that the cross-coefficients in equations (22) and (23) cannot be equal to zero in a system where trapping occurs because then $\Delta\mu_B < 0$ but $J_B > 0$. In other words, the jumping of A and B atoms across the interface cannot be independent of each other if there is trapping. This situation can occur only due to the cross-terms and they may thus be regarded as representations of the co-operative motion of the two components across the interface. On the other hand, it may be useful to treat the migration and the inter-diffusion as two independent processes and thus to use two independent mobilities for them. As emphasized at the end of Section 3, equation (2) already assumes that there are no cross-terms for J^m and J^t. This can be accomplished by keeping the cross-terms for J_A and J_B in equations (22) and (23) and give them proper values. If we modify the L parameters in the following way

$$L_{AA} V_m = M_{AB} x_A^{\gamma/\alpha} + M_{AA}\left(x_A^{\gamma/\alpha}\right)^2 \qquad (35)$$

$$L_{AB} V_m = L_{BA} V_m = M_{AA} x_A^{\gamma/\alpha} x_B^{\gamma/\alpha} \qquad (36)$$

$$L_{BB} V_m = M_{AB} x_B^{\gamma/\alpha} + M_{AA}\left(x_B^{\gamma/\alpha}\right)^2 \qquad (37)$$

then we would find

$$J^m V_m = (M_{AA} + M_{AB})\left(x_A^{\gamma/\alpha} \Delta\mu_A + x_B^{\gamma/\alpha} \Delta\mu_B\right)$$

$$= (M_{AA} + M_{AB}) D^m \qquad (38)$$

$$J^t V_m = x_A^{\gamma/\alpha} x_B^{\gamma/\alpha} M_{AB}(\Delta\mu_A - \Delta\mu_B)$$

$$= x_A^{\gamma/\alpha} x_B^{\gamma/\alpha} M_{AB} D^t \qquad (39)$$

We have thus found that, under steady-state conditions, the kinetic coefficient L in equations (12)–(14) should be equal to $x_A^{\gamma/\alpha} x_B^{\gamma/\alpha} M_{AB}$ according to the sharp interface model. Equations (38) and (39) justify as physically possible the treatment of the two mobilities, M_{AB} and $M_{AA} + M_{AB}$, as independent. This is the choice made in most applications of the sharp interface model and it is implicit in many arguments in the present review. On the other hand, after assuming that J^m, i.e. $J_A + J_B$, is proportional to D^m, as in equation (38), Kaplan *et al.* [27] combined this with equation (24) and solved for J_A and J_B, which would yield

$$J_A = v_R\left[x_A^\alpha x_A^{\gamma/\alpha} \Delta\mu_A + x_A^\alpha x_B^{\gamma/\alpha} \Delta\mu_B\right] \qquad (40)$$

$$J_B = v_R\left[x_B^\alpha x_A^{\gamma/\alpha} \Delta\mu_A + x_B^\alpha x_B^{\gamma/\alpha} \Delta\mu_B\right] \qquad (41)$$

where v_R is a constant and the x values are those of α and γ in contact with the interface. For small deviations from equilibrium they may be represented by the $\alpha+\gamma$ equilibrium values. Kaplan *et al.* observed that the cross-coefficients, $x_A^\alpha x_B^{\gamma/\alpha}$ and $x_B^\alpha x_A^{\gamma/\alpha}$, are not equal and concluded that Onsager's relation is not satisfied for this model if it is assumed that J^m is proportional to D^m. This is a most surprising result since equation (38) together with equation (39) was derived using equations (33) and (34) which obey Onsager's relation. The explanation is that Kaplan *et al.* did not combine equation (38) with an equation like equation (39) although their previous use of equation (18) implies that J^t is proportional to $\Delta(\mu_A - \mu_B)$ for small values of $\Delta(\mu_A - \mu_B)/RT$. Instead they used equation (24) without realizing that x_B^α in the resulting equations (40) and (41) would be a function of $x_B^{\gamma/\alpha}$, $\Delta\mu_A$ and $\Delta\mu_B$. Thus, $x_A^\alpha x_B^{\gamma/\alpha}$ and $x_B^\alpha x_A^{\gamma/\alpha}$ are not true cross-coefficients. This is closely related to the mistake Baker and Cahn made when deriving an expression for J^t using equation (26) instead of equation (27) (see the discussion at the end of Section 7).

Kaplan *et al.* observed that by inserting $D_{\text{int}}^{\text{chem}}$ from equation (9) as D^m they obtained expressions similar to equations (40) and (41) but with the cross-coefficients $x_A^\alpha x_B^\alpha$ and $x_B^\alpha x_A^\alpha$, which are equal. They concluded that this model would thus be preferable because it obeys Onsager's relation whereas the first model did not, according to their test. They characterized the new model as "without solute drag" because the solute drag term, ΔG^t, was not subtracted from D_{int}^{chem} as in equation (10). Concerning their test of Onsager's relation, it should be mentioned that in a subsequent paper [28] they concluded that Onsager's theorem cannot be used to test a particular model. Thus they repudiated their previous claims regarding their models "with" and "without" solute drag. The same conclusion was drawn by Caroli et al. [25] who emphasized that "an extra relationship (imposed by the experimental conditions) between the values of the thermodynamic forces does not influence the response properties of the interface itself". When testing for Onsager's relation in the interface, it is thus important to avoid including such relationships. Equation (26) may be regarded as an example.

Furthermore, in the "without solute drag" model Kaplan and Aziz used an equation looking like equation (2) but, by inserting a driving force not equal to the one given by equation (10), they actually included a cross-term in the equation for J^m. Thus, it seems that they should also have included a corresponding cross-term in the equation for J^t. However, by using equation (18), which is derived from equation (13), they did not. The "without solute drag" model is thus of doubtful value whereas the "with solute drag" model uses the proper driving force in equation (2) and may be regarded as the substitutional version of the sharp interface model developed for interstitial systems by Hillert [5].

It may be mentioned that Langer and Sekerka [29] considered a very particular system where the two phases have a common Gibbs energy curve with two minima. They derived a sharp interface model by starting with a continuous description using the common Gibbs energy curve and extrapolating the bulk compositions to the middle of the interface, in line with Gibbs' treatment of interfaces. For the extrapolation they used the bulk mobilities. However, when calculating the fluxes across the interface, they assumed that the mobility was lower inside the interface and they were thus able to explain trapping. Their case is a rather special one and is not the case considered in the present work, although diffusionless solidification is rather similar.

9. MODEL OF AN INTERFACE AS A SPECIAL PHASE: THE HOMOGENEOUS INTERFACE MODEL

The picture of an interface as a special phase with its own properties has long been used for modelling various phenomena. A typical example is grain-boundary diffusion, which is modelled by a high diffusivity D^I in an interface region of width δ, although only the product $D^I\delta$ is measured experimentally. If the Gibbs energy of that phase is described with its own function, one can plot its curve in a molar Gibbs energy diagram. It should then be remembered that the equilibrium with a bulk phase is represented by a parallel-tangents construction and not with a common tangent, because it is assumed that the interface contains a constant number of atoms (see Ref. [4], p. 392).

Inspired by the modelling of solute drag in grain-boundary migration, presented by Cahn [30] and Lücke and Stüwe [31] in 1962 and to be discussed in Section 12, Hillert [32] in 1970 examined how useful the homogeneous interface model could be in such applications. He thus modelled the interface with constant thermodynamic and kinetic properties over the whole width, δ, and with discontinuous changes on both sides. It is illustrated in Fig. 1(b). Figure 7 gives the molar Gibbs energy diagram for a grain boundary in a binary alloy with a single bulk phase, γ. When the grain boundary migrates with a constant velocity through the γ phase, material of the initial composition, $x_B^{\gamma 0}$, is flowing through the grain boundary and the arrows along the vertical line at $x_B^{\gamma 0}$ in Fig. 7 illustrate what happens to it.

Starting with the $x_B^{\gamma 0}$ point on the γ curve, an amount of Gibbs energy, ΔG^γ, is dissipated by diffusion inside the γ phase as the material climbs to the top of the spike, $x_B^{\gamma/I}$. Then the material passes through a discontinuity and enters the grain boundary without changing the value of $\mu_B - \mu_A$, i.e. the

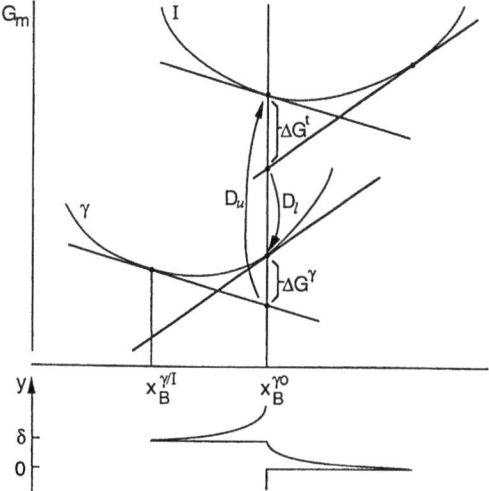

Fig. 7. Molar Gibbs energy diagram from Ref. [32] illustrating the role of diffusion in grain-boundary migration. ΔG^t and ΔG^γ represent the dissipation of Gibbs energy by diffusion inside the boundary and in front of it, respectively, whereas D_u and D_l represent the discontinuous changes of Gibbs energy when material of the representative composition passes into the boundary and out of it, respectively.

slope of the tangent. That requires a supply of driving force D_u defined by the distance between two parallel tangents. The subscript u stands for the upper side of the grain boundary (with the grain boundary moving upwards). Inside the grain boundary the material also climbs to the top of a spike, causing a Gibbs energy dissipation of ΔG^t. Finally, the material passes through a new discontinuity, represented by a second parallel-tangents construction, and enters the growing grain. Some driving force D_l is then recovered. The subscript l stands for the lower side of the grain boundary. In order to move under these conditions, the grain boundary needs a net driving force $D_u - D_l$ and, according to the diagram, it is equal to the total dissipation of Gibbs energy by diffusion, $\Delta G^\gamma + \Delta G^t$.

It should also be noted that for the present application the vertical position of the curve for the grain boundary, relative to the curve for the γ phase, is of no importance because it is only the difference $D_u - D_l$ that matters. The vertical position would represent the interfacial energy, which is not under consideration here. At high velocities the homogeneous interface model needs a drastic modification. That will be discussed in Section 14.

10. USE OF GIBBS ENERGY CURVE FOR THE INTERFACE

In 1989 Ågren [8] showed how one can make calculations for phase transformations based on the sharp interface model by the use of computer software designed for calculations of phase equilibria. At the same time he proposed that one should take into account the composition of the interface in the sharp interface model and he introduced an adjustable parameter to describe that composition. We shall now use the homogeneous interface model to explain his proposal.

Let us start by applying the principles from Fig. 7 to a phase transformation $\gamma \to \alpha$ under steady-state conditions of growth. As for the case discussed earlier, there may be sidewise diffusion inside the parent γ phase and the growing α can differ in composition from the initial γ. However, for the moment we shall not take an interest in the diffusion inside γ or its initial composition. The compositions of main interest will be those of the growing α phase, x_B^α, and of the parent γ phase close to the interface, $x_B^{\gamma/\alpha}$. We shall use a molar Gibbs energy diagram, Fig. 8, to illustrate what happens to material of composition x_B^α as it, due to the migration of the interface at a constant velocity, enters the interface, travels through it and finally deposits on the α phase.

In principle, we know the relative positions and shapes of the Gibbs energy curves for the α and γ phases from the phase diagram information but usually we have no or only a very vague idea about

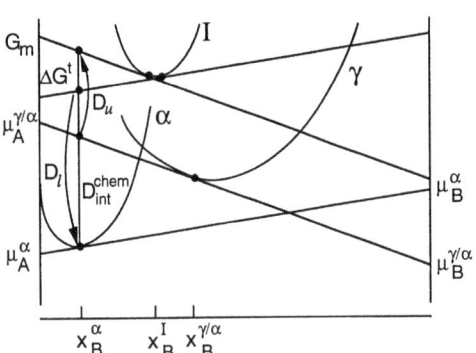

Fig. 8. Molar Gibbs energy diagram illustrating the effect of the composition of the interface on the growth of an α phase with a constant composition according to Ågren's model. The quantity $D_{\text{int}}^{\text{chem}}$ is the driving force over the interface and ΔG^t is the dissipation by trans-interface diffusion. The difference can be used for driving the interface migration. The quantities D_u and D_l represent the discontinuous changes of Gibbs energy for material of the α composition passing into the interface and out of it. Their difference also gives the driving force for the migration of the interface.

the properties of the interface. First we shall place its curve well above the other two, in accordance with the interfacial energy being positive, but horizontally between them. Figure 8 resembles Fig. 7 but now the second phase, α, has been included and the Gibbs energy dissipation in the parent phase, γ, has not been included. The chemical driving force acting over the interface is $D_{\text{int}}^{\text{chem}}$ but, due to the dissipation of Gibbs energy by trans-interface diffusion, it has been reduced and what is left to drive the migration of the interface is

$$D^m = D_l - D_u = D_{\text{int}}^{\text{chem}} - \Delta G^t \qquad (42)$$

In order to simplify the following diagrams, the curve for the interface phase will be lowered until it touches the tangent from $x_B^{\gamma/\alpha}$ (see Fig. 9). That will make the new $D_u = 0$ and the new D_l equal to the old $(D_l - D_u)$. The new D_l may be identified with D^m, according to equation (42), and thus with ΔG^m, and the diagram will resemble Fig. 4, but the composition of the interface, x_B^I, will now play the role of $x_B^{\gamma/\alpha}$ in Fig. 4.

An important difference between Figs 7 and 8 should be emphasized. The quantity $D_l - D_u$ represents the gain in Gibbs energy when the interface migrates. For the phase transformation in Fig. 8 it is positive and can thus provide driving force for the interface migration, $D^m = \Delta G_m$, as demonstrated in Fig. 9. For the grain-boundary migration in Fig. 7 it is negative and the grain boundary cannot move unless a stronger driving force is provided from some other source. What is left of that driving force after subtracting $-(D_l - D_u)$ can be used to drive the migrating interface. The driving force may come from the γ/γ grain boundary being curved or from the parent γ grain being cold worked. Both

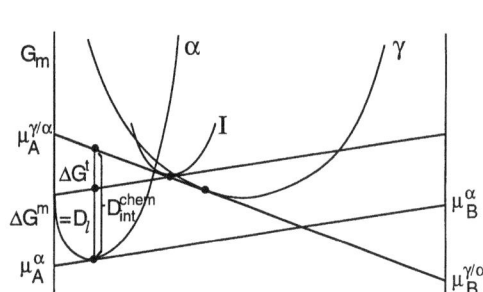

Fig. 9. Molar Gibbs energy diagram similar to Fig. 8 but illustrating that it is more convenient to move the interface curve vertically until it touches the γ tangent. Here $D_u = 0$ and D_l is equal to the Gibbs energy dissipation by the migration of the interface, ΔG^m.

factors would place the Gibbs energy curve for the parent grain above the curve for the growing grain and Fig. 7 could be replaced by a diagram with two γ curves.

11. USE OF REPRESENTATIVE COMPOSITION OF INTERFACE

It was pointed out by Ågren [8] that the essential feature of the Gibbs energy curve for the interface is its horizontal position, i.e. where it falls on the composition axis, and, in particular, the composition where the two tangents to the interface curve intersect. That may be regarded as its "representative composition". Accepting his strategy we shall only need to mark that composition. It will be denoted x_B^I (see Fig. 9). However, one should realize that its exact position depends not only on the positions of the three Gibbs energy curves but also on the value of x_B^α, which is not known in advance. It is also interesting to note from Fig. 9 that $D^m = \Delta G_m$ can no longer be evaluated from the vertical distance of the γ/α point from the α tangent, as it was for the sharp interface model in Figs 3 and 4.

Independent of the exact physical definition of x_B^I, it can be used as a model parameter to simulate various cases of interactions between solute and interface. More physically based modelling can be based on any of the continuous interface models in Figs 1(c)–(e). From such models one can evaluate ΔG^t and will generally find that it differs from ΔG^t according to the sharp interface model. If it is smaller, the situation illustrated in Fig. 9 is obtained. By drawing a line parallel to the α tangent, the value of x_B^I will be found, which would produce the same predictions when inserted in Ågren's model. Owing to its great simplicity, Ågren's model may prove to be a convenient tool for studying various cases but may not replace more ambitious treatments based on continuous interface models. It would rather be based on such models for the estimate of reasonable vaues of x_B^I.

Equations for Ågren's model can be obtained directly from equations (8) and (10) by inserting the representative interface composition

$$\Delta G^t = (x_B^I - x_B^\alpha)\left[\left(\mu_A^{\gamma/\alpha} - \mu_A^\alpha\right) - \left(\mu_B^{\gamma/\alpha} - \mu_B^\alpha\right)\right]$$
(43)

$$D^m = \Delta G^m = x_A^I\left(\mu_A^{\gamma/\alpha} - \mu_A^\alpha\right) + \left(\mu_B^{\gamma/\alpha} - \mu_B^\alpha\right) \quad (44)$$

D^m from equation (44) could be inserted in equation (2) or (17). On the other hand, x_B^I does not replace $x_B^{\gamma/\alpha}$ in $(x_B^{\gamma/\alpha} - x_B^{\gamma_0})$ in calculations of diffusion in the γ phase. Thus, one could let x_B^I approach $x_B^{\gamma_0}$ and even let it fall on the other side of $x_B^{\gamma_0}$ without directly affecting the diffusion in the γ phase. The position of x_B^I depends primarily on the properties of the interface, in particular on its horizontal position. The sharp interface model is recovered with $x_B^I = x_B^{\gamma/\alpha}$. If nothing is known about the properties of the interface, x_B^I could be placed anywhere and with the choice $x_B^I = x_B^\alpha$ one obtains $\Delta G^t = 0$ and that case could be characterized as "without solute drag", but it is not similar to the model Aziz called "without solute drag" because he did not use $\Delta G^t = 0$.

An important difference between Ågren's model and the sharp interface model should be noticed. Without question, $\Delta(\mu_A - \mu_B)$ is the driving force for the exchange of A and B by trans-interface diffusion and $x_B^{\gamma/\alpha} - x_B^\alpha$ is the extent of diffusion required per mole of phase transformation. According to equation (4), it is thus equation (8) that describes the Gibbs energy dissipation if the trans-interface diffusion can be treated as an elementary process, i.e. if the interface is sharp. It is evident that this model completely neglects what happens inside the interface, e.g. if there is a strong tendency for segregation. Such effects are considered in Ågren's model, as shown by the presence of x_B^I in equation (43), and that is such an important difference that his model should not be regarded as a mere modification of the sharp interface model.

Ågren [8] first presented results of numerical calculations using $x_B^I = x_B^{\gamma_0}$ for simplicity, but he clearly stated that this was an arbitrary choice. However, this choice made $\Delta G^t = 0$ for the partitionless case, where $x^\alpha = x^{\gamma_0}$, and thus predicted that all partitionless transformations are diffusionless. That limitation was removed by Jönsson and Ågren [33] by using $x_B^I = (x_B^{L/s} + x_B^s)/2$ when modelling the transition to diffusionless solidification. They later used the same approach when modelling the massive transformation [34]. It was also used by Liu and Ågren when modelling the γ to α transformation in the Fe–C system [10].

In order to model any tendency for segregation it would seem interesting to test a generalized expression,

$$x_B^I = f x_B^{\gamma/\alpha} + (1-f) x_B^\alpha \qquad (45)$$

With f values much larger than 1, one could model cases with a strong tendency for segregation to the interface. However, in partitionless transformations it is conceivable that $x_B^{\gamma/\alpha}$ approaches $x_B^{\gamma_0}$, i.e. x_B^α, at increasing velocities and even decreases further. The choice of x_B^I according to equation (45) would then predict that the segregation disappears and even turns negative if a constant f value is chosen. This may not be an attractive model. To let f vary according to some principle would seem to take this modelling too far. A different method of modifying Ågren's model is thus required in order to allow it to treat partitionless cases with segregation to the interface. A solution to the problem was found by Suehiro et al. [13]. Their main idea can be described as follows.

Let x_B^I be the composition in the middle of the interface and evaluate ΔG^t as the sum of contributions from the two halves of the interface, with the representative compositions $(x_B^{\gamma/\alpha} + x_B^I)/2$ and $(x_B^I + x_B^\alpha)/2$. Then one could introduce a thermodynamic parameter, representing the tendency for segregation, into the calculation of x_B^I, which must be carried out for various velocities. In that case, the segregation would not disappear as $x_B^{\gamma/\alpha}$ approaches x_B^α, but only when the calculated x_B^I approaches x_B^α.

Actually, Suehiro et al. even went one step further and developed the model to include three zones in the interface, each one with its own representative composition. That model is illustrated in Fig. 10 and it may appear that the same result could be obtained with Ågren's original model by choosing a value of x_B^I in the same range of composition (see the dashed arrow). However, as already explained, that model would not work as the growth is approaching partitionless conditions, unless one finds a way to describe how x_B^I varies with the velocity. It may be mentioned that Liu later simplified their model by reducing the number of zones in the interface to two, when modelling the $\gamma \to \alpha$ transformation in the Fe–Mo–C system [12].

12. SOLUTE DRAG IN GRAIN-BOUNDARY MIGRATION

In 1962 Cahn [30] as well as Lücke and Stüwe [31] modelled a grain boundary in a one-phase material as a region of a definite thickness, δ, and with thermodynamic properties varying continuously as a wedge-shaped energy curve from one grain to the other [Fig. 1(c)]. They solved the diffusion equation for a series of constant velocities. The resulting composition profile was asymmetric and they could calculate the net attraction between the grain boundary and the solute atoms by integrating over all the solute atoms present in the energy gradients in the two halves of the grain boundary. Lücke and Stüwe gave an equation equivalent to

$$P_{s.d.} V_m = -\int_{-\infty}^{\infty} x_B \left(\frac{dE}{dy}\right) dy \qquad (46)$$

where y is the coordinate in the growth direction. This net attraction will act as a force dragging a local enrichment of the solute along with the migrating grain boundary, and an opposite force will act on the grain boundary and oppose its migration. $P_{s.d.}$ was originally called "impurity drag", now "solute drag", and according to equation (46) it originates completely from inside the grain boundary because the energy E is constant in both grains. In order to simplify numerical calculations Cahn chose to subtract a term $x_B^0 \int dE$, which is zero when integrated over the grain boundary, and he obtained an equation equivalent to

$$P_{s.d.} V_m = -\int_0^\delta (x_B - x_B^0) \left(\frac{dE}{dy}\right) dy \qquad (47)$$

The net result is the same, of course.

Cahn then added factors that would represent the friction in grain-boundary migration in a pure element, calling them "intrinsic drag" and using the notation P_0. The available driving force would thus yield a rate of migration such that $(P_{s.d.} + P_0)V_m$ is equal to the driving force, expressed in J/mol, which may come from the curvature of the grain boundary or from the parent grain if it has been cold worked. It is interesting to note that, although this is a continuous interface model as far as the calculation of the solute drag is concerned, it is still a sharp interface model as far as the intrinsic drag and its driving force are concerned.

The main result of this treatment is that the solute drag first increases proportionally to the velocity. However, at higher velocities the time for dif-

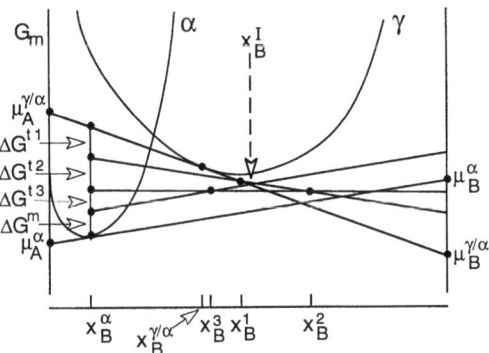

Fig. 10. Ågren's model with three zones, each one with its own representative composition, x_B^1, x_B^2 and x_B^3. The corresponding dissipations of Gibbs energy are also indicated. A single representative composition x_B^I would have the same effect.

fusion inside the grain boundary gets less and the segregated solute atoms find it more and more difficult to keep pace with the migrating boundary. The amount of segregation will thus decrease, and the previous increase of the solute drag will slow down and even stop. The solute drag will thus reach a maximum and then decrease and finally vanish when segregation has become negligible. If the driving force could be increased continuously in an experiment, one would find that the velocity first increases proportionally but then even faster and there may be a critical driving force where the velocity suddenly changes to much higher values where there is practically no diffusion. Thus, the grain boundary has broken away from the enrichment of solute atoms.

It is interesting to note that Schoen and Owen in 1971 [35] applied this approach to model a solute drag by carbon atoms on the migration of a martensitic interface. Instead of the wedge-shaped energy function, they derived theoretically the attraction between carbon atoms and the dislocations in the martensitic interface as a function of distance and their treatment may thus be regarded as a diffuse interface treatment. They did not consider the difference in thermodynamic properties of the two phases except as a source for driving force. Formally, their model thus describes a case of grain-boundary migration in a one-phase material rather than phase interface migration in a two-phase material.

From the definition of the solute drag it is evident that it does not directly describe a dissipation of Gibbs energy but rather a force, a driving force with a negative sign. Thus, it is not completely surprising that it can be evaluated from two different integrals [see equations (46) and (47)], whose integrands vary in different ways through the interface. Of course, there must be a corresponding dissipation of Gibbs energy when a driving force is applied and makes the boundary migrate. Exactly where the dissipation takes place will be discussed in Section 13. On the other hand, exactly where the solute drag, as a net force of attraction, is attached cannot be answered.

13. RELATION BETWEEN SOLUTE DRAG AND GIBBS ENERGY DISSIPATION

It is not a trivial question how the definition of solute drag, as the net force of attraction between solute atoms and interface, can be generalized to the interface in a phase transformation. It turns out that almost all studies, that claim to concern solute drag, actually consider dissipation of Gibbs energy. An exception is a study by Brechet and Purdy [36] who based their work on equation (46). In a second paper [37], they evaluated the solute drag by using instead equation (47). However, equations (46) and (47) are not equivalent for phase transformations

where the term $x_B^0 \int dE$ is not zero when integrated over the interface from one phase to the other. It is necessary to examine which one of equations (46) and (47), if any, should be used for phase transformations.

It should be emphasized that the action of a phase interface on a phase transformation was described by considering the dissipation of Gibbs energy before the concept of "solute drag" was defined (see Ref. [5], for instance). When Hillert [32] explored the homogeneous interface model, he proposed that the dissipation of Gibbs energy due to diffusion is equal to the solute drag as defined by Cahn and by Lücke and Stüwe. The equivalence was then proved under certain conditions by Hillert and Sundman in 1976 [38] starting from the expression for the dissipation of Gibbs energy due to diffusion. Counted per mole of atoms it is

$$\Delta G^{\text{diff}} = \left(\frac{V_m}{v}\right) \int_{-\infty}^{\infty} J_t \left[\frac{d(\mu_B - \mu_A)}{dy}\right] dy \quad (48)$$

Suppose steady-state conditions have been established and has resulted in a partitionless transformation where there is no sidewise diffusion. Then the interdiffusional flux, J_t, can be expressed for each point, yeilding

$$J_t = \left(\frac{v}{V_m}\right)(x_B - x_B^\alpha) \quad (49)$$

$$\Delta G^{\text{diff}} = \int_0^\infty (x_B - x_B^\alpha) \left[\frac{d(\mu_B - \mu_A)}{dy}\right] dy \quad (50)$$

where $x_B^\alpha = x_B^{\gamma_0}$. The first integration limit was here put to $y = 0$, the side of the α phase, because $x_B - x_B^\alpha = 0$ everywhere in the growing α phase under steady-state conditions. It may be convenient to divide the integration into two parts, the interface yielding ΔG^t and the parent γ phase yielding ΔG^γ. One can always write

$$\mu_B - \mu_A = {}^\circ G_B(y) - {}^\circ G_A(y) + f(y, x_B) \quad (51)$$

where $f(y, x_B)$ is the difference in partial Gibbs energy of mixing for B and A, respectively. ${}^\circ G_B(y)$ and ${}^\circ G_A(y)$ are the values for pure A and B, respectively. Since they vary only within the interface, one obtains

$$\Delta G^{\text{diff}} = \int_0^\delta (x_B - x_B^\alpha) \left[\frac{d({}^\circ G_B - {}^\circ G_A)}{dy}\right] dy$$

$$+ \int_0^\infty (x_B - x_B^\alpha) \left(\frac{df}{dy}\right) dy \quad (52)$$

The first term is identical to equation (47) from Cahn's treatment of the solute drag except for the fact that he denoted $({}^\circ G_B - {}^\circ G_A)$ by E. Hillert and Sundman primarily considered the case where the ideal solution model applies everywhere. In that case $(df/dy)dy$ can be replaced by $(\partial f/\partial x_B)_y dx_B$ and the second integral is zero for a partitionless transformation because it is integrated between two limits with the same composition, $x_B^\alpha = x_B^{\gamma_0}$. The same holds for all cases where f is not a function of y. In all such cases the two approaches thus yield the same net result. When f varies with y, then equation (47) must be modified because the force acting on the atoms is no longer given simply by dE/dy, but equation (50) would still apply. It should be emphasized that when equation (50) is used, it is necessary to make a numerical integration only inside the interface where the properties vary, i.e. for the evaluation of ΔG^t. The properties of the γ phase are constant and for partitionless growth the dissipation in γ, ΔG^γ can be obtained directly as

$$\Delta G^\gamma = x_A^\alpha \left(\mu_A^{\gamma_0} - \mu_A^{\gamma/\alpha} \right) + x_B^\alpha \left(\mu_B^{\gamma_0} - \mu_B^{\gamma/\alpha} \right) \quad (53)$$

In the derivation by Hillert and Sundman the properties of the two phases were not defined and the derivation thus holds for grain-boundary migration as well as for partitionless phase transformations. In the former case, equations (46) and (47) are equivalent and are both supported by the dissipation approach. The interesting result is that for phase transformations, where equations (46) and (47) are no longer equivalent, the dissipation approach supports equation (47) but not equation (46). The question how to apply the solute drag theory to phase transformations is thus that it should be done by the use of Cahn's equation (47). However, as we have just seen, when f varies with y, equation (47) must be modified in order to give the same net result as equation (50). It will be postulated that $P_{s.d.} V_m$ should be equal to ΔG^{diff} even when f varies with y. It will also be postulated that such a modification of equation (47) can be made, but it will not be discussed further here.

An important difference between the solute drag approach and the dissipation approach should be emphasized. The integrand of equation (47) or the first term of equation (52) is zero within both phases because E or $(^\circ G_B - ^\circ G_A)$ is constant there. The solute drag is thus located completely inside the interface. That is why the integration limits in equation (47) and for the first term in equation (52) were defined as 0 and δ (the width of the interface). On the other hand, the integrand in equation (48) or equation (50) does not disappear in the parent γ phase if the composition varies, which it does if the solute has piled up in front of the interface. The dissipation of Gibbs energy thus has an important contribution from the spike inside the parent phase, γ. The exact location of the dissipation of Gibbs energy by diffusion can hardly be questioned, but from the solute drag approach it is less evident where to place the dissipation of Gibbs energy. Already when going from equation (46) to (47), one redistributes the contributions within the interface but that is only a mathematical operation and could not have any effect on the physical realities.

Evidently, there must be some coupling between the forces on individual atoms exerted by the interface. Even the solute atoms in the spike in the γ phase are exposed to forces, in this case coming from the gradient in chemical potential. The conclusion should be that there is some coupling between all these forces resulting in a distribution of the Gibbs energy dissipation in exactly the way described by equation (48), based on diffusion. Otherwise, the postulated equality of $P_{s.d.} V_m$ and ΔG^{diff} is incorrect.

The situation may be illustrated with a molar Gibbs energy diagram for a partitionless transformation (Fig. 11), where the dissipation of Gibbs energy in the γ spike, denoted ΔG^γ, has been included and the driving force is evaluated for the whole reaction, D_{total}^{chem}, including diffusion in the γ phase. The total Gibbs energy dissipation due to diffusion is $\Delta G^{diff} = \Delta G^t + \Delta G^\gamma$ and, as we have seen, it is equal to $P_{s.d.} V_m$, when f is independent of y, and it was postulated that the equality holds also when f depends on y. The driving force for the migration of the interface is thus obtained as

$$D^m = \Delta G^m = D_{total}^{chem} - \Delta G^{diff}$$
$$= D_{total}^{chem} - \Delta G^t - \Delta G^\gamma = D_{total}^{chem} - P_{s.d.} V_m \quad (54)$$

After evaluating the solute drag by applying equation (47) or a modification of that equation, it is thus important not to compare with the chemical driving force D_{int}^{chem} evaluated from the composition at the interface, $x_B^{\gamma/\alpha}$, as done in the previous figures. An amount corresponding to ΔG^γ would then be missed. That seems to have happened in the second paper by Purdy and Brechet [37] which, however, was based on the appropriate equation for phase transformations, namely equation (47).

According to the solute drag approach, as well as the dissipation approach, the composition profile must first be calculated. Then the integration is carried out with two different methods that should yield the same result. It may seem unimportant

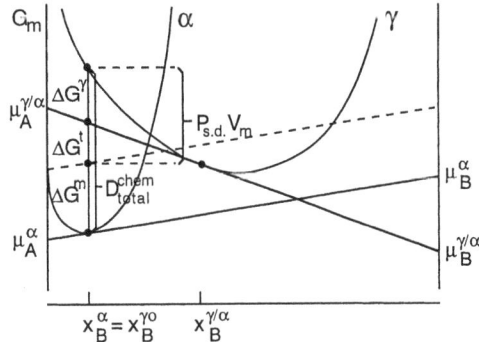

Fig. 11. Molar Gibbs energy diagram for partitionless phase transformation. The effect of solute drag corresponds to $\Delta G^\gamma + \Delta G^t$. Notice that D_{total}^{chem} is different from D_{int}^{chem} in Figs 4 and 9.

what method is used but the dissipation approach has the advantage that it can be used to calculate the loss inside the interface, ΔG^t, and in the spike, ΔG^γ, separately whereas the solute drag approach only yields $\Delta G^t + \Delta G^\gamma$. Furthermore, the solute drag approach cannot, without modification of equation (47), be used for cases where f in equation (51) varies with y, as already discussed. On the other hand, complications may also arise if the transformation is not partitionless. That case will be discussed in Section 18.

It should be mentioned that Shirley [39] has considered the coupling between the interfacial structure and the composition in front of the interface. He also evaluated the solute drag effect by analysing the dissipation of Gibbs energy by diffusion, but he only studied the contribution from the spike in front of the interface (the "extrinsic part"). He paid most attention to the effect of a strain field originating from the structure inside the interface but reaching out in front of the interface. It may be argued that this part of the parent phase should then be regarded as part of the interface. In any case, it should be included in the integrations of equations (46) and (47).

14. HOMOGENEOUS INTERFACE MODEL AT HIGH VELOCITIES

The application of the homogeneous interface model to grain-boundary migration has already been described in Section 9 and Fig. 7. It gives essentially the same result as the wedge-shape model at low velocities but does not predict a decrease at high velocities that makes it possible for the grain boundary to break away from the local enrichment of solute atoms. The reason is evident from Fig. 7. Once the spikes, shown in the lower part of the diagram, have developed to maximum height, the Gibbs energy dissipation is independent of the widths of the spikes according to this construction. It only depends on the height of the spikes. However, that is not in agreement with the solute drag treatment and it may not seem reasonable that there should be any dissipation of Gibbs energy by diffusion if the spikes are much thinner than the interatomic distances. It was thus proposed by Hillert [32] that the top parts of the spikes should be neglected by omitting the last $\delta/4$ of each spike. That modification would decrease ΔG^t from the value given by equation (8) and also decrease ΔG^γ. A similar effect could be expected for phase transformations. It should be emphasized that the transition from diffusion-controlled to partitionless transformation was successfully performed with the sharp interface model without any modification (see Section 4). In that model the dissipation of Gibbs energy by diffusion decreases naturally at high growth velocities. In that respect, the introduction of a homogeneous interface was not an improvement over the sharp interface model.

It is interesting to note that Shirley [39] also did not predict a breakaway from the local enrichment of solute atoms at high velocities and he seems to favour truncation of the spike in order to correct for this shortcoming. He seems to suggest that the solute drag approach has the same shortcoming because he states that "the neglect of the spike-truncation procedure in the Cahn–Lücke–Stüwe model is equivalent to the neglect of the gradient energy". Maybe it should be emphasized that this usually would have no harmful effect if a continuous model is chosen, e.g. the wedge-shaped interface, but an interesting aspect will be discussed in Section 20.

Compared to the wedge-shaped interface model, the homogeneous interface model has some advantages that may sometimes balance the disadvantages:

1. Since the interface is regarded as a separate phase of constant volume, it may be included in molar Gibbs energy diagrams, as already demonstrated, and it has inspired the use of such diagrams to illustrate the sharp interface model, as well. Consequently, the homogeneous interface model should not be regarded as a simple exchange of the wedge-shaped function for a square-well, nor should it be regarded as a modification of the sharp interface model because the representative composition of the interface plays an important role.
2. At constant rates of migration the diffusion equation has a simple analytical solution and it is easy to evaluate the dissipation of Gibbs energy by diffusion.

15. MODELLING OF PHASE TRANSFORMATIONS WITH A CONTINUOUS INTERFACE MODEL

Hillert and Sundman [38] started by applying the wedge-shape model to grain boundaries and phase transformations when evaluating the dissipation of Gibbs energy instead of the solute drag. In order to make the model more flexible they truncated the wedge [see Fig. 1(d)], introducing a central zone with constant properties. A particularly interesting result was the observation that an increased diffusivity in the centre of a grain boundary could decrease the magnitude of the solute drag considerably. It thus appears very difficult to predict solute drag without detailed information on the properties of the interface. For a phase transformation, all the dissipation of Gibbs energy occurs in the γ spike at low velocities and it is interesting to see how its role is gradually taken over by the three zones inside the interface at higher velocities if the diffusivity is higher there. However, in this case the introduction of a higher diffusivity in the interface was found to delay the decrease of the dissipation at higher velocities and may even increase it.

In 1977 Hillert and Sundman [40] applied their treatment to solidification in binary alloys by modelling the interface as a single zone with gradually varying properties from solid to liquid. By considering sidewise diffusion in front of a dendrite tip they were able to predict a spontaneous transition from diffusion-controlled to partitionless solidification. They obtained a diagram very similar to Fig. 5, for the $\gamma \rightarrow \alpha$ transformation in Fe–C alloys, but the critical alloy composition for the onset of partitionless solidification came close to the T_0 line. It may be added that they did not consider a finite interface mobility because they felt that the mobility of the solid/liquid interface is sufficiently high not to have a decisive influence on the transition to partitionless growth, which should occur at a velocity controlled by the rate of diffusion in front of the tip of a dendrite.

It has already been stressed that the predictions may be very sensitive to exactly how the properties of the interface are being modelled. For solidification the predictions may be safer than for solid–solid phase transformations because it seems reasonable to assume that the properties of the interface vary monotonously in this particular case. In solid–solid phase transformations there may very well be a tendency for segregation and there is certainly a higher diffusivity in the middle of the interface.

16. DIFFUSE INTERFACE MODELS

All the interface models described so far are based on the assumption of diffusivities and thermodynamic properties as functions of distance, and the width of the interface thus enters into the definitions of those functions. Except in the homogeneous interface model, those functions vary continuously between the values of the two phases. In the phase field method [41] one has gone one step further by defining a variable ϕ which is 0 for one phase and 1 for the other. The properties change as functions of ϕ instead of distance and the variation of ϕ with distance through the interface is determined by a step-wise, local minimization of the Gibbs energy in a finite element procedure. The shape and position of the interface and its width will thus be results of the calculation. In order to limit the width it is essential to include gradient terms in the Gibbs energy as functions of ϕ and those terms must be chosen in such a way that the predicted width is realistic. However, the width will not be well defined and the interface may be characterized as diffuse [Fig. 1(f)].

The phase field method has been remarkably successful in predicting the development of the shape of growing crystals [42]. However, for the detailed description of the properties of the migrating interface it is critical how the exact form of the Gibbs energy function is chosen. The previous continuous interface models are less predictive and, at the same time, less sensitive to various choices. For instance, they do not critically depend on the inclusion of gradient energies because the width and the shape of the property functions are chosen *a priori*. This is the reason why gradient energies have not generally been included in those models but an attempt to include the gradient energy in the interface model for phase transformations was recently made by Liu *et al.* [43].

With the phase field method one could, at least in principle, simulate the real physics more closely and that method could thus be very predictive, maybe even more predictive than one would like. It may be difficult to make all the predictions come out close to physical realities.

In the previous interface models the pressure difference between the phases is presumed to act in a sharp fashion at a certain position, usually at the point between the interface and the growing phase. Such a pressure difference could be due to the interfacial tension when the interface is curved or due to the friction opposing the migration. With the phase field method those actions are spread out over the entire interface in a way determined basically by the functions of ϕ. This method thus yields the first real continuous model. The possibility of modelling interface migration in alloys with the Monte Carlo technique [44,45] should also be mentioned.

17. PARTITIONLESS GROWTH

Much interest has been focused on the question of where in a phase diagram a massive transformation can be expected, i.e. partitionless growth. It has been pointed out [32,46] that the chemical driving force is exactly sufficient to compensate for the dissipation of Gibbs energy by diffusion in a well-developed spike in the parent phase if the alloy composition falls on the boundary of the one-phase field of the new phase. Hillert [47] has thus claimed that this boundary should be regarded as the natural limit for the massive type of transformation. On the other hand, Perepezko [48] and Menon *et al.* [49] have proposed that the natural limit is the T_0 line. A crucial question is under what conditions the diffusion-controlled growth process is able to break through the barrier caused by a spike, well developed at low velocities? There should be a critical value of the alloy composition inside the two-phase field where this starts to happen.

There have been several attempts to answer this question by modelling the growth process using the compositions at the interface, x_B^α and $x_B^{\gamma/\alpha}$, and the velocity as unknown quantities. Such treatments have shown that growth will spontaneously develop into the partitionless mode if the supersaturation is high enough. Figure 5 presented the results of such a calculation for the transition from edgewise growth of Widmanstätten ferrite to martensitic growth, using the sharp interface model. In that particular case the model predicted that a spontaneous transition would occur if $x_C < 0.0020$, which was close to the phase boundary at $x_C = 0.0017$.

With their model Jönsson and Ågren [34] showed that the critical composition can be modelled to fall anywhere between the phase boundary and the T_0 line, the exact position depending on how the model parameters were chosen. The important factor seems to be the diffusivity in the interface. Already Hillert and Sundman [40] predicted that diffusionless solidification, which is also a partitionless transformation, can occur rather close to the T_0 line. In that case the diffusivity in the interface was assumed to be lower than in the liquid by varying from a high value on the liquid side to a low value on the solid side.

Another important factor is the tendency for segregation to the interface. This was demonstrated by Hillert and Sundman [38] using the truncated wedge model and by Suehiro et al. [13] using a modification of Ågren's model. These calculations predict that the critical limit for massive transformation may move out of the two-phase field and into the one-phase field if the tendency for segregation is strong.

Another strategy for treating partitionless transformations is to assume from the beginning that the growth is partitionless. After the calculation, one can examine (a) whether the chemical driving force is strong enough to balance the various Gibbs energy dissipations, which is necessary in order for that kind of growth to proceed, and (b) whether the partitionless growth will be practically diffusionless or some diffusion will be involved. Even though less complete, results of such modelling may be very instructive which will now be demonstrated. Figure 12 is an example from a recent attempt by Hillert and Schalin [50] to model the partitionless transformation with the wedge-shape model. The horizontal line represents the total chemical driving force evaluated as illustrated in Fig. 11. A part of that force must be used for driving the interface with the velocity v, e.g. according to equation (2), and the remaining part (reduced driving force) would go to zero at some critical velocity. The intersection between this curve and the curve for the Gibbs energy dissipation by diffusion represents posssible conditions for partitionless growth. However, as emphasized by Hillert and Schalin, those conditions are not stable. A small positive fluctuation of the velocity would make the driving force stronger than the dissipation and the growth would speed up and finally reach the end-point of the curve for reduced driving force. The critical question is how the growth process can reach the

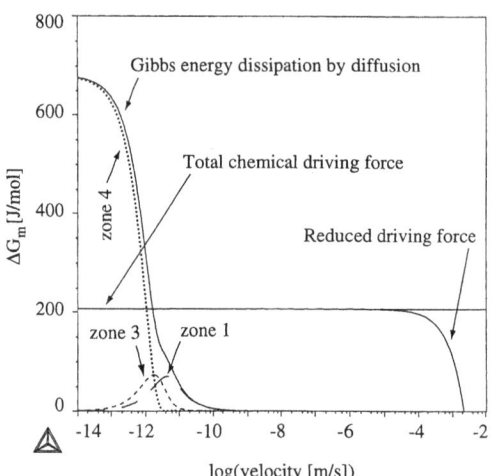

Fig. 12. Dissipation of Gibbs energy by diffusion as a function of velocity, according to the wedge-shape interface model, assuming a constant diffusivity $D = 8 \times 10^{-22}$ m²/s. Zone 4 is from the spike in the parent phase (from Ref. [50]).

point of intersection when the driving force for partitionless growth is much too low to surmount the barrier formed at low velocities, as it is in Fig. 12. Normally, diffusion-controlled growth would be expected in such a case. The only chance to get partitionless transformation would be that the growth in some way attains a velocity higher than about 10^{-12} m/s and the question is if diffusion-controlled growth can do that.

The cause of the difficulty is that the growth velocity of 10^{-12} m/s at the point of intersection would make the width of the spike equal to about $D^\gamma/v = 8 \times 10^{-22}/10^{-12} = 8 \times 10^{-10}$ m. In order to reach that velocity under an initial diffusion-controlled growth, where some solute must diffuse sidewise in order to get out of the way of the new phase, it is necessary that the edge of a growing plate has a radius of the same order of magnitude and that would by itself require a driving force of about $\Delta P V_m = \sigma V_m/r = 1 \times 10^{-5}/8 \times 10^{-10} \approx 10^4$ J/mol in the present case. This is a very high value and could rarely be provided by the chemical driving force.

The calculation presented in Fig. 12 was carried out at 900 K for Fe–6%Ni, which is well inside the two-phase field. The total chemical driving force is only 200 J/mol, which is far from the required value of 10^4 J/mol. It seems that spontaneous development of diffusion-controlled growth into partitionless growth is possible only if the alloy composition is such that the driving force is almost as strong as the dissipation of Gibbs energy at low velocities. That occurs when the alloy composition falls almost on the phase boundary of the new phase. Figure 12 was calculated for a composition half-way between the phase boundary and the T_0 line. It demonstrates that, in order to get partitionless growth well inside the two-phase field, it is necessary to have some kind of dynamic nucleation by which there is no time for a spike to develop [5,38].

Figure 12 was calculated with a constant diffusivity and almost no tendency for segregation. Figure 13 was obtained with a diffusivity varying logarithmically from 8×10^{-22} m^2/s in the parent grain to 5×10^{-12} m^2/s in the middle of the interface and with a stronger tendency for segregation (with a partition coefficient of 1.66). In this case it would be even more difficult for partitionless growth to get started. If it does get started by dynamic nucleation, it will become stabilized at the intersection between the curve for dissipation by diffusion and the curve for reduced driving force close to the lower, right corner. It is evident that partitionless growth would not be quite diffusionless in this case. Furthermore, the hump in the dissipation curve, which is caused by segregation, may stop the development of the growth process after dynamic nucleation much sooner.

18. PHASE TRANSFORMATIONS WITH PARTITIONING AND UNDER NON-STEADY-STATE CONDITIONS

Figure 11 illustrates for partitionless growth how one can evaluate ΔG^m, which is also equal to D^m, from $D_{\text{total}}^{\text{chem}}$. On the other hand, Fig. 9 illustrates that one can evaluate the same quantity for all steady-state conditions from $D_{\text{int}}^{\text{chem}}$. By comparing the two diagrams it may be concluded that D^m can be calculated from $D_{\text{total}}^{\text{chem}}$ even for non-partitionless growth by subtracting ΔG^t, evaluated by integration over the interface, and a hypothetical value of ΔG^γ, evaluated from equation (53) under the assumption that the process is partitionless.

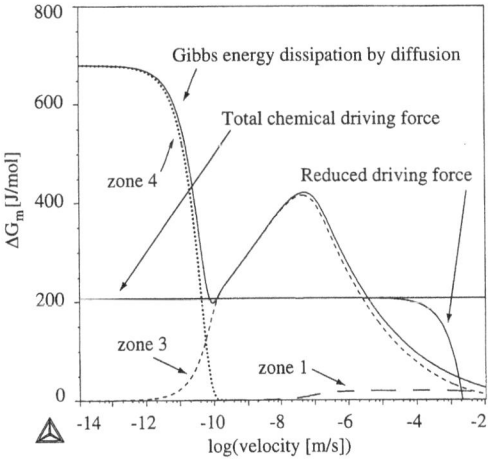

Fig. 13. Dissipation of Gibbs energy by diffusion as a function of velocity, according to the wedge-shape interface model, assuming a variable diffusivity from 8×10^{-22} to 5×10^{-12} m^2/s and a tendency for segregation. Zone 4 is from the spike in the parent phase (from Ref. [50]).

Owing to the fact that interfaces are extremely thin, changes in the local composition under non-steady-state conditions would have a negligible effect and such cases can thus with good accuracy be treated as if there are approximately steady-state conditions as far as the interface is concerned. However, for the rest of the system it may be important and the progress of the phase transformation cannot be properly described without considering the real conditions in the γ phase.

In any case of phase transformation in alloys one can describe the role of the interface by first guessing values of x_B^α and v. Then the composition profile and, in particular, $x_B^{\gamma/\alpha}$ can be calculated under the condition or assumption of partitionless growth, and the dissipation of Gibbs energy by diffusion inside the interface, ΔG^t, is evaluated from equation (48) or equation (50) between the integration limits 0 and δ. The chemical driving force over the interface is evaluated using equation (9) and their difference yields D^m to be inserted in equation (2) to yield the velocity. By iteration one must find the start value of v that agrees with the calculated one. The only independent variable would thus be x_B^α or v. It is interesting to note that this evaluation is independent of any knowledge of what happens inside the shrinking grain. By iteration on a higher level the x_B^α value, which satisfies conditions defined for the α and γ phases, whether there are steady-state conditions or not, can finally be found. For a diffusion-controlled case one must solve the two- or three-dimensional diffusion in the γ phase in order to relate the velocity to the initial γ composition but not in order to evaluate D^m.

Using the postulate that $P_{\text{s.d.}}V_m$ is always equal to $\Delta G^t + \Delta G^\gamma$ for partitionless growth and assuming that it only depends on the conditions inside the interface, it should be possible to evaluate $P_{\text{s.d.}}V_m$ for any case without lateral diffusion in the interface from the real ΔG^t and the hypothetical ΔG^γ. However, it is questionable whether $P_{\text{s.d.}}V_m$ would be particularly useful. As already emphasized, D^m is obtained already from $D_{\text{int}}^{\text{chem}}$ and ΔG^t. A case where the solute drag approach would be useful would be when there is a strong transient and the approach of steady-state conditions inside the interface is slow.

19. ALLOYING EFFECTS IN FE–M–C

When studying the compositions of the three phases in the transformation $\gamma \rightarrow \alpha +$ cementite (cementite being essentially Fe$_3$C) in Fe–M–C alloys, Hultgren [51] found that the substitutional alloying element M is sometimes redistributed between the two new phases but sometimes it is not and then the transformation is regarded as partitionless. He believed that the same phase transformation in Fe–C alloys would occur by local equilibrium at the migrating phase interfaces and proposed that local equilibrium would also hold in Fe–M–C alloys in those cases where M is redistributed. For the partitionless transformation he proposed that it occurs under a restricted local equilibrium at the interfaces, characterized by local equilibrium for the very mobile, interstitial element C but not for the sluggish alloying element M. That restricted local equilibrium he called "paraequilibrium" (PE). Hillert [52] later pointed out that it is conceivable that a partitionless transformation occurs behind a spike in the parent phase and under full local equilibrium (LE). That case he called "false paraequilibium" and later "quasi-paraequilibrium" (Ref. [4], pp. 358–361).

The principles are illustrated in Fig. 14 for a simple case that could be Fe–Mn–C. The dashed lines represent paraequilibrium and the corresponding tie-lines are horizontal because the composition variables u_C and u_{Mn} are used. They are related to the ordinary mole fractions by

$$u_j = \frac{x_j}{(x_{Fe} + x_{Mn})} \quad (55)$$

Thus, it is not necessary to draw such tie-lines. For full equilibrium only one tie-line has been drawn. It represents the full local equilibrium if the α phase forms with the same Mn content as the initial γ, i.e. $u_{Mn}^\alpha = u_{Mn}^0$. The thin line represents an isoactivity line for carbon in γ, going through the $\gamma/\alpha+\gamma$ point. It thus represents the carbon activity at the interface. It has been extrapolated into metastable γ down to the level of u_{Mn}^0. In order to get quasi-paratransformation it is necessary that the initial carbon content in γ, i.e. u_C^0, is lower than in this point. It thus defines a critical point on the u_{Mn}^0 horizontal. By connecting such points for various values of u_{Mn}^0

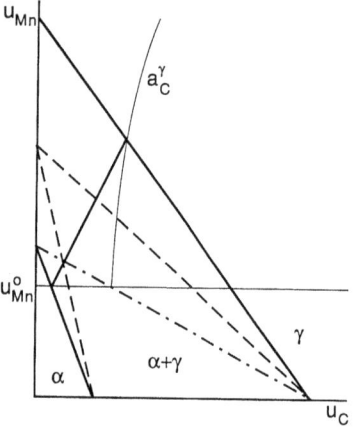

Fig. 14. Schematic, isothermal phase diagram showing the $\alpha+\gamma$ two-phase field in the Fe–Mn–C system at 900 K. The dashed lines represent the two-phase field $\alpha+\gamma$ under paraequilibrium (PE). The dash-dotted line is the upper limit for the partitionless growth, $u_{Mn}^\alpha = u_{Mn}^0$ under full local equilibrium (LE). The thin line is an isoactivity line for carbon in γ, extrapolated into the metastable range of γ.

one gets a critical line for the alloy composition (dash-dotted line in Fig. 14). Quasi-paratransformation is possible below (i.e. to the left of) that line. On the other hand, paratransformation is thermodynamically possible as soon as the initial γ phase is chosen inside the $\alpha+\gamma$ two-phase field according to paraequilibrium. Furthermore, already to the left of the γ boundary, according to full equilibrium, transformation can occur but only by α growing with a different Mn content. Thus there are three critical lines. Each kind of growth can occur below its critical line but its region is not restricted by the next critical line. Below two or three of the critical lines we need a detailed analysis to decide what kind of growth will prevail. In this section it will be shown that intermediate cases may then play an important role.

Starting from a full set of diffusion equations for the multi-component case, Kirkaldy [53] and his group [54] also arrived at the concept of partitionless transformation under full local equilibrium as a limiting case.

Intermediate cases may arise as a consequence of Gibbs energy dissipation by diffusion inside the migrating interface. In order to explain the principles some simplifications will now be made. First it will be assumed that the carbon activity has the same value in the whole region of the interface including the spike of M in the γ grain. That is usually a very good approximation for Fe–M–C alloys when the velocity is controlled by carbon diffusion over much larger distances. Next, the interface will be regarded as planar and it will be assumed that steady-state conditions have been established. In Section 18 it was argued that this is a good approximation for the interface even if the phase transformation does not occur under steady-state conditions and the interface is curved.

Through the assumption of a constant carbon activity, calculations for a ternary system will be very similar to the corresponding calculation for a binary system. We may thus get a diagram very similar to Fig. 12 but now we should forget the horizontal line because the chemical driving force depends on the carbon activity at the interface, a_C^I. The curve should be rather independent of the value of a_C^I if there is not a strong interaction between carbon and the alloying element in the interface. We shall thus look upon the curve in Fig. 12 as if it holds for a range of carbon activities. Any point on the curve would then represent possible growth conditions if there is a value of the carbon activity in the interface that would give the correct value for the driving force and the next question is whether the correct velocity, as controlled by the carbon activity difference within γ, can be obtained with a suitable value of u_C^0 and a suitable geometry. Points on the almost horizontal part in the upper left corner would represent transformations under quasi-paraequilibrium (LE).

Points in the lower right corner would represent partitionless transformations under paraequilibrium (PE). Points in between would represent intermediate cases.

Figure 12 was obtained for a constant diffusivity and a weak tendency for segregation. Figure 13 was an example of the effect of a variable diffusivity and a stronger tendency for segregation. According to the curve in the lower right corner, the growth under paraequilibrium has now been replaced by growth under some Gibbs energy dissipation by diffusion. It would not represent pure paratransformation. Furthermore, it would require very high growth rates, which can hardly be accomplished by growth controlled by carbon diffusion in the γ phase.

The concepts of solute drag, paraequilibrium (PE) and full local equilibrium (LE) have been used extensively for discussions of experimental information on the growth velocity of α in various Fe–M–C alloys. However, in order to make more definite interpretations or predictions, it would be necessary to make a full analysis of the process, including a treatment of the rate of carbon diffusion in the γ phase. It has been attempted by Liu and Ågren [9] who generalized Ågren's interface model to a ternary system using $u_k^I = (u_k^\alpha + u_k^{\gamma/\alpha})/2$ for k denoting Mn and C. They obtained the same kind of result as those in Fig. 5. Then Liu [12] modelled the formation of α in the Fe–Mo–C system using two representative compositions (see Section 11). He was thus able to test the idea that there is a strong tendency for segregation of Mo to the interface. However, one should improve this treatment by using a continuous interface model but, in order to make predictions, it is then necessary to have a reasonable estimate of the diffusivity for trans-interface diffusion.

Purdy and Brechet [37] have also modelled the α formation in Fe–M–C systems. They used the wedge-shape model of an interface and also considered the case where the alloy element is attracted preferentially to the interface. They considered planar growth starting from a surface with a high velocity but decelerating as carbon is piling up in front of the advancing interface. They evaluated how the solute drag would vary as the velocity decreased. A recent contribution by Enomoto [55], using essentially the same method, should also be mentioned.

20. CONTRIBUTION FROM THIN SPIKES

In Fig. 7 it was demonstrated that the homogeneous interface model results in two spikes which can be very thin at high velocities. All the Gibbs energy dissipation is concentrated to these spikes and is not predicted to decrease at high velocities as long as those spikes exist mathematically. On the other hand, it may not seem reasonable that diffusion in such spikes should give rise to Gibbs energy dissipation if they are so thin that they fall well below interatomic distances. In a practical sense, the problem was solved by the truncation procedure. The higher parts were simply omitted by cutting off $\delta/4$ of the width [32]. Calculations based on the wedge shape later gave the impression that this model does not cause the same problem and that was considered as a great advantage. However, an examination of Fig. 13 now reveals that the wedge-shape model does not predict that the contribution from the spike should vanish until the velocity is so high that the width of the spike, estimated as D^γ/v, is very small, about $8 \times 10^{-22}/8 \times 10^{-11} = 10^{-11}$ m according to Fig. 13. It thus seems that there are cases where the wedge-shape model also needs to be modified. One could ask if the truncation procedure could again be applied but it would be more satisfactory if a more physical method could be found. A natural method would be to include gradient energy in the description of the thermodynamic properties and thus to limit the height when the spike gets very thin. That method could probably improve the homogeneous model, also.

On the other hand, knowing x_B^α, being equal to $x_B^{\gamma_0}$ for a partitionless transformation, and choosing a value for the velocity, one can calculate the composition numerically step by step from the side of the α phase and through the interface. With the wedge-shape model that should be straightforward. In the above example problems do not arise until the calculation enters into the γ phase and only if it has an unrealistically steep spike. However, already from the composition profile inside the interface one can evaluate the dissipation inside the interface and the chemical driving force over the interface. Thus everything is known about the interface. In addition one could calculate the total chemical driving force and the difference between the two chemical driving forces must represent the Gibbs energy dissipation in the γ phase, ΔG^γ. As a consequence, that dissipation is independent of the width of the spike. It seems that the same amount of Gibbs energy will dissipate in the γ phase, whether it occurs by diffusion in the spike or by some other process as the atoms leave the γ phase and enter the interface. It thus seems that one can trust the curve presented in Fig. 13.

It should be emphasized that the above discussion does not lead to the conclusion that any thin spike, calculated mathematically, can be used for the evaluation of Gibbs energy dissipation. It seems reasonable that it cannot be applied if there is a sharp discontinuity in the thermodynamic properties. When using such a model it seems advisable first to introduce the gradient energy in the calculation of the composition profile.

21. DIGM OR CIGM

In so-called discontinuous precipitation, a grain boundary between two matrix grains may sometimes migrate together with a new, precipitating phase. Sulonen [56,57] observed that the grain boundary sometimes bulges out and leaves the new phase behind. Hillert and Lagneborg [58] observed that this phenomenon could occur simultaneous to discontinuous precipitation in the same specimen. They discussed possible mechanisms to apply to both phenomena and, in particular, the following two: (a) the change in composition will give rise to coherency stresses and that will provide a driving force for the migration of a boundary into the stressed region, as already proposed for discontinuous precipitation by Sulonen [56]; (b) if the grain-boundary migration is sufficiently rapid, the composition profile in the shrinking grain will be so steep that there is a deviation from local equilibrium between the grain boundary and the shrinking grain, as already proposed for discontinuous precipitation by Hillert [32]. Den Broeder [59] also observed the bulging out of grain boundaries and related it to discontinuous precipitation. He proposed a vacancy mechanism. Hillert and Purdy [60] observed the migration of grain boundaries by grain-boundary diffusion to the surface and evaporation. In order to account for their results they developed the above proposal (b) quantitatively by assuming that the temperature is so low that there is practically no diffusion inside the grains. Their explanation was as follows.

If one of the grains meeting at a boundary is growing, it should at each depth from the surface get a constant composition x_B^γ, controlled by equilibrium with the grain boundary. The shrinking grain should normally have the same composition at its contact with the grain boundary but, further inside, it has the initial composition $x_B^{\gamma 0}$. There will thus be a composition spike and a dissipation of Gibbs energy by diffusion, ΔG^γ, exactly corresponding to the chemical driving force, D^{chem}, due to the exchange of γ with a composition $x_B^{\gamma 0}$ in the shrinking grain for γ with a composition x_B^γ in the growing grain. The migration would be impossible unless driven by some other driving force. However, if the diffusivity is very low in the γ phase, there would be practically no diffusion inside the shrinking grain, ΔG^γ would be zero and the chemical driving force would not dissipate by diffusion. Hillert and Purdy proposed that it would then drive the migration and proposed that the phenomenon should be called CIGM (Chemically Induced Grain-boundary Migration). Once the migration has started, it would no longer be dependent on the other driving force.

Cahn *et al.* [61] objected to this conclusion, referring to a thought experiment where the solute, diffusing from the surface, would be an isotope of the only element in the material. They could see no reason why this would cause the grain boundary to migrate even though it would give the growing grain a lower Gibbs energy than the shrinking grain due to the mixing of isotopes in the growing grain. In order to emphasize their point of view they proposed that the phenomenon should be called DIGM (Diffusion Induced Grain-boundary Migration) because they believed that the grain-boundary diffusion process played a decisive role in the creation of a driving force possibly by unequal grain boundary diffusivities of solute and solvent atoms when they are different chemical species. Anyway, the strength of their argument can hardly be denied but it must be discussed how D^{chem}, which would ordinarily be dissipated by diffusion in the γ phase, would be dissipated if diffusion in γ were prevented by a low bulk diffusivity.

In support of the objection by Cahn *et al.* one may refer to Section 20 where it was concluded that the dissipation of Gibbs energy, corresponding to a spike, would occur in the γ phase even if the spike is too thin to exist. It should make no difference for the migrating grain boundary whether dissipation in the shrinking grain occurs by diffusion or some other mechanism. However, the question still remains how the chemical Gibbs energy is dissipated in the shrinking grain if there is no time for diffusion.

A phenomenon closely related to DIGM has been observed by Yoon and Huppmann [62] when the grain boundary was replaced by a thin liquid film, LFM. It may be argued that there are cases where the assumption of a discontinuity between the liquid and a solid grain should not be too bad. By trusting Cahn's argument one should again ask what happens to the driving force that was previously dissipated by diffusion in the spike in the shrinking γ grain. It must be dissipated by some other process and it was recently suggested that the other process may be adsorption on the solid surface (Ref. [4], pp. 106–108).

It should be mentioned that, accepting Cahn's argument, Hillert in 1983 [63] developed proposal (a) quantitatively by evaluating the part of the chemical driving force, coming from the Gibbs energy that would normally be dissipated by diffusion in the grain, but could be stored as coherency strain energy in the composition gradient of the spike. Its release would be coupled to the migration of the grain boundary and could thus give a driving force for migration. If this is the explanation of the phenomenon, CIGM may be a more appropriate name than DIGM because the migration would be driven by part of the chemical driving force. Alternatively, C in CIGM could be interpreted as standing for "coherency".

Any dissipation of Gibbs energy caused by a process must be paid for by the driving force for the process. Cahn's important lesson to us is that a dissipation of Gibbs energy, caused by a process, will not necessarily act as a driving force for that process. One may compare with a human experience. If one buys something, one must certainly pay for it but if one pays for it, one cannot be sure it will be delivered.

Rabkin [64] has pointed out that one should add gradient energy to the coherency strain energy. Furthermore, together with Gust and Fournelle [65] he has proposed that one should subtract a solute drag effect from the driving force for DIGM. However, the latter proposal does not seem justified because what they define as solute drag is just part of the diffusional dissipation of chemical driving force. It is lost but should not be subtracted from the part of the chemical driving force that is not lost. There should be no need to pay for it once more.

Finally, a recent modelling of DIGM by Kajihara and Gust [66] should be mentioned. They did not accept Cahn's objection but evaluated the driving force as the sum of the contribution from coherency and the absence of dissipation by diffusion in the spike when it is too thin. It is also interesting to note that they in effect only considered grain-boundary diffusion of the solute from the surface and neglected the diffusion of the solvent to the surface which, as a first approximation, may have the same magnitude.

22. DISCONTINUOUS PRECIPITATION

Discontinuous precipitation has also been called recrystallization precipitation because the precipitation of a new phase is accompanied by the growth of one parent grain into another. The growing grain and the new phase usually grow together as a lamellar aggregate. If the new phase is richer in the solute, it grows by drawing solute atoms from the edges of the neighbouring lamellae of the growing parent phase. This occurs by grain-boundary diffusion in a way very much resembling DIGM, the only difference being that the distance is much shorter to the neighbouring lamellae of the new phase than to the surface. As already mentioned, the two phenomena have even been observed side by side in the same specimen [58] and the same mechanisms for the grain-boundary migration have thus been proposed. They do not need to be discussed again. The fact that the diffusion takes place by grain-boundary diffusion is demonstrated by the composition of the growing grain dropping discontinuously to a new value. This is why this phase transformation is called discontinuous precipitation.

It may be added that Sundquist [67] has proposed that some solute drag may be involved. Of course, if the diffusivity varies continuously from a low value in the grain to a high value in the middle of the boundary, there could be some Gibbs energy dissipation due to diffusion in the transverse direction even if the velocity is controlled by the rate of diffusional transportation by lateral boundary diffusion over long distances. That dissipation would come from the chemical driving force and may have an effect on how much would go into the coherency strain energy and then couple to the migration. However, it seems that this would be a minor effect. The main question is not how the chemical driving force is dissipated, but what part of it can avoid being dissipated by diffusion. It seems that the proposal by Rabkin *et al.* [65] is closely related to Sundquist's.

The rate of grain boundary diffusion plays an essential role in controlling the velocity of discontinuous precipitation and it depends on composition differences in the lateral direction in the boundary. Ordinarily, the solute content in a grain boundary depends on the content in the bulk and on a segregation factor (partition coefficient). Rabkin *et al.* [68] have recently pointed out that the segregation factor may decrease considerably when the grain boundary is migrating. They have estimated a "dynamic segregation factor" by applying Cahn's solute drag treatment. This may be essentially the same effect that Brice, and later Aziz, described with equation (19). However, in view of the fact that the velocity, v, is determined by lateral diffusion over distances much longer than the width of the boundary, it may seem that this should be a minor effect. On the other hand, it may be worth while examining this effect and Sundquist's with respect to the variation of the diffusivity across the boundary.

23. CONCLUDING REMARKS

It is evident that many phenomena in alloys, involving the migration of interfaces, cannot be understood without analysing the processes inside the interface and in front of it. In principle, a treatment of these processes would apply to all such phenomena.

Solute trapping is a widely used term when a solute has a higher chemical potential in the new phase than in the parent phase. Usually, the term is used only if the increase has occurred inside the interface. If there is local equilibrium across the interface, there would be no solute trapping according to that definition but the same increase would be observed, now occurring in the spike in front of the interface. However, sometimes the spike is very thin and it may be practically impossible to decide where the increase has occurred. In most cases where there really is trapping due to non-equilibrium at the interface, some of the increase might have occurred in the spike. The term "solute trapping" should thus be used with caution.

This is emphasized by the fact that the solute drag is connected with dissipation of Gibbs energy by diffusion that takes place partly inside the interface and partly in the spike, even though the solute drag as a force can be evaluated from the composition profile inside the interface. It thus seems safer to consider the Gibbs energy dissipation by diffusion because there is no doubt where it occurs. In any case, it seems advisable to distinguish between the two concepts "solute drag" and "Gibbs energy dissipation by diffusion" although they are closely related.

According to the modelling, an enhanced diffusivity in the interface has pronounced effects. For grain-boundary migration it may decrease the dissipation of Gibbs energy by diffusion drastically but for a phase transformation it may increase the dissipation and delay its decrease to higher velocities. It would thus be very important to clarify if the enhanced diffusivity, which is well established for lateral diffusion in interfaces, is also active in the transverse direction. If it is, it may be important to know if the enhanced diffusivity concerns a wider or thinner region than the changes in thermodynamic properties.

When comparing the various interface models, illustrated in Fig. 1, one finds that the sharp model is easy to apply but important aspects will be lost. The homogeneous model is also fairly easy to apply. It has the advantage of inviting graphical representation but its great disadvantage is the need to introduce a truncation procedure when composition gradients get too steep. Among the continuous models the smooth model may be most realistic, but has not yet been examined. The truncated wedge seems flexible enough for practical purposes except in case the sharp breaks in properties at the two sides will prove disadvantageous. The pure wedge shape does not seem to be much inferior but does not offer any special advantages in ease of application. Finally, Ågren's model reproduces some essential features of the more advanced models in the form of the sharp interface model. It is flexible and easy to use except that one must learn how to estimate the representative composition of the interface.

In some situations all continuous models, possibly with the exception of the smooth model, may yield composition spikes that are thinner than interatomic distances. However, this does not necessarily imply that the Gibbs energy dissipation, calculated from them, is incorrect. However, it would be advisable to introduce gradient energy in the description of the thermodynamic properties of the interface as well as the nearest region of the parent phase. That method could even be used to improve the homogeneous model.

Finally, diffuse interface models are important if one wants to model the properties of an interface and, in particular, its width. However, it may be less useful for modelling the role of the interface in phase transformations.

Acknowledgements—This paper is based on the author's Hume–Rothery Lecture 1999. The manuscript was afterwards greatly improved through intensive discussions with Professor John Ågren and a lively correspondence with Professor Michael Aziz.

REFERENCES

1. Gibbs, J. W., *The Collected Works Vol. I*. Yale University Press, New Haven, CT, 1948.
2. Turnbull, D., *Trans. Am. Inst. Min. Engrs*, 1951, **191**, 661.
3. Jackson, K. A., *Can. J. Phys.*, 1958, **36**, 603.
4. Hillert, M., *Phase Equilibria, Phase Diagrams and Phase Transformations—Their Thermodynamic Basis*. Cambridge University Press, Cambridge, 1998.
5. Hillert, M., Internal Report, *The Growth of Ferrite, Bainite and Martensite*. Swedish Institute for Metal Research, Stockholm, 1960.
6. Hillert, M., *Jernkont. Ann.*, 1957, **141**, 757.
7. Zener, C., *Trans. Am. Inst. Min. Engrs*, 1946, **167**, 550.
8. Ågren, J., *Acta metall.*, 1989, **37**, 181.
9. Liu, Z.-K. and Ågren, J., *Acta metall.*, 1989, **37**, 3157.
10. Liu, Z.-K. and Ågren, J., in *Proc. Int. Conf. on Martensite Transformations*, ed. C. M. Wayman and J. Perkins. Monterey Institute of Advanced Studies, Carmel, CA, 1993.
11. Liu, Z.-K., *Acta mater.*, 1996, **44**, 3855.
12. Liu, Z.-K., *Metall. Mater. Trans.*, 1997, **28A**, 1625.
13. Suehiro, M., Liu, Z.-K. and Ågren, J., *Acta metall.*, 1996, **44**, 4241.
14. Aziz, M. J., *Appl. Phys. Lett.*, 1983, **43**, 552.
15. Aziz, M. J. and Kaplan, T., *Acta metall.*, 1988, **36**, 2335.
16. Brice, J. C., in *The Growth of Crystals from the Melt*. North-Holland, Amsterdam, 1965, p. 65.
17. Jackson, K.A., Beatty, K.M. and Blackmore, K.A., Unpublished work, University of Arizona, 1999
18. Olson, G. B., Bhadeshia, H. K. D. H. and Cohen, M., *Acta metall.*, 1989, **37**, 381.
19. Ivantsov, G. P., *Dokl. Akad. Nauk. SSSR*, 1947, **58**, 567.
20. Olson, G. B., Bhadeshia, H. K. D. H. and Cohen, M., *Metall. Trans.*, 1990, **21A**, 805.
21. Mujahid, S. A. and Bhadeshia, H. K. D. H., *Acta metall.*, 1993, **41**, 967.
22. Salwén, A., *Metall. Trans.*, 1993, **24A**, 1507.
23. Krielaart, G.P., Primary ferrite formation from supersaturated austenite, Doctorate thesis, Technische Universiteit Delft, 1995
24. Baker, J. C. and Cahn, J. W., in *Solidification*, ed. T. J. Hughel and G. F. Bolling. ASM, Metals Park, OH, 1971, p. 23.
25. Caroli, B., Caroli, C. and Roulet, B., *Acta metall.*, 1986, **34**, 1867.
26. Onsager, L., deGroot, S. R. and Mazur, P., *Non-Equilibrium Thermodynamics*. North-Holland, Amsterdam, 1969.
27. Kaplan, T., Aziz, M. J. and Gray, L. J., *J. Chem. Phys.*, 1989, **90**, 1133.
28. Kaplan, T., Aziz, M. J. and Gray, L. J., *J. Chem. Phys.*, 1993, **99**, 8031.
29. Langer, J. S. and Sekerka, R. F., *Acta metall.*, 1975, **23**, 1225.
30. Cahn, J. W., *Acta metall.*, 1962, **10**, 1.
31. Lücke, K. and Stüwe, H., in *Recovery and Recrystallization of Metals*, ed. L. Himmel. Interscience, New York, 1963, p. 11.
32. Hillert, M., in *The Mechanism of Phase Transformation in Crystalline Solids*, Institute of

Metals Monograph and Report Series No. 33. Institute of Metals, London, 1969, p. 231.
33. Jönsson, B. and Ågren, J., *J. less-common Metals*, 1988, **145**, 153.
34. Jönsson, B. and Ågren, J., *Acta metall.*, 1990, **38**, 433.
35. Schoen, F. J. and Owen, W. S., *Metall. Trans.*, 1971, **2**, 2431.
36. Brechet, Y. J. M. and Purdy, G. R., *Scripta metall. mater.*, 1992, **27**, 1753.
37. Purdy, G. R. and Brechet, Y. J. M., *Acta metall. mater.*, 1995, **43**, 3763.
38. Hillert, M. and Sundman, B., *Acta metall.*, 1976, **24**, 731.
39. Shirley, C. G., *Acta metall.*, 1978, **26**, 391.
40. Hillert, M. and Sundman, B., *Acta metall.*, 1977, **25**, 11.
41. Gunton, J. D., Miguel, M. S. and Sahni, P. S., in *Phase Transformations and Critical Phenomena*, Vol. 8, ed. C. Domb and J. L. Lebowitz. Academic Press, New York, 1983, p. 267 (see for a general review)
42. Wang, Y., Chen, L.-Q. and Khachaturyan, A., in *Solid → Solid Phase Transformations*, ed. W. C. Johnsson, J. M. Howe, D. E. Laughlin and W. A. Soffa. TMS, Warrendale, PA, 1994, pp. 245–265.
43. Liu, Z.-K., Ågren, J. and Suehiro, M., *Mater. Sci. Engng*, 1998, **A247**, 222.
44. Jackson, K. A., Gilmer, G. H., Temkin, D. E., Weinberg, J. D. and Beatty, K. M., *J. Cryst. Growth*, 1993, **128**, 127.
45. Beatty, K. M. and Jackson, K. A., *J. Cryst. Growth*, 1997, **174**, 28.
46. Karlyn, D., Cahn, J. W. and Cohen, M., *Trans. Am. Inst. Min. Engrs*, 1969, **245**, 194.
47. Hillert, M., *Metall. Trans.*, 1984, **15A**, 411.
48. Perepezko, J. H., *Metall. Trans.*, 1984, **15A**, 437.
49. Menon, E. S., Plichta, M. R. and Aaronson, H. I., *Acta metall.*, 1988, **36**, 321.
50. Hillert, M. and Schalin, M., *Acta mater.*, in press.
51. Hultgren, A., *Trans. Am. Soc. Metals*, 1947, **39**, 915.
52. Hillert, M., *Paraequilibrium, Internal Report*. Swedish Institute of Metals Research, Stockholm, 1953.
53. Kirkaldy, J. S., *Can. J. Phys.*, 1958, **36**, 907.
54. Gilmour, J. B., Purdy, G. R. and Kirkaldy, J. S., *Metall. Trans.*, 1972, **3**, 1455.
55. Enomoto, M., private communication, 1998
56. Sulonen, M., *Acta Polytechniqa Scand.*, 1964, Ch. 28
57. Sulonen, M., *Z. Metallk.*, 1964, **55**, 543.
58. Hillert, M. and Lagneborg, R., *J. Mater. Sci.*, 1971, **6**, 208.
59. den Broeder, F. J. A., *Acta metall.*, 1972, **20**, 319.
60. Hillert, M. and Purdy, G. R., *Acta metall.*, 1978, **26**, 333.
61. Cahn, J. W., Pan, J. D. and Balluffi, R. W., *Scripta metall.*, 1979, **13**, 503.
62. Yoon, D. N. and Huppmann, W. J., *Acta metall.*, 1979, **27**, 73.
63. Hillert, M., *Scripta metall.*, 1983, **17**, 237.
64. Rabkin, E., *Scripta metall. mater.*, 1994, **30**, 1413.
65. Rabkin, E., Gust, W. and Fournelle, R. A., *Interface Sci.*, 1998, **6**, 105.
66. Kajihara, M. and Gust, W., *Scripta mater.*, 1998, **38**, 1621.
67. Sundquist, B. E., *Metall. Trans.*, 1973, **4**, 1919.
68. Rabkin, E., Gust, W. and Estrin, Y., *Scripta mater.*, 1997, **37**, 119.

Credits

© Nature: M. Hillert, Nuclear Reaction Radiography, Nature 168 (1951) 39.

© Maney publishing:

M. Hillert and V.V. Subba, Grey and White Solidification of Cast Iron in "The Solidification of Metals" ISI P110, Iron Steel Inst. (1968) 204.

M. Hillert, The Role of Interfaces in Phase Transformations in "The Mechanism of Phase Transformations in Crystalline Solids", Inst. Metal Monograph No. 33 (1969) 231.

© AIME/TMS/Met. Trans.:

M. Hillert, The Formation of Pearlite in "Decomposition of Austenite by Diffusional Processes", Eds. Zackay and H.I. Aaronson, AIME (1962) 197.

M. Hillert, On the Theories of Growth during Discontinuous Precipitation, Met. Trans. 3 (1972) 2729.

M. Hillert, Diffusion and Interface Control of Reactions in Alloys, Met. Trans. 6A (1975) 5.

M. Hillert, The Uses of Gibbs Free Energy-Composition Diagrams in "Lectures on the Theory of Phase Transformations", Ed. H.I. Aaronson, AIME Warrendale (1975) 1.

M. Hillert, An Analysis of the Effect of Alloying Elements on the Pearlite Reactionin "Solid-Solid Phase Transformations", Eds. H.I. Aaronson, D.E. Laughlin, R.F. Sekerka and C.M. Wayman, AIME (1982) 789.

M. Hillert, Thermodynamics of the Massive Transformations, Met. Trans. 15A (1984) 411.

Reprinted with permission from Elsevier:

M. Hillert, Thermodynamics of Martensitic Transformations, Acta Met. 6 (1958) 122.

M. Hillert, A Solid Solution Model for Inhomogeneous Systems, Acta Met. 9 (1961) 525.

M. Hillert, On the Theory of Normal and Abnormal Grain Growth, Acta Met. 13 (1965) 227.

M. Hillert, Diffusion Controlled Growth of Lamellar Eutectics and Eutectoids in Binary and Ternary Systems, Acta Met. 19 (1971) 769.

S. Björklund, L. Donaghey and M. Hillert, The Effect of Alloying Elements on the Rate of Ostwald Ripening of Cementite in Steel, Acta Met. 20 (1972) 867.

M. Hillert and B. Sundman, A Treatment of the Solute Drag on Moving Grain Boundaries and Phase Interfaces in Binary Alloys, Acta Met. 24 (1976) 731.

M. Hillert and G.R. Purdy, Chemically Induced Grain Boundary Migration, Acta Met. 26 (1978) 333.

G.R. Purdy and M. Hillert, On the Nature of the Bainite Transformation in Steel, Acta Met. 32 (1984) 823.

M. Hillert, Solute Drag, Solute Trapping and Diffusional Dissipation of Gibbs Energy, Acta Mater. 47 (1999) 4481.

A. Borgenstam and M. Hillert, Massive Transformation in the Fe-Ni System, Acta Mater. 48 (2000) 2765.

© D.R.:

M. Hillert, Paraequilibrium published as an Internal Report, Swedish Institute for Metals Research (1953).

M. Hillert, Pressure-Induced Diffusion and Deformation during Precipitation, Especially Graphitization, Jernkont. Ann. 141 (1957) 11.

M. Hillert, The Growth of Ferrite, Bainite and Martensite published as an Internal Report, Swedish Institute for Metals Research (1960).

M. Hillert and L.I. Staffansson, The Regular Solution Model for Stoichiometric Phases and Ionic Melts, Acta Chem. Scand. 24 (1970) 10.

www.ingramcontent.com/pod-product-compliance
Ingram Content Group UK Ltd.
Pitfield, Milton Keynes, MK11 3LW, UK
UKHW051525180426
11947UKWH00018B/1582